AF478558

Excess Electrons in Dielectric Media

Editors

Christiane Ferradini
Professor
Laboratoire de Chimie-Physique
Université René Descartes
Paris, France

Jean-Paul Jay-Gerin
Professor
Département de Médecine
Nucléaire et de Radiobiologie
Université de Sherbrooke
Sherbrooke, Québec, Canada

CRC Press
Boca Raton Ann Arbor Boston London

Library of Congress Cataloging-in-Publication Data

Excess electrons in dielectric media/edited by Christiane Ferradini
  and Jean-Paul Jay-Gerin.
      p.   cm.
    ''July 1990.''
    Includes bibliographical references and index.
    ISBN 0-8493-6962-2
    1. Dielectrics. 2. Excess electrons. I. Ferradini, Christiane.
  II. Jay-Gerin, Jean-Paul.
  QC585.E92   1991
  537'.24--dc20   91-9830
                                                                                       CIP

This book represents information obtained from authentic and highly regarded sources. Reprinted material is quoted with permission, and sources are indicated. A wide variety of references are listed. Every reasonable effort has been made to give reliable data and information, but the author and the publisher cannot assume responsibility for the validity of all materials or for the consequences of their use.

Direct all inquiries to CRC Press, Inc., 2000 Corporate Blvd., N.W., Boca Raton, Florida 33431.

© 1991 by CRC Press, Inc.

International Standard Book Number 0-8493-6962-2

Library of Congress Card Number 91-9830
Printed in the United States

Si ce livre me fasche, j'en prens un autre.

(Michel de Montaigne)

PREFACE

This book brings together in a single volume state-of-the-art reviews of many different aspects of the rapidly expanding field of the study of excess electrons in dielectric media. Scientific conferences are an essential forum for the communication and discussion of new work but their published proceedings, usually without the discussion, tend to be of ephemeral interest and to languish on library shelves unread. I am confident that this will not be the fate of the present volume. The authors of these reviews have presented in coherent form, not only their own special contributions to the field but a balanced and critical review of all relevant work right up to the year of publication. Despite the wide range of interest in different chapters, from purely theoretical methods of calculating the fate of the excess electrons to industrial applications in dielectric breakdown of insulating films, semiconductor electronics and the use of the electron as a probe to study liquid and solid structure they are united by the properties of the excess electrons and there is much cross fertilization in both theoretical and experimental methods.

The growing importance of this field is reflected by the rapid increase in publications over the past two decades and this book will be a valuable stimulus to further work for the authors clearly indicate both what is known and what is not yet known about the behavior of excess electrons in dielectric media.

John W. Boag
Sutton, United Kingdom

FOREWORD

This book is intended as an in-depth account of present knowledge and current problems concerning the physical and chemical interactions of excess electrons in various media.

In addition to their well-known role in chemical bonding, electrons play key roles in many other processes as a result of their ability to exist as relatively free particles. For instance, electrons are universally produced as intermediates upon exposure of matter to high energy radiation and they may also be produced by photoionization or by electrolysis. Furthermore, electron transfer is one of the key reactions of biology, both in the catalysis of oxidation-reduction reactions and in the conversion of sources of energy to usable forms for chemical transformations. It is, therefore, most important to understand the many physicochemical facets of the behavior of excess electrons in matter.

In the past few years, we have become aware of the need for a central reference describing the recent exciting developments in this field, for use by both young investigators and researchers working in related areas. Such considerations were our primary motivation leading to the present volume.

This book is divided into 13 chapters, with a rough balance between theoretical and experimental accounts. Each chapter is self-contained and therefore, can be read independently. The numerous references given in each chapter will greatly facilitate the reader's access to the original literature.

The scope of the included subjects is broad; the volume begins with a description of the primary interactions, transport, and relaxation of low energy (≤ 10 eV) electrons in a variety of dielectric solids (Chapters 1 and 2). The theme of thermalization of subexcitation electrons in molecular solids and liquids is discussed in Chapter 3. Chapter 4 is devoted to the problems of slow electron thermalization, localization, and recombination in polar liquids. The study of the energetics and transport properties of excess electrons in nonpolar dielectric liquids is developed in Chapters 5 and 6. Chapter 7 gives an overview of the quantum simulation methods that have emerged in the last few years, and indicates how these powerful techniques have been applied to study the equilibrium and dynamical behavior of excess electrons in disordered media. A review of the problem of electron solvation in polar liquids is presented in Chapter 8. Chapter 9 deals with excess electrons trapped in polar matrices at low temperature. Chapter 10 is concerned with the study of hot electron transport in silicon dioxide, a subject of particular importance for microelectronic device applications. The fate of excess electrons created in polar dielectric liquids by photoelectrochemical methods or by cathodic generation is discussed in Chapter 11. Studies dealing with excess electron production and decay in organic microheterogeneous systems, such as micelles, vesicles, and microemulsions, are presented in Chapter 12. Finally, the last chapter is devoted to the study of intramolecular long-distance electron-transfer processes in peptides and proteins. These processes are of obvious importance in photosynthesis, respiration, and enzyme catalyzed redox reactions and are currently under very active investigation. We regret, however, that the current state of research did not allow us to include a chapter associated with the problem of the migration of electrons in DNA. The possibility that DNA might act as an electron conductor has, in fact, been postulated frequently, but evidence is still conflicting.

We hope that some of the fascination of the field is conveyed in these chapters and that they will be an effective means for stimulating discussions and for suggesting possible avenues for further investigation.

Christiane Ferradini
Jean-Paul Jay-Gerin

THE EDITORS

Christiane Ferradini, Ph.D., is Director of the Laboratoire de Chimie-Physique and Professor of Physical Chemistry at the Université René Descartes, Paris, France, a position she has held since 1973.

Dr. Ferradini graduated from the Université Paul Sabatier, Toulouse, France, and obtained her Doctorat d'Etat in electrochemistry at the Institut du Radium, Paris, in the laboratory of Irène Joliot-Curie. Following the completion of her thesis in 1955, she became involved in the study of chemical effects of ionizing radiations. In 1973, Dr. Ferradini established, with Dr. J. Pucheault, the Laboratoire de Chimie-Physique of the Université René Descartes which is now devoted to radiolytic studies of radical-mediated mechanisms of biological interest.

Dr. Ferradini was appointed Adjunct Professor in the Department of Nuclear Medicine and Radiobiology, Faculty of Medicine, Université de Sherbrooke, Québec, Canada, in 1986.

Dr. Ferradini has been the author or co-author of several scientific books, and has published over 150 articles.

Jean-Paul Jay-Gerin, Ph.D., is Professor in the Department of Nuclear Medicine and Radiobiology and member of the MRC Group in the Radiation Sciences of the Université de Sherbrooke, Sherbrooke, Québec, Canada. Dr. Jay-Gerin graduated from the Université Joseph Fourier, Grenoble, France in 1967 and received a Doctorat de 3e cycle in physics from the Université de Grenoble in 1970. During the next two years, Dr. Jay-Gerin was employed as a research assistant at the Eaton Electronics Research Laboratory, McGill University, Montréal, Québec, Canada. In 1975, he obtained his Doctorat d'Etat in condensed matter physics from the Université de Grenoble. After post-doctoral work at the Department of Physics, Université de Sherbrooke, he was appointed Assistant Professor in the Faculty of Medicine, Université de Sherbrooke in 1979. He became an Associate Professor in 1983, and Professor in 1988.

Dr. Jay-Gerin has published more than 100 papers; his current research interests are centered on the theoretical aspects of the primary mechanisms of radiation action, the transport of excess electrons in solids and dielectric liquids, electron solvation in polar solvents, and Monte Carlo simulations of the physical and chemical interactions of radiation with matter.

ACKNOWLEDGMENTS

The editors wish to thank all contributing authors for their collaboration during the preparation of this book. We are particularly grateful for the generous help in the review process which we received from many of our colleagues, including G. Ascarelli, J. K. Baird, R. N. Barnett, A. Bernas, P. D. Burrow, G. V. Buxton, J. Casanovas, L. G. Christophorou, M. H. Cohen, J. H. Fendler, D. K. Ferry, J. Goodisman, H. B. Gray, N. J. B. Green, A. Herzenberg, R. A. Holroyd, N. S. Hush, M. Inokuti, S. S. Isied, E. Keszei, L. Kevan, N. V. Klassen, R. M. Marsolais, A. Mozumder, M. Nishikawa, P. Pfluger, C. J. Powell, P. J. Rossky, M. Sprik, I. T. Steinberger, J. K. Thomas, M. Tougaard, N. Ueno, and R. Voltz.

Special thanks are also due to Pauline Grenier, Francine Lussier, and Jacqueline Després whose efficient secretarial assistance greatly facilitated our work.

CONTRIBUTORS

Annette Bernas
C.N.R.S.
Laboratoire de Photophysique Moléculaire
Université Paris-Sud
Orsay, France

Bruce J. Berne
Professor
Department of Chemistry
Columbia University
New York, New York

Anatol M. Brodsky
Professor
School of Arts and Sciences
Department of Chemistry
University of Pennsylvania
Philadelphia, Pennsylvania

Eduard A. Cartier
IBM Research Division
T. J. Watson Research Center
Yorktown Heights, New York

David F. Coker
Professor
Department of Chemistry
Boston University
Boston, Massachusetts

Donelli J. DiMaria
IBM Research Division
T. J. Watson Research Center
Yorktown Heights, New York

Moshe Faraggi
Professor
Department of Chemistry
Nuclear Research Centre-Negev
Beer-Sheva, Israel

Christiane Ferradini
Director
Laboratoire de Chimi-Physique
University René Descartes
Paris, France

Massimo V. Fischetti
IBM Research Division
T. J. Watson Research Center
Yorktown Heights, New York

Dora Grand
C.N.R.S.
Laboratoire de Physico-Chimie des
  Rayonnements
Université Paris-Sud
Orsay, France

Simone Hautecloque
C.N.R.S.
Laboratoire de Physico-Chimie des
  Rayonnements
Université Paris-Sud
Orsay, France

Jean-Paul Jay-Gerin
Professor
Groupe CRM en Sciences des Radiations
Département de Médecine Nucléaire et de
  Radiobiologie
Université de Sherbrooke
Sherbrooke, Québec, Canada

Michael H. Klapper
Professor
Biological Chemistry Division
Department of Chemistry
The Ohio State University
Columbus, Ohio

Richard M. Marsolais
Groupe Surfaces
Secteur Génie des Matériaux
Centre de Recherches de Voreppe-
  Péchiney
Voreppe, France

Raúl C. Muñoz
Division PPE
CERN
Geneva, Switzerland
Presently with
Department of Physics
University of Alabama
Tuscaloosa, Alabama

Masaaki Ogasawara
Associate Professor
Faculty of Engineering
Hokkaido University
Sapporo, Japan

Peter Pfluger
Centre Suisse d'Electronique et de
 Microtechnique S.A.
Neuchâtel, Switzerland

Léon Sanche
Professor
Group CRM of Sciences and Radiation
Department of Nuclear and Radiobiologic
 Medicine
University of Sherbrooke
Sherbrooke, Québec, Canada

Robert Schiller
Professor
Department of Chemistry
Central Research Institute for Physics
Budapest, Hungary

Werner F. Schmidt
Hahn-Meitner-Institut Berlin GmbH
Abteilung Strahlenchemie
Berlin, Germany

René Voltz
Professor
Université Louis Pasteur
Centre de Recherches Nucléaires
Strasbourg, France

TABLE OF CONTENTS

Chapter 1

PRIMARY INTERACTIONS OF LOW ENERGY ELECTRONS IN CONDENSED MATTER

Léon Sanche

TABLE OF CONTENTS

I. INTRODUCTION

The scattering of an electron with an atom or a molecule can be described in terms of forces derived from the potential acting between them. There are basically three types of force between an electron and an isolated atom or molecule:[1] (1) the electrostatic force which is between the projectile electron and the constituent elementary particles of the target; (2) the exchange force, which reflects the requirement that the electron-target system wave function must be antisymmetric under pairwise electron interchange; and (3) the induced polarization attraction which is due to the distortion of the target orbitals by the electric field of the projectile electron. Solution of the Schrödinger equation with an energy dependent potential, taking exactly into account the effects of polarization, static and exchange forces, would necessarily provide the wave functions necessary to fully describe the scattering process at various electron energies. Although such calculations may be feasible for very simple atoms and molecules, in most cases they are considered impractical and the description of the scattering phenomenon is usually provided in terms of approximations appropriate for specific conditions. Even when a complete calculation is possible, the choice of the most adequate approximation often helps elucidate the essential physics of the problem.

A. RESONANT SCATTERING

For understanding electron scattering at low incident energies (0 to 30 eV), it has been found desirable to divide the process into resonant and nonresonant.[2,3] Physically this division is meaningful, since in a resonance, there are always one or more partial waves of the incoming electron wave function which undergo constructive interference within the target (i.e., they are resonant) while the other waves do not. So, with this picture, the electron scattering may be seen as a competition between resonant and nonresonant phenomena. Depending on electron energy, elastic and inelastic scattering can proceed by both of these mechanisms, but often the vibrational and electronic cross sections are considerably affected by the presence of resonances. Resonances occurring in electron impact often enhance inelastic cross sections by orders of magnitude and cause the energy dependence of these cross sections to exhibit oscillatory structure or isolated peaks. Nonresonant cross sections generally show a smooth, slowly varying, energy dependence.

Electron resonances are well described in the literature and many reviews contain information relevant to this scattering phenomenon.[2-7] Resonances occur when the scattered electron resides for a much longer time than the usual scattering time in the neighborhood of the target atom or molecule. From an atomic or molecular orbital perspective a resonant state may be considered as a negative ion formed by an electron which temporarily occupies an orbital of the target. This concept leads to the definition of two major types or categories of resonances or transient anions.[3] If the additional electron occupies a previously unfilled orbital of the target in its ground state, the transitory state is referred to as a single particle resonance. The term "shape" resonance applies more specifically when temporary trapping of the electron is due to the shape of the electron-molecule potential. When the transitory anion is formed by two electrons occupying previously unfilled orbitals, the resonance is called "core-excited" or may be referred to as a two-particle, one-hole state.

When the projectile electron is temporarily captured by the target, it has an increased interaction time. This causes additional distortion of the target whose magnitude depends on the lifetime of the resonance. Consequently, the effect of a resonance in enhancing inelastic cross sections is dependent on the lifetime of the resonance. For example, long-lived resonances ($\Delta t > 10^{-14}$ s) cause a significant displacement of the nuclei of a molecule when the additional electron occupies a strongly bonding or antibonding orbital. When the electron leaves the molecule, nuclear motion is initiated toward the initial internuclear distance, causing excitation of many overtones of the molecule, due to the strong overlap

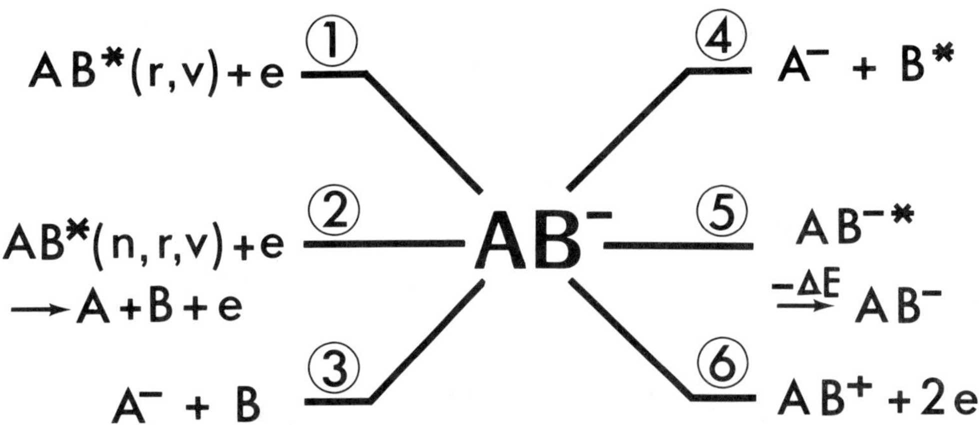

FIGURE 1.   Decay channels of a temporary anion AB^-.

between the nuclear wave function of the resonant state and that of many vibrational states of the ground state of the molecule. On the other hand, when the lifetime is much smaller than a typical vibrational period ($\Delta t \ll 10^{-14}$ s), the nuclei are not significantly displaced. In this case, overlap between the nuclear wave function of the resonant state and that of the vibrational levels of the ground state occurs only between the first few energy levels. Thus, for short resonance times only the lower vibrational levels become excited with considerable amplitude.

Because of the uncertainty principle the transient state has a width in energy ($\Gamma \cdot \Delta t \simeq \hbar$ where Γ is the energy width of the resonance and Δt its lifetime) which serves to characterize and identify the process in the energy dependence of the scattering cross sections or excitation functions. Thus, when resonances are short lived ($\Delta t \ll 10^{-14}$ s) they produce broad peaks in their decay channels (i.e., in the specific excitation functions where they appear). Long-lived resonances ($\Delta t \geqslant 10^{-14}$ s) in atoms produce sharp peaks in elastic and electronic excitation and ionization cross sections. Additionally, in molecules, they also do so in the energy dependence of vibrational excitation cross sections, when they are derived from dissociating states.

Transient molecular anions have more decay channels than their atomic counterparts due to the additional degrees of freedom introduced by nuclear motion. Figure 1 illustrates the possible decay channels of a diatomic transient anion AB^-. The departing electron may leave the molecule in a rotationally, vibrationally (Reaction 1 in Figure 1), or electronically (Reaction 2 in Figure 1) excited state. If the resulting electronically excited neutral state is dissociative, ground state or excited fragments can be produced (2). If the lifetime of the resonance is, at least, of the order of a vibrational period, the AB^- state is dissociative in the Frank-Condon (FC) region and one of the possible fragments has a positive electron affinity, then the anion may dissociate into a stable anion and a neutral fragment in the ground (3) or an excited (4) state. This process is called dissociative attachment (DA). If during its lifetime, the transient anion transfers energy to another system (e.g., to another molecule by collisional interaction) it can become stable when the parent molecule has a positive electron affinity (5). Finally, when the transient anion is formed at energies above the ionization potential, two-electron emission is possible (6). Excluded from the reactions of Figure 1 is the spontaneous emission of a photon by the transitory anion (i.e., $AB^{-*} \rightarrow AB^- + h\nu$), since emission of electromagnetic radiation is rarely faster than electron emission.

The decay of transient anions into these various channels may be better understood by considering the hypothetical potential energy diagram of a diatomic anion AB^- shown in

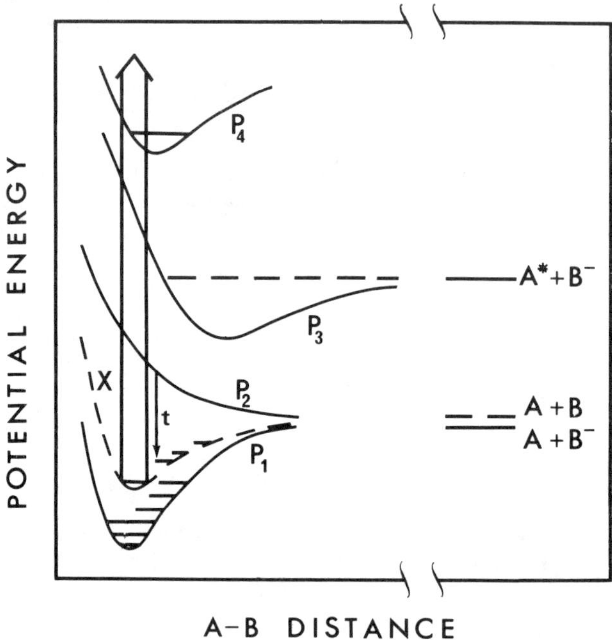

FIGURE 2.   Hypothetical potential energy curves of a molecular anion
AB⁻. The dashed curve represents the ground state of the neutral molecule
AB.

Figure 2. The dashed curve (X) represents the ground state of the molecule AB while the
others are those of the anion system. Curve P_1 represents the ground state of AB⁻ which
is stable for vibrational energy levels below the state X. Here, reaction 5 of Figure 1 may
be illustrated by noting that when an electron is captured in a vibrational level of AB⁻ in
state P_1 lying above the ground state (X) of the molecule, energy transfer from the anion
to another system can cause the electron to be captured in another vibrational state lying
below the ground state X. The anion states P_2 or P_3 may dissociate into fragments A + B⁻
or A* + B⁻, respectively, where A* represents an excited state, if the lifetime of the
resonance is sufficiently long. If not, the electron will be reemitted (e.g., transition [t] in
Figure 2) leaving the target in vibrational and rotational excited states. When the transitory
anion state lies above electronically excited states of the molecule, this latter can acquire
electronic energy in addition to vibrational and rotational motion. Of course, if the state is
not dissociative in the FC region (e.g., P_4) no stable anion fragment can result from that
particular anion and decay is only possible into channels 1, 2, 5, and 6 of Figure 1.

For further information on the mechanism of transient anion formation and its effects
in isolated electron-atom and electron-molecule collisions the reader is referred to the review
articles by Schulz[2] and others.[3-7]

B. NONRESONANT SCATTERING

When the time-dependent amplitude of the projectile electron wave function does not
increase significantly in the vicinity of the target, the scattering process is considered nonresonant. In this case, insight into the physical phenomenon may still be gained from analysis
of the interaction potential, in trying to determine the leading term (or terms) of its expansion.

Consider, as an example, the electrodynamic interaction potential acting between a
molecule and an electron outside a molecule.[9] This potential can be written as

$$V = \frac{\mu_e}{r^2}\, P_1\,(\hat{r} \cdot \hat{R}) - \frac{Q_e}{r^3}\, P_2\,(\hat{r} \cdot \hat{R}) - \frac{\alpha e^2}{2r^4} - \frac{\alpha' e^2}{2r^4}\, P_2\,(\hat{r} \cdot \hat{R}) - \ldots \tag{1}$$

where r is the distance of the incident electron from the molecule and R is the internuclear separation; μ_e is the electric dipole moment. The term containing μ_e is necessarily absent in homonuclear diatomic molecules. The second term involving Q_e, the quadrupole moment, characterizes the quadrupole interaction. These two terms are the "electrostatic" terms in that they pertain to the interaction of the incident electron and the unperturbed molecule. The next two terms, involving α, the spherically symmetric and α', the nonspherical part of the polarizability, are "dynamic" terms. The latter involve the polarization of the molecule by the incident electron. In Equation 1, $\hat{r}$ and $\hat{R}$ are unit vectors in directions r and R, respectively and P_n are Legendre polynomials. The parameters α, α', μ_e, and Q_e are functions of R. It can be seen from Equation 1 that, by estimating the magnitude of the various terms, it may be possible to sort out the dominant scattering mechanism. Furthermore, the potential of Equation 1 can be expanded around the equilibrium internuclear distance R_e as

$$V = V (R - R_e) + (R - R_e) \frac{\partial V}{\partial R}|R = R_e + \ldots \qquad (2)$$

When only these two terms are considered and the molecule is assumed to be a harmonic oscillator, solving the problem within the Born approximation (usually corresponding to small momentum transfer) leads to the optical selection rule $\Delta v = 1$ for vibrational transitions.[10] Thus, within this most restrictive approximation, the electron behaves like electromagnetic radiation. From this analysis, we can expect the electron-molecule potential described by Equation 2 to be responsible for the magnitude of the differential scattering cross sections which are large only for the excitation of the first vibrational energy level ($v = 1$) from the ground state of the molecule.

C. INTERACTION OF LOW ENERGY ELECTRONS (0 to 30 eV) WITH CONDENSED MATTER

When the electron wavelength is short in comparison with the "diameter" of the elementary constituents of condensed matter, the electron can be considered to approximately interact independently with each atom or molecule. Thus, the scattered amplitude within or outside a solid or a liquid is considered the sum of the individually scattered waves. This concept is no longer valid at low energies, where the electron wavelength is of the order of the interatomic or intermolecular distances. In this case, the electron interacts "collectively" with many targets and the scattered intensity must be derived from the sum of the interactions between the electron and each of the elementary constituents of condensed matter. Solving the Schrödinger equation with such a potential is a much more tedious task than for the case of the single-electron target system. It is, therefore, more desirable to describe electron scattering in condensed phases with models of the most prominent interaction. Recent papers by Fano and co-workers outline the relevant concepts and theoretical procedures required to describe the action of slow electrons in condensed media.[11,12] Earlier attempts are to be found in the theoretical work of Fröhlich,[13] Fröhlich and Platzman,[14] and Magee and Helman.[15]

Description of the scattering process in terms of intramolecular resonant and nonresonant mechanisms is an approach which has proven successful in describing the interaction of low energy electrons with *molecular solids*. With this approach, it has often been possible to explain structures in the energy dependence of an inelastic cross section (or a signal proportional to that cross section) by invoking the formation, at specific energies, of transient anions within the solid or near its surface. Comparison with gas-phase data is most useful in identifying the resonant state and in investigating the modifications to the characteristics of the isolated transient anion induced by the presence of neighboring targets. Other nonresonant features in the energy dependence of the cross sections can usually be explained by specifying which part of the interaction potential is dominant.

Low energy (0 to 30 eV) electron scattering within and at the surface of *rare-gas solids and molecular solids* is reviewed in this chapter. Discussion of the phenomena involved is provided with examples from *electron beam experiments* described in the next section. It is shown that while the results of some of these experiments can only be discussed in terms of solid-state concepts, the majority can be interpreted by invoking intraatomic and intra-molecular resonant and/or nonresonant scattering concepts developed from gas-phase experiments. Elastic scattering is discussed in Section III. Data from inter- and intramolecular vibrational excitation and electronic excitation of molecular solids by slow electrons are presented and discussed in Section IV. The subject of molecular dissociation induced by electron impact at low energies, which leads to the production of neutral positive and negative fragments, is reviewed in Section V. The role of resonant processes in electron scattering from molecules physisorbed or chemisorbed in submonolayer amounts on metal surfaces is mentioned in the concluding remarks (Section VI). Methods used to obtain electron elastic and inelastic mean-free paths in dielectrics from beam or photoinjected electron experiments are discussed by Marsolais, Cartier, and Pfluger in Chapter 2. For a review of electron mobilities in dense gases and liquids, the reader is referred to a recent article by Christophorou[16] and Chapter 6.

Molecules condensed in monolayer or submonolayer amounts on conductive surfaces are not considered to form a molecular solid, and consequently, the subject of electron scattering from such targets is not reviewed in this chapter. In these experiments, the electron beam is *usually* utilized as a *probe* to obtain *spectroscopic information* which is related to the geometrical and chemical state of molecules physisorbed or chemisorbed on a single-crystal solid surface.[17,18] In surface spectroscopy,[17] the probed electrons are those which are scattered in a narrow lobe near the specular direction with little momentum transfer parallel to the surface. The interaction takes place via the dynamical dipole moment term of Equation 2 and the $\Delta v = 1$ selection rule applies. Thus, the electron essentially replaces photons in these experiments in order to obtain surface spectroscopic information.

Scattering from atomic and molecular clusters is another aspect of low-energy electron interaction with condensed matter which is not reviewed in this chapter. The information presently available in this field has been obtained from measurement of positive and negative ions formed by crossing a supersonic molecular or atomic beam with an electron beam.[19] Measurements of the electron energy dependences of anion yields has been particularly helpful in elucidating the role of transient anions in the scattering process. This topic has recently been reviewed by Märk.[19]

II. EXPERIMENTS

The interaction of low energy electrons with rare-gas and molecular solids can be investigated by allowing monoenergetic electrons to impinge on a thin multilayer film grown in an ultra-high vacuum system by the condensation of gases or organic vapors on a clean metal substrate held at cryogenic temperatures (15 to 100 K). Depending on the type of apparatus, it is possible to measure the dependence on primary electron energy of the current transmitted through,[20] trapped in,[21] or reflected[22,23] from the film or that of the positive ion,[24] negative ion,[25] and neutral species fluxes[26] emanating from its surface. As a general rule, the film thickness must be larger than the total electron mean-free path in order to minimize effects of the metal substrate. Evidence that a given process occurs within the bulk of the material arises from the presence of multiple scattering effects in the data.[22,27,28] It can also be provided by specific experiments such as matrix isolation[22,29] or overlayer coverage of a film by another substance.[22] A brief description of the different types of experiment is given in this section.

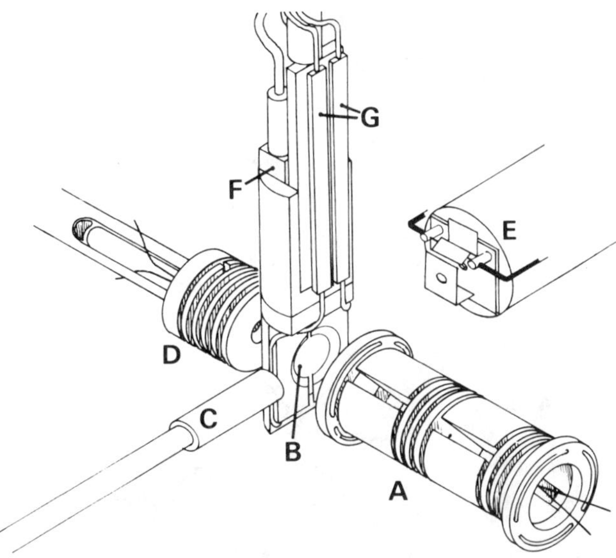

FIGURE 3.   Diagram of typical low energy electron transmission (LEET) spectrometer with mass spectrometer: (A) trochoidal electron monochromator; (B) rotatable target; (C) gas doser; (D) high current electron gun; (E) mass spectrometer; (F) cryostat; and (G) electrical leads.

A. LOW ENERGY ELECTRON TRANSMISSION (LEET) SPECTROSCOPY

A drawing of the type of apparatus used to record LEET spectra[20] as well as electron stimulated desorption (ESD)[30,31] and degradation[32] yields is shown in Figure 3. It consists of a high current electron gun (D), a quadrupole mass spectrometer (E), a gas introduction doser (C), a cooled target (B), and a high resolution trochoidal electron monochromator (A).[33] Depending on the instrument, the target (B) can be rotated and may be cooled down to 15°K (cryostat [F] in Figure 3). Components of the apparatus are housed in a UHV system reaching pressures below 10^{-10} torr. The magnetically collimated electron beam leaving the monochromator (A) with resolution of about 40 meV full width at half maximum (FWHM) impinges on the film condensed on a metal substrate (B) (i.e., the electron collector). The latter is electrically isolated from the cryostat by a sapphire disk and connected to electrical leads (G). LEET spectra are obtained by measuring the current $I_t(E)$ arriving at the metal substrate as a function of incident electron energy. In these experiments, I_t is in order of a few nA and the absolute electron energy scale is calibrated within ± 0.15 eV with respect to the vacuum level by measuring the onset of electron transmission through the films. The metal substrate is usually a polycrystalline metal sheet which can be cleaned by resistive heating (via G). The condensed films are grown using a gas-volume expansion dosing procedure[20] which can be calibrated by monitoring the quantum size effect (QSE) features observed for ultra-thin films.[34,35] With this calibration, the film thicknesses (10 to 500 Å) can usually be estimated with an accuracy equal or better than 30% assuming a layer-by-layer growth.

B. HIGH RESOLUTION ELECTRON ENERGY LOSS (HREEL) SPECTROSCOPY

Energy losses by electrons scattered within thin films are measured by a HREEL spectrometer[22] of a type shown schematically in Figure 4. This apparatus consists essentially of two concentric hemispherical deflectors with appropriate electron optics and a closed-cycle refrigerated cryostat. Electrons leaving the monochromator are focused on the film

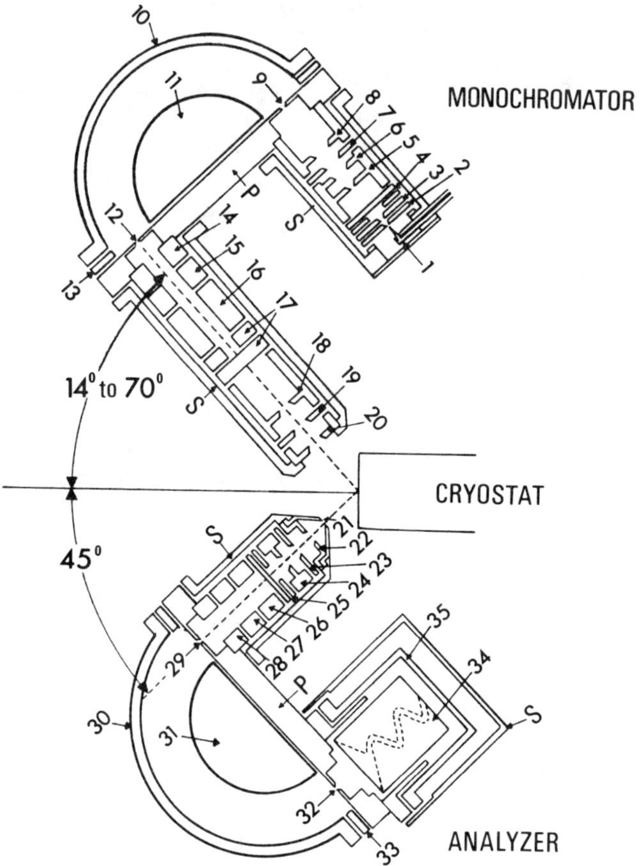

FIGURE 4.   Schematic diagram of hemispherical electron energy loss spectrometer. Electrons emitted from filament (1) are focused at the entrance (9) of the deflector region (10 and 11) by the zoom lens (6, 7, and 8). Energy dispersion is produced by applying a potential of about 1 volt between two concentric hemispheres (10, 11, 30, and 31). Imaging and energy control of the electron beam is achieved by means of electrostatic zoom lenses with cylindrical (14 to 16, 26 to 28) and aperture (6 to 8, 23 to 25) electrodes.

condensed on a metal substrate secured by a press fit to the cold end of the cryostat. The angle of incidence θ_i can be varied from 14 to 70° from the film normal. Electrons reflected at $\theta_r = 45°$ from the film are energy analyzed by another hemispherical deflector. LEET spectra can also be recorded with this type of apparatus. This allows the absolute electron energy scale to be calibrated within ±0.15 eV.

Energy loss spectra are recorded by sweeping the energy of either the monochromator or the analyzer. The energy dependence of the magnitude of a given energy loss event (i.e., the excitation function) is obtained by sweeping the energy of both deflectors with a potential difference between them corresponding to the probed energy loss. HREEL spectra are usually recorded with overall resolutions ranging from 6 to 20 meV FWHM with corresponding currents lying in the 10^{-10} to 10^{-9} Å range.

C. ELECTRON STIMULATED DESORPTION (ESD) OF IONS AND NEUTRALS

A portion of the positive and negative ions produced by electron impact on the film target (B) (Figure 3) can be measured by placing a mass spectrometer (E) near the film

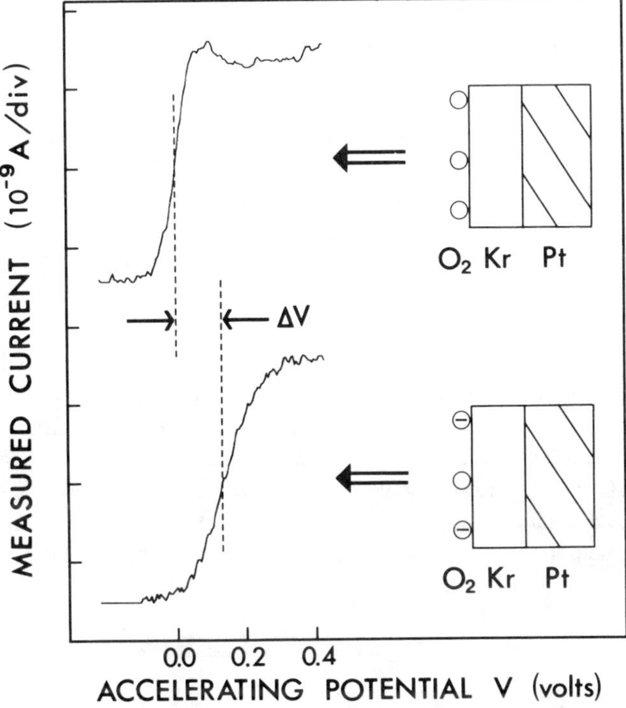

FIGURE 5.   Current transmitted through an uncharged (top) and a charged (bottom) Kr film covered with 0.1 monolayer (ML) of O_2 as a function of the accelerating potential V of the incident electron beam.

surface.[24,31] In the experimental arrangement of Figure 3, the cryostat C is rotated toward E for this measurement. Ions emerging from the film are focused by ion lenses located in front of the entrance of the mass spectrometer. Grids can be inserted between the lenses and the mass spectrometer in order to analyze the ion energies by the retarding potential method. The apparatus can be operated in two modes:[31] (1) the ion-yield mode in which ions of a selected mass are detected as a function of incident electron energy and (2) the ion-energy mode in which the ion current at a selected mass is measured for a fixed electron energy as a function of the retarding potential. The first derivative of the curve obtained in this latter mode represents the kinetic energy (KE) distribution of the ions studied at the chosen incident electron energy. Using the ionization source of the mass spectrometer[36] or a laser resonance-ionization source,[37] the desorption of neutrals can also be analyzed in this type of apparatus. With a standard electron ionization source background ionization can be discriminated using beam modulation lock-in techniques.

D. ELECTRON TRAPPING

The number of charges accumulated in a dielectric film resulting from bombardment with a low energy electron beam of a specific energy can also be measured by LEET spectroscopy.[21,38,39] When electrons from monochromator (A) (Figure 3) have just enough energy to enter a multilayer film deposited on the substrate B a sharp rise, termed the "injection curve" (IC), is seen in the LEET spectrum. The IC for an uncharged film is represented by the upper curve of Figure 5. When the same film is charged by the electron beam, the IC is shifted by ΔV to a higher accelerating potential (bottom curve, Figure 5) since the incoming electrons must then possess additional KE to overcome the negative potential barrier. The IC is also broadened due to the effect of the spatial charges and current density distributions. Such measurements are usually performed in conjunction with LEET, HREEL, and ESD experiments to make sure that the films do not charge significantly during the time of the experiment. However, if the film is allowed to charge by a significant potential ΔV, this latter can be related to the trapping cross section by treating the dielectric film as a charged capacitor.[21] The potential barrier ΔV is related to the charge density $\sigma(t)$, which has accumulated after bombardment time (t), by the relation $\Delta V(t) = \sigma(t)\,h/\epsilon$, where $\sigma(t) = \sigma_o\,(1 - e^{-\beta t})$ and $\beta = \rho\,J_o/e$; ϵ is the dielectric constant of the film, h is its thickness, σ_o is the initial (t = 0) trap density, ρ the trapping cross section, J_o the incident current density and e the unit charge. In the limit $t \rightarrow 0$ a charging coefficient $A = d\Delta V/dt$ directly proportional to the trapping cross section can be expressed as $d\Delta V/dt|_{t=0} = (h\sigma_o\,J_o/\epsilon e)\,\rho = A$.

The experiment is performed as follows. The IC of a freshly deposited multilayer film is first recorded rapidly (e.g., during 0.1 s) to avoid any significant charging. The film is then bombarded at a given voltage (V) applied between the monochromator and the film for a much longer period (e.g., 25 s) with the same incident current (i.e., $I_o \simeq 5 \times 10^{-9}$ A). Afterwards, the IC is again rapidly recorded and the shift ΔV determined by comparison with the initial IC. Such a cycle can be repeated many times on the same film with the same V to obtain the time dependence of the process. To measure the thickness and energy (i.e., V) dependence, a new film has to be deposited for each data point.

III. ELASTIC SCATTERING

Elastic scattering[40] of low energy electrons by multilayer rare-gas and molecular solid films has been investigated by LEET,[20,34,35,41-56] photoinjection,[57-60] and elastic reflection[27,28,56] experiments. In the latter experiments the spectrometer shown in Figure 4 is adjusted to measure electron elastically scattered from the film at particular incident and scattered angles θ_i and θ_r. Then, the amplitude of the elastic peak is measured as a function of electron energy. Since in these experiments, both the incident and outgoing electron momentum are specified, features due to interferences of the electron waves are prominent in the spectra. In well-ordered films, the diffraction structure is dominated by long-range order[56] whereas, in amorphous substances, variation in the structure factor due to short-range order can be detected in the energy dependence of the elastic reflectivity.[27,28]

LEET spectra are especially sensitive to elastic scattering below the electronic-excitation energy threshold, where the transmitted current is purely or vibrationally elastic. However, in transmission, only the incident beam has a well-defined momentum since current which has been scattered into all angles is measured at the metal substrate. Furthermore, when the film is highly disordered, electrons are scattered in all possible directions near the surface,[42] so that, the penetrating momentum is also unspecified. Under these conditions, LEET spectra are expected to represent a directionally averaged band structure.[42,44]

Elastic scattering from rare-gas and molecular solids (i.e., Ar, Kr, Xe, N_2, CO, O_2, CH_4 and several organic molecules)[27,34,42,43,46,50,55,56,61-63] has been found to be nonresonant. The structure in the energy dependence of elastically transmitted or reflected currents of molecular films can be explained by invoking multiple scattering of the electron waves between the potentials of the individual atoms or molecules, as if intramolecular multiple scattering was irrelevant.[56] This is surprising since in gases electron resonances cause easily identifiable structures in the elastic cross section.[2] For example, the N_2^- ($^2\Pi_g$) anion state increases elastic scattering by a factor of 3 in gaseous N_2 around 2 eV.[2] In condensed N_2, no change can be perceived in the elastic cross section although the vibrational excitation cross sections are increased up to two orders of magnitude by the $^2\Pi_g$ state.[61,64] From a purely classical phenomenological point of view, it can be shown[65] that multiple elastic and inelastic scattering between sites tends to decrease the apparent contribution of purely elastic

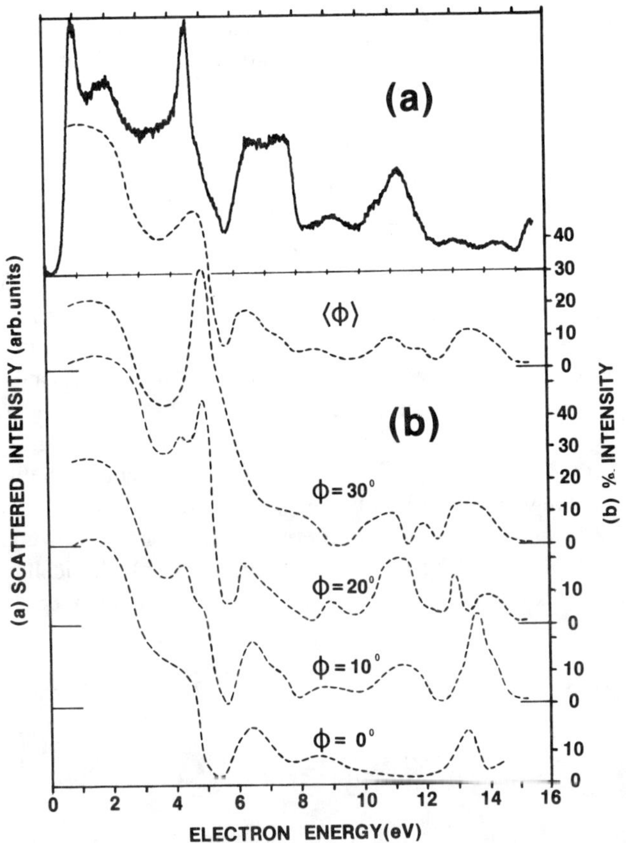

FIGURE 6. (a) Experimental specular reflectivity at incidence angle θ_i = 45° of a 50-ML Ar film; (b) calculated specular reflectivity for the (111) surface of semiinfinite crystal of Ar at θ_i = 45° and at incident azimuthal angles ϕ = 0, 10, 20, 30° and averaged over the azimuth (⟨⟨φ⟩⟩).

scattering with respect to the other cross sections. This would diminish the visibility of resonances in the total and vibrationally elastic cross sections but the actual magnitude of this effect is difficult to estimate until a full quantum mechanical treatment of the problem becomes possible.

An example of the role of nonresonant multiple potential scattering is provided by the data[56] of Figure 6. Part (a) exhibits the experimental elastic reflectivity of a partially ordered 50-monolayer (ML) Ar film recorded at incidence and scattered angles of 45°. The film was grown on a platinum substrate composed of a large number of microfacets having the (111) orientation parallel to the plane of the surface and strong azimuthal disorder.[56] Very good agreement exists between the experimental results in (a) and the results of a low energy electron diffraction (LEED) calculation of the specular (0,0) beam intensity at θ_i = θ_r = 45° averaged over all azimuthal angles (top curve in [b]). Except for the very low (<2 eV) and high (>12 eV) energy portion, every feature of the experimental spectrum is reproduced within 0.5 eV by the ⟨φ⟩ curve. Above 12 eV no particular correction for the additional electron inelastic losses to electronic excitations was considered.[56] The high reflectivity between 0 and 5 eV is also reproduced including the fine structure at 4.7 eV. A simple kinematical calculation at ϕ_i = 45° including the average polarization potential of 1.6 eV predicts a Bragg peak at 4.76 eV. This Bragg peak should lie, therefore, through multiple scattering, at the origin of a band gap and the consequent high reflectivity of the crystal in this energy range.

The energy dependence of exchange and polarization, which is usually considered important in evaluating low energy electron single-target scattering amplitudes,[5] has not been taken into account to produce the specular reflectivity curves of Figure 6 (b). Despite this neglect, the LEED calculation can still describe the essential features of the energy dependence of the elastic intensity in the 2 to 12 eV range. This may be due to the fact that when multiple scattering is taken into account, the short-range "core" region of the potential which is represented by the muffin tin model of LEED theory,[66] dominates the electron-solid interaction.

In the case of amorphous films, no sharp structure is observed in elastic reflectivity curves because the electron coherence and interference is only short range. The electron energy dependence of the elastically scattered intensity from amorphous materials can be expressed via multiple scattering theory[27,28] as an ensemble averaged scattering cross section per scatterer. The energy dependence of such a cross section for elastic scattering is given by the curves at the top of Figure 7a.[28] The values were obtained from the analysis of elastic reflectivity data[27] of a 30-ML amorphous H_2O film. The other curves in this figure represent absolute inelastic cross sections to be discussed in section IV. The absolute elastic cross section of Figure 7a exhibits a rise at low energy and two broad structures whose maxima are located at 6 and 14.5 eV. The cross section for electrons elastically scattered from a disordered medium is expressed[67] in the single event approximation as

$$Q = \frac{1}{\hbar} \left(\frac{m}{2\pi\hbar^2}\right) \int_o^\pi \int_o^{2\pi} |V(|\vec{K}|)|^2 \, S(|\vec{K}|) \, \sin\theta \, d\theta \, d\phi \tag{3a}$$

$$|\vec{K}| = [2(1-\cos\theta)]^{1/2} |\vec{k}_o| \tag{3b}$$

and θ is the angle between the incident $\vec{k}_o$ and scattered $\vec{k}$ electron wave vectors. In this expression, the scattered amplitude is the coherent sum of amplitudes originating from individual sites, whereas the sum of amplitudes multiply scattered from different sites is neglected.[68] $S(\vec{K})$ is a structure factor related to the static correlation between sites in a condensed medium.[69] Expressions 3a and 3b can be applied to explain the results on top of Figure 7a.

Between 1 to 18 eV incident energy, the electron MFP is 3 to 10 times larger than the de Broglie wavelength of the electron, and the modulus of the maximum wave vector transfer $|\vec{K}_{max}| = 2|\vec{k}_o|$ lies from 1 Å^{-1} to 4.3 Å^{-1}, respectively. In this latter range, the structure factor of amorphous ice[70] at 10 K is characterized by two well-developed structures located at $|\vec{K}|_1 = 2 \, \text{Å}^{-1}$ and $|\vec{K}|_2 = 3.1 \, \text{Å}^{-1}$. In the limit where $v(|\vec{K}|)$ can be considered a constant (i.e., for an isotropic scattering cross section of a site), expression 3a simplifies to

$$Q = \frac{2\pi}{\hbar} \left(\frac{m}{2\pi\hbar_2}\right) |V|^2 \int_o^\pi S(|\vec{K}|) \, \sin\theta \, d\theta$$

By performing the angular integration as function of the incident energy, a maximum at energy E_{oi} and E_{oj} should occur in Q, whenever we have for features located at $|\vec{K}|_i$ and $|\vec{K}|_j$ in $S(|\vec{K}|)$, the ratio

$$|\vec{K}|_i/|\vec{K}|_j \simeq (E_{oi}/E_{oj})^{1/2} \tag{4}$$

Taking the energy ratios in Equation 4 from Figure 7a, one obtains $(E_{o1}/E_{o2})^{1/2} = 0.643$ in close agreement with $|\vec{K}|_1/|\vec{K}|_2 = 0.645$ from $S(|\vec{K}|)$; indicating once again that, at least qualitatively, structure in the elastic cross section can be explained by invoking intermolecular interference of the electron waves.

Elastic scattering within thin dielectric films has also been investigated by analyzing the elastic contribution to the LEET current.[34,35,42-50,52,54-56] However, even when elastic scattering dominates, such an analysis is not as straightforward as that of the elastic reflectivity since the LEET current is not simply the complement of the currents arising from addition of the specular and diffracted beam intensities. Other currents arise from scattering at the boundary of the microcrystals and scattering by defects, imperfections and phonons. Figure 8 shows, as an example, LEET spectra for Kr films of different thicknesses and temperatures, deposited on a polycrystalline platinum substrate.[41] Curve (a) of Figure 8 was recorded for the clean metal, whereas curves (b), (c), (d), and (g) represent the energy dependence of the current transmitted through Kr films of 1, 10, 100, and 500 ML, respectively, deposited and held at 17 K. Curves (e) and (f) were recorded at a 100 ML coverage. For curve (e), the temperature was kept constant at 31 K during the deposition and the experiment. Curve (f) was recorded at 17 K after deposition at 31 K. As the Kr film becomes thicker, features 1 to 4 progressively disappear and features 5 to 7 become sharper. The maxima in these latter features correspond within ± 0.2 eV with the position of the first two excitonic levels (n = 1 and 2) of the J = 1/2 and J = 3/2 series of solid krypton[71] which are represented by vertical bars in the figure. In the two-stream approximation,[45] the inelastically scattered current is predicted to be independent of thickness at large thicknesses, whereas, under the same conditions, the elastic current becomes inversely proportional to the thickness. This behavior is observed for structures 5 to 7 and 1 to 4, respectively. An elastic behavior below 10 eV is expected since only acoustic phonons can produce inelastic events in this range.[72] Structures 1 to 4 are also found to depend on the ordering of the Kr film. The films deposited at 17 K (curve d) are the most disordered ones (structural disorder). In the film deposited at 31 K (curve e), the structural disorder is smaller but there is greater thermal disorder (phonons) than in the 17 K deposited films. The films deposited at 31 K and cooled to 17 K (curve f) are the most ordered ones: low structural disorder because of the high deposition temperature and low thermal disorder, since at 17 K the phonon population is greatly reduced (Debye temperature: 64 K).[45] The structures below 10 eV are, therefore, closely related to the ordering of the films.

More detailed analysis of the LEET spectra of Ar, Xe, and CH_4 films according to the Fermi golden rule has been compared with respective electronic conduction band density of states (DOS) calculations.[42,46,49,50] These comparisons clearly indicate a relationship between elastic LEET currents and the DOS of the solid as shown for the case of Ar films[50] in Figure 9. Here, the DOS calculated by the linear analytic tetrahedron method[73] is represented by the solid line. The DOS extracted from the LEET spectrum is represented by the dashed line.[50] Further correlation between the electronic conduction band DOS and "elastic" features in LEET spectra has been achieved[43] by comparing these latter with structure in secondary electron spectra which were previously shown[74,75] to reflect changes in the DOS. Smearing out of LEET features near the melting point of long chain alkane films[43] was attributed to smoothing of the DOS due to breakdown of intermolecular symmetry (e.g., phase transitions) induced by thermal excitations. Changes of the electron effective mass with energy has also been invoked by Plenkiewicz et al[42] to explain the energy dependence of the mean-free paths at low energies.

For thin well-ordered films in which scattering by defects is not sufficiently intense to redistribute electrons in random directions and where energy losses to phonons and vibrations are small, a portion of the penetrating electrons are capable of conserving a specific momentum during their residence in the film. Then, constructive and destructive interferences of the electron wave can evolve between the film-vacuum and film-substrate boundaries.[34,35] These interferences, called quantum size effects (QSE), modulate the usual transmission

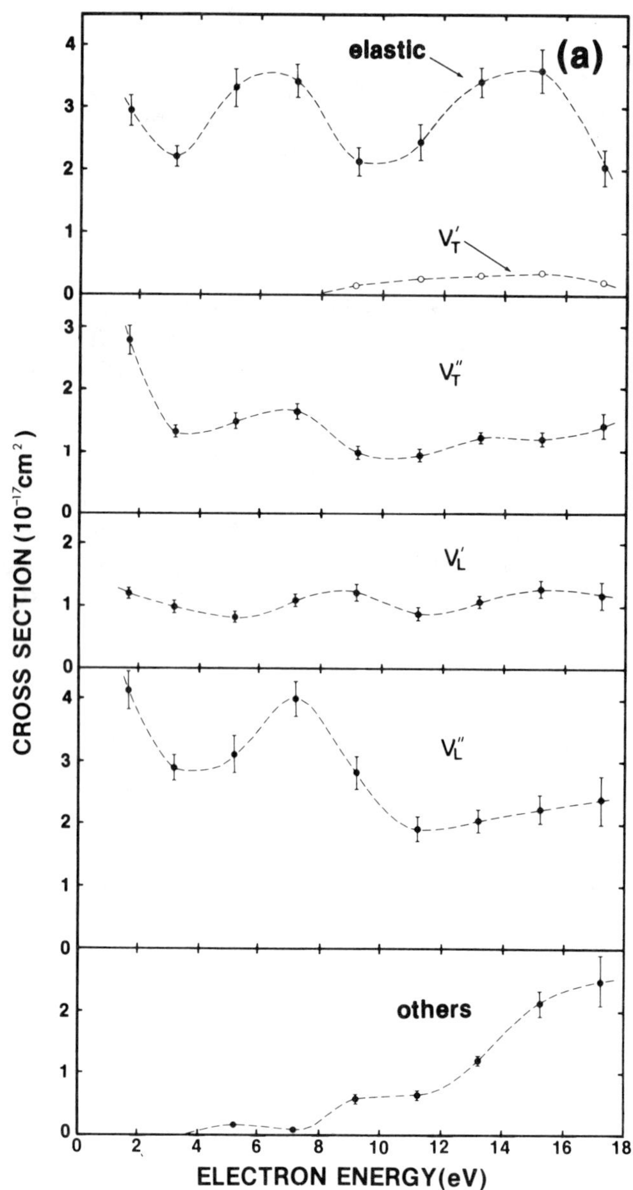

FIGURE 7.   Energy dependence of the elastic and vibrational electron scattering cross sections per scatterer in amorphous ice derived from multiple scattering analysis of energy loss spectra. The bottom curve in (a) represents the cross section ascribed mainly to the sum of electronic excitations.

features emerging from the bulk DOS. This is shown in the LEET spectra of 3 to 6 layers of Ar and CH$_4$ in Figure 10 where the QSE features are indicated by vertical arrows.[35]

IV. INELASTIC SCATTERING

A. INTER- AND INTRAMOLECULAR VIBRATIONAL EXCITATION

By scattering within molecular solids and at their surfaces low energy electrons can excite, with considerable cross section, phonon modes of the lattice[22,28,78,80] and individual

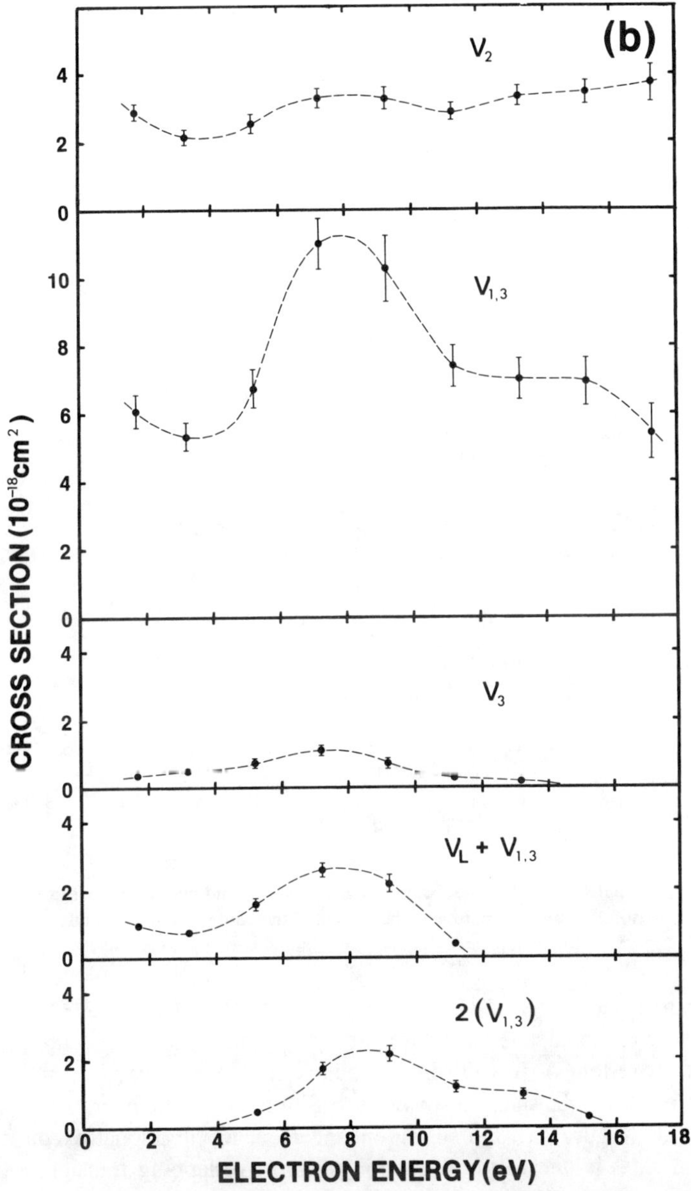

FIGURE 7B.

vibrational levels of the molecular constituents[22,28,61,63,76,78-84] of the solid. These modes can be excited either by nonresonant or by resonant scattering depending on the mechanism prevailing at specific energies.

Examples of resonant excitation of the intramolecular modes of CO and O_2 in the condensed phase are given in Figures 11 and 12, respectively.[22,62] These electron energy loss spectra were recorded for electrons of energies E_i = 2.5, 5, and 20 eV and θ_i = 14° and θ_r = 45°. Those in the inset of Figure 11 were recorded in the specular direction (i.e., θ_i = 45°). The vertical gains in each curve or portion of a curve are referenced to the elastic peak. Each energy-loss peak in Figure 11 can be ascribed to vibrational excitation of ground state CO.[22] Additionally, in O_2 films,[62] some vibrational progressions can be ascribed to intramolecular vibrational excitation of the states $a^1\Delta_g$ and $b^1\Sigma_g^+$ (Figure 12). It can be seen

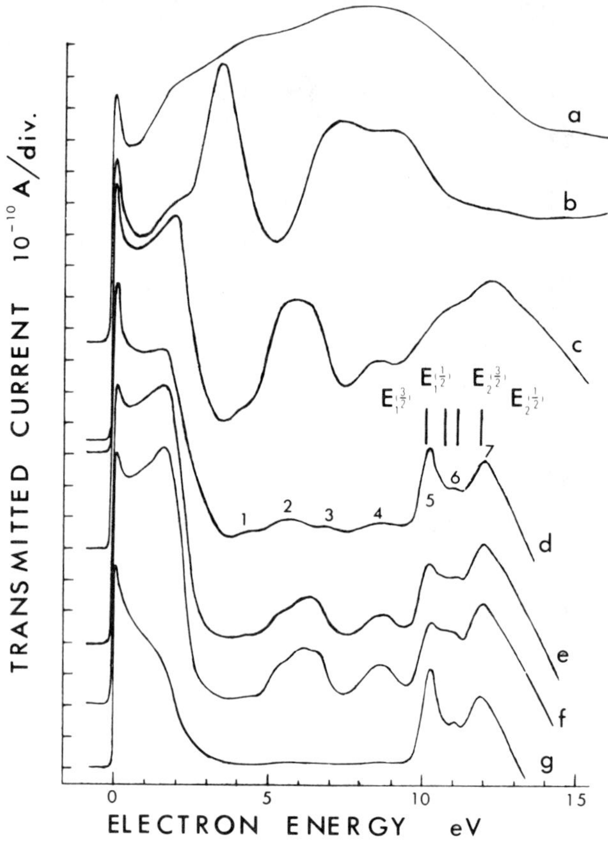

FIGURE 8.   LEET spectra of single layer (b) and multilayer Kr films (c to g). Curve (a) represents the current at the bare platinum substrate. The vertical lines represent the energies of the lowest excitons.

that at certain impact energies (E_i = 2.5 eV and 9 eV for CO and O_2, respectively) the intensities of vibrational energy losses are greatly increased (i.e., up to two orders of magnitude for overtones). In CO films, production of overtones at E_i = 2.5 and 20 eV is attributable to $^2\Pi$ and $^2\Sigma$ shape resonances.[63] Similarly, in the energy loss spectra of multilayer O_2 films (Figure 12), the strong enhancement in vibrational excitation of O_2 in the configuration $^3\Sigma_g^-$ is due to the formation of two overlapping transient anions:[63] the $^2\Pi_u$ and $^4\Sigma_u^-$ states of O_2^-. Only the $^2\Pi_u$ anion decays to the $^1\Delta_g$ and $^1\Sigma_g^+$ states due to spin conservation.

Electron resonances are not present at E_i = 5 eV in CO and the strongest part of the interaction leading to vibrational excitation is expected to arise mainly from interaction with the permanent dipole of CO and polarization (i.e., the first, third and fourth terms in Equation 1). These interactions are mainly effective to produce excitation of one vibrational quantum. In the specular direction, the interaction is restricted to the terms of Equation (2) and the Δv = 1 selection rule applies. As seen in the lower curve of the inset in Figure 11 only the v = 1 level is excited at E_i = 5 eV. At E_i = 2.5 eV, in the specular beam both resonant and dipolar scattering is possible and strong excitation of v = 1 is accompanied by excitation of a few overtones due to the formation of the $^2\Pi$ temporary state of CO$^-$ (upper inset, Figure 11).

Resonances can be best identified by the structures they produce in the energy dependence of the cross sections of their decay channels (i.e., in the excitation functions). The excitation

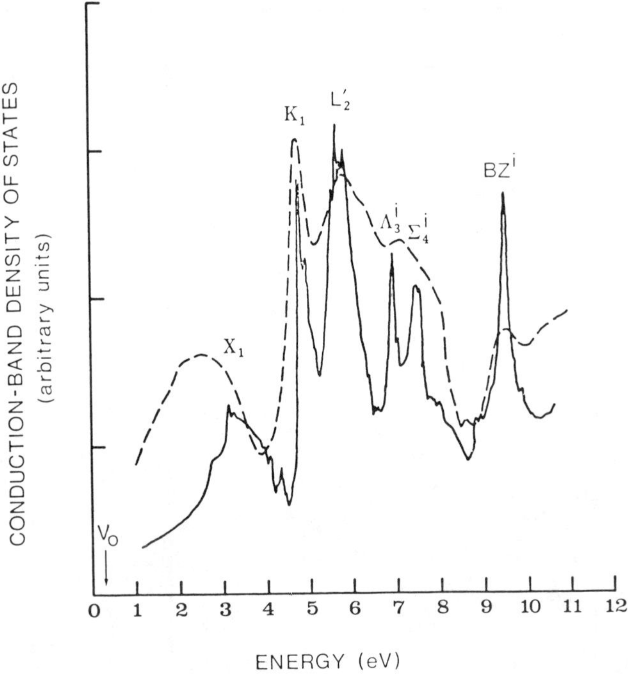

FIGURE 9. Electron conduction-band density of states of solid argon:[50] calculated (solid line) and determined from analysis of LEET data for solid argon recorded below the onset of the first excitonic excitation at about 20 K (dashed line). The zero of energy is that of the vacuum level. V_o is the energy of the bottom of the conduction band measured with respect to the vacuum level. It is equal to 0.25 eV for solid Ar.[144]

function of the v = 1 level of N_2 between 0 and 30 eV recorded with a 50-Å N_2 film is shown in Figure 13a. The four maxima can be ascribed to transient anion formation.[76] The symmetries of the 2, 12, and 19 eV anions are $^2\Pi_g$, $^2\Delta_g$ and $^2\Sigma_g{}^+$ but the configuration of the 8 eV state is still undetermined in both the gas and solid phase. The lowest energy feature in Figure 13a is displayed on an extended scale in Figure 13b (i.e., v = 1) along with the excitation functions of the v = 2 and v = 3 vibrational levels of condensed molecular nitrogen in the 0 to 5 eV region. The oscillatory structure, which modulates the broad peak in each excitation function, results from vibrational motion of the $^2\Pi_g$ transient N_2^- state.[76] The vertical bars in Figure 13b represent the energies of the first eight "vibrational peaks" of that state as recorded in the v = 1 decay channel in gaseous N_2.[85] The average spacing between the first six of these vibrational peaks is 0.27 eV as compared to the corresponding value of 0.29 eV in the solid.[76] From the shift of the first vibrational peak in the upper curve of Figure 13b in going from 1.93 eV in the gas to 1.23 eV in the film, we find a value of 0.7 eV for the relaxation energy. This lowering of the energy of the $^2\Pi_g$ anion is due to electronic polarization of the surrounding N_2 molecules by the temporarily localized charge.

If the lifetime of the anion was long compared to a vibrational period of a N_2 molecule, the oscillatory structure in Figure 13b would form nonoverlapping, well-defined peaks corresponding to vibrational levels of N_2^-. On the other hand, if the lifetime was much shorter than a vibrational period, no structure would be observed. In the intermediate case (i.e., lifetime of the order of the vibrational period) overlapping oscillatory structure is observed. Thus, the structure in Figure 13b does not truly represent vibrational levels but indicates that the lifetime of the resonance is of the order of the vibrational period of N_2

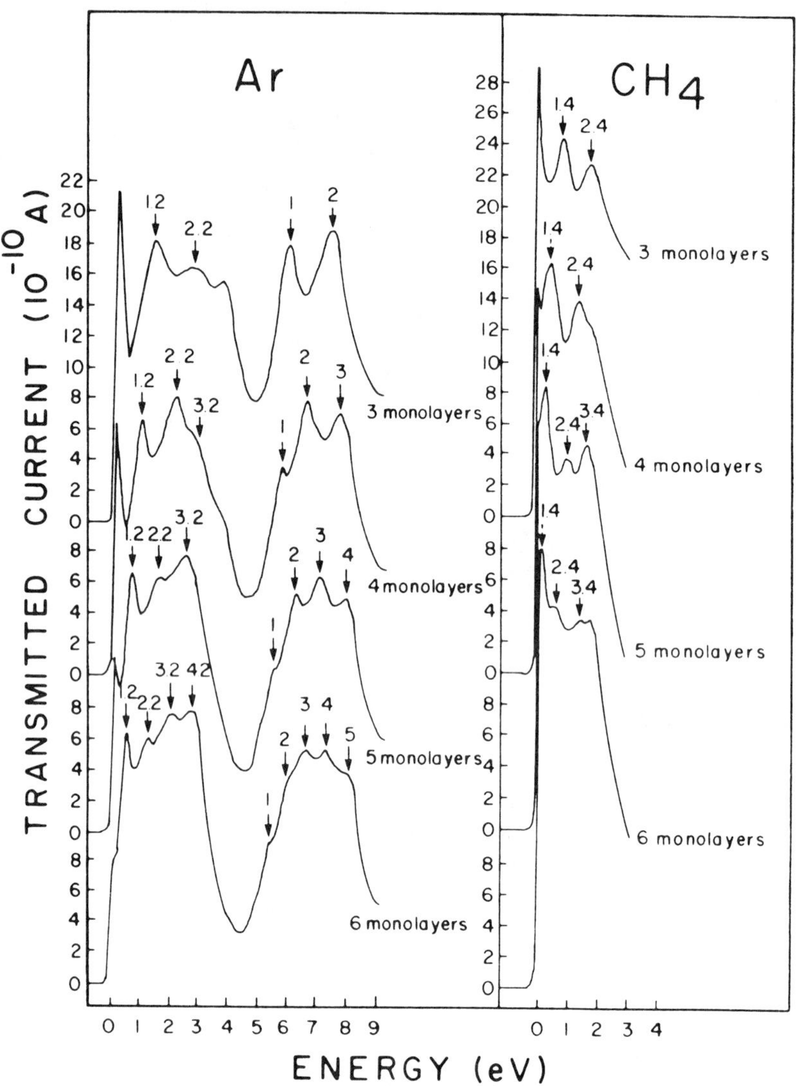

FIGURE 10.   Transmitted current in Ar and CH$_4$ as a function of incident electron energy for film thicknesses ranging from 3 to 6 ML. The arrows point to the interference structure caused by multiple reflection of the electron waves between the film-vacuum and film-substrate interfaces. The numbers refer to the assigned values of the interference order.

($\simeq 10^{-14}$ s). An estimate of the energy width (Γ) and lifetime of the resonance can be obtained by measuring the half width at half maximum on the low energy side of the first vibrational peak in each excitation function of Figure 13b. This measurement yields an approximate energy width (FWHM) of 620 meV which compares with the gas-phase value of 190 meV and corresponds to a lifetime of 3×10^{-15} s. A more precise value may be derived from a theoretical analysis of the vibrational amplitudes.[86] This result indicates that the combination of Coulomb, exchange, and centrifugal forces which forms a trapping mechanism for the additional electron[2] is only slightly perturbed by the neighboring molecules, whereas the long-range polarization part of the scattering potential is lowered by 0.7 eV by the surrounding media.

As a general rule resonances have a tendency to dominate the electron-impact vibrational excitation cross sections in molecular solids. However, for solids composed of molecules

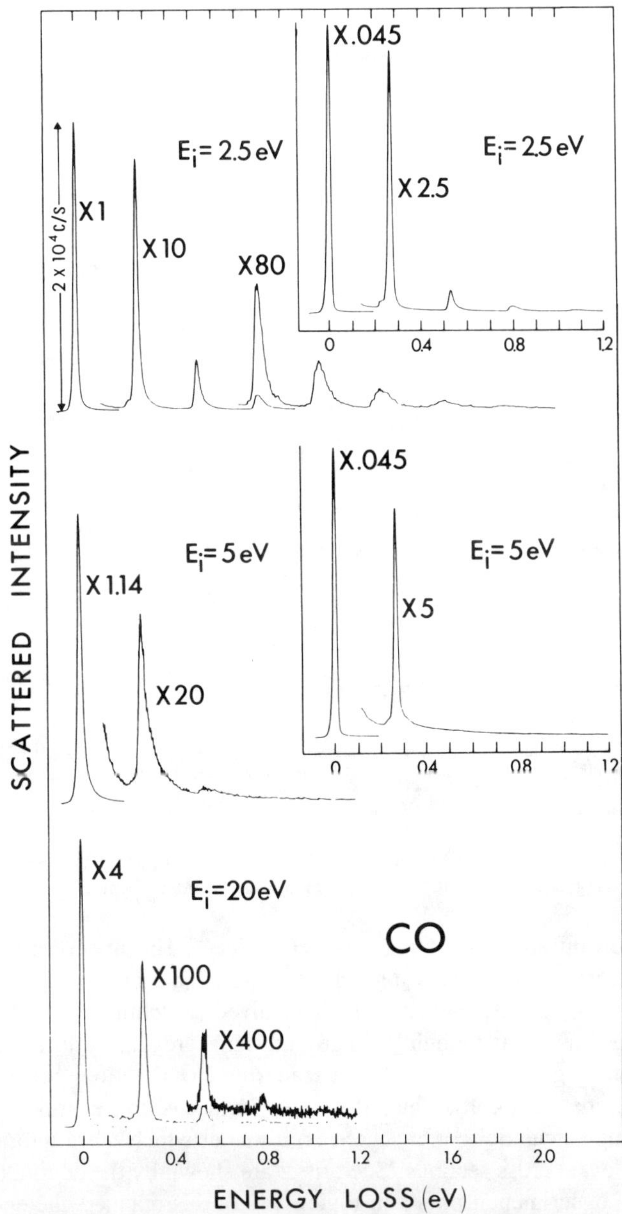

FIGURE 11. High resolution electron energy loss (HREEL) spectra for 2.5-, 5-, and 20-eV electrons impinging on a 50 Å CO film. The curves shown in the inserts were recorded in the specular direction with $\theta_i = \theta_r = 45°$. The others were recorded with a beam incident at $\theta_i = 14°$ from the surface normal and $\theta_r = 45°$.

having a strong dipole moment, nonresonant scattering may be strong enough to excite at all energies with considerable amplitude, the lowest-energy lattice and molecular vibrational modes. Recent results obtained with amorphous H_2O films provide a good example of the competition between resonance and dipolar scattering in producing inter- and intramolecular vibrational excitation.[27,28,78]

HREEL spectra[78] for a 30-ML film of amorphous ice recorded at $\theta_i = 14°$ and $\theta_r = 45°$ and for incident energies of 7.2 and 9.2 eV are shown in Figure 14. These spectra can

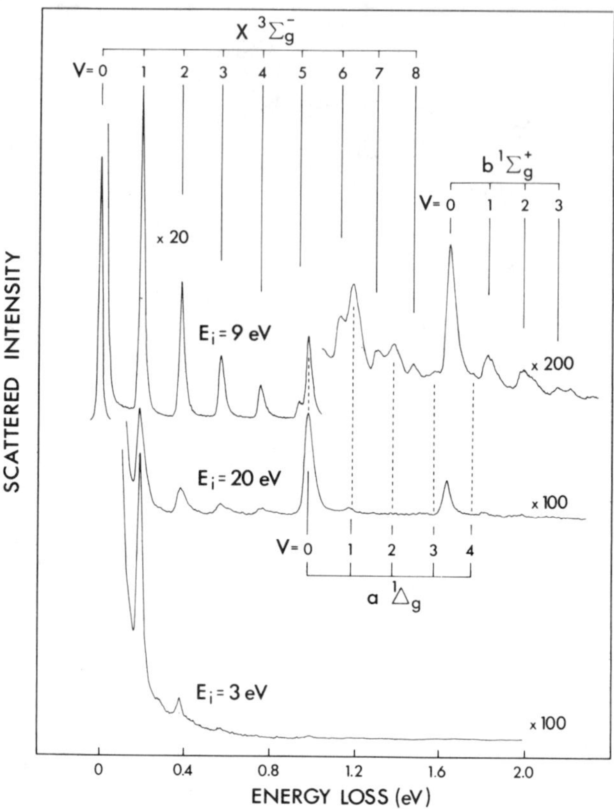

FIGURE 12.   HREEL spectra for electrons of 9-, 20-, and 3-eV primary energies incident at $\theta_i = 14°$ on a multilayer disorderd O_2 film.

be compared with infrared and Raman spectra of ice.[87] The major energy-loss peaks are identified as a translational mode (ν_T) at 24 meV, librational modes (ν_L) at 62 and 95 meV, the bending mode (ν_2) at 205 meV and the unresolved stretching modes ($\nu_{1,3}$) at 425 meV. The remaining features along with the background signal are mainly attributed to overlapping and multiple scattering.[78] In order to extract scattering cross sections per scatterer from such spectra, HREEL spectra recorded at different incident electron energies must be fitted to multiple scattering calculations[27,28,88] based on a model which takes into account the magnitudes of the relevant cross sections. Such fits[78] are shown by the continuous curves which pass through the experimental points in Figure 14. In this manner, the energy dependence of the cross sections for the major energy losses can be generated,[28] as shown in Figure 7.

The absolute electron scattering cross sections for exciting the translational ($\nu_T'\nu_T''$) and librational (ν_L' and ν_L'') modes of amorphous ice in the 2 to 18 eV range are shown in Figure 7a. The bottom curve labeled "others" essentially represents contributions to the inelastic cross section arising from electronic excitation and ionization.[28] Cross sections for exciting the bending (ν_2) and stretching (ν_1 and ν_3) modes are shown in Figure 7b. The latter two are found at two slightly different frequencies giving rise to the curve labeled $\nu_{1,3}$ which represents the cross section for an unresolved peak containing energy-loss contributions from both the symmetric and asymmetric stretching modes. The 2 to 18 eV behavior of the absolute cross section for exciting the strongest combination mode ($\nu_L + \nu_{1,3}$) and the most intense of the overtones [$2(\nu_{1,3})$] are shown at the bottom of Figure 7b. The ν_T'' and ν_L'' lattice modes are clearly characterized by a peak around 7 eV having $\simeq 2$ eV FWHM and a rise at low energy. The $\nu_{1,3}$ modes exhibit a broader ($\simeq 4$ eV FWHM) feature around 8 eV with a smaller rise at low energy.

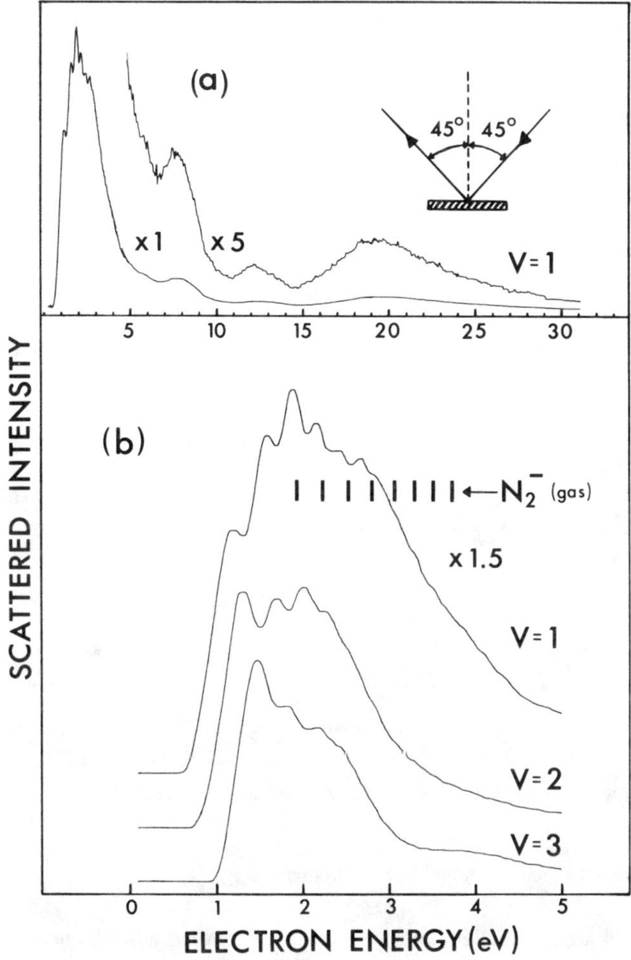

FIGURE 13. (a) Energy dependence of vibrational excitation of the first vibrational level of ground-state N_2 in $\simeq$ 50-Å multilayer molecular nitrogen film. The four maxima are interpreted as due to the formation of temporary N_2^- states. (b) Excitation functions for the first three vibrational levels of N_2 in the 0- to 5-eV region. The oscillatory structure is due to nuclear motion within the temporary N_2^- state.

In the curves of Figure 7, any increase toward lower energies has been ascribed to nonresonant scattering (i.e., electron-dipole and electron-polarization interaction) which is present throughout the spectra and possibly accounts for the constant portion in the magnitude of the cross sections.[28] Note that no constant portion exists in the cross section shown in the bottom curve of Figure 7b. In the absence of a large quadrupole moment, the magnitude of the cross section for overtones arises only via resonance scattering since the dipolar interaction term, which is dominant in Equation 1, can only produce $\Delta v = 1$ excitations.

From comparison of the energies and widths of the features of Figure 7 with those found in the gas-phase cross sections the strong enhancement of the amplitude of the $v_{1,3}$ mode near 8 eV can be ascribed to the formation of a $^2B_2$ H_2O^- state whereas the broad peak near 7 eV in the v_L'' cross section correlates to a $^2B_1$ anion state.[78] While the excitation of an intramolecular vibrational mode by a resonance process can be inferred from gas-phase mechanisms, the excitation of lattice vibrational modes by the longer lived $^2B_1$ state requires a new mechanism. Because of the overlapping of electron-molecule interaction potentials

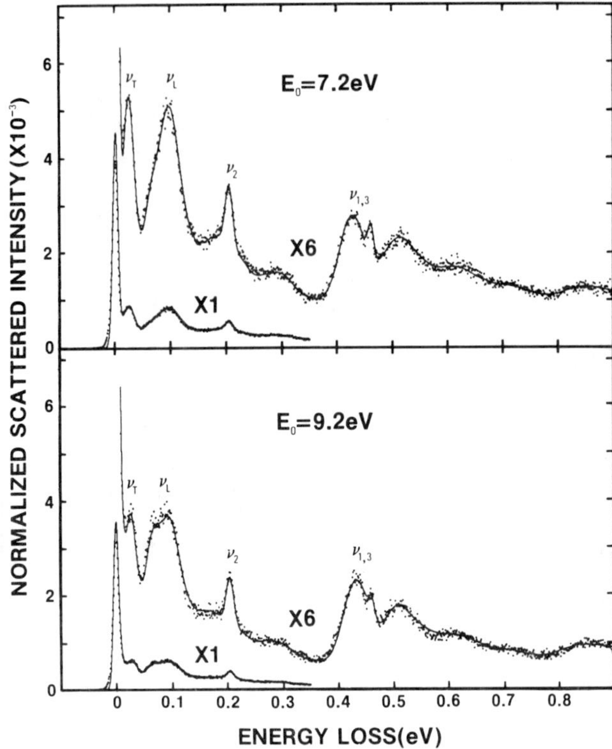

FIGURE 14.   HREEL spectra of a 30-ML ice film recorded at $\theta_i = 14°$ and $\theta_r = 45°$ (dotted curves). The superimposed continuous curves result from multiple-scattering calculations of the electron backscattered intensities. E_o represents the incident electron energy.

in the condensed phase, a transient anion state is also coupled with the nuclear and electronic polarization modes of the surrounding medium. When an electron is localized in the vicinity of an H_2O molecule a force is exerted between the H_2O^- state and the permanent dipoles of the surrounding molecules which pull these latter toward the anion initiating translational and rotational motion. Calculations with a classical harmonic oscillator mode[78] indicate that a resonance whose lifetime is greater than 10^{-14} s can transfer a significant amount of energy to librational motion by this mechanism, which is also effective in producing translational motion for resonance times of the order or larger than 10^{-13} s.

B. ELECTRONIC EXCITATION

Electronic excitation by low energy electrons in multilayer atomic and molecular films condensed on metallic substrates has been investigated by both LEET[20,41,44,52,53,89-94] and electron energy loss[29,62,83,93,95-97] spectroscopies. The inelastic features in LEET spectroscopy usually appear as broad maxima resulting from a convolution of inelastically scattered current distributions created by the much reduced energy of electrons having produced excitons and band-to-band transitions within the film[20,93] Excitation of a bound electron in a molecular film to an unoccupied state by an incoming electron scatters this latter to a lower state in the conduction band. Near the electronic excitation threshold, such energy loss electrons "fall" to the lowest "conduction" level V_o and their transmission probability becomes unity when the band edge (i.e., V_o) lies below the vacuum level.[20] As the electron energy is increased excitation of the same transition produces electrons of correspondingly increasing energy up to a point where they have nonzero probability to escape in the vacuum. From

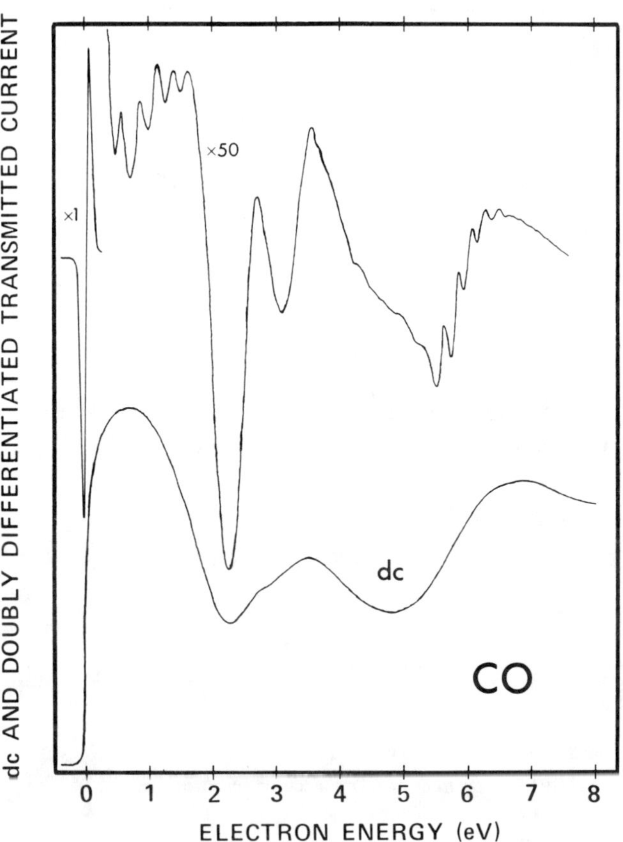

FIGURE 15. LEET spectrum (bottom curve) of CO on Pt together with its doubly differentiated (DD) LEET spectrum (top curve). The film thickness and temperature were 10 ML and 20 K, respectively.

then on transmission diminishes progressively. This process, first postulated by Hiraoka and Hamill[89] is referred to as the "V_o-mechanism".[93] It causes an increase in transmission followed by a tailing decrease as the electron energy is swept across the energy of the electronic transition. As shown in recent calculations,[94] this maximum in transmission occurs near the energy of the electronic transition causing the current increase. LEET spectra are therefore most sensitive to electronically excited states having strong excitation cross sections near threshold. This is shown in Figure 8 where transmission maxima above 9 eV are found at the energies of the lowest excitons indicated by vertical lines.[41] Inelastic features 5, 6, and 7 are not appreciably modified by ordering of the film compared to those lying at lower energy since their magnitude does not depend on the band structure of Kr in first order.

By recording singly differential[51,89,90] or doubly differential (DD)[20,91,93,98] LEET spectra, sharp features in transmission become more apparent. As shown in a recent model simulation of LEET spectra,[94] sharp features can arise from an abrupt change in the total cross section caused by a large increase in electronic energy losses near an excitation threshold or by the formation of a long-lived anion at a specific energy. The effect of the large threshold cross section of the $a^3\Pi$ state of CO on the DD-LEET spectrum of a 5-ML CO film[93] is shown in Figure 15. The DD transmitted current shows a series of sharp peaks between 5.5 and 6.5 eV corresponding to the vibronic levels of $a^3\Pi$ state of isolated CO.[99] All these levels are lowered by about 0.3 eV, compared to the gas-phase values, due to the additional KE the electron can transfer to electronic excitation as a consequence of the lowering of its potential energy in the solid (i.e., $V_o \simeq -0.3$ eV).

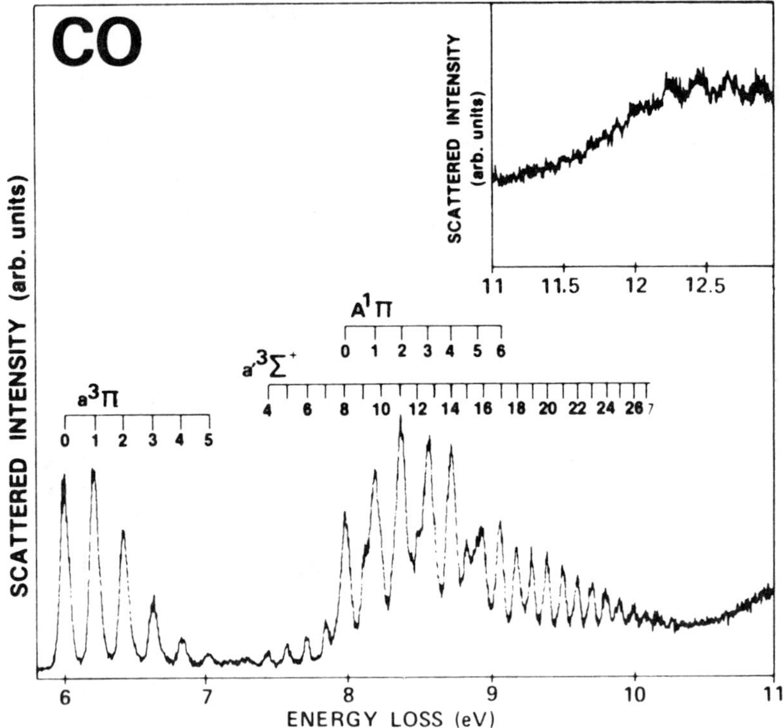

FIGURE 16.   HREEL spectrum of CO condensed on Pt in the range 5.9 to 13 eV. The incident energy is 19 eV; $\theta_i = \theta_r = 45°$.

Current theoretical models to explain the origin of strong threshold effects, such as those shown in Figure 15, bear on two mechanisms. One model[100] is based on a distortion of the scattered S wave, which would enhance the electron wave function in the neighborhood of a molecule because of the existence of a virtual bound state, whose capability to bind an electron depends on the internuclear distance. According to this model resonances can appear when vibrations are included. In the other model[101] the long-range dipole potential plays a crucial role in the enhancement of the wave function. The latter is greatly modified in the presence of a critical dipole potential, namely, a dipole just about to bind an electron. This model can predict a sharp peak at threshold which has a counterpart in the elastic channel. However, because of the *screening* of the long-range portion of the electron-molecule potential in molecular solids, it is unlikely that a mechanism which depends critically on a long-range interaction would be responsible for the strong threshold cross section in condensed CO. Hence, a virtual bound state mechanism must be favored to explain condensed phase data.

From LEET experiments in N_2, CO,[93] and solid rare gases films,[44,45] the threshold behavior of the electronic excitation cross sections appears to be similar in both solid and gas phases. In fact, comparison of low energy HREEL spectra indicate that, in general, valence states have about the same amplitude in these two phases.[29,93] An example is shown in Figure 16 which displays HREEL spectra[93] for 15 ML of CO on Pt recorded at 19 eV incident energy and $\theta_i = \theta_r = 45°$. The intensity of energy loss peaks are about three orders of magnitude smaller than the elastic one. Below 10.5 eV these spectra are similar to those recorded in the gas phase at the same primary energy and at about 90° from the primary beam.[102] Above 10.5 eV, in a region where an abundance of Rydberg states appears in the gas phase, only a few weak peaks are present.

The data in Figure 16 further indicate that the position of all the peaks is shifted down by a few tens of meV with respect to the gas-phase values.[99,102] The shift is more important (>25 meV) for high values of the vibrational quantum number. The energy resolution of the electron beam allows detection of broadening of the energy levels whose actual width can be easily estimated if we assume that the observed FWHM results from the convolution of two Gaussian line shapes. For well-resolved vibronic peaks in CO, the estimated FWHM is 30, 14, and 31 meV for the $a^3\Pi$, $a'^3\Sigma^+$, and $A^1\Pi$ states, respectively.[93] Peak broadening is expected to be due to multiple scattering, faster decay of the excitation, and dispersion. However, the absence of significant dependence of the position of the peaks on the angle of incidence,[93] along with the small broadening of the vibronic structure, indicates little dispersion of these excitons.

Sharp features in DD-LEET spectra due to long-lived, two-particle, one-hole resonances (i.e., transient anions which involve electronic excitation of the target) can be distinguished from those caused by threshold effects because they exhibit different line shapes and, particularly, because they occur at energies much lower than those of their parent neutral excited state.[20,98] In a two-particle, one-hole (i.e., core-excited) resonance the electron is captured essentially by the positive electron affinity of an excited electronic state of an atom or a molecule. Thus, the energy of such a temporary anion in condensed matter will be lowered from that of the neutral parent molecules, from which it is derived, by a value which depends on both the electronic polarization of the parent excited state and that of the solid. Core-excited resonances can, therefore, be identified by correlating their energies with that of the parent electronic excited state of the solid taking into account the relatively large energy shift introduced by the electron affinity of the solid plus that of the excited parent state. An example is provided by sharp features 2 to 9 in the DD-LEET spectrum of 1,3,5 hexatriene film[20] in Figure 17. The energies of these features can be correlated with those of valence excited states of the solid found at 1.6 ± 0.2 eV higher energy. Structures 2 and 3 are associated with electron trapping by the $1\ ^3B_u$ and $1\ ^3A_g$ electronically excited states of hexatriene,[103] whereas structures 4 to 7 correlate with vibronic levels[103] of electronic configuration $1\ ^1B_u$.

Although similar correlations could be found in the DD-LEET spectra of 14 hydrocarbon films having strongly polarizable π bonds,[20,98] the presence of these sharp features could not be detected by HREEL spectroscopy.[104] Sharp core-excited resonances have a small magnitude, usually only a few percent or less of the total electronic excitation cross section, and their sharpness makes them difficult to observe in experiments where spatial fluctuations of potential are present due to disorder. This may account for their visibility being restricted to high resolution DD-LEET experiments in which only sharp features are amplified.

Another explanation which has been advanced to explain these observations pertains to the fundamental mechanism responsible for the formation of core-excited resonances in molecular solids. In such a solid, the electronically excited state which binds the additional electron can move with finite wave vector, and so does the temporarily captured electron, if charge transfer to another molecule is favored instead of electronic deexcitation by electron emission. Furthermore, if the additional electron remains with the exciton until the electron-exciton complex reaches the metal substrate, the trapped electron cannot be reemitted to vacuum. In this case, electron-exciton complexes contribute exclusively to the LEET current. The existence of electron-exciton complexes in organic molecular solids was postulated theoretically[105] before the observation[98] of resonant LEET structures. Recent models[106-109] of the propagation of electron-exciton complexes in organic thin films now account for the small magnitude of these effects and the long lifetimes required for their detection at the film-metal interface.[109]

Fragmentary data, on exciton formation by electron impact on molecular solids, have also been obtained from experiments involving the detection of emitted metastable atoms

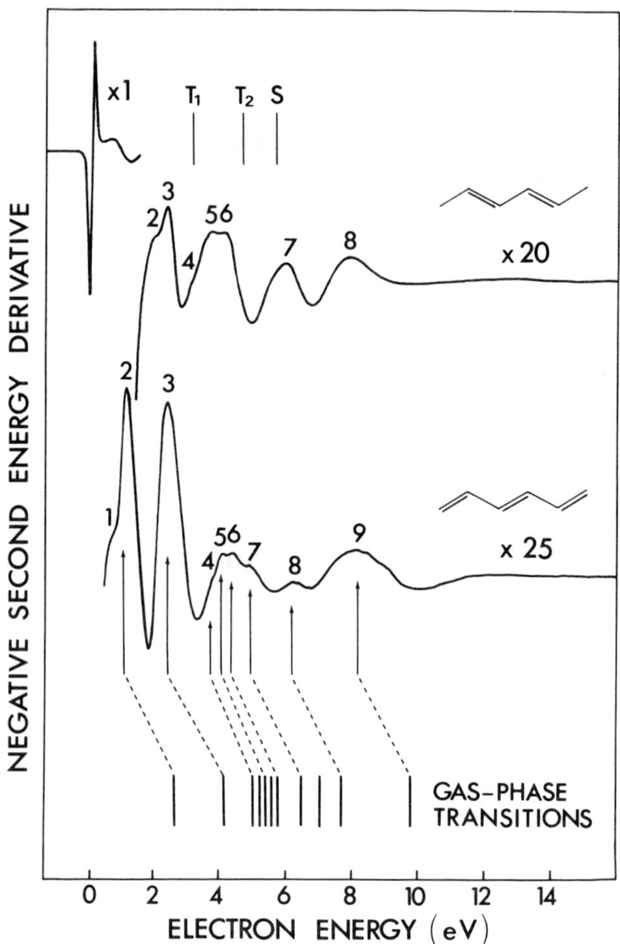

FIGURE 17.   The upper curve represents the DD-LEET spectrum of condensed trans-2-, trans-4-hexadiene. It is divided into two segments. The one near 0 eV is the injection peak. The remaining portion of the spectrum has its vertical amplitude scaled by a factor of 20. The vertical bars on top represent the energies of the known electronic transitions for this molecule. The lower curve is the DD-LEET spectrum of condensed 1,3,5-hexatriene. Structures 2 to 9 coincide with the gas phase electronic transitions of this molecule which are indicated by vertical bars at the bottom of the figure. This coincidence is achieved by increasing the energy scale of the DD-LEET spectrum by 1.6 eV. No such correlation is possible with the hexadiene data.

and molecules[26] and monitoring the emission of photons of particular wavelengths as a function of incident energy.[110] In these latter experiments, peak intensities in the luminescence excitation functions could be correlated to LEET features corresponding to exciton production near threshold.[110]

In conclusion, electronic excitation in molecular solids induced by slow electrons has been found to be similar to the gas-phase below the band gap. Above this energy, Rydberg transitions disappear as well as core-excited resonances associated with such states. Electronic transitions have larger widths and slightly different energies. As expected from the strength of the exchange interaction at low energies, spin forbidden states are prominent in this range. In solids composed of molecules having strongly polarizable π bonds, the formation of long-lived electron-exciton complexes must be postulated to explain DD-LEET and HREEL observations.

V. MOLECULAR DISSOCIATION

Above a certain energy threshold, electrons impinging on isolated or condensed molecules can cause disruption of internuclear bonds which leads to fragmentation. Depending on the intermediate state involved and charge exchange processes, these fragments can be either neutral, positively, or negatively charged. In molecular film experiments, dissociation products can be measured by the ESD techniques of the types described in Section II.C, when they possess sufficient KE to overcome the polarization forces they induce in the solid. Charged particles unable to leave the solid can be monitored by measuring the trapping cross section as described in Section II.D.

A. POSITIVE IONS AND NEUTRALS

Although most of the results on ESD of positive ions from molecular solids have been obtained at energies beyond the scope of the present article, part of the recorded yield functions often lie below 30 eV.[112-119] Some of these experiments, reviewed in this section, provide insight into the desorption mechanism.

Compared to other types of slow electron beam experiments the energy dependence of cation ESD signals (i.e., the yield functions) receding from surfaces partially or completely covered with molecules, does not exhibit appreciable structure.[24,111-117] This can be seen in Figure 18, where the H^+ ESD yield in the 20 to 50 eV range is displayed for 0.7-, 1.2-, and 16-ML H_2O coverage (θ) of a Ni (111) surface.[112] The threshold region is enlarged in the inset. Additional information on the ESD process can be obtained by analyzing the energy of the positive ions for different electron impact energies, as shown in Figure 19. This latter displays the KE distribution of H^+ ions receding from a 16-ML H_2O film deposited on Ni (111).[113] The KE analysis was performed at 25-, 30-, 35-, and 45-eV incident electron energies. From such analysis and calculations of the potential energy curves of the H_2O excited states involved in H^+ fragmentation from H_2O clusters, Noell et al.[113] suggested specific electronic states corresponding to possible desorption pathways. The dissociating states are, in general, not dipole allowed and involve two electron excitations. Near threshold, these states correspond to excited states of the neutral rather than those of the ionized configuration. So, dipolar dissociation (DD) is the effective mechanism in this limit (e.g., for a diatomic molecule AB, DD is represented by $e + AB \rightarrow AB^* + e \rightarrow A^+ + B^- + e$). Predissociation and shake-up involving excitations of the atomic oxygen orbital 2s contribute to the H^+ yield above 28 eV.[113]

The predominance of positive ion emission from neutral excited states is in good agreement with the results of Sambe et al.[118] who found that, contrary to the gas phase in which 0^+ formation by electron impact arises predominantly via O_2^+ intermediates, the 0^+ signal from single and multilayer films of 0_2 on Pt arises from DD. This difference can be explained by considering the role of the image force in enhancing the signal from neutral dissociating states in comparison to that arising from O_2^+ states.[118] The image charge in the metal substrate and/or in the dielectric film lowers the energy of the final O_2^+ state leading to $0 + 0^+$ products but does not significantly change that of an initial O_2^* state; the energy difference between the initial state and a given $0 + 0^+$ dissociation limit is therefore larger for neutral O_2^*. Hence, 0^+ ions from O_2^* states have more KE and can, therefore, more easily escape the image force of the solid. The absence of charge transfer and the amount of configuration space further contribute in promoting dissociation from neutrals.[118]

Cation ESD from hydrocarbons [e.g., C_6H_6, C_6H_{12} $(CH_3)_3$ CH; $(CH_3)_4$ C, $(CF_3)_3$ CH, $C_6H_2F_4$, C_4H_9F] and deuterated hydrocarbons was studied in detail by Kelber and Knotek,[115,116,119] who emphasized the role of multivalence hole states. They argued that such states resulting from the decay of a C 2s hole give rise to both H^+ and CH_3^+ desorption and are localized within methyl or other functional groups of organic solids containing only

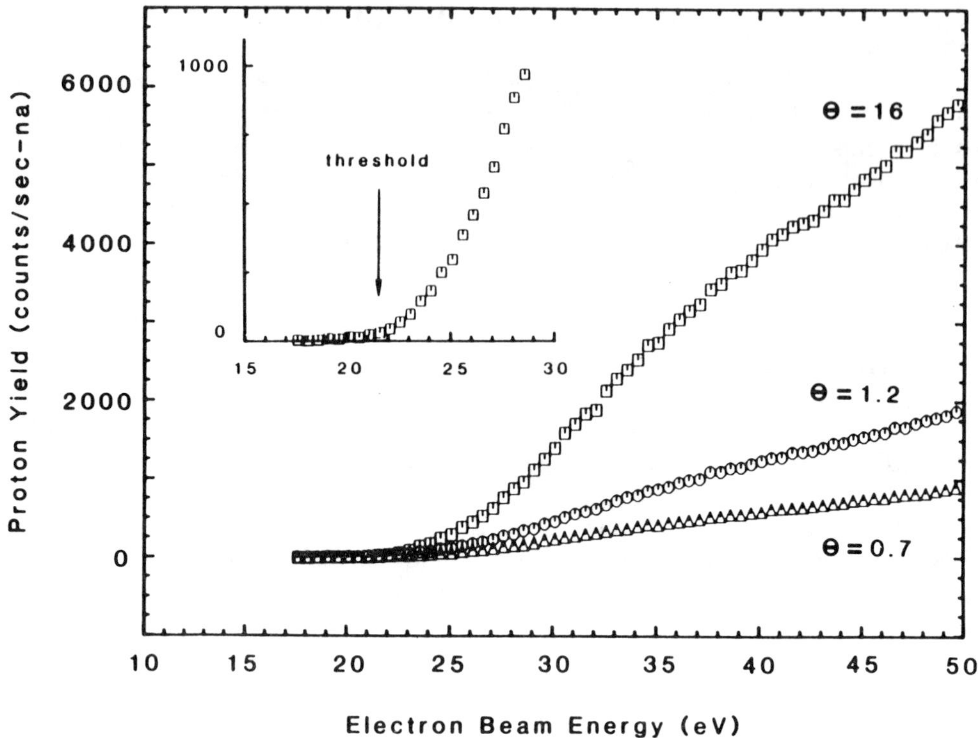

FIGURE 18. H^+ desorption yield as a function of incident electron beam energy for three coverages (θ in ML) of H_2O on Ni(111).[113] The inset is an enlargement of the threshold region for $\theta = 16$. Each yield curve is normalized to the incident electron beam current.

hydrogen and carbon.[116] With fluorinated hydrocarbons having separate H and F bonding sites, the F^+ and CF^+-threshold potentials were correlated with excitation from the F 2s level.[119] Further comparison between the appearance potentials of H^+ and F^+, in fluorinated hydrocarbons with common and separate H and F bonding sites, indicated that the final state excitations leading to cation desorption from these solids are localized within functional groups but not necessarily with individual C–H or C–F bonds of the functional group.[115]

ESD of neutrals arising from dissociative states has been recorded primarily from species chemisorbed on metal surfaces[37] and it is only recently that such measurements have been extended to multilayer molecular films;[36] namely, amorphous ice. The 1- to 30-eV yield function has been recorded for both molecular and atomic hydrogen. The H_2 signal onsets at 6.3 ± 0.3 eV and exhibits two broad peaks at 15 and 28 eV. The effective cross section for H_2 ESD from a 4-ML H_2O film was found to lie within 10^{-18} to 10^{-17} cm^2 in the incident energy range 10 to 30 eV.

B. NEGATIVE IONS

Anion emergence into vacuum from electron bombarded molecular solids[25,30,31,118,120-126] can proceed via DA and/or DD. This latter process produces a smooth continuous signal which, beyond threshold, increases monotonically with electron energy. The threshold energy for dissociation within the solid corresponds to the dissociation energy of the isolated positive and negative species screened by the polarization induced by the products in the solid. Below the energy threshold for DD, ESD of negative ions can only proceed via DA.[25] This process constitutes a particular channel for the decay of molecular transient anions (Figure 1) formed within molecular solids and near their surfaces. As explained in Figure 2, ESD of anions

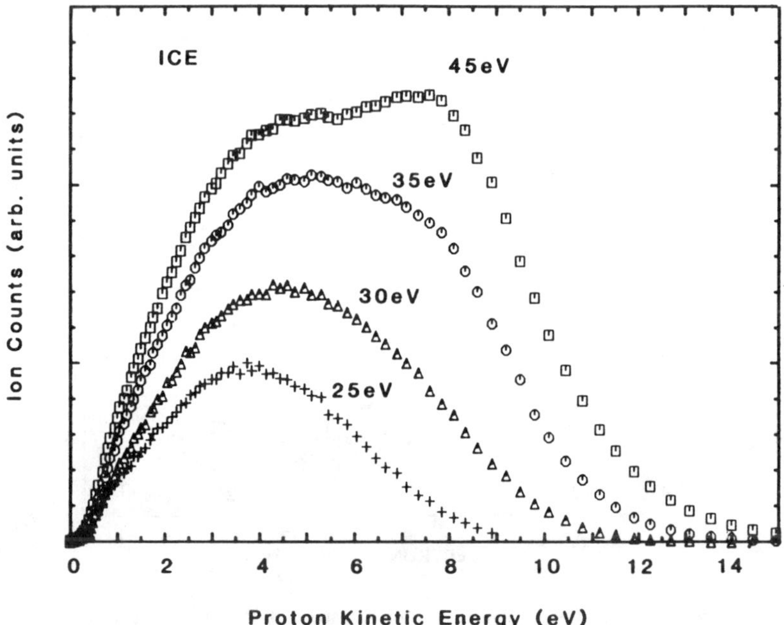

FIGURE 19.   The H^+ kinetic energy distribution for θ = 16 ML in Figure 18 at four separate electron beam energies.[113] The curves are not normalized to each other.

by DA arises from transient states which dissociate before autoionization. Since a given molecular configuration of a transient anion appears at a well-defined energy (Figure 2) each peak in the electron-energy dependence of the anion ESD yield identifies the energy of a particular resonant state.

Anion ESD by both DA and DD processes is exemplified in Figure 20 by the energy dependence of H^- yields recorded with multilayer films of condensed C_nH_{2n+2} (n = 1,2,4,5,6) molecules.[123] All curves exhibit single peak, whose maximum is located around 10 eV, and a rise which onsets at higher energy (16 to 18 eV). This latter is due to DD. The 10-eV peak is caused by the DA reaction

$$e^- + C_nH_{2n+2} \rightarrow (C_nH_{2n+2})^- \rightarrow H^- + C_nH_{2n+1}$$

Similar yield functions are observed for CH_3^-, CH_2^-, and CH^- ions, in order of decreasing intensity.[123] These anions also arise from the decay of transitory $(C_nH_{2n+2})^-$ states. DA features have also been observed in the O^-, C^-, Cl^-, and H^- ESD yield functions measured from molecular solids formed by condensing O_2, NO, CO, Cl_2, N_2O, H_2O, and hydrocarbon molecules (e.g., Figure 20) on a metal substrate.[30,31,118,120-124] More recently, the anion-complexes $Ar \cdot O^-$ and $Ar \cdot Cl^-$ were produced by ESD from Ar matrices containing N_2O and Cl_2, respectively.[125] O^-, $Ar \cdot O^-$, $Kr \cdot O^-$, O_3^-, (or $O_2 \cdot O^-$) and O_2^- were also produced by ESD from solid rare-gas matrices containing O_2.[125,126] The ESD yield functions from O_2 isolated on solid rare-gas surfaces have also been reported.[118,122]

Analysis of the electron energy dependence of these anion yields revealed that the gas-phase DA mechanism was perturbed in the condensed phase by: (1) multiple electron scattering prior to electron attachment to a molecule; (2) post-dissociation interaction (PDI) of the resulting anion with the solid; and (3) changes in the symmetry of the single-electron molecule potential introduced by the presence of neighboring targets. This latter change in symmetry led to the observation of $\Sigma^- \leftrightarrow \Sigma^+$ transitions which are forbidden by electron

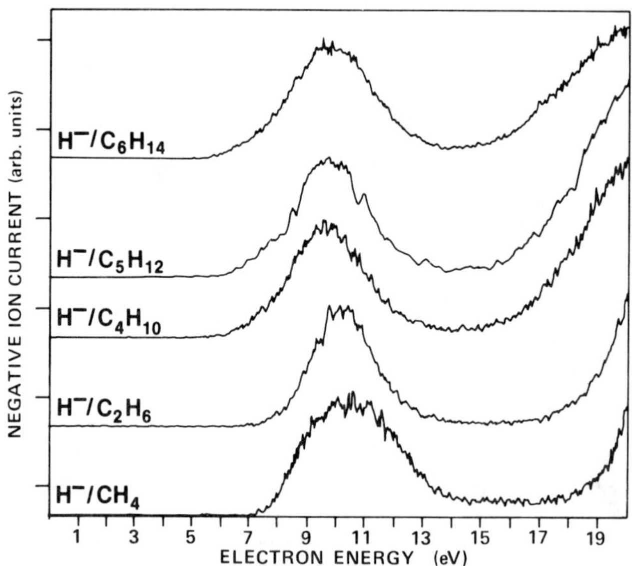

FIGURE 20.   Energy dependence of H^- signal desorbed by electrons impinging on a 6-ML film of C_nH_{2n+2} molecules deposited on a Pt substrate.

impact in the gas-phase.[127] In addition, coherent scattering within solids was found to enhance ESD yields.[122] By analyzing the energy distributions of the desorbing anions as a function of incident electron energy,[31,120,121] it has been possible to distinguish between the contribution to the total yield arising directly from DA at a molecular site and the other two portions of the signal which involve multiple scattering and PDI. It was concluded that the lattice was not involved in the dissociation dynamics and that, therefore, the energy and momentum conservation laws of the free molecule could be applied to solid state DA reactions.[31,121]

Since much of our present understanding of DA at or near the surface of molecular solids has been derived from experiments with condensed O_2, the results of these latter are discussed in the rest of this section to illustrate the mechanics of this resonant process in the presence of neighboring molecules.

In Figure 21, the electron energy dependences of the O^- ESD yield from a 3-ML O_2 film (a and b)[120] and a 20-ML Ar film containing 10 and 2% volume O_2 (c and d, respectively)[126] are compared with the O^- yields obtained by electron impact in gaseous O_2 (e).[128] For Kr and Xe matrices containing 10% O_2, the results are similar to those obtained in Figure 21 (c).[126] Figure 21 (b) was recorded with a retarding potential $V_R = 1.5$ V applied to the grids at the entrance of the mass spectrometer.[120]

The results of Figure 21 can be explained by considering the potential energy curves of the O_2^- system lying below 15 eV.[129-131] Although a multitude of molecular O_2^- states can be formed from the interaction of 0 and O^- fragments [e.g., 24 from the ground state $O(^3P)$ and $O^-(^2P)$,[129] only a few of these possess the proper characteristics to produce substantial decay into the DA channels. Briefly, the state involved must have a multiplicity of 2 or 4, be repulsive in the Frank-Condon region, possess a lifetime of the order or longer than the O_2 vibrational period and be accessible in the energy range of interest. The multiplicity condition is dictated by spin conservation and the long lifetime is necessary to avoid electron detachment or autoionization before electron stabilization on one of the separating oxygen atoms.

Molecular orbital analysis of the O_2^- states joining the two lowest separated atom limits [i.e., $O(^3P) + O^-(^2P)$ and $O(^1D) + O^-(^2P)$] indicate that below 15 eV only the three states,

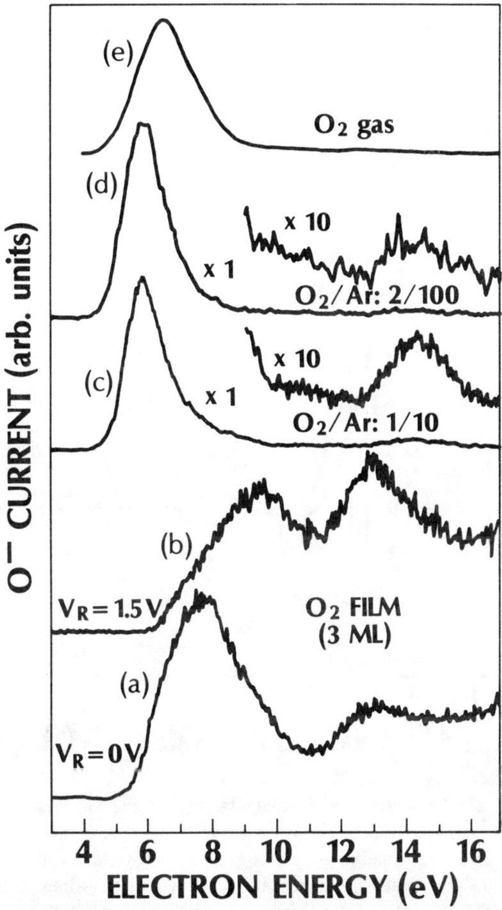

FIGURE 21. Incident electron energy dependence of O^- stimulated desorption yields from (a) and (b) 3-ML O_2, films; (c) and (d) 20-ML Ar films containing 10 and 2% volume O_2, respectively. Curve (e) represents the gas-phase yields.[128] V_R is a potential retarding the desorbing anions.

shown in the potential energy diagram of Figure 22, have the proper characteristics to produce O^- via DA with a considerable cross section.[130] The dissociation limits of the molecular states of O_2^- are known from Wigner-Witmer rules.[131] According to these latter, the lowest $^2\Sigma_g^+$ state, shown in Figure 22, correlates with the $O(^1D) + O^-(^2P)$ dissociation limit. The correlation of that state of the $O(^3P) + O^-(^2P)$ limit shown in Figure 22 is the result of curve crossing with another O_2^- state of the same symmetry.[129] Experimental evidence from ESD measurements from pure O_2 multilayer films[120] and in CO and N_2 matrices containing O_2[121] corroborate the theoretical analysis, indicating the existence of three O_2^- states located around 6, 9, and 13 eV, which contribute to the O^- signal in the 4 to 16 eV range. Analysis of ion energies at various electron energies within this range[120] indicates that the $^2\Sigma_g^+$ states dissociate predominantly into the lowest limit whereas the upper state dissociates predominantly into the second dissociation limit.

Out of the states shown in Figure 22, the Σ^+ configurations cannot be formed by electron scattering from a single O_2 molecule in the $^3\Sigma_g^-$ ground state,[127] since in the single-target-electron frame of reference, the one-electron wave function must be σ^+. This selection rule excludes $\Sigma^- \leftrightarrow \Sigma^+$ transitions. So, it turns out that in gaseous experiments only the $^2\Pi_u$ state produces O^- with considerable amplitude. This can be seen from curve (e) in Figure 21 which exhibits only one broad peak with a maximum at 6.5 eV arising from the reaction

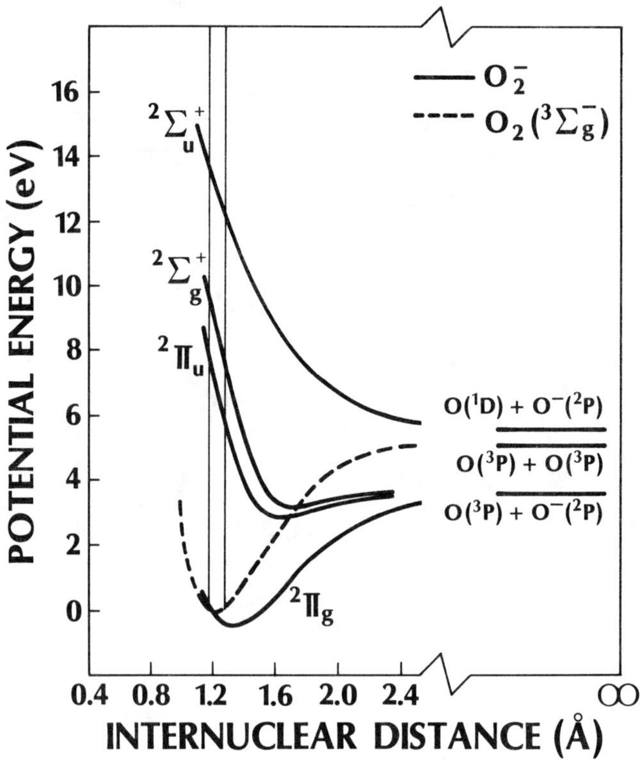

FIGURE 22.   Potential energy diagram[129,130] of the O_2^- system (solid curves) and ground state O_2 (dashed curve). The states shown are only those which are expected to produce significant O^- yields via dissociative attachment. The curves are correlated to the dissociation limit giving the most abundant signal as determined from energy analysis of desorbing O^- ions.[120]

$$e + O_2(^3\Sigma_g^-) \rightarrow O_2^-(^2\Pi_u) \rightarrow O^-(^2P) + O(^3P)$$

The same reaction dominates ESD from O_2-doped rare gas matrices (e.g., curves c and d in Figure 21) where, as in the gas phase, the yield function consists essentially of a single peak having a maximum at a slightly lower energy (i.e., 6.2 eV). This is expected since in matrices electrons attach principally to O_2 molecules which are isolated from each other by rare gas atoms. However, as the concentration of O_2 molecules increases from 2 (curve d) to 100% (curve a) another prominent structure appears at 13 eV in the yield function and the 6.2 eV peak distorts, broadens significantly and shifts to 7.6 eV. These modifications are amplified when only higher energy ions (i.e., >1.5 eV, curve b) are allowed to reach the mass spectrometer. In both curves a and b the appearance of the new features is attributable to breakdown of the $\Sigma^- \leftrightarrow \Sigma^+$ selection rule, due to the change in electron-target symmetry in condensed O_2. The peaks at 9.5 and 13 eV are therefore ascribed, respectively, to the $^2\Sigma_g^+$ and $^2\Sigma_u^+$ states shown in Figure 22. Kinetic energy analysis of the ions contained in the weak 14.5 eV peak of Figure 21c indicates that the majority arise from multiple electron scatterings prior to formation of the $^2\Pi_u$ O_2^- symmetry-allowed state.[126]

Thus, the absence of the 9.5 and 13 eV peaks in curves c and d indicates that, contrary to the O_2–O_2 interaction which perturbs the single-target symmetry, the presence of Ar atoms near O_2 molecules has essentially no perturbative effect. Nevertheless, this does not exclude the possibility that a much smaller portion of the anion signal desorbing from the Ar matrix

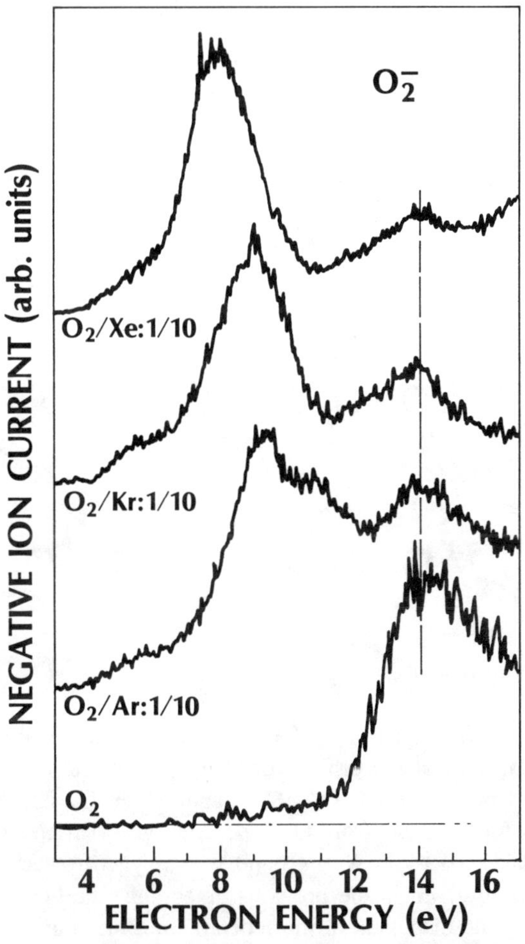

FIGURE 23. Electron energy dependence of O$_2^-$ yields desorbing from 5-ML O$_2$, 20-ML 10% O$_2$ Ar, 20-ML 10% O$_2$ Kr, and 20-ML 10% O$_2$ Xe films, respectively.

could arise from sites where two or more O$_2$ molecules are in close proximity. Such a signal has been shown to arise from isolated O$_2$ dimers at low O$_2$ concentration.[126]

In conclusion, the type of atoms or molecules surrounding a transient anion has been found to have a drastic effect on the symmetries of the states involved in the DA process. Yield functions of oxygen anions other than O$^-$ desorbing from O$_2$-doped rare-gas matrices are also different from those of the pure O$_2$ films[126] as shown for the case of O$_2^-$ ESD in Figure 23. In this figure, the O$_2^-$ yield function of a pure 5-ML O$_2$ film is compared with those of 20-ML 10% O$_2$ Ar, Kr, and Xe matrix films. Similar results have been obtained for the O $\cdot$ O$_2^-$ and/or O$_3^-$, Ar $\cdot$ O$^-$ and Kr $\cdot$ O$^-$ ESD signals arising from matrix films.[126] With the exception of the O$^-$ signal, all ESD yield functions from dilute O$_2$-doped matrices can be explained by postulating that electrons interact with O$_2$ dimers according to scheme proposed in Figure 24. The electron first attaches to an O$_2$ molecule within a dimer. Following electron capture, the O$_2 \cdot$ O$_2^-$ unit can react electrostatically with a matrix atom M to form the state M $\cdot$ O$_2 \cdot$ O$_2^-$ which then dissociates into M $\cdot$ O$^-$ + O $\cdot$ O$_2$ or M $\cdot$ O + O$_2 \cdot$ O$^-$ fragments. If the anion dissociates only within the dimer (i.e., within O$_2 \cdot$ O$_2^-$) the receding O$^-$ ion can remain electrostatically bonded to O$_2$ producing ESD of the anion complex O$_2 \cdot$ O$^-$. The O$_2 \cdot$ O$_2^-$ transient state can also dissociate into the limit O$^-$ + O $\cdot$ O$_2$ giving

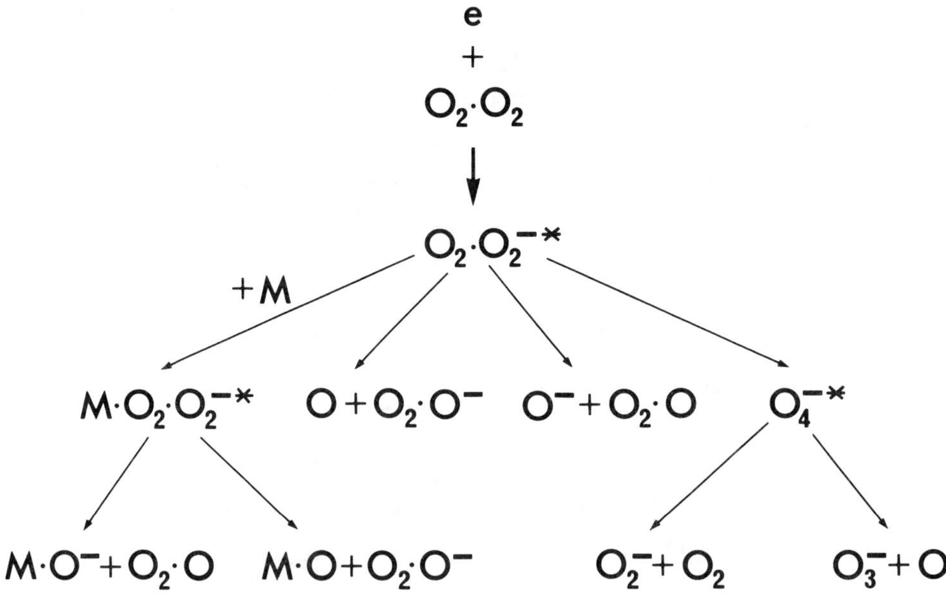

FIGURE 24.   Possible decay channels of an O_2^- transient ion within the dimeric configuration. M represents an atom or molecule of the matrix containing O_2.

rise to a relatively small O^- signal from forbidden Σ^+ states. Finally, O_3^- and O_2^- yields can arise from hybridization of the orbitals of $O_2 \cdot O_2^{-*}$ into an electronically excited O_4^- state which dissociates into the limits $O_2^- + O_2$ and $O_3^- + O$. The higher energy portion of the yield functions for O_2^- ESD from O_2-doped rare-gas matrices and pure O_2 films can also arise from PDI of $O^-(^2P)$ ions with ground state oxygen molecules.[126]

With the information given in the previous paragraph, the behavior of the O_2^- signal shown in Figure 23 can be discussed in more detail. In dilute matrices isolated dimers are present and the reaction pathway (Figure 24)

$$e + O_2 \cdot O_2 \rightarrow O_2 \cdot O_2^- \rightarrow O_4^- \rightarrow O_2^- + O_2$$

is possible. It follows from momentum and energy conservation that starting around 4-eV incident electron energy O_2^- can be produced by this reaction with sufficient KE to overcome the polarization potential of the solid.[126] Isolated O_2 dimers are absent from pure O_2 films, a situation which leads to the inhibition of the electronic rearrangement $O_2 \cdot O_2^- \rightarrow O_4^-$. Thus, in this case the O_2^- signal onsets around 9 eV, an energy which allows O_2^- propulsion into vacuum via PDI of O^- produced from dissociating O_2^- states (i.e., $O^- + O_2 \rightarrow O_3^- \rightarrow O_2^- + O$). In pure solid O_2, only the $^2\Sigma_u^+$ state near 14 eV is efficient in producing O_2^-, whereas in the matrices initial formation of all O_2^- states (i.e., $^2\Pi_g$, $^2\Sigma_g^+$ and $^2\Sigma_u^+$) contributes to the O_2^- signal.

When an anion is produced by DA, it does not necessarily escape in vacuum, but can remain trapped in the solid or at its surface. Surface charging by DA for O_2 molecules deposited at the surface of a 20-ML Kr film has been measured with the method described in Section II.D. The surface charging coefficient A_s, for such a film, is represented by the middle curve in Figure 25, as a function of the energy of the electron beam causing the charge.[38] This result (Figure 25c) is comparable to the energy dependence of the anion yields derived from the gas-phase attachment rate coefficient for stable O_2^- production (Figure 25a)[132] and the gas-phase DA cross section (Figure 25b).[2] No signal has been reported

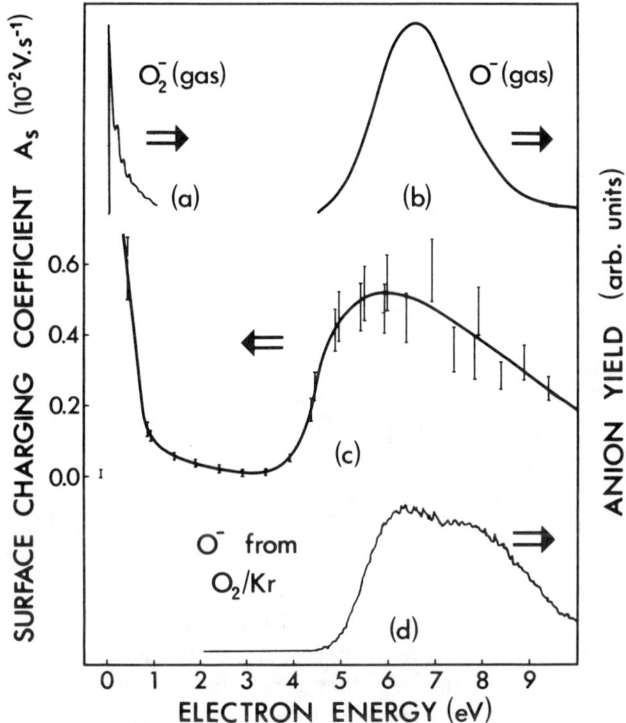

FIGURE 25. Anion yields produced by 0 to 10 eV electron impact on gaseous O_2 (a and b) and on a 20 ML Kr film covered with 0.1 ML O_2 (d). The electron energy dependence of the surface charging coefficient A_s for 0.1 ML O_2 on a 20-ML Kr film is shown in (c).

between 1.2 and 4.5 eV in the gas-phase.[2] Curve (d) represents the ESD signal from a Kr film covered with 0.1 ML of O_2.[122] This coverage corresponds to the O_2 coverage of the charging experiment from which curve (c) was obtained.[38]

From the correspondence between curves (b), (c), and (d) charge trapping at the surface of the Kr film has been ascribed to dissociation of the $^2\Pi_u$ state of O_2^- into the limit $O(^3P)$ + $O^-(^2P)$.[38] According to the potential energy curves of the O_2^- system shown in Figure 22, desorbing $O^-(^2P)$ ions could arise from either the $^2\Pi_u$ and the $^3\Sigma_g^+$ state of O_2^-. But, only a small fraction of the signal probably arises from the Σ state since this latter must be formed near another O_2 molecule in order to break the $\Sigma^- \leftrightarrow \Sigma^+$ selection rule. The maximum in the charging coefficient A_s is found at about 0.6 eV below the maximum in the gas-phase O^- yield due to lowering of the $^2\Pi_u$ state by the polarization potential of the Kr surface. The DA maximum in the ESD yield arises from the same O_2^- surface state as that in the charging signal. However, the DA maximum lies at higher energy, since in order to overcome the polarization force of the Kr film, desorbing O^- ions must arise on the average, from a higher energy portion in the FC region of the O_2^- $^2\Pi_u$ state.

In the 0 to 1.2 eV range, temporary electron attachment to gaseous O_2 leads to the formation of the $^2\Pi_g$ resonance state of O_2^- whose potential energy curve is shown in Figure 22. This occurs in vibrational levels v $\geqslant$4, since the v <4 levels lie below ground state O_2. However, when during the lifetime of the $^2\Pi_g$ anion vibrational energy is transferred to another molecule in a third body collision, the v<4 levels can be reached and the electron becomes permanently attached to the molecule[2] (process 5 in Figure 1). Vibrational energy transfer may also occur when O_2^- ($^2\Pi_g$) ions are formed within van der Waals clusters which exist in high pressure gases[133,134] and in supersonic molecular beams.[18] The comparison in

Figure 25 suggests that a similar stabilization process is effective at the Kr surface and responsible for surface charge accumulation in the range 0 to 2 eV. The electron can either lose energy to vibrationally excite O_2 via the $^2\Pi_g$ O_2^- state and afterwards stabilize at a Kr trapping site or the $^2\Pi_g$ state may itself stabilize by energy transfer to phonons. The estimated lifetime of the $^2\Pi_g$ O_2^- state within clusters is of the order of 10^{-12} s[135] which is comparable to the phonon vibrational period of the Kr lattice $(5.10^{-13}$ s).[135] Hence, the additional electron resides a sufficiently long time at an O_2 site to polarize the phonon modes of the Kr lattice and thereby transfer energy to lattice vibrations.

VI. CONCLUDING REMARKS

The mechanisms responsible for the elastic and inelastic interactions of electrons with molecular solids have been reviewed in this chapter along with those processes which lead to the formation of positive and negative ions near the surface of these solids. These mechanisms have been described in terms of resonant and nonresonant scattering of the electron waves. This division, first introduced to describe electron scattering with gaseous atoms and molecules,[2] can also be extended to other fields of research concerned with the interaction of electrons with matter.

In fact, resonant electron scattering has also been observed from isolated molecules physisorbed[22,30,76,83,84,118,136] and chemisorbed[137,138] on single-crystal and polycrystalline metal surfaces. Even though transient anions have their symmetries and lifetimes greatly perturbed by metal surfaces,[139,140] in certain cases, they persist for sufficiently long times to enhance vibrational excitation[22,76,83,84,136,138] and produce anions via DA.[30,118,122] Furthermore, when the probed molecule is well oriented on a single-crystal surface, the angular distribution of vibrational[141] and electronic[142] electron energy loss cross sections can provide information on the surface orientation of the molecule.[136,138] Description of electron scattering from molecules physisorbed or chemisorbed on metal surfaces in terms of dipoles, "impact" and resonance scattering is to be found in a recent review article by Gadzuk.[143]

ACKNOWLEDGMENTS

The author is indebted to Miss Johanne Provencher and Mr. Luc Parenteau for their valuable assistance in the preparation of this article. Thanks are also extended to Mr. Marc Michaud, Dr. Paul Rowntree and Dr. C. Ferradini for suggestions and helpful comments. This work was sponsored by the Medical Research Council of Canada.

REFERENCES

1. **Burke, P. G.,** *Potential Scattering in Atomic Physics,* Plenum Press, New York, 1977; **Brown, S. C.,** *Electron-Molecule Scattering,* John Wiley & Sons, New York, 1979.
2. **Schulz, G. J.,** Resonances in electron impact on atoms, *Rev. Mod. Phys.,* 45, 378, 1973; Resonances in electron impact on diatomic molecules, *Rev. Mod. Phys.,* 423, 1973.
3. **Allan, M.,** Studies of triplet states and short-lived negative ions by means of electron impact spectroscopy, *J. Electr. Spectr. Rel. Phenom.,* 48, 219, 1989.
4. **Shimamura, I. and Takayanagi, K.,** *Electron-Molecule Collisions,* Plenum Press, New York, 1984.
5. **Hinze, J.,** *Electron-Atom and Electron-Molecule Collisions,* Plenum Press, New York, 1983.
6. **Christophorou, L. G.,** *Electron-Molecule Interactions and their Applications,* Vol. 1 and 2, Academic Press, Orlando, 1984; **Christophorou, L. G.,** The lifetime of metastable negative ions, *Adv. Electr. Electron Phys.* 46, 55, 1978.

7. **Mott, N. F. and Massey, H. S. W.,** *The Theory of Atomic Collisions,* Clarendon, Oxford, 1965; **Inokuti, M., Itikawa, Y., and Turner, J. E.,** Inelastic collisions of fast charged-particles with atoms and molecules: the Bethe theory revisited, *Rev. Mod. Phys.,* 50, 23, 1978; **Inokuti, M.,** Inelastic collisions of fast charged-particles with atoms and molecules: the Bethe theory revisited, *Rev. Mod. Phys.,* 43, 297, 1971; **Massey, H. S. W.,** *Negative Ions,* University Press, London, 1976.

8. **Sanche, L. and Schulz, G. J.,** Electron transmission spectroscopy: resonances in triatomic molecules and hydrocarbons, *J. Chem. Phys.,* 58, 479, 1973.

9. **Takayanagi, K.,** Scattering of slow electrons by molecules, *Prog. Theor. Phys. Suppl. Jpn.,* 40, 216, 1967.

10. **Itikawa, Y.,** Electron-impact vibrational excitation of H_2O, *J. Phys. Soc. Jpn.,* 36, 1127, 1974.

11. **Fano, U. and Stephens, J. A.,** Slow electrons in condensed matter, *Phys. Rev. B,* 34, 438, 1986; **Fano, U.,** Short-range and long-range interactions of slow-electrons in condensed matter: effects on reflection and transmission, *Phys. Rev. A,* 36, 1929, 1987.

12. **Fano, U.,** Studies of slow electron action on condensed media, *Radiat. Phys. Chem.,* 32, 95, 1988.

13. **Fröhlich, H.,** Electric breakdown in ionic crystals, *Proc. R. Soc. Ser. A,* 160, 230, 1937.

14. **Fröhlich, H. and Platzman, R. L.,** Energy loss of moving electrons to dipolar relaxation, *Phys. Rev.,* 92, 1152, 1953.

15. **Magee, J. L. and Helman, W. P.,** Energy loss of electrons in random motion, *J. Chem. Phys.,* 66, 310, 1977.

16. **Christophorou, L. G.,** Gas/liquid transition: interphase physics, in *The Liquid State and Its Electrical Properties,* Plenum Press, New York, 1988.

17. For a review of surface analysis by high resolution electron energy loss spectroscopy see **Ibach, H. and Mills, D. L.,** *Electron Energy Loss Spectroscopy and Surface Vibrations,* Academic Press, New York, 1982.

18. **Demuth, J. E. and Avouris, P.,** Surface spectroscopy, *Phys. Today,* 36, 62, 1983.

19. **Märk, T. D.,** Cluster ions: production, detection and stability, *Int. J. Mass Spectrom. Ion Proc.,* 79, 1, 1987.

20. **Sanche, L.,** Transmission of 0—15 eV monoenergetic electrons through thin-film molecular solids, *J. Chem. Phys.,* 71, 4860, 1979; **Sanche, L., Bader, G., and Caron, L.,** Transmission of 0—15 eV monoenergetic electrons through aliphatic and alicyclic hydrocarbon films, *J. Chem. Phys.,* 76, 4016, 1982.

21. **Marsolais, R. M., Deschênes, M., and Sanche, L.,** Low energy electron transmission method for measuring charge trapping in dielectric films, *Rev. Sci. Instr.,* 60, 2724, 1989.

22. **Sanche, L. and Michaud, M.,** Interaction of low-energy electrons (1—30 eV) with condensed molecules. II. Vibrational-librational excitation and shape resonances in thin N_2 and CO films, *Phys. Rev. B,* 30, 6078, 1984.

23. **Cloutier, P. and Sanche, L.,** A trochoidal spectrometer for the analysis of low-energy inelastically back-scattered electrons, *Rev. Sci. Instrum.,* 60, 1054, 1989.

24. For a review of the data on ESD of cations and the techniques used for measuring positive ion formation by electron impact on molecules or atoms adsorbed or chemisorbed on surface see **Tolk, N. H., Traum, M. M., Tully, J. C., and Madey, T. E.,** *Desorption Induced by Electronic Transitions,* Springer Ser. in Chem. Phys. Vol. 24, Springer-Verlag, New York, 1983.

25. For a review of anions formation by electron impact on condensed molecular solids see **Sanche, L.,** Dissociative attachment in ESD from condensed molecules, in *Desorption Induced by Electronic Transitions DIET III,* Springer Ser. in Surface Sci. Vol. 13, Stulen, R. H. and Knotek, M. L., Eds, Springer-Verlag, Berlin, 1988, 78.

26. **Leclerc, G., Bass, A. D., Michaud, M., and Sanche, L.,** Angle-resolved electron stimulated desorption of metastable atoms from solid Argon, *J. Electr. Spectr. Rel. Phenom.,* 52, 725, 1990; **Alvey, M. D., Dresser, M. J., and Yates, J. T.,** Metastable angular distributions from electron-stimulated desorption, *Phys. Rev. Lett.,* 56, 367, 1986.

27. **Michaud, M. and Sanche, L.,** Total cross sections for slow-electron (1—20 eV) scattering in solid H_2O, *Phys. Rev. A,* 36, 4672, 1987.

28. **Michaud, M. and Sanche, L.,** Absolute vibrational excitation cross sections for slow-electron (1—18 eV) scattering in solid H_2O, *Phys. Rev. A,* 36, 4684, 1987.

29. **Sanche, L. and Michaud, M.,** Electron energy-loss vibronic spectroscopy of matrix-isolated benzene and multilayer benzene films, *Chem. Phys. Lett.,* 80, 184, 1981.

30. **Sanche, L.,** Dissociative attachment in electron scattering from condensed O_2 and CO, *Phys. Rev. Lett.,* 53, 1638, 1984.

31. **Azria, R., Parenteau, L., and Sanche, L.,** Dynamics of dissociative attachment reactions in electron stimulated desorption: Cl^- from condensed Cl_2, *J. Chem. Phys.,* 87, 2292, 1987.

32. **Leclerc, G., Cui, Z., and Sanche, L.,** Effective dissociation cross section for the low-energy (0.5—31 eV) electron impact on solid n-hexane thin films, *J. Phys. Chem.,* 91, 6461, 1987.

33. **Stamatovic, A. and Schulz, G. J.**, Characteristics of the trochoidal electron monochromator, *Rev. Sci. Instrum.*, 41, 423, 1970.

34. **Perluzzo, G., Sanche, L., Gaubert, C., and Baudoing, R.**, Thickness-dependent interference structure in the 0—15 eV electron transmission spectra of rare-gas films, *Phys. Rev. B*, 30, 4292, 1984.

35. **Perluzzo, G., Bader, G., Caron, L. G., and Sanche, L.**, Direct determination of electron band energies by transmission interference in thin films, *Phys. Rev. Lett.*, 55, 545, 1985.

36. **Vezina, C., Leclerc, C., and Sanche, L.**, Détermination de la section efficace effective de degadation de l'eau en phase solide par l'etude de la desorption de 1_2^+ par stimulation electronique, in abstracts of the *Sième Journées d'Etudes sur la Chimie sous Rayonnement*, Sherbrooke, Quebec, Canada, 1990.

37. **Burns, A. R., Stechel, E. B., and Jennison, D. R.**, *Rovibrational Laser Spectroscopy of ESD Neutrals from Chemisorbed Species;* **Stechel, E. B., Jennison, D. R., and Burns, A. R.**, Dynamics in neutral DIET from chemisorbed molecules, in *Desorption Induced by Electronic Transitions DIET III*, Springer Ser. in Surface Sci. Vol. 13, Stulen, R. H. and Knotek, M. L., Eds., Springer-Verlag, Berlin, 1988.

38. **Sanche, L. and Deschênes, M.**, Mechanisms of charge trapping at a dielectric surface: resonance stabilization and dissociative attachment, *Phys. Rev. Lett.*, 61, 2096, 1988.

39. **Sambe, H., Ramaker, D. E., Deschênes, M., Bass, A. D., and Sanche, L.**, Enhancement of dissociative electron attachment cross section in O_2 condensed on a Kr film, to be published.

40. **Sanche, L.**, Primary Interactions of Low-energy electrons in condensed matter, *Excess Electrons in Dielectric Media*, Jay-Gerin, J.-P. and Ferradini, C., Eds., CRC Press, Boca Raton, FL, 1991, chap. 1.

41. **Sanche, L., Perluzzo, G., Bader, G., and Caron, L. G.**, Temperature and thickness dependence of the 0—15 eV electron transmission spectra of rare gas films, *J. Chem. Phys.*, 77, 3285, 1982.

42. **Plenkiewicz, B., Plenkiewicz, P., Perluzzo, G., and Jay-Gerin, J.-P.**, Analysis of low-energy electron transmission experiments through thin solid xenon films in the elastic scattering region, *Phys. Rev. B.*, 32, 1253, 1985; **Plenkiewicz, B., Plenkiewicz, P., and Jay-Gerin, J.-P.**, Energy dependence of the mean free path of excess hot electrons in solid xenon in the elastic scattering region, *Phys. Rev. B*, 33, 5744, 1986.

43. **Ueno, N., Sugita, K., Seki, K., and Inokuchi, H.**, Low-energy electron transmission and secondary-electron emission experiments on crystalline and molten long-chain alkanes, *Phys. Rev. B*, 34, 6386, 1986.

44. **Bader, G., Perluzzo, G., Caron, L. G., and Sanche, L.**, Structural-order effects in low-energy electron transmission spectra of condensed Ar, Kr, Xe, N_2, CO and O_2, *Phys. Rev. B*, 30, 78, 1984; **Caron, L. G., Perluzzo, G., Bader, G., and Sanche, L.**, Electron transmission in the energy gap of thin films of argon, nitrogen, and n-hexane, *Phys. Rev. B*, 33, 3027, 1986.

45. **Bader, G., Perluzzo, G., Caron, L. G., and Sanche, L.**, Elastic and inelastic mean-free-path determination in solid xenon from electron transmission experiments, *Phys. Rev. B*, 26, 6019, 1982.

46. **Jay-Gerin, J.-P., Plenkiewicz, B., Plenkiewicz, P., Perluzzo, G., and Sanche, L.**, Electron mean free path and conduction-band density-of-state in solid methane as determined from low-energy electron transmission experiments, *Sol. State Comm.*, 55, 1115, 1985; **Keszei, E., Jay-Gerin, J.-P., Perluzzo, G., and Sanche, L.**, Quasielastic hot-electron transport in solid N_2 films, *J. Chem. Phys.*, 85, 7396, 1986.

47. **Goulet, T. and Jay-Gerin, J.-P.**, Theoretical study of the transmission of low-energy (0—10 eV) electrons through thin-film organic molecular solids: benzene, *Radiat. Phys. Chem.*, 27, 229, 1986; **Goulet, T., Pou, V., and Jay-Gerin, J.-P.**, A procedure for determining low-energy ($<$10 eV) electron mean free paths in molecular solids: benzene, *J. Electr. Rel. Phenom.*, 41, 157, 1986; **Goulet, T., Jay-Gerin, J.-P., and Patau, J.-P.**, Monte Carlo simulations of low-energy ($<$10 eV) electron transmission and reflection experiments: application to solid xenon, *J. Electr. Rel. Phenom.*, 43, 17, 1987.

48. **Goulet, T., Keszei, E., and Jay-Gerin, J.-P.**, Probabilistic description of particle transport. I. General theory of quasielastic scattering in plane-parallel media, *Phys. Rev. A*, 37, 2176, 1988; **Keszei, E., Goulet, T., and Jay-Gerin, J.-P.**, Probabilistic description of particle transport. II. Analysis of low-energy electron transmission through thin solid Xe and N_2 films, *Phys. Rev. A*, 37, 2183, 1988.

49. **Plenkiewicz, P., Jay-Gerin, J.-P., Plenkiewicz, B., and Perluzzo, G.**, Electron mean free path and conduction-band density of states in solid argon, *Sol. State Comm.*, 57, 203, 1986.

50. **Plenkiewicz, P., Plenkiewicz, B., and Jay-Gerin, J.-P.**, Conduction-band density of states of solid argon, *Sol. State Comm.*, 65, 1227, 1988.

51. **Kiraoka, K. and Nara, M.**, Conduction-band formation through the temporary anion state in organic-solids, *Bull. Chem. Soc. Jpn.*, 57, 2243, 1984; **Hiraoka, K. and Nara, M.**, Conduction band structure of solid n-alkanes studied by electron-transmission spectra, *Chem. Phys. Lett.*, 94, 589, 1983.

52. **Harrigan, M. E. and Lee, H. J.**, Theory of current characteristics of a thin dielectric film under slow electron impact, *J. Chem. Phys.*, 60, 4909, 1974.

53. **Huang, J. T. J. and Magee, J. L.**, On transmission of low-energy electrons in alkane thin films, *J. Chem. Phys.*, 61, 2736, 1974.

54. **Cheng, I. Y. and Funabashi, K.**, On the low energy event in Hiraoka-Hamill experiments, *J. Chem. Phys.*, 59, 2977, 1973.

55. **Perluzzo, G., Bader, G., Caron, L. G., and Sanche, L.,** Temperature dependence of some structures in electron transmission spectra of Xe solid films, *Phys. Rev. B, 26,* 3976, 1982.

56. **Michaud, M., Sanche, L., Gaubert, C., and Baudoing, R.,** Low-energy electron reflection and transmission on Ar films condensed on polycrystalline platinum, *Surf. Sci., 205,* 447, 1988.

57. **Chang, Y. C. and Berry, W. B.,** Electron range studies in solid hydrocarbon films at 77 K, *J. Chem. Phys., 61,* 2727, 1974; **Hino, S., Sato, N., and Inokuti, H.,** Electron escape depths of organic solids, *J. Chem. Phys., 67,* 4139, 1977; **Grechov, V. V.,** Low-energy (≤ 2 eV) electron escape depths of tetracene films, *Chem. Phys. Lett., 96,* 237, 1983.

58. **Pfluger, P., Zeller, H. R., and Bernasconi, J.,** Hot-electron transport in polymeric dielectrics, *Phys. Rev. Lett., 53,* 94, 1984.

59. **Bernasconi, J., Cartier, E., and Pfluger, P.,** Hot-electron transport through thin dielectric films: Boltzmann theory and electron spectroscopy, *Phys. Rev. B, 38,* 12567, 1988; **Cartier, E. and Pfluger, P.** Experimental determination of energy dependent inelastic and elastic scattering rates of hot-electrons in large bandgap insulators, *Phys. Scri., T23,* 235, 1988.

60. **Marsolais, R. M., Cartier, E. A., and Pfluger, P.,** Hot electron transport in condensed organic dielectrics, *Excess Electrons in Dielectric Media,* Jay-Gerin, J.-P. and Ferradini, C., Eds., CRC Press, Boca Raton, FL, 1991, chap. 2.

61. **Sanche, L. and Michaud, M.,** Vibrational excitation via shape resonances in electron scattering from N_2 multilayer films, *Chem. Phys. Lett., 84,* 497, 1981.

62. **Sanche, L. and Michaud, M.,** Resonance-enhanced vibrational excitation in electron scattering from O_2 multilayer films, *Phys. Rev. Lett., 47,* 1008, 1981.

63. **Sanche, L. and Michaud, M.,** Vibrational-librational excitation and shape resonances in electron scattering from condensed N_2, CO, O_2 and NO, in *Resonances in Electron-Molecules Scatttering van der Waals Complexes and Reactive Chemical Dynamics,* ACS Symp. Ser. no. 263, Truhlar, D. G., Ed., 1984, 211.

64. **Fano, U., Stephens, J. A., and Inokuti, M.,** Absence of resonances in the elastic-scattering of electrons in molecular solids, *J. Chem. Phys., 85,* 6239, 1986.

65. **Michaud, M.,** Private communication. This can be shown, for example, by calculating energy-loss intensities using multiple scattering formula (6) in Reference 63.

66. **Pendry, J. B.,** *Low-energy Electron Diffraction,* Academic Press, New York, 1971; **Clarke, L. J.,** *Surface Crystallography: An Introduction to Low Energy Electron Diffraction,* John Wiley, New York, 1985.

67. **Davis, T. H., Schmidt, L. D., and Minday, R. M.,** Kinetic theory of excess electrons in polyatomic gases, liquids, and solids, *Phys. Rev. A, 3,* 1027, 1971.

68. **Van Hove, L.,** Correlations in space and time and Born approximation scattering in systems of interacting particles, *Phys. Rev., 95,* 249, 1954.

69. **Duke, C. B., Anderson, J. R., and Tucker, C. W., Jr.,** The inelastic collision model. II. Second-order perturbation theory, *Surf. Sci., 19,* 117, 1970; **Duke, C. B. and Laramore, G. E.,** Quantum field-theory of inelastic diffraction. I. Low-order perturbation theory, *Phys. Rev. B, 3,* 3183, 1971.

70. **Narten, A. H., Venkatesh, C. G., and Rice, S. A.,** Diffraction pattern and structure of amorphous solid water at 10 and 77 K, *J. Chem. Phys., 64,* 1106, 1976.

71. **Resca, L. and Rodriguez, S.,** Excitation states in solid rare-gases, *Phys. Rev. B, 17,* 3334, 1978.

72. **Schwentner, N.,** Mean-free path of electrons in rare-gas solids derived from vacuum-UV photoemission data, *Phys. Rev. B, 14,* 5490, 1976; **Klein, M. L. and Koehler, T. R.,** Lattice dynamics of rare gas solids, in *Rare Gas Solids,* Vol. I, Klein, M. L. and Venables, J. A., Eds., Academic, New York, 1976, 301.

73. **Rath, J. and Freeman, A. J.,** Generalized magnetic susceptibilities in metal: application of the analytic tetrahedron linear energy method to sc, *Phys. Rev. B, 11,* 2109, 1975.

74. **Fujihira, M. and Inokuchi, H.,** Photoemission from polyethylene, *Chem. Phys. Lett., 17,* 554, 1972.

75. **Ueno, N. and Sugita, K.,** Secondary electron emission spectroscopy of solid and liquid high-density polyethylene, *Sol. State Comm., 34,* 355, 1980.

76. **Sanche, L. and Michaud, M.,** Vibrational structure in the N_2^- ($^2\Pi_g$) electron resonance of N_2 films, *Phys. Rev. B, 27,* 3856, 1983.

77. **Sanche, L., Perluzzo, G., and Michaud, M.,** Electron transmission spectroscopy of matrix-isolated N_2, *J. Chem. Phys., 83,* 3837, 1985.

78. **Michaud, M. and Sanche, L.,** Opening of new decay channels for core-excited resonances, *Phys. Rev. Lett., 59,* 645, 1987.

79. **Sanche, L. and Michaud, M.,** Vibrational excitation via shape resonances in electron scattering from the NO dimer, *J. Chem. Phys., 81,* 257, 1984.

80. **Thiel, P. A., Hoffmann, F. M., and Weinberg, W. H.,** Monolayer and multilayer adsorption of water on Ru(001), *J. Chem. Phys., 75,* 5556, 1981.

81. **Sakurai, M., Okano, T., and Tuzi, Y.,** Vibrational excitation of physisorbed CO_2 on a Ag(111) surface, *J. Vac. Sci. Technol. A, 5,* 431, 1987.

82. **Hoffman, F. M., Felter, T. E., Thiel, P. A., and Weinberg, W. H.,** The adsorption of cyclic hydrocarbons on Ru(001). II. Cyclohexane, *Surf. Sci.,* 130, 173, 1983.

83. **Demuth, J. E., Schmeisser, D., and Avouris, P.,** Resonance scattering of electrons from N_2, CO, O_2 and H_2 adsorbed on a silver surface, *Phys. Rev. Lett.,* 47, 1166, 1981; **Schmeisser, D., Demuth, J. E., and Avouris, P.,** Electron-energy-loss studies of physisorbed O_2 and N_2 on Ag and Cu surfaces, *Phys. Rev. B,* 26, 4857, 1982.

84. **Bader, G., Chiasson, J., Caron, L. G., Michaud, M., Perluzzo, G.,and Sanche, L.,** Absolute scattering probabilities for subexcitation electrons in condensed H_2O, *Rad. Res.,* 114, 467, 1988.

85. **Ehrhardt, H. and Willmann, K.,** Die Winkeläbhangigkeit der Resonanz-streuung niederenergetischer Elektronen, *Z. Phys.,* 204, 462, 1967.

86. **Birtwistle, D. T. and Herzenberg, A.,** Vibrational excitation of N_2 by resonance scattering of electrons, *J. Phys. B,* 4, 53, 1971; **Dubé, L. and Herzenberg, A.,** Absolute cross sections from the "boomerang model" for resonant electron-molecule scattering, *Phys. Rev. A,* 20, 194, 1979.

87. **Hagen, W., Tielens, A. G. G. M., and Greenberg, J. M.,** The infrared spectra of amorphous solid water and ice IV between 10 and 140 K, *Chem. Phys.,* 56, 367, 1981; **Wong, P. T. T. and Whalley, E.,** Raman spectrum of ice-8, *J. Chem. Phys.,* 64, 2359, 1976.

88. **Michaud, M. and Sanche, L.,** Interaction of low-energy electrons (1—30 eV) with condensed molecules. I. Multiple scattering theory, *Phys. Rev. B,* 30, 6067, 1984.

89. **Hiraoka, K. and Hamill, W. H.,** Characteristic energy losses by slow electrons in organic molecular thin-films at 77 K, *J. Chem. Phys.,* 57, 3870, 1972; **Hiraoka, K. and Hamill, W. H.,** Characteristic energy-losses by slow electron impact on thin-film alkanes at 77 K, *J. Chem. Phys.* 59, 5749, 1973.

90. **Hiraoka, K.,** Determination of energies of quasifree electron state V_o in organic solids from electron transmission spectra, *J. Phys. Chem.,* 85, 4008, 1981; **Hiraoka, K. and Nara, M.,** Transmission of low-energy electrons through thin-films of benzene and hexane at 80 K, *Bull. Chem. Soc. Jpn.,* 54, 1589, 1981.

91. **Leclerc, G., Goulet, T., Cloutier, P., Jay-Gerin, J.-P., and Sanche, L.,** Low-energy (0—10 eV) electron transmission spectra of multilayer tryptophan films, *J. Phys. Chem.,* 91, 4999, 1987.

92. **Ashby, C. I. H.,** Electronic excitation in electron bombardment enchancement of chemical reactions, *Appl. Phys. Lett.,* 43, 609, 1983.

93. **Marsolais, R. M., Michaud, M., and Sanche, L.,** Near-threshold electronic excitation by electron impact of multilayer physisorbed N_2 and CO, *Phys. Rev. A,* 35, 607, 1987.

94. **Marsolais, R. M. and Sanche, L.,** Mechanisms producing inelastic structures in low-energy electron transmission spectra, *Phys. Rev. B,* 38, 11118, 1988.

95. **Merkel, P. B. and Hamill, W. H.,** Energy loss spectra of low-energy electrons scattered from thin solid molecular films, *J. Chem. Phys.,* 55, 1409, 1971.

96. **Avouris, P. and Persson, B. N. J.,** Excited states at metal surfaces and their nonradiative relaxation, *J. Phys. Chem.,* 88, 837, 1984.

97. **Avouris, P., Schmeisser, D., and Demuth, J. E.,** Nonradiative relaxation of electronically excited N_2 on Al(111). Comparison with nonlocal optical theory, *J. Chem. Phys.,* 79, 488, 1983.

98. **Sanche, L.,** Resonance transmission of 0—15 eV electrons in solid benzene and pyridine, *Chem. Phys. Lett.,* 65, 61, 1979; **Sanche, L.** Interaction of 0—15 eV electrons with molecular solids: formation of charged excitonic complexes, *J. Phys. C,* 13, L677, 1980.

99. **Herzberg, G.,** *Spectra of Diatomic Molecules,* van Nostrand, NY 1966.

100. **Gauyacq, J. P. and Herzenberg, A.,** Nuclear-excited Feshbach resonances in electron + HCl scattering, *Phys. Rev. A,* 25, 2959, 1982.

101. **Ohno, M. and Domcke, W.,** Theory of resonance and threshold effects in the electronic excitation of molecules by electron impact, *Phys. Rev. A,* 28, 3315, 1983.

102. **Daviel, S., Wallbank, B., Comer, J. and Hicks, P. J.,** Electron energy-loss spectroscopy of carbon-monoxide using a new position-sensitive multidetector spectrometer. I. The energy region 6—10.9 eV, *J. Phys. B,* 15, 1929, 1982.

103. **Flicker, W. M., Mosher, O. A., and Kupperman, A.,** Low-energy variable angle electron-impact excitation of 1,3,5-hexatriene, *Chem. Phys. Lett.,* 45, 492, 1977; **Frueholz, R. P. and Kuppermann, A.,** Vibronic structure of the second triplet state of 1,3,5-hexatriene, *J. Chem. Phys.,* 69, 3433, 1978.

104. **Michaud, M. and Sanche, L.,** unpublished results.

105. **Mavroyannis, C.,** Excitation spectrum and electromagnetic properties of electron-exciton bound-states in molecular-crystal, *Physics* 77, 343, 1974; **Nagaev, E. L. and Sokolova, E. B.,** State of conduction electrons interacting with Frenkel excitons, *Phys. Status Solidi B,* 64, 441, 1974.

106. **Dalidchik, F. I. and Slonim, V. Z.,** Autoionization bands of electron-exciton complexes in periodic structures of reduced spatial dimension, *JETP Lett.,* 31, 112, 1980; **Agranovitch, V. M. and Zakhidov, A. A.,** Interaction of excitons with charge carriers in organic solids: charged excitonic complexes, *Chem. Phys. Lett.,* 68, 86, 1979.

107. **Singh, J.,** Formation of a complex charge carrier due to the interaction between a free Frenkel exciton and excess electron in molecular-crystal, *Phys. Status Solidi B,* 103, 423, 1981.

108. **Bader, G., Caron, L., and Sanche, L.,** Electron-exciton complex formation in organic solids, *Sol. State Comm.,* 38, 849, 1981.

109. **Bader, G. and Caron, L. G.,** Resonance of electronic and electron-exciton complex modes in molecular crystals, *J. Chem. Phys.,* 77, 3166, 1982; **Caron, L. G., and Bader, G.,** Electron-exciton complexes and the maxima in low-energy electron transmission spectroscopy, *J. Chem. Phys.,* 80, 119, 1984.

110. **Hiraoka, K., Nara, M., and Iijima, Y.,** Luminescence emission and excitation spectra of benzene thin film under slow electron impact at 77 K, *J. Phys. Chem.,* 87, 3959, 1983.

111. **Stulen, R. H. and Knotek, M. L.,** *Desorption Induced by Electronic Transitions DIET III,* Springer Ser. in Surface Sci., Vol. 13, Springer-Verlag, Berlin, 1988.

112. **Stulen, R. H. and Thiel, P. A.,** Electron-stimulated desorption and thermal desorption spectrometry of H_2O on Ni (111), *Surf. Sci.,* 157, 99, 1985.

113. **Noell, J. O., Melius, C. F., and Stulen, R. H.,** Mechanisms of electron-stimulated desorption of protons from water: gas, chemisorbed and ice phases, *Surf. Sci.* 157, 119, 1985.

114. **Stockbauer, R., Bertel, E., and Madey, T. E.,** The origin of H^+ in electron stimulated desorption of condensed CH_3OH, *J. Chem. Phys.,* 76, 5639, 1982.

115. **Kelber, J. A. and Knotek, M. L.,** Electron-stimulated desorption from partially fluorinated hydrocarbon thin films: molecules with common versus separate hydrogen and fluorine bonding sites, *Phys. Rev. B,* 30, 400, 1984.

116. **Kelber, J. A. and Knotek, M. L.,** Electron-stimulated desorption of condensed branched alkanes, *Surf. Sci.,* 121, L499, 1982.

117. **Rosenberg, R. A., Rehn, V., Knotek, M. L., and Stulen, R. H.,** Photon-stimulated desorption and electron-stimulated desorption of H^+ from solid H_2O and NH_3, *J. Vac. Sci. Technol. A,* 5, 1085, 1987.

118. **Sambe, H., Ramaker, D. E., Parenteau, L., and Sanche, L.,** Image-charge effects in electron-stimulated desorption: O^- from O_2 condensed on Ar films grown on Pt, *Phys. Rev. Lett.,* 59, 236, 1987.

119. **Kelber, J. A. and Knotek, M. L.,** Electron-stimulated desorption in organic molecular solids, *J. Vac. Sci. Technol. A,* 1, 1149, 1983.

120. **Azria, R., Parenteau, L., and Sanche, L.,** Dissociative attachment from condensed O_2: violation of the selection rule $\Sigma^- \leftrightarrow \Sigma^+$, *Phys. Rev. Lett.,* 59, 638, 1987.

121. **Azria, R., Parenteau, L., and Sanche, L.,** Mechanisms for O^- electron stimulated desorption via dissociative attachment in condensed CO, *J. Chem. Phys.,* 88, 5166, 1988; **Azria, R., Sanche, L. and Parenteau, L.,** O^- electron stimulated desorption from O_2 and CO and N_2 matrices. *Chem. Phys. Lett.,* 156, 606, 1989.

122. **Sambe, H., Ramaker, D. E., Parenteau, L., and Sanche, L.,** Electron-stimulated desorption enhanced by coherent scattering, *Phys. Rev. Lett.,* 59, 505, 1987.

123. **Sanche, L. and Parenteau, L.,** Ion-molecule surface reactions induced by slow (5—20 eV) electrons, *Phys. Rev. Lett.,* 59, 136, 1987.

124. **Sanche, L. and Parenteau, L.,** Dissociative attachment in electron-stimulated desorption from condensed NO and N_2O, *J. Vac. Sci. Technol. A,* 4, 1240, 1986.

125. **Sanche, L. and Parenteau, L.,** Production of anion-atom complexes by electron stimulated desorption, *J. Chem. Phys.,* 90, 3402, 1989.

126. **Sanche, L. and Parenteau, L.,** Dissociative attachment reactions in electron stimulated desorption from condensed O_2 and O_2-doped rare-gas matrices, *J. Chem. Phys.,* 91, 2664, 1989.

127. **Sambe, H. and Ramaker, D. E.,** The σ^- selection rule in electron attachment and autoionization of diatomic molecules, *Chem. Phys. Lett.,* 139, 386, 1987.

128. **Rapp, D. and Briglia, D. D.,** Total cross sections for ionization and attachment in gases by electron impact. II. Negative-ion formation, *J. Chem. Phys.,* 43, 1480, 1965.

129. **Krauss, M., Neumann, D., Wahl, A. C., Das, G., and Zemke, W.,** Excited electronic states of O_2^-, *Phys. Rev. A,* 7, 69, 1973.

130. **Sambe, H. and Ramaker, D. E.,** Forbidden electron attachment in O_2, *Phys. Rev. A,* 40, 3651, 1989.

131. **Krupenie, P. H.,** The band spectrum of molecular oxygen, *J. Phys. Chem. Ref. Data,* Vol. 1, No. 2, 423, 1971.

132. **Spence, D. and Schulz, G. J.,** Three-body attachment in O_2 using electron-beams, *Phys. Rev. A,* 5, 724, 1972.

133. **Shimamori, H. and Fessenden, R. W.,** Thermal electron attachment to oxygen and van der Waals molecules containing oxygen, *J. Chem. Phys.,* 74, 453, 1981.

134. **Hatano, Y.,** Electron attachment to van der Waals molecules, in *Electronic and Atomic Collisions,* Lorentz, D. C. et al., Eds., Elsevier, New York, 1986.

135. **Kern, K., Zeppenfield, P., David, R., and Comsa, G.,** Observation of adsorbate-substrate vibrational coupling in physisorbed Kr films on Pt(111), *Phys. Rev. B,* 35, 886, 1987.

136. **Palmer, R. E., Rous, P. J., Wilkes, J. L., and Willis, R. F.,** Determination of adsorbate molecular-orientation from resonance electron-scattering angular-distributions, *Phys. Rev. Lett.,* 60, 329, 1988.

137. **Kesmodel, L. L.,** Resonance scattering of electrons from benzene chemisorbed on Pd(100), *Phys. Rev. Lett.,* 53, 1001, 1984.

138. **Jones, T. S. and Richardson, N. V.,** Temporary negative-ion formation in chemisorbed species: the energy and angular dependence of vibrational losses for electron scattering studies of HCOO/Ni(110), *Phys. Rev. Lett.,* 61, 1752, 1988.

139. **Gadzuk, J. W.,** Resonance vibrational excitation in electron-energy-loss spectroscopy of adsorbed molecules, *Phys. Rev. B,* 31, 6789, 1985; **Gadzuk, J. W.,** Shape resonances, overtones, and electron energy loss spectroscopy of gas phase and physisorbed diatomic molecules, *J. Chem. Phys.,* 79, 3982, 1983.

140. **Gerber, A. and Herzenberg, A.,** Resonance scattering of electrons from N_2 adsorbed on a metallic surface, *Phys. Rev. B,* 31, 6219, 1985.

141. **Davenport, J. W., Ho, W., and Schrieffer, J. R.,** Theory of vibrationally inelastic electron scattering from oriented molecules, *Phys. Rev. B,* 17, 3115, 1978.

142. **Nagano, S., Luo, Z. P., Metiu, H., Huo, W. M., Lima, M. A. P., and McKoy, V.,** On the possibility of using differential cross section measurements for the electronic excitation of adsorbates by an electron beam to determine the adsorbate orientation, *J. Chem. Phys.,* 85, 6153, 1986.

143. **Gadzuk, J. W.,** Excitation mechanisms in vibrational spectroscopy of molecules on surfaces, in *Vibrational Spectroscopy of Molecules on Surfaces,* J. T. Yates, Jr. and T. E. Madey, Eds., Plenum Publishing, New York, 1987, 49.

144. **Schwentner, N., Himpsel, F.-J., Saile, V., Skibowski, M., Steinmann, W., and Koch, E. E.,** Photoemission from rare gas solids: electron-energy distributions from the valence bands, *Phys. Rev. Lett.* 34, 528, 1975.

Chapter 2

HOT ELECTRON TRANSPORT IN CONDENSED ORGANIC DIELECTRICS

Richard M. Marsolais, Eduard A. Cartier, and Peter Pfluger

TABLE OF CONTENTS

I. INTRODUCTION

Transport and relaxation of excess nonthermalized carriers is a topic of major importance for engineering materials subjected to high electrical fields. Organic insulators, and in particular polymers, are used in a large number of applications ranging from power generation equipment to microelectronic devices. The requirements on lifetime and reliability of the organic insulators are stringent, and reliable accelerated test methods allowing the prediction of possible failures are still in their infancy. Major difficulties arise from the fact that a large variety of instabilities may occur under the combined electrical, mechanical, chemical, and thermal stresses to which these materials are subjected in practice. A qualitative review of the major prebreakdown and breakdown mechanisms has recently been given by Zeller and coworkers,[1,2] and earlier work is summarized in books by O'Dwyer[3] and Bartnikas and McMahon.[4]

Among all the mechanisms leading to aging and breakdown in organic solids, we shall be concerned in the present article only with processes of electronic origin. It was suspected already for decades that avalanching induced by impact ionization of hot carriers is a decisive factor for ultimate breakdown; in contrast, it was recognized only recently that excitations induced by hot carriers unavoidably lead to aging processes in organic dielectrics subjected to electrical field strengths exceeding some critical value F_c.[5,6] The key question pertains to the quantification of this "electronic aging", i.e., the relative importance of this effect in comparison to the other aging processes.

In electronic aging, the detrimental role of the electrical field is to drive mobile electrons into harmful high energy states. As long as $F < F_c$ anywhere in the dielectric, excess electrons which always exist accidentally (e.g., due to cosmic rays, light, heat, etc.) rapidly relax and dissipate their energy in the dielectric (Figure 1a): the carriers are quickly thermalized and retrapped in defect states. On the other hand, if the average energy gain in the electric field exceeds the losses due to the interactions of the carriers with its host solid, a hot carrier population is sustained in energy levels characterized by a high carrier mobility (Figure 1b). A schematic energy distribution for the excess electrons under a given field $F > F_c$ is shown in Figure 2. We distinguish four regions on the energy scale. In the lowest energy region, the carriers interact strongly with the trap states of the system and their transport is defect-controlled. Above E_μ, which describes a mobility threshold, the carriers are quasi-free in the conduction states of the dielectric. In large bandgap insulators the mobility edge E_μ often lies in the vicinity of the vacuum energy level. Depending on the material, E_μ can be found below the vacuum level as well as above. Beyond an energy value $E^{(1)}_{crit}$, it is known that the energy relaxation of the carriers partly contributes to microscopic damage of the material by breaking bonds and creating defect states.[5] Finally, above $E^{(2)}_{crit}$, which is of the order of the gap energy of the dielectric, the electrons contribute to impact ionization which may eventually lead to the so-called "intrinsic" breakdown, i.e. avalanching. It is, therefore, essential that the electron population above the critical damage energies be assessed as a function of electrical field conditions for any given material.

The determination of N(E,F), which represents the steady-state energy distribution of the hot electrons at a given electrical field strength F, is undoubtedly involved. It was one of the last years' milestones to recognize that it can be expressed simply as follows:[7,8]

$$N(E,F) = \sqrt{E} \cdot \exp\left\{ -\frac{3}{2e^2F^2} \int_{E_\mu}^{E} m^*(E') \cdot s^2_{eff}(E')\, dE' \right\} \tag{1}$$

Here, $m^*(E)$ is a (direction) averaged effective mass for electron states at an energy $E > E_\mu$, and $s_{eff}(E)$ is an effective scattering function describing all the relevant excitations at that

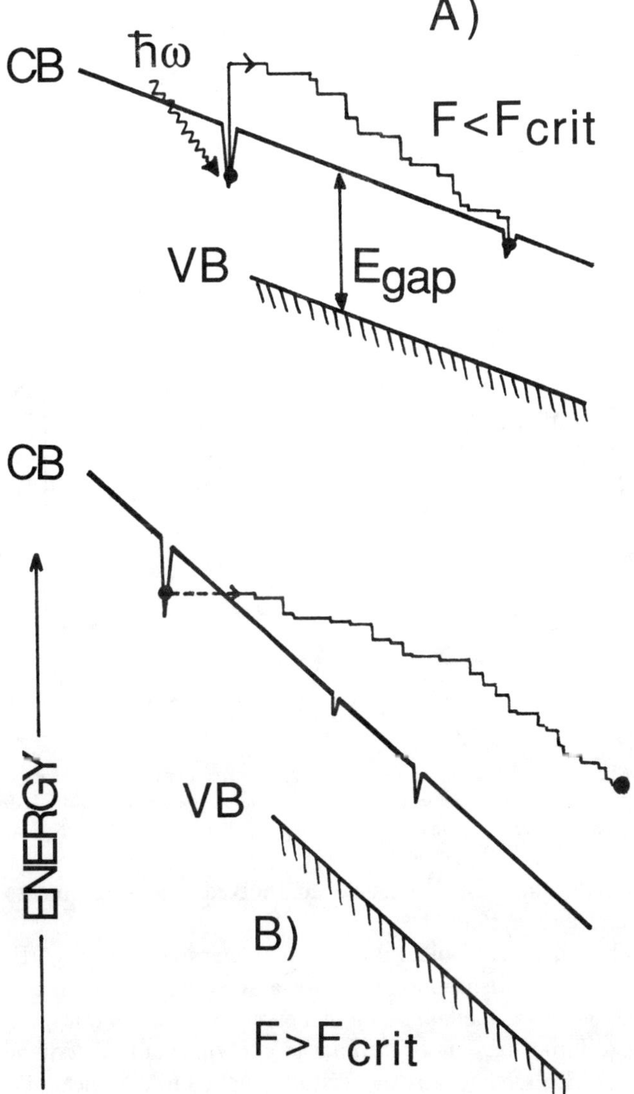

FIGURE 1. Schematic behavior of electrons in the conduction energy levels of a dielectric under field. The electrical field strength is indicated by the slope of the conduction band edge. (A) Under subcritical conditions, excited electrons (e..g, by light), are quickly retrapped; (B) Under overcritical fields electrons injected into the conduction band (exemplified here is a tunneling process) are kept in high mobility states.

energy. E_μ is the energy delimiting carriers in trapped energy states ($E < E_\mu$) from mobile carriers in quasi-free states ($E > E_\mu$), and $s_{eff}(E)$ is defined by

$$s_{eff}(E) = [s_u(E) \cdot s_p(E)]^{1/2} \tag{2a}$$

where $s_u(E)$ and $s_p(E)$ are the energy and the momentum relaxation rates (per unit time), respectively. On the other hand, $s_{eff}(E)$ can, at the same time, be generally described as a convoluted function of all the individual microscopic scattering functions:

$$s_{eff}(E) = f[s_j(E)] \tag{2b}$$

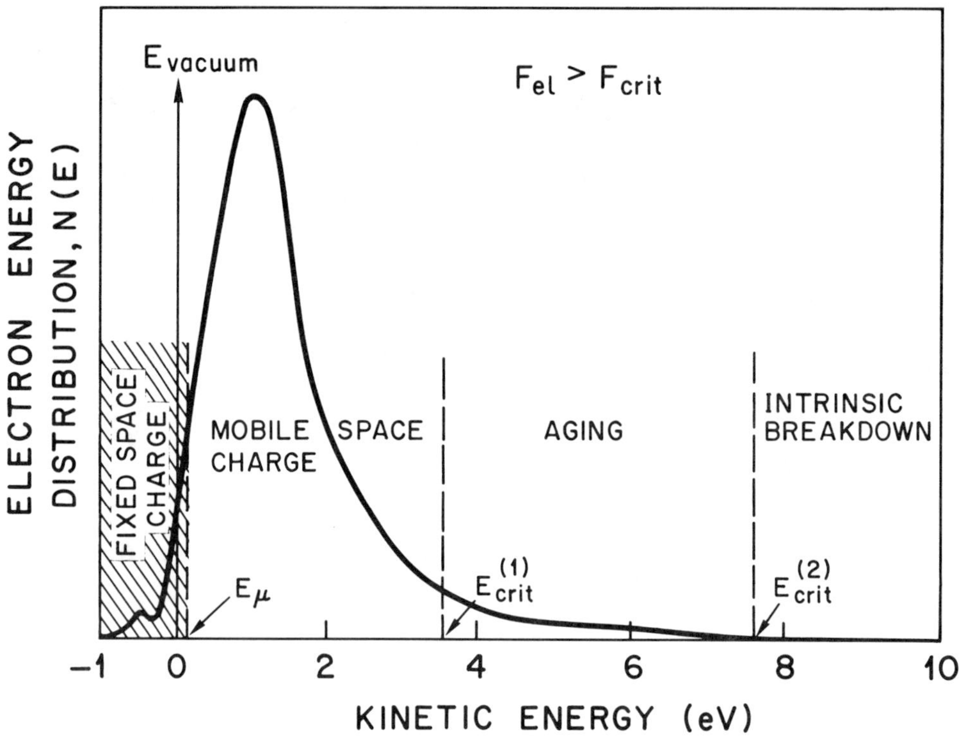

FIGURE 2.   Schematic representation of the electron energy distribution in a large bandgap insulator under an electric field $F > F_{crit}$. The shape of the curve is not fully realistic and was chosen so for instructive reasons. The transition points between the different regimes are those known for saturated hydrocarbon organic dielectrics.

wherein j runs over all the elastic, inelastic, and mixed scattering processes which are of relevance to the dielectric system.

The problem has thus been boiled down to the determination of $m^*(E) \cdot s_{eff}^2(E)$. Once this quantity is known, the distribution $N(E)$ for a given field strength F can be calculated according to Equation 1. The method of internal photoemission for transport analysis (IPTA) has been developed in our laboratory in order to experimentally determine this product in organic dielectrics.[9] This method was successfully applied to the linear alkane hexatriacontane, $n\text{-}C_{36}H_{74}$, which was chosen for the experiments because of its model character for polyethylene, to tetratetracontane, $n\text{-}C_{44}H_{90}$, to liquid phases of these materials, and to SiO_2.[9-11]

However, the decomposition of $s_{eff}(E)$ into individual scattering rates remains a particularly difficult task, even more so that Equation 2b leads to sufficiently simple forms only under special conditions. This is the case for example, when the energy is predominantly dissipated by the excitation of just one vibrational mode of an organic dielectric (e.g., a C–H stretch mode), all other collisions leading essentially to elastic scattering whereby only momentum is transferred, and when no energy is gained by the excess carriers. Owing to circumstances close to these conditions, our earlier analyses of the IPTA experiments yielded physically reasonable and meaningful results.

The purpose of the present chapter is thus threefold: first, we present a more sophisticated treatment of Equation 2b and, accordingly, a more rigorous analysis of the IPTA experiments. Second, and more importantly, we use a set of physical scattering models and parameters consistent with the previous analysis as an input in Monte Carlo simulations. It will be shown that we thereby, indeed, reproduce the experimental data. The consequence is that the experiments, the assumptions and parametrization of the underlying microscopic mechanisms, the analysis method and the simulations all form a complete self-consistent description of the problem. Finally, a side aspect of the present work is to have a more complete theoretical framework, verified and calibrated by the experiments, which will render a significant part of the experimental IPTA efforts unnecessary in the future.

This chapter is organized as follows: in Section II, we recall the techniques for acquiring IPTA data and analyzing them; Section III is devoted to a discussion of the physical scattering mechanisms involved in the subsequent analysis and to the introduction of their specific parametrized expressions; Section IV briefly describes the Monte Carlo simulation procedure while a more detailed description of s_{eff} (Equation 2b) follows in Section V; and computer-generated "IPTA-spectra" are presented and discussed in the last section.

II. THE TECHNIQUE OF INTERNAL PHOTOEMISSION FOR TRANSPORT ANALYSIS

The IPTA method is based on the injection of hot electrons into the conduction band of insulators at a metal-dielectric interface and the measurement of energy distributions of electrons transmitted through insulating films of varying thickness. The changes of the energy distributions contain information about the scattering events inside the dielectric. The energy level diagram of the internal photoemission experiments for transport analysis is shown in Figure 3 and a typical data set is given in Figure 4.

A model is required to extract the scattering rates (or lengths) from the spectra. For this purpose, the transport problem of hot quasi-free electrons through thin planar dielectric films was initially considered in detail by solving the steady-state Boltzmann equation for the electron current density $J(E, \theta, x)$, where θ is the direction cosine and where x describes the position within the film of thickness d.[11] This treatment is akin to the type of approach used in the interpretation of Low-Energy Electron Transmission (LEET) experiments.[12] This extensive study showed that the complexity of the problem does not allow for a thorough analysis of the experimental spectra. Nevertheless, it was realized that the equations describing the evolution of an energy interval ΔE at the top of the energy distribution E_{max} (see Figures 3 and 4), were greatly simplified. In this part of the spectrum, no down-scattering from higher energy states can occur and an exact formal solution could be obtained. A "procedure", which has become known as the "topmost interval method", was then introduced to usefully exploit the experimental data. It consisted of integrating the intensity of the emission in an energy interval ΔE at the top of the energy distribution, E_{max}, and in plotting it as a function of the thickness of the film. The thickness dependence thus obtained could then be traced back to elastic and inelastic scattering rates per unit length, γ_e and γ_i, respectively. These scattering rates are simply related to the scattering rates per unit time, s_e and s_i, via the relation $s(E) = \gamma(E) \cdot v(E)$, $v(E)$ being the electron velocity within the solid dielectric.

For the case of the topmost energy interval $[E_{max} - \Delta E, E_{max}]$ the following expressions were obtained for the intensity of the electrons transmitted at $E = E_{max}$ through a dielectric layer of thickness d:[11]

$$I(E_{max}, d) = I_0 \cdot \frac{2\gamma_{eff}(E_{max})}{\Gamma^+ \cdot \exp[\gamma_{eff}(E_{max}) \cdot d] - \Gamma^- \cdot \exp[-\gamma_{eff}(E_{max}) \cdot d]} \tag{3}$$

where

$$\Gamma^{\pm} = \gamma_e + 2\gamma_i \pm \gamma_{eff} \tag{3a}$$

and

$$\gamma_{eff} = 2 \cdot \sqrt{\gamma_i(\gamma_i + \gamma_e)} \tag{3b}$$

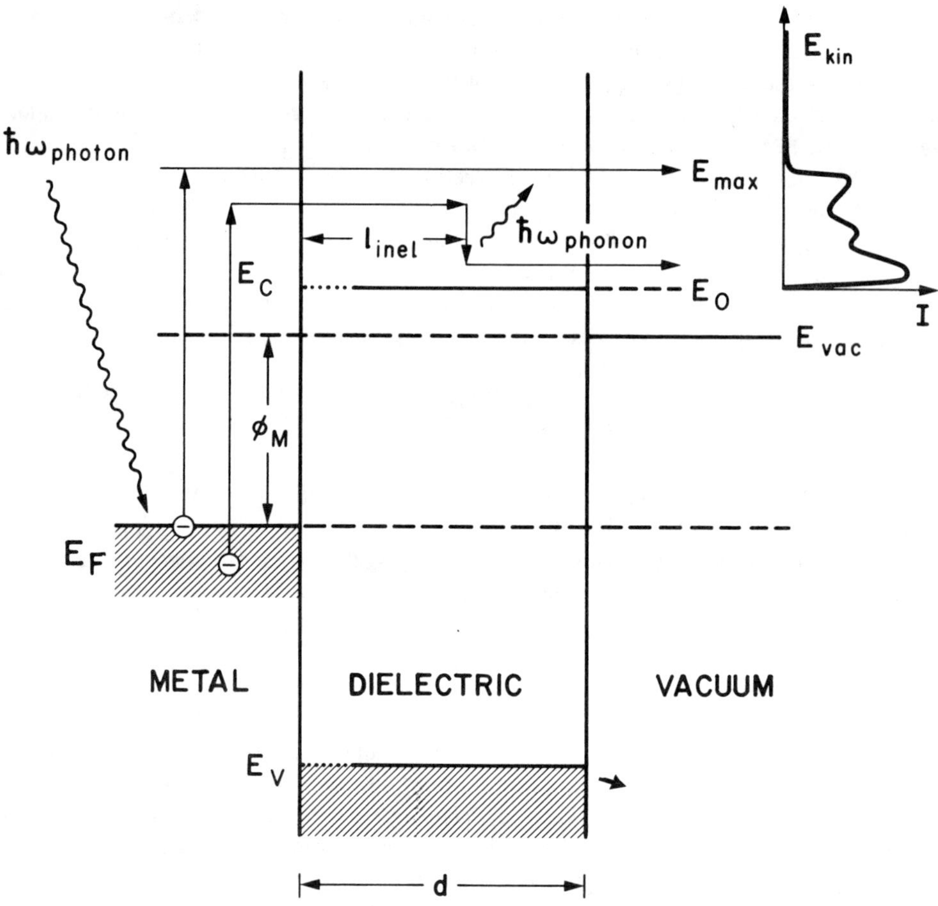

FIGURE 3.   Energy-level diagram of the IPTA (internal photoemission for transport analysis) experiment. In most cases, polycrystalline Pt was used as electron emitter and long chain paraffines (n-$C_{36}H_{74}$, n-$C_{44}H_{90}$) as well as sputtered SiO_2 films were used as dielectric layers. (From Cartier, E. and Pfluger, P., *Phys. Rev. B*, 34, 8822, 1986. With permission.)

Equation 3 reduces to an exponential decay function

$$I(E_{max}, d) = I_0 \frac{2\gamma_{eff}}{\Gamma^+} \cdot \exp\left[-\gamma_{eff}(E_{max}) \cdot d\right] \tag{4}$$

for $d \gg 1/\gamma_{eff}$, and to an over-exponential decrease of the zero-loss intensity for $d \ll 1/\gamma_{eff}$:

$$I(E_{max}, d) = I_0 \cdot \frac{1}{1 + [\gamma_e(E_{max}) + 2\gamma_i(E_{max})] \cdot d} \tag{5}$$

For the case of the large thicknesses, we can further distinguish the limiting cases of purely elastic and purely inelastic scattering, and Equation 3 then becomes

$$I(E_{max}, d) = I_0 \cdot \frac{1}{1 + \gamma_e(E_{max}) \cdot d} \qquad \text{when} \qquad \frac{\gamma_i}{\gamma_e} \longrightarrow 0 \tag{6a}$$

and

$$I(E_{max}, d) = I_0 \cdot \exp\left[-2\gamma_i(E_{max}) \cdot d\right] \qquad \text{when} \qquad \gamma_e = 0 \tag{6b}$$

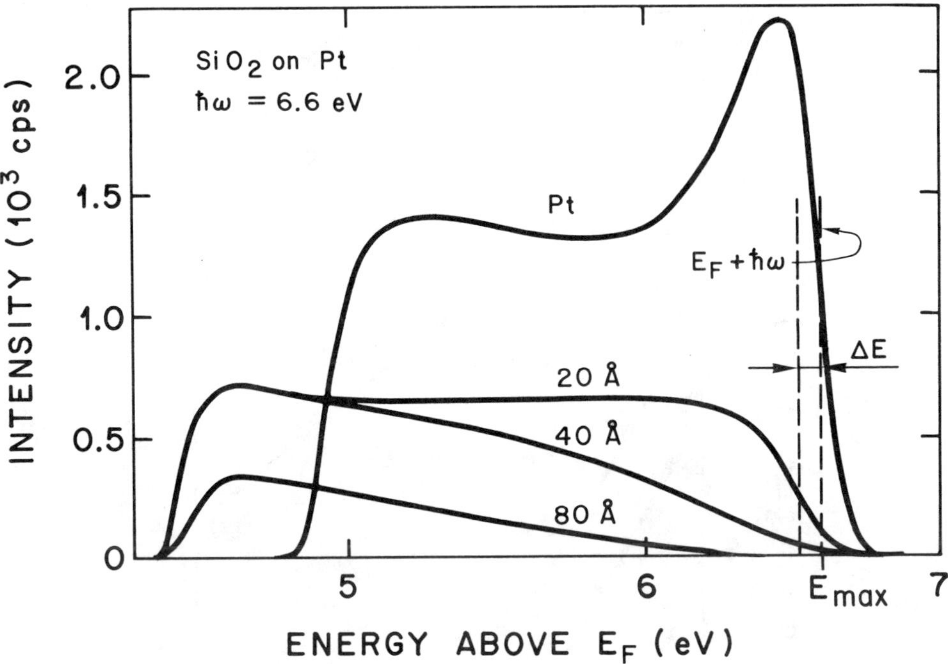

FIGURE 4. Energy distribution curves of electrons photoemitted from Pt after traversing SiO$_2$ films of various thickness (From Pfluger, P., Cartier, E., and Dersch, H., *Proc. of the IEEE Int. Symp. on Electrical Insulation, Cambridge, MA, 1988, 135. With permission.*)

In this description, the physics is all contained in overall elastic and inelastic scattering rates without detailed specification of the corresponding microscopic mechanisms. Even so, the above formulas are not completely model independent and certain properties of the scattering processes are assumed. To give just one example, Equation 3b implies that the elastic scattering is implicitly taken to be isotropic. If anisotropy is taken into account, Equation 3b becomes

$$\gamma_{eff} = 2 \cdot \sqrt{\gamma_i \left[\gamma_i + 2 (1 - \alpha) \cdot \gamma_e \right]},$$

where α is the fraction of forward-scattered electrons. We do not attempt here to give a complete description of which formula holds under which conditions, but rather would like to emphasize that the interpretation of the measured intensities is quite delicate. We also point out that Equation 3b involves an angular integration procedure of the escaping electrons over all directions, explaining the factor of 2 and the fact that this equation does *not* simply reduce to $\gamma_{eff} = \gamma_i$ in the case $\gamma_e \to 0$, i.e., vanishing elastic scattering. The latter situation does occur in the expression for the angle-resolved transmitted intensity I(E, θ, d) when the emission angle is normal to the overlayer. The consequence for the experimentalists is that the detailed interpretation of the measured intensities depends on the acceptance angle of the electron spectrometer, on eventual bias voltages used to collect electrons in lower energy ranges, etc.

If we trust that these technical issues are well under control, then the energy dependence of the scattering lengths can be scanned by experimentally varying the energy position of the topmost energy interval, denoted E$_{max}$ in Figures 3 and 4. In practice, this is done by varying the energy of the incident light $\hbar\omega_{photon}$. The effective scattering length, $1/\gamma_{eff}$, is then determined for each energy by the exponential thickness dependence of the intensity at E$_{max}$ for large film thickness according to Formula 4. Figure 5 illustrates the procedure

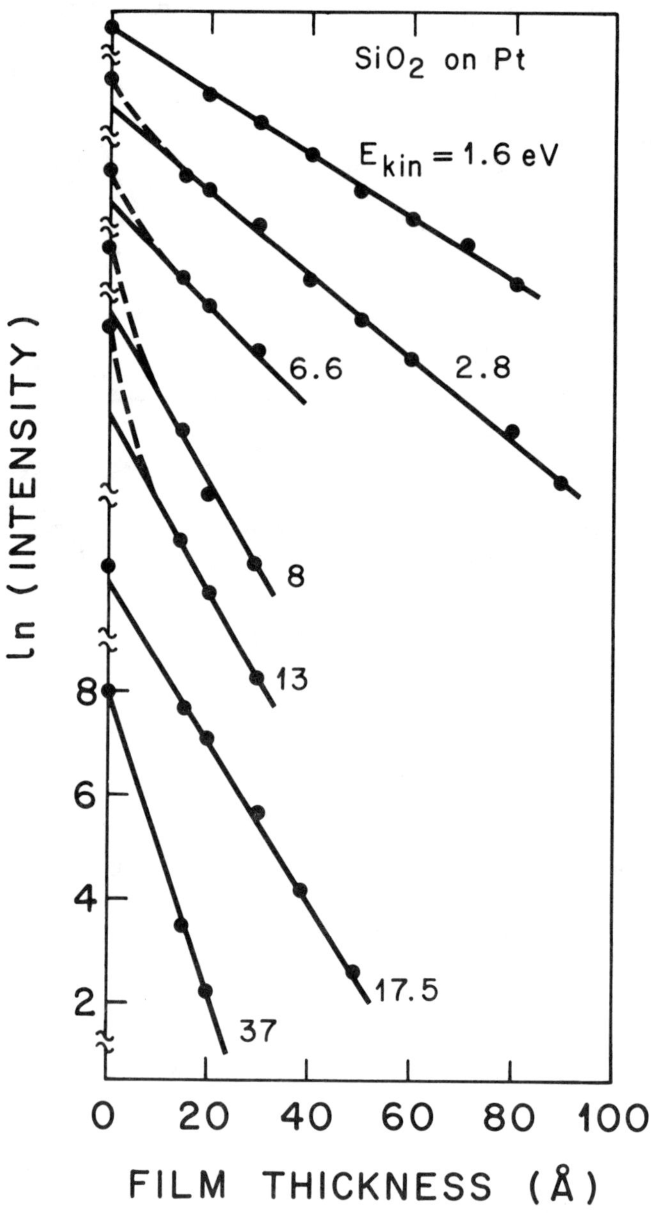

FIGURE 5.   Damping of photoelectrons injected at various topmost energy levels into SiO$_2$ as a consequence of the scattering processes in the overlayers.

with a data set on SiO$_2$, including the spectra of Figure 4. The various dampings of the transmitted intensities in the topmost energy intervals as a function of film thickness are clearly shown. The different slopes reflect the variation of the scattering rates (lengths) for the electrons as a function of their energy in the conduction states of the dielectric. It should be pointed out that the so-called large thickness limit makes sense only so long as d $\gg$ 1/γ_{eff} and so long as the scattering is highly inelastic. Otherwise, the thickness dependence is closer to what Equations 5 or 6a would predict, and the slope yields only an approximate value for γ_{eff}. These effects are further detailed in Section V. Moreover, and more importantly, the finite experimental resolution scatters the data and a seemingly straight line may hide a slight curvature.

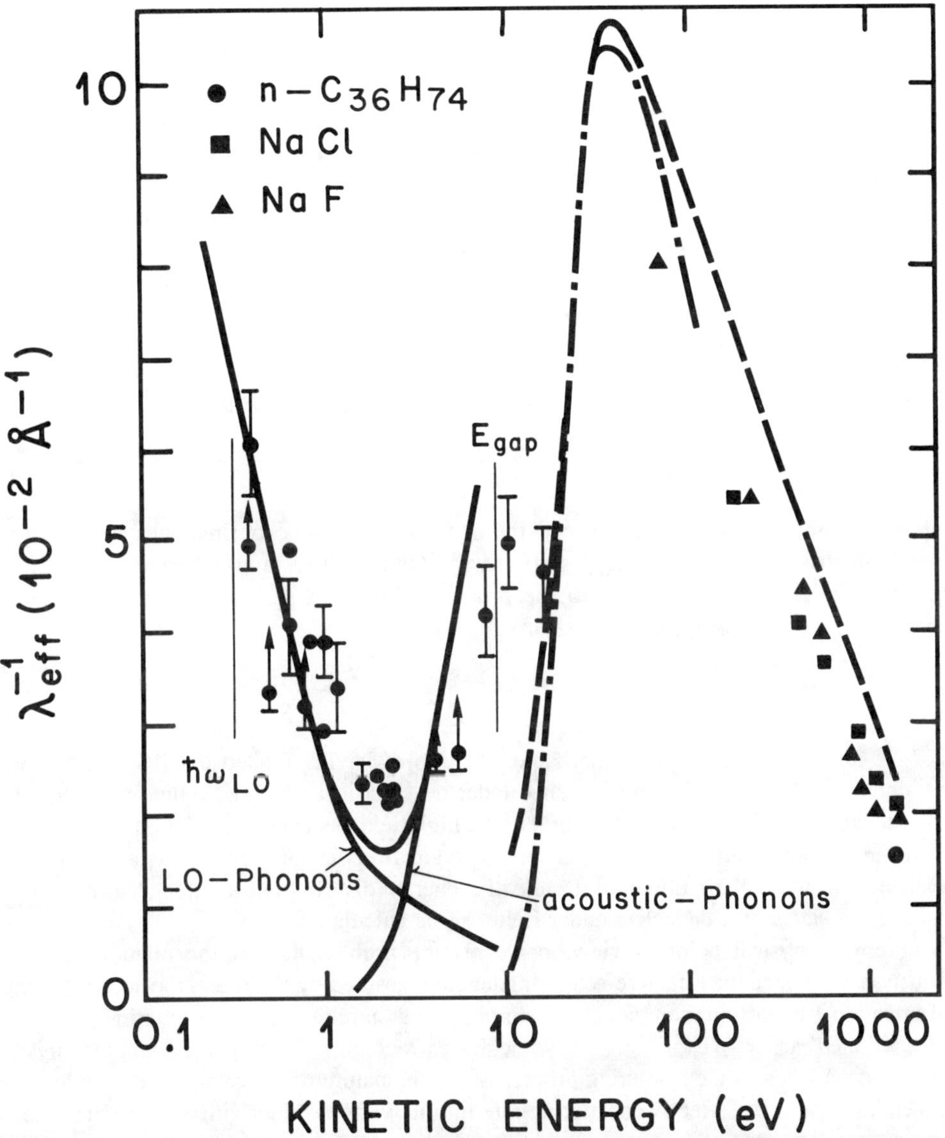

FIGURE 6. Effective inverse scattering rate per unit length, γ_{eff} ($= 1/\lambda_{eff}$), in the organic insulator n-C$_{36}$H$_{74}$ as a function of electron kinetic energy ($\bullet$). For comparison, results on NaCl ($\blacksquare$) and NaF ($\blacktriangle$) from the literature are included (see Reference 9 and references therein for details). The pseudo-universal curve derived from measurements in inorganic insulators (dashed) and the calculated inverse effective scattering length for electron-electron scattering in polyethylene (dash-dotted) are shown. The solid lines show the inverse effective scattering length in NaCl calculated from energy and momentum relaxation frequencies.[7] The individual contributions from optical and acoustic phonons are indicated. (From Cartier, E. and Pfluger, P., *Phys. Rev. B*, 34, 8822, 1986. With permission.)

In Figure 6, we return to the material class of central interest in this paper, *viz.*, organic insulators; the figure presents the experimentally determined values of the effective scattering rate per unit length, γ_{eff}, for the organic model dielectric n–C$_{36}$H$_{74}$, as obtained from an extensive set of IPTA experiments. A discussion of this curve and its implications on dielectric breakdown and aging is found in earlier papers and shall not be repeated here.[6,9] What is of central interest in the present work is the correspondence of γ_{eff} to γ_i and γ_e, and to the microscopic scattering processes. However, there are some complications in the separation of the elastic and inelastic rates. We first recall that the electrons primarily scatter by emission and absorption of phonons. Since the thermal occupation probability of vibrational states quickly decreases with energy, optical phonons with typically $\hbar\omega_{ph} \sim 100$ to 360 meV in hydrocarbons,[13] as compared to $k_BT \sim 25$ meV, are predominantly emitted. On the other hand, the low energy acoustic modes are emitted and absorbed in roughly equal amounts. Consequently, the interaction of the electrons with the acoustic phonons produces a quasi-elastic motion resulting from an unceasing up and down in energy space, while their interaction with optical phonons leads to large changes in energy preferentially downward, thus to mostly inelastic behavior. This simple identification is further supported by the fact that the experimental energy resolution is typically 50 to 150 meV. Hence, a single scattering event with optical phonons is sufficient for the electron to leave the detector energy window whereas several events with acoustic phonons are necessary to produce the same effect. Based on these attributions, the effective scattering rate may be written as

$$\gamma_{eff} = 2\,[\gamma^{opt}(\gamma^{opt} + \gamma^{ac})]^{1/2} \tag{7}$$

where γ^{opt} and γ^{ac} are the optical and the acoustic phonon scattering rates per unit path length of carriers with energy E_{max}, respectively. On the other hand, it can easily be shown that $\gamma_{eff}\,(E_{max})$ is simply related to the product $m^*(E_{max}) \cdot s_{eff}^2\,(E_{max})$ which appears in Expression 1 for N(E) by the expression[5-10]

$$m^*(E_{max}) \cdot s_{eff}^2\,(E_{max}) = \frac{1}{2}\,\gamma_{eff}^2\,(E_{max}) \cdot E_{max} \tag{8}$$

We are, therefore, in the remarkable situation of having related the IPTA data which are obtained from thin film experiments under no electrical field and which yield $\gamma_{eff}\,(E)$, with the key ingredient in the calculation of the high field electronic population N(E) relevant for dielectric aging and breakdown. From a pragmatic point of view this is all we need to evaluate the integral in Equation 1, and the energy distributions of hot electrons in the conduction states of a dielectric under field can be calculated.

From a physical point of view, however, the problem is that the relation given by Equation 8 between the effective-scattering length measured by IPTA and the needed product of Equation 1 is only simple under very simplifying assumptions on the scattering processes, as mentioned earlier. We thus ought to somehow extract information on the dominant physical mechanisms, such as the scattering processes and the nature of the carrier, i.e., we look for a method to split the effective scattering rate into its various underlying microscopic mechanisms. As a matter of fact, the purpose of such an effort is not only to understand the physics for advanced applications but also to acquire a handy procedure for selecting possible processes and evaluate their strength on the basis of the topmost interval method only, i.e., without resorting to involved calculations on the full spectra.

III. RELAXATION PROCESSES

A. GENERAL REMARKS

In order to define the basis for a more sophisticated treatment, we now discuss in this chapter some basics of the scattering physics for nonthermal carriers in the conduction state regions of an organic dielectric and we attempt to identify the major scattering processes of importance. The reader, unfamiliar with some of the solid state theoretical concepts presented below, will find the necessary background in any sufficiently advanced textbook on theoretical solid state physics, such as J. M. Ziman, O. Madelung, or C. Kittel.[14-16]

First, a few remarks are needed on the nature of electrons in such an environment. An electron injected in a dielectric material propagates as a charge carrier, the nature of which is a function of energy. At sufficiently high energies, the carrier behaves as a free electron, apart from small perturbations due to the environment, such as a slight anisotropy as well as a slightly different mass and a shifted kinetic energy caused by virtual excitations and long range interactions. The range of energies of interest in our problem ($E < 10eV \approx E_g$, the gap energy) however, is such that the carrier is strongly influenced by the excitation spectrum and possesses a variable character.[17] As the energy decreases, the carrier is more and more influenced by the band structure, i.e., it is strongly perturbed by coherent multiple elastic scattering, so as to develop strongly renormalized properties, and eventually may even propagate as a polaron. Below some energy close to the so-called bottom of the conduction band E_c, this coherent character is lost and the propagation proceeds through hopping or tunneling between localized states. Despite this variable character, we adopt the point of view that the carrier motion can always be described as a quasi-electron of (anisotropic) effective mass m* scattered by a "lossy" medium. This picture undoubtedly breaks down when $E < E_c$ or when the scattering rate is too large, as we shall see below, but is an instructive starting point in our analysis.

The propagation of an electron in a dielectric medium is governed by the complex elementary excitations spectrum and by the electron-excitation interactions. The thermalization process results, in a complicated manner, from the real excitations, i.e., those allowed by conservation laws and selection rules, and is described as a series of impact scattering events leading to the creation of annihilation of elementary excitations. The motion of the quasi-electron thus obtained will be approximately that of a free electron of anisotropic effective mass m* and kinetic energy $E - E_c$, where E_c is said to be the bottom of the conduction band but is more appropriately referred to as (minus) the electron affinity. The corresponding dispersion relation is:

$$E(\vec{k}) = E_k + E_c \equiv \frac{\hbar^2 k_1^2}{2\,m^*_1} + \frac{\hbar^2 k_2^2}{2\,m^*_2} + \frac{\hbar^2 k_3^2}{2\,m^*_3} + E_c \tag{9}$$

It is obvious, though, that this simple representation does not take into full account the band structure properties and fails to properly handle the propagation of the electron in localized states below E_c.

Let us now examine, in order of decreasing energy transfer, the thermalization mechanisms most relevant to the problem of breakdown in polymers. Notice that the quantity obtained from theory and needed in a Monte Carlo simulation program is the scattering probability per *unit time*, $s(E) = 1/\tau(E)$. This quantity relates directly to the energy broadening and thus to the unstationary character of the perturbed energy level. The scattering lengths deduced from the slopes in the IPTA experiments correspond to a scattering probability per *unit length*, $\gamma(E)$. It is directly related to the mean-free path by $\lambda(E) = 1/\gamma(E)$. We recall that the two quantities are related through the electron's velocity: $\gamma(E) = \tau(E)/v(E)$, where $v(E)$ is the instantaneous speed of the electron as given by the dispersion relation.

We now proceed to the discussion of the different relevant microscopic scattering processes.

B. ELECTRON-HOLE PAIR CREATION AND EXCITONS

Electron-hole pair creation is a precursor of the ultimate short-term electronic breakdown since avalanche effects under strong electric fields soon follow. It can occur when the kinetic energy of the electron is of the order of or exceeds the gap energy, which is in the vicinity of 10 eV in fully saturated polymers. The threshold is pronounced, the creation rate going like a power of the excess energy.[18,19] In polymers, it would be more appropriate to regard these excitations as Wannier or Frenkel type excitons.[20] Once excited to an energy level close to the bottom of the conduction band, the electron will easily be driven into the band by strong fields where it propagates like its progenitor. Rice et al. have shown the existence of such levels below E_c in chain compounds.[21]

Other electronic excitations of lower lying energy are important in polymer degradation. The weakest bonds will indeed be broken at energies of the order of a few electron volts. For the alkane family, the threshold has been found to lie in the vicinity of 3.5 eV.[5] The dissociative electronic excitations, to use the molecular spectroscopy terminology, induce defects and convey aging effects in polymers, but are not always detrimental. They may be beneficial when they induce cross links. They are harmful, though, when they produce defects which can lead to trapping, charge injection, and lowering of the bottom of the conduction band.

Owing to the large energies involved compared to room temperature, the mechanisms discussed so far are genuinely inelastic.

C. OPTICAL MODES

Moving on to excitations involving lower energies, the interaction of low energy electrons with vibrational modes is a central issue in molecular solids. It is usually not easy to decide whether the best description of this problem is in terms of localized modes or in terms of phonons. It is much more important, however, to assess whether the interaction is perturbative, i.e., accessible to a first Born approximation calculation, or resonant, i.e., a local temporary enhancement of the electron wave function accompanied by virtual trapping which demands a multiple-scattering calculation. Since this latter problem is already extensively discussed elsewhere in this book,[17] we shall confine our derivation to the idealized problem of phonon scattering and to the first Born approximation for both the optical and acoustic modes.

The optical mode phonons have been calculated for crystalline polyethylene.[22] Most of the modes below 175 meV are strongly dispersed, the width of the bands being around 30 meV, while the modes above 175 meV have a quite flat dispersion. The localization of the modes entailed by disorder should not change this picture significantly since most of the strong modes correspond to intrachain deformations and are only weakly influenced by neighboring chains. In that matter, electron energy loss (EEL) spectroscopy will help quantify these effects (see later). As a starting point, we shall assume that the modes are dispersionless in the calculation. Using the first Born approximation, the emission and absorption rates are given by[23]

$$\frac{1}{\tau} = \frac{e^2 \epsilon_{LO}}{8\pi^2 \epsilon_0} \left(\frac{1}{\epsilon_\infty} - \frac{1}{\epsilon_s} \right) \int d^3k' \frac{1}{q^2} \left(N_q + \frac{1}{2} \pm \frac{1}{2} \right) \delta \left(E_k - E_{k'} + \hbar\omega_{LO} \right) \qquad (10)$$

where ω_{LO} is the angular frequency of the longitudinal optical mode (the only modes to contribute in normal processes), ϵ_0 is the permittivity of vacuum (in SI units), ϵ_∞ and ϵ_s are the relative dielectric functions above and below the mode frequency, respectively, $\vec{q}$ is the momentum transfer as given by $\vec{k}' = \vec{k} \mp \vec{q}$, N_q is the thermal occupation probability of the emitted or absorbed phonon, and E_k is the kinetic energy of the scattered electron. Hereafter, we will assume that, in such a notation as $\pm$ or $\mp$, the top sign corresponds to emission and the bottom sign to absorption processes. It is straightforward to show for a quadratic electronic dispersion relation and for a flat phonon dispersion relation that the rates read

$$\frac{1}{\tau} = \frac{\pi e^2 m^* \omega_{LO}}{\epsilon_0 h^2} \left(\frac{1}{\epsilon_\infty} - \frac{1}{\epsilon_s} \right) \left(N_{LO} + \frac{1}{2} \pm \frac{1}{2} \right) \frac{1}{2k} \ln \left(\left[\frac{\sqrt{E} + \sqrt{E \mp \hbar\omega}}{\sqrt{E} - \sqrt{E \mp \hbar\omega}} \right]^2 \right) \qquad (11)$$

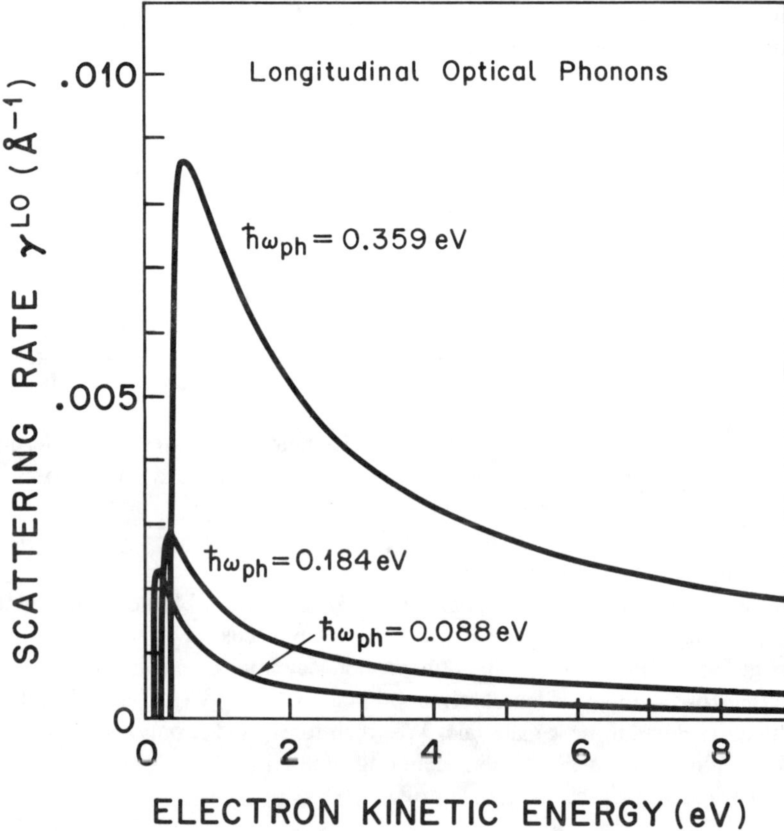

FIGURE 7.   Scattering rates per unit length γ as a function of electron energy, for electrons coupling to three different longitudinal optical phonons ("Fröhlich scattering"). The curves were calculated according to Equation 11.

Figure 7 illustrates the energy dependence of the scattering rate per unit length, $\gamma(E)$, resulting from this equation for a realistic case. We consider the emission from the three phonons which are known to be relevant for polyethylene and linear alcanes,[13,22] namely $\hbar\omega = 0.088eV$, $0.184eV$, and $0.359eV$. The absorption rate is not shown because it is by far, smaller than the emission rate, except at, or below, threshold.

The strength of the electron coupling constant to the LO mode is governed by three separate quantities: (1) the phonon dispersion relation, (2) the electron effective mass and (3) the dielectric function. The former, which was assumed flat in the above derivation, will not significantly change our results for the emission rate if dispersion is taken into account because these phonon levels are so scarcely populated that the occupation term reduces to identity. The absorption rate, on the other hand, will predominantly favor the phonons corresponding to the minima in the dispersion curves. Indeed, the occupation term reduces to $\exp(-\hbar\omega/k_BT)$, giving rise to the same result as before with the occupation term replaced by $\exp(-\hbar\omega/k_BT)$. The effective mass poses a more delicate problem in the case of anisotropic materials. A study of the integral (Equation 10) suggests that the scattering is governed by the largest effective mass, although altogether somewhat reduced when the ratio of the effective masses is large. Finally, the dielectric function can be determined experimentally. The IR reflectivity on bulk polyethylene and alkane films was measured and essentially the same spectra were obtained for both cases.[24] The signal-to-noise ratio being better in the data on polyethylene, we used it to extract the dielectric contribution to the coupling constant and obtained $(1/\epsilon_\infty - 1/\epsilon_s) \approx 0.015$, 0.005, and 0.004 for the main three phonons, respectively. The curves in Figure 7 were drawn according to these results, assuming the same effective mass for all three of them $(m^* = m_e)$.

The angular probability distribution of the scattered electrons for LO phonon excitations follows directly from consideration of momentum and energy conservation.[25] With respect to the direction of the wave vector prior to scattering, $\vec{K}$, the azimuthal angle is random while the polar angle obeys the following differential angular distribution for both emission and absorption events:[25]

$$d\,P\,(\theta) = \frac{2\sqrt{E_k E_{k'}}\,\sin(\theta)\,d\theta}{(E_k + E_{k'} - 2\sqrt{E_k E_{k'}}\,\cos(\theta))\,\ln\left(\dfrac{E_k + E_{k'} + 2\sqrt{E_k E_{k'}}}{E_k + E_{k'} - 2\sqrt{E_k E_{k'}}}\right)} \tag{12}$$

whereby E_k and $E_{k'}$ are the electron kinetic energies before and after the scattering event, respectively. This distribution favors the forward direction, and even more so as the ratio $E/\hbar\omega$ increases.

D. ACOUSTIC PHONON SCATTERING

The scattering by acoustic phonons is a classical problem in Monte Carlo simulation of hot electron transport in semiconductors which has been discussed by several authors.[26] The validity of the usual framework of the deformation potential model is clearly dubious for kinetic energies of the order of the deformation potential energy, E_1, which is of the order of a few electron volts in most materials. We, therefore, used it only as a starting ground in our investigation of acoustic phonon scattering in polymers.

The scattering is given by the Golden rule

$$\frac{1}{\tau} = \sum_{k'} W\,(\vec{k},\,\vec{k}) = \frac{2\pi}{\hbar} \sum_{k'} |M\,(\vec{k},\,\vec{k})\,|^2\,\delta(E\,(\vec{k}') - E(\vec{k})) \tag{13}$$

where the transition matrix element $M\,(\vec{k},\vec{k}')$ for the deformation potential interaction reads

$$M\,(\vec{k},\,\vec{k}') = |\,\langle\,\vec{k} + \vec{q},\,n_q \pm 1\,|\,H'\,|\vec{k},\,n_q\rangle| \tag{14}$$
$$= \frac{\hbar q^2 E_1^2}{2V\rho\omega_q}\left\{N_q + \frac{1}{2} \pm \frac{1}{2}\right\}$$

with V the volume, ρ the density, ω_q the phonon angular frequency and N_q the Bose-Einstein distribution. Consideration of the conservation of energy and momentum in normal processes yields at sufficiently high energies $(k \gg m^*\,c_s/\hbar$, where c_s is the phase speed of sound)

$$\frac{1}{\tau} = \frac{m^* E_1^2}{4\pi\hbar^2\rho}\int_0^{2k} dq\,\frac{q^3}{\omega_q}\left\{N_q + \frac{1}{2} \pm \frac{1}{2}\right\} \tag{15}$$

In the case of a linear phonon dispersion, $\omega_q = c_s\,q$, and at sufficiently high temperature $(k_B T \gg \hbar\omega_q)$, the rates reduce to the following k dependence:

$$\frac{1}{\tau} = \frac{m^* E_1^2}{\hbar^2\rho c_s}\left\{\frac{k_B T}{\hbar c_s} \pm \frac{2}{3}k\right\}k \tag{16}$$

The dominant term is therefore linear in k, i.e., it goes like the square root of the kinetic energy when a parabolic electronic dispersion curve is assumed. Extensions of this formula are worked out in the review paper by Jacoboni and Reggiani.[26]

Polyethylene departs from this simple situation in at least two important ways. First, the dispersion curves are anisotropic, as already mentioned, and the speed of sound is much larger along the chains ($c_s \approx 4 \times 10^4$ m/s) than perpendicular to the chains ($c_s \approx 1.5 \times 10^3$ m/s), even in the crystalline phase.[22] The acoustic scattering rate is, therefore, expected to be dominated by the phonons normal to the chains in some complicated manner. The case of semiconductors suggests that the resulting scattering rate exhibits a similar energy dependence with m* replaced by an average $\langle m^* \rangle^3 \approx m_{\parallel}^* \cdot m_{\perp}^{*2}$ when the bands are parabolic.[26] Second, the phonon dispersion curves differ greatly from the ideal case. For instance, theoretical calculations for crystalline polyethylene show that two main branches coexist and couple below 100 meV, which can be viewed as a dispersion made of one flat branch ($\hbar\omega \approx 0.025$ eV $\approx$ constant) and one inverted parabolic branch (with $\hbar\omega \approx 0$ at the center and at the edge of the Brillouin zone and reaching a maximum of 0.07 eV near the center of the zone).[22] Different energy dependences might therefore be expected; a specific model would be needed to assess the relevance of these effects. Nevertheless, Equations 15 and 16 provide a first estimate of the scattering rates. According to the above considerations, the dominant contribution to the normal processes comes from the smaller effective mass, $m^*/m_e \approx 0.05$,[27] and $c_s \approx 1.5 \times 10^3$ m/s. Assuming E_1 to be isotropic, and using furthermore $E_1 \approx 5$ eV, $\rho \approx 1$ g/cm^3, and $E_k = 1$ eV, the order of magnitude of the scattering rate is then $1/\tau \approx 3 \times 10^{14}$ s$^{-1} \approx 0.014$ Å^{-1}.

Other mechanisms related to vibrations can lead to an increase of the absolute scattering rate with energy, and also eventually to an overall functional dependence

$$\frac{1}{\tau} \sim E^\alpha \tag{17}$$

with $\alpha > 0.5$ as compared to Equation 16. Phonon dispersion effects were already mentioned above. Effective mass effects would originate from the change in the average effective mass with increasing energy. Indeed, the electron frees itself from the band structure and gains isotropy as the energy increases. Finally, resonances are well documented in low energy electron spectroscopies and are known to enhance the cross section by orders of magnitude.

Concerning the angular probability distribution, Jacoboni and Reggiani describe how to take into account the thermal population in the determination of the deflection angle and of the energy transfer on the basis of energy and momentum conservation.[26] For instance, assuming a linear phonon dispersion ($\omega_q = c_s q$) and isotropy, the distribution is

$$P(k,q)\, dq = \frac{m^* E_1^2}{4\pi\rho c_s \hbar^2 k} \left\{ N_q + \frac{1}{2} \pm \frac{1}{2} \right\} q^2 dq \tag{18}$$

and q is chosen with the aid of some computerized mathematical technique. The polar deflection angle results from this procedure while the azimuthal angle is uniformly distributed since we assumed isotropic symmetry.

E. UMKLAPP PROCESSES

Umklapp processes can occur as soon as $\vec{q}$ is outside the first Brillouin zone. This is bound to happen when k is larger than half of the smallest dimension of the zone. The lattice parameters for polyethylene are a = 7.36 Å and b = 4.92 Å normal to the chains, and c = 2.54 Å along the chains.[28] This corresponds to edges of the first Brillouin zone at π/a = 0.43 Å^{-1}, π/b = 0.64 Å^{-1} and π/c = 1.24 Å^{-1}. Assuming for the effective masses $m_{\parallel}^* = 0.05\, m_e$ and $m_{\perp}^* = 1.5\, m_e$,[27] the corresponding kinetic energies are 0.47 eV, 1.04 eV, and 117 eV, respectively. These values are obtained from a mere extrapolation of the curvature of the bottom of the conduction band. It is obvious that the bands flatten near the

Brillouin zone edge; hence, the value of 177 eV in a more realistic treatment should rather be in the range of 10 to 20 eV. In any case, our conclusion is that umklapp processes will, therefore, start to play a role at energies as low as 0.5 eV, but will be confined to the a*b* plane for energies lower than 10 eV.

Acoustic and optical phonon scattering are likewise submitted to umklapp processes, although, owing to the relative flatness of the optical branch, the latter is only marginally affected. The effect of the folding back to the first Brillouin zone implicit in umklapp processes will definitely be more pronounced in the case of acoustic phonons, resulting most often in an increase of the scattering rates. Moreover. whereas the normal processes usually allow interactions only with the longitudinal modes, the umklapp processes let the transverse modes contribute to the scattering, increasing the acoustic phonon scattering rate even more. Only specific calculations can provide energy dependences for the scattering rates involving umklapp processes, such as the one reported in Reference 19.

F. DEFECT SCATTERING

By nature, polymer agglomerates are full of defects: the chains have finite lengths and are not straight, bonds can be broken and crosslinks may be present. The rate of scattering on such (usually neutral), defects normally decreases with energy. For energies not exceeding a quarter of the ionization energy, Ziman reports the following energy dependence assuming that the scattering center behaves like a neutral hydrogen atom,[14]

$$\gamma \propto \frac{\epsilon \hbar^2}{m^* e^2} \cdot \frac{1}{k} \tag{19}$$

We consider the scattering due to the finiteness of the chains or to disorder as the scattering of the wave function by potential barriers. In a simple reasoning, if R is the intensity reflection coefficient on such a barrier and if ℓ is the mean distance between them, then the change of intensity per unit length is roughly

$$\frac{dI}{dx} = -\frac{RI}{\ell} \tag{20}$$

such that the scattering rate is $\gamma = R/\ell$. Since the reflection coefficient on a barrier decreases with energy, a similar behavior is expected from such a kind of defect scattering. Assuming a square potential, the energy dependence is $\gamma \sim \Delta^2/E^2$ when $E \gg \Delta$, where Δ is the height of the barrier. For energies in the vicinity of Δ, this coefficient saturates to about unity and trapping defects play a dominant role.

G. ENERGY BROADENING

The electron is, by construction, assumed to propagate as a long-lived quasi-particle with well-defined properties. This assumption should be reexamined when the scattering rate is large, i.e., the total scattering rate should in principle always be such that the imaginary energy shift of the chosen quasi-particle should be small compared to its kinetic energy.[29,30] The equation

$$\frac{1}{\tau(\vec{k})} = \frac{2\text{Im}\{\Delta E_k\}}{\hbar} \tag{21}$$

where $1/\tau(\vec{k})$ is the inverse total scattering rate per unit time and ΔE_k is the imaginary energy shift of the quasi-electron due to all interactions, illustrates that the spread in energy actually changes $\tau(\vec{k})$ itself in a self-consistent manner. This self-energy effect can be handled separately and should be built in the theoretical scattering equations. These considerations are essential when the electric field is turned on.[31]

IV. MONTE CARLO SIMULATION PROCEDURE

A crucial step in the analysis of the data consists in relating *in a quantitative way* the thickness evolution of the IPTA spectra described in Section II to the underlying microscopic processes discussed in some detail just before. As mentioned earlier, this difficult task was initially tackled with the help of a simplified version of the Boltzmann equation where the motion of the carriers was handled as currents and where only pure elastic and inelastic processes contributed to the scattering. Explicit analytical expressions were obtained for the thickness dependence of the current at the topmost energy in various limits.[11]

For a deeper investigation we chose to use Monte Carlo simulation methods to analyze both, the data in the topmost energy interval and the full energy and thickness dependences of the transmission spectra. Narrowed down to our problem, the Monte Carlo method of calculation consists in simulating the propagation and eventual faith of a very large number of electrons in a probabilistic way. The trajectories are independent from each other since the normal current densities are extremely low and since the state of the system is only marginally affected by the electron trajectories for small current densities. Such assumptions would clearly have to be relaxed for large current densities in real bulk systems when several electrons travel simultaneously near breakdown and when significant degradation or local heating can take place. The specific Monte Carlo procedure used here is extensively described in the literature.[26]

In the simulation program, we considered the very simple geometry of two ideal parallel planes separating three idealized media: the metal substrate on one side, the vacuum on the other side and a thin dielectric film sandwiched between the two. The frame of reference is the film: the y-z plane is on the metal-film interface and the x axis is perpendicular to the two interfaces and points toward vacuum (see Figure 8). The metal-film interface is in the x = 0 plane and the film-vacuum interface is in the x = d plane. An electron is first given initial conditions (namely a position, an energy, and a momentum) at the film-metal interface, i.e., it is injected from the metal. Its propagation is then simulated on the basis of probability distributions until it escapes the film either to the metal or to vacuum. The procedure is repeated a number of times, sufficient to ensure a satisfactory signal-to-noise ratio.

The propagation of an electron is decomposed into free traveling times t interrupted by scattering events, both of which are governed by the scattering processes operative in the system analyzed, i.e., all open channels. The scattering mechanisms fed into the Monte Carlo programs include optical and acoustical phonon, pure elastic and pure inelastic processes. The first two obey the specific functions discussed in the previous section. The pure elastic case has been introduced to reproduce the scattering on defects, which we furthermore assumed to be isotropic. As a matter of fact, we also used it to represent trap-controlled mobility knowing very well that this description needed some justification.[32] The pure inelastic case represents electronic excitation processes as long as its energy dependence and the resulting angular deflection are theoretically sound. We finally introduced a hybrid inelastic case which, in addition, allowed for a description of the so-called superelastic events (up and down of the electrons on the energy scale). This latter facility was used to investigate the effects of pseudo-elastic and pseudo-inelastic events (see below). The Monte-Carlo procedure is sufficiently flexible to let us turn on and off any set of scattering channels we need. The real difficulties are in fact to decide on which channels to use, to determine their theoretical probability functions and to have enough computer power to handle it all.

The state of the electron (and of the system) at the beginning or just after a scattering

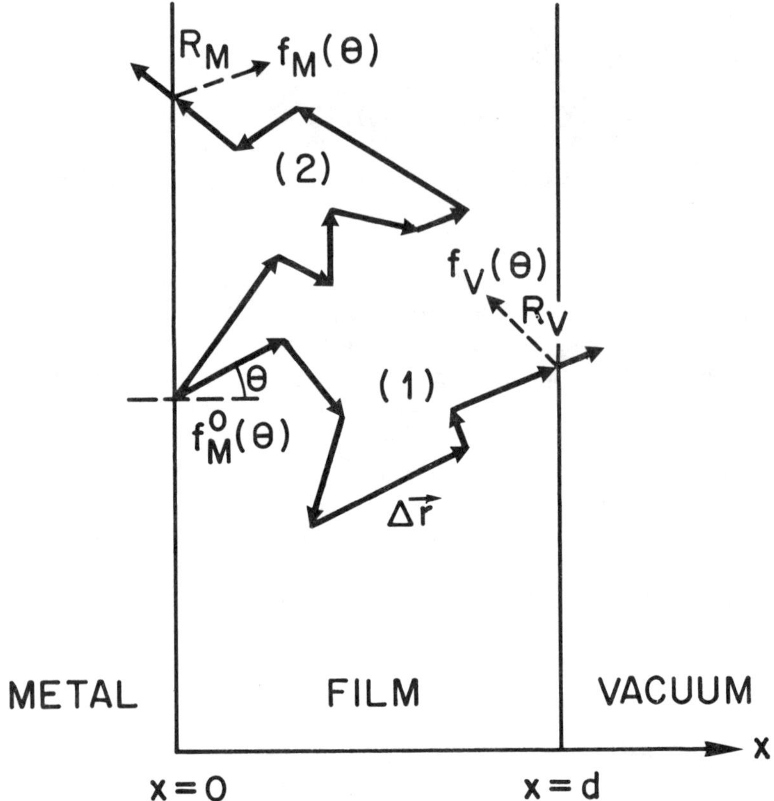

FIGURE 8.   Monte Carlo procedure adapted in our study. Two hypothetical trajectories are shown. The initial direction of motion is selected according to an angular distribution $f_M^O(\theta)$. Scattering events may be elastic or inelastic. Interfaces are traversed with probabilities 1-R_M and 1-R_v respectively. In case of reflection, the functions $f_M(\theta)$ or $f_v(\theta)$ govern the new direction.

event (at interfaces or in the bulk) and the free traveling time between collisions are all subject to a probabilistic description, according to specific techniques such as the rejection technique.[26]

At fixed energy, the free traveling time is simply governed by the exponential probability distribution:[33]

$$P(t)\, dt = \frac{dt}{\tau_{tot}} \exp\left(-\frac{t}{\tau_{tot}}\right) \qquad (22)$$

where $1/\tau_{tot}$ is the total scattering rate. The procedure we adopted to determine the actual path length is standard and makes use of the so-called real and self-scattering events.[33] The free-flight terminates when a real scattering event occurs. If the electron is still in the film, a scattering channel is selected using the technique for a discrete distribution and the outcome of the scattering (angular deflection, final energy) is subsequently selected randomly according to appropriate probability distributions and conservation laws. Otherwise, other probability distributions determine whether the electron is reflected back into the film or transmitted to the metal or to vacuum. In the event of self-scattering, the propagation of the electron resumes unaltered. A trajectory is generated by repeating the procedure until the electron has left the film. We maintain the use of self-scattering events even in the absence of electric field in order to implement a thickness parameter which induces more scattering when the film is thin. This parameter is unity at large thicknesses but increases like $\gamma_{tot} \cdot d$ at small thicknesses, whereby γ_{tot} is the sum of the scattering rates over all the open channels.

The semi-classical equation which describes the motion of an electron between collisions (i.e., scattering events) is:

$$\vec{x}' = \vec{x} + \vec{v} \cdot t \tag{23}$$

where $v = \hbar^{-1} \cdot \partial E(k)/\partial k \approx \hbar k/m^*$ is one component of the group velocity of the electron. We assume an isotropic behavior and a band dispersion relation, $E(k)$, which is reduced to that of a parabolic band with an energy-dependent effective mass m^*. This over-simplification, which can easily be changed in the program, is acceptable owing to the strong angular average resulting from multiple scattering in the film and at the interfaces, especially in disordered media.

The interfaces are looked upon as scattering centers and act as potential steps between two homogeneous media. The result of interface scattering will be either transmission or reflection, but in any case an elastic process. This elastic scattering can be diffuse or specular. In the *specular* case, the change in wave vector is determined by the conservation of energy alone and is not probabilistic, similar to Snell's law for the refraction of light. In the *diffuse* case, one has to supply a specific angular distribution for the scattered electron, which we write generally as:[34]

$$p(\theta) = (n + 1) \sin (\theta) \cos^n (\theta) \tag{24}$$

where θ is given with respect to the x axis and $n = 0$ corresponds to an isotropic distribution. For the transmission coefficients through the interfaces, both the classical and the quantum mechanical cases were considered.[35]

V. ANALYSIS OF THE EFFECTIVE SCATTERING RATE

Before we come to the analysis of the full energy dependence of the transmission spectra in Section VI, we first proceed to a more profound analysis of the topmost energy interval data and the effective scattering rates extracted from this experimental body. To start, we comment on a few specific important features from the Boltzmann analysis. In particular the large thickness exponential decay of the emitted current (Equation 4) is found to be related *not* to the inelastic scattering rate but rather to a *convolution over both the elastic and the inelastic rates*. It is obvious, though, that the effective rate is dominated by the inelastic contribution unless γ_e clearly exceeds γ_i. It turns out, moreover, that the exponential decay is only an asymptotic, or steady state, regime which takes place beyond some thickness in the vicinity of $d \approx 1/\gamma_{eff}$, a thickness which can be extremely large when $\gamma_i \ll \gamma_e$. The small thickness limit exhibits a $1/(1 + \gamma d)$ type of decay just like the large thickness limit of the pure elastic case, albeit the exact thickness dependence of this initial decay is also a function of the entrance angular distribution at the film-metal interface. This transitory regime is caused by the randomization process of the average direction of motion due to the elastic scattering channel (corresponding to low energy excitations in a real system).

Equation 3b is a key element in our analysis because it stipulates how inelastic and elastic processes combine to yield the measurable effective scattering rate per unit length. Yet, it is invalidated by the approximations and simplifying assumptions made to derive Equations 3 to 6, such as the two-stream approximation. This question has been thoroughly discussed by Bernasconi et al.[11] and by other authors.[12,36-39] We shall only point out here that the results of our Monte Carlo simulations demonstrate the usefulness of these equations as a starting point in the analysis of the data. A major difficulty is, however, that Equation

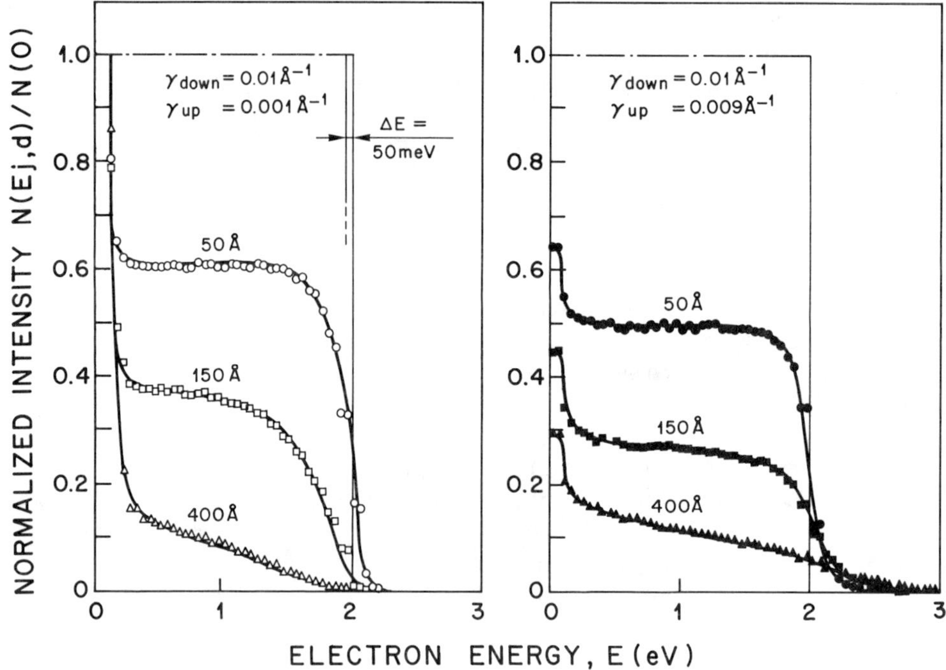

FIGURE 9. Monte Carlo simulation of IPTA spectra assuming an energy-independent emission from the substrate and a single inelastic scattering channel with energy transfer $\hbar\omega = 110$ meV. The series on the left assumes weak phonon absorption ($\gamma_{up} = 0.1\,\gamma_{down}$), corresponding to a quasi-inelastic case, while the series on the right assumes strong absorption ($\gamma_{up} = 0.9\,\gamma_{down}$), representing a quasi-elastic situation. The symbols correspond to the calculated emission from each energy slot (50 mV wide), while the full lines are just an aid to the eye.

3b is also invalidated by the fact that both γ_e and γ_i are ill defined in real systems since most scattering processes lead to both emission and absorption events, each one occurring at a rate different from the other. A channel is therefore never either pure elastic or pure inelastic, although one can still distinguish the so-called quasi-elastic and quasi-inelastic cases. The main consequence of the possibility of up and down transitions is a gradual energy spreading which confers a mixed character to the channel. The quasi-elastic case will indeed look partially inelastic since the energy window is effectively depleted (as soon as the energy spread exceeds the energy width ΔE), while the quasi-inelastic case will look partially elastic since the window is, to some extent, replenished by absorption events.

We have examined this problem in more detail with the aid of our Monte Carlo simulation program. The electrons counted in the topmost energy window correspond to an average stream of propagating carriers constantly depleted, but only partly replenished, by scattering events i.e., the propagating electrons step up and down in energy at each scattering event, moving in and out of the window. This description goes beyond the preliminary analysis described in Section II, for which we recall that only step-down processes in energy were considered.

A full account of this investigation will be given later and here we shall discuss only the simplest possible case, namely a medium wherein a propagating electron can either lose or gain an energy quantum at two different rates, γ_{down} and γ_{up}, respectively, the scattering being assumed isotropic.[40] Figure 9 illustrates the effects on the spectra of the imbalance between up and down scattering rates and shows the differences between scattering of mostly inelastic or mostly elastic character. The series of spectra on the left side corresponds to the quasi-inelastic case ($\gamma_{up} \ll \gamma_{down}$) whereas the series on the right corresponds to the quasi-elastic case ($\gamma_{up} \approx \gamma_{down}$). The energy channel width was taken to be 0.05 eV, the energy transfer per scattering event was 0.110 eV, 50,000 trajectories were allowed per channel, and all scattering events were assumed isotropic. At a sufficiently small thickness, the spectra maintain their flatness in the center, while they are rounded up at the upper edge due to depletion and spreading. The difference in spreading between the quasi-elastic case and the quasi-inelastic case shows up quite early and, as expected, the spreading is asymmetric in the quasi-inelastic case. The peak developing just above the zero of energy results from the fact that we assume $E_0 = 0$ and forbid transitions to below E_0. An increasing fraction of the spectrum is affected by the energy spreading as the thickness of the film increases. When the logarithm of the intensity at any given energy is plotted as a function of thickness, a downward kink appears when the spread reaches that energy. (This feature will actually occur in experimental spectra, as we shall see later.)

Figure 10 exhibits the Monte Carlo results at the topmost energy for two different angular distributions of the injected electrons: isotropic and normal-specular. Here, "normal-specular" means that the injected electrons all have an initial velocity normal to the surface of the metal, without any angular spread whatsoever. All interface reflection coefficients were set to zero in this example. The two extreme cases, namely the genuine elastic case ($\gamma_{up} = \gamma_{down}$) and the pure inelastic case ($\gamma_{up} = 0$), behave as predicted by Equations 6a and b, to within a geometrical factor for the scattering rates. This factor arises from the entrance angular distribution and from the drastic reduction to one dimension done in the double-stream approximation.[11] The realistic, intermixed cases display, as expected, a gradual evolution from quasi-elastic to pure inelastic.

The differences between the cases of isotropic and normal injection are remarkable. For the more practical situations, corresponding to predominantly, but not purely, inelastic behavior, the slopes for large thicknesses are quite similar in both injection modes, while major differences occur at small thicknesses. Clearly, within the distance needed to randomize the initial angular distribution, the intensity reduction of the injected population is a very sensitive quantity to angular effects. These two very different cases were calculated to obtain, from comparison with experimental data, an indication for a reasonable choice regarding the entrance angular distribution in the Monte Carlo procedures.

Most theoretical investigations invoke a combination of pure elastic and pure inelastic processes to explain the data.[11,12,39] The effects of these two extremes on the spectra are very different. It is clear though that some processes cannot be classified in either of these two categories. Moreover, most microscopic theoretical models yield γ_{up} and γ_{down} while most simple theoretical models (as well as our intuition) call for γ_e and γ_i. It is therefore essential to know how γ_{up} and γ_{down} merge to form γ_e, γ_i and, consistently, γ_{eff}. To try to answer this question , at least partly, we have performed computer simulations in the isotropic case, which is more realistic than the normal-specular case.[11] The effective scattering rate is then found to obey a law very close to

$$\gamma_{eff} = \sqrt{2}\,|\,\gamma_{down} - \gamma_{up}\,| \tag{25}$$

Inspired by analytical results from the double-stream, we sought an effective scattering rate of the form of Equation 3b where γ_i and γ_e are now given by

$$\gamma_i = \gamma_{up} + \gamma_{down} - f\,(\gamma_{up}, \gamma_{down}) \qquad \text{and} \qquad \gamma_e = f\,(\gamma_{up}, \gamma_{down}) \tag{26}$$

where $f(\gamma_{up}, \gamma_{down})$ is an admixture function, which is clearly a function of all the properties of the system, such as the angular distribution at the entrance and the directionality of the various scattering events.[40] This admixture is a measure of the elastic- and inelastic-like aspects developing from the imbalance between gain and loss events. The above equations yield

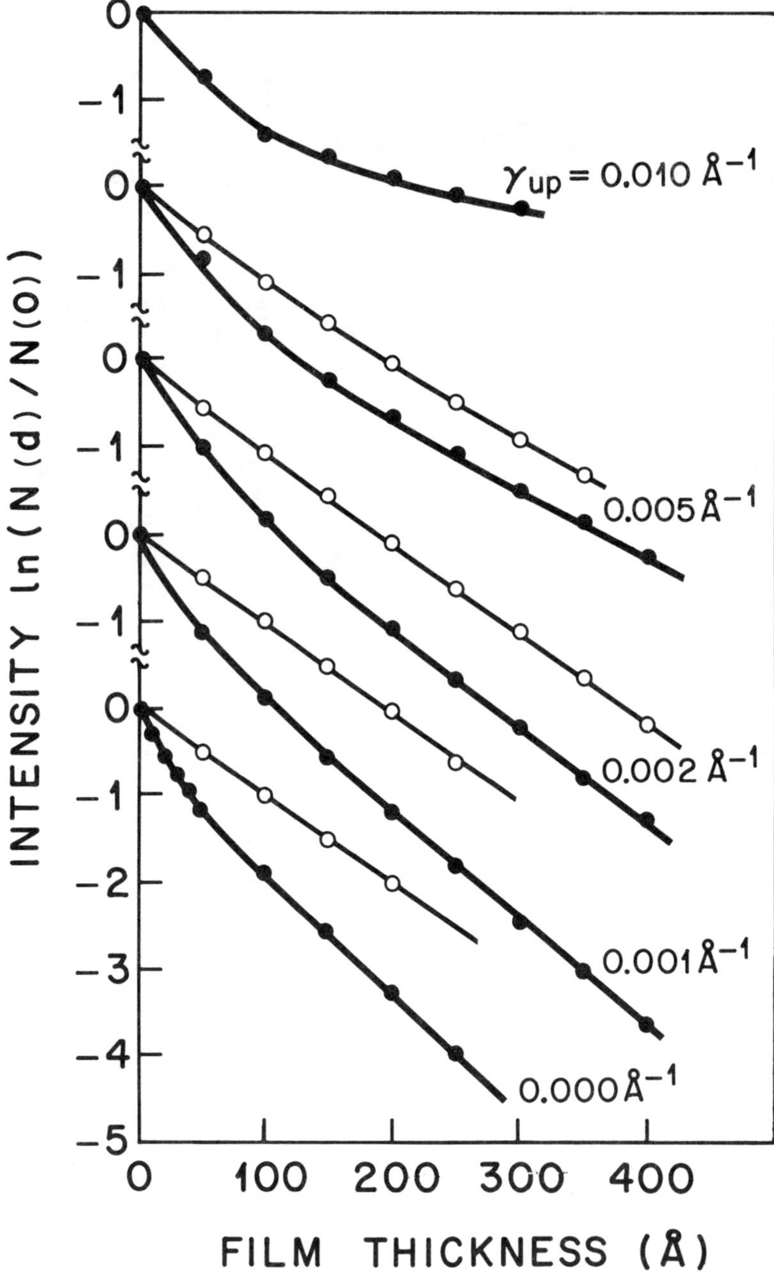

FIGURE 10.   Results for the topmost energy internal obtained from Monte Carlo simulations like those of Figure 9. Results are given for different absorption rates, while the emission rate was kept fixed and equal to $\gamma_{down} = 0.01\ \text{Å}^{-1}$. The slot width is 0.05 eV. The open dots represent the cases with normal electron injection, the full dots those with isotropic injection conditions.

$$f(\gamma_{up}, \gamma_{down}) = \frac{4\,\gamma_{up}\gamma_{down}}{\gamma_{up} + \gamma_{down}} \tag{27}$$

This expression fulfills the intuitive requirement that γ_i vanishes when $\gamma_{up} = \gamma_{down}$. With such definitions, the prefactor which was a 2 in Equation 3b now is a $\sqrt{2}$ and in fact Equation 3b is extended by the system of Equations 25 to 27.

Other factors, such as the finite width of the energy window, also contribute to mix up elastic and inelastic characters, as mentioned above. The effect of the width is in fact to confer an elastic component to an inelastic channel when $\hbar\omega < \Delta E$ since two or more loss events are necessary to leave the layer. We shall only point out that this effect is especially important when the loss mechanism involves losses of the order of the integration window ($\hbar\omega < 100$ meV), as in the presence of umklapp processes.

VI. SIMULATION OF IPTA SPECTRA

Finally, we use Monte Carlo simulations of the full IPTA spectra to test both the method and our interpretation of the data, and also in view of future investigations of electrical field effects. A thorough analysis is quite complicated and may be misleading if all mechanisms are included at once without preliminary investigations and/or sufficient knowledge of the physical processes. We judged it more enlightening to proceed in an interative fashion, adding more ingredients and corrections as the analysis improved and as experimental data became available. We also want to put forth an iterative analysis procedure using the transformation of microscopic scattering processes into effective scattering rates as the first iteration loop, i.e., to seek a handy tool for *first selecting possible processes and for adjusting their individual strength so as to reproduce the energy dependence of the effective scattering rate*. This turned out to be quite a powerful approach. To illustrate our procedure, let us concentrate on the series of IPTA spectra for liquid tetratetracontane n-$C_{44}H_{90}$,[41,42] taken at 92°C on a platinum substrate, and shown in Figure 11.[11]

The very first step consists in qualitatively studying the experimental spectra. The presence of a strong and narrow feature near the zero of energy should be associated with the existence of a bottom of a conduction band. Physically, this band is not necessarily a true conduction band, at least not in three dimensions, but definitely corresponds to the lowest energy levels allowed for the film + electron system. The lack of fine structure above zero supports the idea of a weakly coherent motion for the injected electron. We thus assume the electronic dispersion relation to be isotropic, with the proviso that the effective mass m* might not be unity owing to some averaging process and to renormalization (due to the electron-phonon interaction for example). Note, in passing, that strong anisotropy is possible and has been observed recently on hexatriacontane films in the crystalline phase ($m_{\parallel}^{*} \approx 0.05\ m_e$ and $m_{\perp}^{*} \approx 1.5\ m_e$).[27] Figure 12, which summarizes several topmost analyses on liquid tetratetracontane films, is also a source of information. The upturn at the lowest thicknesses indicates, by comparison with Figure 10, that the injection from the metal substrate is not specular (in the sense defined above), although a bit more so near 0.6 eV. Let us consider it to be isotropic at all energies. Moreover, the existence of truly exponential decays, i.e., straight lines in the semilog scale, gives evidence for a substantial contribution from inelastic processes at all energies. The occurrence of the change of regime in the vicinity of 50 Å further indicates that reasonable values for γ_{eff} are in the neighborhood of 0.02 Å^{-1}. This observation is confirmed by the γ_{eff} values deduced from the slopes at larger thicknesses.[9]

The next step is to decompose γ_{eff} into its constituents. In order to unravel the dominant mechanisms, we first include only two, definitely relevant and intrinsically different, mechanisms, namely pure elastic scattering, whatever it means, and optical phonon scattering, which is basically inelastic. The energy dependence of the optical contribution is assumed to obey the Fröhlich formula,[7,8] with the strength given initially by the infrared reflectivity data, as described in Section III. We then reckon the necessary elastic contribution using Equation 3b with $\sqrt{2}$ as a prefactor, Equations 25 to 27, and assuming that $\gamma_e = \gamma_e^{\mathrm{pure}} + \gamma_e^{\mathrm{optic}}$ and $\gamma_i = \gamma_i^{\mathrm{optic}}$. The result, shown in Figure 13 together with the data, deserves a few comments. First, $\gamma_e(E)$ and $\Sigma\ \gamma_i(E)$ differ from the functions appearing in Reference

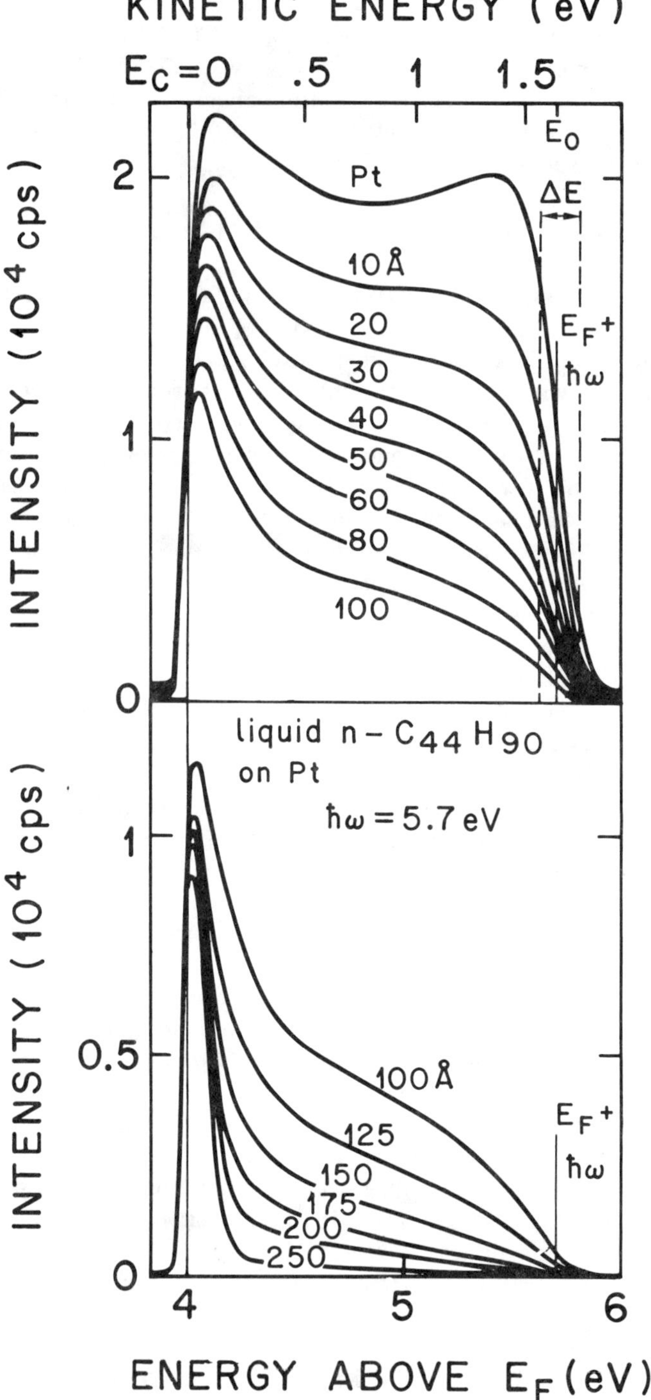

FIGURE 11.   Experimental energy distributions of hot electrons after transport through liquid n-$C_{44}H_{90}$ films of various thicknesses, obtained by the IPTA technique. (From Bernasconi, J., Cartier, E., and Pfluger, P., *Phys. Rev. B,* 38, 12567, 1988. With permission.)

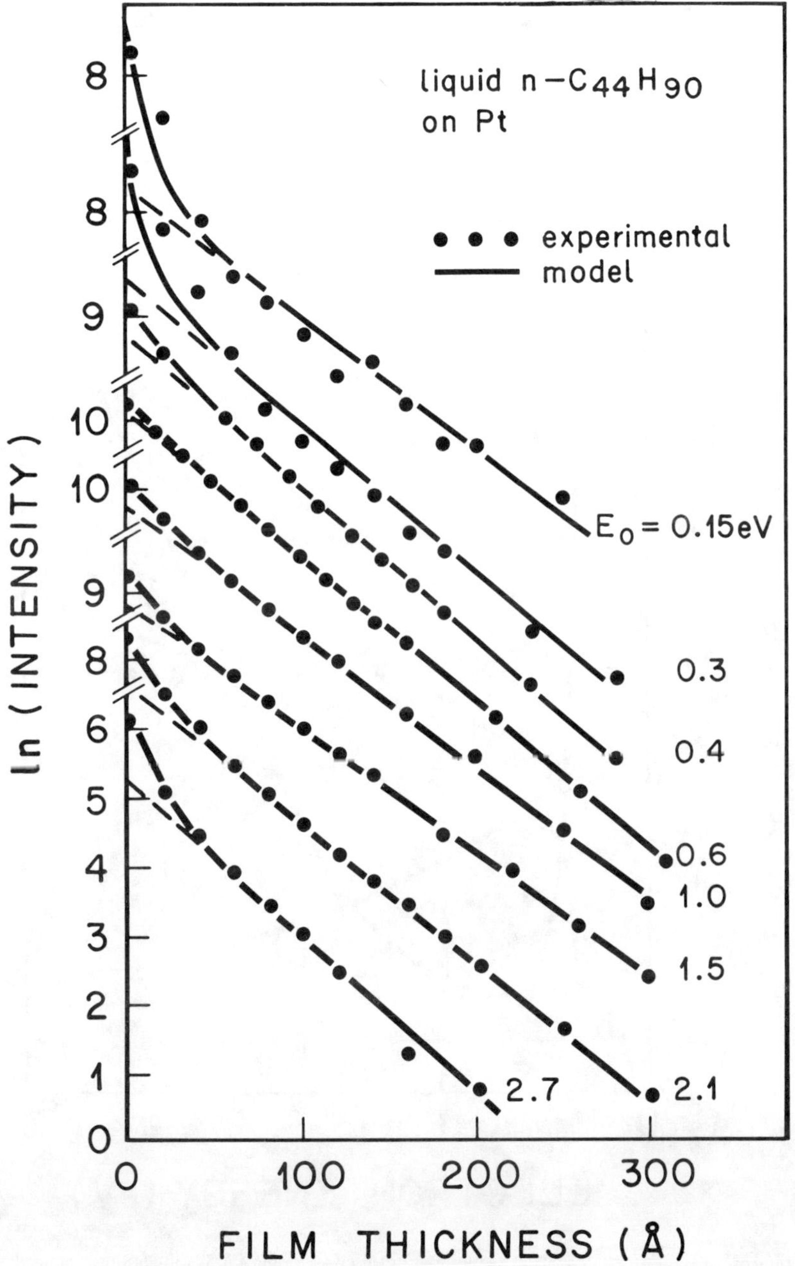

FIGURE 12.   Thickness dependence of the intensity transmitted through liquid n-$C_{44}H_{90}$ films in the topmost energy interval. The solid lines correspond to the simple analytical description according to Equation 3. (From Bernasconi, J., Cartier, E., and Pfluger, P., *Phys. Rev. B*, 38, 12567, 1988. With permission.)

11, but correspond to one plausible decomposition of $\gamma_{eff}(E)$. Second, the slightly structured effective scattering rate results from vastly different constituents. Finally, the structure in the optical phonon scattering rate reemerges in the effective rate, and expectedly in the calculated IPTA spectra. This suggests smoother thresholds than predicted by the Fröhlich formula, as though the bottom of E_c was not sharp. Indeed, the lack of sharpness caused by defects, impurities, or different chain lengths, is inherent to imperfect solids like the

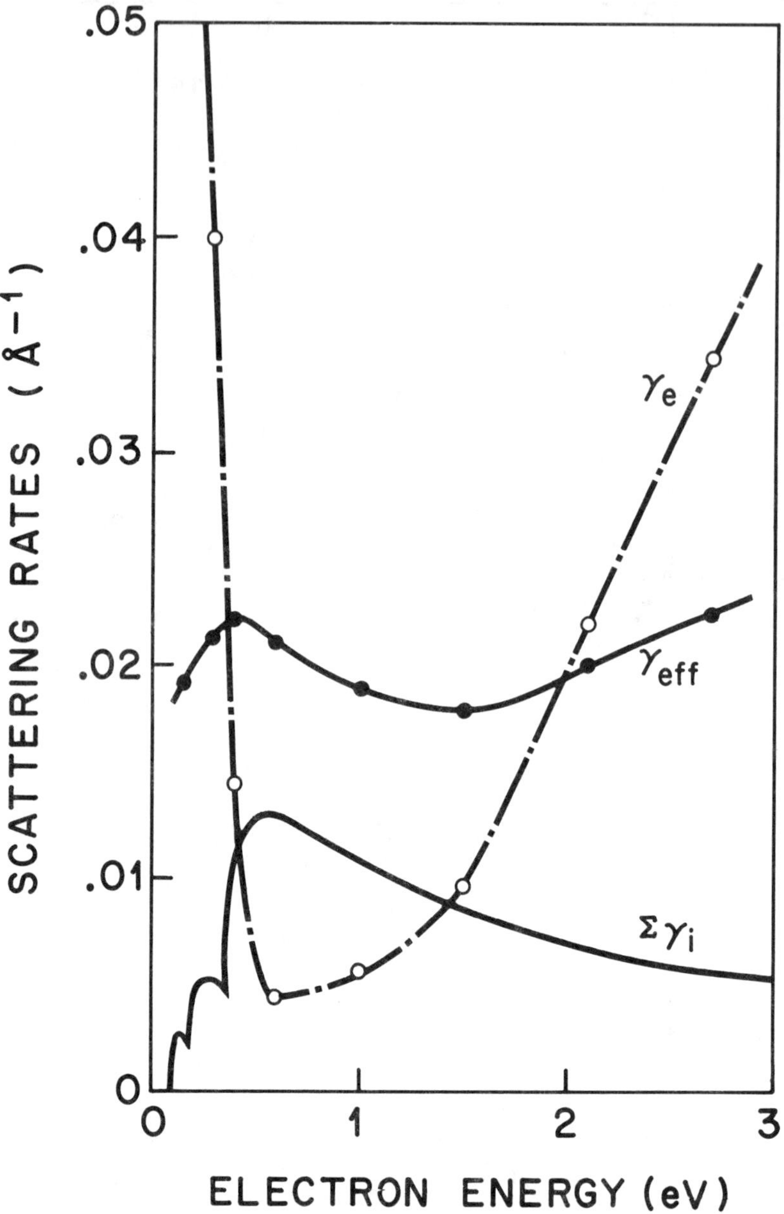

FIGURE 13.   Energy-dependent scattering rates used in Monte Carlo simulations, obtained from a decomposition based on the experimental data of Figure 12. $\Sigma\gamma_i$ is the sum of the emission scattering rates for all three optical phonon channels shown in Figure 7; γ_e is the pure elastic scattering rate extracted from γ_{eff} and $\Sigma\gamma_i$.

long chain alcanes. This problem has been discussed elsewhere in the case of simpler systems.[17,32,35] Furthermore, the vanishing of the optical phonon scattering rates toward E_c, our zero of electron energy, necessitates unreasonably large elastic rates, so large that the electron energy level is enormously broadened. Even if the inelastic rate possessed a low energy tail, as suggested above, this would still imply an increasingly large elastic rate as the energy decreases below 0.4 eV, which intuitively is attributed to defect scattering. Still, the elastic scattering rate may be large enough to doubt the validity of the free electron approach altogether, suggesting hopping or tunneling between localized states instead. In

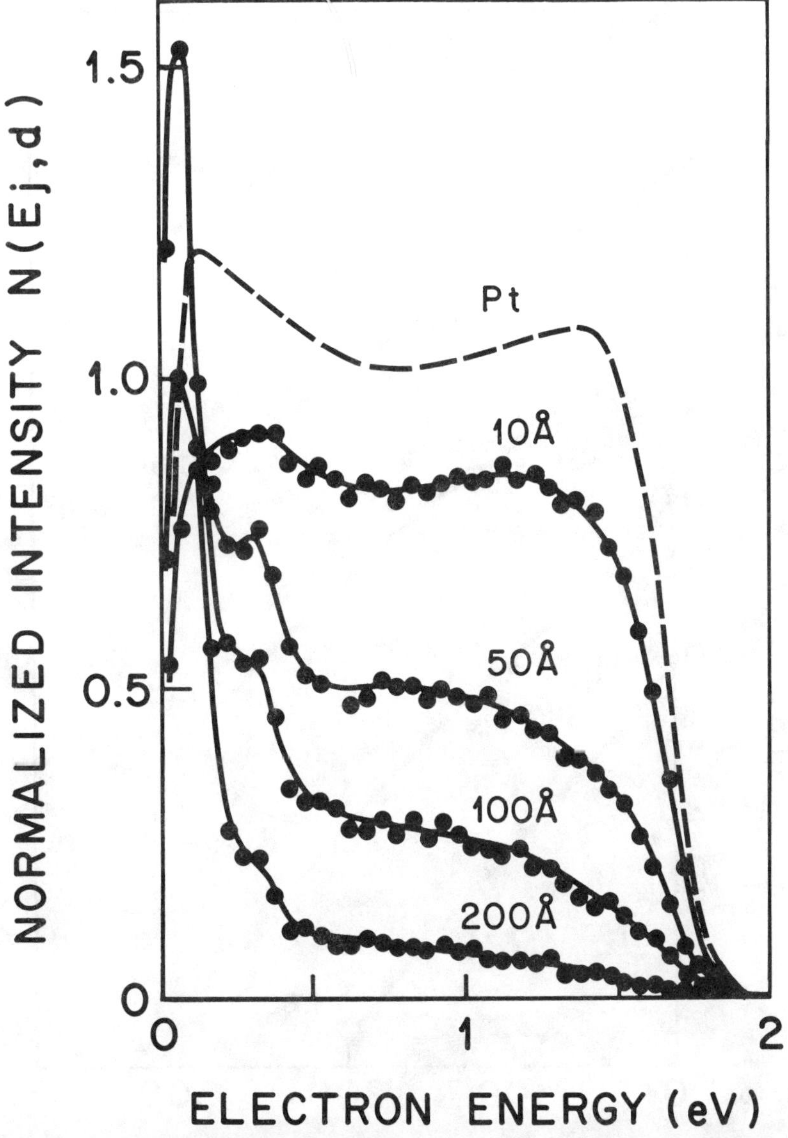

FIGURE 14. Monte Carlo generated IPTA spectra based on the scattering functions of Figure 13. A total of 5000 trajectories was simulated in each channel. The dashed line corresponds to the experimental spectrum of the platinum substrate. The dots are the data generated by Monte Carlo simulation. The full lines are only guides to the eye.

such an eventuality, our free-carrier picture breaks down and we should, in principle, implement hopping processes. Caron et al.[32] have investigated the consequences of such effects on the transmission of monoenergetic low energy electrons and found that it gives rise to a thickness dependence like a pure elastic channel with the hopping probability playing the role of the scattering length. Finally, the sizeable increase in the elastic rate above 1 eV cannot be fully accounted for by classical (normal or umklapp) acoustic phonon scattering alone.

Figures 14 and 15 exhibit the result of a Monte Carlo simulation performed with the scattering rates of Figure 13. The only other input has been a fit to the clean platinum spectrum to generate the entrance probability from the metal. The entrance angular probability

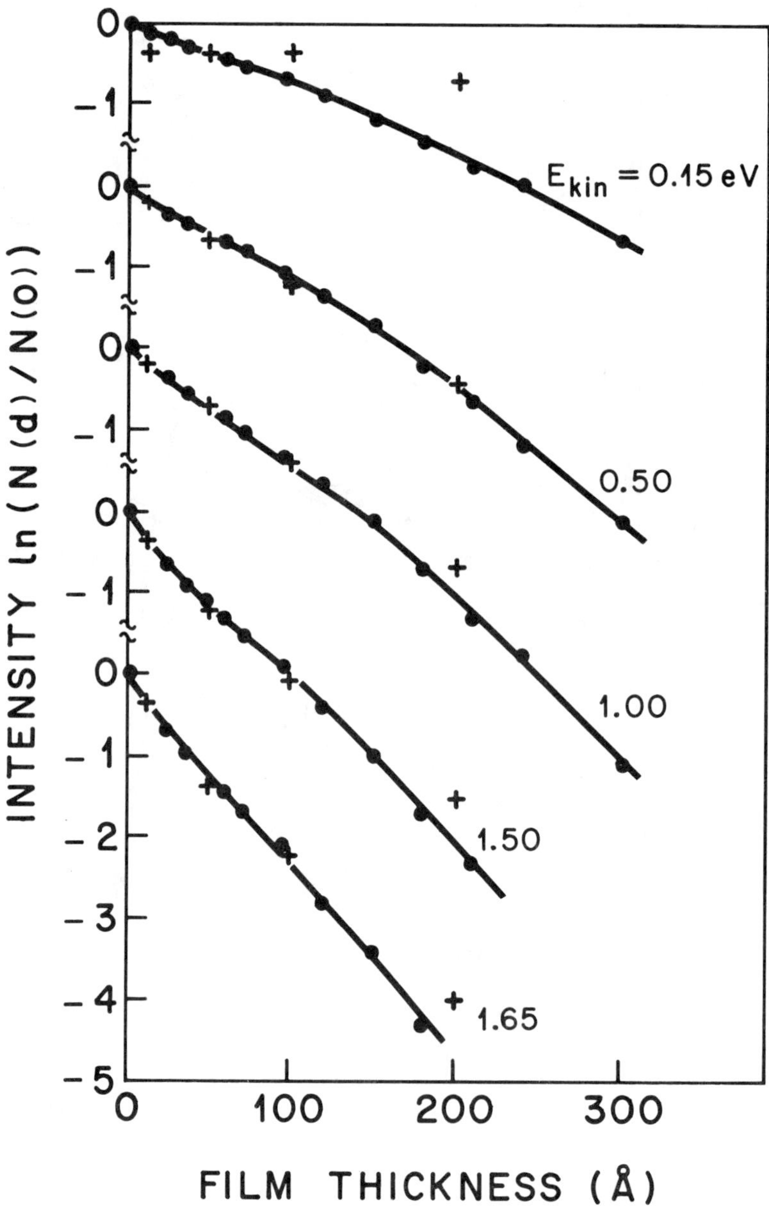

FIGURE 15.   Thickness dependence of the intensity transmitted through n-$C_{44}H_{90}$ films at various electron kinetic energies outside (below) the topmost energy interval, normalized to the emission intensity from the Pt substrate. The dots represent the experimental data taken from Figure 11, the crosses correspond to the Monte Carlo simulations. The full lines are only guides to the eye.

distribution was isotropic. The reflectivity of the interfaces was set equal to zero and the effective mass was taken equal to unity (we needed to renormalize the film thicknesses by a constant factor 1.2 to match the data of Figures 11 to 13, the former originating from a series of measurements in which the thicknesses of the films had been calibrated on an absolute scale with only ± 15% accuracy). The Monte Carlo generated spectra of Figure 14 should be compared to the data of Figure 11 and the agreement is quite satisfactory for the first iteration. Indeed, a consistent fit at all thicknesses and energies of IPTA spectra is basically impossible without proper account of all major processes. Two discrepancies have nonetheless appeared and are best observed in Figure 15 which shows the reduction of the transmitted intensity at several fixed energy values in the spectra. First, the experimental decrease at large thicknesses departs from the initial exponential decay, an effect not reproduced by our simulation. This departure may simply originate from the spreading effect mentioned in Section V. This effect clearly appears in the single-channel calculation shown in Figure 9, where it was seen that the spread extends to lower and lower energies as the thickness increases, entailing a departure from the ''intrinsic'' exponential decrease at a given energy when the tail of the spread reaches that energy. Second, the emission below about 0.3 eV electron energy is not well handled. Especially the height of the low energy peak is not well reproduced. This height, which results from a piling up of downcoming electrons, would definitely decrease if the thresholds of the optical emission rate were not so sharp, since some of the intensity would be transferred to lower energies. Some of the scattered electrons could also be transferred to below the vacuum level, where the electrons could not escape the film any longer, unless absorption events would bring them back above vacuum. The fine structure appearing in the spectra at energies below 0.5 eV (Figure 14) is entirely due to the strong maximum of the optical rate, as shown in Figure 13. Notice that, according to our analysis, the optical rate could not take on much larger values. It could take on smaller values, but this decrease should be accompanied by proper modifications of the effective mass and/or should be confirmed by experimental data, e.g., angular resolved EEL spectra. Other side effects can also be invoked, such as a nonzero reflectivity of the film-vacuum and film-substrate interfaces near zero energy.[33] Clearly, a more correct calculation will demand a smoothing of the curve to take into account the finite resolution of the apparatus, a variation of the absolute and relative strength of the optical modes, possible changes of the energy levels or other parameters in the film as the film grows.

To sum up, the preceding calculations have shown that the physics of the electron propagation in liquid tetratetracontane is dominated by strong optical phonon scattering on the one hand and highly elastic processes on the other hand. The same is true for solid alcanes, where hot electrons are concerned. Differences between solid and liquid hydrocarbons are only prominent in the transport of relatively slow electrons, *viz.*, those in the vicinity of the mobility threshold E_μ. The scattering by optical phonons turned out to be consistently described by the Fröhlich equation with the exception of the low energy behavior, near threshold, which cannot be as sharp as predicted. This discrepancy is most likely due to the lack of sharpness of the bottom of the conduction band and/or to its anisotropy. An averaging process can thus spread out the thresholds. The increase of the elastic channel with energy is ascribable to acoustic phonon scattering, either through normal or umklapp processes, or a combination of the two. The quasi-elastic behavior (and the spreading out effect) that the data seem to suggest can be provided by acoustic processes, although the effective energy transfer $\hbar\omega$ involved in these processes is not well known, especially since it results from an average over the acoustical phonon dispersion curves. The low energy divergence of the elastic channel hints at other physical mechanisms (defects, trapping, etc.), as discussed earlier.

The most interesting aspect of these results is the striking consistency between topmost and full IPTA spectra analyses. This is even more remarkable since we have verified that small changes in the mechanisms already have drastic consequences on the corresponding IPTA spectra. It should be emphasized at this point that the values for the transport parameters calculated above are just one possibility, the very first step of a series of iterations. We do not pretend to have found their proper and final values, because the model and assumptions are too crude at present.

VII. CONCLUSION AND OUTLOOK

The foregoing has demonstrated the scientific validity and the self-consistency of our approach englobing the IPTA experimental technique for hot electron transport studies in organic insulators, the modeling of the scattering processes, and the analysis technique of the IPTA spectra. Furthermore, the way was paved for a more sophisticated Monte Carlo analysis of IPTA spectra. Promising directions have been successfully identified, such as the implementation of a quasi-elastic channel on top of the pure elastic channel or a smoothing of the threshold of the optical channels. Some delicate parameters such as reflection coefficients, angular distributions, or anisotropic effective masses will need more attention in the future. This type of investigation will unavoidably call for some sort of systematic sensitivity study to evaluate the statistical error on the fit and the accuracy of the values obtained for the transport parameters.

An interesting step worth attempting, even in the absence of a full understanding of the IPTA spectra, consists in turning on the electric field. Owing to its importance for high-switching speed microelectronics, such a study has been carried along for SiO_2.[43] A particular difficulty in the case of polymers originates from the low kinetic energy propagation which is dominated by defect scattering and trapping mechanisms. The output of such a study would clearly be a knowledge of the time evolution of the kinetic energy for any given electron. Consequently, the kinetic energy distribution at, or out of equilibrium could be assessed. Degradation rates and breakdown properties could be determined by computer experiments rather than doing extensive IPTA studies and going through Equation 1. Such a study is now being tackled by one of the authors (E. A. Cartier).

At high fields the lifetime of an electron in any given state can be very short. Moreover, the scattering processes are definitely affected by high fields since the energy of an electron changes significantly during scattering. These considerations directly reemerge in the basic assumptions of the Monte Carlo procedure which, as described here, is essentially a semiclassical approximation. However, theoretical investigations have shown that quantum-mechanical effects can be taken into account reliably.[44]

ACKNOWLEDGMENTS

This work concludes the basic research activity on hot electron transport in solid and liquid organic dielectric films at Asea Brown Boveri. It is our pleasure to thank H. R. Zeller for initiating this activity 7 years ago, for his active support throughout this period, and his key contributions in the early stages of the project. During the course of this activity, important theoretical contributions were carried out by J. Bernasconi. The present study relies on these contributions and we gratefully acknowledge his invaluable comments and suggestions.

We also thank G. Blatter for the numerous discussions we had on theoretical matters and P. Brüesch for the measurement and interpretation of IR spectra. We greatly appreciated the help of S. Thuillier who performed some of the computer simulations presented in Section V.

REFERENCES

1. **Zeller, H. R., Baumann, Th., Cartier, E., Dersch, H., Pfluger, P., and Stucki, F.,** The physics of electrical breakdown and prebreakdown in solid dielectrics, *Advances in Solid State Physics (Festkörperprobleme),* 27, 223, 1987.
2. **Pfluger, P., Zeller, H. R., and Bernasconi, J.,** Hot-electron transport in polymeric dielectrics, *Phys. Rev. Lett.,* 53, 94, 1984.

3. **O'Dwyer, J.**., *The Theory of Electrical Conduction and Breakdown in Solid Dielectrics,* Oxford University Press, London, 1973.

4. **Bartnikas, R. and McMahon, E. J., Eds.,** *Engineering Dielectrics,* Vol. II, Electrical Properties of Solid Insulating Materials, ASTM, Philadelphia, 1979.

5. **Cartier, E. and Pfluger, P.,** Detection of hot-electron induced radiation damage in organic dielectrics by exoelectron emission from thin films, *IEEE Trans. on Electrical Insulation,* EI-22, 123, 1987.

6. **Pfluger, P., Cartier, E., and Dersch, H.,** Electron transport in thin insulating films; physics and effects on dielectric aging, *Proc. of the IEEE Int. Symp. on Electrical Insulation,* Cambrigde MA, 1988, 135.

7. **Sparks, M., Mills, D. L., Warren, R., Holstein, T., Maradudin, A. A., Sham, L. J., Loh, E., Jr., and King, D. F.,** Theory electron-avalanche breakdown in solids, *Phys. Rev. B,* 24, 3519, 1981.

8. **Zeller, H. R., Pfluger, P., and Bernasconi, J.,** High-mobility states and dielectric breakdown in polymeric dielectrics, *IEEE Trans. on Electrical Insulation,* EI-19, 200, 1984.

9. **Cartier, E. and Pfluger, P.,** Transport and relaxation of hot conduction electrons in an organic dielectric, *Phys. Rev. B,* 34, 8822, 1986.

10. **Cartier, E. and Pfluger, P.,** Experimental determination of energy dependent inelastic and elastic scattering rates of hot electrons in large bandgap insulators, *Physica Scripta,* T23, 235, 1988.

11. **Bernasconi, J., Cartier, E., and Pfluger, P.,** Hot-electron transport through thin dielectric films: Boltzmann theory and electron spectroscopy, *Phys. Rev. B,* 38, 12567, 1988.

12. **Bader, G., Perluzzo, G., Caron, L. G., and Sanche, L.,** Elastic and inelastic mean-free-path determination in solid xenon from electron transmission experiments, *Phys. Rev. B,* 26, 6019, 1982.

13. **Pireaux, J. J., Thiry, P., Caudano, R., and Pfluger, P.,** Surface analysis of polyethylene and hexatriacontane by high resolution electron energy loss spectroscopy, *J. Chem. Phys.,* 84, 6452, 1986.

14. **Ziman, J. M.,** *Electrons and Phonons,* Clarendon Press, Oxford, 1967.

15. **Madelung, O.,** *Introduction to Solid-State Theory,* Springer-Verlag, Berlin, 1978.

16. **Kittel, C.,** *Quantum Theory of Solids,* Wiley & Sons, New York, 1963, 1987.

17. **Sanche, L.,** Primary interactions of low-energy electrons in condensed matter, *Excess Electrons in Dielectric Media,* Jay-Gerin, J.-P., Ferradini, C., Eds., CRC Press, Boca Raton, FL, 1991, chap. 1.

18. **Kane, E. O.,** Electron scattering by pair production in silicon, *Phys. Rev.,* 159, 624, 1967.

19. **Blatter, G. and Bäriswyl, D.,** High-field transport phenomenology: hot electron generation at semiconductor interfaces, *Phys. Rev. B,* 36, 6446, 1987.

20. **Davydov, A. S.,** *Theory of Molecular Excitons,* Plenum Press, New York, 1971.

21. **Rice, M. J. and Phillpot, S. R.,** Polarons and bipolarons in a model tetrahedrally bonded homopolymer, *Phys. Rev. Lett.,* 58, 937, 1987.

22. **Tasumi, M. and Shimanouchi, T.,** Crystal vibrations and intermolecular forces of polymethylene, *J. Chem. Phys.,* 43, 1245, 1965.

23. See for example: **Conwell, E. M.,** *High Field Transport in Semiconductors, Solid State Phys.,* Suppl. 9, Academic Press, New York, 1967.

24. **Brüesch, P.,** unpublished results.

25. **Fitting, H. J. and Friemann, J. U.,** Monte-Carlo studies of the electron mobility in SiO_2, *Phys. Stat. Sol. (a),* 69, 349, 1982.

26. **Jacoboni, C. and Reggiani, L.,** The Monte Carlo method for the solution of charge transport in semiconductors with applications to covalent materials, *Rev. Mod. Phys.,* 55, 645, 1983; **Reggiani, L.,** *Hot Electron Transport in Semiconductors,* Topics in Applied Physics Vol. 58, Springer, Berlin, 1985; and references therein.

27. **Rei Vilar, M., Blatter, G., Pfluger, P., Heymann, M., and Schott, M.,** Monoenergetic and directed electron emission from a large-bandgap organic insulator with negative electron affinity, *Europhys. Lett.,* 5, 375, 1988.

28. **Fox, D., Labes, M. H., and Weissberger, A.,** *Physics and Chemistry of the Organic Solid State,* Vol. 1, Interscience, Wiley, New York, 1966.

29. **Heitler, W.,** *Quantum Theory of Radiation,* Clarendon Press, Oxford, 1970.

30. **Merzbacher, E.,** *Quantum Mechanics,* Wiley, New York, 1970, chap. 18.

31. **Di Maria, D. J. and Fischetti, M. V.,** Hot electron transport and trapping in silicon dioxide, *Excess Electrons in Dielectric Media,* CRC Press, Boca Raton, FL, 1991, chap. 10.

32. **Caron, L. G., Perluzzo, G., Bader, G., and Sanche, L.,** Electron transmission in the energy gap of thin films of argon, nitrogen, and n-hexane, *Phys. Rev. B,* 33, 3027, 1986.

33. **Rees, H. D.,** *Phys. Lett. A,* 26, 416, 1968; **Boardman, A. D., Fawcett, W., and Rees, H. D.,** Monte Carlo calculation of the velocity field relationship for galium arsenide, *Solid State Commun.,* 6, 305, 1968.

34. **Goulet, T., Keszei, E., and Jay-Gerin, J.-P.,** Probabilistic description of particle transport. I. General theory of quasielastic scattering in plane-parallel media, *Phys. Rev. A,* 37, 2176, 1988.

35. **Marsolais, R. M. and Sanche, L.,** Mechanisms producing inelastic structures in low-energy electron transmission spectra, *Phys. Rev. B,* 38, 11118, 1988.

36. **Chandrasekhar, S.,** *Radiative Transfer,* Clarendon, Oxford, 1950.

37. **Mahan, G. D.,** Hot electrons in one dimension, *J. Appl. Phys.,* 58, 2242, 1985; **Mahan, G. D., and Canright, G. S.,** Hot electrons in one dimension. II. Backscattering, *Phys. Rev. B.,* 35, 4365, 1987.

38. **Chantry, J., Phelps, A. V., and Schulz, G. J.,** Theory of electron collision experiments in intermediate and high gas densities, *Phys. Rev.,* 152, 81, 1966.

39. **Michaud, M. and Sanche, L.,** Interaction of low-energy electrons (1—30eV) with condensed molecules. I. Multiple scattering theory, *Phys. Rev. B,* 30, 6067, 1984.

40. **Marsolais, R. M.,** unpublished results.

41. The phase transitions of n-$C_{44}H_{90}$ were easily detectable in spectral features, particularly those at very low electron kinetic energy. The melting point of the thin films was found to lie within 2°C of the melting point of bulk n-$C_{44}H_{90}$.

42. **Ueno, N., Sugita, K., Seki, K., and Inokuchi, H.,** Low-energy electron transmission and secondary-electron emission experiments on crystalline and molten long-chain alkanes, *Phys. Rev. B,* 34, 6386, 1986.

43. **Fischetti, M. V., Di Maria, D. J., Brorson, S. D., Theis, T. N., and Kirtley, J. R.,** Theory of high-field electron transport in silicon dioxide, *Phys. Rev. B,* 31, 8124, 1985.

44. **Reggiani, L., Lugli, P., and Jauho, A. P.,** Quantum kinetic equation for electronic transport in nondegenerate semiconductors, *Phys. Rev. B,* 36, 6602, 1987.

Chapter 3

THERMALIZATION OF SUBEXCITATION ELECTRONS IN DENSE MOLECULAR MEDIA

René Voltz

TABLE OF CONTENTS

I. INTRODUCTION

Ionization of molecular matter by high energy electromagnetic or charged particle radiation invariably produces electrons with an energy spectrum far from the thermally equilibrated Maxwell-Boltzmann distribution. Whether primary or secondary, the electrons are continuously slowed down by creating electron-hole pairs and vibrational excitations of the medium. When their kinetic energy has dropped below the lowest electronic excitation energy of the molecules, further moderation can only take place by transferring energy to the nuclear degrees of freedom. As stressed by Hart and Platzman[1] a long time ago, the electrons which accumulate in this "subexcitation" energy domain should be treated as a distinct, separate part of the degradation spectrum, in which no additional electronic activations are produced and which is characterized by greatly reduced energy loss rates. Due to their high mobility and their protracted existence before thermalization, such subexcitation electrons can explore large spatial zones and initiate various observable activation or reaction processes, even with components present in small concentrations.[1] They are also conspicuous in determining the geminate ion pair separations and the spatial structure of the spurs,[2] which govern many of the permanent radiation effects in dense molecular matter.

In the reviews concerning the fundamental processes in radiation chemistry and physics[1-6] or the rapidly developing field of photoelectronic phenomena in molecular materials,[6-9] the urgent need of a better knowledge of the role and properties of subexcitation electrons in the intermediate stages of radiation action has been repeatedly noted. This clearly requires a more precise understanding of the inelastic and quasi-elastic scattering processes, together with the corresponding energy loss and transport properties which govern the dynamical behavior of the electrons prior to thermalization. Then reliable discussions of practically significant quantities, like "thermalization time", "thermalization length" and thermalized electron distributions around the hole in geminate pairs, should be easier.

Most of the recent studies of subexcitation electrons in molecular media were focused on the gas phase.[9,10] But even for the dense phases (with some exceptions, such as Magee's contribution[11]) the properties of individual molecules have ordinarily been used to investigate the interaction mechanisms with the vibrational modes. The fundamental limits of such a viewpoint have been experimentally demonstrated by Sanche[12] in recent studies of scattering of slow electrons in solid molecular films. If one considers, for example, the molecular resonant states, involving the formation of transient negative ions under isolated molecule conditions, the role of such local resonances in dense matter should instead be viewed as forming a tight-binding conduction band. The most natural description of the electron dynamics in the subexcitation energy domain is then in terms of quasi-free states of the electrons endowed with a proper effective mass and interacting with the molecular and lattice vibrations via the characteristic deformation potential mechanisms.[13-19] Such a representation, standard in the treatment of charge-carrier transport[14-17] and hot-electron relaxation[18] in metals and semiconductors, is less developed in similar studies of condensed molecular materials; this is notably the case for hot-electron thermalization.

This chapter is devoted to the theme of thermalization of subexcitation electrons in molecular solids and liquids; it is presented in the conceptual framework developed in contemporary condensed matter physics to account for charge carrier relaxation and transport. The emphasis will be on as simple and complete a presentation as possible of the entire thermalization process. An effort is made to derive the theoretical expressions required for realistic numerical estimates of the experimentally significant quantities, but the referred experimental data serve only as an illustrative purpose. No exhaustive review of the experimental information, directly or indirectly relevant to the theme, is attempted.

Section II is an overview of the physical system under consideration: the moving electrons and their interaction partners. In the subexcitation energy domain, the electrons generally act as electronic polarons and interact with the molecular vibrations, the optical and acoustic lattice modes and, eventually, the orientational polarization field in dipolar systems. Special attention is given to the specific properties of the molecular media, related to the high frequency molecular vibrations: the dynamical manifestations of the corresponding vibronic interactions underscore a demarcation between the supravibrational and subvibrational energy domains where the carriers behave, respectively, as electronic and vibronic (small) polarons.

In Section III, we analyze the dynamics of the electrons coupled to the intramolecular and lattice vibrational modes from a common point of view based on the deformation potential method. We must, therefore, begin by specifying the nature and form of the deformation potential, taken to represent, in each case, the electron-vibration interaction. For the molecular and optical lattice phonon modes, the connection with the linear vibronic interaction energy terms, more familiar in molecular spectroscopy, must be established. We then consider both the scattering and bound states of each of the interacting electron-phonon systems, as well as the associated electron energy loss and self-trapping properties. In the discussion, a central role is envisaged for the vibronic ''relaxation energy'', taken to characterize the mechanism under consideration. Rather than presenting a detailed and lengthy quantum theoretical derivation of the essential results, we prefer a more qualitative description, combining simple, classical arguments with the necessary quantum constraints such as the uncertainty principle. It is hoped that a heuristic approach, as suggested by a work of Rose,[19] helps the understanding, both simple and comprehensive, of the basic physics involved.

With the help of the results thereby obtained on the electron energy loss rates and on the inelastic and elastic scattering probabilities, we are then in a position to discuss in some detail the dynamical behavior of the subexcitation electrons during the thermalization process. This is attempted in Section IV, where we examine the evolution, on the dimensions of energy and space, of the electrons during the two main stages of their relaxation. In the upper part of the subexcitation energy domain, the numerical estimates confirm the dominant role of intramolecular vibrational excitations in efficient moderation, and of the acoustic phonons in spatial diffusion of the hot electrons. The kinetic description is borrowed from ''age theory'' and enables one to properly define the energy-loss time scale and the diffusion length that characterize this supravibrational stage of the process (Section IV.A). The final stage of electron moderation toward thermal equilibrium is treated in Section IV.B: it concerns the subvibrational energy domain, where the electrons generally have an increased effective mass and interact only with the lattice phonons by scattering events involving small energy loss and gain; relaxation is then best described as a Fokker-Planck process, with diffusion and streaming down the energy axis, together with acoustic phonon controlled spatial diffusion. The derived characteristic thermalization times and lengths agree with the existing experimental estimates in hydrocarbon liquids and crystals.

II. ELECTRONS AND PHONONS IN MOLECULAR MEDIA

In dense molecular matter, a moving charged particle interacts with electronic (exciton) and nuclear (phonon) modes which may be classified according to the nature of the interaction. One class includes the various polarization modes that are coupled to the charge carrier by its *long-range Coulomb interaction field*. These are, in principle, associated with the electronic, ionic and orientational polarization contributions, which generally have well-separated spectral and temporal response properties. The other important class is that of the intramolecular and lattice vibrational modes interacting with the charge carrier via a *short-range deformation potential* mechanism.[13-15]

A. FREE ELECTRONS AND POLARONS
The dynamical consequences of the interaction of the moving charge carrier with one of the aforementioned electronic or vibrational modes of the medium are usually described

in the framework of an *adiabatic separation* of the motions of the interacting subsystems. The general picture then depends critically on the comparison of two characteristic times. The first represents the time of interaction of the charged particle with a molecular site in the medium, given by

$$\tau = a/v = \hbar/B \tag{1}$$

where a is the lattice constant or mean spacing between molecular neighbors and v is the velocity of the charge carrier which can eventually be expressed in terms of an associated energy band width B. For the other interacting subsystem, the natural unit for the temporal response is taken as ω^{-1}, if $\hbar\omega$ is the energy of the electronic or vibrational excitation considered. For a charged particle of kinetic energy so high that

$$v > a\omega \text{ or } \omega\tau < 1 \tag{2}$$

the particle is the fast subsystem; an impulsive excitation of the slow modes with the concomitant *energy loss of the fast particle* is the interesting dynamical feature. On the other hand, for sufficiently slow charge carriers, when

$$v < a\omega \text{ or } \omega\tau > 1 \tag{3}$$

the electronic or vibrational mode of the medium is the fast subsystem, which instantaneously follows the slow moving particle; here no real energy exchanges take place, but the charged particle, dressed by the interacting modes, has increased self-energy and effective mass, becoming what is called a "polaron".[13-15]

"Subexcitation electrons" are defined as satisfying the low energy condition (Equation 3) for the electronic polarization fields ($\omega \equiv \omega_e$). These correspond to the characteristic plasmon and exciton modes of the medium, but, in molecular matter, the lowest frequency to be considered is generally that of the lowest triplet exciton state, with energies $\hbar\omega_e$ of several electron-volts and frequencies $\omega_e \approx 10^{15}$ to 10^{16} s^{-1}. According to the foregoing arguments, the moving subexcitation electrons are "dressed" by the electronic polarization clouds. They must, therefore, be regarded as "electronic polarons" which, compared to the bare electron, have a higher effective mass m* and an additional self-energy.[13-15,20,28] Crude estimates of such stabilizing polarization energies are often made using familiar models.[21,22] Following Born's approach, for example, one may consider the difference of the free energies of charging an ion in the vacuum and in the dielectric medium, respectively, obtaining:

$$P_e = \frac{e^2}{2a}\left(1 - \frac{1}{\epsilon_\infty}\right) \tag{4}$$

Here ϵ_∞ is the high frequency, optical dielectric constant and *a* is the effective radius of the ion, being of the order of the lattice constant. In crystalline and liquid nonpolar hydrocarbons ($\epsilon_\infty = 2$), values of P around 1 to 2 eV are commonly considered.[4,6,7] According to general polaron theory,[13,14,20,28] the self-energy and the effective mass are related to the energy of the coupled exciton field, $\hbar\omega_e$, and to the bare electron mass m, by

$$P_e = \alpha\hbar\omega_e, \ (m^*/m) = 1 + \alpha/6$$

where α is a characteristic polaron coupling constant. In view of the above mentioned numerical values for P_e and $\hbar\omega_e$, one generally has $\alpha < 1$; as a consequence, only a weak mass renormalization is expected to take place for the electronic polarons in molecular media ($m^* \approx m$).

B. MOLECULAR AND LATTICE VIBRATIONS

The energy losses of the subexcitation electrons can only occur by excitation of the slower vibrational motions. In molecular media, the intermolecular forces are so weak that, to a first approximation, the vibrational modes can be classified according to their parentage in the free molecular motions. Since each molecule contributes its own vibrations plus three rotations and three translations, we shall consider, together with the internal molecular vibrations, the lattice rotational (librational) and translational modes.[23,24]

The *intramolecular vibrations* ($\omega \equiv \omega_M$) involve various bond stretchings and bendings.[23,24,43] The stretching modes are more important in coupling with electrons as demonstrated by the vibrational progressions in the typical vibronic spectra. These progressions indicate that the electronically excited states and, by analogy, the negative ionic states have a slightly expanded nuclear skeleton relative to the ground state. In hydrocarbon media the wave numbers of such molecular stretching motions range typically between 1000 cm^{-1} for the C–C bonds and 3000 cm^{-1} for the C–H bonds. Such values correspond to frequencies ω_M of the order of 10^{14} s^{-1} and to Debye temperatures $\Theta_D = \hbar\omega_M/k_B$ (k_B: Boltzmann constant) higher than 10^3 K. The thermal population number given, following the Bose-Einstein distribution, by

$$\langle n \rangle = [\exp(\hbar\omega_M/k_BT) - 1]^{-1} \tag{5}$$

is therefore negligible for such high frequency modes. The interaction of the intramolecular modes with the moving electron is naturally local and short ranged. It involves the linear electron-vibration interaction terms responsible for molecular vibronic coupling and may be treated by introducing an appropriate "optical" deformation potential.[13,14,17-19] (see Section II.A).

Among the *lattice vibrational modes*, there are always three *acoustic modes* ($\omega \equiv \omega_a$), which have zero energy for zero wavevector of the corresponding waves. All the other modes are termed "optical" and have frequencies ($\omega \equiv \omega_L$) approximately independent of the wave number. The energies are generally of the order of 100 sm^{-1}; for organic crystals, for example, the values are between 40 and 140 cm^{-1} for the different librational phonon branches, and between 30 and 70 cm^{-1} at the zone edges for the translational, acoustic modes.[24] As usual, the long wavelength acoustic phonon energies are taken as linear functions of the wave number q, $\hbar\omega_a = \hbar c_s q$, where c_s is the sound velocity in the dense medium ($c_s \approx 10^5$ cm·s^{-1} in organic materials).[18,23] Phonon wave numbers of ≈ 100 cm^{-1} correspond to $\Theta_D \approx 145°$K, which shows that the acoustic and optical lattice vibrations at room temperature have populations (Equation 5) which satisfy equipartition of energy:

$$\langle n_a \rangle = k_BT/\hbar\omega_a, \quad \langle n_L \rangle = k_BT/\hbar\omega_L \tag{6}$$

The totality of these lattice modes, to which hydrogen bond bending and stretching motions (50 to 200 cm^{-1}) must eventually be added, are responsible for continuous energy exchanges with the electron during the whole thermalization process. The appropriate coupling mechanisms are represented by the short-range acoustic or optical deformation potentials.[13-15,17-19] (cf., Section III).

Paying attention only to nonionic materials, we shall not consider the role of the longitudinal optical (LO) phonons associated with ionic polarization fields.[13-16] By contrast, the effect of orientational polarization is known to be of essential importance for electrons in *dipolar media* (liquid water, hydroxylic solvents, etc.). The energy loss mechanism is qualitatively different from the foregoing electron-phonon interaction processes.[25] It involves the macroscopically described long-range coupling to the dipolar relaxation modes as represented in the simplest cases, by the Debye form of the dielectric spectral response function:

$$\epsilon(\omega) = \epsilon_\infty + (\epsilon_o - \epsilon_\infty)/(1 - i\omega\tau_D) \qquad (7)$$

where, in addition to the high frequency dielectric constant ϵ_∞, the static dielectric constant ϵ_o is also incorporated. τ_D is Debye's time constant for relaxation of the dipoles by rotational diffusion. As an example, we may mention the values for liquid water:[26] $\epsilon_o \approx 80$, $\epsilon_\infty = 2$, $\tau_D = 8$ ps.

C. INTERACTION ENERGY AND TIME SCALES

The dynamical behavior of subexcitation electrons in molecular matter takes place between definite limits on the energy and time scales, which are set by the response times of the coupled molecular or lattice motions. Two energy ranges in general must be distinguished.[11,27]

In the *high energy range,* with electron energies E higher than the intramolecular vibrational quanta ($\hbar\omega_M \approx 0.1$ to 0.3 eV),

$$\hbar\omega_e > E > \hbar\omega_M \qquad (8)$$

the electrons act as "light" electronic polarons, with a weakly renormalized effective mass m_e^*. Such "supravibrational" electrons exchange energy with practically all the vibrational modes of the medium. But here the high frequency, intramolecular vibrations are more conspicuous because they are responsible for the efficient energy loss processes with a fast cascading down to the lower limit $\hbar\omega_M$. All the while losing energy by emission of the intramolecular phonons, the electrons also interact with the low frequency, thermally populated lattice modes. This causes quasi-elastic scattering with emission and absorption of relatively small vibrational quanta, resulting in continuous energy exchanges with the lattice together with spatial diffusion of the electron (cf., Section IV.A).

At later times, in the lower velocity range, with *subvibrational electron energies* such that

$$\hbar\omega_M > E \qquad (9)$$

the electrons become the slow subsystem which is followed by the faster molecular vibrations. They must then be regarded as moving vibronic compound systems or "heavy" *vibronic polarons.* The corresponding effective masses m_{ev}^* are related to the electronic masses m_e^* by means of a vibrational factor, $m_{ev}^*/m_e^* \approx S_v^{-2}$ where S_v is the intramolecular vibrational overlap integral between the zero-point molecular neutral and negative ion states. It is given by $S_v = \exp(-S/2)$, where S is the "Huang-Rhys factor" characteristic for vibronic coupling.[13,14,24] With $S > 1$, a notable increase of the electron effective mass ($m_{ev} > 3\ m_e^*$) is thus generally expected. These heavy subvibrational electrons are coupled to the low frequency optical and acoustic modes of the medium in quasi-elastic collision processes with small energy exchanges. They behave as Brownian particles on the energy dimension as well as in ordinary space, relaxing, according to diffusion equations, to thermally equilibrated Boltzmann-Maxwell distributions involving eventually the self-trapped states associated with the various lattice motions.[29,30] This will be considered in more detail in Section IV.B.

In the case of polar materials one has to consider, in addition to the previous short-range, electron-phonon interactions, the long-range coupling of the moving electrons to the permanent molecular dipole moments in the medium. The dynamical behavior of the electrons in the orientational polarization field is scaled by the characteristic Debye relaxation time τ_D introduced before. For charged particles with velocities so high that $v > a/\tau_D$ (i.e., $\tau < \tau_D$), the coupling to the dipolar relaxation modes results in continuous energy transfer to the dielectric medium with rates that may be comparable to those of the aforementioned mechanisms.[11,25] At lower velocities, when $v < a/\tau_D$ ($\tau > \tau_D$) the dynamical influence of the medium is better expressed in terms of the "dielectric friction" processes found to be important for slowly moving ions.[31,32] But for electrons, the concerned kinetic energies are generally too low for being of some practical importance.

III. ELECTRON-PHONON INTERACTION MECHANISMS IN MOLECULAR MEDIA

The dynamical features of electron-phonon interactions in dense matter must, in principle, be described by a quantum mechanical treatment of the electron-phonon scattering processes in the momentum space.[13-19] As already indicated, we shall, instead, present the essential features of the interaction of electrons with the molecular and lattice vibrations more qualitatively by combining classical arguments in the real space with the necessary quantal constraints.[19] Such a heuristic approach is convenient to focus on the basic physics common to all the possible energy loss and trapping mechanisms for moving electrons in molecular matter. The results have validity in only an order of magnitude, consistent with the uncertainty principle, but are easily tested by comparison with the "exact" results in the literature of solid state theory.[13-19]

Among the quantum physical conditions which must be kept in mind, we will of course regard the energy exchanges between electrons and vibrations as taking place via emission or absorption of intramolecular and lattice vibrational quanta, $\hbar\omega_M$ and $\hbar\omega_L$. Less trivial perhaps are the limits placed by the uncertainty principle upon the localizability of a particle: for an electron with maximum momentum m*v, the smallest wave packet that can be formed is of the order of

$$R = \hbar/m*v. \tag{10}$$

This length may therefore be visualized as the "effective size" of the electron.[19]

In each of the cases, concerned either with intramolecular vibrations or with optical and acoustic lattice phonons, we have to consider the scattering states of the moving electrons with the collisional energy exchanges, as well as the bound states of the electron-phonon system, responsible for electron localization and trapping. In the former case, the electrons are in "extended" states with wave functions extending over many molecular sites in the crystal or liquid; in the latter case, the electronic wave function is rapidly decreasing over a distance of the order of the lattice constant a. A convenient general form is

$$\psi(r) = \psi_o \exp(-\lambda r/a), \tag{11}$$

where the parameter λ represents the degree of spatial localization. It varies between $\lambda = 1$ for an electron completely localized at a molecular site, and $\lambda = O$ for an extended state; in order to account properly for the limit of delocalized quasi-free electron states, ψ_o can be taken as N exp(ikr), where N is an appropriate normalization constant.

A. OPTICAL PHONON MODES

In the description of the electron-phonon interactions in nonionic molecular media, we first consider the intramolecular vibrations and the nonpolar optical lattice modes presented in the previous section.

1. The Localized Oscillator Model

It is enlightening to begin the analysis of electron-phonon coupling processes with the elementary model of the electron interacting with a single oscillator. In molecular matter, this applies directly to the intramolecular vibrations or to the localized optical phonon modes of the lattice. The model not only serves to introduce the basic dynamical features of interest, but also provides a useful guide for the following more general treatments concerned with elastic continua.

Under the familiar adiabatic approximation separating the molecular vibrational and electronic motions, the kinetic energy of the oscillator is neglected. In the absence of interaction, the energy of the electron-oscillator system is then taken as

$$E_e^o(x) = E_e + \frac{1}{2} \mu\omega^2 x^2$$

where E_e denotes the energy of the electron, μ, ω, and x, respectively, represent the mass, angular frequency, and coordinate of the oscillator, located at some site r in the medium. When the electron interacts with the oscillator, the latter is displaced by a small amount δ_e, which results in an energy change given, in the linear approximation, by $\delta\epsilon = -\mu\omega^2\delta_e x$. This change can then be taken as the effective potential seen by the electron interacting with the oscillator, *viz.*, $V_e(x) = -(\mu\omega^2\delta_e)x$. The total energy of the interacting system is thus of the form:

$$\epsilon_e(x) = E_e + \frac{1}{2} \mu\omega^2 x^2 - (\mu\omega^2\delta_e)x \tag{12}$$

where the successive terms, respectively, represent the electronic energy, the elastic energy of the vibration, and the linear vibronic interaction potential.

Considering first the bound states of the system, when the electron is localized ($\lambda = 1$, in Equation 11) at the oscillator, we may look for the most stable vibronic configuration. Minimization of the expression (Equation 12) with respect to x shows that it corresponds to $x_o = \delta_e$, with an energy

$$\epsilon_e(x_o) = E_e - E_R, \quad E_R = \frac{1}{2} \mu\omega^2\delta_e^2 \tag{13}$$

where E_R is called the vibronic "relaxation energy".

The other limiting case corresponds to the scattering states of the system when the electron is in an extended state. Of interest here is the transfer of energy from the fast, quasi-free electron to the oscillator. In the heuristic approach, we begin with the classical picture of a point-like swift particle interacting with the oscillator through the linear vibronic potential. If the oscillator is subjected to the transient force $-[\partial V_e(x)/\partial x]$ for the brief interaction time τ, defined by Equation 1, the momentum and energy transferred are given by

$$\Delta p_{cl} = -(\partial V_e/\partial x)\tau = (2E_R)^{1/2}\mu^{1/2}\omega\tau$$

$$\Delta E_{cl} = \Delta p_{cl}^2/2\mu = E_R\omega^2\tau^2$$

respectively. This enables one to define a classical form for the energy transfer rate:

$$(dE/dt)_{cl} = \Delta E_{cl}/\tau = E_R\omega^2\tau$$

In a second step, the classical results must be corrected to account for the spatial uncertainty of the electron, which can be visualized as having an effective radius given by Equation

10. Hence, the interaction with the localized oscillator lasts for a time $\tau_e = R/v$, during which the probability of presence of the electron at the molecular site is not unity, but approximately $(a/R)^3$. The expression for the actual amount of energy transferred to the oscillator must include this probability factor and the new interaction time, so that

$$\Delta E = E_R \omega^2 \tau_e^2 (a/R)^3$$

The corresponding *energy transfer rate* then becomes:

$$(dE/dt) = \Delta E/\tau_e = E_R \omega^2 (a/R)^2 \tau \tag{14}$$

where τ and R are defined by Equations 1 and 10. It will be noted later that this simple result may be a useful approximation to the exact expressions derived by more rigorous arguments.

2. The Elastic Continuum Model

While the localized oscillator approach is certainly the most natural to account for the influence of molecular vibrations, the electron-lattice interaction processes are more frequently described by using the elastic continuum model for the phonons.[13-16,33-35] The two approaches are, of course, intimately related by the familiar rules connecting discrete and continuous oscillator fields. In practice, the electron-phonon interaction is represented by short-range optical deformation potentials.

a. Optical Deformation Potential

In the case of the optical phonons, associated either with intramolecular or lattice modes, the vibrational state is characterized by the local oscillator amplitude $q(r)$ at each point r of the continuum. The elastic energy density is then taken as $\rho\omega^2 q^2/2$ where ρ is the oscillator mass density and ω the angular frequency of the optical mode. Following arguments similar to those for the localized oscillator, the energy of an electron in the medium depends upon the vibrational state according to a linear term which is then taken as the electron-phonon interaction potential density in the form $-D_{op} q(r)|\psi(r)|^2$. Here $|\psi(r)|^2$ accounts for the local electron density and D_{op} is the characteristic *optical deformation parameter*.

We thus write the total energy of the electron-phonon system as

$$\epsilon[\psi,q] = \int dr\ \psi^*(r)\left(-\frac{\hbar^2}{2m^*}\nabla^2\right)\psi(r) + \frac{1}{2}\rho\omega^2 \int dr|q(r)|^2 - D_{op}\int dr|\psi(r)|^2 q(r) \tag{15}$$

augmented by the lattice kinetic energy, which is again neglected in the adiabatic approximation.[13,14,33-35] This energy has the same general form as Equation 12, but here it is a functional of the electronic and vibrational fields, $\psi(r)$ and $q(r)$. The first term describes the kinetic energy $< p^2/2m^* >$ of the electron with effective mass m^* in a state which may be either extended or localized, as represented for example by Equation 11. The two other terms represent the elastic lattice energy, and the linear electron-lattice interaction energy, respectively.

b. Self-trapped and Quasi-free Electron States

For each given electronic state with fixed $\psi(r)$, we can investigate the most stable electron-phonon states in a more comprehensive way as before, looking for the existence of stable localized or self-trapped states, together with the extended or free states of the electron coupled to the optical phonon field. Minimization of Equation 15 with respect to $q(r)$ yields the relation between the electronic state $\psi(r)$ and the resulting vibrational state $q_o(r)$ of the medium, *viz.*

$$q_o(r) = D_{op}|\psi(r)|^2/\rho\omega^2 \tag{16}$$

The associated energy of the system is then obtained as

$$\begin{cases} \epsilon[\psi] = E_e[\psi] - E_R[\psi] \\[2ex] E_e[\psi] = \int dr\psi^*(r)(-\dfrac{\hbar^2}{2m^*}\nabla^2)\psi(r) \\[2ex] E_R[\psi] = \dfrac{1}{2}\dfrac{D_{op}^2}{\rho\omega^2}\int dr|\psi(r)|^4 \end{cases} \tag{17}$$

where $E_e[\psi]$ has been defined before. $E_R[\psi]$ may be regarded as the vibronic relaxation energy of the lattice in presence of the electron in state $\psi(r)$. Remembering the general expression (Equation 11) of the electronic wave function, one notes that the energies (Equation 17) depend upon the degree of localization λ: $\epsilon(\lambda) = E_e(\lambda) - E_R(\lambda)$.

If one takes up the case $\lambda = 1$, corresponding to the limiting situation of an electron completely localized at a molecular site, the previous results of the localized oscillator model must be recovered. An exact evaluation of Equation 17 depends on the exact form of the electronic wave functions, but the results are well represented by the following expressions

$$\begin{cases} E_e(\lambda = 1) = (\hbar^2/2m^*)(\pi/a)^2 \equiv E_e \\[2ex] E_R(\lambda = 1) = D_{op}^2/2\mu\omega^2 \equiv E_R \end{cases} \tag{18}$$

where $\mu = 4\pi a^3 \rho/3$ is the oscillating mass per molecular site. We may note that for electrons with effective mass m^* in a conduction band, the energy E_e thus defined is of the order of the conduction band width B. A comparison of Equations 13 and 18 shows that the optical deformation parameter of the elastic continuum is given, in terms of the localized oscillator properties as

$$D_{op} = \mu\omega^2\delta_e \tag{19}$$

In the expression (Equation 17) of the total energy of the system, E_e is the positive kinetic energy term which represents the delocalized, quasi-free aspects of the electronic behavior. The negative relaxation energy term E_R stands for the opposite effect of binding to the local oscillators with formation of trapped states. The occurrence of free and trapped electronic states is thus determined by the ''electron-phonon coupling constant''[13,33]

$$g_{op} = E_R/E_e \simeq E_R/B \tag{20}$$

When $g_{op} > 1$, a stable state, with the electron trapped at a molecular site, exists; such a system is called a ''small or molecular polaron''. If $g_{op} < 1$, the molecular polaron cannot be stable and the electron remains quasi-free.

To complete the analysis beyond the limiting case $\lambda = 1$, we now consider the influence of the degree of localization of the electron ($0 < \lambda < 1$) as defined in Equation 11. It is easy to verify that the expressions (Equation 17) become $E_e(\lambda) = E_e\lambda^2$ and $E_R(\lambda) = E_R\lambda^3$ in terms of the energies defined by Equation 18. The total energy then reads

$$\epsilon(\lambda) = E_e\lambda^2 - E_R\lambda^3 \tag{21}$$

Inspection of this function shows that the condition for $\epsilon(\lambda) < 0$, i.e., for the existence of bound electron-phonon states, is more generally expressed as $g_{op}\lambda > 1$. Among these trapped states, the molecular polaron characterized by maximal localization ($\lambda = 1$) is the most stable. On the other hand, when $g_{op}\lambda < 1$, one has $\epsilon(\lambda) > 0$, which corresponds to untrapped, quasi-free and extended electronic states. Here, the positive kinetic energy term $E_e(\lambda)$ overcomes the local relaxation energy $E_R(\lambda)$ of the electron, thus preventing self-trapping. It is interesting to note that for $\lambda^*_{op} = 2/3\, g_{op}$, the energy ϵ is maximum and equal to $\epsilon^* = 4E_e/27\, g^2_{op}$. During the course of a localization process, when an electron evolves from an extended state [$\lambda = 0$, $\epsilon(\lambda) = 0$] to the limiting self-trapped molecular polaron state ($\lambda = 1$, $\epsilon(\lambda) = E_e - E_R$), it thus encounters a positive potential barrier ϵ^*_{op} at an intermediate state characterized by λ^*_{op}.[33-35]

c. Energy Loss and Scattering of Electrons

The *rate of energy loss* of quasi-free electrons to the elastic continuum can be derived by generalizing the heuristic approach outlined before for the localized oscillator. The optical deformation potential representing the short-ranged coupling of the electron with the effective size R, defined by Equation 10, is taken as:

$$V_{op}(r) = \begin{cases} -D_{op}q(r), & (r < R) \\ \\ 0 & (r > R) \end{cases} \tag{22}$$

The material continuum in contact with the electron is thus submitted to the transient force $-\nabla V(r)$ for the interaction time $\tau_e = R/v = (R/a)\tau$. Then, the resulting transfers of momentum and energy are given by $\Delta p = D_{op}\tau_e$ and $\Delta E = \Delta p^2/2M$, where $M = \rho(4\pi R^3/3)$ is the effective mass of the medium. The energy transfer rate, $(dE/dt)_{op} = \Delta E/\tau_e$ is then easily found to be formally identical to Expression 14 for the localized oscillator model. By introducing the wave-number $k = m^*v/\hbar = R^{-1}$, one has

$$(dE/dt)_{op} = E_R\omega^2(ka)^2\tau$$

where E_R is the local vibronic relaxation energy, defined by Equation 18.

Using the Equations 1 and 18, with $v = (2E/m^*)^{1/2}$, we can express the rate as

$$\left(\frac{dE}{dt}\right)_{op} = \frac{3}{4}\frac{D^2_{op}}{2^{1/2}}\frac{m^{*3/2}}{\pi\hbar^2\rho}E^{1/2} \tag{23}$$

from which we may deduce the expressions of the mean-free time, $t_{op} = \hbar\omega(dE/dt)^{-1}_{op}$, and of the *mean-free path* $\ell_{op} = (vt_{op})$ between the events where the quanta $\hbar\omega$ are transferred to the optical phonon field,

$$\frac{1}{t_{op}} = \frac{3}{4}\frac{D^2_{op}}{2^{1/2}\pi\hbar^3}\frac{m^{*3/2}}{\rho\omega}E^{1/2} \tag{23A}$$

$$\frac{1}{\ell_{op}} = \frac{3}{4}\frac{D^2_{op}}{2\pi\hbar^3}\frac{m^{*2}}{\rho\omega} \tag{23B}$$

These results, obtained from qualitative arguments, are very close to the exact expressions derived by quantum mechanical transition rate theory. Equation 23A, for example, may be compared to the probability per unit time of spontaneous emission of an optical phonon by a fast electron[36]

$$W_{op}^{(+)} = \frac{D_{op}^2 \, m^{*3/2}}{2^{1/2}\pi\hbar^3\rho\omega} \, (E - \hbar\omega)^{1/2} \tag{24}$$

This shows that the Expression 23 underestimates the actual values by a factor 3/4 (unimportant for the accuracy here required) and applies best for high electron energies, i.e., if $E \gg \hbar\omega$.

When an electron interacts with the optical modes, it can also absorb phonons with a transition probability given by the companion formula of Equation 24, i.e.,

$$W_{op}^{(-)} = \frac{D_{op}^2 \, m^{*3/2}}{2^{1/2}\pi\hbar^3\rho\omega} \, (E + \hbar\omega)^{1/2} \tag{25}$$

In the presence of an optical phonon population n_{op}, the energy exchanges with the moving electrons involve spontaneous and induced emission together with absorption of the phonons. This may be characterized by the probability per unit time of the electron being scattered out of its state k:[36]

$$\frac{1}{\tau_{op}} = (1 + n_{op}) \, W_{op}^{(+)} + n_{op} \, W_{op}^{(-)}$$

In the important case of the low frequency lattice modes, Equation 6 applies with $<n_{op}> \gg 1$; also $\hbar\omega \ll E$, meaning that $W^{(-)} \approx W^{(+)} = t_{op}^{-1}$. As a consequence, one obtains for such situations

$$(\tau_{op})^{-1} = 2(k_B T/\hbar\omega) \, t_{op}^{-1}$$

We can also introduce the mean-free path of the electron between scattering by the low frequency optical modes given by

$$(\lambda_{op})^{-1} = 2(k_B T/\hbar\omega) \, \ell_{op}^{-1} \tag{26}$$

According to Equation 23B (corrected for the spurious 3/4 factor), we obtain more explicitly

$$\lambda_{op} = \pi\hbar^4\rho\omega^2/m^{*2}D_{op}^2(k_B T) \tag{27}$$

It is interesting to note that the mean-free paths ℓ_{op} and λ_{op} are constants independent of the energy of the electron.

B. ACOUSTIC LATTICE PHONON MODES

Besides coupling to the various polar modes, the electrons in molecular media also interact with the *longitudinal acoustic* modes of the lattice. Originally introduced by Shockley and Bardeen, the treatment follows the familiar acoustic deformation potential approach.[13-15,17-19]

1. Acoustic Deformation Potential

The short-range deformation interaction potential is attributed here to the strain of the elastic continuum caused by the acoustic waves. The state of the oscillating lattice is represented, at each point r, by the local dilatation $\Delta(r)$, i.e., the fractional change in volume when the lattice is expanded in the neighborhood of r. Then, the elastic energy density is written as $C \Delta^2/2$, where C is the elastic modulus given by

$$C = \rho c_s^2 \tag{28}$$

in terms of the mass density ρ of the medium and the longitudinal sound velocity c_s. Since the energy $E_e(k)$ of the electron with wave vector k depends on the local strain of the lattice, one expects an energy change linear in the dilatation. Accordingly, this can be taken as the electron-acoustic phonon interaction potential density, $-D_a\Delta(r)|\psi(r)|^2$, where D_a is the characteristic *acoustic deformation potential* parameter.

The parameter D_a repesents the change of the electronic energy $E_e(k)$ upon local dilatation $\Delta(r)$, *viz.*, $D_a = dE_e/d\Delta$. Its magnitude in molecular media can be evaluated, following Toyozawa,[33] by taking the energy of the electron near the bottom of a conduction band using the appropriate tight-binding approximation, i.e.,

$$E_e = e(a) - \nu t(a) \tag{29}$$

where the dependence on the lattice constant must be explicitly considered. In Equation 29 $e(a)$ represents the negative molecular ion energy in the lattice, where the free molecular value is corrected by the electronic polarization energy of the medium $P_e(a)$, taken most simply as the Born expression (Equation 4). The second term in Equation 29 is of the order of the band width B, where

$$B = \nu t(a) \tag{30}$$

It contains the energy $t(a)$ of intermolecular electronic transfer to the ν nearest neighbors separated by the lattice constant a, and is given by the intersite electronic transfer integrals

$$t(a) = t_o \exp(-\alpha a) \tag{30A}$$

where α^{-1} is of the order of the molecular electronic orbital radii. Considering the change of the electronic energy (Equation 29) with the dilatation $\delta\Delta = 3\,\delta a/a$, one then obtains

$$\frac{dE_e}{d\Delta} = -\frac{a}{3}\frac{dP_e}{da} - \frac{\nu a}{3}\frac{dt}{da}$$

Each of these two terms represents a contribution to the deformation potential, which is easily derived from Expressions 4 or 30A. The results are

$$D_a = D'_a + D''_a, \quad D'_a = P_e/3, \quad D''_a = \alpha a B/3 \tag{31}$$

where D'_a and D''_a account for intrasite ("diagonal") and for intersite ("nondiagonal") interactions, respectively: their magnitude can accordingly be deduced from the electronic properties respectively P_e and B of the molecular material.

In an equivalent way, the change of electronic energy with dilatation can be related to the change, with the molecular number density N, of the energy V_o of the electron at the conduction band bottom, relative to the energy in the vacuum: $dE_e/d\Delta = -(dV_o/dN)/N$. Since V_o is the sum of the mean polarization energy U_p and of the electronic kinetic energy T, one recovers, likewise the separate contributions D'_a and D''_a. This is, in fact, the viewpoint adopted in many of the discussions on the mobility of electrons in nonpolar liquids.[37-40]

2. Self-trapped and Quasi-free Electrons

In the same way as for the interaction with the optical phonons, we may analyze the energy of the system of the electron coupled to the acoustical phonon field $\Delta(r)$. Neglecting the kinetic energy of the lattice, this energy is given by a functional of the electronic and vibrational fields $\psi(r)$ and $\Delta(r)$ of a form similar to Equation 15

$$\epsilon[\psi,\Delta] = \int dr\psi^*(r)\left(-\frac{\hbar^2}{2m^*}\nabla^2\right)\psi(r) + \frac{1}{2}\rho c_s^2 \int dr|\Delta(r)|^2 - D_a \int dr|\psi(r)|^2\Delta(r) \qquad (32)$$

where the terms represent the electronic, elastic, and electron-phonon interaction energies, respectively.

Fixing the electronic state $\psi(r)$, and minimizing the functional $\epsilon[\psi,\Delta]$ with respect to $\Delta(r)$, yields the relationship describing the influence of the electronic state on the strain $\Delta_o(r)$ in the most stable electron-phonon configuration, *viz.*

$$\Delta_o(r) = D_a|\psi(r)^2/\rho c_s^2 \qquad (33)$$

The corresponding energy is, similarly to Equation 17, obtained as the sum of the pure electronic energy and an electron-phonon relaxation energy

$$\epsilon(\psi) = E_e[\psi] + E_{aR}[\psi]$$

where $E_e[\psi]$ is defined as in Equation 17. The *acoustic phonon relaxation energy* is given by

$$E_{aR}[\psi] = [D_a^2/(2\rho c_s^2)]\int dr|\psi(r)|^4 \qquad (34)$$

If we envisage Expression 11 of the electronic wave function, we recover, for the energy $\epsilon[\psi]$, a form exhibiting the same dependence on the degree of electronic localization as before [cf., Equation 21], i.e.,

$$\epsilon(\lambda) = E_e\lambda^2 - E_{aR}\lambda^3 \qquad (35)$$

Here E_e is still given by Equation 18, but E_{aR} is defined as

$$E_{aR} = \left(D_a^2/2\rho c_s^2\right)/\left(\frac{4}{3}\pi a^3\right) = D_a^2 q^2/2\mu\omega^2 \qquad (36)$$

if the characteristic acoustic phonon relationship $\omega = c_s q$ is used.

To analyze the conditions of occurrence of trapped and untrapped states of the electrons in presence of the acoustical phonon field, we may apply the aforementioned arguments for the optical phonons. Introducing the acoustic phonon-electron coupling constant

$$g_a = E_{aR}/E_e = E_{aR}/B \qquad (37)$$

the bound electron-phonon states can exist whenever $g_a\lambda > 1$. The most stable self-trapped configuration corresponds to $\lambda = 1$, i.e., to the "small polaron" where the electron is localized around a molecular site. In the present case, it should be kept in mind that small polaron binding results from intermolecular as well as intramolecular mechanisms as mentioned above. Untrapped electrons are encountered when $g_a\lambda < 1$, i.e., for sufficiently extended electronic states and sufficiently small electron-phonon coupling constants. Noteworthy still is the positive energy barrier $\epsilon_a^* = 4\,E_e/27\,g_a^2$ separating the self-trapped and the free electron states at the localization parameter $\lambda_a^* = 2/3\,g_a$. As stressed by Toyozawa[13,33] Rashba,[35] and others,[34,41] the co-existence of stable self-trapped and metastable quasi-free electrons separated by a potential barrier is, in fact, a specific consequence of the short-range electron-phonon interactions. It is, therefore, a common feature for the acoustic and optical deformation potential mechanisms and may have nontrivial dynamical consequences.[35,41,42]

3. Scattering of Electrons

To derive the energy loss and scattering rates of the electrons interacting with the acoustical modes of the lattice, the interaction potential is taken in a form similar to Equation 22

$$V_a(r) = \begin{cases} -D_a\Delta(r) & (r < R) \\[2mm] 0 & (r > R) \end{cases} \tag{38}$$

where the two factors are however different in nature from those of the optical deformation potential (Equation 22). If we denote by $u(r)$ the displacement of the point r in the vibrating medium, the local dilatation is given by $\Delta(r) = \nabla u(r)$. In the presence of an acoustic wave with wave vector q, this becomes $\Delta(r) = i\,q\cdot u$; as a consequence, the transient force experienced by the contact volume of radius R is here obtained as $-\nabla V_a(r) = iD_a\,q$. Taking into account the interaction time $\tau_e = R/v$ and the mass $M = \rho(4\pi R^3/3)$ of the interaction volume, the energy transfer from the electron to the acoustical phonon field thus becomes $\Delta E = D_a^2 q^2 \tau_e^2 / 2M$. Then the energy loss rate, considered as $\Delta E/\tau_e$, can be written in the general form similar to Equation 14 as

$$(dE/dt)_a = E_{aR}\omega^2(a/R)^2\tau$$

where E_{aR} is the acoustic relaxation energy of the medium defined by Equation 36.

The counterparts of Expression 23, for the energy loss rate, mean-free time and mean-free path, are easily derived as

$$\left(\frac{dE}{dt}\right)_a = \frac{3}{4}\,\frac{D_a^2 q^2 m^{*3/2}}{2^{1/2}\pi\hbar^2\rho}\,E^{1/2} \tag{39}$$

$$\frac{1}{t_a} = \frac{3}{4}\,\frac{D_a^2 q^2 m^{*3/2}}{\pi\hbar^3\rho\omega}\,E^{1/2} \tag{39A}$$

$$\frac{1}{\ell_a} = \frac{3}{4}\,\frac{D_a^2 q^2 m^{*2}}{2\pi\hbar^3\rho\omega} \tag{39B}$$

Apart from the unimportant factor 3/4, already mentioned before, they agree closely with the expressions established by more rigorous theoretical approaches.[14-19]

In Equations 39, 39A and 39B, the values of the phonon momentum are related to the electron energy by the condition of momentum conservation. Qualitatively, this is taken into account by noting that the highest frequency acoustic phonon that can be emitted is given by the momentum conservation condition $\hbar q = 2\,m^*v$, which can be used in Expressions 39, 39A, and 39B. The *rate of energy loss* to acoustic phonons then becomes:

$$\left(\frac{dE}{dt}\right)_a = 6\,\frac{D_a^2 m^{*5/2}}{2^{1/2}\pi\hbar^4\rho}\,E^{3/2} \tag{40}$$

which is a factor 3/2 larger than the "exact" result derived by Conwell.[36]

In view of their low frequencies, the acoustic phonons ordinarily have non-negligible occupation numbers given by the energy equipartition formula (Equation 6). Hence, the processes of phonon absorption and emission by the moving electrons are equally important. The resulting quasi-elastic scattering phenomena are therefore best described by the temperature dependent mean-free times τ_a and lengths λ_a, obtained following the same arguments as for the low frequency optical modes of the lattice, *viz.*,

$$\tau_a^{-1} = 2(k_B T/\hbar\omega)t^{-1}, \quad \lambda_a^{-1} = 2(k_B T/\hbar\omega)\ell_a^{-1}$$

The *mean-free path for quasi-elastic scattering* of the electrons by acoustic phonons is, according to Equation 39B, of the form:

$$\lambda_a = \pi\hbar^4\rho c_s^2/m^{*2}D_a^2(k_B T) \tag{41}$$

which can be compared to λ_{op} given by Equation 22. Here again, the mean-free path is independent of the electron velocity.

IV. RELAXATION OF SUBEXCITATION ELECTRONS

In nonpolar molecular media, exemplified by organic liquids or crystals, the energy relaxation and spatial transport properties of hot electrons are completely determined by the energy loss rates and mean-free path lengths of the inelastic and quasi-eleastic electron-phonon scattering processes described in Section III. We shall attempt to assess the importance of each of the possible contributions, considering in turn the "supravibrational" and "subvibrational" electron energy ranges defined in Section II.C.

A. HIGH ENERGY SUBEXCITATION ELECTRONS

Such electrons have energies in the supravibrational range (Equation 8), where an effective mass near the mass of the bare particle may be assumed, i.e., $m_e^* = m$ (cf. Section II.C.).

1. Excitation of Intramolecular Vibrations

The high frequency molecular vibrations are expected to play the major role for the energy loss of the relatively fast subexcitation electrons. They behave as optical modes according to the description in Section III.A. As already indicated, we shall denote by ω_M and E_{MR} the angular frequency and vibronic relaxation energy that characterize such vibrations.

Ordinarily, the symmetric modes that retain the molecular symmetry in the allowed electronic transitions arc thc most important. In the case of aromatic molecules (benzene, naphtalene, anthracene...) the importance of the totally symmetric C–C stretching modes is well established from theoretical and experimental spectroscopic studies;[7,23,43] here the relaxation energies are generally of the order of the vibrational energy, so that the choice $E_{MR} = \hbar\omega_M = 0.1$ eV appears appropriate for numerical estimates. Similar information for other systems is scarce, but molecular photoemission and ultraviolet absorption spectroscopy generally indicate that "Huang-Rhys numbers" ($S_M = E_{MR}/\hbar\omega_M$) higher than unity could be envisaged in many instances.

The *energy loss rate* of the electrons in exciting the molecular vibrations is given, following Equation 14, by

$$(dE/dt)_M = E_{MR}\omega_M(\omega_M\tau)(ka)^2 \tag{42}$$

With $v = (2E/m)^{1/2}$ and $k = mv/\hbar$, one finds the numerical values $v = 6 \times 10^7 \times E^{1/2}$ cm $\cdot$ s^{-1}, $k = 0.5 \times E^{1/2}$ Å^{-1}, where E is given in eV units. Taking a = 5 Å for the intermolecular spacing, one then obtains $\tau = 0.8 \times 10^{-15} \times E^{-1/2}$ s and ka $= 2.5 \times E^{1/2}$. With $\hbar\omega_M = E_{MR} = 0.1$ eV ($\omega_M = 1.5 \times 10^{14}$ s^{-1}), one further gets $\omega_M\tau = 1.25 \times 10^{-1} \times E^{-1/2}$, $E_{MR}\omega_M = 1.5 \times 10^{13}$ eV s^{-1} and finally, $(dE/dt)_M = 1.2 \times 10^{13} \times E^{1/2}$ eV $\cdot$ s^{-1}. However as stressed in discussing Expressions 23 and 24, it is the excess energy $E - \hbar\omega_M$, rather than E, that must be considered. We should, therefore, write

$$(dE/dt)_M = 10^{13}(E - \hbar\omega_M)^{1/2} \text{ eV} \cdot \text{s}^{-1}, E > \hbar\omega_M \tag{43}$$

In the case of stronger vibronic coupling, with Huang-Rhys numbers higher than unity, the loss rate is accordingly higher.

It is interesting to note that the rough estimates often made,[7,44] by taking $(dE/dt)_M = (\hbar\omega_M)\omega_M$, indeed yield the right order of magnitude; the accord must, however, be regarded as more accidental than really justified. In his analysis,[11] Magee does not take into account the known properties of molecular vibronic coupling, but treats all the p intramolecular vibrations equally in terms of some effective deformation potential, characterized by an interaction energy taken to be, without further justification, of the order of 1 eV. The corresponding electron energy loss rate is then obtained as $(dE/dt)_M - 1.4 \times 10^{12}p$ eV $\cdot$ s^{-1}; for polyatomic molecules with p > 10 (p $\approx$ 50, for hydrocarbons in the hexane class), values comparable to Equation 43 are then found as well. The model adopted to describe the interaction of the electron with the molecular modes, and the result indicating an influence of the total number p of the nuclear degrees of freedom in the molecule, appear, however, as problematic.

The *mean-free path* ℓ_M of the electron between two intramolecular vibrational excitation events is given by an expression of the form in Equation 23B. It is more conveniently written as

$$a/\ell_M = (E_{MR}/\hbar\omega_M)(\omega_M\tau)^2(ka)^2 \tag{44}$$

Making use of the above-mentioned numerical values, we then obtain the simple estimate

$$\ell_M = 10(\hbar\omega_M/E_{MR})a$$

In the cases of relatively small vibronic coupling, like solid or liquid aromatic hydrocarbons, when $(E_{MR}/\hbar\omega_M) \lesssim 1$ is known to apply, we thus expect values so high that $\ell_M \gtrsim 10a \approx 50$ Å. For solid benzene, Goulet and Jay-Gerin[45,46] find $\ell_M = 80$ Å by analyzing the experimental data of low energy electron transmission through thin benzene films. Stronger vibronic coupling, with larger Huang-Rhys numbers, means smaller mean-free paths; values around $(E_{MR}/\hbar\omega_M) \approx 5$ and therefore, $\ell_M \approx 10$ Å could well be encountered in still unexplored molecular systems.

2. Interaction with the Lattice Modes

At usual temperatures, practically all the low frequency acoustic and optical lattice vibrations are thermally populated, as noted in Section II.B. The hot electrons are then involved in all the possible phonon absorption and emission processes. These may be regarded as *quasi-elastic*, because the individual relative energy changes of the electron are generally small. Since all the lattice phonons, up to those with the maximum wave number $q_{max} = \pi/a$, are thermally excited, the momenta of the electron and the phonons are comparable. This means that the direction of motion of the electrons may change considerably in each of the quasi-elastic collisions. These may, therefore, be regarded as approximately isotropic.

a. Optical Phonons

These lattice modes are described by angular frequency ω_L and vibronic relaxation energy E_{LR}. The vibronic coupling strength may be characterized by Huang-Rhys numbers $S_L = E_{LR}/\hbar\omega_L$ ranging from 2 to 10 in organic crystals;[47] for the numerical estimates, $\hbar\omega_L = 0.01$ eV and $E_{LR} = 5\hbar\omega_L$ may thus be appropriate.

Electron energy loss via spontaneous emission of such lattice phonons is represented by expressions of the loss rate, $(dE/dt)_L$, and mean-free path between phonon emission, ℓ_L, having the same form as in the Equations 42 and 44 for the intramolecular modes. For electrons of the same velocity, one then has

$$\begin{cases} (dE/dt)_L/(dE/dt)_M = E_{LR}\omega_L^2/E_{MR}\omega_M^2 \\[2ex] \ell_L/\ell_M = E_{MR}\omega_M/E_{LR}\omega_L \end{cases} \tag{45}$$

Consideration of the abovementioned values shows that $(dE/dt)_L \approx 5 \times 10^{-3}(dE/dt)_M$ and $\ell_L \approx 20\,\ell_M$, which confirms a negligible role of the optical lattice phonons in the retardation of fast subexcitation electrons.

For the elastic scattering processes determined essentially by the temperature dependent, induced phonon emission and absorption processes, the mean-free path length λ_L may be evaluated under the form of Equation 26, i.e., $\lambda_L = (\hbar\omega_L/2k_BT)\ell_L$. This means that $\lambda_L \approx 0.2\,\ell_L = 4\,\ell_M$, so that relatively large values ($\lambda_L \approx 200$ Å) are generally expected for the mean-free path lengths of quasi-elastic scattering with optical modes.

b. Acoustic Phonons

Their angular frequencies are denoted by ω_a. The strength of coupling to the electrons is characterized by the acoustic relaxation energy E_{aR} defined, according to Equation 36, as $E_{aR} = D_a^2/2\,\mu c_s^2$, where $\mu = 4\,\pi\,a^3\rho/3$. With the characteristic values already mentioned for hydrocarbons (a $\approx$ 5 Å, $\rho \simeq 1$ g $\cdot$ cm^{-3} and $c_s \simeq \cdot 10^5$ cm $\cdot$ s^{-1}), one obtains $\mu = 5 \times 10^{-22}$g and $\mu c_s^2 = 3$ eV. The deformation potential is evaluated from Expression 31: for the intrasite contribution, $D'_a = P_e/3$, a value of 0.5 eV is in order; for the contribution arising from the intermolecular exchange integrals, we may take $D''_a \approx B = 1$ eV, corresponding to a reasonable order of magnitude for the band width of quasi-free electrons in molecular solids; we thus obtain $D_a \simeq 1.5$ eV.

The energy loss rate by emission of acoustic phonons, $(dE/dt)_a$, at the electron energy E is given by Equation 40. It can be compared to Expression 23, for optical phonon emission: considering the intramolecular vibrational contribution and setting $D_{op}^2 = 2(\mu\omega^2)E_R$ (cf., Equation 18), one has

$$\begin{cases} (dE/dt)_a/(dE/dt)_M = 4(m/\mu)(D_a/\hbar\omega_M)^2(E/E_{MR}) \\[2ex] (dE/dt)_a/(dE/dt)_L = 4(m/\mu)(D_a/\hbar\omega_L)^2(E/E_{LR}) \end{cases} \tag{46}$$

for comparison to the losses via intramolecular and lattice optical modes, respectively. Adopting the numerical values already noted, the estimates obtained are: $(dE/dt)_a \approx (1.5$

$\times$ 10^{-2}) $(dE/dt)_M$ and $(dE/dt)_a \approx (3E)(dE/dt)_L$, if E is written in eV units. Although as efficient as the optical lattice phonons, the acoustic modes cannot significantly compete with the intramolecular vibrations in slowing down the moving electrons.

The quasi-elastic scattering processes with the acoustic modes are characterized by the energy independent mean-free path λ_a of Equation 41. Its importance is conveniently compared to that for the optical lattice phonons, λ_L, given by Equation 27, knowing that

$$\lambda_a/\lambda_L = 2 \, \mu c_s^2 \cdot E_{LR}/D_a^2 \tag{47}$$

Inserting the foregoing numerical estimates, one finds that $\lambda_a = 1.3 \times 10^{-1}\lambda_L$, which indicates a dominant contribution of the acoustic modes in the quasielastic scattering processes. The comparison with the inelastic scattering processes, described by mean-free path ℓ_M, is easy by using the relationship

$$(\lambda_a/\ell_M) = (\mu c_s^2 E_{MR}\hbar\omega_M)/D_a^2 \cdot k_B T \tag{48}$$

The estimate thus obtained is $\lambda_a = 0.5 \, \ell_M$; Equation 48 more generally shows that $\lambda_a \lesssim \ell_M$ could be justified in most of the molecular materials at room temperatures. In the already mentioned study of solid benzene,[45,46] a quasi-elastic scattering mean-free path of 10 Å was found together with the inelastic scattering length $\ell_M = 80$ Å: this agrees with the present arguments, which attribute quasi-elastic scattering essentially to the acoustic phonons.

3. Energy Loss and Spatial Transport Properties

These are governed by the energy loss rates and the momentum relaxation rates of the quasi-free electrons, derived in the preceding sections. For approximate isotropic quasi-elastic scattering, it would be good to consider the spatial distribution of the electrons as spherically symmetric.[29,30,48]

a. Energy Loss Times and Lengths

The most important dynamical feature of the electrons in the supravibrational energy range (Equation 8) is fast energy relaxation by excitation of the intramolecular vibrations, described by the loss rates (Equations 42 and 43). Such processes of cascading down the energy axis, from an initial level E_o, until the lower limit $E = \hbar\omega_M$ is reached, take place over an *energy loss time* defined by[29]

$$\theta_M(E_o) = \int_{\hbar\omega_M}^{E_o} dE'(dE'/dt)_M^{-1} \tag{49}$$

This time scale can be taken as characteristic of the first stage of energy relaxation of relatively high energy subexcitation electrons. From Expression 43, we easily find that

$$\theta_M(E_o) = 2(E_o - \hbar\omega_M)|(dE/dt)_M^{-1}|_{E \, = \, E_o} \tag{49A}$$

In view of the aforementioned numerical estimates, this means that $\theta_M(E_o) = 2 \times 10^{-13}$ $(E_o - \hbar\omega_M)^{1/2}$s, where E_o is in eV. Characteristic cascading times $\theta_M(E_o)$ are therefore, ordinarily expected on the 10^{-13} to 10^{-12} s time scales.

Spatial transport of the subexcitation electrons in the higher energy range (Equation 8) is strongly influenced by the quasi-elastic, but momentum relaxing collisions with the lattice phonon modes. Among these, the acoustic phonons are generally expected to be by far the most efficient, as noted in Section III.A.2. Another notable contribution to momentum relaxation may however, arise from the inelastic scattering with the intramolecular modes.

Therefore, the *momentum relaxation times* τ_r and mean-free path lengths λ_r should be, in general, taken as

$$\tau_r^{-1} = \tau_a^{-1} + t_M^{-1}, \quad \lambda_r^{-1} = \lambda_a^{-1} + \ell_M^{-1}$$

In terms of these quantities, spatial transport of the electrons is then represented by the energy dependent diffusion coefficient

$$D(E) = v^2(E)\tau_r(E)/3 = \lambda_r v(E)/3 \tag{50}$$

From this, we may estimate the total *diffusion length* $b(E_o)$ of the electrons during the whole cascading process as[29]

$$b^2(E_o) = 6 \int dt D = 2 \int_{\hbar\omega_M}^{E_o} dE'(dE'/dt)_M^{-1} v^2 \tau_r \tag{51}$$

Making use of $(dE/dt)_M = \hbar\omega_M/t_M(E)$, $\ell_M = v(E)t_M(E)$ and $\lambda_r = v(E)\tau_r(E)$, and bearing in mind the previously noted energy independence of the mean path lengths ℓ_M and λ_a, one thus obtains the simple result

$$b^2(E_o) = 2\ell_M\lambda_r \frac{(E_o - \hbar\omega_M)}{\hbar\omega_M} \tag{51A}$$

where the last factor represents the number of cascade steps. In the representative example of solid benzene,[45,46] with $\ell_M = 80$ Å and $\lambda_a = 10$ Å, it is found that $\lambda_r \approx \lambda_a$, and that $b(E_o) \simeq 40[(E_o - \hbar\omega_M)/\hbar\omega_M]^{1/2}$ Å.

b. Hot Electron Distribution Functions

A more explicit description of the temporal and spatial behavior of the electrons is obtained by considering the distribution function of the electrons created initially with the energy $E_o > \hbar\omega_M$, at the origin $r = 0$. Assuming isotropic transport properties, the distribution function $f(E,r;t)$ represents the probability of finding, at time t, the electron energy and distance from the origin in the ranges $(E, E + dE)$ and $(r, r + dr)$, respectively. It obeys the transport equation

$$\frac{\partial}{\partial t} f = D(E)\nabla_r^2 f + \left(\frac{dE}{dt}\right)_M \frac{\partial f}{\partial E} + \delta(r)\,\delta(t)\,\delta(E - E_o) \tag{52}$$

where the three terms account for the previously mentioned processes of isotropic diffusion, of energy loss by exciting molecular vibrations, and initial preparation of the hot electrons, respectively.

When the interest is restricted to energy relaxation, it is sufficient to use the reduced distribution function

$$g(E,t) = \int dr\, f(E,r;t) \tag{53}$$

and the corresponding simplified equation

$$\frac{\partial}{\partial t} g = \left(\frac{dE}{dt}\right)_M \frac{\partial g}{\partial E} + \delta(t)\,\delta(E - E_o) \tag{53A}$$

It is conveniently transformed by the introduction of the new variable[48]

$$\theta(E,E_o) = - \int_{E_o}^{E} dE'(dE'/dt)_M^{-1} \tag{54}$$

We obtain

$$\frac{\partial}{\partial t} g = - \frac{\partial}{\partial \theta} g + \delta(t)\delta(\theta)/\left|\left(\frac{dE}{dt}\right)_M\right|_{E_o}$$

The solution of this equation is readily obtained as

$$g(E,t) = \left(\frac{dE}{dt}\right)_M^{-1}\bigg|_{E_o} \delta[t - \theta(E,E_o)] \tag{55}$$

This shows that the relaxing electrons reach the lowest limit of the energy range, $E = \hbar\omega_M$, exactly at the time $t = \theta(\hbar\omega_M,E_o)$ equal to the energy loss time $\theta_M(E_o)$ defined by Equation 49.

The complete picture, including energy relaxation and spatial diffusion, is of course reflected by the solution of Equation 52. We shall write it in the form

$$f(E,r;t) = g(\theta,t)h(\theta,r) \tag{56}$$

remembering that θ is related to E by Equation 54. When Equation 56 is substituted in Equation 52, one gets the equation for $h(\theta,r)$

$$D(\theta)\nabla_r^2 h - \frac{\partial}{\partial \theta} h = - \delta(r)\delta(\theta)$$

With the change of variable

$$\sigma(E,E_o) = \int_o^{\theta} d\theta' D(\theta') = - \int_{E_o}^{E} dE'(dE'/dt)_M^{-1} D(E') \tag{57}$$

we obtain

$$\nabla_r^2 h - \frac{\partial h}{\partial \sigma} = - \delta(r)\delta(\sigma)$$

The solution has, therefore, the form of the Green function of the diffusion equation, *viz.*,

$$h(E,r) = [4\pi\sigma(E,E_o)]^{-3/2} \exp[-r^2/4\sigma(E,E_o)] \tag{58}$$

Together with the temporal factor (Equation 55), Equation 58 completely determines the electron distribution function (Equation 56). One easily verifies that, at time $t = \theta_M(E_o)$, when the electrons reach the lowest level $E = \hbar\omega_M$ of the energy range of interest, the spatial distribution is a Gaussian characterized by the mean square displacement $b^2(E_o) = 6\,\sigma(\hbar\omega_M,E_o)$, with the diffusion length $b(E_o)$ defined by Equations 51 and 51A.

It should be noted that all the foregoing results are subject to the conditions of validity for using a diffusion equation to describe spatial transport. The distribution function (Equation

58) in particular, must be slowly varying over distances of the order of the mean-free path λ_r: $(\partial h/\partial r)/h < \lambda_r^{-1}$, i.e., $r < (2\ b/^2\lambda_r) = (2\ \ell_H/3)\ (E_o - \hbar\omega_M)/\hbar\omega_M$. In view of the afore-mentioned numerical values, such critical distances should be around some hundreds of angstroms. At larger distances, where the Gaussian function (Equation 58) predicts negligibly small densities, the actual density is in fact determined by those electrons reaching the distance r after a limited number of collisions only. The number of such electrons is pro-portional to $\exp(-r/\lambda_r)$, so that the Gaussian must then be replaced by the exponential distribution $\exp(-r/\lambda_r)$.[48]

B. SUBVIBRATIONAL ELECTRONS

When the relaxing hot electrons cross the limit $E = \hbar\omega_M$ and enter the subvibrational energy domain (Equation 9), they become the slow subsystem, when compared to the intramolecular vibrational modes, and are converted in the vibronic molecular polaron states, considered in Section III.A. The strength of molecular vibronic coupling, represented by the Huang-Rhys number $E_{MR}/\hbar\omega_M$, is then reflected by the extent of renormalization of the effective mass of the charge carriers

$$(m^*/m) = \exp(E_{MR}/\hbar\omega_M) \tag{59}$$

The corresponding band widths are reduced by the same factor, *viz.*, $B^*/B = m/m^*$.

1. Interaction with the Lattice Phonons

Coupling of the subvibrational electrons with the optical and acoustic phonon modes of the lattice leads to real energy exchanges with rates given by the same expressions as for the supravibrational electrons in Section IV.A. But the renormalized effective mass m^* must be properly taken into account: velocity, wave number, and interaction time (Equation 1) of the "heavy" electrons become $v^* = (2E/m^*)^{1/2}$, $k^* = m^*v^*/\hbar$ and $\tau^* = a/v^*$, respectively. As a consequence, the numerical values for the "light" electrons of Section IV.A.1 must be rescaled as: $v^* = 6 \times 10^7 \times (mE/m^*)^{1/2}$ cm $\cdot$ s^{-1}, $k^* = 0.5 \times (m^*E/m)^{1/2}$ Å^{-1}, τ^* $= 0.8 \times 10^{-15}$ $(mE/m^*)^{-1/2}$s.

In the case of the *optical phonons*, the energy loss rate and the mean-free path between phonon emission by the subvibrational electrons are governed by expressions similar to 42 and 44

$$\begin{cases} (dE/dt)_L = E_{LR}\omega_L(\omega_L\tau^*)(k^*a)^2 \\[2mm] a/\ell_L = (E_{LR}/\hbar\omega_L)(\omega_L\tau^*)^2(k^*a)^2 \end{cases} \tag{60}$$

where the aforementioned values of k^* and τ^* must be used. Taking $S_M = E_{MR}/\hbar\omega_M = 1$ (which has already been indicated as typical for hydrocarbon systems) implies that $(m^*/m) = e = 2.718$. Using the numerical values introduced previously ($\hbar\omega_L = 0.01$ eV, $E_{LR} = 5\ \hbar\omega_L$), then leads to

$$\begin{cases} (dE/dt)_L = 6 \times 10^{10}(m^*/m)^{3/2}E^{1/2} = 2.7 \times 10^{11}E^{1/2}\text{eV} \cdot \text{s}^{-1} \\[2mm] (a/\ell_L) = 5 \times 10^{-3}(m^*/m)^2 = 3.5 \times 10^{-2} \end{cases}$$

For $E = 0.1$ eV, this means that $(dE/dt)_L \approx 10^{11}$ eV $\cdot$ s^{-1}. The energy independent mean-free path length ℓ_L is of the order $\ell_L \approx 30$ a. In practice the mean-free path length is more significant for quasi-elastic scattering with the optical phonons, due to the totality of the phonon emission and absorption processes; it is of the form of Equation 26, $\lambda_L = (\hbar\omega_L/2 k_B T)\ell_L$, i.e., $\lambda_L = 0.2\ \ell_L$, at room temperature. As a result of the higher effective mass, the values thus expected, $\lambda_L \approx 6a$ (≈ 30 Å) are strongly reduced compared to those previously derived for the "light" supravibrational electrons.

For the *acoustic phonon interactions* of the subvibrational electrons, we must reevaluate the deformation potential parameter D_a, given by Equation 31. Among the two contributions, the first, D'_a, related to the electronic polarization energy is unchanged with respect to the value derived for the light supravibrational electrons ($D'_a = 0.5$ eV, cf., Section A.2.b). But the second term, D''_a, related to the electron band width must be reduced by the factor (m^*/m); instead of the foregoing value ($D_a = 1.5$ eV), we shall therefore use $D^*_a = 1$ eV for the deformation potential of the heavy subvibrational electrons. The corresponding energy loss rate and quasi-elastic scattering mean-free path are then easily evaluated, by comparison to $(dE/dt)_L$ and λ_L, following the relationships in Equations 46 and 47 with the appropriate modifications, *viz.*,

$$\begin{cases} (dE/dt)_a/(dE/dt)_L = 4(m^*/\mu)(D^*_a/\hbar\omega_L)^2(E/E_{LR}) \\[2ex] \lambda_a/\lambda_L = 2\mu c_s^2 E_{LR}/D_a^{*2} \end{cases} \tag{61}$$

Numerically, we find that $(dE/dt)_a = 4E \times (dE/dt)_L = 10^{12}E^{3/2}$ eV $\cdot$ s^{-1}: when $E = 0.1$ eV, $(dE/dt)_a \simeq 4 \times 10^{10}$ eV $\cdot$ s^{-1}. For the subvibrational electrons, the interactions with the acoustic and optical phonons provide comparable energy loss rates. Regarding the mean-free paths, one gets $\lambda_a = 0.3\ \lambda_L$ ($\approx 2a$), which indicates that the acoustic phonons should usually (but not always) be dominant in the quasi-elastic collisions that govern the spatial transport properties of subvibrational electrons.

2. Energy Relaxation and Thermalization Times

By contrast to the higher energy range, where energy relaxation of the electrons is dominated by spontaneous emission of the high frequency molecular vibrational quanta, the subvibrational electrons are submitted to the totality of the phonon absorption and emission processes with the low frequency lattice modes. The influence of such small and continuous energy gain and loss events upon the behavior of the electron in energy space can be described, in terms of the distribution function (Equation 53), by a Fokker-Planck type equation[30,49,50]

$$\frac{\partial g}{\partial t} = -\frac{\partial}{\partial E}\left(\frac{\langle \Delta E \rangle}{\Delta t}\right)g + \frac{1}{2}\frac{\partial^2}{\partial E^2}\left(\frac{\langle \Delta E^2 \rangle}{\Delta t}\right)g$$

where $\langle \Delta E \rangle$ and $\langle \Delta E^2 \rangle$ represent the first and second moments of the probability functions for stochastic energy exchanges. For the first moment, one clearly has

$$\langle \Delta E \rangle/\Delta t = -\mathscr{R}(E); \quad \mathscr{R}(E) = (dE/dt)_L + (dE/dt)_a \tag{62}$$

where $\mathscr{R}(E)$ is the total energy loss rate to the optical and acoustic phonons, governed by Equations 60 and 61. The second moment defines the diffusion coefficient in energy space of the electron

$$\mathscr{D}(E) = \frac{1}{2}\frac{\langle \Delta E^2 \rangle}{\Delta t} \tag{63}$$

The transport equation can also be expressed in terms of a probability flux $J(E)$, as

$$\frac{\partial g}{\partial t} = -\frac{\partial}{\partial E} J, \quad J(E) = -\mathcal{R}(E)g + \frac{\partial}{\partial E}(\mathcal{D} \cdot g) \qquad (64)$$

This can be equivalently written in the form

$$J(E) = V(E)g - \mathcal{D}(E)\frac{\partial}{\partial E}g, \quad V(E) = -\mathcal{R}(E) - \frac{d}{dE}\mathcal{D} \qquad (64A)$$

thereby introducing the directed velocity on the energy axis, $V(E)$.

Thermal equilibrium is represented by the Maxwell-Boltzmann distribution function

$$g_o(E,T) = \rho(E)\exp(-E/k_BT), \quad \rho(E) = C\,E^{1/2} \qquad (65)$$

where ρ is the electron state density. It must obviously satisfy the condition

$$J_o = V(E)g_o - \mathcal{D}(E)(dg_o/dE) = 0 \qquad (66)$$

from which we derive a relation between the diffusion coefficient and the directed velocity,[30] *viz.*,

$$V(E)/\mathcal{D}(E) = (1/2\,E) - (1/k_BT)$$

Together with Equations 62 and 64A, this relation enables one to derive the form of $\mathcal{D}(E)$ and thereby to completely determine the Fokker-Planck equation that governs the energy relaxation and approach to thermal equilibrium of the electrons.[30,50,51] Exact solutions of the nonlinear equation thus derived are, however, difficult to obtain in general.

When the electrons are still far from thermal equilibrium, so that $E \gg k_BT$, one easily verifies the simpler approximate relation

$$V(E) = -\mathcal{R}(E), \quad \mathcal{D}(E) = kT\mathcal{R}(E), \quad E \gg k_BT \qquad (67)$$

which we shall use to analyze the essential nonequilibrium properties of the subvibrational electrons. The corresponding transport equation for the distribution function $g(E,t)$ is then given by the counterpart of Equation 53A, i.e.,

$$\frac{\partial}{\partial t}g = \mathcal{D}(E)\frac{\partial^2}{\partial E^2}g + \mathcal{R}(E)\frac{\partial}{\partial E}g + \delta(t)\delta(E - E_o) \qquad (68)$$

describing the evolution, on the energy axis, of the electrons initially having the energy E_o.

The function $g(E,t)$ is most simply determined under the assumption of negligible energy variations of the transport coefficients $\mathcal{D}(E)$ and $\mathcal{R}(E)$. Taking, for these quantities, the constant values at the initial energy

$$\mathcal{D}_o = \mathcal{D}(E_o), \quad \mathcal{R}_o = \mathcal{R}(E_o) \qquad (69)$$

Equation 68 becomes the familiar differential equation for Green's function of diffusion with streaming. One therefore has

$$g(E,t) = (4\pi\mathcal{D}_o t)^{1/2}\exp[-(E - E_o + \mathcal{R}_o t)^2/4\,\mathcal{D}_o t] \qquad (70)$$

a result also noted by Kurosawa.[30] For an assessment of the validity of the assumption made, we must compare the relative energy variations of $\mathcal{R}(E)$ and $\mathcal{D}(E)$ to that of the distribution function (Equation 70), i.e., $g'/g = [(E - E_o)/2\,\mathcal{D}_o t] + (1/2\,k_B T)$. Remembering that $\mathcal{R}(E) = \alpha\,E^n$ ($n = 1/2$ and $3/2$ for optical and acoustic phonon interactions, respectively), one has $\mathcal{D}'/\mathcal{D} = \mathcal{R}'/\mathcal{R} = n/E$; this shows that the energy variations of the transport coefficients in Equation 68 are indeed negligible under the condition $E \gg k_B T$ already assumed to establish the transport equation by means of Equation 67.

According to Equation 70, the dynamical features of the subvibrational electrons in energy space are displayed by the quantities

$$
\begin{cases}
\langle E \rangle = \int dE'\,E'\,g(E',t) = E_o - \mathcal{R}_o t \\[2mm]
\langle \Delta E^2 \rangle = \int dE'\,(E' - \langle E \rangle)^2\,g(E',t) = 2\mathcal{D}_o t
\end{cases}
\tag{71}
$$

describing a uniform drift together with dispersion on the energy axis. One readily verifies that, in the limit of vanishing diffusion, one recovers a distribution of the same form as Equation 55, i.e.,

$$
g(E,t) = \mathcal{R}_o^{-1}\,\delta[t - \theta(E,E_o)], \quad \theta(E,E_o) = (E_o - E)/\mathcal{R}_o
$$

Similar to the energy loss time $\theta_M(E_o)$ in the supravibrational energy range, the lattice phonon induced relaxation of the subvibrational electrons down to the thermal energy $3\,k_B T/2$, can be characterized by the *thermalization time*, defined as follows

$$
\theta_L^*(E_o) = (E_o - 3\,k_B T/2)/\mathcal{R}_o
\tag{72}
$$

The actual significance of this simple result depends on how far the arguments leading to Equation 70 are valid in the low energy part of the epithermal range; the errors introduced should not be very important.

For numerical estimates, we use the values noted in the preceding section for the energy loss rates to the optical and acoustic lattice phonons ($\mathcal{R}_o \approx 10^{11}$ eV $\cdot$ s^{-1}) for the electrons entering the subvibrational domain with the energy $E_o \simeq 0.1$ eV. The thermalization times (Equation 72) are then found around $\theta_L^* = 10^{-12}$ s. This agrees with the values derived by Warman[39,52] for liquid hydrocarbons, by analyzing experimental results on the electric field dependence of electron mobility.

3. Spatial Transport and Thermalization Length

For the hot subvibrational electrons, spatial transport is characterized by a diffusion coefficient $D^*(E)$ and a drift mobility $\mu(E)$, suitable for quasi-free charge carriers

$$
D^*(E) = v^*(E)\lambda_r/3, \quad \mu(E) = (e/m^*)\tau_r(E)
\tag{73}
$$

Here λ_r and $\tau_r(E)$ are the mean-free path and time between the momentum relaxing collisions with the lattice phonons. Most of the experimental information on the nature of the transport and the phonon scattering processes comes from the drift mobility studies of thermally equilibrated electrons in nonpolar liquids and crystals. Under such conditions, the co-existence of the quasi-free electron states with the much less mobile self-trapped states, described in Sections III.A.2 and III.B.2, must be taken into account. In the liquid hydrocarbons,[37-40] the observed mobilities have indeed been analyzed as a weighted sum of the quasi-free and self-trapped electron mobilities, by considering the two states in thermal equilibrium. Their quantum levels are separated by an energy found to be of the same order as the aforementioned lattice relaxation energies E_{LR} and E_{aR} (from 0.1 to 0.2 eV in the hydrocarbons). For the free electron mobility in such systems, momentum relaxation via acoustic deformation interactions is considered to be dominant.[38,39] In crystalline naphtalene, the data on the drift mobilities have likewise been described in terms of acoustical phonon limited motion of quasi-free electrons.[53] In accord with the estimates in Section IV.B.1, we may thus generally set $\lambda_r \approx \lambda_a$ and $\tau_r(E) \approx \tau_a(E)$ in Equation 73. Numerically, if one takes $\lambda_a = 10$ Å, one finds $D^*(E) \simeq 0.1$ cm^2 s^{-1} at energy $E \simeq 0.1$ eV.

The distribution function f(E,r;t) which accounts for the full dynamical behavior of the hot subvibrational electrons, in space and energy, is obtained from the transport equation which generalizes Equation 68

$$\frac{\partial}{\partial t} f = D^*(E)\nabla_r^2 f + \mathscr{D}(E)\frac{\partial^2}{\partial E^2} f + \mathscr{R}(E) \frac{\partial}{\partial E} f + \delta(t)\delta(E - E_o)\delta(r) \qquad (74)$$

Here D*(E) is given by Equation 73. Proceeding in the same way and under conditions of validity similar to those discussed for the transport equation (68), we neglect the energy variation of the coefficient D*(E) as well as for $\mathscr{D}(E)$ and $\mathscr{R}(E)$. Taking again the values for $E = E_o$ and writing $D^*(E_o) = D_o^*$ we then obtain the solution under the simplified form as follows

$$f(E,r;t) = g(E,t)h(r,t), \quad h(r,t) = (4\pi D_o^* t)^{-3/2}\exp(-r^2/4D_o^* t) \qquad (75)$$

where g(E,t) is given by Equation 70.

This enables us to evaluate the spatial width of the electron distribution when the subvibrational electrons, initially created with energy E_o at the origin $r = 0$, reach the thermal energy at the time $\theta_L^*(E_o)$ defined by Equation 72. It can be represented by the thermalization length defined as

$$b^*(E_o) = [6D^*(E_o)\theta_L^*(E_o)]^{1/2} \qquad (76)$$

Considering the values previously mentioned as typical for the 0.1 eV electrons in hydrocarbon systems ($D^* = 0.1$ cm^2 s^{-1}, $\theta_L^* = 10^{-12}$ s), we may expect thermalization lengths of the order of 80 Å. This compares well with the experimental findings.[6] In the case of crystalline anthracene, Chance and Braun[54] have measured $b^* \approx 60$ Å; comparable lengths were found by Holroyd et al.[55] in various liquid hydrocarbons: 36 Å for *n*-pentane and *n*-hexane, 70 Å for cyclopentane and 2-Methylpentane; other values reported are 180 Å for tetramethylsilane and 130 Å for neopentane.[46]

In practice, the most significant quantities are of course the *total thermalization times and lengths,* that include the contributions from the two successive energy ranges of the subexcitation electrons. When these are prepared with a supravibrational energy $E_o > \hbar\omega$, the total time θ_{th} for approach of thermal equilibrium is the sum of the periods spent in each of the energy intervals given by Equations 8 and 9, so that

$$\theta_{th}(E_o) = \theta_M(E_o) + \theta_L^*(\hbar\omega_M) \qquad (77)$$

where Expressions 49A and 72 must be used. Following the aforementioned estimates, no considerable difference is expected between the two time periods, even if the contribution in the subvibrational range should generally be the longest ($\sim$picoseconds). The total thermalization length b_{th} is related to the spatial width of the electron distribution function obtained as the convolution of the Gaussian functions with the widths $b(E_o)$ and $b^*(\hbar\omega_M)$, that respectively characterize the transport of the relaxing electrons in the supra- and subvibrational energy ranges. One thus gets

$$b_{th}^2(E_o) = b^2(E_o) + b^{*2}(\hbar\omega_L) \tag{78}$$

in terms of the lengths given by Equations 51 and 76. According to the foregoing results, comparable contributions from the two energy ranges may generally be expected. This disagrees with the usual suggestion that the thermalization distance of the subexcitation electrons is mainly due to the subvibrational contribution.[6,9,27] The arguments and estimates presented in this chapter indicate that a situation where $b^*(\hbar\omega_L) \gg b(E_o)$ should be the exception rather than the rule for the subexcitation electrons in dense molecular media.

V. CONCLUDING REMARKS

Before closing this chapter, two additional comments are in order. *First,* the chapter is almost exclusively devoted to the electron-phonon interaction processes in nonpolar media. In the polar materials, the moving electrons are subject to an additional energy loss mechanism due to their coupling with the dielectric polarization field represented by the polarization vector $P(r,t)$.[13,14,26] In order to describe the linear dielectric response of the medium to the longitudinal field generated by the electron, one introduces the displacement $D(r,t)$ as the source of polarization, writing[31]

$$P(r,t) = \frac{1}{4\pi} \int_o^\infty dt' \, \gamma(t')D(r,t - t')$$

where the temporal response function $\gamma(t)$ is the Fourier transform of $\gamma(\omega) = [1 - \epsilon(\omega)^{-1}]$, given in terms of the dielectric function (Equation 7). It contains two terms:

$$\gamma(t) = \left(1 - \frac{1}{\epsilon_\infty}\right)\delta(t) + \frac{1}{\widetilde{\epsilon}} \frac{1}{\tau_L} e^{-t/\tau_L}$$

with

$$\frac{1}{\widetilde{\epsilon}} = \frac{1}{\epsilon_\infty} - \frac{1}{\epsilon_o}, \text{ and } \tau_L = \frac{\epsilon_\infty}{\epsilon_o} \tau_D \tag{79}$$

The first term describes instantaneous electronic polarization, related to the mass renormalization of the electronic polaron (cf., Section II.A). The second term represents the influence of orientational polarization, showing that the corresponding coupling strength is determined by the dielectric function $\widetilde{\epsilon}$ and that the relevant dielectric relaxation time is the "longitudinal time" τ_L.

The self-trapped and quasi-free state properties of the electrons interacting with the long-range polarization field can be analyzed as in the interacting systems involving the short-range elastic phonon fields described in Sections III.A.2 and III.B.2. Self-trapping is found to occur even in extended states (i.e., $\lambda \ll 1$ in Equation 11) forming the well-documented "large", dielectric polarons.[13,14,41] The most stable configuration corresponds to maximal localization ($\lambda = 1$) and serves to define the characteristic polarization relaxation energy E_{PR}. It is derived following the same lines as for the optical and acoustic phonon energies (Equations 18 and 36), with the result

$$E_{PR} = e^2/\widetilde{\epsilon}\, a \tag{80}$$

When the electron moves, it does work on the lattice and the absorbed energy is thereby converted into heat during the dipolar relaxation of the permanent dipole moments. Relaxation thus takes place down the energy axis toward the lowest energy level of the system given by Equation 80. The rate limiting dynamical component is the longitudinal polarization field, which relaxes exponentially with the longitudinal relaxation time τ_L of Equation 79. One therefore expects an electron energy loss rate given by

$$(dE/dt)_P = E_{PR}/\tau_L = (e^2/a)(\epsilon_o - \epsilon_\infty)\epsilon_\infty^2 \tau_D \qquad (81)$$

which agrees with the result derived more carefully by Fröhlich and Platzman.[25]

In the case of water, the values mentioned in Section II.B yield $\widetilde{\epsilon}^{-1} = 0.54$ and $\tau_L = 0.2$ ps, so that $(dE/dt)_p = 7 \times 10^{12}$ eV $\cdot$ s^{-1} (with a = 5 Å). Other numerical estimates are:[56] $\widetilde{\epsilon}^{-1} = 0.54$, $\tau_L = 3.3$ ps and $(dE/dt)_p = 4 \times 10^{11}$ eV $\cdot$ s^{-1} for methanol; $\widetilde{\epsilon}^{-1} = 0.50$, $\tau_L = 9.8$ ps and $(dE/dt)_p = 1.5 \times 10^{11}$ eV $\cdot$ s^{-1} for ethanol. These electron energy and effective mass independent loss rates can indeed compete with the above-derived energy losses $(dE/dt)_M$ and $\mathcal{R}(E)$ to the molecular and lattice vibrations. They must be added to these quantities in the kinetic Equations 53A and 68. As a consequence, the moderation times $\theta_M(E_o)$, $\theta_L{}^*(\hbar\omega_M)$ and therefore $\theta_{th}(E_o)$, as well as the moderation lengths $b(E_o)$, $b^*(\hbar\omega_M)$ and $b_{th}(E_o)$, are all decreased with an extent easily evaluated in the individual cases. *Second*, in this chapter, we used a representation of the electrons based on concepts ordinarily taken as well defined in ordered solids. Their extension to disordered or liquid systems may, therefore, appear as questionable. In the short times (<1 ps) involved in subexcitation electron dynamics, the liquid structure can, in fact, be regarded as frozen, so that the liquids behave as a statically disordered system.[57] It is known that, under such conditions, the picture of extended, quasi-free electron states in a conduction band is still appropriate for energies higher than the "mobility edge," E_c.[58] Disorder may be responsible for additional momentum and energy relaxation processes, only when the fluctuations of the local energy and the interaction energy terms in the tight-binding model (cf., Equation 29) have correlation lengths less than the de Broglie wavelength R, given by Equation 10. Only minor modifications are thereby expected. In the energy domain ranging from the bottom of the band to the limit E_c, the electron states are localized.[58] When they reach such low energies, in the very last period of their thermalization process in disordered systems, the excess electrons may therefore relax by hopping between localized states rather than by energy transfers from extended states.

ACKNOWLEDGMENT

The author wishes to express his appreciation to the reviewer for his very helpful comments.

REFERENCES

1. **Hart, E. J. and Platzman, R. L.,** Radiation chemistry, in *Mechanisms in Radiobiology,* Forssberg, A. and Errera, M., Eds., Academic Press, New York, 1961, 63.
2. **Mozumder, A.,** Charged particle tracks and their structure, in *Advances in Radiation Chemistry,* Vol. 1, Burton, M. and Magee, J. L., Eds., Wiley, New York, 1969, 1.
3. **Kaplan, I. G. and Miterev, A. M.,** Interaction of charged particles with molecular medium and track effects in radiation chemistry, in *Advances of Chemical Physics,* Vol. 68, Prigogine, J. and Rice, S. A., Eds., J. Wiley, Chichester, 1987, 255.
4. **Lipsky, S.,** Ionization and excitation in non-polar organic liquids, *J. Chem. Educ.,* 58, 93, 1981.
5. **Voltz, R.,** Comments on the activated species in irradiated organic media, in *Progress and Problems in Contemporary Radiation Chemistry,* Teply, J., Ed., Acadeimia Prague, 1971, 139.

6. **Pope, M. and Swenberg, C. E.,** *Electronic Processes in Organic Crystals,* Clarendon Press, Oxford, 1982, Chaps. 3 and 6.
7. **Silinsh, E. A.,** *Organic Molecular Crystals,* Springer, Berlin, 1980, chap. 2.
8. **Yakovlev, B. S. and Lukin, L. V.,** Photoionization in non-polar liquids, in *Advances of Chemical Physics,* Vol. 60, Prigogine, I. and Rice, S. A., Eds., J. Wiley, Chichester, 1985, 99.
9. **La Verne, J. A. and Mozumder, A.,** Energy loss and thermalization of low-energy electrons, *Radiat. Phys. Chem.,* 23, 637, 1984.
10. **Inokuti, M., Kimura, M., and Kowari, K.,** Energy spectra of subexcitation electrons, *Chem. Phys. Lett.,* 152, 504, 1988.
11. **Magee, J. L.,** Electron energy loss processes at subelectronic excitation energies in liquids, *Can. J. Chem.,* 55, 1847, 1977.
12. **Sanche, L.,** Interaction of slow electrons with atoms and molecules in solid films, *Radiat. Phys. Chem.,* 32, 269, 1988.
13. **Nakajima, S., Toyozawa, Y., and Abe, R.,** *The Physics of Elementary Excitations,* Springer, Berlin, 1980, chaps. 6 and 7.
14. **Davydov, A.,** *Théorie du Solide,* Mir, Moscow, 1980, chap. 7.
15. **Kittel, C.,** *Quantum Theory of Solids,* Wiley, New York, 1963, chap. 7.
16. **Pines, D.,** *Elementary Excitations in Solids,* Benjamin, New York, 1964, chap. 2.
17. **Shockley, W.,** *Electrons and Holes in Semiconductors,* Van Nostrand, Princeton, 1950, chap. 17.
18. **Conwell, E. M.,** High field transport in semiconductors, in *Solid State Physics,* Supplement 9, Seitz, F., Turnbull, D., and Ehrenreich, H., Eds., Academic Press, New York, 1967.
19. **Rose, A.,** The acoustoelectric effects and the energy losses by hot electrons, *RCA Rev.,* 27, 600, 1966.
20. **Toyozawa, Y.,** Theory of the electronic polaron, *Prog. Theor. Phys.,* 12, 421, 1954.
21. **Born, M.,** Volumen und Hydratationswärme der Ionen, *Z. Physik,* 1, 45, 1920.
22. **Mott, N. F. and Gurney, R. W.,** *Electronic Processes in Ionic Crystals,* Clarendon Press, Oxford, 1948.
23. **Munn, R. W. and Siebrand, W.,** Theory of electron-phonon interactions in organic crystals, *Disc. Faraday Soc.,* 51, 17, 1971.
24. **Druger, S. D.,** Theory of charge transport processes in organic molecular solids, in *Organic Molecular Photophysics,* Vol. 2, Birks, J. B., Ed., Wiley, New York, 1975, 313.
25. **Fröhlich, H. and Platzman, R. L.,** Energy loss to dipolar relaxation, *Phys. Rev.,* 92, 1152, 1953.
26. **Davies, M.,** in *Dielectric Properties and Molecular Behavior,* Hill, N. E., Vaughan, W. E., Price, A. H., and Davies, M., Eds. Van Nostrand-Reinhold, New York, 1969, chap. 5.
27. **Mozumder, A. and Magee, J. L.,** Theory of radiation chemistry. VIII. Ionization of nonpolar liquids by radiation in the absence of external electric field, *J. Chem. Phys.,* 47, 939, 1967.
28. **Wang, S., Mahutte, C. K., and Matsuura, M.,** Electron polarization effect on electron states in insulators, *Phys. Stat. Sol. (B),* 51, 11, 1972.
29. **Episov, S. E. and Levinson, Y. B.,** The temperature and energy distribution of photoexcited hot electrons, *Adv. in Phys.,* 36, 331, 1987.
30. **Kurosawa, T.,** Notes on the theory of hot electrons in semiconductors, *J. Phys. Soc. Jpn,* 20, 937, 1965.
31. **Zwanzig, R.,** Dielectric friction on a moving ion, *J. Chem. Phys.,* 38, 1063, 1963.
32. **Wolynes, P. G.,** Molecular theory of solvated ion dynamics, *J. Chem. Phys.,* 68, 473, 1978.
33. **Toyozawa, Y.,** Exciton-Lattice interaction, in *Vacuum Ultraviolet Radiation Physics,* Koch, E., Haensel, R., and Kunz, C., Eds., Pergamon-Vieweg, Braunschweig, 1974, 317.
34. **Emin, D.,** On the existence of free and self-trapped carriers in insulators, *Adv. Phys.,* 22, 57, 1973.
35. **Rashba, E. I.,** Self-trapping of excitons, in *Excitons,* Rashba, E. I. and Sturge, M. D., Eds., North-Holland, Amsterdam, 1982, chap. 13.
36. **Conwell, E. M.,** in *Solid State Physics,* Supplement 9, Seitz, F., Turnbull, D., and Ehrenreich, H., Eds., Academic Press, New York, 1967.
37. **Schiller, R., Vass, Sz., and Mandics, J.,** Energy of the quasi-free electrons in liquid hydrocarbons, *Int. J. Radiat. Phys. Chem.,* 5, 491, 1973.
38. **Berlin, Y. A., Nyikos, L., and Schiller, R.,** Mobility of localized and quasi-free excess electrons in liquid hydrocarbons, *J. Chem. Phys.,* 69, 2401, 1978.
39. **Warman, J. M.,** The dynamics of electrons and ions in non-polar liquids, in *The Study of Fast Processes and Transient Species by Electron Pulse Radiolysis,* Baxendale, J. M. and Busi, F., Eds., Reidel, Dordtrecht, 1982, 433.
40. **Schmidt, W. S.,** Electron mobility in non-polar liquids: the effect of electron structure, temperature and electric field, *Can. J. Chem.,* 55, 2197, 1977.
41. **Mott, N. F. and Stoneham, A. M.,** The lifetime of electrons, holes and excitons before self-trapping, *J. Phys. C10,* 3391, 1977.
42. **Emin, D.,** Temporal decay of the optically induced properties of a small-polaronic solid, *J. de Phys.,* 42, C4-535, 1981.
43. **Duke, C. B.,** Localization in molecular solids, *Mol. Cryst. Liq. Cryst.,* 50, 63, 1979.

44. **Silinsh, E. A., Kolesnikov, V. A., Muzikante, I. J., and Balode, D.,** On charge carrier photogeneration mechanisms in organic molecular crystals, *Phys. Stat. Sol. (B)* 113, 379, 1982.

45. **Goulet, T., Pou, V. and Jay-Gerin, J. P.,** A procedure for determining low-energy (<10 eV) electron mean free paths in molecular solids: benzene, *J. Electron Spectrosc. Relat. Phenom.,* 41, 157, 1986.

46. **Goulet, T. and Jay-Gerin, J. P.,** Theoretical study of the transmission of low-energy (0—10 eV) electrons through thin-film organic molecular solids: benzene, *Rad. Phys. Chem.,* 27, 229, 1986.

47. **Matsui, A., Mizuno, K., and Kobayashi, M.,** Exciton dynamics in organic molecular crystals, *J. de Phys.,* 46, C7-19, 1985.

48. **Alkhieser, A. and Peletminski, S.,** *Les Méthodes de la Physique Statistique,* Mir, Moscow, 1980, chap. 1.4.

49. **Van Kampen, N. G.,** *Stochastic Processes in Physics and Chemistry,* North-Holland, Amsterdam, 1983, chap. 8.

50. **Holway, H. L. and Fradin, D. W.,** Electron avalanche breakdown by laser radiation in insulating crystals, *J. Appl. Phys.,* 46, 279, 1975.

51. **Conwell, E. M.,** *Solid State Physics,* Suppl. 9, Seitz, F., Turnbull, D., and Ehrenreich, H., Eds., Academic Press, New York, 1967.

52. **Warman, J. M.,** Estimates of electron thermalization times for some dielectric liquids from drift velocity data, *Rad. Phys. Chem.,* 17, 21, 1981.

53. **Duke, C. B.,** The localization of injected charges in polymers and molecular solids, *Can. J. Chem.,* 63, 236, 1985.

54. **Chance, R. R. and Braun, C. L.,** Temperature dependence of intrinsic carrier generation in anthracene single crystals, *J. Chem. Phys.,* 64, 3573, 1976.

55. **Holroyd, R. A., Dietrich, B. K., and Schwarz, H. A.,** Ranges of photoinjected electrons in dielectric liquids, *J. Phys. Chem.,* 76, 3794, 1972.

56. **Rips, I. and Jortner, J.,** Dynamic solvent effects on outer-sphere electron transfer, *J. Chem. Phys.,* 87, 2090, 1987.

57. **Cohen, M. H.,** Electrons in fluids: the role of disorder, *Can. J. Chem.,* 55, 1906, 1977.

58. **Mott, N. F., Davies, E. A., and Street, R. A.,** States in the gap and recombination in amorphous semiconductors, *Phil. Mag.,* 32, 961, 1975.

Chapter 4

EXCESS ELECTRON LIFE HISTORY BY DIELECTRIC RELAXATION

Robert Schiller

TABLE OF CONTENTS

I. INTRODUCTION

Electric charge carriers and dielectric substances attract each other. The phenomenon, known as dielectric polarization, is understood as the result of the formation and orientation of molecular dipoles under the action of the charge carriers' electric field. The attractive force to build up takes some time, particularly if the substance is a polar dielectric, that is if it consists of permanent dipoles. In such substances, polarization is mainly due to partial alignment of the molecular dipoles parallel to the electric field.

The great role dielectric polarization plays in the life of a charge carrier can easily be seen, if one considers the process on a macroscopic scale. Fill one kilogram of electrons into a spherical vessel of one liter volume and put it near to a water surface. The force which acts between vessel and water is about fifty times the force which keeps the earth on its orbit around the sun.

This chapter deals with some aspects of the polarization interaction between polar substances and low energy electrons. Electrons with energies below the lowest vibrational level of the substance are considered; special attention is paid to time-dependent processes. The aim of this chapter is to give some insight into the contemporary thinking about thermalization and localization of excess electrons and about the dynamics of ion-electron recombination. In order to achieve this, we have to start with an admittedly cursory and incomplete overview of polarization and solvation phenomena.

II. SOME ASPECTS OF DIELECTRIC POLARIZATION

A. DIELECTRIC SUBSTANCE AS A CONTINUUM

The extent of polarization is determined by the density of the charge carriers, $\rho(r,t)$, and by the relative permittivity (dielectric constant) of the medium, ϵ. For isotropic and homogeneous substances ϵ is a scalar quantity which usually varies with time, $\epsilon = \epsilon(t)$, whereas generally speaking, ϵ is a tensor. The discussions throughout this chapter will be limited to isotropic systems.

Charge density is usually described in terms of the displacement vector, $\mathbf{D}$, as

$$\text{div } \mathbf{D} = 4\pi\rho \tag{1}$$

The field strength, $\mathbf{E}$, which prevails inside the medium is given by $\mathbf{D}$ and ϵ as

$$\mathbf{D} = \epsilon \, \mathbf{E} \tag{2}$$

The electric energy of a system, U, is defined by the amount of spatial distribution of the charges and by the field strength, through the volume integral

$$U = (1/8\pi) \int_V \mathbf{D} \cdot \mathbf{E} \, dV \tag{3}$$

Whereas Equation 3 is of general validity, the linear relationship, Equation 2, is a material equation which, although most often found true, might break down at high field strengths. Nevertheless, we will assume the linear relationship to hold.

The extent of polarization is characterized by the polarization vector, $\mathbf{P}$, as

$$\mathbf{D} - \mathbf{E} = 4\pi\mathbf{P} \tag{4}$$

In view of Equation 1 the divergence of $\mathbf{P}$ gives the density of the polarization charges.

For those areas of the medium where no charge carrier is present Equation 1 reduces to the expression

$$\text{div } \mathbf{D} = 0 \tag{5}$$

Electrodynamics teach us that, if charge motion is slow enough to make magnetic fields negligible, the electric field has a scalar potential or, in other words, is free of rotation.

$$\text{rot } \mathbf{E} = 0 \tag{6}$$

(For these basic laws of electrodynamics, see References 1 to 3.)

Due to the sluggish response of the molecular dipoles to a change in charge density or in field strength, $\mathbf{D}$ and $\mathbf{E}$ do not vary synchronously. Moreover, the time lag depends on whether polarization build-up is controlled by $\mathbf{D}$ or by $\mathbf{E}$. These problems are treated in terms of the macroscopic theory of dielectric relaxation.

There is a certain fraction of polarization charges that forms instantaneously upon a change in either $\mathbf{D}$ or $\mathbf{E}$. For that fraction the two quantities differ only in a constant factor, $\mathbf{D}_{inst}(t) = \epsilon_\infty \mathbf{E}(t)$, where ϵ_∞ is called high frequency relative permittivity. Instantaneous response or high frequency means, in the present context, the scale of infrared frequencies. The slowly forming contribution of $\mathbf{D}$ is to be expressed as a convolution of $\mathbf{E}(t)$ and of a material function, $\alpha(t)$, the latter characterizing noninstantaneous polarization. The time dependence of $\mathbf{D}$ is to be written as[3]

$$\mathbf{D}(t) = \epsilon_\infty \mathbf{E}(t) + \int_{-\infty}^{t} \mathbf{E}(u)\alpha(t - u)du \tag{7}$$

The mathematically simplest assumption for $\alpha(t)$, which fortunately enough has been found to hold for a number of real systems including water, is an exponential function,

$$\alpha(t) = \frac{\epsilon_s - \epsilon_\infty}{\tau} e^{-t/\tau} \tag{8}$$

Here τ is called the Debye relaxation time and ϵ_s is the static relative permittivity. If neither $\mathbf{D}$ nor $\mathbf{E}$ varies in time, the two static quantities are related as

$$\mathbf{D} = \epsilon_s \mathbf{E} \tag{9}$$

The combination of Equations 7 to 9 is often given in a differential form as

$$\frac{d(\mathbf{D} - \epsilon_\infty \mathbf{E})}{dt} = -\frac{\mathbf{D} - \epsilon_s \mathbf{E}}{\tau} \tag{10}$$

This differential equation clearly reflects the asymmetry in $\mathbf{E}$ and $\mathbf{D}$.

Fröhlich[3] demonstrated in the particular case of a plate condenser that if an $\mathbf{E}$ jump is applied to the plates, polarization builds up with the relaxation time, τ, whereas a $\mathbf{D}$ jump makes polarization increase with the usually much shorter time constant, ϑ, given as

$$\vartheta = \frac{\epsilon_\infty \tau}{\epsilon_s} \tag{11}$$

Later it has been shown, first in the context of ion-electron recombination processes,[4,5] and afterwards in more general terms,[6] that whenever the total amount of charges is kept constant, the relaxation time of polarization always equals ϑ. Hence, ϑ is often called constant charge relaxation time.

If there is a jump in the field strength, i.e. $E(t)$ is a step function, $\epsilon(t)$ can be written as

$$\epsilon_E(t) = \epsilon_s - (\epsilon_s - \epsilon_\infty)e^{-t/\tau} \tag{12}$$

whereas, if polarization is controlled by a **D** jump, relative permittivity reads as

$$\frac{1}{\epsilon_D(t)} = \frac{1}{\epsilon_s} - \left(\frac{1}{\epsilon_s} - \frac{1}{\epsilon_\infty}\right) e^{-t/\vartheta} \tag{13}$$

More recently Hubbard and Onsager[7] discovered a seemingly different meaning of τ and ϑ. The polarization vector, **P**, as any other vector, can be split into a longitudinal component, $\mathbf{P_L}$, which points to the direction of the electric field, and into a transversal component, $\mathbf{P_T}$, which is perpendicular to the field.[8] Elementary vector analysis shows that the divergence and the rotation of **P** can be written as

$$\text{div } \mathbf{P} = \text{div } \mathbf{P_L}; \text{ rot } \mathbf{P} = \text{rot } \mathbf{P_T} \tag{14}$$

The combination of Equations 4 to 6 and 10 shows that the components of P relax with different rates, *viz.*,

$$\frac{d \text{ rot } \mathbf{P_T}}{dt} = -\frac{\text{rot } \mathbf{P_T}}{\tau} \tag{15}$$

and

$$\frac{d \text{ div } \mathbf{P_L}}{dt} = -\frac{\text{div } \mathbf{P_L}}{\vartheta} \tag{16}$$

In view of these expressions ϑ is often called longitudinal relaxation time and is denoted by τ_L.

The physical content of the difference between the two relaxation times has been the subject of some thinking. Friedman[9] visualized the two time constants by devising appropriate thought experiments. Newton and Friedman[10] have shown that the identity of the constant charge and the longitudinal relaxation times are not just incidental, neither is that of the constant field and the transversal relaxation times. As they states, ''We can identify $4\pi\mathbf{P_L} = -\mathbf{E}$ and $4\pi\mathbf{P_T} = \mathbf{D}$, so the relaxation of **E** is a relaxation of the longitudinal part of the polarization vector field while a relaxation of **D** is a relaxation of the transverse part of the polarization field.'' A chemical kinetic analogy between the two relaxation processes and the effect of pressure or volume jumps on an equilibrium chemical system was also suggested.[10] Kivelson and Friedman,[11] understanding the difference between τ and ϑ in terms of a molecular model, proposed that small fluctuations in the orientation of the molecular dipoles might be detected in ϑ rather than in τ.

By considering mechanical motions within a piece of polar substance under the influence of external electric fields, τ was seen to control rotation, since it was found that

$$-\frac{d\omega}{dt} = \frac{\omega}{\tau} + \text{const.} \tag{17}$$

holds where ω denotes angular velocity. Similarly, ϑ controls translation as

$$-\frac{dv}{dt} = \frac{v}{\vartheta} + const. \tag{18}$$

where v is translational velocity. Thus the two relaxation times are inversely proportional to the corresponding coefficients of mechanical friction.[12]

The exponential form of $\alpha(t)$ (Equation 8), called Debye relaxation, occurs frequently, particularly among liquids,[13] but it is not of general validity. In some cases $\alpha(t)$ can be given as a sum of two or three exponentials. More generally, however, a distribution function of relaxation times, $f(\tau)$, can be defined and $\alpha(t)$ be given as

$$\alpha(t) = \int_o^\infty f(\tau)e^{-t/\tau}\, d\tau \tag{19}$$

Rips and Jortner[14] demonstrated that the averages of the constant field and the constant charge relaxation times are correlated according to Equation 10.

$$\langle\vartheta\rangle = \frac{\epsilon_\infty}{\epsilon_s}\langle\tau\rangle \tag{20}$$

where the average for τ is defined as

$$\langle\tau\rangle = \int_o^\infty \tau\, f(\tau)d\tau \tag{21}$$

B. ELEMENTS OF MOLECULAR THEORIES

The microscopic theory of dielectric polarization dates back to the 1870s in the works of Clausius and Mosotti, an achievement "a century ahead its time",[13] because it treated the phenomenon as a many-body problem. Following these lines Debye developed the statistical mechanical treatment of an ensemble of noninteracting dipoles embedded in a continuum of an unstructured insulator. As a result, he evaluated the orientational contribution to the polarization vector, P_o, finding

$$\frac{4\pi}{3}P_o = \frac{\epsilon_s - 1}{\epsilon_s + 2} - \frac{\epsilon_\infty - 1}{\epsilon_\infty + 2} = \frac{4\pi}{9}N\frac{\mu^2}{kT}, \tag{22}$$

where μ is the moment, N is the density of the dipoles, k is Boltzmann's constant and T is absolute temperature.[13,15] In the spirit of statistical mechanical thinking, P_o was evaluated as the expectation value of the dipole moments in an ensemble of unit volume.

More recent theories on the relative permittivity of polar liquids retain this basic idea whereas they try to remedy the original model in two respects. The structure of the insulator is taken into consideration in terms of some theory of the liquid state (e.g., Reference 16) and the electric interactions between the molecular dipoles are also allowed. Wertheim[17] applied the method of mean spherical approximation (MSA), a very powerful tool in liquid state theory, to the present task. MSA regards molecules of a liquid as hard spheres held together by some attracting potential which acts only outside the hard sphere radius. An important notion of the theory is the Ornstein-Zernike direct correlation function, $c(r)$. This function gives the probability of two molecules to be at a distance, r, from each other, provided they are not perturbed by any other molecule of the system. According to MSA $c(r)$ and the intermolecular potential, $V(r)$, are proportional to each other,

$$c(r) = -\frac{V(r)}{kT} \tag{23}$$

Finally, the pair distribution function, g(r), which gives the probability of two molecules to be at a distance, r, from each other, by allowing for the effects of the surrounding molecules, can be evaluated in terms of the liquid state theory by Ornstein and Zernike.

Wertheim introduced the classical dipole-dipole interaction potential (a nonspherical potential) for V(**r**) if r > R prevails, R being the hard sphere radius, and thus evaluated g(**r**) for an ensemble of hard spheres, each one of which carries a point dipole in its center. In analogy to Equation 22 his result for P_o can be given as

$$\frac{4\pi}{3} P_o = \frac{4\pi}{9} N \frac{\mu^2}{kT} [1 - F(g(\mathbf{r}))] \tag{24}$$

where F is a somewhat complicated but tractable function. The static relative permittivity can be found by Equation 24 as

$$\epsilon_s = \epsilon_\infty \frac{(1 + 4\xi)^2 (1 + \xi)^4}{(1 - 2\xi)^6} \tag{25}$$

Here ξ is defined by N and g(**r**) its value varying between 0 (nonpolar liquids, $\epsilon_\infty = \epsilon_s$) and $^1/_2$ (ϵ_s diverges). A most interesting feature of the final result is its independence of the hard sphere radius, R, showing that orientational polarization stems exclusively from the point dipoles.

C. IONS IN POLAR LIQUIDS AT EQUILIBRIUM

The energy of an ion in a dielectric material is lower than in vacuum. Obviously so, charge carrier and insulator attract each other. The free energy change, F_s, of the process in which an ion of radius R_i and of charge q is transferred from vacuum into a liquid can be calculated as

$$F_s = - (F_{vac} - F_{diel}) = -\frac{1}{2} (1 - \frac{1}{\epsilon_s}) \frac{q^2}{R_i} \tag{26}$$

F_s is called the free energy of solvation. The above equation, named after Born, is a straightforward consequence of Equations 2 and 3 by writing $D = q/r^2$. (For an explanation why the term free energy is preferred to energy in this context, see Reference 3.)

The Born expression, although it agrees remarkably well with a great number of observations, can be regarded only as a first approximation, since an ion among molecules is certainly different from a charged macroscopic sphere embedded in a structureless continuum. It would be impossible to review here even the most important results of contemporary theories of solvation. The interested reader might consult, e.g., the articles in the volumes edited by Dogonadze et al.[18] One simple trend of these developments, which seems to be directly relevant to excess electron solvation dynamics, will be selected here.

In the wake of Wertheim's work quoted above, Chan et al.[19] treated a mixture of ions and molecular dipoles in terms of the mean spherical approximation. In this theory ions are regarded as hard spheres of radius R_i with a point charge of magnitude q in the center of each sphere, whereas solvent molecules are taken as hard spheres of radius R_d with a point dipole of moment μ in the center of each sphere. Ion-ion, ion-dipole, and dipole-dipole interactions are considered separately and pair correlation functions are evaluated. These enable one to compute the local dipole densities around the ions, by virtue of which the free energy of solvation reads as

$$F_s = -\frac{1}{2}(1 - \frac{1}{\epsilon_s})\frac{q^2}{R_i + 2R_s} \tag{27a}$$

with

$$R_s = \left(\frac{1}{2} - \frac{3\xi}{1 + 4\xi}\right) R_d \tag{27b}$$

where ξ is the same quantity which figures in and ϵ_s is given by Equation 25. The above expression can be interpreted as a Born equation corrected both for the finite size of the molecules and for molecular interactions. The continuum expression, Equation 26, turns out to be a good approximation for large ions in liquids of high relative permittivity.

D. SOLVATION DYNAMICS

Solvation does not set in instantaneously. In the spirit of the Born equation the dynamics of solvation can be described as a process of dielectric relaxation in a structureless continuum, around a charged sphere. If the substance obeys Debye law, Equation 10, solvation is an exponential process with a single relaxation time, ϑ. The time dependence of the solvation free energy, $F_s(t)$, can be calculated by inserting $\epsilon_D(t)$ of Equation 13 in the place of ϵ_s in Equation 26.

Theories, which are based on more realistic models, consider the insulator as a liquid of molecular structure, much in the same way as do theories of relative permittivity and equilibrium solvation. Wolynes[20] developed a model with the excess electron in mind, since he regarded this species as an appropriate microscopic probe of polarization.

The author accepted the three-stage model of the excess electron life history. Ionization brings about energetic electrons, delocalized in the liquid and the electron imparts its excess energy to the medium. After that the electron gets localized in a shallow trap due to some random fluctuation of the medium. This localized species starts the polarization process, deepening its own trap, until finally the equilibrium solvation shell is formed. Thermalization, trapping and polarization are thought to be temporally distinct processes. The theories to be dealt with now, discuss the third stage only, in terms of the mean spherical approximation. This implies the nonobvious assumption that the initially localized electron can be regarded as a hard sphere.

By applying MSA to a nonequilibrium situation R_s, the correction term in Equation 27 turns out to depend on time. This is due to the time dependence of relative permittivity, $\epsilon(t)$. A detailed calculation which, among other things, assumes Equation 8 to prevail, results finally in a complicated kinetic picture. The portion of the liquid which lies far from the trapped electron can be treated as a continuum, hence it gets polarized with the constant charge relaxation time, ϑ. At smaller distances, however, the microscopic structure of the liquid and the finite size of the ion become perceptible and this results in a relaxation time, τ_G, longer than ϑ, given as

$$\tau_G = \frac{1 + \dfrac{R_i}{R_d}(\epsilon_\infty + 3)}{1 + \dfrac{R_i}{R_d}(\epsilon_s + 3)}\, \tau \tag{28}$$

Thus one cannot describe solvation as a simple exponential process, although dielectric relaxation obeys an exponential law. The result agrees with earlier numerical calculations[21]

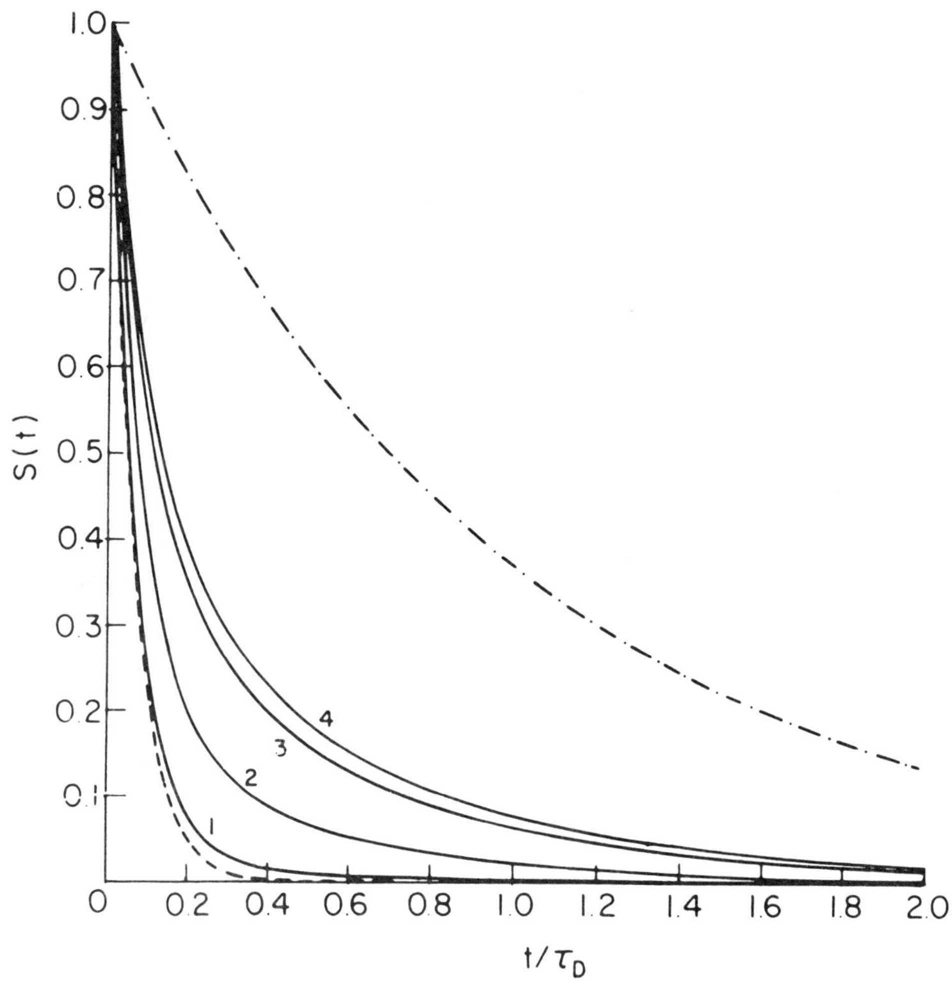

FIGURE 1.   Solvation time correlation function, S(t), with $\epsilon_s = 30$ and $\epsilon_\infty = 2$. The solid lines correspond to different R_d/R_i values: 0.1 for (1), 1 for (2), 10 for (3), 100 for (4). Dashed curve continuum result with constant charge relaxation time, ϑ; dashed-dotted curve continuum result with constant-field relaxation time, τ. (From Rips, I., Klafter, J., and Jortner, J., *J. Chem. Phys.*, 88, 3246, 1988. With permission.)

and with a remark made by Onsager[22] proposing "that the $\gg$ snowball $\ll$ forms outside in," i.e., ϑ characterizes the rate of polarization only in those liquid regions which lie far from the ion, whereas the process is slower in the immediate vicinity of the ion.

The analytical treatment of Wolynes' model by Rips et al.[23] supports Onsager's prediction. The solvation time correlation function, S(t), which is essentially a normalized, linear function of $F_s(t)$, is seen to have a complicated time dependence. The parameter R_d/R_i, the solvent-to-ion size ratio, decides the actual time dependence (Figure 1). If R_d is much smaller than R_i the continuum approximation holds and solvation is of exponential kinetics with time constant ϑ. In the opposite case, if $R_d \gg R_i$ prevails, the time constant of the process is as large as about τ. Intermediate size ratios correspond to intermediate solvation rates.

Experiments on solvation kinetics support the above findings.[24-27] The measurements, referring partly to the solvation of abruptly formed dipoles and not of ions, show that solvation times usually fall between τ and ϑ. Whereas this is true for heavy ions or dipoles, it seems to contradict the observations of electron solvation. This point will be set forth in Section III.B.

Loring and Mukamel[28] suggested a different model for solvation dynamics by regarding the liquid as a lattice of interacting point dipoles. They were thought to rotate freely, whereas their translational motion was forbidden. Equations for $\epsilon_E(t)$ and $\epsilon_D(t)$ were derived for this model showing that both of the relative permittivities depend not only on time but also on the distance from the polarizing ion. This renders the rigorous definition of τ and ϑ impossible. The authors are at variance with Onsager[21] maintaining that relaxation is fastest at a distance twice the lattice spacing, polarization building up slower both outside and inside of this critical value.

The above authors assume that orientational polarization is due to rotational motion only, in agreement with Debye's original ideas.[15] Chandra and Bagchi,[29,30] however, strongly argue in favor of the importance of translational displacement. Moreover, Onsager's suggestion on the "snowball which forms outside in" breaks down if translation contributes significantly to polarization relaxation.[33] This question brings one back to the physical interpretation of τ and ϑ. Although detailed microphysical models were seen just now which blur the difference, one might feel tempted to agree with Fleming[31] who states that "the ratio $\epsilon_\infty/\epsilon_s$ gauges the number of molecules required to cooperate to give the ϑ response, rather than the τ response which, roughly speaking, corresponds to a single-molecule reorientation." This picture has been attributed to τ already by Debye. Friedman[32] has, however, recently pointed out that there is a difference between the two relaxation times even in ionic crystals, where molecular rotation is nonexistent. One might think, in keeping with Reference 12, that τ refers to the overall rotational motion of the substance. This stems either from the rotation of individual molecular dipoles or from the macroscopic torsion of the sample. Dilute solutions or low pressure gases realize the first possibility, ionic crystals the second one. Based on a statistical mechanical calculation which takes dipole-dipole interaction into account in a rigorous way, Kivelson and Friedman[11] could show that the longitudinal (i.e., parallel-to-the-field) component of the collective dipole density is much smaller than the transverse one. The relaxation of the small longitudinal component is much faster than the reorientation of a single molecule. That is thought to be the mechanism behind the difference between ϑ and τ.

III. EXCESS ELECTRON DYNAMICS

A. MODERATION OF LOW ENERGY ELECTRONS

Excess electrons, whether produced by ionizing radiation, photochemically or through injection from a solid electrode, are not in thermal equilibrium with their environment. The net result of the energy exchange between faster-than-thermal electrons and the medium consists in electron moderation. The modes of energy exchange depend on the energy range. With an actual energy below the lowest vibrational level of the medium, the only ways an electron can impart its energy to the environment are dielectric polarization and elastic collision. Since this latter process, due to the large disparity between the masses of an electron and a molecule, contributes but little to the overall energy transfer, it is the electric interaction which is solely responsible for electron thermalization in the so-called subvibrational range.

The electron loses as much kinetic energy, K, as electric energy is gained by the medium. Thus the task is to define the rate of the electric energy change due to electron motion. Formally this can be done by Equations 2 and 3 finding

$$-\frac{dK}{dt} = \frac{dU}{dt} = \frac{1}{4\pi} \int \mathbf{E} \frac{\partial \mathbf{D}}{\partial t} \, dV \tag{29}$$

In order to evaluate this integral one has to know the space and time dependence of $\mathbf{D}$, i.e., the charge density of the electron as it traverses the medium, and the dielectric function which correlates **D** and **E**. As to the latter, Fröhlich and Platzman,[34] who were the first to tackle the problem in terms of Equation 29, assumed the medium to obey Debye's law, Equation 10. The electron is regarded as a point charge at $r_e(t)$, hence D at any other point r is given as

$$\mathbf{D}(r,t) = - e \frac{\mathbf{r} - \mathbf{r}_e(t)}{|\mathbf{r} - \mathbf{r}_e(t)|^3} \tag{30}$$

where e is the elementary charge. Since **D** diverges as **r** approaches r_e, mathematical expediency requires exclusion of a certain volume in the vicinity of the point charge from the range of integration. This is usually taken to be either a cylinder around the path[34] or a sphere around the site[35,36] of the electron, both regions being thought to be of molecular dimensions. The last information needed is the electron path, $r_e(t)$. Two extreme cases were investigated, a straight line path and that of random motion. The numerical work of Magee and Helman[33] and the analytical treatment of Tachiya and Sano,[36] in good agreement with each other and with the earlier paper by Fröhlich and Platzman,[34] show that either the type of the electron path, or the shape of the excluded volume has little if any bearing on the rate of moderation. The rate reads as

$$- \frac{dK}{dt} = \frac{\pi e^2}{4a} \frac{\epsilon_s - \epsilon_\infty}{\tau n^4} \tag{31}$$

where n is the index of refraction at infrared frequencies and a is the radius of the excluded cylinder.

Inserting the properties of liquid water and estimating a as 3×10^{-8} cm one finds $-(dK/dt) \approx 10^{13}$ eVs^{-1}, whereas for near-melting point ice it is about 10^7 eVs^{-1}. The large difference is mainly due to the big change in τ at the melting point.

Exponential polarization build-up is not the only possible choice. Jonscher[37] treated the problem of moderation by employing a non-Debye relaxation function of the form $\alpha(t) = At^{-n}$ with $0 < n < 1$. This law seems to occur frequently with solid dielectrics. His results, being presented in a somewhat qualitative manner, are difficult to compare with previous works.

Fröhlich and Platzman[34] estimated the upper bound of the subexcitation energy range to be about 10 eV in water. Recent experiments, however, indicate lower values. Michaud and Sanche[38,39] determined the vibrational cross section of amorphous ice at 14 K for 1 to 20 eV electrons. With these data at hand Goulet and Jay-Gerin[40] performed a Monte Carlo calculation, computing electron thermalization times and distances. Averaging over all energies below 7.4 eV and considering the energy distribution characteristic to irradiated liquid water,[41] they found an average thermalization distance of about 13 nm and an average thermalization time, $t_{th} = 0.7 \times 10^{-13}$ s. The latter value agrees with the estimate based on Equation 31; the mechanism upon which the calculation rests is, however, vibrational excitation and not dielectric relaxation. That shows that two mechanisms of energy degradation prevail in the several eV range and that it is only below 0.1 eV where dielectric phenomena take over the full brunt of thermalization.

Recently Mozumder[42] summarized a series of experimental results on thermalization. By extrapolating the microwave conductivity date of Warman et al.[43] on dry and humid air, he obtained for liquid water $t_{th} = 4.5 \times 10^{-15}$ s. The drift velocity measurements in water vapor by Christophorou et al.[44] resulted in $t_{th} = 2 \times 10^{-14}$ s. The data of Sanche et al.[45] referring to electron transmission and energy loss across frozen layers of H_2O indicate $t_{th} = 2.4 \times 10^{-14}$ s. The differences of the data are larger than the experimental error, nevertheless, one has to consider that all the calculations depend on the initial energies and energy spectra of the subexcitation electrons, only known incompletely. Mozumder[42] thinks that $t_{th} = 2.4 \times 10^{-14}$ s is a reasonable estimate within a factor of two.

Photoelectrochemical methods have provided radiation chemists with a new, powerful tool for the investigation of low energy electron processes. A metal or semiconductor, immersed into a polar liquid, polarized as a cathode and illuminated with visible or near UV light, injects electrons of controlled energy into the liquid.[46-48] Recently, Kreitus et al.[49] and Konovalov et al.[50] measured electron thermalization lengths in liquid water by making use of photoelectrochemical techniques. The results of the latter group for 2.5 eV electrons agree reasonably well with theoretical expectations.[40] Konovalov et al. interpret their results as a proof of thermalization to proceed by a nonlocal mechanism. This statement is in keeping with dielectric relaxation as a means of thermalization, since this mechanism involves an ensemble of molecules.

B. PREFORMED TRAPS OF SELF-TRAPPING

A good number of students of excess electron physical chemistry share the view, first launched by Tewari and Freeman,[51] that shallow potential wells, which are present in liquids either as a consequence of their equilibrium structure or of near-equilibrium fluctuations, play a central role in excess electron stabilization. As it was already mentioned in Section II.4, these shallow traps are assumed to localize the thermalized, quasi-free electrons, a stage which is followed by stabilization through solvation. Since the experimental basis of this idea was summarized by Kenney-Wallace and Jonah[52] and some further aspects were reviewed recently,[53] the forthcoming discussion will be limited to two basic questions only: whether liquid structure theories support and whether solvated electron formation kinetics demand the existence of preformed traps.

A substance consisting of molecular dipoles may localize electrons without any structural rearrangement or collective phenomena. It has been shown[54] that if the dipole moment of a solitary molecule is larger than $\mu = 1.625$ debye (1 debye $= 10^{-18}$ esu) the electron can form a bound state in the field of the stationary dipole. Eventual molecular rotation does not influence this limit markedly.[55] The aggregate effect of two collinear dipoles secures a more stable binding.[56] Given the dipole moment of an H_2O molecule $\mu(H_2O) = 1.83$ debye, electron localization in the field of one single water molecule seems to be feasible. Nevertheless, mass spectrometry has not revealed the existence of stable H_2O^- ions, although resonant virtual negative ion states of nearly zero energy have been demonstrated by electron scattering.[57,58] The appearance of negatively charged water clusters, $(H_2O)_n^-$ with n >8, in high pressure water vapor[57] clearly indicates that electron localizes in an ensemble of water molecules.

These findings, together with fast kinetic spectrometry of solvated electron formation, which indicates localization to be an extremely fast process,[52,59] prompted some theoretical studies on preformed traps. Tachiya and Mozumder[60] computed the electric fields of tetrahedrally and octahedrally located dipoles with fixed relative positions but of uncorrelated random orientation and solved the Schrödinger equation of the electron for those fields. The calculation shows that shallow traps, which are thought to be effective in electron trapping, are present in a density high enough to make preformed traps play an important role in electron localization. In a later paper the same authors[61] further developed the above model in order to describe the temporal evolution of orientational polarization. By that they conceived a theory for the dynamics of electron solvation, and using some admittedly arbitrary assumptions, they evaluated the period of time which is needed for a localized electron to form an equilibrium solvation shell. The estimated values are found to always be shorter than the experimental data.

Schnitker et al.[62] used an approach much different from the previous one, discussing the structure of liquid water in a most rigorous way by making use of molecular dynamics simulation, whereas the electron was simply taken as a point charge. Nevertheless, the potential energy distribution at different sites in water was found to be somewhat similar to that obtained by Tachiya and Mozumder.[60] The main finding of this work is the statement that potential traps, present in water by virtue of configurational fluctuations, are of the right depth and occurrence probability to render localization in preformed traps "a very plausible mechanism". The authors are much aware of the need to perform an explicit quantum mechanical calculation, in order to see whether the applied crude approach of the electron properties influences their results.

A quantum statistical treatment of the excess electron in liquid water was performed by Jonah et al.[63] Their results were discussed in more detail in a later paper.[64] It was demonstrated by numerical calculations that, in the vicinity of a localized electron, the water molecules are aligned in such a way that it is not the molecular dipole moment vectors but the O–H bonds that point toward the electron. And, more important in the present context, potential traps were found only in those regions where the electron polarizes its environment. Thus, based admittedly on a limited number of molecular configurations only, the authors do not favor the idea of preformed traps.

Experimental results, however, render indirect evidence of trapping prior to solvation. The analysis of optical absorption spectra of the excess electron in low temperature liquid alcohols and in water "right upon formation", i.e., at a time scale much shorter than ϑ, indicates, according to Houée-Levin and Jay-Gerin,[65] the existence of trapped states with an energy below the free electron continuum. This finding seems to be in agreement with the predictions of Schnitker, et al.[62] Photoionization experiments[66] also indicate that ejected electrons "immediately" find a microscopic environment where they can localize prior to solvation.

High time resolution kinetic spectrometry indeed indicates pretrapping will proceed. It was established both by chemical[67] and spectroscopic[68-74] methods that there exists a well-defined, short-lived state of the excess electron which forms "promptly" upon irradiation and which relaxes "slowly" into an equilibrium state. Gauduel and co-workers[73,74] established that the early species, which absorb in the infrared, appear with a characteristic time of 100 fs and relax toward solvated electron spectrum with a relaxation time of 240 fs. The obvious idea was to assign the fast forming species to a localized electron without an equilibrium solvation shell. Keszei and Jay-Gerin[75] reevaluated the experimental data of References 73 and 74 establishing an even shorter time scale of the same order of magnitude.

The very fast appearance of an IR spectrum seems to be a valid argument in favor of the preformed trap idea. Nevertheless, some caution may be in order. Wiesenfeld and Ippen[70] construed their measurements, which showed fully solvated electrons to appear in laser illuminated water sooner than 0.3 ps and in *n*-butanol after some 10 ps, in terms of preformed traps. However, Zusman and Helman[76] offered a theoretical explanation based solely on electron induced dielectric relaxation. They described the transient optical absorption in terms of the time-dependent density matrix of the electron states which were influenced by the energy fluctuations of the medium. The fluctuations were not taken to be spontaneous but to be controlled by the electric field of the electron, according to Equations 3 and 10. By this token it was shown that the correlation function of the medium energy fluctuations, C(t), decays exponentially as

$$C(t) \;=\; Ae^{-|t|/\vartheta} \qquad\qquad (32)$$

It is C(t) that defines the time dependence of the optical spectrum. Thus, according to this rigorous theory, it is again the constant charge relaxation time, ϑ, that defines the time scale of the process. The agreement between the calculated curves and the ps time scale experiments in Reference 70 is convincing.

The apparent contradiction between fast electron localization and slow dielectric relaxation was also a subject of some consideration in the past. This, however, was due to the fact that relaxation was erroneously characterized by the constant field relaxation time, τ, instead of ϑ. But in certain cases solvated electron spectra were seen to appear even earlier than ϑ. This was found for some low temperature[77] and high temperature alcohols and water[78] (see also Reference 52) and particularly for ice.[79-83] These observations launch a real problem.

Let us recall what contemporary theories state about solvation times (cf., Section II.4). It is clearly stated that a classical ion gets solvated with time constants falling between τ and ϑ. Thus, if these theories are correct, the species that solvate faster than ϑ must be different from a classical ion. By the same argument it must also differ from an electron in a preformed trap, since this obeys the same solvation kinetics as a classical ion does. The fact that electron solvation is often faster than ϑ, whereas solvation of a trapped electron is predicted to proceed slower than ϑ, casts serious doubts upon the role of preformed traps in electron solvation. Some earlier calculations have already indicated that it is the non-classical nature of the electron and its spread-out charge distribution, that offers a clue to the problem of fast solvation.

In the semicontinuum model of Fueki et al.[84,85] the one-electron Schrödinger equation is solved for the field of four or six discrete dipoles as nearest neighbors which are embedded in a continuum of the dielectric liquid. The electron polarizes its molecular-and-continuous environment whereas it stays in its actual ground state throughout the process.

In another approach[86,87] the liquid is regarded as a dielectric continuum which relaxes according to Debye's law, Equation 10, whereas the electron moves along in the medium as a wave packet. The charge density of the packet depends on space and time, $\rho = \rho\,(\mathbf{r},t)$. Solvation is thought to be completed, as the extent of polarization attains its equilibrium value at the actual site of the wave packet. In other words, both polarization and electron motion must be considered. Equilibrium sets in sooner than ϑ because, as the electron moves on, its high density parts quickly polarize the medium for the oncoming low density tail. The mathematical condition of equilibrium solvation to be attained is given by the inequality

$$\rho(\mathbf{r},o) + \int_0^t \frac{\partial\rho(\mathbf{r},u)}{\partial u}\, e^{u/\vartheta}\, du \leq 0 \tag{33}$$

Equilibrium is attained by moment t for those portions of the wave packet for which Equation 33 holds.

Using the over simplified model of a one-dimensional, nonspreading wave packet which travels with constant velocity, a quick estimate for the solvation time, t_s, can be given by Equation 33 as

$$t_s = \left(\frac{8^{1/2}h}{3k}\right)^{1/2} \left(\frac{\vartheta}{T}\right)^{1/2} = 6.75 \times 10^{-6} \left(\frac{\vartheta}{T}\right)^{1/2} \tag{34}$$

if ϑ is expressed in seconds and T in Kelvin.

Both the semicontinuum and the above-described continuum model render good estimates for the observed solvation times.[68] Recently it was shown[53] that Equation 34 gives a good estimate also for electron localization in near melting point ice. In this system with ϑ (ice) = 650 ns one finds by Equation 34 t_s(calc) = 332 ps in good agreement with the experimental value[79-83] t_s(exp) = 400 ps. Thus t_s is shown to be orders of magnitude smaller than ϑ: the solvation time of a spread out wave packet might be much shorter than dielectric relaxation time is, without assuming any other process but dielectric relaxation. A theory by Tachiya and Watanabe[88] is based on ideas partly similar to those of References 86 and 87 with the important difference that these authors consider the unsolvated electron to be in a localized state. Perhaps this is why they predict solvation to take a period of time of several ϑs.

C. A UNIFIED MODEL OF RECOMBINATION, THERMALIZATION, AND SELF-TRAPPING

The results and considerations on thermalization and self-trapping of slow electrons clearly indicate that it is the same effect, dielectric relaxation, with time constant ϑ, which is operative in both processes. Thus one might reasonably think that thermalization does not precede localization but they proceed more or less simultaneously. Probably the same holds for recombination.

Radiolysis or photolysis produces ions and electrons in pairs. Ion-electron distances and electron energies are such that a considerable fraction of pairs can be taken as independent of any neighboring charge carriers.[89,90] The model to be described in the following passage concentrates upon ion-electron pairs instead of solitary excess electrons.

Transport of ions in their mutual Coulomb field has been dealt with in much detail in order to understand ion recombination. It is clear that there is a finite probability of an ion pair recombing, partly due to Coulombic attraction, partly to random reencounter. If G_o denotes the yield of primary ionizations, and W_{esc} the probability of escape for an ion from its geminate partner, i.e. $(1 - W_{esc})$ is the probability of recombination, the yield of excess electrons, $G_{e_s^-}$, equals

$$G_{e_s^-} = G_o\, W_{esc} \tag{35}$$

After some early attempts Onsager[91] solved the steady state diffusion-migration (Fokker-Planck) equation for a pair of classical ions. He assumed the relative permittivity to be constant, that means the calculations referred to nonpolar media. Mozumder[92,93] developed Onsager's method to polar systems by including Equation 13 in the expression of interionic forces. Another approach[4,5] made use of thermodynamic arguments together with Equation 10 in order to estimate W_{esc}. The recombination of the electron as a nonclassical entity, i.e., the problem of quantum Brownian motion, has been dealt with only recently.[94]

A simple model is now being proposed for the description of electron moderation, localization, and partial recombination. Let us regard an ion-electron pair as a ground state hydrogen-like atom embedded in a relaxing dielectric continuum.[95] One can safely assume that this "atom" is always in its ground state although Coulomb potential varies with time, since dielectric relaxation is slow in comparison to electron transitions. A similar assumption was used recently for the quantum molecular dynamics simulation of excess electron transport.[96] The energy of such an entity is given by elementary quantum mechanics, corrected for the presence of the dielectric substance, as

$$E = -\frac{me^4}{2\hbar^2\,\epsilon_D(t)} \tag{36}$$

where m denotes electron mass. The energy decreases as $\epsilon_D(t)$ increases in time according to Equation 13, whereas the electron is thought to remain in its ground state throughout the process. The charge density (the absolute square of the eigenfunction) is, by the same token, also time dependent of the form

$$\rho(r,t) = \frac{\alpha^3}{8\pi}\, e^{-\alpha r} \tag{37}$$

with

$$\alpha = \frac{2me^2}{\hbar^2\, \epsilon_D(t)} \tag{38}$$

That is, the charge distribution becomes broader as time goes by.

The kinetic energy of the electron, K, is

$$K = -E \tag{39}$$

The rate of thermalization, $-(dK/dt)$, can be evaluated by differentiating Equation 36, finding for the initial rate an expression similar to Equation 29 (cf., Reference 34), *viz.*,

$$-\left.\frac{dK}{dt}\right|_{t=0} = \frac{me^4}{\hbar^2}\,\frac{\epsilon_s - \epsilon_\infty}{\epsilon_\infty^3\, \tau} \tag{40}$$

This equation yields for liquid water a value of about 2×10^{12} eV/s which, with an initial energy of about 0.1 eV, gives an estimate for thermalization time as $t_{th} = 5 \times 10^{-14}$ s in reasonable agreement with the data quoted in Section III.1.

The electron cloud, polarizing the medium, brings about an energy well the depth of which, $U_p(t)$, varies with time. In order to evaluate $U_p(t)$ one has to consider the change of $\rho(t)$ and the polarization of the medium as simultaneous processes. The time dependence of $U_p(t)$ can be given in a good approximation as

$$U_p(t) \propto -\left(\frac{1}{\epsilon_D(t)} - \frac{1}{\epsilon_D^2(t)}\right) \tag{41}$$

The depth of the well is considerably larger than the electron energy, $|U_p| \gg |E|$ holds throughout the process.

The form of Equation 41 reveals an important fact. It shows that polarization brings about a deep potential trap instantaneously $[\epsilon_D(t = 0) = \epsilon_\infty]$ which becomes shallower with elapsing time. The present model shows self-trapping as effective at the very first moment of ion pair formation. The energy of the electron, trapped in its own polarization field at this very early stage is in the order of 0.1 eV, thus self-trapping is seen to precede thermalization. There is no pressing need by the kinetics to assume the effect of preformed traps. The instantaneously formed potential well relaxes to an equilibrium solvation shell as ϵ_D approaches ϵ_s.

A most important test of the model would be its comparison with the femtosecond time scale spectroscopy of the solvated electron. This work has not been carried through for the time being. Apart from technical difficulties the model in its present form deals with the ground state of the electron but says nothing about its excitations.

The theory was tested against solvated electron yields, measured in a series of polar liquids and in H_2O ice. Electrons can escape from any attracting field if only some energy is imparted to them from the surrounding. The energy transport from the medium to the electron can be calculated in terms of thermodynamic energy fluctuations. The coupling of fluctuations in the solvent and fluctuations of the electronic state was already suggested in a work on quantum simulation of hydrated electron structure and dynamics.[97] Such calculations show immediately that U_p is much too deep for any fluctuation to delocalize the electron. Medium energy fluctuations can make only localized electrons leave the field of the positive ion. Thus electrons first localize and only localized electrons may or may not escape recombination.

Energy fluctuation, which enhances escape, must be at least as large as the potential energy of the electron in the field of the positive ion, $-2E$. Combining this requirement with the Gaussian distribution of fluctuations one finds the probability of electron escape to be

$$W_{esc} = \frac{1}{2} \, erfc \left(-\frac{2E}{\sigma} \right) \tag{42}$$

where $\sigma^2 = kT^2 c_v$, c_v is the constant volume specific heat of the medium and erfc denotes the complementary error function. The argument depends on relative permittivity and specific heat as

$$-\frac{2E}{\sigma} \propto \frac{1}{\epsilon_D^2(t) \, c_v^{1/2}} \tag{43}$$

The above expressions describe the kinetics of recombination controlled by the gradual increase of relative permittivity. In order to evaluate the final escape probability one has to replace $\epsilon_D(t)$ by ϵ_s in Equation 43. The calculated curve of the solvated electron yield is compared with experimental data[98] in Figure 2. The only free parameter used in curve fitting was G_0 in view of Equation 35. The agreement between theory and experiment looks reasonable. The phenomenological correlation between $G_{e_s^-}$ and ϵ_s, observed by Jay-Gerin and Ferradini,[98] has now been theoretically substantiated in terms of Equations 42 and 43. Also the very low stabilized electron yield of near melting point ice can be understood by this model which seems to describe solvated electron yields successfully in terms of two macroscopic material properties.

Recent experimental results lend some indirect evidence to the validity of the ion-electron pair model, discussed above. Long et al.[99] demonstrated that the solvation dynamics of an electron depends on whether the electron has been ejected from a Cl^- ion or from a water molecule. This shows that the presence or absence of the geminate positive ion markedly influences the time dependence of the excess electron processes.

IV. SUMMARY AND CONCLUSIONS

The aim of this chapter is to give an overview of some ideas relevant to the understanding of the early life history of excess electrons in polar liquids. The main attention being paid to physical principles and to contemporary work, experimental findings, and results of historical importance were mentioned only where they were needed by the argument. Additional material can be found elsewhere.[53]

After summarizing some basic ideas in the macroscopic theory of polarization and certain aspects of the molecular theories of relative permittivity, dielectric relaxation, and solvation, the problem of slow electron thermalization, localization, and recombination was discussed. The question, whether localized electrons are self-trapped entities or their localization is predestined by trapping sites which are present in the liquid, irrespective of the electric field of the electrons, was treated in some detail. Having summarized the present status of the problem one is inclined to prefer the model of self-trapping. On the one hand, recent liquid theories, although in a somewhat equivocal way, do not seem to support the idea of preformed traps. On the other hand, electron solvation kinetics was observed to be much too fast to be attributed to the stabilization of an electron localized in a preformed trap. This is a somewhat ironical situation. The idea of preformed traps was conceived in order to understand the fast localization of the excess electrons. By now it has turned out that a localized electron would form an equilibrium solvation shell slower than the constant charge relaxation time ϑ, whereas experiments show that solvation is usually faster than ϑ. The solution lies probably with the fact that an excess electron cannot be regarded as a classical heavy ion.

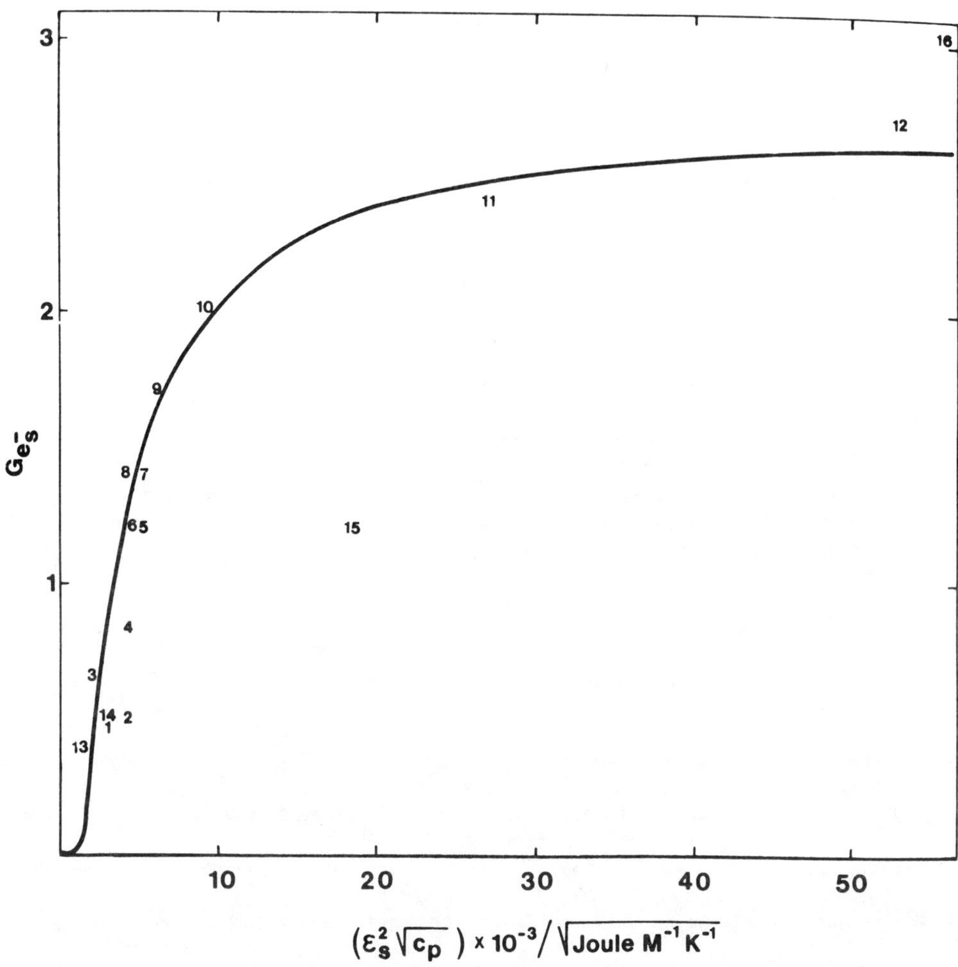

FIGURE 2. Dependence of solvated electron yield on static relative permittivity and specific heat. Continuous curve calculated by Equations 35, 42, and 43 with $G_o = 5.4$ $(100 \text{ eV})^{-1}$. Experimental points: (1) 1-pentanol; (2) isobutanol; (3) tert-butanol; (4) 2-butanone; (5) acetone; (6) 2-propanol; (7) 1-butanol; (8) 1-propanol; (9) ethanol; (10) methanol; (11) hydrazine; (12) H_2O; (13) pyridine; (14) ethylenediamine; (15) ethylene glycol; (16) D_2O. For experimental data, see Reference 94. (From Schiller, R., *J. Chem. Phys.*, 92, 5527, 1990. With permission.)

Stressing the importance of the nonclassical nature of the electron, with reference also to earlier work, a model was suggested by which thermalization, localization, and recombination can be described as simultaneous processes. The model, consisting of an ion-electron pair treated as an H-like atom which is embedded into a relaxing polar medium, enables one to understand instantaneous localization as a self-trapping process. Also, it gives a reasonable estimate to the rate of thermalization and predicts solvation kinetics. In order to test the theory, solvated electron yields were calculated as a function of static relative permittivity and specific heat of the medium, obtaining fair agreement with experimental data. It is strongly suggested that the destiny of a slow electron rests with dielectric polarization.

ACKNOWLEDGMENTS

The author expresses his thanks to Dr. A. Bernas, Professor H. L. Friedman, Dr. E. Keszei, and Dr. M. Tachiya for prepublication information, comments and criticism. Particular thanks are due to Dr. L. Nyikos for critical reading of the manuscript.

REFERENCES

1. **Jackson, J. D.,** *Classical Electrodynamics,* Wiley, New York, 1962.
2. **Landau, L. D. and Lifshitz, E. M.,** *Electrodynamics of Continuous Media,* Addison-Wesley, Oxford, 1960.
3. **Fröhlich, H.,** *Theory of Dielectrics,* Clarendon, Oxford, 1950.
4. **Schiller, R.,** The behavior of ion spurs and tracks in irradiated dipolar systems, *J. Chem. Phys.,* 43, 2760, 1965.
5. **Schiller, R,** Thermal electron escape in irradiated dipolar systems, *J. Chem. Phys.,* 47, 2278, 1967.
6. **Schiller, R.,** The dielectric constant of dipolar substances in the environment of ions produced by irradiation, *Chem. Phys. Lett.,* 5, 176, 1970.
7. **Hubbard, J. and Onsager, L.,** Dielectric dispersion and dielectric friction in electrolyte solutions. I., *J. Chem. Phys.,* 67, 4850, 1977.
8. **Morse, P. M. and Feshbach, H.,** *Methods of Theoretical Physics, Part II,* McGraw-Hill, New York, 1953, 1759.
9. **Friedman, H. L.,** Fast response of a dielectric to the motion of a charge, *J. Chem. Soc. Faraday Trans.,* 2, 79, 1465, 1983.
10. **Newton, M. D. and Friedman, L. H.,** Green function theory of charge transfer processes in solution, *J. Chem. Phys.,* 88, 4460, 1988.
11. **Kivelson, D. and Friedman, H.,** Longitudinal dielectric relaxation, *J. Phys. Chem.,* 93, 7026, 1989.
12. **Schiller, R.,** Macroscopic friction and dielectric relaxation, *IEEE Trans. Elec. Ins.,* 24, 199, 1989.
13. **Jonscher, A. K.,** *Dielectric Relaxation in Solids,* Chelsea Dielectric Press, London, 1983, 285.
14. **Rips, I. and Jortner, J.,** Dynamic solvent effects on outer-sphere electron transfer, *J. Chem. Phys.,* 87, 2090, 1987.
15. **Debye, P.,** *Polar Molecules,* Dover, New York, 1948, chap. V.
16. **Croxton, C. A.,** *Liquid State Physics — A Statistical Mechanical Introduction,* Cambridge University Press, London, 1974.
17. **Wertheim, M. S.,** Exact solution of the mean spherical model for fluids of hard spheres with permanent electric dipole moments, *J. Chem. Phys.,* 55, 4291, 1971.
18. **Dogonadze, R. R., Kálmán, E., Kornyshev, A. A., and Ulstrup, J., Eds.,** *The Chemical Physics of Solvation,* Parts A,B,C, Elsevier, Amsterdam, 1985, 1986, 1988.
19. **Chan, D. Y. C., Mitchell, D. J., and Ninham, B. W.,** A model for solvent structure around ions, *J. Chem. Phys.,* 70, 2946, 1979.
20. **Wolynes, P. G.,** Linearized microscopic theories of nonequilibrium solvation, *J. Chem. Phys.,* 86, 5133, 1987.
21. **Calef, D. F. and Wolynes, P. G.,** Smoluchowski-Vlasov theory of charge solvation dynamics, *J. Chem. Phys.,* 78, 4145, 1983.
22. **Onsager, L.,** Comment to the paper by Jortner, J. and Gaathon, M., Effects of phase density on ionization processes and electron localization in fluids, *Can. J. Chem.,* 55, 1819, 1977.
23. **Rips, I., Klafter, J., and Jortner, J.,** Dynamics of ion solvation, *J. Chem. Phys.,* 88, 3246, 1988.
24. **Castner, E. W., Maroncelli, M., Jr., and Fleming, G. R.,** Subpicosecond resolution studies of solvation dynamics in polar aprotic and alcohol solvents, *J. Chem. Phys.,* 86, 1090, 1987.
25. **Maroncelli, M. and Fleming, G. R.,** Picosecond solvation dynamics of coumarin 153: the importance of molecular aspects of solvation, *J. Chem. Phys.,* 86, 6221, 1987.
26. **Simon, J. D. and Shyh-Gang Su,** Intramolecular electron transfer and solvation, *J. Chem. Phys.,* 87, 7016, 1987.
27. **Nagarajan, V., Brearly, A. M., Kang, T. J., and Barbara, P. F.,** Time-resolved spectroscopic measurements on microscopic solvation dynamics, *J. Chem. Phys.,* 86, 3183, 1987.
28. **Loring, R. F. and Mukamel, S.,** Molecular theory of solvation and dielectric response in polar fluids, *J. Chem. Phys.,* 87, 1272, 1987.
29. **Chandra, A. and Bagchi, B.,** The role of translational diffusion in the polarization relaxation in dense polar liquids, *Chem. Phys. Lett.,* 151, 47, 1988.
30. **Bagchi, B. and Chandra, A.,** Solvation of an ion and of a dipole in a dipolar liquid: how different are the dynamics?, *Chem. Phys. Lett.,* 155, 533, 1989.
31. **Fleming, G. R.,** Comment, in *Solvation. Faraday Disc. Chem. Soc.,* No. 85, Faraday Div. Royal Soc. Chem., London, 1988, 231.
32. **Friedman, H. L.,** Personal communication, 1989.
33. **Chandra, A. and Bagchi, B.,** Breakdown of Onsager's conjecture on distance dependent polarization relaxation in solvation dynamics, *J. Chem. Phys.,* 91, 2594, 1989.
34. **Fröhlich, H. and Platzman, R. L.,** Energy loss of moving electrons to dipolar relaxation, *Phys. Rev.,* 92, 1152, 1953.

35. **Magee, J. L. and Helman, W. P.,** Energy loss of electrons in random motion, *J. Chem. Phys.,* 66, 310, 1977.
36. **Tachiya, M. and Sano, H.,** Energy loss of electrons in random motion. Analytical result, *J. Chem. Phys.,* 67, 5111, 1977.
37. **Jonscher, A. K.,** Movement of charge carriers in lossy dielectrics, *J. Phys. D: Appl. Phys.,* 13, L137, 1980.
38. **Michaud, M. and Sanche, L.,** Total cross section for slow electron (1 to 20 eV) scattering in solid H_2O, *Phys. Rev.* A, 36, 4672, 1987.
39. **Michaud, M. and Sanche, L.,** Absolute vibrational cross sections for slow-electron (1 to 18 eV) scattering in solid H_2O, *Phys. Rev.* A, 36, 4684, 1987.
40. **Goulet, T. and Jay-Gerin, J.-P.,** Thermalization distances and times for subexcitation electrons in solid water, *J. Phys. Chem.,* 92, 6871, 1988.
41. **Kaplan, I. G. and Mitverev, A. M.,** Interaction of charged particles with molecular medium and track effects in radiation chemistry, in *Advances in Chemical Physics,* Vol. LXVIII, Prigogine, I. and Rice, S. A., Eds., Wiley, New York, 1987, 255.
42. **Mozumder, A.,** Conjecture on electron trapping in liquid water, *Radiat. Phys. Chem.,* 32, 287, 1988.
43. **Warman, J. M., Zhou-lei, M., and van Lith, D.,** Electron thermalization in nanosecond pulse-ionized dry and humid air, *J. Chem. Phys.,* 81, 3908, 1984.
44. **Christophorou, L. G., Gant, K. S., and Baird, J. K.,** Slowing-down of subexcitation electrons in polyatomic gases, *Chem. Phys. Lett.,* 30, 104, 1975.
45. **Bader, G., Chaison, J., Caron, L. G., Michaud, M., Perluzzo, G., Sanche, L., Green, A., and Mann, A.,** Personal communications to M. Mozumder, cited in Ref. 40.
46. **Barker, G. C.,** Electrochemical effects produced by light-induced electron emission, *Ber. Bunsenges.,* 75, 728, 1971.
47. **Brodskii, A. M. and Gurevich, Yu. Ya.,** *Theory of electron emission from metals (in Russian),* Nauka, Moscow, 1973.
48. **Gurevich, Yu. Ya., Pleskov, Yu. V., and Rotenberg, Z. A.,** *Photoelectrochemisty,* Consultants Bureau, New York, 1980, chap. 1.
49. **Kreitus, I. V., Benderskii, V. A., and Tiliks, Y. E.,** Photoelectron emission from metal into concentrated electrolytes. Part II. Thermalization path-length of low-energy electrons and e^-_{aq} bulk reactions, *J. Electroanal. Chem.,* 140, 311, 1982.
50. **Konovalov, V. V., Raitsimring, A. M., and Tsvetkov, Yu. D.,** Thermalization lengths of "subexcitation electrons" in water determined by photoinjection from metals into electrolyte solutions, *Radiat. Phys. Chem.,* 32, 623, 1988.
51. **Tewari, P. H. and Freeman, G. R.,** Radiolysis of hydrocarbons: dependence of the free-ion yield on molecular structure, *J. Chem. Phys.,* 49, 954, 1968.
52. **Kenney-Wallace, G. A. and Jonah, C. D.,** Picosecond spectroscopy and solvation clusters. The dynamics of localizing electrons in polar fluids, *J. Phys. Chem.,* 86, 2572, 1982.
53. **Schiller, R.,** Slow electrons in relaxing dipolar media, *Radiat. Phys. Chem.,* 34, 61, 1989.
54. **Crawford, O. H.,** Bound states of a charged particle in a dipole field, *Proc. Phys. Soc.,* 91, 279, 1967.
55. **Garrett, W. R.,** Critical binding of an electron to a non-stationary electric dipole, *Chem. Phys. Lett.,* 5, 393, 1970.
56. **McAloon, B. J. and Webster, B. C.,** A molecular orbital study of a dimer model for the hydrated and ammoniated electron, *Theor. Chim. Acta,* 15, 385, 1969.
57. **Armbruster, M., Haberland, H., and Schindler, H. G.,** How can one make negatively charged water clusters (hydrated electrons)?, in *Electron and Ion Swarms,* Christophorou, L. G., Ed., Pergamon Press, New York, 1981, 203.
58. **Seng, G. and Linder, F.,** Vibrational excitation of polar molecules by electron impact II. Direct and resonant excitation of H_2O, *J. Phys. B: Atom. Molec. Phys.,* 9, 2539, 1976.
59. **Jay-Gerin, J.-P. and Ferradini, C., Eds.,** Electron solvation in polar liquids, in *Excess Electrons in Dielectric Media,* CRC Press, Boca Raton, FL, 1991, chap. 8.
60. **Tachiya, M. and Mozumder, A.,** Model of pre-existing traps for electrons in polar glasses, *J. Chem. Phys.,* 60, 3037, 1974.
61. **Tachiya, M. and Mozumder, A.,** Solvation process of the trapped electron in polar liquids, *J. Chem. Phys.,* 63, 1959, 1975.
62. **Schnitker, J., Rossky, P. J., and Kenney-Wallace, G. A.,** Electron localization in liquid water: a computer simulation study of microscopic trapping sites, *J. Chem. Phys.,* 85, 2986, 1986.
63. **Jonah, C. D., Romero, C., and Rahman, A.,** Hydrated electron revisited via the Feynman path integral route, *Chem. Phys. Lett.,* 123, 209, 1986.
64. **Jonah, C. D., Bartels, D. M., and Chernovitz, A. C.,** Primary processes in the radiation chemistry of water, *Radiat. Phys. Chem.,* 34, 145, 1989.

65. **Houée-Levin, C. and Jay-Gerin, J.-P.,** Interpretation of the absorption spectra of incompletely relaxed electrons in polar liquids as a direct measure of the energetic distribution of band-tail localized states, *J. Phys. Chem.,* 92, 6454, 1988.

66. **Goulet, T., Bernas, A., Ferradini, C., and Jay-Gerin, J.-P.,** On the electronic structure of liquid water: conduction-band tail revealed by photoionization data, *Chem. Phys. Lett.* 170, 492, 1990.

67. **Hamill, W. H.,** A model for the radiolysis of water, *J. Phys. Chem.,* 73, 1341, 1969.

68. **Chase, W. J. and Hunt, J. W.,** Solvation time of the electron in polar liquids. Water and alcohols, *J. Phys. Chem.,* 79, 2835, 1975.

69. **Hunt, J. W.,** Early events in radiation chemistry, in *Advances in Radiation Chemistry,* Burton, M. and Magee, J. L., Eds., Wiley, New York, 1976, 185.

70. **Wiesenfeld, J. M. and Ippen, E. P.,** Dynamics of electron solvation in liquid water, *Chem. Phys. Lett.,* 73, 47, 1980.

71. **Huppert, D., Kenney-Wallace, G. A., and Rentzepis, P. M.,** Picosecond infrared dynamics of electron trapping in polar liquids, *J. Chem. Phys.,* 75, 2265, 1981.

72. **Nikogosyan, D. N., Oraevsky, A. A., and Rupasov, V. I.,** Two-photon ionization and dissociation of liquid water, *Chem. Phys.,* 77, 131, 1983.

73. **Gauduel, Y., Martin, J. L., Migus, A., Yamada, N., and Antonetti, A.,** Femtosecond study of electron localization and solvation in pure water, in *Ultrafast Phenomena V,* Fleming, G. R. and Siegman, A. E., Eds., Springer, Berlin, 1986, 308.

74. **Migus, A., Gauduel, Y., Martin, J. A., and Antonetti, A.,** Excess electrons in liquid water: first evidence of a pre-hydrated state with femtosecond lifetime, *Phys. Rev. Lett.,* 58, 1559, 1987.

75. **Keszei, E. and Jay-Gerin, J.-P.,** Molar extinction coefficient of the incompletely relaxed excess electron in liquid water, *Radiat. Phys. Chem.,* 33, 183, 1989.

76. **Zusman, L. D. and Helman, A. B.,** Time-resolved spectroscopy of solvated electrons, *Chem. Phys. Lett.,* 114, 301, 1985.

77. **Baxendale, J. H. and Wardman, P.,** Electrons in liquid alcohols at low temperatures, *J. Chem. Soc. Faraday Trans. I.,* 69, 584, 1973.

78. **Gilles, L., Aldrich, J. E., and Hunt, J. W.,** Solvation time of the electron in liquid alcohols and water at room temperature, *Nature (London) Phys. Sci.,* 243, 70, 1973.

79. **Verberne, J. M., Loman, H., Warman, J. M., de Haas, M. P., Hummel, A., and Prinsen, L.,** Excess electrons in ice, *Nature (London),* 272, 343, 1978.

80. **Warman, J. M. and Jonah, C. D.,** Electron solvation in crystalline ice on a subnanosecond timescale, *Chem. Phys. Lett.,* 79, 43, 1981.

81. **Warman, J. M., de Haas, M. P., and Verberne, J. B.,** Decay kinetics of excess electrons in crystalline ice, *J. Phys. Chem.,* 84, 1240, 1980.

82. **Warman, J. M., Kunst, M., and Jonah, C. D.,** Subnanosecond time-resolved optical absorption studies of electron solvation in ice, *J. Phys. Chem.,* 87, 4292, 1983.

83. **de Haas, M. P., Kunst, M., Warman, J. M., and Verberne, J. B.,** Nanosecond time-resolved conductivity studies of pulse-ionized ice. 1. The mobility and trapping of conduction-band electrons in H_2O and D_2O ice, *J. Phys. Chem.,* 87, 4089, 1983.

84. **Fueki, K., Da-Fei Feng, and Kevan, L.,** Application of the semicontinuum model to temperature effects on solvated electron spectra and relaxation rates of dipole orientation around an excess electron in liquid alcohols, *J. Phys. Chem.,* 78, 393, 1974.

85. **Fueki, K., Feng, D.-F., and Kevan, L.,** Relaxation time of dipole orientation around a localized excess electron in alcohols, *J. Phys. Chem.,* 80, 1381, 1976.

86. **Schiller, R.,** Self-trapping of radiation produced electrons, *Nature (London),* 217, 1141, 1968.

87. **Schiller, R. and Vass, Sz.,** Calculation of electron solvation times for several dipolar liquids, *Int. J. Radiat. Phys. Chem.,* 6, 223, 1974.

88. **Tachiya, M. and Watanabe, H.,** Solvation processes of the electron in polar liquids. II. Continuum model, *J. Chem. Phys.,* 66, 3056, 1977.

89. **Hummel, A.,** Ionization in nonpolar molecular liquids by high-energy electrons, in *Advances in Radiation Chemistry,* Vol. 4., Burton, M. and Magee, J. L., Eds., Wiley, New York, 1974, 1.

90. **Hummel, A.,** Single-pair diffusion model of radiolysis of hydrocarbon liquids, in *Kinetics of Nonhomogeneous Processes,* Freeman, G. R., Ed., Wiley, New York, 1987, 215.

91. **Onsager, L.,** Initial recombination of ions, *Phys. Rev.,* 54, 554, 1938.

92. **Mozumder, A.,** Neutralization of isolated ion pair in polar media I. Theory for the yield of solvated electrons, *J. Chem. Phys.,* 50, 3153, 1969.

93. **Mozumder, A.,** Neutralization of isolated ion pair in polar media II. Evolution of the neutralization process, *J. Chem. Phys.,* 50, 3162, 1969.

94. **Tachiya, M.,** Personal communication, 1990.

95. **Schiller, R.,** Ion-electron pairs in condensed polar media treated as H-like atoms, *J. Chem. Phys.,* 92, 5527, 1990.

96. **Schnitker, J. and Rossky, P. J.,** Excess electron migration in liquid water, *J. Phys. Chem.,* 93, 6965, 1989.
97. **Rossky, P. J. and Schnitker, J.,** The hydrated electron: quantum simulation of structure, spectroscopy and dynamics, *J. Phys. Chem.,* 92, 4277, 1988.
98. **Jay-Gerin, J.-P. and Ferradini, C.,** On the variation of the free ion yield with the static dielectric constant in the radiolysis of liquids, *Radiat. Phys. Chem.,* 33, 251, 1989.
99. **Long, F. H., Lu, H., and Eisenthal, K. B.,** Femtosecond studies of electron photodetachment of simple ions in liquid water: solvation and geminate recombination dynamics, *J. Chem. Phys.,* 91, 4413, 1989.

Chapter 5

ELECTRONIC ENERGY LEVELS IN NONPOLAR DIELECTRIC LIQUIDS

Werner F. Schmidt

TABLE OF CONTENTS

I. INTRODUCTION

The study of the electronic properties of dielectric nonpolar liquids represents an important subject of the physics of insulators. While electronic conduction and excitation in solid insulators has received intense attention already for many decades the investigation of such processes in nonpolar liquids is of more recent origin. The simplest class of liquids are the liquefied rare gases (LHe, LNe, LAr, LKr, LXe) for which a detailed picture of the electron energetics and dynamics has been developed. Another class of liquids is represented by hydrocarbons and their derivatives. In very pure samples, electronic transport can be observed and electron mobilities, as a function of temperature and electric field strength, have been measured for more than 75 different liquids. Two recent reviews give overviews of the results obtained.[1,2]

These two classes of liquids are, in the pure state, perfect insulators. Thermal excitation of charge carrier pairs can be neglected under most circumstances. This property allows the injection and observation of small concentrations of charge carriers. Radioactive sources, accelerators, lasers, and other light sources are used for the injection of electrons. Electrical and optical detection methods are employed. These investigations lead to a deeper understanding of electronic charge carrier transport in liquids and other disordered systems. Furthermore, they promote a deeper understanding of the primary radiation- and photochemical processes and of the reaction kinetics of excess electrons with chemical solutes. This knowledge is a prerequisite for a better understanding of such diverse fields as photo- and radiation chemistry, electrical insulation, radiobiology, polymer physics, or detectors for ionizing radiation.

Recently, increased interest in electronic conduction in nonpolar liquids has been spurred by their employment as detection media in liquid ionization chambers and other detection systems in high energy physics.[3,4]

The liquids we are dealing with are comprised of atoms or molecules with ionization potentials of 8 to 24.6 eV. Upon condensation certain shifts in energy of the electronic levels take place. These liquids exhibit valence electron levels and electronic conduction levels which are well separated by a forbidden gap. Measurements of the photoconductivity and the photoelectric effect are the main experimental tools for the study of these shifts. Once the excess electron is injected into the liquid it may become localized. This state is connected with a thermally activated drift mobility. Special conditions are met for liquid helium where in the bulk and on the surface localized electrons have been observed. Liquid neon represents a borderline case as the existence of delocalized and localized electrons has been reported. Information on the energetics of hot electrons comes from mobility measurements at high electric field strength. The transport theory based on the Boltzmann collision equation allows the calculation of the mean electron energy as a function of the electric field strength. Scintillation studies provide additional calibration points for the energy levels involved. The properties of the positive hole levels are less well understood. A short summary of their physical properties must suffice. The topic of electron attachment/detachment equilibria cannot be treated within the framework of this chapter although the pertinent rate constants are determined by the energetics of the electrons and scavenger molecules in the liquid.

The theoretical treatment of the energy levels of excess electrons in liquid hydrocarbons and in most liquified rare gases is still in its infancy. Important progress has been made during the last few years with respect to the description of the electronic conduction level. A more refined theoretical understanding exists for excess electrons in the bulk and on the surface of liquid helium.

The energy levels we are concerned with in this article are depicted in Figure 1. Usually the energy of an electron in the vacuum is set to be zero. In the liquid, the conducting state

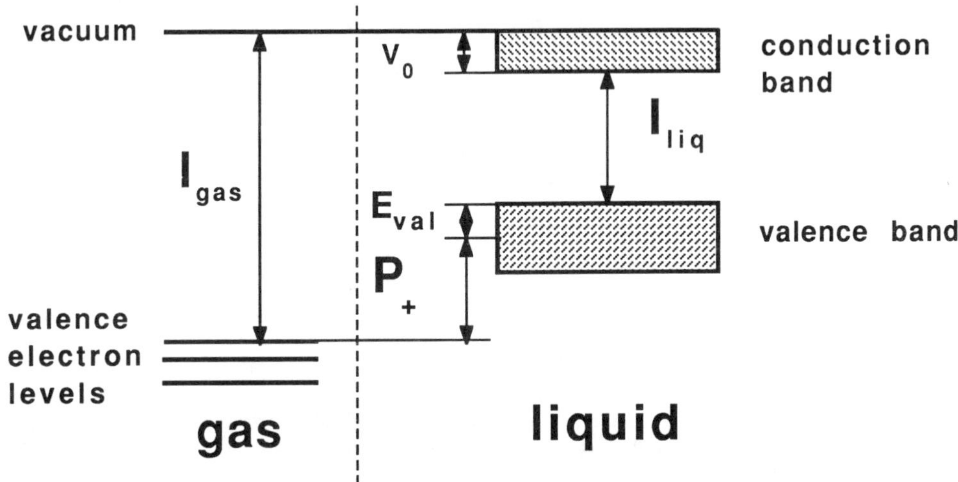

FIGURE 1.   Electronic energy levels of atoms or molecules in the gas and liquid phase.

is shifted by a value V_0 below or above the vacuum level. The uppermost valence electron level is shifted towards the vacuum by the polarization energy of the positive hole or ion, P_+, and by a broadening of the valence electron levels due to the increased interaction of the electronic shells of the atoms or molecules. The following relationship gives the shift in energies,

$$I_{liq} = I_{gas} + P_+ + V_0 + E_{val} \tag{1}$$

P_+, and E_{val} are negative, V_0 may be either positive or negative. While V_0 and I_{liq} have been measured, the order of magnitude of E_{val} can only be estimated from measurements of photoelectron spectra in the solid phase. Values for P_+ have been obtained, but the theoretical description of the structure of P_+ has not progressed very far beyond Born's equation.[5]

Restrictions in space prevented an extensive coverage of all these topics. Each subject will be discussed with respect to its basic aspects leaving the more detailed description to the specialized literature. References included have been selected from the point of view of giving the reader a foot in the door. With the use of modern data banks he will have no difficulty in extracting more information on a particular topic. The chapter is organized as follows: in Section II we discuss experimental methods and results of the study of energy levels; in Section III, theoretical models are presented for the delocalized electron level, for electron bubbles and for positive holes. Models for all other subjects are treated in Section II together with the experimental results.

II. EXPERIMENTAL METHODS AND RESULTS OF THE STUDY OF ELECTRONIC ENERGY LEVELS

A. PHOTOCONDUCTIVITY
1. Photoelectric Effect and Electronic Conduction Level

An important quantity in the discussion of excess electrons in dielectric liquids is given by the energy V_o of the electronic conduction level in the liquid with respect to the vacuum. It is the energy of the delocalized excess electron state. This energy can be either positive or negative depending on the strength of interaction between the excess electron and the molecules of the liquid. A positive V_o means that energy has to be spent in order to bring

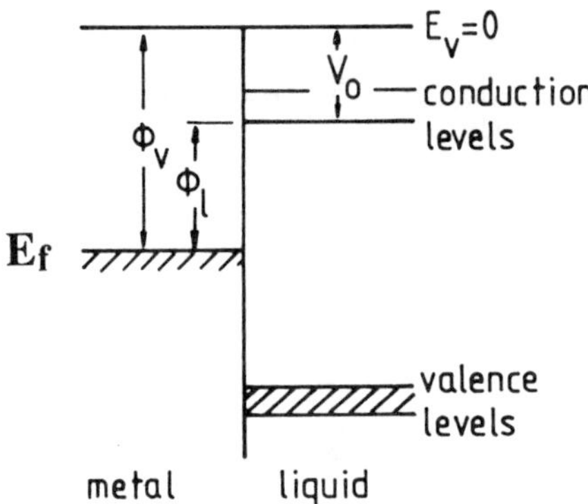

FIGURE 2.   Electronic energy levels at the metal/liquid interface; (Φ_v) Work function of the metal in vacuum; (Φ_l) work function in liquid; (E_f) Fermi level.

an electron from the vacuum into the bulk of the liquid, a negative value of V_o means that energy is released during this process. This energy can be determined by means of the photoelectric effect.[6]

The work function of a metal cathode in contact with a liquid Φ_{liq} is given by its vacuum work function Φ_{vac} and V_o as,

$$\phi_{liq} = \phi_{vac} + V_0 \tag{2}$$

The relations are depicted schematically in Figure 2.

For the yield of photoelectrons from a metal as a function of the photon energy, Fowler[7] derived the following dependence,

$$\frac{i}{I} = \alpha\ A\ T^2\ F(x) \tag{3}$$

where i denotes the current of photoelectrons, I is a measure of the light intensity, α is the quantum yield of the process, T denotes the absolute temperature, and A is a constant. The function $F(x)$, also known as Fowler function is defined as

$$F(x) = e^x - \frac{e^{2x}}{2^2} + \frac{e^{3x}}{3^2}\cdots \tag{4}$$

for $x < 0$, and

$$F(x) = \frac{x^2}{2} + \frac{\pi^2}{6} - \left(e^{-x} - \frac{e^{-2x}}{2^2} + \frac{e^{-3x}}{3^2}\cdots\right) \tag{5}$$

for $x > 0$. The dimensionless quantity x is defined as,

$$x = \frac{h\nu - \phi}{k_B T} \tag{6}$$

k_B is the Boltzmann constant.

It follows from Equations 5 and 6 that the yield of photoelectrons at constant temperature is a function of $h\nu - \Phi$, only. Equation 5 represents a universal curve which can be plotted as log $(i/I)_{Fow}$ vs. x. The measured data log $(i/I)_{meas}$ are plotted vs. $h\nu$. The experimental curve is shifted horizontally with respect to the universal curve by $\Phi/(k_B T)$. The vertical shift represents the apparent quantum yield of the emission process. An experimental example is given in Figure 3.[8] Near the work function threshold, i.e., for x <1, Equation 5 can be approximated by the first term which leads to the well-known dependence of the normalized photocurrent on the photon energy as,

$$(i/I)^{1/2} \propto (h\nu - \phi)$$

In order to obtain V_0, measurements of the work function of a metal cathode in vacuum and in contact with the liquid have to be performed. Experimental work demonstrated that the V_0 values obtained depended on the type of metal used for the measurement.[9,10] Two effects influence the work function measurement. First, oxide layers may form a barrier for electron injection due to the formation of a double layer.[10] Second, even with a completely clean metal, surface states in the metal-liquid interface may be filled with electrons from the metal also giving rise to the formation of a double layer. This is schematically depicted in Figure 4. The strength of this double layer depends on the concentration of occupied surface states q_{ss} (surface charge density, C/m^2) and on the thickness δ.[10] The work $\Delta\Phi$ for an electron to overcome this double layer is given as,

$$\Delta\phi = e_0 \frac{q_{ss}}{\epsilon_r \epsilon_0} \delta \tag{7}$$

ϵ_r is the relative dielectric constant of the liquid, ϵ_0 is the permittivity of free space, $\epsilon_0 = 8.85 \times 10^{-12}$ F/m, $e_0 = 1.6 \times 10^{-19}$ C. Although the details of the influence of these double layers on V_0 are not understood, it can be stated that measurements with gold cathodes give more negative V_0 values as compared to zinc or magnesium cathodes which suffer from oxide layers.[9] Values of V_0 obtained for different liquids are compiled in Table 1. Conduction band energies of hydrocarbons as a function of density were measured by Nishikawa and his co-workers.[11] Examples are shown in Figure 5. Measurements of V_0 in liquid argon, krypton, xenon, and methane as a function of density have been carried out by Reininger et al.[12—14] Some of their results are shown in Figure 20A and B of Section III.A.1.

2. Single Photon Ionization of Solutes and Solvents

Generally, a reduction of the ionization energy occurs during condensation (see Equation 1) due to the fact that the polarization energy, P_+ amounts to -1 to -3 eV depending on the dielectric constant, ϵ_r of the liquid and on the ionic radius, R. An estimate can be obtained from Born's formula,[5]

$$P_+ = -\frac{e_0^2}{8 \pi \epsilon_0 R} (1 - \frac{1}{\epsilon_r}) \tag{8}$$

V_0 may be either positive or negative but it varies between $+0.2$ eV and -0.7 eV, only. Exceptions are liquid helium and liquid neon where the absolute value of the polarization energy is <1 eV due to the low dielectric constant but the V_0 values are positive, $+0.67$ eV in the case of LNe and $+1.05$ eV in the case of LHe. In liquid helium the ionization energy I_{liq} is greater by more than 0.5 eV than the gas-phase ionization energy I_{gas}. This leads to a unique behavior in experiments concerned with charge transfer/at the metal/liquid interface under conditions of high electric field strength and electric breakdown.[15]

The quantity I_{liq} can be determined by photoconductivity measurements. The ionization

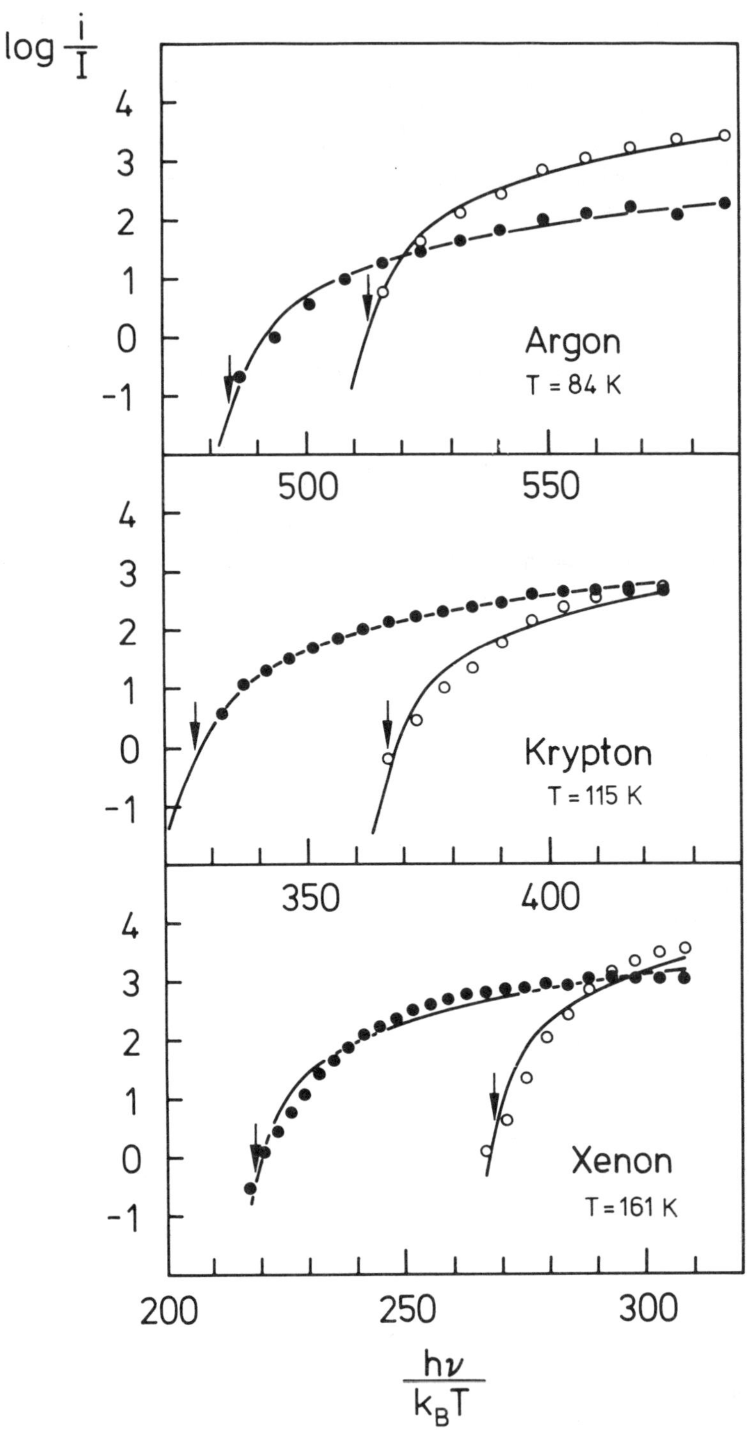

FIGURE 3.   Fowler curves for photoinjection of electrons in LAr, LKr, and LXe: (○) Vacuum curve; (●) liquid curve. (From Tauchert, W. and Schmidt, W. F., *Z. Naturforsch.*, 30a, 1085, 1975. With permission.)

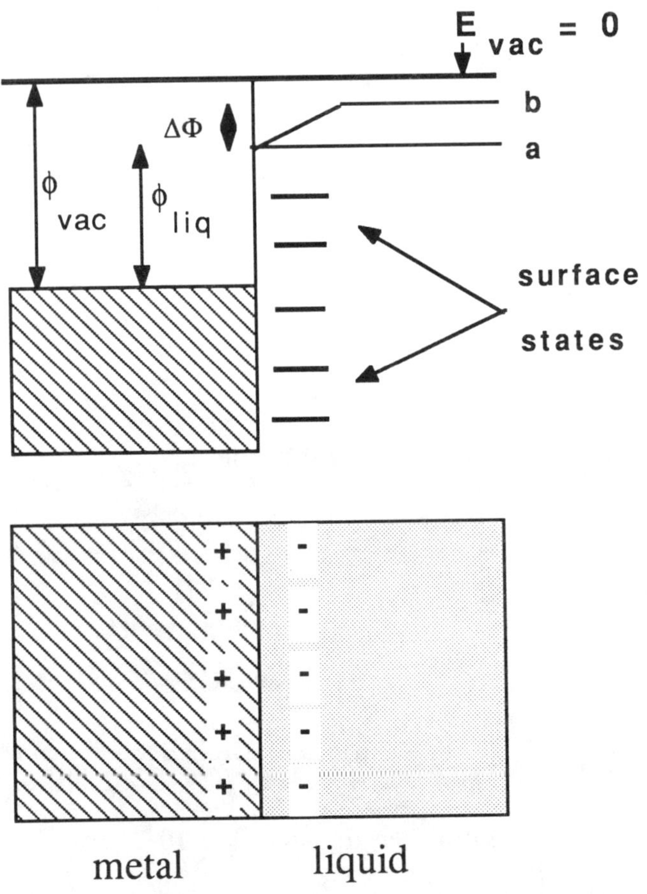

FIGURE 4.   Schematic representation of surface states at the metal/liquid interface; upper figure — energetic conditions, lower figure — schematic representation of the double layer.

TABLE 1
V_0 Values for Various Liquids[6,8-11]

Liquid	T(K)	V_0(eV)
Helium	4.2	+1.05
Neon	25	+0.67
Argon	84	−0.20
Krypton	116	−0.40
Xenon	161	−0.67
Methane	100	−0.25
Ethane	100	+0.20
n-Pentane	296	0.0
n-Hexane	296	+0.04
Neopentane	296	−0.43
Neohexane	296	−0.15
Isooctane	296	−0.18
Tetramethylsilane	296	−0.62
Tetramethyltin	296	−0.7

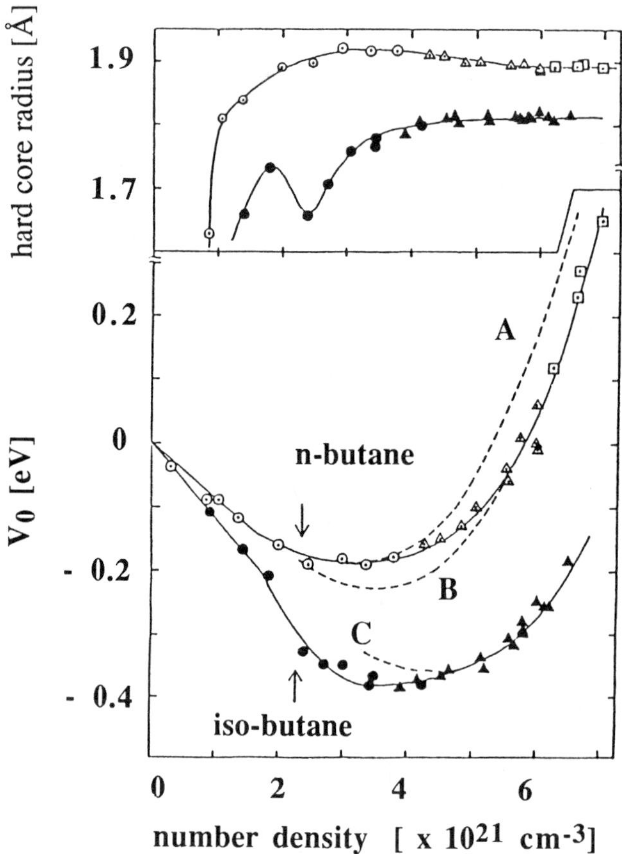

FIGURE 5. V_0 as a function of density for *n*-butane and iso-butane. Results of calculations by the SJC-model are shown by curves A (a = 0.1918 nm), B (a = 0.1892 nm), and C (a = 0.1812 nm). The solid line above 1.5×10^{21} cm^{-3} was obtained with the density dependent hard core radii given in the upper figure (From Nakagawa, K., Itoh, K., and Nishikawa, M., *IEEE Trans. El. Insul.*, EI23, 509, 1988. With permission.)

energy corresponds to photons of the vacuum ultraviolet spectrum where light sources deuterium lamps, rare gas discharge lamps and synchrotron radiation are available. Since the VUV light is strongly absorbed by the liquid, window electrodes of suitable material as LiF or MgF_2 are necessary. Usually, they are covered with a thin layer of gold. A typical set-up is depicted in Figure 6.[16] A photoconductivity spectrum measured in a solution of TMPD in neopentane is shown in Figure 7.[17] As the wavelength of the light is reduced, the photo currents exceed the dark current when the photon energy is high enough to ionize TMPD molecules dissolved in the neopentane. Below 180 nm the neopentane begins to absorb the light thus preventing further photoionization of TMPD in the bulk of the solution. At 140 nm the photocurrent begins to rise again due to the ionization of neopentane. After reaching a maximum the photocurrent falls down to zero again. This drop is due to the increase of absorption of the optical window material in this region. Taking into account the lamp spectrum, the transmission of the monochromator and the absorption spectrum of the gold covered LiF or MgF_2, respectively, the true variation of the photo conductivity with photon energy can be derived. Figure 8A shows the measured photocurrent of 2,2,4,4-tetramethylpentane while in Figure 8B the normalized photocurrent is shown.[18] In order to determine I_{liq} an extrapolation is necessary. Usually, the data are plotted vs. photon energy

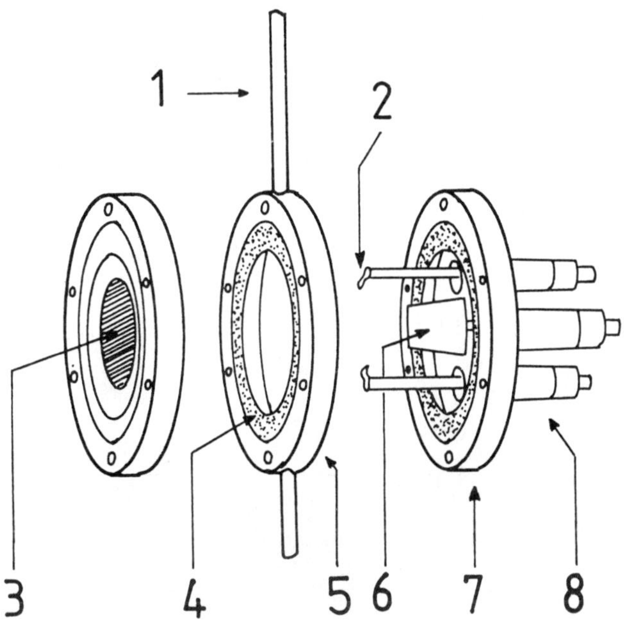

FIGURE 6. Photoconductivity cell for VUV measurements: (1) Filling tube; (2) contact spring; (3) gold coated LiF window; (5) double-sided flange; (6) back electrode; (7) flange; (8) electric feedthrough. (From Böttcher, E. H. and Schmidt, W. F., *J. Chem. Phys.*, 80, 1353, 1984. With permission.)

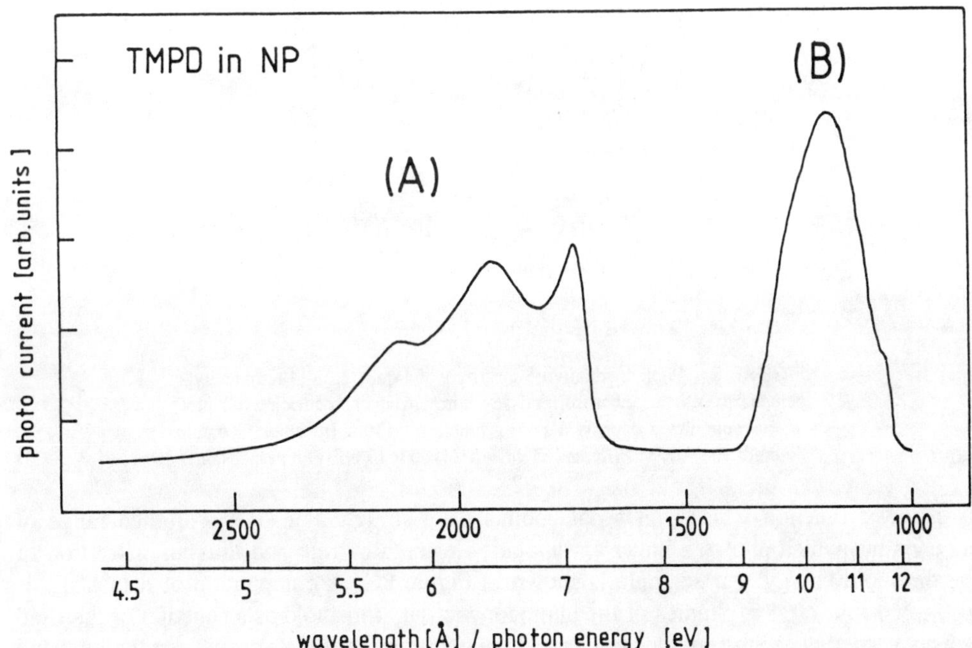

FIGURE 7. Photoconductivity spectrum of a solution of *N,N,N′,N′*-tetramethyl-*p*-phenylene diamine (TMPD) and 2,2-dimethylpropane (neopentane, NP): (A) photoionization of TMPD; (B) Photoionization of NP. (From Böttcher, E. H. and Schmidt, W. F., *5th Tihany Symp. Radiat. Chem.*, 427, 1982. With permission.)

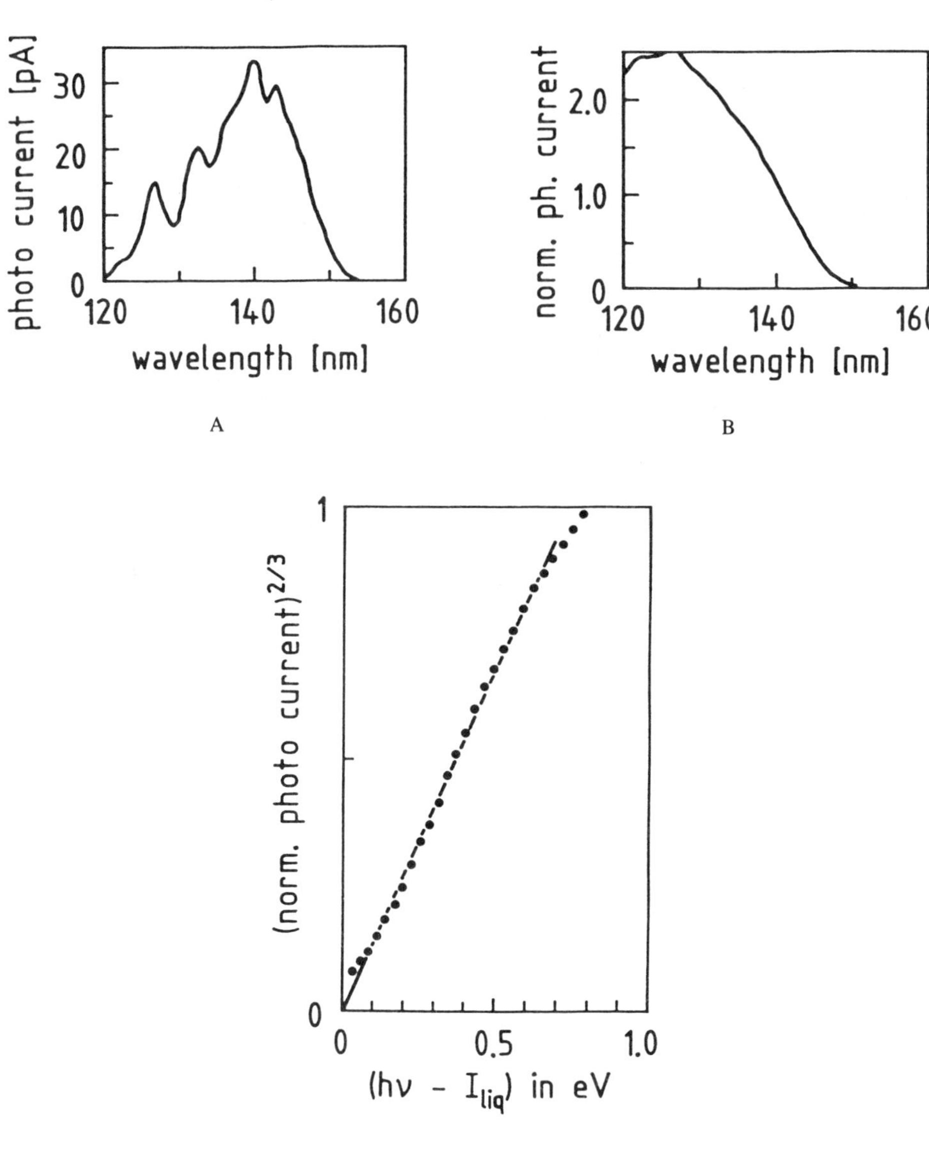

FIGURE 8.   Photoconductivity spectrum of liquid 2,2,4,4-tetramethyl-pentane; (A) actual photocurrent, (B) normalized photocurrent, (C) plot of the normalized data as a power function. (From Buschick, K. and Schmidt, W. F., *IEEE Trans. El. Insul.*, 24, 353, 1989. With permission.)

as a power function which yields data points on a straight line over a limited range of energies but which allows a linear extrapolation to zero current and thus an estimation of the threshold energy. An example is shown in Figure 8C. A compilation of data of I_{liq} is given in Table 2.[16,18-24] Studies of the photoconductivity threshold as a function of the fluid density carried out with xenon demonstrated the evolution of the electronic conduction band with increasing density.[22] The threshold, as a function of number density, is shown in Figure 9. The decrease is due to the combined effect of V_0 and P_+ which both decrease with density in the range studied here.

Single photon induced conductivity of solutes in nonpolar solvents yields information on the reduction of the ionization potential of the solute molecule due to the effect of V_0

TABLE 2
Ionization Energies I_{liq} of Neat Liquids[16,18-22,25]

Liquid	T(K)	I_{liq}(eV)
Xenon	161	9.2
Krypton	121	11.56
Tetramethylsilane	294	8.1 ± 0.05
		8.05 ± 0.05
Tetramethylgermanium	294	7.6 ± 0.05
Tetramethyltin	294	6.9 ± 0.1
2,2-Dimethylpropane (Neopentane)	294	8.85 ± 0.05
		8.55 ± 0.05
Methylbutane (Isopentane)	294	9.15 ± 0.05
n-Pentane	294	9.15 ± 0.1
		8.85 ± 0.05
2,2-Dimethylpropane (Neohexane)	294	8.73 ± 0.05
		8.50 ± 0.05
n-Hexane	294	8.6 ± 0.05
3-Methylpentane	294	8.85 ± 0.1
2,2,4-Trimethylpentane (Isooctane)	294	8.3 ± 0.05
2,2,4,4-Tetramethylpentane	295	8.2
n-Tridecane	294	9.25 ± 0.05
Tetrakis(dimethylamino)ethylene	295	3.77 ± 0.02
n-Pentene-1	294	8.33 ± 0.05
Tetramethylethylene	294	6.80 ± 0.05
Cyclopentane	294	8.80 ± 0.05
Cyclohexane	294	8.75 ± 0.1
		8.43 ± 0.05
Hexamethyldisilane	294	6.75 ± 0.1
Triethylsilane	294	8.25 ± 0.1
Bis(trimethylsilyl)ethane	295	6.85 ± 0.05
Polydimethysiloxane (50cSt)	296	7.29 ± 0.05

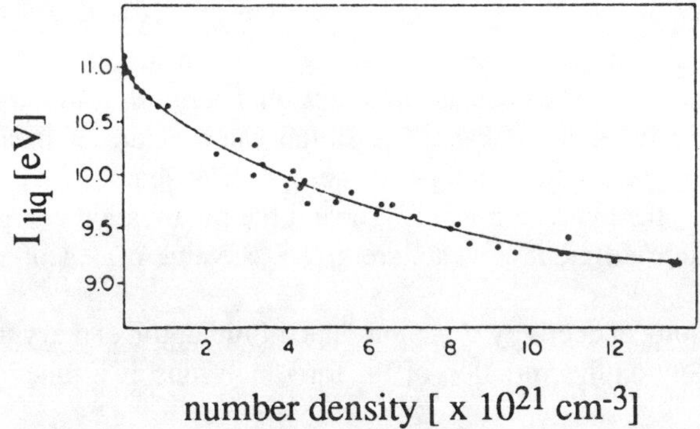

FIGURE 9. I_{liq} of LXe as a function of number density. (From Reininger, R., Asaf, U., and Steinberger, I. T., *Phys. Rev.*, B28, 3193, 1983. With permission.)

and P_+. Under these conditions a broadening of the valence levels of the isolated molecule due to interaction with the molecules of the liquid should be negligible. The ionization energy of a solute molecule or atom I_{sol}(liq) in solution is given as

$$I_{sol}(liq) = I_{sol}(gas) + P_+ + V_0 \qquad (9)$$

I_{sol}(gas) is the gas-phase ionization energy.

TABLE 3
Ionization Energies I_{solute} of Various Solutes in Different Solvents at Room Temperature[26,27a]

Solute	Solvent	I_{gas}	I_{solute}	P_+
1,2 Benzanthracene	2,2,4 TMP	7.47	6.16	−1.07
	Neopentane		6.15	−0.87
Azulene	2,2,4 TMP	7.42	6.06	−1.12
	TMSi		5.78	−1.09
Perylene	2,2,4 TMP	7.00	5.78	−0.98
	TMSi		5.47	−0.98
DABCO	2,2,4 TMP	7.20	5.81	−1.15
Triphenylamine	2,2,4 TMP	6.86	5.66	−0.96
Pyrene	Neopentane	7.44	6.20	−0.81
α-Methylnaphthalene	Neopentane	7.96	6.20	−1.33

[a]   All values in eV.

TABLE 4
Ionization Energies I_{solute}, V_0, and P_+ in eV for Anthracene in Different Hydrocarbons at Room Temperature

	I_{solute}	V_0	P_+
Gas	7.47		
2,2,4 Trimethylpentane	6.14	−0.24	−1.09
2,2,4,4 Tetramethylpentane	6.07	−0.36	−1.04
Tetramethylsilane	5.87	−0.55	−1.06

From Holroyd, R. A., Preses, J. M., and Zevos, N., *J. Chem. Phys.*, 79, 483, 1983. With permission.

Aromatic molecules were studied in a variety of hydrocarbon solvents.[25] Some results are summarized in Table 3. The decrease in ionization energy in solution is due to the effect of $|P_+|$ which is always greater than $|V_0|$. If the same solute is measured in different hydrocarbons, approximately the same polarization energy should result since ϵ_r is approximately 2 for most of the hydrocarbons. In Table 4 the polarization energies of the positive anthracene ion in three different solvents are given. A value of −1.06 eV is obtained for all three solvents.

The extrinsic photoconductivity of xenon-doped fluid argon and krypton was studied by Reininger et al.[28] Data on the variation of P_+ for xenon ions as a function of fluid density were obtained.

The availability of photons of the X-ray region at electron storage rings allows the excitation of electrons from inner shells of suitable solute molecules. First reports on the photoconductivity of organometallics in hydrocarbon solution are due to Sham.[29]

3. Multiphoton Effects

Ionization of liquid molecules or solutes can also be achieved by multiphoton absorption. Generally, the sum of the photon energies should exceed the ionization energy, I_{liq}. Two cases have to be distinguished. First, no excited state exists which matches the photon energy of the exciting light. The ionization is effected by the simultaneous absorption of two or more photons and the charge carrier yield depends on the light intensity to the second or higher power. Second, there exists an excited state with a certain lifetime, ionization is

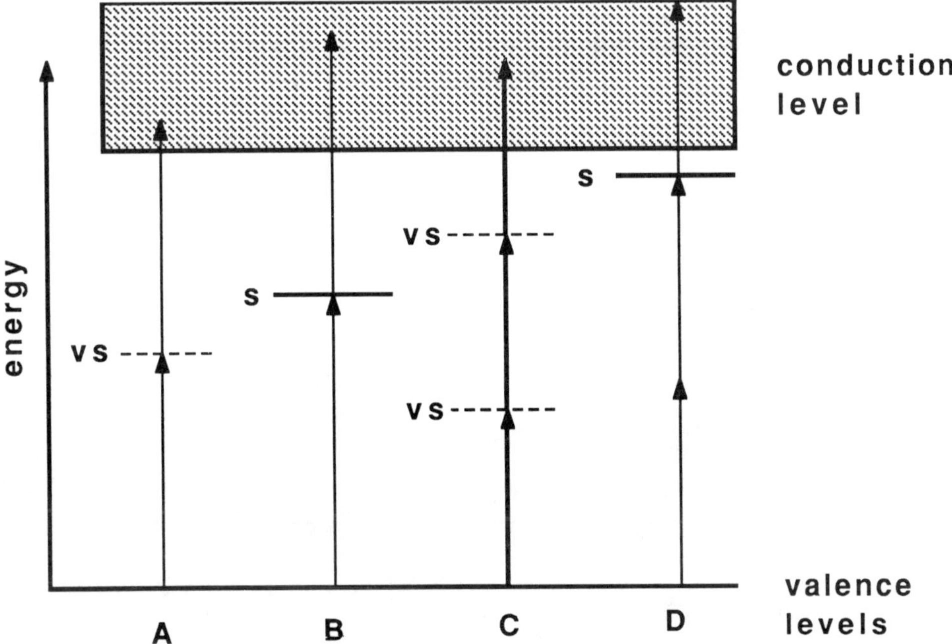

FIGURE 10.   Multiphoton ionization pathways: vs — virtual state; s — singlet state.

effected then by ionizing this state with the second or third photon. Some of the possibilities are depicted in Figure 10.

There seem to be only a few studies on the ionization by multiphoton absorption of pure liquids. Biphotonic ionization of liquid tetramethylsilane at different wavelengths was reported by Böttcher and Schmidt[30] who used an excimer laser. A quadratic dependence of the charge carrier yield on the photon energy of the exciting light was found for 249 nm (4.98 eV), 308 nm (4.03 eV), and 353 nm (3.51 eV). The quantum yield for charge carrier production decreased by more than two orders of magnitude from 249 nm to 353 nm. The photon energy of 3.51 eV is already below the ionization energy of liquid tetramethylsilane (8.05 eV). In the photoionization of solute molecules in liquid hydrocarbons similar effects have been reported. It was found that with biphotonic ionization lower thresholds were obtained. A possible explanation may be the participation of short-lived excited states. Multiphoton ionization of neat benzene was investigated by Scott et al.[31] They found that three-photon processes were observed with an excited singlet state at 6.99 eV with a lifetime of 10 ns or longer formed by two photon absorption.

More studies have been carried out with solutions. Early work is due to Pillof and Albrecht[32] who studied the biphotonic ionization of TMPD (N,N,N′N′ tetramethyl-p-phenylene diamine) in 3-methylpentane. They found the square dependence on the light intensity. Siomos et al.[33,34] investigated the two photon induced ionization threshold of several aromatic molecules in n-pentane. The values obtained were lower than the corresponding thresholds measured in single photon ionization experiments. In a later paper, Faidas and Christophorou[35] attributed these lower values to three-photonic processes near the threshold which were not identified in the earlier work. For fluoranthene in tetramethylsilane they report a threshold of 5.7 eV. In another more complete study, two-photon ionization onsets were reported for several aromatic molecules in various hydrocarbons.[36] Some of their results are compiled in Table 5. Within the variation of the relative dielectric constant ϵ_r and the limits of error, the polarization P_+ energy is constant.

TABLE 5
Two-Photon Ionization Thresholds of Azulene at Room Temperature

Solvent	I_{liq}[a]	V_0	P_+
n-Tridecane	6.28	+0.21	−1.35
n-Pentane	6.12	+0.01	−1.31
2,2,4,4 Tetramethylpentane	5.70	−0.33	−1.39
Tetramethylsilane	5.45	−0.55	−1.42
Tetramethyltin	5.33	−0.75	−1.34
Gas phase	7.42		

[a]   All in eV.

From Faidas, H. and Christophorou, L. G., *J. Chem. Phys.*, 86, 2505, 1987.

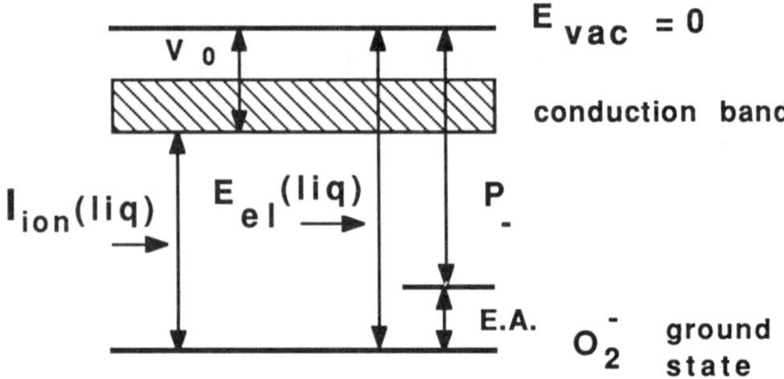

FIGURE 11.   Energetic conditions at photodetachment of electrons from negative oxygen ions in a liquid hydrocarbon.

4. Photodetachment from Negative Ions, Photoassisted Electron Diffusion

Defined negative ions can be generated by electron attachment to electronegative solutes. If we set the energy of the vacuum as zero, $E_{vac} = 0$, then the energy of the attached electron with respect to vacuum is given as,

$$E_{el} \text{ (gas)} = \text{E.A.} \tag{10}$$

where E.A. denotes the electron affinity. E.A. is negative if energy is released due to the attachment. It is also known as the ionization energy of the negative ion. If this ion is placed into a nonpolar liquid then an energy E_{el} (liq) is released which is the sum of the polarization energy of the negative ion and the solvation energy of the neutral solute molecule. Since the heat of solution of neutral organic molecules in hydrocarbons is much smaller than P_-, E_{el} (liq) may be written as,

$$E_{el} \text{ (liq)} = \text{E.A.} + P_- \tag{11}$$

where P_- is given by the Born formula (see Equation 8). Note that P_- is always negative. If the electron is to be detached from the negative ion but is allowed to remain in the liquid, an ionization energy I_{ion}(liq) is necessary which is given as

$$I_{ion}(\text{liq}) = \text{E.A.} + P_- - V_0 \tag{12}$$

V_0 is the energy of the electronic conduction level. These energetic conditions are shown schematically in Figure 11. I_{ion}(liq) can be measured, for instance, as a photo conductivity

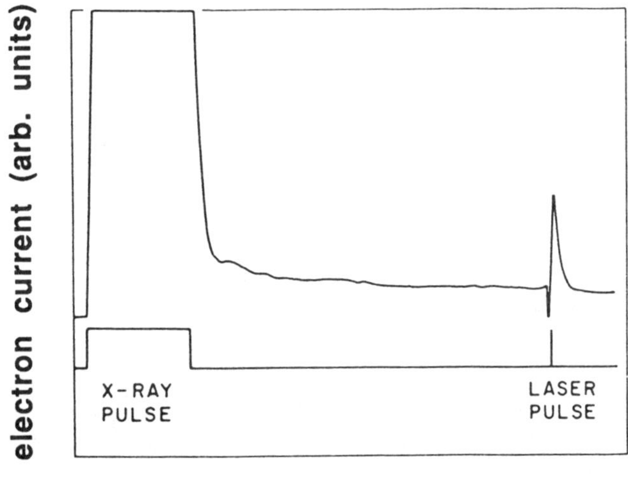

FIGURE 12.   Oscilloscopic trace of electron detachment from O_{2^-} in tetramethylsilane. (From Sowada, U. and Holroyd, R. A., *J. Chem. Phys.*, 70, 3586, 1979. With permission.)

threshold. Such measurements have been described for the system O_2/liquid hydrocarbons.[37,38] Negative ions of O_2 were produced by irradiation of a solution with a pulse of high energy electrons or X rays. Ionization of the molecules of the liquid yields positive ions and electrons,

$$M \longrightarrow M^+ + e^-$$

the electrons react with O_2 to give,

$$e^- + O_2 \longrightarrow O_{2^-}$$

If the concentration of O_2 is high enough then all electrons will react during the radiation pulse and after the pulse positive and negative ions will be present in the liquid which will start to recombine. A light pulse of sufficiently high photon energy leads to photo detachment of electrons and due to their much greater mobility to an observable current peak which decays rapidly in time due to reattachment. An example is shown in Figure 12. Variation of the photon energy with a tunable laser allows the determination of the photoconductivity threshold. In the gas phase the photodetachment cross section was found to follow a dependence on the photon energy E, given as

$$\sigma = BE \, (E - E_{th})^{3/2} \tag{13}$$

The same law was found to hold for the detachment in the liquid solutions, too. In the gas phase the electron affinity of O_2^- has been determined to be 0.44 eV. The results obtained in the hydrocarbons are summarized in Table 6. From a comparison of the magnitude of the photodetachment cross section in the gas phase and in the liquid an estimate of the effective mass of the electron in the liquid can be obtained. Baird[39] derived a relationship connecting the constant B and the effective mass of the electron in the gas and liquid phases, respectively. From measured data he derived as the electron effective mass in LAr (87K) $m^* = 0.18 \, m_0$; and $m^* = 0.27 \, m_0$ for liquid tetramethylsilane, 2,2,4-trimethylpentane,

TABLE 6
Photoconductivity Threshold of O_2- in Nonpolar Liquids

Liquid	T(K)	E_{th}(eV)	V_0(eV)	Ref.
Tetramethylsilane	296	2.08	-0.5	38
		2.30		37
	200	2.32	-0.47	38
2,2-Dimethylpropane	296	2.02	-0.43	38
2,2,4 Trimethylpentane	296	2.55	-0.18	38
		2.40		37

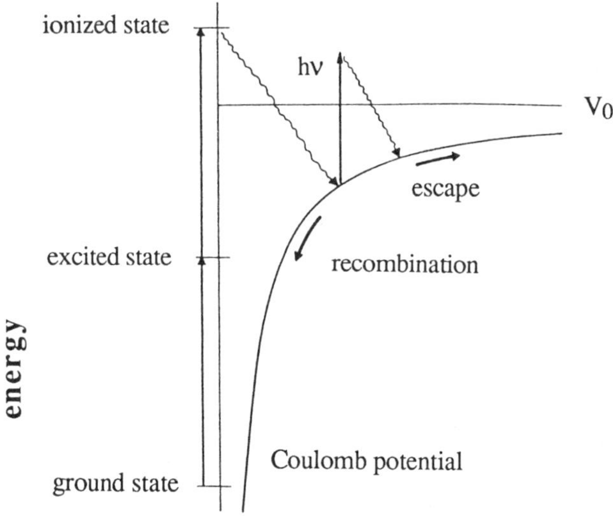

FIGURE 13.   Model of photodetrapping.

and 2,2,-dimethylbutane at 296 K, respectively (m_0 free electron rest mass). In a recent paper, Baird and Schulman[40] extended the model to include the temperature dependence of the photodetachment threshold.

Localized electrons can also be considered as temporary negative ions states. The electron can leave its trap either by absorption of phonons (thermal activation) or by absorption of photons (photoassisted diffusion). The latter process requires an energy of the order of 1 eV as can be deduced from the optical absorption spectrum of localized electrons in *n*-hexane. The process of photoassisted diffusion has been demonstrated experimentally by Balakin et al.[41,42] Electrons were produced by photoionization of anthracene in *n*-hexane or 2,2,4-trimethylpentane by a laser pulse of 347 nm. Subsequent illumination of the solution with a pulse of 694 nm led to a temporary increase in the photocurrent due to the liberation of trapped electrons. The process responsible for this increase can be envisaged as follows,

$$h\nu + e_{tr} \longrightarrow e_{hot} \longrightarrow e_{th} \longrightarrow e_{tr}$$

where, e_{tr} = trapped electron, e_{hot} = superthermal quasi-free electron, e_{th} = thermalized quasi-free electron. This effect has an additional consequence for the yield of free electron/ion pairs. Electrons which are localized within the Onsager radius r_c from their positive ion can escape into the bulk of the liquid by photoassisted diffusion (see Figure 13). As a

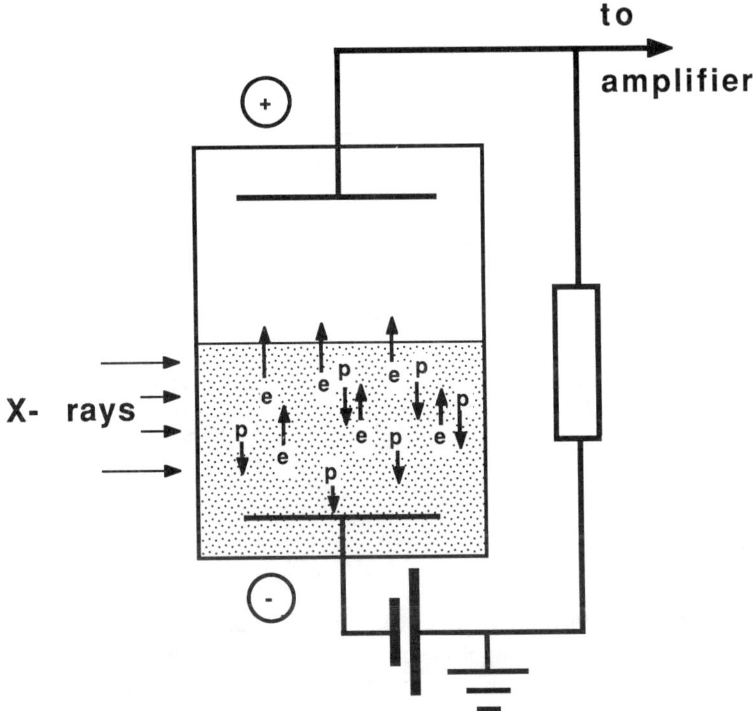

FIGURE 14.   Experimental set up for the study of electron emission

consequence the free ion yield increases when measured under the conditions of simultaneous illumination by 347 nm and 694 nm light.[43,44]

The wavelength dependence of the photoinduced diffusion coefficient was compared to the optical absorption spectrum of the trapped electron theoretically by Funabashi.[45] It was assumed that the absorption spectrum of the electron is due to electron-transfer transitions with a distribution of transfer distances. At constant light intensity, the diffusion coefficient is shown to exhibit a peak at shorter wavelengths than the absorption peak.

B. ELECTRONS AT INTERFACES
1. Electron Emission from Liquids

Electron emission from the liquid into the vapor phase should occur since for delocalized electrons (also called quasi-free electrons) the barrier for escape from the liquid is given by V_0. The energy of localized electrons (also called trapped electrons) may also be above or below the vacuum level). Two cases can be distinguished, emission of delocalized electrons, for instance from liquid argon or liquid xenon, or emission of localized electrons, for instance electrons from liquid helium, liquid neon, or liquid n-hexane. A schematic representation of an experimental set-up is shown in Figure 14. Electrons are produced in the liquid either by ionization with X- or γ-rays from the outside or by a radioactive α source deposited on the cathode.

Emission from electron bubbles in liquid helium was observed experimentally by Schoepe and Rayfield.[46] The total energy of the electron bubble is several tenths of one electron volt above the vacuum level (see Section III.B.1). When the bubble approaches the LHe surface ($z = 0$, $z > 0$ in the liquid) it comes under the influence of its repulsive image charge ($\epsilon_r(\text{LHe}) > \epsilon_r(\text{vac})$) and of the homogeneous applied electric field E. The total potential energy $V(z)$ is given as

$$V(z) = A/z + e_0 E z \tag{14}$$

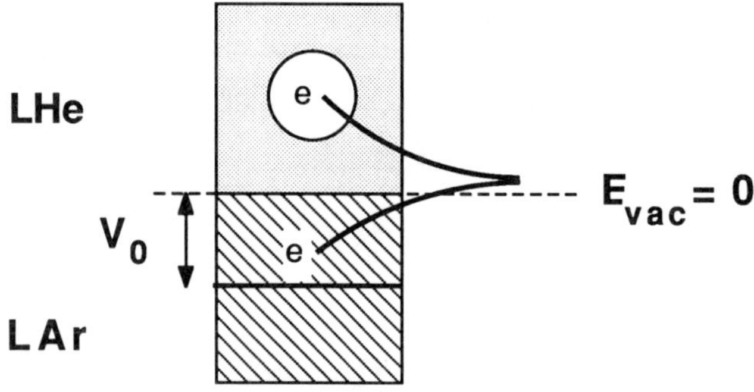

FIGURE 15.   Energetic conditions of electron emission: (a) from electron bubbles (LHe); (b) from delocalized states (LAr).

with

$$A = e_0^2 \, \frac{(\epsilon_r - 1)}{16 \, \pi\epsilon_0\epsilon_r \, (\epsilon_r + 1)} \tag{15}$$

ϵ_r denotes the relative dielectric constant of the liquid while the constant of the vapor was set equal to 1. A potential minimum occurs at around 200 Å below the surface with an applied field of 150 V/cm. Here, the electron bubble is trapped and the electron leaves the bubble by tunneling through this liquid layer into the vapor space. The same effect was observed by Bruschi et al.[47] in liquid neon. No detailed studies are available for emission of trapped electrons in hydrocarbons.

A different situation exists for delocalization electrons in the heavier rare-gas liquids. Here, $V_0 < 0$ and the barrier can be overcome by electric field assisted thermal emission. If the electric field strength is high enough the emission probability approaches 1. In the case of liquid argon ($V_0 = -0.20$ eV), a few kV/cm at the liquid/vapor interface are sufficient.[48] The two energetic situations described before are summarized in Figure 15. Electron emission from some liquid hydrocarbons has been reported by Boriev et al.[49]

Recently, new interest has arisen in this phenomenon since novel liquid photocathodes can be realized which are based on the photoionization of suitable solute molecules in a hydrocarbon and subsequent extraction of the electron and detection in the vapor space above.[50]

2. Two-dimensional Electrons

For liquids with a positive value of V_0 electron transfer from the vacuum into the liquid is hampered by this barrier. Especially, the large value of $V_0 = 1$ eV of liquid He leads to the formation of electron surface states.[51,52] An electron above the liquid surface experiences an image potential which is given by

$$V_{image} (z) = - f_\epsilon \, \frac{e_0^2}{4\pi\epsilon_0 \, z} \tag{16}$$

where z denotes the coordinate perpendicular to the surface (see Figure 16A) and f_ϵ depends on the relative dielectric constant ϵ_r,

$$f_\epsilon = \frac{\epsilon_r - 1}{4(\epsilon_r + 1)} \tag{17}$$

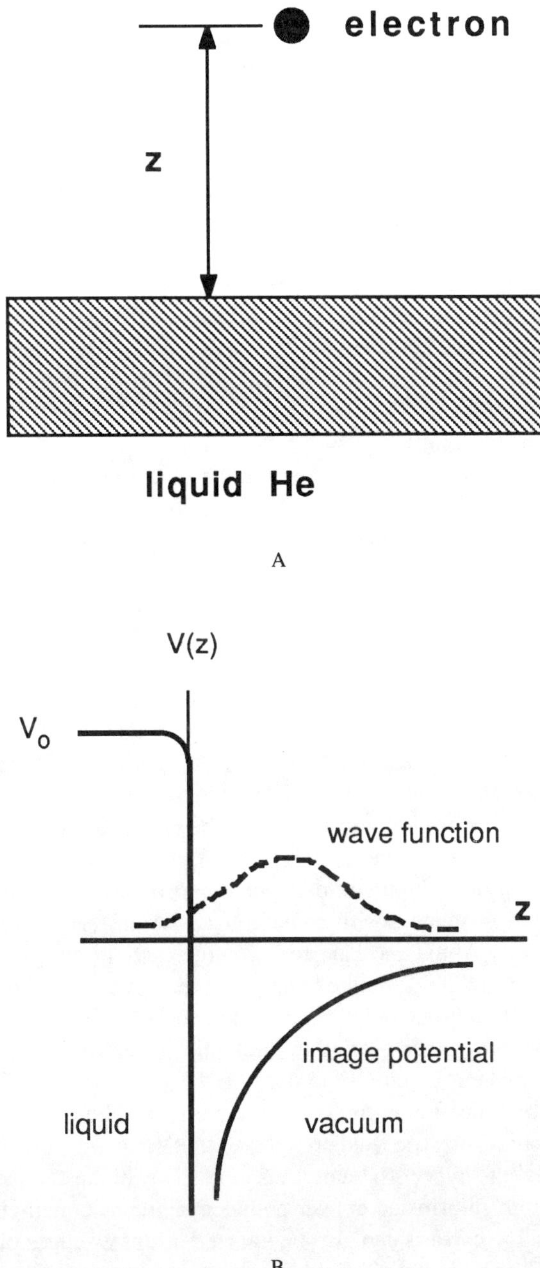

FIGURE 16.   Electron on a LHe surface: (A) geometry; (B) potential.

Under the influence of this potential the electron is attracted toward the surface. On the other hand, the overlap of its wave function with the wave functions of the He electrons leads to a strong repulsion. As a result the excess electron forms a slight depression on the liquid He surface. Parallel to the surface it has a high mobility. For the vapor/liquid surface of liquid He^4 the factor f_ϵ is approximately 1/144. The attractive potential gives rise to hydrogen-like wave functions with the Bohr radius becoming a_B/f_ϵ. The energy levels correspond to a Rydberg series given as,

$$E_n = \frac{f_\epsilon^2}{2\,n}\,Ry \qquad (18)$$

Transition between the energy levels corresponds to frequencies of 125.9 GHz for $n = 1$ to $n = 2$ and 148.6 GHz for $n = 1$ to $n = 3$. The ground state energy ($n = 1$) is -9 K while the first excited state is at -2 K.[53,54] The potential is schematically depicted in Figure 16B. The Bohr radius for the ground state (where the probability density has a maximum) is approximately 76 Å from the surface, the expectation value is at 114 Å. These electrons move around the surface as a classical two-dimensional Coulomb gas.

At higher surface densities of electrons collective effects become important and the electrons form a regular lattice connected with a depression of the liquid surface.[55] Electrons in surface states have also been observed with liquid hydrogen and liquid neon.[56]

3. Photoassisted Hole Injection

Very little work seems to have been done on the injection of positive charge carriers at an electrode in contact with a nonpolar liquid. In principle, electrons from the valence levels of the atoms or molecules of the liquid near the electrode could be excited by light to singlet or triplet states which are above the Fermi level (see Figure 2). If the electrode is on positive potential the excited electron could tunnel into the metal and a positive charge carrier would be injected into the liquid. So far, this process has not been observed in pure nonpolar liquids.

Injection of positive charge carriers at a semiconductor electrode in contact with a hydrocarbon was observed with the participation of molecules with a lower ionization potential than the hydrocarbon. Reinis and Cressman[57] used tertiary alkylamines in dodecane. Positive charge carriers and electrons were produced in a selenium layer. The holes were driven to the semiconductor/liquid interface where they reacted with the amines (see Figure 17, upper part). Under the influence of an electric field of 10^5 V/cm positive amine ions were injected the mobility of which could be measured by a time of flight experiment. An extensive study on the sensitized positive charge transfer at a semiconductor/dodecane interface was carried out by Ahuja and Hauffe.[58] Charge carrier pairs were produced by light absorption in metal free phthalocyanine, copper phthalocyanine, and chlorinated copper phthalocyanine. The holes migrated to the semiconductor/ liquid interface where they reacted with electron donating compounds present in the liquid. Over 10 different electron donating compounds were investigated. Electric field strengths between 10^4 and 10^5 V/cm were applied which led to measurable injection currents of positive ions. The overall quantum efficiency determined with an applied electric fieldstrength of 43 kV/cm was low, of the order of 10^{-8} to 10^{-5} depending on the type of donor molecule. The highest injection currents were observed with the system chlorinated copper phthalocyanine/hexamethyl phosphor triamide/ dodecane. Positive charge carriers can also be injected at tips or edges on positive potential, under the influence of the high electric field strength (see Figure 17, lower part). The details of this process are still not well understood.

C. HIGH ELECTRIC FIELDS
1. Electron Drift Velocity and Attachment

Measurements of the electron drift velocity as a function of the applied electric field strength are carried out in order to determine the low field electron mobility μ_0. In liquids with a value of $\mu_0 > 10$ cm^{2}V^{-1}s^{-1}, a field dependent mobility is observed if the electrical field strength exceeds a critical value. This effect has been explained as being due to a heating up of the electron distribution by the electric field.[59] Electrons take up more energy from the electric field between two subsequent collisions with the molecules or atoms of the liquid than they can lose. Analysis of the drift velocity/electric field strength data in

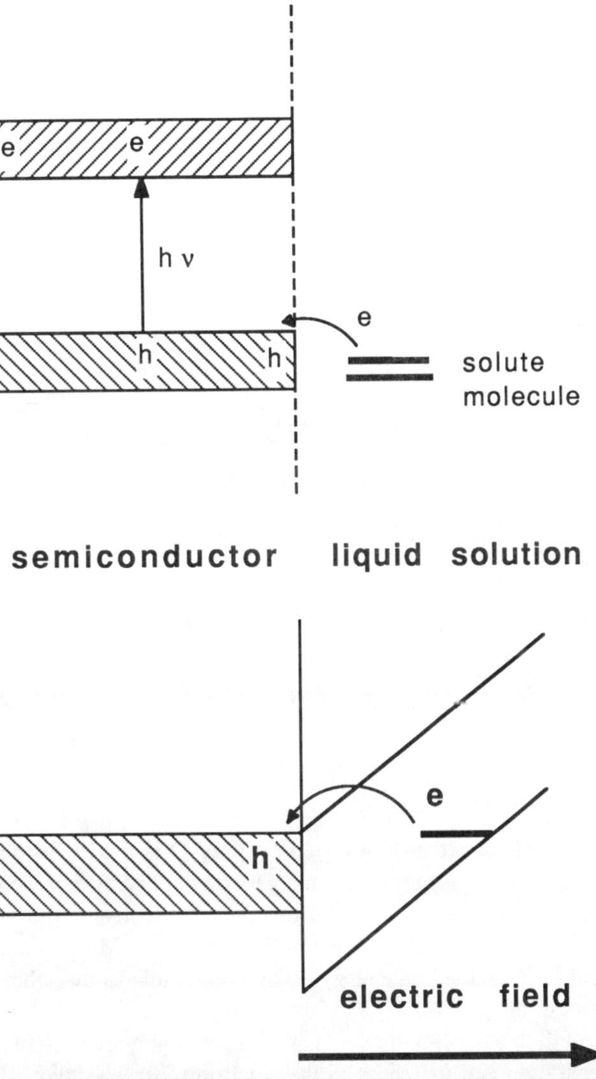

FIGURE 17.   Energetic conditions for injection of positive holes; upper figure — hole injection at a photoconductor/liquid interface; lower figure — hole injection by tunneling at high electric field strength.

liquid argon and xenon by means of the Cohen Lekner theory[60,61] yielded the electron mean energy as a function of the electric field strength. Since at low electric field strength the electrons are in thermal equilibrium with the atoms of the liquid, their mean energy is given by $k_B T$, at higher electric field strength F, an increase of electron energy occurs and the mean energy is given then as $k_B T + \epsilon(F)$. An example of such an analysis by Gushchin et al.[61] is shown in Figure 18. Further increase of the electric field strength above 10^5 V/cm is possible in the vicinity of thin wires or tips. When the electrons enter this region their mean energy increases rapidly and it becomes sufficient for the excitation of light (about 7 eV) in LXe and at higher field strengths ionization of Xe atoms in the liquid by collisions (above 9 eV) takes place.[62,63] From these findings it seems to be apparent that in LAr, LKr and LXe the conduction band extends over several electron volts in energy.

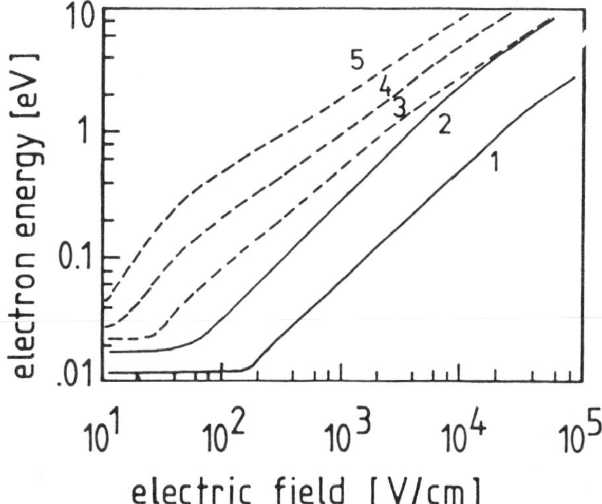

FIGURE 18.   Electron mean energy as a function of the applied electric field strength: (1) LAr, 85 K; (2) LAr, 140 K; (3) LXe, 165 K; (4) LXe, 200 K; (5) LXe, 230 K. (From Gushchin, E. M., Kruglov, A. A., and Obodovskii, I. M., *Sov. Phys. JETP,* 55, 650, 1982. With permission.)

The electron mean energy extracted from drift velocity data was corroborated by measurements of the electron diffusion coefficient by Shibamura et al.[64] Adapting the Townsend method for diffusion measurements in gases they were able to determine the electron mean energy

$$\langle \epsilon \rangle = e_0 D/\mu \tag{19}$$

as a function of electric field strength up to 10^4 V/cm. D is the diffusion coefficient, μ is the electron mobility. Additional support for the general dependence of electron mean energy on electric field strength can be obtained from measurements of the attachment rate of specific scavengers as a function of the electric field strength. Measurements with SF_6, O_2, and N_2O in LAr and LXe yielded an energy scale comparable to that obtained by Gushchin et al.[61] or Shibamura et al.[65]

Measurements in molecular liquids necessitate much higher electric field strengths which lead to electric breakdown and increases in the electron mean energy of 0.05 eV to 0.1 eV only have been achieved.[66,67]

2. Recombination Fluorescence

Ionization of solute molecules in nonpolar solvents leads to geminate positive solute ion/electron pairs which will undergo recombination. The recombination may lead to excited states of the solute molecules which decay by fluorescence to the electronic ground state. This recombination fluorescence is influenced by the application of an external electric field. Increase of the field strength leads to a decrease of the fluorescence light intensity. The electric field helps germinate electron/ion pairs to escape their mutual attraction. This effect has been observed in the photoionization of TMPD in various hydrocarbons[68,69] and also in the liquefied rare gases where recombination leads to fluorescence of the rare gas atoms.[70]

D. LOCALIZED ELECTRONS
1. Optical Absorption of Electrons

Electrons exhibit an optical absorption in several liquids. The optical absorption spectra of solvated electrons in polar liquids as water, alcohols, and liquid ammonia are well known.[71]

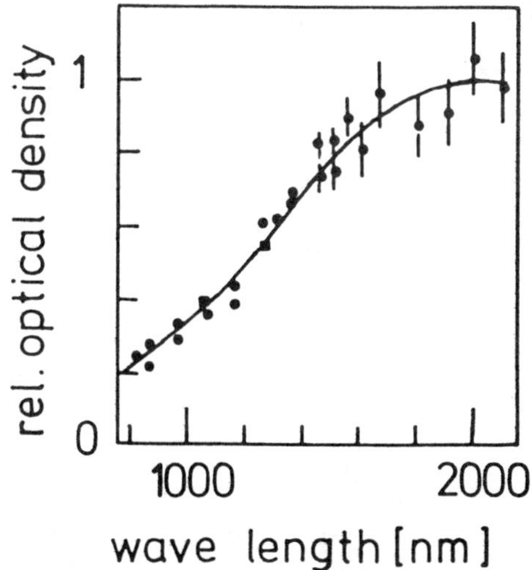

FIGURE 19. Optical absorption of localized electrons in liquid propane
at −185°C. (From Gillis, H. A., Klassen, N. V., Teather, G. G., and
Lokan, K. H., *Chem. Phys. Lett.*, 10, 481, 1971. With permission.)

The electron is trapped by the oriented dipolar molecules. The absorption spectrum shows
a maximum in the red part of the visible spectrum. Optical absorption due to localized
electrons in liquid hydrocarbons has been reported by Baxendale et al.[72-74] and by Gillis et
al.[75] The electron trap is thought to be formed by a preexisting random orientation of
molecules. The optical absorption starts in the red part of the visible spectrum and rises
toward the infrared exhibiting a maximum at around 2 μm (0.62 eV). It is probably due to
a transition from a bound state to a delocalized state. At higher temperatures the spectrum
shifts further toward the red. Thermal motion of the molecules forming the trap leads to a
reduction of the trapping energy. Figure 19 shows the absorption spectrum of trapped
electrons in liquid propane at −185°C.[75]

In hexafluorobenzene (HFB), van den Ende et al.[76] observed the optical absorption
spectrum of an electron temporarily trapped at one HFB-molecule or at a group of several
HFB-molecules. The absorption maximum was found to be at 675 nm (1.8 eV). Theoretical
models for solvated and localized electrons have been developed by a number of authors.
Details can be found in more recent reviews.[77,78]

2. Positronium Formation

Studies of excess electron properties in nonpolar liquids are complemented by investigations on the properties of positronium in these liquids. Positronium is formed by the
reaction of positrons with electrons and its formation manifests itself in an abrupt change
of the positron annihilation rate due to the fact that positronium is much more stable.
Positronium may become trapped in dense media by density fluctuations. In LHe positronium
forms a self-trapped state, a positronium bubble comparable to that of the electron. A more
detailed discussion of these states must be left to the specialized literature.[79]

III. THEORETICAL MODELS FOR ELECTRONIC ENERGY
LEVELS

A. MODELS FOR THE GROUND STATE ENERGY OF ELECTRONS
1. Wigner-Seitz Model

For the calculation of the ground state energy of a delocalized electron in a nonpolar liquid the Wigner Seitz model has been applied.[80] Each molecule or atom is replaced by a sphere of radius r_s which is given as,

$$r_s = \sqrt[3]{\frac{3}{4\pi n_0}} \tag{20}$$

where n_0 denotes the number density. Within this sphere, the potential V is assumed to be spherically symmetric,

$$V = V(r) \tag{21}$$

Translational symmetry is assumed,

$$V(r + 2\,r_s) = V(r) \tag{22}$$

For the calculation of the energy levels the Schrödinger equation has to be solved,

$$\left[-\left(\frac{h^2}{8\pi^2 m}\right)\nabla^2 + V(r) \right]\Psi_0(r) = V_0\Psi_0(r) \tag{23}$$

with the boundary condition

$$\nabla\Psi_0(r)\,\big|_{r=r_s} = 0 \tag{24}$$

The potential $V(r)$ consists essentially of two parts. First, the electron experiences a repulsive potential V_a by the electrons of the molecule or atom on account of the Pauli principle, and second, the electron experiences an attractive potential U_p on account of polarization effects,

$$V(r) = V_a + U_p \tag{25}$$

V_a only has a short range and it does not depend on the aggregate state. U_p has a long range and depends on the number density. The repulsive part V_a can be approximated by

$$V_a = \infty \qquad \text{for} \qquad r < a \tag{26a}$$

$$V_a = 0 \qquad \text{for} \qquad r > a \tag{26b}$$

where a denotes the scattering length of the repulsive potential. It is taken as the hard core radius of the molecule or atom. For the polarization term U_p the following expression was proposed,

$$U_p = -\frac{3\,\alpha\,e^2}{2r_s^4}\left[\frac{8}{7} + \frac{1}{1 + \frac{8}{3}\pi\alpha\,n_0}\right] \tag{27}$$

α is the polarizability. For the solution of Equation 23 Springett et al.[80] obtained

$$\Psi_0 = \frac{\sin[k_0(r - a)]}{r} \tag{28}$$

where k_0 is given by

$$k_0 = \left[\left(\frac{8\pi^2 m}{h_2}\right)(V_0 - U_p)\right]^{1/2} \tag{29}$$

k_0 is the wave vector of the electron in its ground state and it is obtained from Equation 28 and Equation 24

$$\tan\left[k_0\left(r_s - a\right)\right] = k_0 r_s \tag{30}$$

The ground state energy of the electron V_0 is given as the sum of the kinetic and potential energy of the electron,

$$V_0 = T + U_p \tag{31}$$

$$T = \frac{h^2 k_0^2}{8\pi^2 m} \tag{32}$$

Equations 29 and 30 can now be used either to calculate V_0 with hard core data or measured scattering lengths, or they can be employed to determine values of ''a'' from measured V_0 data.

Improvements of the Springett Jortner Cohen model[80] (SJC) depend on the choice of a suitable pseudopotential for the repulsive part V_a of the potential $V(r)$ (see Equation 25). First calculations of pseudopotentials for excess electrons in water and methane have been performed by Ishimaru and Fukui.[81] They assumed for the calculation that the electron is residing in a trap which is formed by four molecules with one of the OH or CH bonds directed toward the cavity center, respectively. The results demonstrated that the electron is localized in liquid water while it will remain quasi-free in liquid methane. Calculations of the ground state energy V_0 by means of a pseudopotential were carried out for LAr and LCH_4 by Plenkiewicz et al.[82,83] They concluded that the pseudopotential should be based on a proper atomic potential which satisfactorily reproduces the experimental gas-phase low energy electron scattering data. By following the Wigner-Seitz theory within the SJC model, the total potential of an electron at position $\vec{r}$ in the Wigner-Seitz sphere is given as

$$V(|\vec{r} - \vec{r}_0|) = V_{opt}(|\vec{r} - \vec{r}_0|) + 4\pi n \int_{r_{WS}}^{\infty} V_{opt}(r)\, g(r)\, F(r)\, r^2 dr \tag{33}$$

where r_0 and r_{WS} are the center and the radius of the Wigner-Seitz sphere, respectively, while V_{opt} denotes the electron/atom or electron/molecule interaction potential. $g(r)$ is the pair correlation function of the liquid and $F(r)$ is a screening function which takes into account that a dipole induced in a given atom or molecule by the electron is affected by the other dipoles. $F(r)$ was approximated by the same function as was used by Springett et al. and $g(r)$ was obtained from a solution of the Percus-Yevick equation for hard spheres. The main improvement of the JSC-model occurred in the calculation of V_{opt} which is composed of three terms,

$$V_{opt}(r) = V_{st}(r) + V_{ex}(r) + V_{pol}(r) \tag{34}$$

where V_{st} is the static potential, V_{ex} is the potential due to exchange interactions, while V_{pol} stands for the polarization potential. Calculations of V_0 gave excellent agreement with experimental results by Asaf et al.[12,13] (see Figures 20A and 20B). The present approach seems to indicate that the use of a high precision atomic or molecular potentials which describes the low energy electron scattering in the gas phase is the key element in the

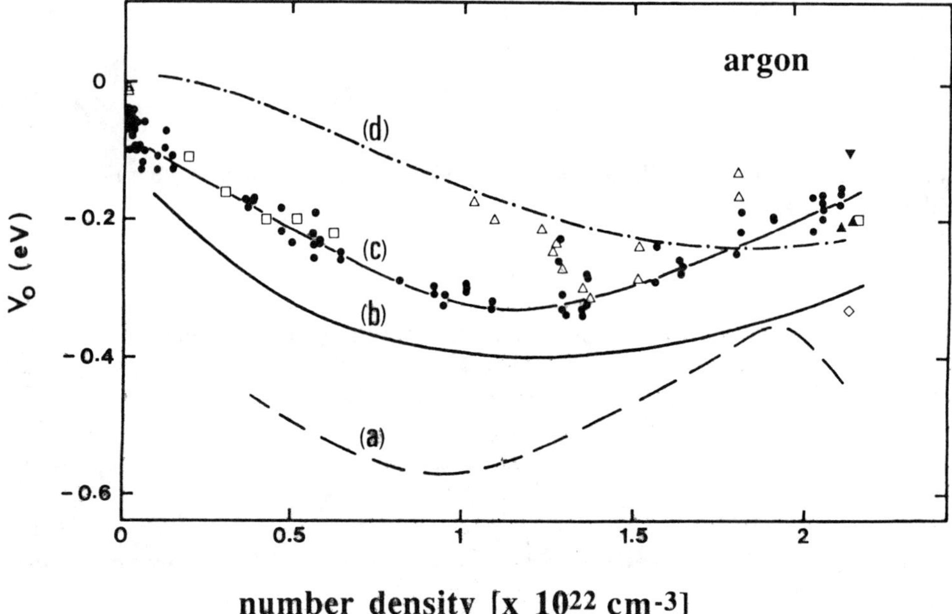

number density [x 10^{22} cm^{-3}]

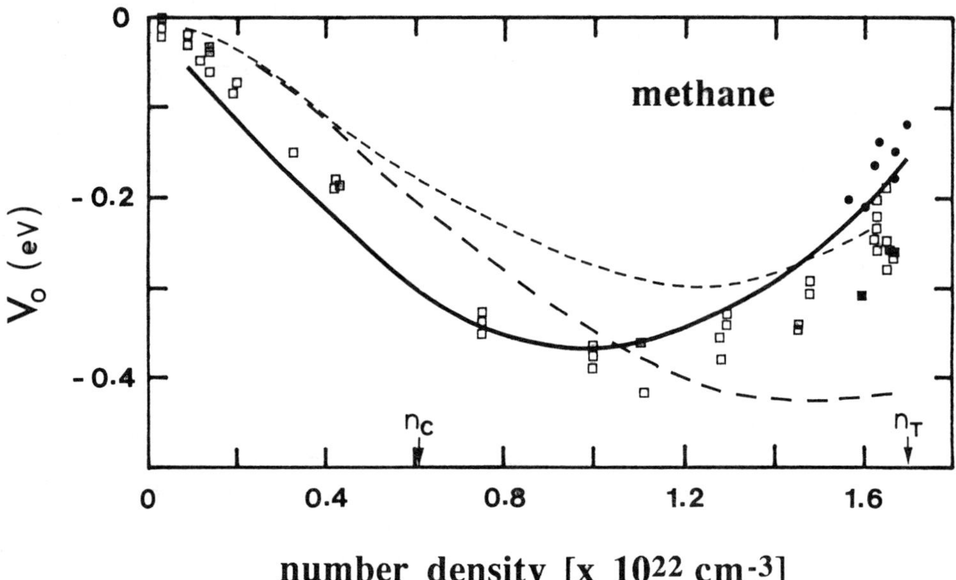

number density [x 10^{22} cm^{-3}]

FIGURE 20. V_0 vs. number density: (A) Argon; (a) theoretical results by Basak and Cohen;[104] (b) calculation by Plenkiewicz et al.;[82] (c) smooth curve representing various experimental data; (d) calculations by Reininger et al.[14] (B) Methane: solid line — calculations by Plenkiewicz et al.;[83] dashed lines — calculations by Asaf et al.[13] according to the SJC model; upper dashed curve a = 0.112 nm, lower dashed curve a = 0.109 nm; experimental data points from various sources.

application of the Wigner-Seitz theory for the treatment of excess electron states in liquids. It should be noted though that the rare gases and methane can be described by spherically symmetric potentials.

An extension of the SJC model to the variation of V_0 of other liquid hydrocarbons with density (i.e., temperature) was carried out by Nakagawa et al.[11] and Yamaguchi et al.[84] Best agreement between theory and experiment was obtained if a variation of the hard core radius with density was assumed (see Figure 5).

2. Semicontinuum Model

Other hydrocarbons might require different theoretical concepts. Kimura et al.[85-87] proposed the concept of a semicontinuum model for the trapped electron with a structured first solvation shell of molecules around the electron. The number of molecules in this shell is taken to be N = 4, 6, or 8. Since the electron/molecule interaction depends on the orientation of the hydrocarbon molecule, in principle, an infinity of energy levels is possible. Kimura et al.[86] performed the calculations for special orientations of the molecules in the first coordination layer. Later, Kimura and Fueki[87] extended these concepts to the calculation of the electron localization process in ethane and propane as the density is increased from the rarefied gas to the liquid. The variation of V_0 and of E_{1s}, the ground state energy of the localized electron with density were calculated for N = 4, 6, or 8 molecules in the first solvation shell, respectively. Localization occurs if $V_0 - E_{1s} > 0$. For liquid ethane this is predicted to occur between 4 and 7.5 $\times$ 10^{21} cm^{-3} with N = 4 or N = 8, respectively. Döldissen et al.[88] observed a sharp decrease of electron mobility over several orders of magnitude as the density exceeded 6 $\times$ 10^{21} cm^{-3}.

3. Quantum Field Theory

A completely different theoretical treatment of the barrier for electron penetration into LHe and LNe was put forward by Tankersley[89] who used quantum field theory in order to calculate the first order correction to the optical potential,

$$V_{opt} = 2\pi \, (h/2\pi)^2 \, n \, a/m \tag{35}$$

Here a is the e-He scattering length, m is the electron mass, and n is the liquid number density. The energy barrier V_0 is obtained as,

$$V_0 = (2\pi \, (h/2\pi)^2 n \, a/m) \left\{ 1 + 2 \, a \, \pi^{-1} \int_0^\infty (1 - S(k))dk \right\} \tag{36}$$

with S(k) the structure factor. V_0 can be conveniently expressed as a sum of ϵ_1 and ϵ_2 which are given as,

$$\epsilon_1 = \alpha \, n \, a \tag{37a}$$

and

$$\epsilon_2 = \beta \, I \, n \, a^2 \tag{37b}$$

with

$$I = \int dk \, (1 - S(k)) \tag{37c}$$

and

$$\alpha = 0.479 \times 10^{22} \text{ eV cm}^3/\text{Å} \tag{37d}$$

and

$$\beta = 0.305 \times 10^{22} \text{ eV cm}^3/\text{Å} \tag{37e}$$

ϵ_1 is the optical potential which is linear in a and ϵ_2 is the first order correction which is quadratic in a. The V_0 values obtained for LHe, LNe, and LAr compared favorably with the measured data.

B. ELECTRON BUBBLES AND 2-D ELECTRONS

Electrons may form microscopic bubbles in liquid helium, neon, and other cryogenic liquids. Usually, a large positive value of V_0 seems to be a prerequisite for localization of an electron in a bubble. The energetic conditions for the formation of an electron bubble in LHe have been treated by a number of authors.[80,90-94] In a simplified manner, we may depict the energetic conditions in the following way. The energy level of a low energy electron in the vacuum is set to be $E_{vac} = 0$. In order to bring it into the liquid as an extended state an energy V_0 is necessary. The creation of a microscopic bubble of radius R in the liquid requires volume and surface work, E_p and E_s, respectively,

$$E_p = \frac{4\pi}{3} R^3 p \tag{38}$$

$$E_s = 4\pi R^2 \gamma \tag{39}$$

where p denotes the external pressure on the bubble and γ denotes the surface tension of the liquid. Introduction of an electron into the bubble leads to localization of the electron in a well with a barrier height given by $(V_0 - P_-)$ where P_- is the polarization energy of an electron confined to a bubble of radius R in a dielectric continuum. P_- can be calculated by means of Born's formula (see Equation 8). The localization of the electron in the well leads to a ground state electronic energy, E_e which can be estimated by solving the Schrödinger equation of the one-dimensional potential well. The total energy, E_t of the electron bubble is then given as,

$$E_t = E_e + E_s + E_p + P_- \tag{40}$$

The bubble is energetically stable if $E_t < V_0$. Decrease of the bubble radius, for instance by external pressure will eventually increase the ground state energy to the value of V_0 and the bubble will cease to exist.

A special case of 2-D electrons on the surface of liquid helium is represented by multiply charged bubbles and drops. The electrons form a 2-D Wigner crystal and in first approximation it was found that the electronic properties of multielectron bubbles and drops coincide.[95]

The electron bubble model has also been proposed for those liquid hydrocarbons where low mobility electrons have been observed.[91,92] Sample calculations by Hammer et al.[92] did not yield stable bubbles in liquid hexane if the liquid is treated as a nonpolar dielectric continuum. The microdipoles of the C–H bonds seem to be responsible for the localization (see also Section III.A.1).

C. POSITIVE HOLES

In most nonpolar dielectric liquids the positive charge carrier mobilities measured were characteristic of ionic species. Exceptionally high positive charge carrier mobilities were found in the six-membered cycloalkanes as cyclohexane, methylcyclohexane, and decalin.[96,97] The mobilities are characterized by a low temperature dependence. This type of behavior is characteristic of hole transport. Electrons from neighboring molecules of suitable orientation jump over to the positive species thus transporting the positive charge. With other hydrocarbon molecules, the holes react under conversion to a positive ion,

$$h_+ (\text{solvent hole}) + A \text{ (molecule)} \xrightarrow{k_{ct}} A_+ (\text{ion}) + M(\text{solvent molecule})$$

Reactions of this type are only possible if the ionization energy I_{solute} of the solute molecule A in the solvent is smaller than the ionization energy I_{liq} of the solvent molecules M. The rate constant k_{ct} for the charge transfer was found to increase dramatically with increasing difference in ionization energy ΔI,

$$\Delta I = I_{liq} - I_{solute} \tag{41}$$

although above 1 eV it seemed to level off.[98,99] The highest values reported for k_{ct} are of the order of 10^{11} $M^{-1}s^{-1}$.[100]

The mechanism of hole transport in the cycloalkane is not yet understood. In molecular crystals, for instance anthracene, hole transport is facilitated by the strong interaction of the molecules in the lattice. The broadening of the valence levels in hydrocarbon liquids, however, is small. Recently, proton migration has been forwarded as a possible explanation of the higher positive charge carrier mobility in cyclohexane.[101]

D. V_0 AND ELECTRON MOBILITY

A correlation between electron ground state energy V_0 and electron mobility in liquid hydrocarbons was proposed by Schiller et al.[102] and Nyikos et al.[103] If electron localization is possible in a particular liquid it modifies the electron mobility since the electron will spend some time immobilized in a trap. Since localization occurs when $V_0 - E_t > 0$ an equilibrium is thought to exist between electrons in the localized state and in the quasi-free state. The probability of localization P, depends on V_0 as follows,

$$P = (2\pi)^{-1/2} \int_{-\infty}^{(V_0 - E_t)/\sigma} \exp(-t^2/2)\, dt \tag{42}$$

σ is a fluctuation parameter, E_t is the energy of the localized state. The electron mobility μ is given as

$$\mu = \mu_f (1 - P) \tag{43}$$

μ_f is the electron mobility in the quasi-free state. The experimental data could be rationalized by means of Equations 42 and 43 by taking $E_t = -0.28$ eV and $\mu_f = 65$ $cm^2V^{-1}s^{-1}$, respectively. Other sets of values E_t and μ_f yield similarly good agreement between this model and experimental data.

Correlations between the mobility in the quasi-free state and V_0 in LAr have been derived by Basak and Cohen[104] in the framework of the deformation potential theory. This approach was extended to the liquid hydrocarbons by Holroyd and Cipolini.[105] Another path for the calculation of electron drift mobilities in liquids was taken by Baird and Rehfeld,[106] who assumed trapped and free electronic states to be in thermodynamic equilibrium. Applying statistical thermodynamics to the equilibrium they derived an activation energy of the drift mobility which linked it to the effective electron mass in the conduction band, the photo-diffusion threshold, the oscillator frequencies, and the volume fraction of traps in the liquid.

IV. CONCLUSION

Only a short overview of the energetic properties of excess electrons in nonpolar dielectric liquids could be given. With increasing application of sophisticated measurement techniques employing multiphoton excitation, high electric field strengths, high pressures, and sensitive optical absorption measurements new exciting results are possible. On the theoretical side, improvement of existing models by use of better experimental input data from other fields and new theoretical concepts are necessary in order to advance the understanding of excess electrons in fluids. The importance of such kinds of studies is underlined by the increasing number of applications nonpolar liquids find in other fields of science and engineering.

REFERENCES

1. **Holroyd, R. A. and Schmidt, W. F.,** Transport of electrons in nonpolar fluids, *Ann. Rev. Phys. Chem.*, 40, 439, 1989.
2. **Freeman, G. R., Ed.,** *Kinetics of Nonhomogeneous Processes*, Wiley, New York, 1987.
3. **Engler, J.,** Status and perspectives of liquid argon calorimeters, *Nucl. Instrum. Meth.*, 225, 525, 1984.
4. **Albrow, M. G., Apsimon, R., Aubert, B., Bacci, C., Bezaguet, A., Bonino, R., Cennini, P., Centro, S. et al.,** Performance of a uranium/tetramethylpentane electromagnetic calorimeter, *Nucl. Instr. Meth.*, A265, 303, 1988.
5. **Born, M.,** Volumen and Hydratationswärme der Ionen, *Z. Physik*, 1, 45, 1920.
6. **Holroyd, R. A. and Allen, M.,** Energy of excess electrons in non-polar liquids by photoelectric work function measurements, *J. Chem. Phys.*, 54, 5014, 1971.
7. **Fowler, R. H.,** The analysis of photoelectric sensitivity curves for clean metals at various temperatures, *Phys. Rev.*, 38, 45, 1931.
8. **Tauchert, W. and Schmidt, W. F.,** Energy of the quasifree electron state in liquid argon, krypton, and xenon, *Z. Naturforsch.*, 30a, 1085, 1975.
9. **Tauchert, W., Jungblut, H., and Schmidt, W. F.,** Photoelectric determination of V_0 values and electron ranges in some cryogenic liquids, *Can. J. Chem.*, 55, 1860, 1977.
10. **Casanovas, J. and Guelfucci, J. P.,** Conduction level and thermalization range parameter of electrons in liquid alkanes, *IEEE Trans. El. Insul.*, 23, 515, 1988.
11. **Nakagawa, K., Itoh, K., and Nishikawa, M.,** The effect of molecular structure on the density dependence of electron mobility and conduction band energy in nonpolar liquids, *IEEE Trans. El. Insul.*, EI23, 509, 1988.
12. **Reininger, R., Asaf, U., and Steinberger, I. T.,** The density dependence of the quasifree electron state in fluid xenon and krypton, *Chem. Phys. Lett.*, 90, 287, 1982; Erratum ibid. 99, 192, 1983.
13. **Asaf, U., Reininger, R., and Steinberger, I. T.,** The energy V_0 of the quasi-free electron in gaseous, liquid and solid methane, *Chem. Phys. Lett.*, 100, 363, 1983.
14. **Reininger, R., Asaf, U., Steinberger, I. T., and Basak, S.,** Relationship between the energy V_0 of the quasi-free electron and its mobility in fluid argon, krypton and xenon, *Phys. Rev.*, B28, 4426, 1983.
15. **Yoshino, K., Ohseko, K., Shiraishi, M., Terauchi, M., and Inuishi, Y.,** Dielectric breakdown of cryogenic liquids in terms of pressure, polarity, pulse width and impurity, *J. Electrostatics*, 12, 305, 1982.
16. **Böttcher, E. H. and Schmidt, W. F.,** Photoconductivity of non-polar liquids induced by vacuum-ultraviolet light, *J. Chem. Phys.*, 80, 1353, 1984.
17. **Böttcher, E. H. and Schmidt, W. F.,** Photoconductivity induced by single photon excitation of tetramethyl-p-phenylenediamine and anthracene in liquid pentanes, *5th Tihany Symp. Radiat. Chem.*, 427, 1982.
18. **Buschick, K. and Schmidt, W. F.,** Vacuum ultraviolet photoconductivity of 2,2,4,4 tetramethylpentane and bis(trimethylsilyl)ethane, *IEEE Trans. El. Insul.*, 24, 353, 1989.
19. **Casanovas, J., Grob, R., Delacroix, D., Guelfucci, J. P., and Blanc, D.,** Photoconductivity studies in some non-polar liquids, *J. Chem. Phys.*, 75, 4661, 1981.
20. **Casanovas, J., Guelfucci, J. P., and Terrissol, M.,** Determination of the ionization quantum yield and thermalization distance of photoelectrons created by VUV photoionization in pure liquid alkanes and tetramethylsilane, *Radiat. Phys. Chem.*, 32, 361, 1988.
21. **Baron, P. L., Casanovas, J., Guelfucci, J. P., and Laou Sio Hoi, R.,** Photoconductivity induced by VUV Photons in polydimethylsiloxane and polymethylphenylsiloxane oils, *IEEE Trans. El. Insul.*, 23, 563, 1988.

22. **Reininger, R., Asaf, U., and Steinberger, I. T.,** Photoconductivity and the evolution of energy bands in fluid xenon, *Phys. Rev.,* B 28, 3193, 1983.
23. **Asaf, U. and Steinberger, I. T.,** Photoconductivity and electron transport parameters in liquid and solid xenon, *Phys. Rev.,* B10, 4464, 1974.
24. **Reininger, R., Asaf, U., Steinberger, I. T., and Laporte, P.,** Evolution of photoconductivity in fluid xenon, *J. Electrostatics,* 12, 123, 1982.
25. **Holroyd, R. A., Ehrenson, S., and Preses, J. M.,** Electron mobility, ion yields, and photoconductivity in liquid tetrakis(dimethylamino)ethylene, *J. Phys. Chem.,* 89, 4244, 1985.
26. **Holroyd, R. A., Preses, J. M., Böttcher, E. H., and Schmidt, W. F.,** Photoconductivity induced by single-photon excitation of aromatic molecules in liquid hydrocarbons, *J. Phys. Chem.,* 88, 744, 1984.
27. **Holroyd, R. A., Preses, J. M., and Zevos, N.,** Single-photon induced conductivity of solutes in nonpolar solvents, *J. Chem. Phys.,* 79, 483, 1983.
28. **Reininger, R., Steinberger, I. T., Bernstorff, S., Saile, V., and Laporte, P.,** Extrinsic photoconductivity in xenon-doped fluid argon and krypton, *Chem. Phys.,* 86, 189, 1984.
29. **Sham, T.-K.,** X-ray absorption studies of liquids: structure and reactivity of metal complexes in solution and x-ray photoconductivity of hydrocarbon solutions of organometallics, in *Topics in Current Chemistry,* Vol. 145, Springer-Verlag, Berlin, 1988, 82.
30. **Böttcher, E. H. and Schmidt, W. F.,** Laser-induced photoconductivity of liquid tetramethylsilane, *Proc. 7. ICDL,* IEEE Publ. 81CH1594-1, Berlin, 1981, 89.
31. **Scott, T. W., Twarowski, A. J., and Albrecht, A. C.,** Multiphoton ionization of liquid benzene: the ionization mechanism, *Chem. Phys. Lett.,* 66, 1, 1979.
32. **Pillof, H. S. and Albrecht, A. C.,** Biphotonic ionization: a flash photoconductivity study of TMPD in 3-methylpentane solutions, *J. Chem. Phys.,* 49, 4891, 1968.
33. **Siomos, K., Kourouklis, G., and Christophorou, L. G.,** Laser two-photon ionization of organic molecules in dielectric liquids, *Chem. Phys. Lett.,* 80, 504, 1981.
34. **Siomos, K. and Christophorou, L. G.,** Laser two-photon ionization spectroscopy of molecules in liquids, *Chem. Phys. Lett.,* 72, 43, 1980.
35. **Faidas, H. and Christophorou, L. G.,** Multiphoton ionization of fluoranthene in tetramethylsilane, *J. Chem Phys, 86, 2505, 1987*
36. **Faidas, H. and Christophorou, L. G.,** Laser multiphoton ionization of aromatic molecules in non-polar liquids, *Radiat. Phys. Chem.,* 32, 433, 1988.
37. **Lukin, I. V. and Yakovlev, B. S.,** Electron photodetachment from O_2 in liquid hydrocarbons, *Chem. Phys. Lett.,* 42, 307, 1976.
38. **Sowada, U. and Holroyd, R. A.,** Laser photodetachment of electrons from O_{2-} in nonpolar liquids, *J. Chem. Phys.,* 70, 3586, 1979.
39. **Baird, J. K.,** Negative ion photodetachment and the electron effective mass in liquids, *J. Chem. Phys.,* 79, 316, 1983.
40. **Baird, J. K. and Schulman, T. P.,** The relationship between gas phase and liquid phase electron photodetachment cross sections in the threshold region: application to anthracene and perfluorobenzene anions, *Radiat. Phys. Chem.,* 32, 493, 1988.
41. **Balakin, A. A. and Yakovlev, B. S.,** Photoionization of trapped electrons in liquid n-hexane, *Chem. Phys. Lett.,* 66, 299, 1979.
42. **Balakin, A. A., Lukin, L. V., Tolmachev, A. V., and Yakovlev, B. S.,** Determination of the mean free path of a quasifree electron prior to localization in liquid isooctane, *High Energy Chem. (USSR),* 15, 123, 1981.
43. **Lukin L. V., Tolmachev A. V., and Yakovlev B. S.,** The photoexcitation of trapped electrons produced in the photoionization of anthracene in liquid n-hexane, *Chem. Phys. Lett.,* 81, 595, 1981.
44. **Lukin L. V., Tolmachev A. V., and Yakovlev B. S.,** Photoionization of anthracene in liquid methyl-cyclohexane: effects of electron-ion pair excitation, *High Energy Chem. (USSR),* 16, 415, 1982.
45. **Funabashi, K.,** Photoinduced diffusion of trapped electrons, *J. Chem. Phys.,* 76, 5519, 1982.
46. **Schoepe, W. and Rayfield, G. W.,** Tunneling from electronic bubble states in liquid helium through the liquid-vapor interface, *Phys. Rev.,* A7, 2111, 1973.
47. **Bruschi, L., Mazzi, G., Santini, M. and Torzo, G.,** Transmission of negative ions through the liquid-vapor surface in neon, *J. Phys.-C: Solid State Phys.,* 8, 1412, 1975.
48. **Dolgoshein, B. A., Lebedenko, V. N., and Rodionov, B. U.,** New method of registration of ionizing-particle tracks in condensed matter, *JETP Lett.,* 11, 351, 1970.
49. **Boriev, I. A., Balakin, A. A., and Yakovlev, B. S.,** Electron emission from nonpolar organic liquids, *High Energy Chem., (USSR),* 12, 16, 1978.
50. **Peskov, V., Charpak, G., Miné, P., Sauli, F., Scigocki, D., Séguinot, J., Schmidt, W. F., and Ypsilantis, T.,** Liquid and solid photocathodes, *Nucl. Instrum. Methods,* A269, 149, 1988.
51. **Cole, M. W.,** Electronic surface states of liquid helium, *Rev. Mod. Phys.,* 46, 451, 1974.

52. **Grimes, C. C.,** Electrons in surface states on liquid helium, *Surf. Sci.,* 73, 379, 1978.

53. **Grimes, C. C. and Brown, T. R.,** Direct spectroscopic observation of electrons in image-potential states outside liquid helium, *Phys. Rev. Lett.,* 32, 280, 1974.

54. **Yu Ming S. and Chia-Wei, W.,** Electronic surface states on liquid helium, *J. Low Temp. Phys.,* 14, 418, 1974.

55. **Williams, F. I. B.,** Collective aspects of charged-particle systems at helium interfaces, *Surf. Sci.,* 113, 371, 1982.

56. **Troyanovskii, A. M. and Khaikin, M. S.,** Electron mobility in two dimensional layer over the surface of solid hydrogen, *Soviet Phys.-JETP,* 54, 214, 1981.

57. **Reinis, G. J. and Cressman, P. J.,** Tertiary alkylamines as sensitizers for (+) charge injection from selenium into n-dodecane and as ionic (+) charge carriers in n-dodecane, *Proc. 6th Intern. Conf. Cond. Breakdown Diel. Liqu.,* Mont-Saint-Aignan, France. Éditions Frontieres, Dreux, France July 1978, 235.

58. **Ahuja, R. C. and Hauffe, K.,** Charge transfer across organic photoconductor/insulating liquid interface. I. Metal free phthalocyanine alkanes, *Ber. Bunsen Ges. Phys. Chem.,* 84, 68, 1980; Part II. Metal free phthalocyanine dodecane containing electron donor/acceptor molecules, *Ber. Bunsen Ges. Phys. Chem.,* 84, 129, 1980; Part III. Copper phthalocyanine, chlorinated copper phthalocyanine/dodecane containing electron acceptor/donor additives, *Ber. Bunsen Ges. Phys. Chem.,* 84, 138, 1980.

59. **Cohen, M. H. and Lekner, J.,** Theory of hot electrons in gases, liquids and solids, *Phys. Rev.,* 158, 305, 1967.

60. **Lekner, J.,** Motion of electrons in liquid argon, *Phys. Rev.,* 158, 130, 1967.

61. **Gushchin, E. M., Kruglov, A. A., and Obodovskii, I. M.,** Electron dynamics in condensed argon and xenon, *Sov. Phys. JETP,* 55, 650, 1982.

62. **Lansiart, A., Seigneur, A., Moretti, J. L., and Morucci, J. P.,** Development research on a highly luminous condensed xenon scintillator, *Nucl. Instrum. Methods,* 135, 47, 1976.

63. **Derenzo, S. E., Mast, T. S., Zaklad, H., and Muller, R. A.,** Electron avalanche in liquid xenon, *Phys. Rev.,* 9, 2582, 1974.

64. **Shibamura, E., Takahashi, T., Kubota, S., and Doke, T.,** Ratio of diffusion coefficient to mobility for electrons in liquid argon, *Phys. Rev.,* 20, 2547, 1979.

65. **Bakale, G., Sowada, U., and Schmidt, W. F.,** Effect of an electric field on electron attachment to SF_6, N_2O, and O_2 in liquid argon and xenon, *J. Phys. Chem.,* 80, 2556, 1976.

66. **Döldissen, W. and Schmidt, W. F.,** Saturation of the electron drift velocity in liquid tetramethylsilane, *Chem. Phys. Lett.,* 68, 527, 1979.

67. **Bakale, G. and Beck, G.,** Field dependent electron attachment in liquid tetramethylsilane, *J. Chem. Phys.,* 84, 5344, 1986.

68. **Bullot, J., Cordier, P., and Gauthier, M.,** Electric field quenching of recombination fluorescence in photoionized nonpolar solutions. I. Free electron quantum yield determination, *J. Chem. Phys.,* 69, 1374, 1978.

69. **Bullot, J., Cordier, P., and Gauthier, M.,** Photoionization in non-polar liquids studied by electric field quenching of recombination fluorescence and photoconductivity, *J. Phys. Chem.,* 84, 1253, 1980.

70. **Shibamura, E., Crawford, H. J., Doke, T., Engelage, J. M., Flores, I., Hitachi, A., Kikuchi, J., Lindstrom, P. J., Masuda, K., and Ogura, K.,** Ionization and scintillation produced by relativistic Au, He, and H ions in liquid argon, *Nucl. Instrum. Methods,* A260, 437, 1987.

71. **Hart, E. J.,** The hydrated electron, in *Actions Chimiques et Biologiques des Radiations,* Vol. 10, Haissinsky, M. Ed., Masson et Cie, Paris, 1966.

72. **Baxendale, J. H., Bell, C., and Wardman, P.,** Pulse radiolysis observations of solvated electrons in liquid hydrocarbons, *Chem. Phys. Lett.,* 12, 347, 1971.

73. **Baxendale, J. H., Bell, C., and Wardman, P.,** Observations on solvated electrons in aliphatic hydrocarbons at room temperature by pulse radiolysis, *J. Chem. Soc. Faraday Trans.,* I, 69, 776 1973.

74. **Baxendale, J. H. and Rasburn, E. J.,** Pulse radiolysis study of the kinetics of electron reactions in liquid n-hexane at room temperature, *J. Chem. Soc. Faraday Trans.,* I, 70, 705 1974.

75. **Gillis, H. A., Klassen, N. V., Teather, G. G., and Lokan, K. H.,** Optical measurements on solvated electrons in pulse irradiated liquid propane, *Chem. Phys. Lett.,* 10, 481, 1971.

76. **van den Ende, C. A. M., Nyikos, L., Warman, J. M. and Hummel, A.,** Mobility, reaction kinetics and optical absorption spectrum of the excess electron in pure C_6F_6 and admixtures with non polar liquids, *Radiat. Phys. Chem.,* 19, 297, 1982.

77. **Feng, D. F. and Kevan, L.,** Theoretical models for solvated electrons, *Chem. Rev.,* 80, 1, 1980.

78. **Kestner, N. R.,** *Electron-solvent and anion solvent interactions,* Kevan, L. and Webster, B. Eds., Elsevier, Amsterdam, 1976, chap. 1.

79. **Iakubov, I. T. and Khrapak, A. G.,** Self-trapped states of positrons and positronium in dense gases and liquids, *Rep. Prog. Phys.,* 45, 697, 1982.

80. **Springett, B. E., Jortner, J., and Cohen, M. H.,** Stability criterion for the localization of an excess electron in a nonpolar fluid, *J. Chem. Phys.,* 48, 2720, 1968.

81. **Ishimaru, S., and Fukui, K.,** Calculations of the pseudopotential for the excess electron in water and methane, *Theoret. Chim. Acta (Berlin),* 39, 103, 1975.
82. **Plenkiewicz, B., Jay-Gerin, J.-P., Plenkiewicz, P., and Bachelet, G. B.,** Conduction band energy of excess electrons in liquid argon, *Europhys. Lett.,* 1, 455, 1986.
83. **Plenkiewicz, B., Plenkiewicz, P., Jay-Gerin, J.-P., and Jain, A.,** Density dependence of the ground-state energy of excess electrons in liquid methane, *J. Chem. Phys.,* 90, 4907, 1989.
84. **Yamaguchi, Y., Nakajima, T., and Nishikawa, M.,** Conduction band energy in dense ethane fluid, *J. Chem. Phys.,* 71, 550, 1979.
85. **Kimura, T., Fueki, K., Narayana, P. A., and Kevan, L.,** A semicontinuum model with a structured first solvation shell for excess electrons in liquid and glassy alkanes, *Can. J. Chem.,* 55, 1940, 1977.
86. **Kimura, T., Fueki, K., and Kevan, L.,** Ground state energy of excess electrons in n-hexane, 2,2,4-trimethylpentane, and tetramethylsilane, *J. Chem. Phys.,* 68, 3945, 1978.
87. **Kimura, T. and Fueki, K.,** Effect of density on excess electron localization in ethane and propane, *J. Chem. Phys.,* 66, 366, 1977.
88. **Döldissen, W., Schmidt, W. F., and Bakale, G.,** Excess electron mobility in ethane: density, temperature, and electric field effects, *J. Phys. Chem.,* 84, 1179, 1980.
89. **Tankersley, L. L.,** Energy barrier for electron penetration into helium, *J. Low Temp. Phys.,* 11, 451, 1973.
90. **Miyakawa, T. and Dexter, D. L.,** Stability of electronic bubbles in liquid neon and hydrogen, *Phys. Rev.,* 184, 166, 1969.
91. **Schiller, R.,** Localization probability and mobility of electrons in liquid hydrocarbons, *J. Chem. Phys.,* 57, 2222, 1972.
92. **Hammer, H., Schoepe, W., and Weber, D.,** Note on a bubble model for excess electrons in liquid hydrocarbons, *J. Chem. Phys.,* 64, 1252, 1976.
93. **Raz, B. and Jortner, J.,** Energy of the quasifree electron state in dense neon, *Chem. Phys. Lett.,* 9, 222, 1971.
94. **Hernandez, J. P.,** Self-trapped states of an electron in a structurally disordered system, *Phys. Rev.,* A7, 1755, 1973.
95. **Artemév, A. A., Khrapak, A. G., and Yakubov, I. T.,** Multi-electron states in small dielectric particles, *Soviet J. Low Temp. Phys.,* 11, 555, 1986.
96. **DeHaas, M. P., Hummel, A., Infelta, P. P., and Warman, J. M.,** The measurement of a positive charge mobility larger than that of the excess electron in irradiated liquid trans-decalin, *J. Chem. Phys.,* 65, 5019, 1976.
97. **DeHaas, M. P., Warman, J. M., Infelta, P. P., and Hummel, A.,** The direct observation of a highly mobile positive ion in nanosecond pulse irradiated liquid cyclohexane, *Chem. Phys. Lett.,* 31, 382, 1975.
98. **Mehnert, R., Brede, O., Bös, J., and Naumann, W.,** Transfer of positive charge in non-polar liquids studied by pulse radiolysis, *J. Electrostatics,* 12, 107, 1982.
99. **Mehnert, R., Brede, O., Bös, J., and Naumann, W.,** Charge transfer from the carbon tetrachloride radical cation to alkyl chlorides, alkanes, alkenes and aromatics, *Ber. Bunsen Ges. Physik. Chem.,* 83, 992, 1979.
100. **Warman, J. M., Infelta, P. P., DeHaas, M. P., and Hummel, A.,** Rate constants for the reaction of positive ions with solutes in irradiated liquid cyclohexane, *Chem. Phys. Lett.,* 43, 321, 1976.
101. **Sauer, M. C. and Schmidt, K. H.,** The decay of the high mobility cation in cyclohexane, *Radiat. Phys. Chem.,* 32, 281, 1988.
102. **Schiller, R., Vass, Sz., and Mandics, J.,** Energy of the quasi-free electrons and the probability of electron localization in liquid hydrocarbons, *Int. J. Radiat. Phys. Chem.,* 5, 491, 1973.
103. **Nyikos, L. and Schiller, R.,** Electron mobility and conduction state energy in hydrocarbon mixtures, *Chem. Phys. Lett.,* 34, 128, 1975.
104. **Basak, S. and Cohen, M. H.,** Deformation-potential theory for the mobility of excess electrons in liquid argon, *Phys. Rev.,* B20, 3404, 1979.
105. **Holroyd, R. A. and Cipolini, N. E.,** Dependence of conduction band energy and electron mobility on fluid density, *Proc. 6th Int. Congr. Rad. Res.,* Tokyo, May 1979, 228.
106. **Baird, J. K. and Rehfeld, R. H.,** Thermodynamics of electron transport in amorphous insulators, *J. Chem. Phys.,* 86, 4090, 1987.

Chapter 6

MOBILITY OF EXCESS ELECTRONS IN DIELECTRIC LIQUIDS: EXPERIMENT AND THEORY

Raúl C. Muñoz

TABLE OF CONTENTS

I. INTRODUCTION

Since the discovery of a conductance transient due to excess electrons in dielectric hydrocarbon liquids,[1,2] the field has grown rapidly. More than 70 liquids have been investigated,[3] and studies have been performed on the behavior of the excess electron mobility as a function of electric field, temperature, pressure, and molecular structure of the liquid. Nevertheless, a comparable development of a coherent theoretical framework for understanding excess electron motion in dielectric liquids is lacking.

The purpose of this chapter is to review the development of this field. However, limitations of space impose that a choice be made between the width of the coverage of different topics and the depth of the presentation. We have chosen to cover in depth some of the most recent findings, at the expense of treating briefly or omitting altogether, some other exciting topics that certainly merit consideration within the field of excess electron mobility in dielectric liquids.

Such is the case of liquids where evidence of charge transport involving *hole* motion has been found; it is also the case of excess electron behavior in liquids where the repulsive part of the potential representing the electron-medium interaction, originating from Pauli's exclusion principle, is strong enough to lead to the formation of a cavity or 'bubble' around the electron, as in the case of He,[4,5] Ne,[6,7] and H_2.[8] Thus, Section IV contains an overview of drift mobility data that excludes both polar liquids and liquids composed of electronegative molecules, as well as liquids where charge transport via hole motion has been reported; it covers very briefly the data concerning electron bubbles. Similarly, Section VI contains an overview of theoretical work, which omits some very successful efforts that led to an understanding of bubble formation;[9,10] it also excludes a considerable body of work that relates to electron solvation. Bubble formation is intimately related to Chapter 5 by W. F. Schmidt in this volume; electron solvation is discussed by J. P. Jay-Gerin and C. Ferradini in Chapter 8.

The organization of this chapter is as follows. Sections II through V are devoted mostly to a discussion of experimental aspects. Section II contains a phenomenological definition of the drift and the Hall mobility. Section III contains a brief description of some of the experimental methods used to measure these quantities. Section IV contains an overview of drift mobility data. For clarity of presentation we include in Section IV, together with the data involving electron bubbles, a short qualitative discussion of why these entities are formed in some liquids and not in others; a detailed discussion of bubble theory is beyond the scope of this work. Section V is devoted to an overview of Hall mobility data. The remainder of this chapter (Sections VI through VIII) focuses mainly on theoretical aspects. Section VI contains an overview of theoretical work concerning electron mobility. Section VII explores the significance of the Hall mobility and of the mobility ratio (the ratio of the Hall to the drift mobility). This chapter finishes with Section VIII where the evidence concerning some of the fundamental problems that remain to be solved is summarized. The references are representative of the main work done in the field, but are by no means comprehensive; they are simply intended as a guide to help the interested reader dig further into the literature.

II. HALL MOBILITY AND DRIFT MOBILITY: A PHENOMENOLOGICAL DEFINITION

When excess electrons are present in an insulating material under the influence of an electric field, $\mathbf{E}$, and a magnetic field, $\mathbf{B}$, the average electron velocity $\langle \mathbf{v} \rangle$ is given by [11-13]

$$\langle \mathbf{v} \rangle = \mu_D \, \mathbf{E} + \mu_H \, \mu_D \, \mathbf{E} \times \mathbf{B} \tag{1}$$

where μ_D stands for the drift mobility and μ_H stands for the Hall mobility.

Dimensional analysis indicates that the constant of proportionality μ_H defined by Equation 1 has units of $[B^{-1}]$. That μ_H has, indeed, the same units as μ_D and therefore is a mobility, follows from the Lorentz equation describing the force acting on a *free* particle that moves under the influence of an electric field $\mathbf{E}$ and a magnetic field $\mathbf{B}$, $\mathbf{F} = q(\mathbf{E} + \mathbf{v} \times \mathbf{B})$, which implies that $[E] = [v\,B]$ or $[B^{-1}] = [v\,E^{-1}]$. Equation 1 is inspired on this *itinerant electron gas* picture, where the charge carriers move about in the sample undergoing occasional scattering events.

III. EXPERIMENTAL METHOD

A. DETERMINATION OF μ_D

Measurements of μ_D have been carried out mainly using two different methods: the measurement of a conductance transient due to electrons after a short ionizing radiation pulse, and the measurement of the electron time of flight.

1. The Electron Conductance Transient (ECT) Method

The ECT method has been employed in cells with a parallel plate geometry. It is based on the relationship between the current, i, and the charge density, n, in a parallel plate cell, where a voltage, V, is applied across two electrodes of area, A, separated by a distance, d,

$$i = n\,q\,(v_D{}^+ + v_D{}^-)\,A = n\,q\,(\mu_D{}^+ + \mu_D{}^-)\,V\,\frac{A}{d} \sim n\,q\,\mu_D{}^-\,V\,\frac{A}{d} \tag{2}$$

where $v_D{}^+$, $v_D{}^-$, $\mu_D{}^+$, and $\mu_D{}^-$ stand for the drift velocity and drift mobility of positive ions and electrons respectively, and q is the electron charge. The approximation indicated on the right hand side of Equation 2 is normally a very good one, since usually $\mu_D{}^- \gg \mu_D{}^+$. A typical experimental set up is shown schematically in Figure 1a; a typical ECT signal is illustrated in Figure 1b.

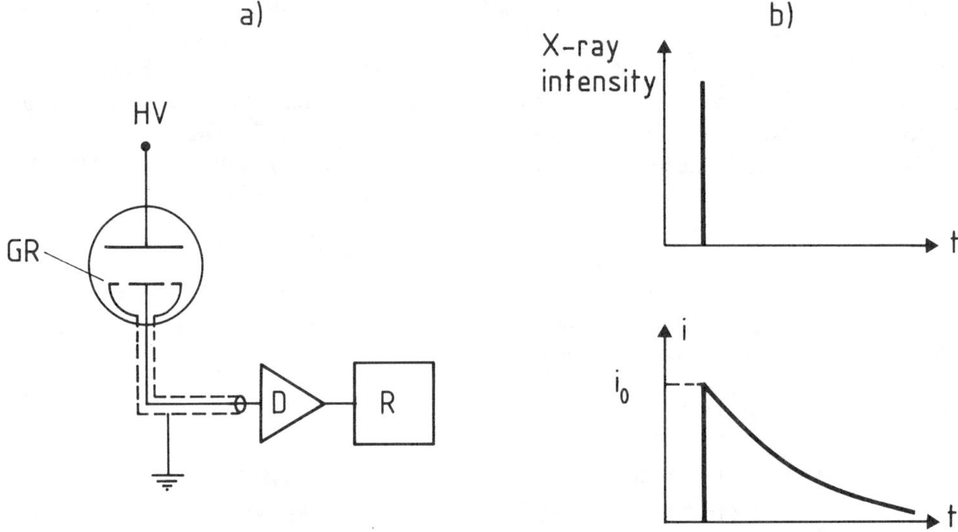

FIGURE 1.   (a) Typical cell configuration used in the determination of μ_D according to the ECT method. HV — high voltage electrode; GR — guard ring; D — detecting circuit that produces an output voltage proportional to the input current; R — oscilloscope or transient recorder. (b) Typical signal obtained after illuminating the sample uniformly with a short X-ray pulse;[16] i_0 — initial electron current.

Once the initial value i_0 of the current in the sample has been measured, then in order to calculate μ_D^- from Equation 2 it is necessary to accurately measure n and the cell constant A/d. The latter can readily be determined by measuring the conductance of an ionic liquid of known conductivity.[14] The former involves a different experiment, specifically designed to determine the number of electron ion pairs G_{fi} (E) that escape initial recombination*, a quantity that has been found to depend on the strength of the electric field E.[3,15] Once the free ion yield G_{fi} is known, a simultaneous measurement of i_0 and of the radiation dose allows the determination of μ_D^-.

In the ECT method for i_0 to faithfully represent the initial value of the electronic current in the sample, it is necessary that the two processes that tend to reduce this current — namely, electron loss to ions via electron-ion bulk recombination, and electron loss to impurities via electron attachment — play a negligible role in a time scale of the order of the response time of the detecting circuit or of the width of the ionizing radiation pulse, whichever is the longest. Otherwise, corrections must be applied.[16]

2. The Time of Flight (TOF) Method

The TOF method relies solely on the accurate determination of the drift time t_D. In a parallel plate geometry, setting B = 0 in Equation 1 leads to v = (d/t_D) = $\mu_D(V/d)$, therefore μ_D = $d^2/(V\,t_D)$. The contribution to the signal arising from the motion of the positive ions has again been neglected, and the superscript indicating the sign of the carrier has been dropped. The experimental set-up is similar to that of the ECT method (Figure 1a). Typical signals are shown in Figure 2 for different experimental conditions.

For this method to give accurate results, the electronic current flowing in the sample must remain unaffected by processes of electron attachment to impurities or electron-ion bulk recombination, during a time of the order of t_D.

---

*   A method known as the ''charge clearing field'' has been developed to determine $G_{fi}(0)$. As stated in Reference 17, it consists of ''ionizing the liquid by a short pulse of radiation, then immediately applying a clearing field of such magnitude that all the free ions are drawn to the electrodes before any appreciable amount of homogeneous recombination has had time to occur.''

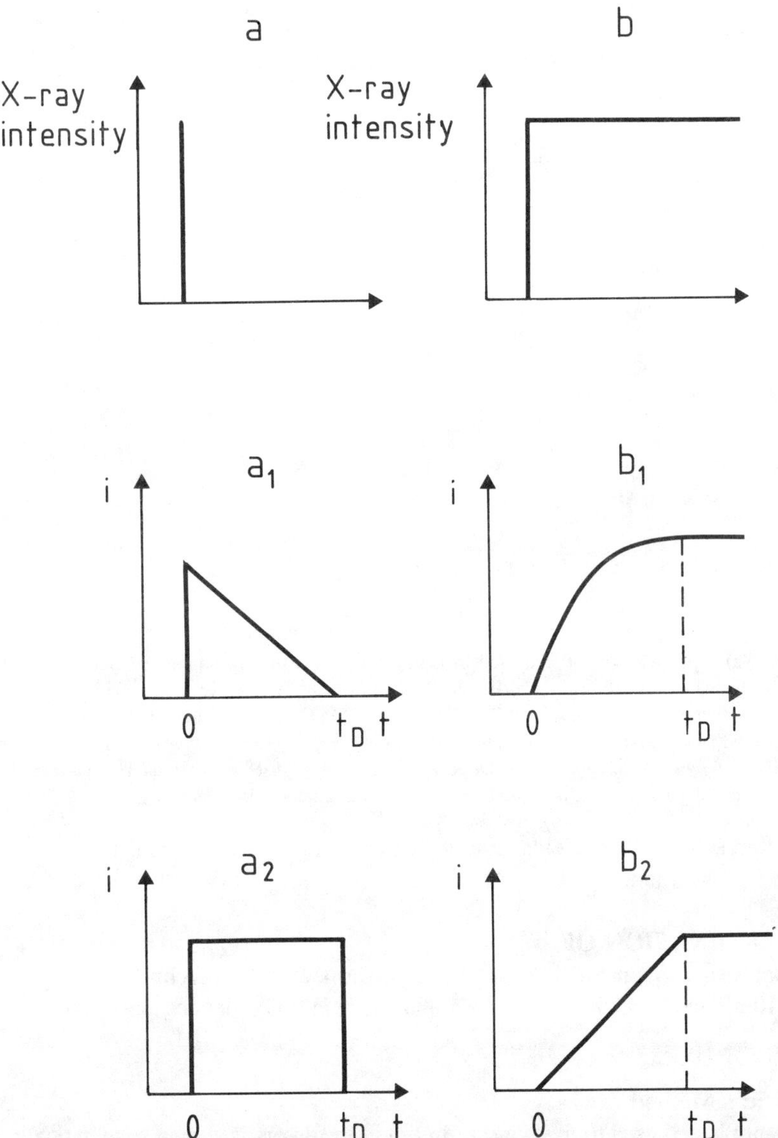

FIGURE 2. Determination of μ_D according to the TOF method, using a parallel plate cell similar to that shown schematically in Figure 1a: (a) δ function excitation; (a₁) response of the detecting circuit D after uniform illumination of the sample;[150] (a₂) response of D after irradiating a thin slab close to the HV electrode; (b) step function excitation; (b₁) response of the detecting circuit D after uniform illumination of the sample;[2] (b₂) response of D after irradiating a thin slab close to the HV electrode.[2]

A comparison of these two experimental techniques indicates that the TOF method has the disadvantage that it demands a sample purity (with respect to electron attaching impurities) which is, at least, one order of magnitude higher than that required when using the ECT method. Also, to avoid electron-ion bulk recombination, the permissible dose when using the TOF method is normally one order of magnitude lower than that which is acceptable with the ECT method. The advantage of the TOF technique is that no separate experiment is required to measure G_{fi}, neither is it necessary to perform accurate dosimetry. Conse-

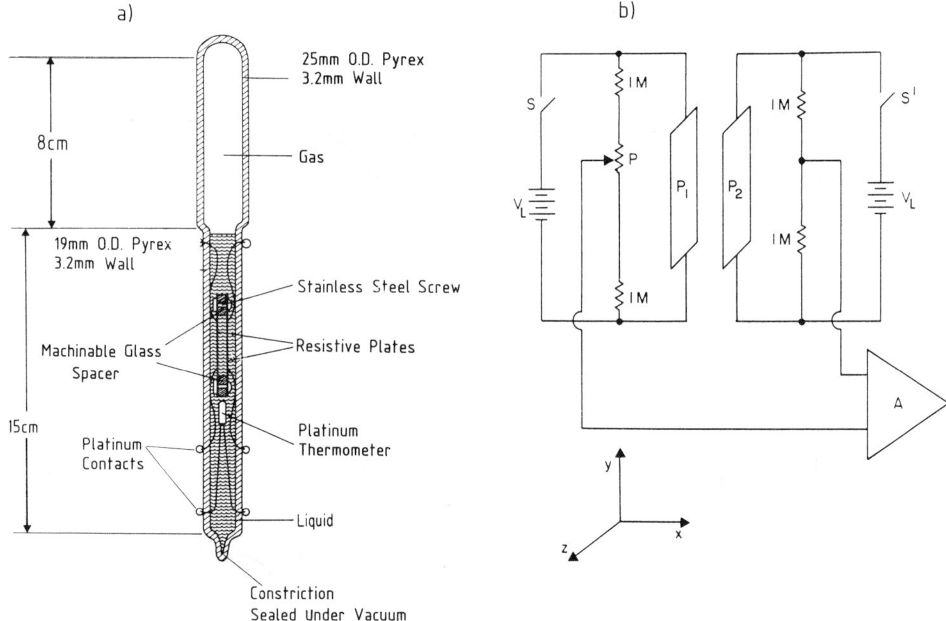

FIGURE 3.   (a) Schematics of the cell used to measure μ_H in insulating liquids; (b) schematics of Redfield's technique as applied to insulating liquids: P_1 and P_2 are metal oxide resistive plates, with the metal oxide film deposited on an aluminum oxide substrate; P is a miniature 100 kΩ, 10-turn potentiometer linear to 0.25%; S and S' represent field switching relays (for simplicity, the field reversing relays have been omitted); V_L represents a 300 V battery; A represents the *differential* charge sensitive amplifier used to sense the Hall current. The magnetic field **B** is oriented along z. For further details, see References 12 and 13.

quently, the TOF method yields results that are more accurate than those obtained when using the ECT method.

B. DETERMINATION OF μ_H

Recent measurements of μ_H have been performed using a technique[12] originally developed by Redfield,[18] which was adapted and modified in order to apply it to liquid insulators.[12,19]

1. Redfield's Method

The sample cell used to measure μ_H in liquid insulators is shown schematically in Figure 3a, and the fundamental aspects of Redfield's technique, as applied to liquids, are illustrated in Figure 3b. The electric field is set up on the sample by means of a voltage applied to each of the resistive plates P_1 and P_2. A longitudinal electric field $E_L = V_L/L$ (along the y axis) arises in the sample owing to the continuity of the tangential component of the field across the resistive plate-liquid interface. The action of the above field on excess electrons produces a longitudinal current i_L. According to Equation 1, a magnetic field, **B**, oriented along the z axis will produce a small rotation of this current; it will produce a transverse current i_H flowing between P_1 and P_2. A voltage difference ΔV_t applied between the center of P_1 and the center of P_2 by means of a voltage divider (Figure 3b), will cancel the effect of the magnetic field and will provide a measure of the Hall field $E_H = \mu_H B E_L$. For a cell comprising two electrodes, such as that pictured in Figure 3a, μ_H can be calculated from*

---

*   For a cell with more than two electrodes, this relationship is not necessarily valid, and the correction to the right-hand side of Equation 3 must be empirically found. The reason is that, in a cell with more than two electrodes, charges moving in the sample will induce a current not only in electrodes P_1 and P_2 but also in the additional electrodes; these extra currents are not considered in the derivation of Equation 3 (see Postscript).[12,19]

$$\mu_H = \frac{1}{2B}\left(\frac{\Delta V_t}{t}\right)\left(\frac{L}{V_L}\right) \tag{3}$$

where ΔV_t is the transverse voltage that is necessary to cancel the difference between the signals obtained for $+B$ and $-B$; t is the plate separation and L the plate length.[12,13,18]

The conceptual simplicity of this technique masks some experimental difficulties; these are summarized below.

The first concerns the choice of the detector for i_H. To achieve a signal to noise ratio significantly larger than one, it is necessary to employ a technique developed in nuclear physics to detect small currents; it is necessary to use a *charge sensitive amplifier* rather than a current amplifier.*

If the sample is sufficiently pure, the charge Q_H induced in the detecting circuit by i_H will be $Q_H \sim i_H \times TOF$, which for a typical $i_H \sim 1$ nA and a TOF ~ 100 µs leads to $Q_H \sim 10^{-13}$ C, well above the noise level of $\sim 1.8 \times 10^{-15}$ C(rms) achieved in the detecting circuit[12] and certainly well above the typical noise level of commercial charge amplifiers. Once again, the signal to noise ratio depends, not surprisingly, on sample purity. Typical electron lifetimes of 100 µs or more have been achieved to perform these measurements successfully. Such purity levels are at least one order of magnitude more severe than those necessary to perform measurements of μ_D using the TOF method (for example, an electron lifetime of 100 µs in tetramethyl silane, TMS, corresponds to an equivalent concentration of $O_2 < 3$ *parts per billion*).**[20]

Second, the transverse deflection caused by the magnetic field **B** produces a current that flows, say, *into* plate P_1 and *away* from plate P_2, or vice versa (Figure 3b). Departures of the electric field from a uniform longitudinal field configuration produce, on the contrary, a transverse fringing field that exhibits a high degree of symmetry with respect to a plane parallel to plates P_1 and P_2 located halfway between them.[12,19] In practice, since the ratio t/L of plate separation to plate length is not zero (owing to the finite length of the resistive plates), it is not unusual to have a transverse component of the field *one order of magnitude larger* than E_H.[19] This unwanted field gives rise to a signal that is picked up by the detector even when $B = 0$. To reduce this error and to increase the signal to noise ratio, it is convenient to use as detector for i_H a *differential* charge sensitive amplifier; that is, an amplifier capable of rejecting currents that arrive in phase at its two inputs, and that responds only to signals that arrive 180° out of phase. The effectiveness of this technique is, however, limited by the balance of stray capacitances seen at each input of the preamplifier.[12,19]

The fringing field can be reduced one order of magnitude by adding two additional electrodes perpendicular to P_1 and P_2 biased at the proper potential, at the expense of losing the simple relationship (Equation 3) between μ_H and ΔV_t.[12,19]

Third, errors due to a polarization field that may arise out of charge that remains unneutralized in the cell should be avoided. (Recall that a polarization field of 1 V/cm would already imply an error of the order of *50% or more*!) To minimize these errors — which are two orders of magnitude more important in this experiment than they are in a TOF experiment due to the fact that $E_H = \mu_H B E_L << E_L$ — it is necessary to pulse the longitudinal field with alternating polarity; this makes the decay of any accumulated space charge possible. In samples exhibiting a high electron mobility, such as TMS and neopentane (NP), it was found that in order to eliminate the polarization field it was necessary to irradiate the sample with blank X-ray pulses (beam pulses with *no field* applied to the sample).

---

* A charge sensitive amplifier produces an output voltage pulse that is proportional to $\int_0^t i(t')\,dt'$, where i(t) is the (time dependent) current present at the input of the amplifier; a current amplifier produces an output voltage proportional to i(t).

** Application of a similar purification process—based on the purification method developed during the course of this work—led to the UA1 upgrade at CERN, e. g. the effort of building a calorimeter employing 12,000 liters of 2,2,4,4 tetramethyl pentane.

However, such precautions are not necessary in liquids that exhibit a lower mobility, such as 2,2-dimethyl butane (22-DMB) and 2,2,4-trimethyl pentane (224-TMP).[12] Further details on the experimental technique have been published.[12,13]

IV. OVERVIEW OF DRIFT MOBILITY DATA

Some remarkable features emerge when studying the drift mobility of excess electrons in insulating nonpolar liquids. The first surprise is related to the effect of disorder close to the triple point. We may naively expect that, upon melting, the broken translational symmetry (symmetry which is customarily associated with the existence of extended electronic states or "band", in the terminology of solid state physics) results in a dramatic reduction of μ_D. On the contrary, in cases where the excess electrons do exhibit a high mobility and where a measurement of μ_D has been made both in the solid and in the liquid phase close to the melting point, as in Ar, Kr, Xe,[21] methane,[22] 2,2-dimethyl propane (or neopentane, NP),[23,24] and TMS,[25] the experiment indicates that μ_D decreases upon melting by a modest factor of ~ 2.2. This may be considered as an experimental proof that a band exists also in the liquid close to the melting point, even if the mathematical formalism to describe it — the equivalent of Bloch's theorem in solid state physics — has not yet been developed. We return to this point in Section VI.

The second notable feature is the diversity of values found for μ_D (covering five orders of magnitude) in insulating liquids that otherwise display rather similar physical properties. As an example, consider the electron mobility found in methane, $\mu \sim 400$ cm^2/ V · s at 111 K, and in ethane, $\mu \sim 0.0013$ cm^2/ V · s at the same temperature.[26]

A third remarkable feature is related to the difference in the mobility of excess electrons observed among the rare gas liquids. In liquid He and Ne, the electron mobilities are approximately 0.025 cm^2/ V · s at 4.2 K and 0.001 cm^2/ V · s at 25 K, respectively, whereas in liquid Ar, Kr, and Xe they are 475 cm^2/ V · s at 85 K, 1800 cm^2/ V · s at 117 K, and 1700 cm^2/ V · s at 163 K (see Table 1). The mobilities observed in the former have been interpreted as arising from electrons residing in a cavity or "bubble", those observed in the latter have been interpreted as arising from the motion of quasi-free electrons. For clarity of presentation, data concerning electron bubbles are reviewed at the end of this section.

A. EFFECT OF MOLECULAR STRUCTURE

Table 1 contains data arranged to display the dependence of μ_D on molecular structure. Inspection of Table 1 indicates that a high μ_D correlates with a high degree of molecular sphericity*; this correlation has been discussed recently.[26-31] This trend stands out with particular clarity in the case of the C_6 isomers. In the case of *n*-alkanes, the longer the chain the smaller the μ_D, although this trend levels off beyond *n*-hexane, at about 0.05 cm^2/ V · s. Further information involving data for other hydrocarbon liquids may be found in the literature.[3,15,26,32-35]

B. EFFECT OF TEMPERATURE
1. Liquids Exhibiting a High μ_D

The liquids listed in Table 1 as having spherical or nearly spherical molecules exhibit an electron mobility which display a similar behavior when the density of the liquid is varied, either by changing its temperature and/or its pressure, from densities close to the density at the triple point N_T, to densities close to the critical density N_C. The data displayed in Figures 4 and 5 indicate that μ_D increases as the density decreases below N_T, reaches a maximum that takes place at roughly the same density at which a minimum in the level of the conduction

---

* The exception to this rule is the case of He and Ne (not included in Table 1), and is discussed at the end of this section.

TABLE 1
Excess Electron Mobility in Dielectric Liquids

Liquid	μ_D (cm²/V · s)	Temp (K)	E_a(eV)[g]	Ref.
Spherical molecules				
Ar	475	85[a]	—	21
Kr	1800	117[b]	—	46, 159
Xe	1700	163[c]	—	46
Nearly spherical molecules				
Methane	475	92[d]	—	169
2,2-dimethyl propane (NP)	50	RT[e]	—	66, 67
Tetramethyl silane (TMS)	98	RT	—	174
C_6 isomers				
n-hexane	0.093	295	0.16	32
2-methyl pentane	0.29	293	—	122
3-methyl pentane	0.22	RT	0.2	171
2,3-dimethyl butane	1.1	293	—	122
2,2-dimethyl butane	10	296	0.06	170
n-alkanes				
Methane	475	92[d]	—	169
Ethane	0.0013	111[f]	0.13	26, 172
Propane	0.05	175	0.13	172
n-butane	0.27	293	0.17	172
n-pentane	0.15	295	0.20	32
n-hexane	0.093	295	0.16	32
n-heptane	0.046	295	0.17	32
n-octane	0.040	295	0.18	32
n-nonane	0.047	295	0.19	32
n-decane	0.038	295	0.22	32

[a] Triple point 83.8 K.
[b] Triple point 115.9 K.
[c] Triple point 161.3 K.
[d] Triple point 90.7 K.
[e] RT stands for room temperature.
[f] Triple point 89.8 K.
[g] E_a denotes activation energy measured along the liquid-vapor coexistence line.

band V_0 is observed[36] (the exception to this rule is perhaps the case of Kr, where the mobility maximum occurs at a density somewhat higher than that corresponding to the V_0 minimum — see Figure 4), and then decreases sharply as the density of the fluid decreases further toward N_C. A similar behavior has been found in methane.[37,38]

An interesting effect is observed in the case of Ar, TMS, and NP, where data at pressures higher than those corresponding to the coexistence line are available.[39,40] As shown in Figure 6, it is observed that when the mobility measured at different temperatures and pressures is plotted as a function of density, the apparent complexity of the data is replaced by a simple trend similar to that shown in Figures 4 and 5 described above, with the mobility maxima occurring at approximately the same density, pointing to the fact that density is one of the appropriate variables that should be used to describe changes in electron mobility.

2. Liquids Exhibiting Low and Intermediate μ_D

Except for some recent studies by Muñoz et al. discussed in Section IV. D, all previous measurements of mobility in liquids exhibiting low and intermediate μ_D (i.e., $\mu_D \leq 10$ cm²/ V · s) have been performed along the liquid-vapor coexistence line. Thus, the data and the activation energies that have been reported reflect the composite effect of changes in mobility caused by temperature-induced changes in density (the change in density between the melting

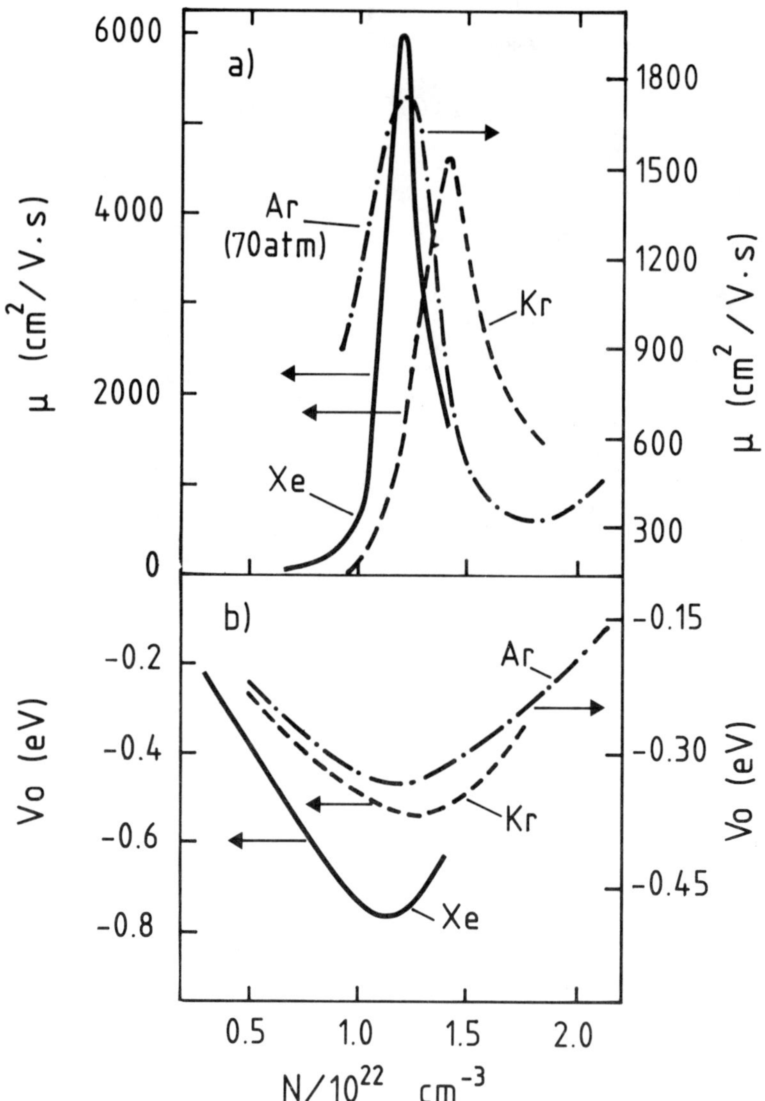

FIGURE 4.   (a) Density dependence of the zero field mobility in liquid Ar, Kr, and Xe.
Data for Ar[39] along the 70 atm. isobar; data for Kr and Xe[46,159] along the liquid-vapor
coexistence line. The densities at the triple and critical points, in units of 10^{22} atoms/cm^3,
are: (2.10, 0.81), (1.75, 0.65), and (1.41, 0.50), respectively. (b) Density dependence of
the position of the bottom of the conduction band V_0 in liquid Ar,[94] Kr, and Xe.[173]

and the critical point, for most liquids considered here, is about a factor of 2.5), as well as
changes induced by temperature in the excess electron kinetic energy and possibly in the
collision and/or the trapping mechanisms at work. Because of this, activation energies
determined in this way are of a rather dubious nature.

The data for most n-alkanes measured in this way (the notable exception being methane)
were found to follow approximately an Arrhenius-type dependence,[26,33] although deviations
from it could be observed. In n-pentane, a higher activation energy is observed at higher
temperatures; in ethane, the opposite behavior occurs. When the temperature dependence
of the mobility is expressed as $\mu(T) = \mu_0 \exp(-E_a/k_B T)$, where k_B stands for Boltzmann's
constant, the values of μ_0 range between 100 and 1000 cm^2/ V · s, and the activation energies
E_a range from ~ 0 to 0.2 eV.

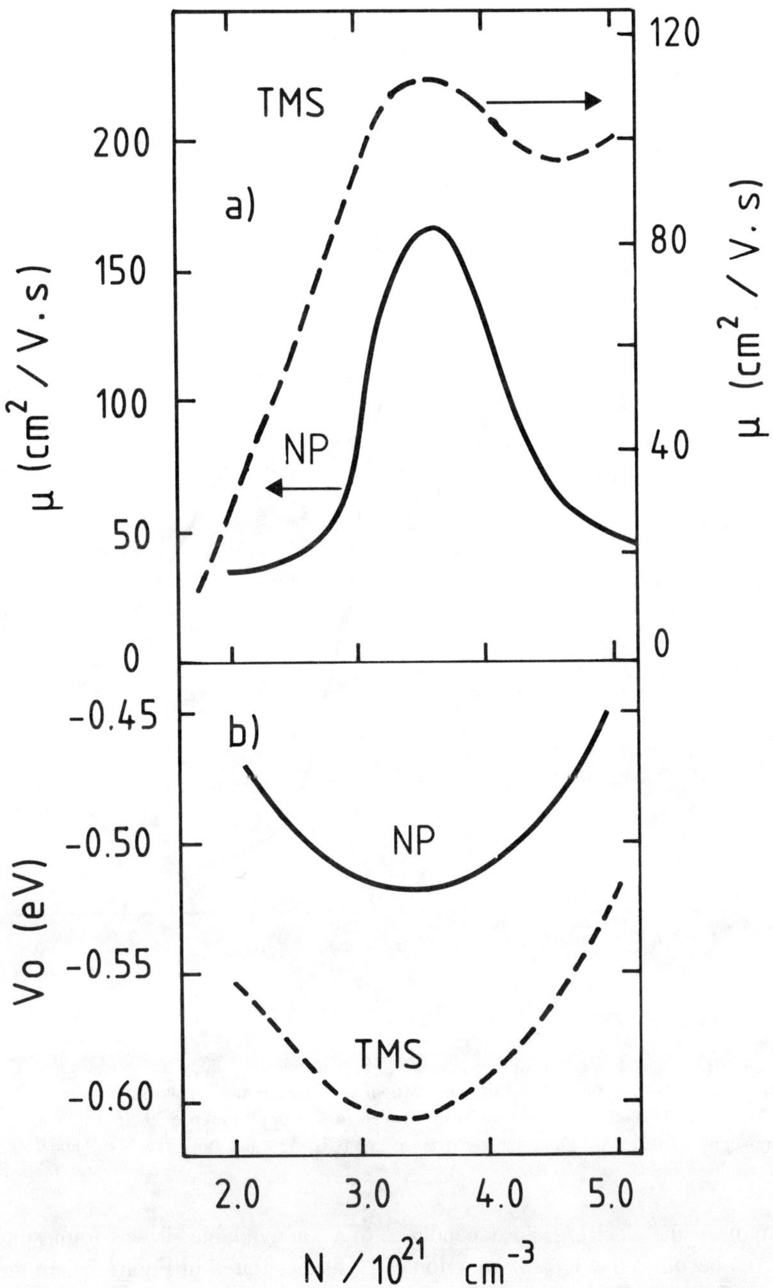

FIGURE 5. (a) Density dependence of the zero field mobility in liquid tetramethyl silane (TMS)[36,174] and neopentane (NP)[66] measured along the liquid-vapor coexistence line. The densities at the melting and critical points, in units of 10^{21} molecules/cm³, are: (5.24, 1.67) and (5.26, 1.94), respectively. (b) Density dependence of the position of the bottom of the conduction band V_0 in TMS and NP.[36]

C. Studies on the Electric Field Dependence of μ_D

At low enough fields, the excess electron drift velocity v_D is proportional to the electric field E. At higher fields deviations from this ohmic behavior have been found. As shown in Figure 7, two distinct patterns have been observed.

In liquids where excess electrons exhibit a high mobility (e.g., $\mu_D > 10$ cm²/ V · s) and

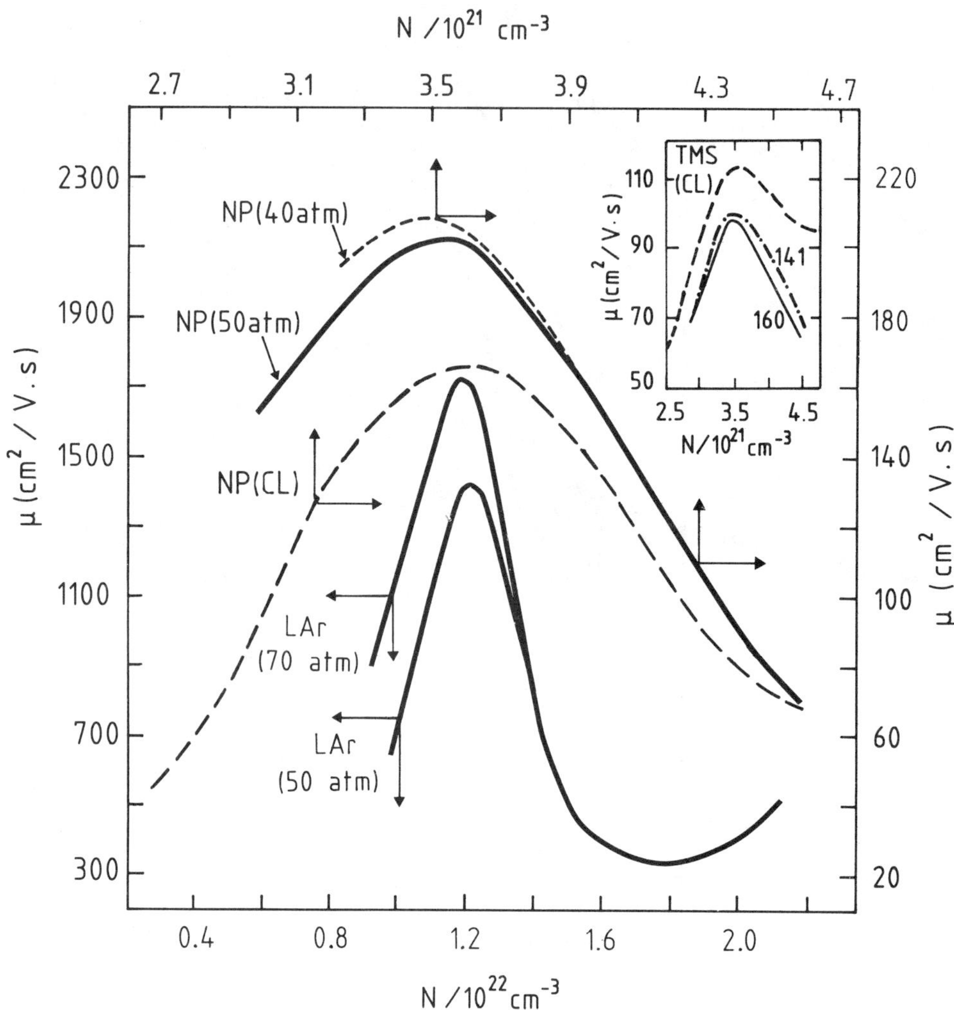

FIGURE 6.   Density dependence of μ_D in neopentane (NP) and liquid Ar along different isobars. Curve labeled NP(CL) denotes μ_D in NP measured along the liquid-vapor coexistence line.[66] Data for the 40 and 50 atm isobar in NP from Reference 40; data for the 50 and 70 atm. isobar in Ar from Reference 39. Insert: Density dependence of μ_D in tetramethyl silane (TMS) along the coexistence line (CL);[36,174] and along the 141°C and 160°C isotherms at pressures ranging from 40 to 1000 bar.[40]

where a study of the electric field dependence of v_D is available, it was found that at large fields, v_D (E) becomes a *sublinear* function of E, as illustrated in Figure 7a. In the rare gas liquids,[41] methane,[42] NP,[43] and TMS,[44] v_D (E) can be described approximately by a $E^{+1/2}$ dependence at intermediate fields, and saturation has been observed at large fields in all but NP. Such field dependence has been interpreted as arising from charge transport via extended states involving hot electron phenomena, a point discussed in more detail in Section VI.

In the rare gas liquids, the addition of a small concentration of molecular solutes was found to lead to an increase in v_D above the saturation value in the pure liquid; v_D either reaches a higher constant value or passes through a maximum at fields >10 kV/cm;[41] this has been interpreted in terms of inelastic collisions between the excess electrons and the impurity molecules.

In liquids where electrons exhibit μ_D <1 cm²/ V · s, at large fields it was found that v_D (E) can be described by a *superlinear* function of E,[26] as shown in Figure 7b. As discussed

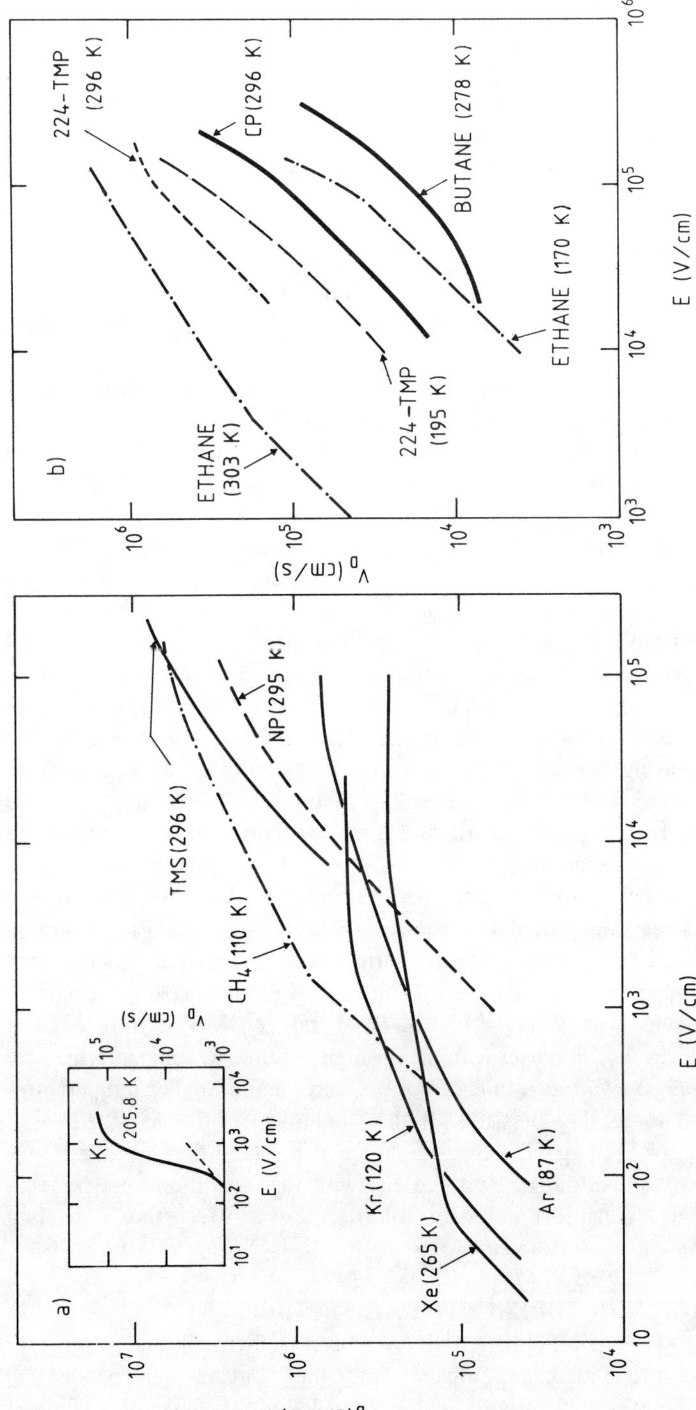

FIGURE 7. (a) Electric field dependence of μ_D in liquids composed of spherical or nearly spherical molecules[41-44]; temperature indicated in the figure. Insert: field dependence of μ_D in liquid Kr at 205.8 K.[159] Dotted line indicates ohmic behavior. (b) Field dependence of μ_D in several hydrocarbon liquids[26,42] at the temperatures indicated.

later (Section VI), such behavior has been interpreted as evidence of a charge transport mechanism involving localized electron states. Note that the *same liquid may exhibit both types of behavior,* a *superlinear* dependence of v_D (E) on E if its density is such that μ_D <1 cm²/ V · s, and a *sublinear* dependence if μ_D > 10 cm²/ V · s (Figure 7b). However, this criteria for separating sublinear from superlinear field dependence is only approximate, for (as shown in the insert in Figure 7a), in Kr at 205.8 K, the observed μ_D in the limit of zero field is μ_0 ~ 5 cm²/V · s. Note that in this case, v_D (E) displays a superlinear field dependence for E <200 V/cm, and sublinear dependence for E >1 kV/cm. Similar behavior is observed in Ar[39,45] and Xe[46] close to the critical point; we return to this point in Section VIII.

D. STUDIES ON THE PRESSURE DEPENDENCE OF μ_D

Recently, Muñoz et al. measured μ_D in several insulating liquids at pressures ranging from 1 to 2500 bar and temperatures between 20 and 120°C.[47-49] Three distinct patterns were observed.

As illustrated in Figure 8a, in TMS the electron drift mobility μ_D decreases at all temperatures with increasing pressure, the decrease being higher at higher temperatures. The isochoric mobility is proportional to $T^{-0.9}$; the predictions of the Basak and Cohen theory[50] are in qualitative agreement with experiment, although the mobility predicted at room temperature and 1 bar is about 2.5 times the mobility found experimentally.[47]

A somewhat different behavior is observed in *n*-hexane, 3-methyl pentane (3-MP), *n*-pentane, and cyclopentane (CP). In these compounds, μ_D also decreases at all temperatures with increasing pressure, the effect being more pronounced at higher temperatures, but the isochoric mobility *increases* with increasing temperature in a manner which is rather well described by an Arrhenius-type dependence. The activation energies describing the isochoric mobilities turn out to be 50 to 70% of the activation energies corresponding to the mobility when measured along the liquid-vapor coexistence line. The isochoric activation energies increase slightly with increasing density in the case of *n*-pentane and *n*-hexane, but are constant in the case of CP and 3-MP.[48,49] The decrease in mobility with increasing pressure at constant temperature may be interpreted as arising from a pressure-induced change in the equilibrium constant, K, for the reaction e_{qf} + trap $\longleftrightarrow$ e_t, where e_{qf}, trap, and e_t stand for quasi-free electrons, *preexisting traps,* and trapped electrons, respectively. The volume change associated with this reaction (calculated from $\Delta V = -RT\ (\partial \ln K/\partial P)_T$, assuming that the molar volume of quasi-free electrons is zero) turns out to be negative; this has been interpreted as arising from electrostriction of the solvent around the localized electron.[49]

A more enigmatic behavior was observed in 224-TMP and 22-DMB, as illustrated in Figure 8b. At low temperatures, μ_D *increases* with increasing pressure. At some intermediate temperature, μ_D (P) *remains constant* as a function of pressure, and at higher temperatures μ_D actually *decreases* with increasing pressure. If the isochoric mobility is fitted to an Arrhenius-type law, i.e., $\mu(N,T) = \mu(N)\ exp\ (-E_a/k_BT)$, it is found that a modest 14% increase in density produces a fivefold decrease in E_a (with the activation energy being smaller than k_BT at the higher densities) and a fourfold decrease in $\mu(N)$, thus indicating the inapplicability of this law to these two cases.

E. ZERO FIELD MOBILITY IN SUPERCRITICAL FLUIDS

Electron mobilities at low electric field strength have been isothermally measured over a wide density range in supercritical methane,[51] ethane,[52] propane,[53] butanes[54] and pentanes.[55] In the limit of dilute gas, the excess electron is scattered by individual molecules, and the scattering rate depends linearly on the density of the fluid. Therefore, the product of the electron mobility μ and the gas density N along an isotherm, when plotted as a function of density, is a constant.

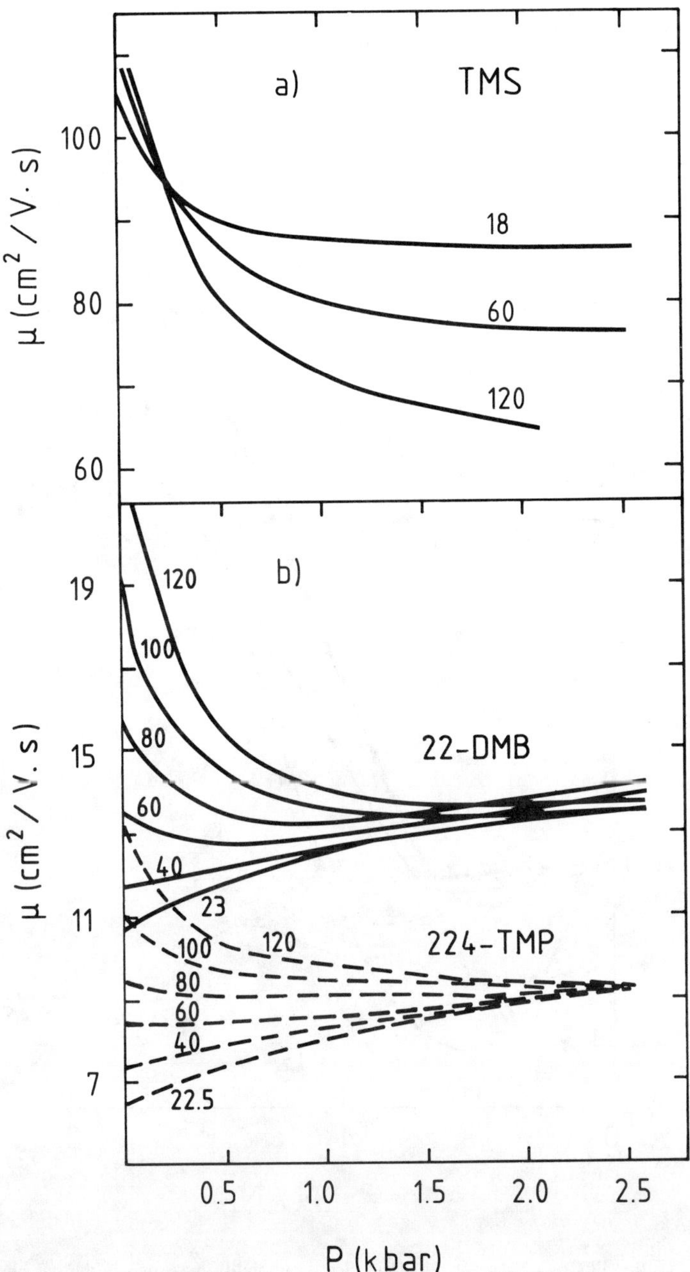

FIGURE 8. (a) Pressure dependence of μ_D in tetramethyl silane (TMS)[48], temperature in °C indicated in the figure. (b) Pressure dependence of μ_D in 2,2,4-trimethyl pentane (224-TMP)[48] and 2,2-dimethyl butane (22-DMB),[49] at the temperature in °C indicated in the figure.

To facilitate comparison between different samples, a plot of the product μN (normalized by the value characteristic of the dilute gas $(\mu N)_{\text{dil gas}}$) as a function of density, is displayed in Figure 9.

Some interesting features are worth mentioning. At very low densities, $N/N_c < 0.05$, the excess electron mobility does exhibit the expected density dependence $\mu(N,T) \sim c(T)/N$, where $c(T)$ is a weakly temperature dependent constant. As the density increases, the product

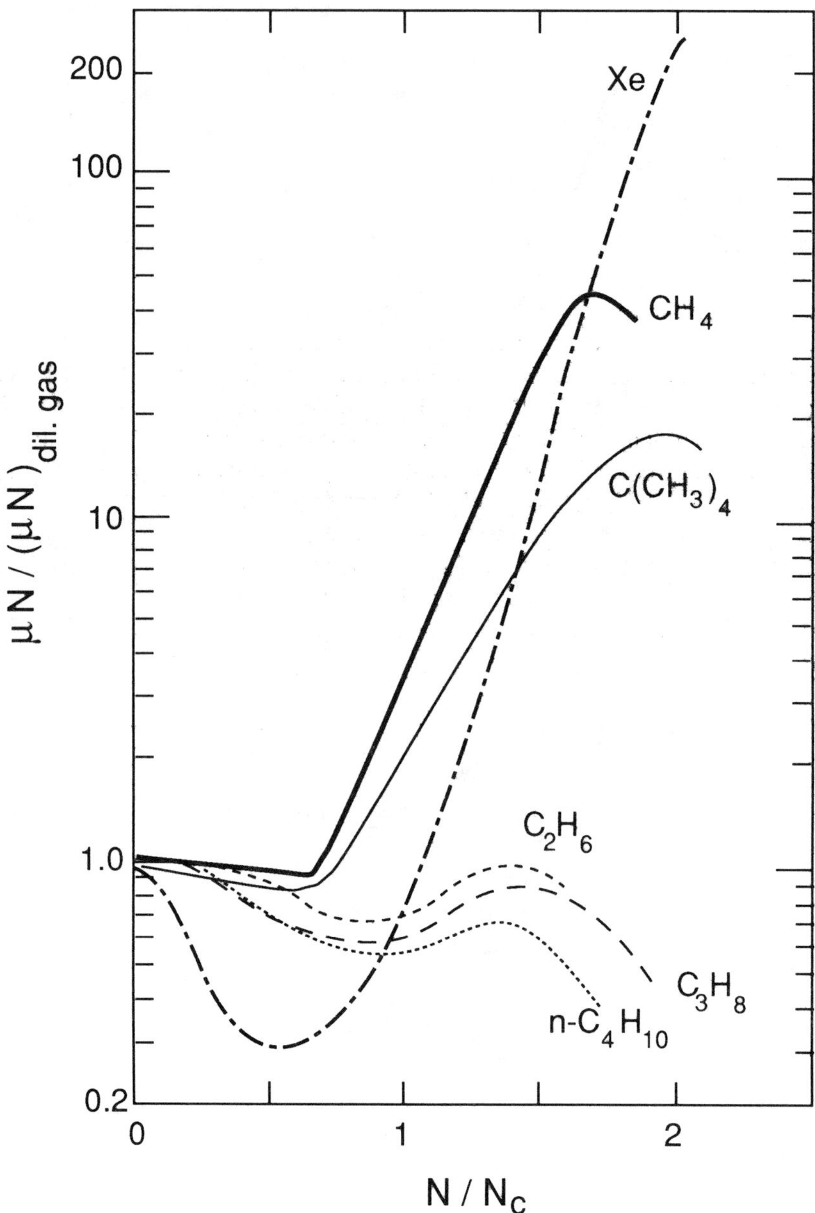

FIGURE 9.   Plot of μN (μ — zero field electron mobility measured in the gas phase, N — gas density) normalized by the values corresponding to the dilute gas limit (μN)$_{\text{dil gas}}$, vs. normalized density N/N$_C$ (N$_C$ — critical density).[46,54,55] The hydrocarbon data correspond to measurements in the supercritical fluid; the Xe data refer to measurements performed at temperatures and pressures not exceeding those of the liquid-vapor coexistence line.

μN drops below its dilute gas limit, except perhaps in methane. As the density increases further toward the critical density N$_C$ and beyond, the density normalized mobility displays a behavior which is markedly different in the gases composed of spherical or nearly spherical molecules (Xe, methane, and NP), as opposed to fluids made up of elongated molecules, such as ethane, propane, and *n*-butane. In the former, as the density increases beyond the critical density, the product μN undergoes an *increase of more than one order of magnitude* over the value characteristic in the dilute gas; in the latter, this product exhibits a local maximum and then continues to decrease with increasing density, but it stays *below* its dilute gas value (Figure 9).

The increase in μN beyond its dilute gas limiting value observed in the gases composed of spherical or nearly spherical molecules, can be interpreted as indicating the onset of conduction band, i.e., the appearance of *extended electron states* encompassing many fluid molecules. In this regime, the excess electrons are no longer scattered by individual molecules, but by density fluctuations that produce local changes in the level of the bottom of the conduction band V_0; this point is discussed in Section VI. The transition from dilute gas $N/N_C < 0.05$ to dense gas $N/N_C > 1$ presumably involves an intermediate region where the excess electrons are no longer scattered by individual molecules, but do not yet occupy states that are "shared" by many fluid molecules; that is, charge conduction is not yet dominated by the appearance of extended electron states. In this region, the electrons undergo multiple scattering from several molecules at a time. Why this multiple scattering results in a decrease of the product μN below its dilute gas limit in some fluids (such as Xe, and perhaps NP — see discussion in Section VIII.C), and not in others (methane), is not yet understood.

Concerning the observed behavior of the density normalized mobility in fluids composed of elongated molecules, the fact that μN does not exceed its dilute gas limit, not even in the dense gas region, can be interpreted as indicating the "absence" of extended electron states. In spite of the proximity of neighboring molecules, orientational disorder prevents the formation of such states; this point is further discussed in Section VI.

F. LIQUID He AND Ne: ELECTRON BUBBLES

The mobility of negatively charged carriers has been measured in fluid He at temperatures at and below 4.2 K[4,5,56-58] and in liquid Ne along the liquid-vapor coexistence line.[6,7,58] The data in fluid He and liquid Ne are displayed in Figure 10.

There is a striking difference between the behavior of excess electrons in fluid He and liquid Ne, when compared with the other rare gas liquids, Ar, Kr, and Xe (Figure 4). The electron mobility in the former is several orders of magnitude smaller than in the latter. Moreover, as shown in Figure 10a, in dense He gas the electron mobility undergoes a dramatic drop of five orders of magnitude when the pressure approaches the saturated vapor pressure.

1. Electron Bubbles or Impurity Anions?

The possibility that the negatively charged species observed in fluid He is an electronegative impurity can be ruled out for several reasons, a few of which are summarized below.

First, the mobility of ions in a classical dielectric liquid is, in first approximation, governed by hydrodynamics. The limiting velocity that a sphere of radius R, carrying a charge q, attains under an applied electric field depends on the viscosity η of the liquid; the ionic mobility is determined by Stoke's law:[4] $\mu = q/(6\pi\eta R)$. Using the known values of the viscosity of liquid He, Meyer et al.[4] calculated a Stoke's radius in ^{4}He I of $R_+ = 4.3$ Å for positive ions, and $R_- = 8.4$ Å for the negative species at 3 K, and $R_+ = 5$ Å and $R_- = 14$ Å in ^{3}He at 3.2 K. The radius of the negative species is unreasonably large for an ion, suggesting instead that the negative charge carrier is a complex entity involving not an impurity anion but several helium atoms.

Second, the measurement of the mobility of negative carriers in He gas has been extended to higher temperatures and pressures.[59,60] The data reported by Jahnke and Silver (JS) reveals a high mobility species coexisting with a low mobility species,[60] the latter is similar to that found by Levine and Sanders at lower temperatures.[5] The high mobility species was interpreted as a nearly free electron colliding with He atoms; the low mobility species was

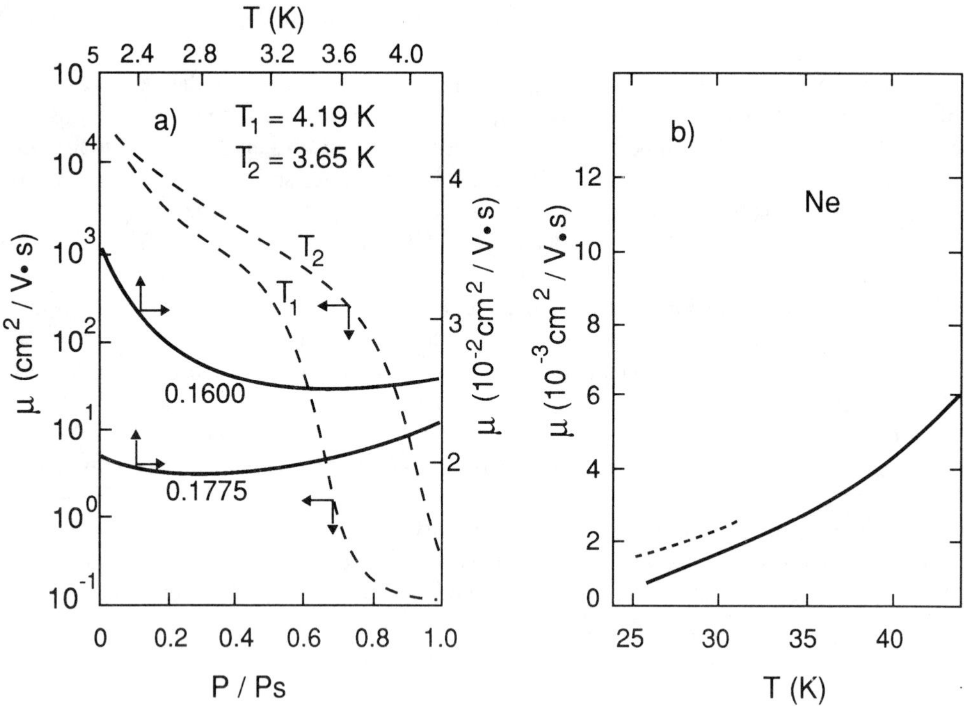

FIGURE 10.   (a) Drift mobility of negatively charged carriers in fluid ^{4}He. Dotted line: isothermal mobility of negative carriers in dense ^{4}He gas, plotted vs. the vapor pressure normalized by the saturated vapor pressure, at the temperature indicated in the figure.[5] Solid line: mobility of negative carriers in ^{4}He I, at the density (g/cm^3) indicated in the figure.[4] (b) Mobility of negative carriers in liquid Ne, measured along the liquid-vapor coexistence line. Solid line, data from Reference 6; dotted line, data from Reference 7.

interpreted as an electron residing in a cavity, or bubble. Interestingly enough, the results obtained by JS with He 99.95% pure were the same as those obtained with He 99.9995% pure. Moreover, if the low mobility species was an impurity anion, then the strength of the signal observed in the low mobility branch should exhibit a linear dependence on the density of the fluid, contrary to experimental observation.[60] These results, together with the dramatic mobility drop observed by Levine and Sanders (Figure 10a), imply that this drop is due to bubble formation in the gas phase, since from the theory of ion mobility in gases, the mobility of negative ions (which could arise out of electronegative impurities present in the He gas) is not expected to exhibit such a drastic density/pressure dependence.[5]

Another argument in favor of the bubble model is provided by completely different experiments. If the electron in the liquid is confined within a bubble, then the transition from a nearly free state to a localized state involves a penalty in energy — the kinetic energy of localization or zero-point energy. An excess electron going from the gas into the liquid should see an energy barrier that has been estimated to be ~1 eV. The existence of this barrier has been confirmed by different experiments: (a) by making a comparison of the work function of a photosensitive material (a Cesium-antimony photocathode) measured under vacuum with that obtained with the photocathode submerged in liquid He;[61] (b) by extracting the charge carriers produced by a ^{210}Po source immersed in liquid He across the liquid-vapor interface, and detecting these carriers in the collector exposed to the He gas above the liquid; the signal observed for negative carriers is 10^4 times larger than that observed for positive carriers.[62]

Finally, we mention what is, in our opinion, the most convincing evidence yet produced confirming the correctness of the bubble model. Northby and Sanders reported the current observed in a TOF spectrometer submerged in liquid He, when illuminated with a Xe lamp.[63] The chosen frequency of the voltage (a square wave) applied between the grids of the spectrometer was such that only charge carriers with a mobility higher than that of electron bubbles could arrive at the collector; carriers were produced by a ^{210}Po source. If electrons are ejected from bubbles when the light is turned on, a collector current should appear, and no current should be observed with the light off. A photocurrent was, indeed, recorded, but no signal was observed when studying currents due to the motion of positive ions.

2. Electron Bubbles in Other Liquids

The search for electron bubbles via measurements of electron mobility has been extended to H_2. The observation of negative carriers displaying a large mobility coexisting with carriers displaying a low mobility has been reported in dense H_2 gas;[8] a low mobility species has been reported in liquid H_2.[58,64]

Concerning the electric field dependence of the drift velocity of these species, it has been found that at large fields, the negative carriers observed in liquid He, Ne, and H_2 all exhibit the superlinear field dependence expected from localized electrons. Moreover, a signal produced by fast negative carriers was also reported in liquid Ne; it was interpreted as nearly free electrons *preceding electron localization in a bubble.*[58]

3. Why Bubbles?

We now turn to a brief, qualitative discussion of why excess electrons form bubbles in some dielectric liquids and not in others. For a detailed discussion of the theory of bubbles, the reader is referred to Reference 15 and the literature quoted therein.

Whenever there are two states available to a system, the condition of stability is that the free energy of the system in the stable state be less than that in the unstable state. For an electron in the bubble state to be stable relative to the quasi-free state, the condition $E_t < V_0$ must be fulfilled. Here V_0 is the energy of the quasi-free electron in the liquid relative to vacuum, and E_t is the reversible work of putting an electron from vacuum into a bubble, i.e.,

$$E_t = 4\pi R^2 \gamma + \frac{4}{3} \pi R^3 P - \frac{\epsilon - 1}{2 \epsilon} \frac{q^2}{R} + E_0 \qquad (4)$$

where R stands for the bubble radius, γ is the surface tension, P is the pressure, ϵ is the dielectric constant of the fluid, q is the electron charge, and E_0 is the ground state energy of the electron in a bubble. In Equation 4, the first and second terms represent the surface and volume work necessary to create a bubble, and the third (negative) term, written in Gaussian units, represents the polarization energy of the bubble, assuming that the charge q is all contained within a sphere of radius R.

Obvious candidates for bubble formation are the liquids that exhibit a large, positive V_0. The values of V_0 determined by photoelectric work function measurements in some selected liquids are listed in Table 2, along with the molecular polarizabilities. Inspection of Table 2 reveals that indeed, He, Ne, and H_2 do exhibit a large, positive V_0; hence bubbles may be expected to be stable in these liquids; this is in contrast to Ar, Kr, and Xe which exhibit a negative V_0. Therefore, in the latter three liquids, we may expect that the quasi-free state will be stable, rather than the bubble state. Calculations of bubble stability have predicted stable bubbles in ^{4}He (R = 16.8 Å at 3 K), Ne (R = 7.5 Å at 25 K), and H_2 (R = 10.7 Å at 19 K).[15] However, the values used in this calculation ($V_0 = +1.3$ eV for ^{4}He, $V_0 = +0.50$ eV for Ne, $V_0 = +0.90$ eV for H_2) are somewhat different from those determined experimentally (Table 2).

The fundamental reason underlying bubble formation arises from Pauli's exclusion prin-

TABLE 2
Conduction Band Energies and
Polarizabilities in Dielectric Liquids

Liquid	V_0 (eV)	Temp (K)	α (Å³)	Ref.
He	+1.05	4.2	0.201[a]	61
Ne	+0.67	25	0.390[a]	175
Ar	−0.19	85	1.62[a]	94
Kr	−0.35	117	2.46[a]	173
Xe	−0.63	163	3.99[a]	173
H_2	>+2.0	15	0.714[b]	175
			1.03[c]	

[a]   Reference 81, page 462.
[b]   Polarizability perpendicular to the molecular axis, Reference 10.
[c]   Polarizability parallel to the molecular axis, Reference 10.

ciple. The potential well giving rise to the bound state of the electron exists because of the repulsion between the excess electron and the electrons belonging to the molecules that constitute the fluid; the bubble is maintained by the zero-point kinetic energy of the electron against the pressure-volume and surface energies, which tend to contract it. The potential characterizing the excess electron/molecule interaction in the dilute gas of a fluid that serves as a host for bubbles, should exhibit a strong short-range repulsive core due to the Hartree interaction, and a weak long range attractive component due to the polarizability of the molecules. In other words, it should exhibit a positive scattering length (the radial part of the electron wave function is "pushed out" from the scatterer). This quantity is, indeed, positive in He, Ne, and H_2, but negative in Ar, Kr, and Xe.[10] (A negative scattering length corresponds to a situation where the radial part of the electron wave function is "pulled into" the scatterer.) The predominance of this repulsive core is responsible for the positive V_0 observed in liquid He, Ne, and H_2, whereas the predominance of the attractive component of the potential (arising from larger atomic polarizabilities due to increased atomic number) determines the negative V_0 observed in liquid Ar, Kr, and Xe.

V. OVERVIEW OF HALL MOBILITY DATA

Muñoz et al. measured μ_H along the liquid-vapor coexistence line in TMS[13,19,65] and NP;[12,19,66,67] Itoh, Muñoz, and Holroyd measured μ_H in 2,2,4,4-tetramethyl pentane (2244-TMP), 224-TMP, and 22-DMB.[68] The temperature dependence of μ_H and of the mobility ratio $r = \mu_H/\mu_D$ in TMS and NP is displayed in Figure 11; that observed in 2244-TMP, 224-TMP, and 22-DMB is shown in Figure 12.

An interesting behavior is observed in NP. At some 30°C below the critical point (T_c = 160°C), μ_D starts to decrease with increasing temperature, and the decrease becomes steeper in the proximity of the critical point. At 140°C, $\mu_D \sim$ 150 cm²/V · s, and at the critical point μ_D reaches a value of about 30 cm²/V · s.[19,66,67] However, in this region μ_H remains nearly constant, dropping from ~220 cm²/V · s at 140°C to about 170 cm²/V · s at T_c. Consequently, the mobility ratio μ_H/μ_D increases from ~1.5 at 140°C to ~5.5 at T_c. This behavior has been interpreted as evidence of *intrinsic traps*.[19,66,67] A more detailed discussion is presented in Section VII.

On the contrary, in TMS (T_c = 175°C), a liquid with physical properties similar to those of NP, it was found that both μ_H and μ_D gradually decrease with increasing temperature above 100°C, from 110 cm²/V · s at 100°C to about 50 cm²/V · s at 164°C, resulting in a

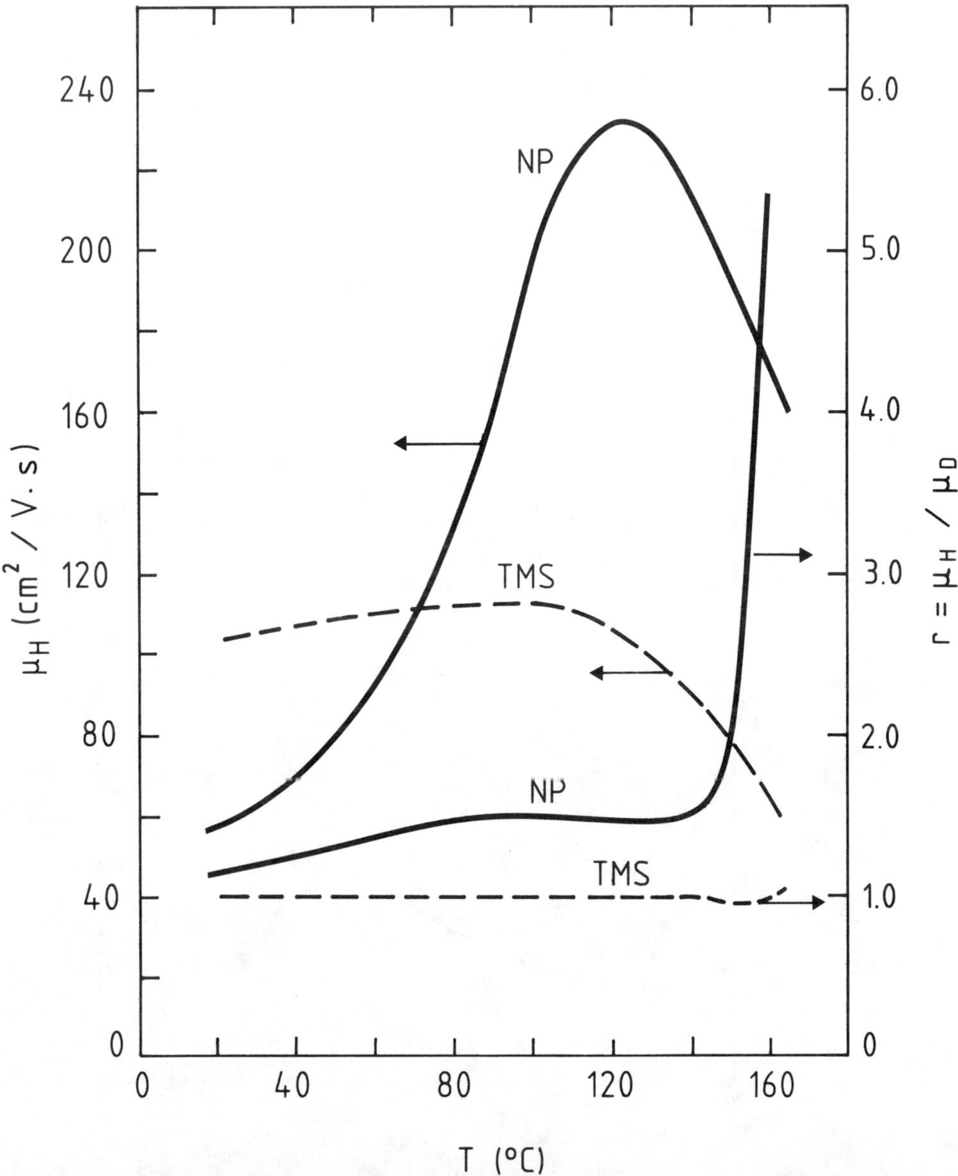

FIGURE 11.   Temperature dependence of μ_H and of the mobility ratio $r = \mu_H/\mu_D$ in tetramethyl silane (TMS) and neopentane (NP). For TMS, μ_H data taken from Reference 13, μ_D data used to calculate r taken from Reference 174. For NP, μ_H taken from Reference 12, μ_D data used to calculate r taken from Reference 66.

mobility ratio which is 1.0 $\pm$ 10%, from room temperature to 164°C.* This has been interpreted as resulting from a scattering mechanism whose effectiveness increases with increasing temperature, and *not from intrinsic traps.*[13]

As displayed in Figure 12, in 22-DMB and 2244-TMP the ratio μ_H/μ_D is approximately

* There is an earlier report of $\mu_H = 125 \pm 10$ cm^2/V · s at room temperature[65] and thus r = 1.25 $\pm$ 10% in TMS; the measurement was performed using a 4-electrode cell. Later measurements of two different samples, using two different 2-electrode cells such as that shown in Figure 3a, yielded $\mu_H = 101 \pm 7$ cm^2/V · s and $\mu_H = 108 \pm 7$ cm^2/V · s,[13] which combined together yield $\mu_H = 104 \pm 5$ cm^2/V · s. The difference between these two sets of measurements is probably due to a systematic error in the measurement of μ_H not reported in Reference 65, stemming from the application of Equation 3 to a 4-electrode cell.[12,13] See References 12 and 13 and the Postscript at the end of this chapter.

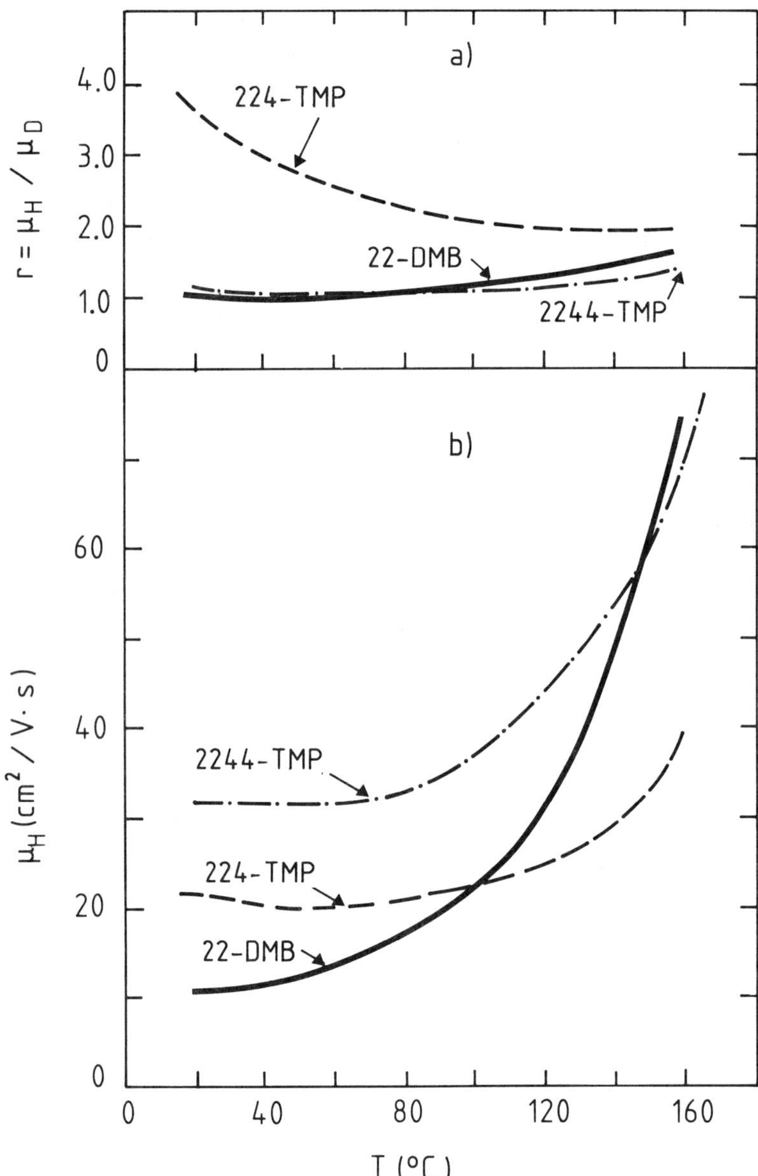

FIGURE 12.   (a) Temperature dependence of the mobility ratio r = μ_H/μ_D in 2,2,4-trimethyl pentane (224-TMP), 2,2,4,4-tetramethyl pentane (2244-TMP), and 2,2-dimethyl butane (22-DMB).[68] (b) Temperature dependence of μ_H in 2244-TMP, 224-TMP and 22-DMB.[68]

1.0 up to $\sim$ 100°C; from there on it increases slightly. Hence, traps probably do not play a significant role in the charge transport process in these two liquids. On the contrary, in 224-TMP, the mobility ratio is significantly larger than 1.0 at all temperatures, being $\sim$3.7 at room temperature and decreasing to $\sim$2.0 at 160°C. As discussed below (Section VII), such behavior is not readily explainable in terms of the scattering mechanisms known in solid state physics. This suggests that in this liquid the charge transport process involves traps within this temperature range, the traps playing a less significant role at higher temperatures. If traps are assumed to be distributed uniformly with a trap density, N_t (independent of temperature), and are characterized by an activation energy E_a, then the Hall mobility data in 224-TMP is consistent with $E_a \sim$ 0.05 eV and $N_t \sim$ 8 $\times$ 10^{18} cm^{-3}.[68] This trap density is to be compared with the density of states available to a quasi-free electron, that is, with the effective density of states $N_0 = 2 \, (2\pi m^* k_B T/h^2)^{3/2}$ that a nearly free electron gas would exhibit at the bottom of the conduction band. If $m^* = m_0$ (m^*: electron effective mass; m_0: free electron mass) then, in 224-TMP, 2.4×10^{19} cm$^{-3} < N_0 < 4.2 \times 10^{19}$ cm^{-3} between 20 and 150°C.[68]

Recently, Itoh, Holroyd, and Nishikawa measured μ_H and μ_D in mixtures of n-pentane and neopentane (NP) up to 30% n-pentane, at temperatures between 20 and 150°C.[69] The addition of n-pentane molecules to NP lowers both μ_H and μ_D. In a 30% n-pentane solution at temperatures below 145°C, the values of μ_H are smaller by a factor of 4 to 6 than those found in pure NP at the same temperature. However, within the range of concentrations and temperatures covered in the experiment, it was found that the ratio μ_H/μ_D falls into the range 0.9 to 1.5. These results were interpreted as indicating that the addition of n-pentane molecules to NP results in a *more efficient electron scattering mechanism rather than in more effective electron traps.*[69]

In cases where evidence of traps has been found (NP close to the critical point, and 224-TMP in the temperature range between 20 and 160°C), the experiment indicates that μ_H and μ_D are of the same order of magnitude. Therefore, the residence time of electrons in the traps is comparable to the relaxation time $\tau = \mu \, (m^*/q)$ estimated by using for μ the value obtained for μ_H. In a sample exhibiting $\mu = 10$ cm^2/V $\cdot$ s and $m^* = m_0$, $\tau \sim 6 \times 10^{-13}$ s. Since the time scale involved in TOF experiments and in the measurement of the Hall effect ranges from a few microseconds to several hundred microseconds, the excess electrons undergo, on the average, a number of scattering and/or trapping events that is in the 10^7 to 10^9 range.

Once again, we find that liquids that otherwise exhibit very similar physical properties, display charge transport mechanisms that are appreciably different.

VI. OVERVIEW OF THEORETICAL WORK

The large variation of electron mobility observed among different liquids, the correlation between mobility and molecular sphericity, and the *change* in the charge transport mode observed in some liquids according to the thermodynamic state of the fluid (the outstanding example being the rare gas liquids), are all phenomena that have yet to be explained. Although we are not currently aware of any theory capable of explaining *all* of these features, a consideration of the role of disorder provides a qualitative answer to some of these questions. It also provides a qualitative understanding of the nature of the trapping mechanisms revealed by the measurement of the Hall effect in liquids such as 224-TMP.

A. ROLE OF DISORDER

It seems natural to begin by asking, *which of the concepts appropriate to crystalline solids can be used in non-crystalline materials,* and particularly in liquids?

The first concept, equally valid for crystalline and non-crystalline materials, is the density of states, $N(E)$.[70] The quantity $N(E) \, dE$ denotes the number of states per unit volume available for an electron with a given spin direction and with energy between E and $E + dE$.

A second concept central to the theory of crystalline solids is the concept of a band; that is, the spectrum of eigenenergies, $\varepsilon(\mathbf{q})$, corresponding to the eigenfunctions, $\varphi_q(\mathbf{r})$, satisfying Schroedinger's equation in a periodic (crystalline) potential. In an *ideal* crystal (i.e., a crystal *lattice without impurities or defects, and without phonons*) solutions of the Schroedinger equation can be found in the form of Bloch's functions; that is, functions of the type $\varphi_q(\mathbf{r}) = \exp(i\mathbf{q} \cdot \mathbf{r}) \, u_q(\mathbf{r})$, where $u_q(\mathbf{r})$ is a function displaying the periodicity of the lattice, and $\mathbf{q}$ is a wave vector of the reciprocal lattice. Bloch's functions are electron states extending throughout the crystal. By virtue of being eigenstates of a *Hamiltonian that* *already contains the periodic potential, an electron* represented by such functions (i.e., an excess electron in a dielectric crystal) is *no longer scattered by the mere presence of the atoms in the lattice; it is scattered by deviations of the potential from a perfect periodic potential.* Another way of saying this, is that the mean-free path of an excess electron in an *ideal* insulating crystal, is determined by the reflections of the electron wave function at the boundary of the crystal, thus leading to a large mobility at zero temperature.*

Whether or not this band concept is applicable to non-crystalline materials, and in particular to liquids, depends on the electron mean-free path ℓ. If ℓ is large, in the sense that $k\ell = (2\pi/\lambda)\ell \gg 1$ (where $\mathbf{k}$ is the wave vector associated with the excess electron, λ stands for its de Broglie wavelength, and $\ell = v_{TH}\,\tau$, with $\mu = |q|\,\tau/m^*$, here τ is the momentum relaxation time, $q = -e_0$ is the electron charge, v_{TH} is the velocity of the excess electron moving with a kinetic energy of the order of the thermal energy $k_B T$), then the band picture may be applied to the liquid. In such a case, the carriers move with well-defined velocities and undergo *occasional* scattering events, therefore, the wave vector $\mathbf{k}$ is still a good quantum number.[70]

Considerable evidence was gathered nearly 30 years ago by Ioffe and Regel,[71] pointing to the fact that in materials where the coordination number is preserved upon melting, the band structure observed in the solid is roughly preserved in the liquid. The evidence comes from a decrease in conductivity by about a factor of 2 upon melting, which has been observed in a number of metals and semiconductors. The decrease in electron mobility by roughly a factor of 2 upon melting, observed in the compounds listed in Table 1 as having spherical or nearly spherical molecules, is consistent with this data. Thus the experiment shows that the mechanisms that limit the electron mean-free path in crystalline materials do not become orders of magnitude more efficient upon melting; *the existence of a band of extended electron states does not require full long range order* (i.e., translational symmetry).[71,72]

The second possibility, $k\ell \sim 1$, implies that "scattering" events can no longer be viewed as occasional, isolated events; the itinerant electron gas picture breaks down, and k is no longer a good quantum number. As discussed below (Section VII), if the effective mass m* is set equal to the free electron mass m_0, then, at room temperature, condition $k\ell \sim 1$ corresponds to an electron mobility $\mu \sim 10$ cm^2/V · s; a larger mobility leads to $k\ell > 1$ and vice versa.

The third alternative, $k\ell < 1$, corresponds to some new phenomena not found in crystalline solids.[70] Condition $k\ell < 1$ implies that the electron mean-free path ℓ is comparable to or smaller than the molecular diameter; therefore, when this condition is satisfied, the *concept of electron mean-free path ℓ breaks down.* The interaction of the excess electron with the atoms making up the sample is strong enough to give rise to localized states at the bottom of the conduction band. As explained by Mott and Davis, "there is nothing unfamiliar about the concept of localized states; they are simply 'traps',... The new concept for amorphous materials is that a continuous density of states, N(E), can exist in which for a range of energies the states are all traps, or in other words localized, and for which the mobility at the zero of temperature vanishes, *even though the wave functions of neighboring states overlap.* Moreover, at the bottom of a conduction band or top of a valence band, such localized states must necessarily occur in a disordered material."[70]

We now turn to the question concerning the correlation between electron mobility and molecular sphericity, and the role of disorder. When molecules are brought together to form a dielectric fluid, the eventual formation of extended electron states requires, as a necessary

---

*   Of course, in the actual insulating crystals, deviations of the crystal lattice from a perfect lattice (arising from impurities, defects or thermal vibrations — phonons), give rise to an electron mean-free path which is much shorter than that determined by the reflections of the electron wave function at the boundary of the crystal. Therefore, the actual value of the electron mobility is determined by *these deviations* and *not* by the size of the crystal.

condition, the nonvanishing overlap of the wave functions representing an excess electron interacting with an isolated molecule; such extended states will only occur if overlap between these wave functions should take place over many neighboring molecules. In the rare gases, the wave function mentioned above is an s-state; in liquids composed of nearly spherical molecules, we expect this wave function to be almost isotropic. On the contrary, in liquids composed of elongated molecules, the corresponding molecular orbital is expected to be highly anisotropic, and a nonvanishing overlap between orbitals centered on two neighboring molecules will take place only if the molecules exhibit the "correct" relative orientation.

In contrast to the case of crystalline solids, neighboring molecules in a fluid where translational symmetry is nonexistent do not all exhibit the same orientation with respect to each other. Thus, in a fluid made up of elongated molecules, the probability of finding many neighboring molecules with the "correct" mutual orientation that would allow an appreciable overlap of these wave functions is small; therefore, in such systems the formation of extended electron states is hindered by orientational disorder. This is in contrast to the case of fluids composed of spherical or nearly spherical molecules, where overlap between these wave functions should occur, once the molecules are brought sufficiently close to each other, irrespective of their orientation.

This qualitative reasoning leads to the prediction that, when the density of a dielectric gas made up of molecules which do not attach electrons, increases toward the liquid density, the product of the low field electron mobility and the gas density will *increase over and above the value corresponding to the dilute gas limit* (thus signaling the appearance of extended electron states) if the molecules that constitute the gas are spherical, or nearly spherical, as in the case of TMS, NP, methane, and, of course, Ar, Kr, and Xe. On the contrary, such an effect *should not take* place if the gas is made up of elongated molecules.

As discussed in Section IV, these crude, qualitative predictions are in agreement with experiment. Therefore, the dependence of electron mobility on molecular sphericity can be understood as a consequence of orientational disorder.[73] The "absence" of extended electron states in liquids made up of elongated molecules can be interpreted (in terms of the concepts developed in the theory of noncrystalline materials) as localized states present at the bottom of the conduction band, which dominate the conduction process. The ground state energy of these traps has been calculated using the semicontinuum model, in which the interaction between the excess electron and the medium is represented by: (1) electron-point dipole and electron-point polarizability interactions with the molecules located in the first shell; and (2) a polarization potential describing the interaction of the electron with the fluid located beyond the first shell.[74-76]

B. ELECTRON TRANSPORT IN LIQUIDS DISPLAYING A HIGH μ_D

Attempts have been made to calculate the density and temperature dependence of the low field drift mobility, $\mu_o(N,T)$ (μ_o is the coefficient of proportionality between the drift velocity and the electric field, for fields low enough for ohmic behavior to be observed), and calculations have been made of the field dependence of the electron drift velocity $v_D(E)$ in liquid Ar, close to the triple point. Two of these theories have played a central role; they both used the Boltzmann transport equation (BTE).

Moreover, μ_H data became available very recently, and only in samples exhibiting a relatively high electron mobility (i.e., $\mu \geqslant 10$ cm^2/V $\cdot$ s). To date, no effort has been made to elucidate the significance of μ_H nor of the mobility ratio r = μ_H/μ_D in insulating liquids, in terms of the details related to the scattering mechanisms at work. Current understanding of μ_H and of the mobility ratio, r, is based on direct application of the standard theoretical framework developed in solid state physics, based upon BTE. Thus, a brief description of the formalism of the Boltzmann equation appears necessary.

1. Boltzmann Transport Equation

When charge carriers occupy extended states such that the mean-free path ℓ is long compared with the de Broglie wavelength λ, we may visualize the motion of these electrons as that of an electron gas characterized by a mass m* (the effective mass).

The effect of switching on the external fields **E** and **B** is to produce a shift in momentum space of the electron distribution function, f (to within a normalization factor, f equals the number of excess electrons per unit volume of phase space), from a Maxwellian centered at the origin of momentum space — that therefore carries no momentum — to another distribution that carries net momentum. Equilibrium is reached when the rate of change $(\partial f/\partial t)_{E,B}$ of the electron distribution function, f, due to the external fields **E** and **B** is equal in magnitude but opposite to the rate of change, $(\partial f/\partial t)_{COLL}$, due to collisions, so that the net rate of change of f is zero. When equilibrium prevails, the momentum and the kinetic energy gained by the electrons owing to acceleration between scattering events is lost via collisions. In this sense, collisions play a randomizing role.[13]

In a homogeneous sample, the rate of change of f due to the external fields can be written as[77,78]

$$\left(\frac{\partial f}{\partial t}\right)_{E,B} = \frac{q}{m^*}\,(\mathbf{E} + \mathbf{v} \times \mathbf{B})\left(\frac{\partial f}{\partial \mathbf{v}}\right) \tag{5}$$

In an insulator, the rate of change of f due to collisions can be written, following Boltzmann, as[78,79]

$$\left(\frac{\partial f}{\partial t}\right)_{COLL} = -\int [f(\mathbf{v})\,T(\mathbf{v},\mathbf{v}') - f(\mathbf{v}')\,T(\mathbf{v}',\mathbf{v})]d^3\mathbf{v}' \tag{6}$$

Here $T(\mathbf{v},\mathbf{v}')\,d^3\mathbf{v}'$ represents the transition probability per unit time that a carrier traveling with a velocity **v** before the collision, will emerge from it traveling with **v'**. As discussed by Overhauser and Huberman,[80] the operator defined by Equation 5 acts continuously on f. Moreover, the entities being scattered are assumed to behave as classical point particles. In order that **v** and **v'** be defined, the transition from **v** to **v'** must take place in a time that is *short* compared with the mean time between scattering events.

2. Effective Mass

At this point it seems appropriate to clarify the similarities and differences that emerge when the formalism outlined above is applied to describe the motion of excess electrons in gases and in liquids. In the dilute gas limit, the excess electron interacts only with one gas molecule at a time, and therefore m* can be replaced by m_0, the free electron mass. In the case of the liquid, since the de Broglie wavelength λ is long compared with the molecular diameter, the electron interacts simultaneously with many liquid molecules. The theoretical description must resolve this conflict between electrons which, on the one hand, display a large mobility (i.e., quasi-free electrons with a mean-free path $\ell > \lambda$), and, on the other hand, interact simultaneously with many molecules that in the gas phase are known to behave as quite effective electron scatterers. The solution to this puzzle resides in the existence of extended electron states, such that the excess electron is "shared" by many molecules.

In solid state physics, the concept of effective mass m* has been introduced to describe the motion of an electron occupying extended states (a band); it incorporates the effects of interference (Bragg reflection) arising from the multiple scattering of the electron off a collection of scatterers arranged in a periodic lattice. As discussed above, the constituents of this (fictitious) gas move about, undergoing *occasional* scattering events, which no longer arise because of the mere presence of the atoms in the lattice, but rather from *deviations* of the crystalline potential from a perfect periodic potential. The multiple scattering of the electrons off the atoms in a perfect lattice (i.e., a lattice without impurities or defects, and without phonons), determines only the *acceleration* that one electron would experience when acted upon by an external force; the ratio between the external force and the acceleration is the *effective mass m**.[81] The actual value of m* (and whether or not m* $= m_0$), depends entirely on the wave vector dependence of the eigenenergies $\epsilon(\mathbf{k})$ corresponding to the eigenfunctions that satisfy the translational symmetry required by the periodic potential, through the relation m* $= \hbar^2/ (\partial^2\epsilon(\mathbf{k})/\partial\mathbf{k}^2)$.[81] The curvature $(\partial^2\epsilon(\mathbf{k})/\partial\mathbf{k}^2)$ depends ultimately on the width of the conduction band; a small band width will give rise to a large effective mass.

In the theories discussed below, the authors borrowed the concept of effective mass m* from solid state physics, but set m* $= m_0$ probably because of lack of experimental information concerning this parameter.

3. Theory of Cohen and Lekner

Cohen and Lekner (CL)[82] developed a solution of BTE in an attempt to explain the low field data of Schnyders et al. at 84 K,[83] and the sublinear dependence of the drift velocity $v_D(E)$ on electric field E reported by Swan in liquid Ar close to the triple point.[84,85] Cohen and Lekner used a standard expansion[79] of the electron distribution function $f(\epsilon)$ in terms of Legendre polynomials $P_n\{\cos[\varphi(\mathbf{p},\mathbf{E})]\}$, of which they retained only the first two terms

$$f(\epsilon) = F_0(\epsilon) P_0 + F_1 (\epsilon) P_1 \{\cos[\varphi(\mathbf{p},\mathbf{E})]\} \tag{7}$$

where ϵ is the electron kinetic energy, and $\varphi(\mathbf{p},\mathbf{E})$ is the angle between the momentum $\mathbf{p}$ of the electron and the external field $\mathbf{E}$. The functions F_0 and F_1 are determined as solutions of a set of differential equations involving the mean-free path Λ_0 and Λ_1 for energy and momentum transfer, respectively; the latter are in turn determined by the structure of the liquid and the scattering cross section $\sigma(\epsilon,\theta)$ stemming from the electron-liquid interaction.

Lekner[86] applied this theory of hot electrons to the motion of excess electrons in liquid Ar. He described the electron-fluid interaction via the potential

$$U_\alpha (R) = - [\frac{\alpha q^2}{2 (R^2 + R_\alpha^2)^2}] g(R) \tag{8}$$

where, when $R_\alpha = 0$, the expression in square brackets represents the polarization energy of a test charge q immersed in a gas composed of molecules exhibiting a polarizability α. The adjustable parameter R_α was introduced to account for exchange and correlation effects; g(R) is a function involving the pair correlation function, introduced to describe the structure of the liquid.

When m* $= m_0$, the CL theory, containing only one adjustable parameter, R_α (the value of which was chosen to fit the scattering data of slow electrons by Ar atoms in the dilute gas limit), gave a calculated mobility only 20% lower than the experimental value at 84 K, and fitted the observed field dependence of the drift velocity reasonably well over the range $10^1 \le E \le 10^4$ V/cm. The sublinear dependence of $v_D(E)$ on E (corresponding to a *decrease* in $\mu_D(E)$ with increasing field E) is attributed to the excess electrons picking up more energy from the field during acceleration between scattering events than they can lose through essentially elastic collisions with the liquid. The net result is an increase of the electron kinetic energy above $(3/2) k_B T$, which in turn leads to stronger electron scattering, therefore to a smaller mobility. Since this theory predicted a low field mobility $\mu_0 \sim [a^2 S(0)]^{-1} \sim (a^2\chi_T)^{-1}$ (where a is the scattering length characterizing the electron-fluid interaction,

$S(0) = N k_B T\chi_T$ is the long wavelength limit of the structure factor, N is the density of the fluid, T is the absolute temperature, and χ_T is the isothermal compressibility), it provided an explanation for the decrease of ~2 observed in μ_D upon melting, as being essentially due to an increase in χ_T. It also predicted a value of $V_0 = -0.46$ eV at the triple point, more than double the value $V_0 = -0.19$ eV found recently (Figure 4b).

However, the apparent success did not prevail under close scrutiny. The largest discrepancy between theory and experiment persisted; at 84 K, and at fields $E > 10^4$ V/cm theory predicted a *decrease* in the drift velocity with increasing field,[86] whereas experiment indicated a nearly constant v_D.[21] The saturation of v_D in the rare gas solids and liquids is now well established, and in the liquids the measurement has been extended to higher fields (Figure 7a). The results reflect a poorly understood scattering mechanism between the excess electron and the liquid, rather than an experimental artifact due to impurities, as Lekner thought.[86]

The other crucial test is the prediction of the density and temperature dependence of the low field mobility μ_0. When new data became available,[39,87] it was clear that the theory failed to reproduce the sharp maximum observed experimentally (Figure 4a).

Subsequent calculations by Jahnke, Holzwarth and Rice,[88] indicated that Lekner's contention that the mobility maxima could be explained by a change of sign of the scattering length a with decreasing density (with a passing through zero at the density at which the mobility maximum occurs)[89,90] was incorrect; the scattering length calculated according to Lekner's method decreases only by about 20% when the liquid density decreases from the density at the triple point to the critical density. Furthermore, Gryko, and Popielawski found that Lekner's calculation was very sensitive to the choice of the potential representing the electron-fluid interaction, and when applied to liquid Kr, it resulted in a severely underestimated mobility.[91]

These calculations indicate that the failure of the CL theory does not arise from the formalism they propose to describe hot electron phenomena at high fields,[82] but rather from the inability of Lekner's effective pseudopotential model to represent the electron-fluid interaction adequately even at low fields, where the formalism provides an answer that is qualitatively correct (e.g., a drift velocity $v_D(E)$ that depends linearly on the strength of the electric field E).

4. Theory of Basak and Cohen

Basak and Cohen (BC)[50] developed a deformation potential theory to explain the density and temperature dependence of the low field mobility data in liquid Ar,[39] taking as a central feature of the theory the assumption that, in liquid Ar, the mobility maximum and the V_0 minimum occur at approximately the same density.* (At the time when the BC paper was published, this correspondence between conduction band minima and mobility maxima had been experimentally verified only in TMS and NP,[36] as shown in Figure 5). BC solved BTE by means of the standard relaxation time approximation, which consists of expressing the collision operator (Equation 6) in the following fashion[78,92]

$$\left(\frac{\partial f}{\partial t}\right)_{COLL} = -\frac{f - f_0}{\tau(\varepsilon)} \tag{9}$$

where f_0 is the equilibrium distribution function (a Maxwellian) when $\mathbf{E} = \mathbf{B} = 0$, and f is the equilibrium distribution function when $\mathbf{E} \neq 0$, $\mathbf{B} \neq 0$; $\tau(\varepsilon)$ is the energy-dependent

---

*   As shown in Figure 4, experiments proved this assumption to be correct not only in Ar but also in Kr and Xe (also in the case of methane), although in liquid Kr the mobility maximum occurs at a density somewhat higher than that corresponding to the V_0 minimum.

relaxation time. If the fields **E** and **B** were suddenly turned off, collisions would cause f to relax exponentially toward f_0, with a time constant equal to the relaxation time $\tau(\varepsilon)$.[78] Although BC did not include the magnetic field **B** in their calculation, we retain the generality of the treatment as the results will be used in the next section to discuss μ_H and the mobility ratio, r.

Within the relaxation time approximation, it can be shown that, in the weak field limit, f is given, to first order in **E** and **B**, by[78]

$$f(\varepsilon) = f_0(\varepsilon) \left[1 + \frac{q\tau}{k_B T} \mathbf{v} \cdot \mathbf{E} + \frac{q^2\tau^2}{m^* k_B T} \mathbf{v} \cdot \mathbf{E} \times \mathbf{B} \right] \qquad (10)$$

and thus the average velocity $\langle \mathbf{v} \rangle$ of the electron gas is[78]

$$\langle \mathbf{v} \rangle = \frac{q}{m^*} \left[\langle \tau \rangle \, \mathbf{E} + \frac{q}{m^*} \langle \tau^2 \rangle \, \mathbf{E} \times \mathbf{B} \right] \qquad (11)$$

where the average $\langle z \rangle$ of a scalar function $z(x)$ is defined by

$$\langle z \rangle = \frac{4}{3\sqrt{\pi}} \int_0^\infty z(x) \, x^{\,3/2} \exp(-x) \, dx \qquad (12)$$

Since $\tau \sim \mu \, (m^*/q)$, from Equation 10 it is apparent that, *for the relaxation time approximation to apply, the change in the electron distribution function f induced by the external fields E and B should be considered as a perturbation,* in the sense that it should be *small* compared to f_0, e. g. $|f - f_0|/f_0 << 1$.[78,79]

When $\mathbf{B} = 0$, Equations 11 and 12 can be written in the equivalent form $\mu = (2q/3m^* k_B T) \langle \tau\varepsilon \rangle_0$ utilized by BC, where the symbol $\langle \ \rangle_0$ indicates an average over a Maxwellian electron population, using as weighting function a density of states $\sim \varepsilon^{1/2}$, appropriate for a free electron gas.[50] To calculate the relaxation time τ, BC proposed an effective Hamiltonian to describe the electron-fluid interaction

$$H_{eff} = \frac{\mathbf{p}^2}{2m^*} + V_0 \left[n\,(\mathbf{r}) \right] \qquad (13)$$

where m^* is the electron effective mass and V_0 is the position of the bottom of the conduction band, which is assumed to depend on the local fluid density n(r). Electron scattering occurs when the electron moves from one region to another where the local density is different, thus resulting in a potential change

$$\Delta V_0\,(\mathbf{r}) = V_0[n(\mathbf{r})] - \langle V_0 \rangle = V_0' \, \Delta\, n\,(\mathbf{r}) + \frac{1}{2} V_0'' \, [\Delta\, n\,(\mathbf{r})]^2 + \frac{1}{6} V_0''' \, [\Delta\, n\,(\mathbf{r})]^3 \qquad (14)$$

where the primes indicate derivatives with respect to density and $\langle V_0 \rangle$ indicates the value of V_0 at the average density.

Using this Hamiltonian, BC derived a transition rate $T\,(\mathbf{v}, \mathbf{v}')$ independent of the kinetic energy, ϵ, of the carrier, and thus a momentum relaxation time inversely proportional to the density of states of a free electron gas, $\tau \sim \epsilon^{-1/2}$. The BC mobility is

$$\mu = \frac{q}{(m^*)^{5/2}} \frac{2\sqrt{2\pi}\hbar^4}{3 \, N^2 \, (k_B T)^{3/2} \, \chi_T \, [(V_0')^2 + f\,(\alpha,\beta,\gamma)]} \qquad (15)$$

where $f(\alpha,\beta,\gamma)$ is a function involving integrals of the structure factor $S(\mathbf{q})$ and derivatives of V_0.

Berlin, Nyikos, and Schiller[93] developed independently a theory along similar lines. When $f(\alpha,\beta,\gamma)$ is set to zero in the BC result, the mobility calculated by Berlin et al. differs from the BC mobility by a factor of $36/\pi^2$.

The detailed calculation of Reininger, Asaf, and Steinberger,[94] incorporating the wave vector dependence of the structure factor $S(\mathbf{k})$ on $\mathbf{k}$, indicates that the BC theory reproduces the main features of the experiment. Going from densities lower than the critical density N_C to densities close to the triple point density, it predicts a mobility which first decreases with increasing density, reaching a minimum in the neighborhood of N_C. As the density increases further, the mobility rises to a maximum, which occurs at approximately the same density at which the minimum in V_0 is observed, reflecting the fact that scattering is the weakest (and the denominator in Equation 15 is minimum) when $V_0' = 0$; a further increase in density produces a decrease of mobility towards the triple point. However, the predicted mobility maximum is much too sharp, and the mobility predicted at the critical density is severely underestimated, owing to the large value of χ_T close to the critical point.

5. Beyond Basak and Cohen

Attempts have been made to rescue the CL theory by improving the effective electron-liquid interaction potential and by including multiple scattering effects.[95] More refined calculations of hot electron behavior in the rare gas liquids near the triple point have been performed using CL formalism.[96,97] A threshold energy of about 10 eV (different for each liquid) for inelastic collisions between the excess electron and the liquid, as well as a detailed dependence of $\Lambda_0(\varepsilon)$ and $\Lambda_1(\varepsilon)$ (the mean-free path for energy and momentum transfer) on the electron kinetic energy ε, valid for collisions below threshold, were introduced into the theory.[97] The results suggest that: (1) the electron distribution function becomes skewed and anisotropic at high fields, such that the transverse diffusion coefficient is markedly larger than the longitudinal diffusion coefficient (in liquid Xe at 100 kV/cm, the predicted ratio of transverse to longitudinal diffusion coefficient is close to 100); (2) at fields of 10 kV/cm, the average electron kinetic energy is already of the order of 1 eV. However, in liquid Ar and Xe, the calculation predicts a *decrease* of the electron drift velocity with increasing field strength, at fields in excess of about 40 kV/cm; in liquid Kr, this decreasing trend sets in at fields larger than 5 kV/cm. Nevertheless, a further increase of electric field beyond 300 kV/cm *reverses* this dependence, leading to a drift velocity that *increases* with increasing electric field. In order to recover a drift velocity that remains constant with increasing field, a (modified) structure factor that depends linearly on the kinetic energy of the electron must be introduced into the theory; such dependence *is in contradiction with experimental data on the structure factor* $S(\mathbf{k})$.[97]

A Monte Carlo simulation of hot electrons in liquid Ar has recently been reported; it also predicts a *decrease* of the electron drift velocity with increasing field at fields E $>$40 kV/cm.[98] A numerical integration of BE in a form similar to that proposed by CL has been reported,[99] introducing an inelastic scattering channel with a threshold energy of about 2 eV; the calculated electron drift velocity as a function of electric field in liquid Ar agrees reasonably well with experiment.

A modified version of the BC theory has been proposed for liquid hydrocarbons, and for liquid Ar[100] and liquid Xe.[101] Changes in V_0 due to density fluctuations are modeled as 3-D square barriers and square wells, whose radii is considered as an adjustable parameter to fit the experimental data. Calculation of the scattering cross section proceeds according to the method of partial waves, in order to circumvent the limitations of the first Born approximation used by BC.[100,101] A substitution of the isothermal by the adiabatic compressibility has also been proposed;[102] it transforms the main term of the BC theory in the mobility that would be expected from acoustic phonon scattering. Concerning the electron mobility near the critical point, calculations of mobility in nonpolar fluids have been reported,[103-107] and the formation of a polaron has been discussed.[108,109]

Calculations of the electron drift velocity in liquid Ar using the method of molecular dynamic simulation have been performed,[110,111] starting from a diametrically opposite point of view, namely, that electron transport does not occur via a conduction band but rather as a collision assisted electron transfer from atom to atom. This approach has been the object of debate;[112,113] it is inconsistent with the evidence discussed above, pointing to the existence of a conduction band in the liquid.

A calculation of electron mobility in liquid Ar that considers the fluid as a mixture of dense regions acting as high mobility channels and less dense regions acting as low mobility channels for electron propagation, has been reported; the mobility maximum is due to regions of density higher than the average and which are stabilized and sustained by the presence of the excess electron due to the attractive electron-fluid interaction.[114]

Recently, an extension of BC theory was developed, and mobility calculations were reported for liquid Ar[115] and liquid Xe.[116] Another mechanism was added to the scattering mechanism arising from static density fluctuations proposed by BC: that of scattering by acoustic phonons; that is, density fluctuations which propagate as a sound wave through the liquid. Calculation of the scattering cross section associated with static density fluctuations proceeds according to the method of partial waves. In the case of Ar, a linear dependence of the effective mass $m^*(N)$ on density N was assumed; in the case of Xe, the effective mass was considered equal to the value $m^* = 0.27\ m_0$ reported near the triple point.[117] In Xe, agreement between theory and experiment seems good at low densities, but discrepancies of about a factor of 2 appear at densities 20% lower than that of the triple point.[116] In Ar the opposite occurs: theory and experiment agree within 20% at densities close to that of the triple point, but theory underestimates the mobility near the critical point.[115]

C. ELECTRON TRANSPORT IN LIQUIDS DISPLAYING LOW AND INTERMEDIATE μ_D

Two main models have been put forward to explain the temperature dependence of the excess electron mobility observed in those liquids where an Arrhenius-type law has been found.

1. The Two-State Model

Minday, Schmidt, and Davis proposed the so-called two-state model.[15,118] According to this model, the excess electrons alternate between extended states and shallow traps. The observed mobility is then given by

$$\mu = \frac{\tau_F}{\tau_F + \tau_T}\ \mu_F + \frac{\tau_T}{\tau_F + \tau_T}\ \mu_T \tag{16}$$

where τ_F denotes the mean time spent in the quasi-free or conducting state between two trapping events, τ_T denotes the mean time spent in a trap; μ_F and μ_T stand for the mobility of the electron in the extended and trapped states, respectively. The assumption that $\tau_F\mu_F >> \tau_T\mu_T$ leads to $\mu \sim \mu_F/[1 + (\tau_T/\tau_F)]$. Assuming further that $\tau_T/\tau_F >> 1$ and that the residence time in a trap is temperature dependent, $\tau_T = \tau_0 \exp(E_A/k_BT)$, leads to

$$\mu = \mu_F \left(\frac{\tau_F}{\tau_0}\right) \exp\left(- E_A / k_BT\right) \tag{17}$$

The mobility μ_F in the extended state may also be temperature dependent. However, over a limited temperature range, the exponential dependence will dominate, thus leading to a straight line in the Arrhenius plot. From measurements of the temperature dependence of the mobility alone, no information other than the preexponential factor and the activation energy E_A can be obtained about the parameters of this model.

Application of an electric field, E, leads to a voltage drop across the trap diameter, d, thus reducing the trapping energy and leading to a superlinear dependence of $v_D(E)$ on E and therefore to an increase in mobility given by[26]

$$\mu(E,T) = \mu(0,T) \exp(q E d / k_B T) \tag{18}$$

where $\mu(0,T)$ is given by Equation 17.

2. Hopping Models

Another model that does not invoke the participation of extended states has been proposed[26] to explain the mobilities following an Arrhenius law. According to this model, charge transport occurs when the excess electron jumps over a barrier separating one trap from another. Denoting by Λ the mean jump distance, and by v, the attempt frequency, the diffusion coefficient is $D = \Lambda^2 v$. Assuming that the attempt frequency is an activated process, i.e., $v = v_0 \exp(-E_A/k_B T)$, and applying the Einstein relation $\mu = q D/k_B T$, leads to

$$\mu(0,T) = \frac{q}{k_B T} \Lambda^2 v_0 \exp(-E_A / k_B T) \tag{19}$$

The mobility increases with electric field according to[26]

$$\mu(E,T) = \mu(0,T) \frac{\sinh(q \Lambda E / 2 k_B T)}{(q \Lambda E / 2 k_B T)} \tag{20}$$

The small polaron model leads also to an Arrhenius-type activated mobility at low fields, and to a field dependent mobility as indicated by Equation 20 at high fields. According to this model (originally developed to explain electron motion in molecular crystals), an excess electron moving through a lattice may become trapped by the cloud of induced polarization which surrounds it; the resulting entity is known as a small polaron (small, because the perturbation of the lattice does not extend far beyond the nearest neighbors). Electron transport occurs when the electron hops to another site that exhibits a molecular configuration similar to that displayed by the original trapping site; this "coincidence" between molecular configurations results from thermal fluctuations. For this model to apply, the residence time of an excess electron in a trap should be short compared with the tunneling time across the barrier separating the two sites, but long compared with the jump time. The low field drift mobility associated with small polaron hopping motion can be formally obtained from that indicated by Equation 19, if the frequency v_0 is replaced by $\sqrt{\pi}|J|^2/[2\hbar(E_a k_B T)^{1/2}]$, where J is the transfer integral between two neighboring sites and E_a is the activation energy of μ_D.[26] This model has been applied to ethane and propane;[26] it is further discussed below (Section VII).

3. Other Models

Schiller and co-workers developed a general theory of electrons in insulating non-polar liquids.[32,93,119,120] For this purpose, the concept of *electron localization due to energy fluctuations* in the liquid was introduced. The probability, P, of finding a region capable of localizing one excess electron is given by

$$P = \frac{1}{\sqrt{2\pi}\Sigma} \int_{-\infty}^{V_0} \exp\left[-(E - E_t)^2 / 2\Sigma^2\right] dE \qquad (21)$$

where E_t is the energy of the localized state, V_0 is the energy of the extended state, and Σ is a parameter describing the distribution of energy fluctuations. The observed drift mobility is then

$$\mu_D = \mu_F (1 - P) \qquad (22)$$

where, as before, μ_F denotes the mobility of the extended state, and the contribution arising from the trapped state has been neglected. According to this model, the prediction of the density and temperature dependence of the low field mobility requires knowledge of the full dependence of V_0 (N) on N, and of E_t (N, T) on N, T. Since P depends on V_0, a correlation between μ and V_0 should exist. Indeed, such a correlation between the values of μ and V_0 measured at room temperature has been empirically derived:[121]

$$\mu = \frac{125}{1 + 360 \exp (15\ V_0)} \qquad (23)$$

However, some data fit within this classification pattern based on the calculation of the localization probability (Equation 21), while others do not.[119,120] The same is true regarding an alternative formulation of this theory proposed by Freeman and co-workers,[122,123] and concerning a *percolation theory* proposed by Kestner and Jortner.[124] Schiller and co-workers have also proposed a percolation theory to explain electron mobility in liquid mixtures;[125,126] the model has been shown to provide reasonable agreement with experimental data.

Nevertheless, the recent measurements of μ_H described in Section V cast doubts on the validity of the assumptions underlying some of these models (involving either nearly free electron states coexisting with trapped states, or regions which are "transparent" to electron propagation coexisting with regions which are "opaque") when applied to the pure liquids, namely: (1) that *all* insulating nonpolar liquids can be described either as a mixture of regions acting as high and low mobility channels for electron propagation,[124] or as an ensemble of trapped and quasi-free electrons;[119,120,122,123] and (2) that the difference between liquids exhibiting high, intermediate, or low electron mobility is mostly due to the relative importance of the regions that are transparent to electron propagation, as compared with those that are opaque (or between the population of the quasi-free and trapped states, in the case of the two state model). Within this theoretical framework, the Hall effect singles out the quasi-free states, since the observed Hall mobility is essentially that characterizing the charge transport process via the high mobility channels or via the conduction band; this point is discussed further in the next section.

According to these models, since *n*-pentane exhibits an electron mobility which is some two orders of magnitude lower than that observed in neopentane (NP)—see Table 1—we would expect that the channels which are opaque to electron propagation play a role which is significantly more important in *n*-pentane than in NP, and that the electron mobility characterizing regions which are transparent to electron propagation is comparable in these two liquids. Therefore, we may expect that the addition of *n*-pentane molecules to NP should result in an increase in the density of traps (or regions that are opaque to electron propagation), over and above those that might already exist in pure NP. Consequently, mixtures of *n*-pentane and NP should exhibit a drift mobility μ_D that decreases with increasing *n*-pentane concentration, but a Hall mobility μ_H that is similar to that observed in pure NP at the same temperature. As mentioned in Section V, the experiment indicates that, on the contrary, addition of *n*-pentane to NP results in a *decrease of both μ_H and μ_D* with increasing *n*-

pentane concentration; the ratio μ_H/μ_D falls in the range of 0.9 to 1.5 at temperatures $20°C < T < 145°C$, in spite of the fact that at 30% n-pentane concentration, the values of μ_H are already a factor of 4 to 6 smaller than those observed in pure NP.[69]

If the charge transport process involves traps as well as extended states in NP and in n-pentane, then the experiment indicates that in n-pentane-NP mixtures these traps play a role which is *similar* to that in pure NP. Indeed, the experimental findings suggest that adding n-pentane to NP results in a *more efficient electron scattering mechanism rather than in more effective electron traps; the excess electrons behave as if they remained in a quasi-free state.*[69]

Moreover, in 2244-TMP, 22-DMB, and TMS, the observed μ_H at room temperature coincides with μ_D, indicating that *traps* (or opaque regions for electron propagation) *do not play a significant role* (Figures 11 and 12). And yet, the electron mobility observed in TMS at room temperature is 10 times that observed in 22-DMB. As discussed below, the mobility at room temperature in 22-DMB is $\mu \sim 10$ cm^2/V · s, a value that is perhaps uncomfortably small for appropriately describing charge transport through extended states or via regions which are transparent to electron propagation.

4. Path Integral Calculations

Recently, a new theoretical approach was developed to study the behavior of excess electrons in fluids.[127-130] This novel technique uses the path integral method proposed by Feynman to calculate the properties of an electron in a polar lattice (polaron).[131] Relevant quantities such as the electron partition function turn out to be isomorphic to a ring polymer where the elements that are nearest neighbors in the ring are connected by harmonic potentials. Calculations on the polymer yield information about the behavior of the excess electron. A significant advantage of this method is that it is able to treat the role of solvent fluctuations self-consistently for both extended and localized electrons, within a single perspective.[128]

Numerical calculations have so far been performed in a fluid of hard spheres of diameter σ and density ρ.[128] The excess electron is assumed to be in equilibrium with the solvent; it is excluded from the region occupied by the hard spheres. The results indicate that as the solvent density increases the electron ring polymer is compressed, and at high enough densities the excess electron becomes localized. The localization transition takes place at fluid densities between $\rho = 0.10$ σ^{-3} and $\rho = 0.15$ σ^{-3}. An increase in temperature produces both an increase of the density at which localization occurs, and a broadening of the transition region. The results of this calculation have been crosschecked by comparing them with a Monte Carlo simulation of an excess electron in a hard sphere fluid;[132] both methods yield results that agree qualitatively.

To calculate the mobility (indeed, the diffusion coefficient), the formalism described above (which is appropriate for the calculation of equilibrium properties) was extended in order to determine the temporal development of the relevant quantities. The slope of the mean square displacement of the excess electron at long times, yields the diffusion coefficient.

Numerical calculations performed in the hard sphere solvent indicate that the electron diffusion constant undergoes a drop of more than two orders of magnitude, approaching the diffusion constant of the hard sphere solvent, when the density of the fluid increases beyond $\rho = 0.3$ σ^{-3}. The effect of increasing temperature is twofold. First, a reduction in the de Broglie wavelength, λ, of the excess electron causes the transition from high to low mobility (diffusion coefficient) to take place at higher densities. Second, at still higher densities (where the mobility has already dropped to a low value) the mobility increases with increasing temperature, a behavior reminiscent of the activated mobility found in a number of hydrocarbon liquids. Unfortunately, detailed comparison between theory and experiment cannot be made until more realistic models of the fluid are used in the calculation.

VII. THEORETICAL INTERPRETATION OF THE HALL MOBILITY AND OF THE MOBILITY RATIO μ_H/μ_D

At this point, we might pause and ask why should μ_H be different from μ_D? The answer is: μ_H is defined as the coefficient of proportionality between the average velocity $\langle v \rangle$ and μ_D $\mathbf{E} \times \mathbf{B}$, while μ_D is defined as the coefficient of proportionality between $\langle v \rangle$ and $\mathbf{E}$ (Equation 1). As pointed out in Section II, μ_H and μ_D have the same dimensions as a consequence of the Lorentz equation. They do, however, represent independent quantities determined by completely different experiments; thus they *do not need to be equal*.

A. LIQUIDS EXHIBITING A HIGH ELECTRON MOBILITY

The theoretical interpretation we present below is based on the validity of the band concept and on the solution of BTE according to a particular perturbation method, the relaxation time approximation. We refer to Equations 1 and 11, from which it follows[133]

$$\mu_D = \frac{q}{m^*} \langle \tau \rangle \tag{24}$$

$$\mu_H = \frac{q}{m^*} \frac{\langle \tau^2 \rangle}{\langle \tau \rangle} \tag{25}$$

$$r = \frac{\mu_H}{\mu_D} = \frac{\langle \tau^2 \rangle}{\langle \tau \rangle^2} \tag{26}$$

The particular case in which the momentum relaxation time, τ, can be described by a simple power law dependence on the kinetic energy, ϵ, of the carriers, $\tau(\epsilon) = \lambda \, \epsilon^\alpha$ (λ and α independent of ϵ), leads to[133]

$$r = \frac{\Gamma(5/2) \, \Gamma \, (2\alpha + 5/2)}{(\Gamma \, (\alpha + 5/2))^2} \tag{27}$$

Among the scattering mechanisms known in solid state physics, there is acoustic phonon scattering,[134,135] and scattering by dislocations of the lattice,[134] which, as in the scattering mechanism proposed by BC, lead to $\alpha = -1/2$ and $r = 3\pi/8 = 1.18$. For scattering by neutral impurities[134] and by nonpolar optical phonons,[135] when $\epsilon << \hbar\omega_0$ (ω_0 being the optical phonon frequency at zero wave vector), $\alpha = 0$ and $r = 1.0$. In the case of scattering by ionized impurities,[134] $\alpha = 1.5$ and $r = (315/512) \, \pi = 1.93$.

However, a word of caution is necessary. These answers have been obtained using BTE; that is, by assuming that the excess electrons undergo occasional scattering events, and thus *the observed changes in μ_H and μ_D reflect changes in the momentum relaxation time τ. The presence of traps may invalidate some of these conclusions.*

If traps are characterized by a mean residence time, τ_T, and scattering events are characterized by a relaxation time, τ, we expect that traps would affect the measurement of μ_D when $\tau_T \gtrsim \tau$. In such a case, BTE may need to be modified, because the operator describing the action of external fields on the electron distribution function f (Equation 5) is assumed to act continuously,[80] while the *trapping and detrapping process introduces of necessity a delay τ_T between the absorption and reemission of the carrier.* This delay is *not* incorporated in BTE (it is incorporated neither into the operator describing the effect of collisions on f, Equation 6, nor in the relaxation time approximation, Equation 9, since collisions are assumed to take place in a time that is *short* compared with τ. We return to this point below.

In the particular case where the liquid is considered as a mixture of regions acting as high and low mobility channels for electron propagation, or as an ensemble of trapped and quasi-free electrons, trapping (or propagation through low mobility channels) *increases* the total transit time of the carriers between the measuring electrodes, and thus, *decreases* the observed μ_D with respect to what would be measured if trapping did not take place. Therefore the *drift mobility μ_D is determined by scattering and trapping processes.*

On the other hand, the transverse deflection produced by the magnetic field **B** (Equation 1) is proportional to μ_H. The contribution to the Hall signal arising from trapped and extended states will be [P μ_H(Tr) μ_T + (1 − P) μ_H(CB) μ_F] **E** × **B**, where P is the probability of electron localization, μ_F and μ_H(CB) are the microscopic and Hall mobility corresponding to the extended state (conduction band), and μ_T and μ_H(Tr) are the microscopic and Hall mobility corresponding to the trapped state. The mobilities μ_F and μ_H(CB) are what would be measured in a time of flight and in a Hall experiment if there were not traps; the mobilities μ_T and μ_H(Tr) are what would be measured in a sample that exhibited only traps and no extended states. Within the two state model, the mechanisms involved in charge transport from trap to trap are assumed to be significantly less efficient than charge transport via the conduction band, therefore P μ_H(Tr) μ_T << (1 − P)μ_H(CB) μ_F. Hence the contribution to the Hall signal reduces to (1 − P) μ_F μ_H(CB) **E** × **B** = μ_D μ_H(CB) **E** × **B**, since (1 − P) μ_F = μ_D is the actual drift mobility (Equation 22).[68] Thus, *in spite of traps, the observed Hall mobility is the Hall mobility characterizing the charge transport process in the conduction band.* A similar argument holds for the model where the liquid is considered as a mixture of transparent and opaque regions for electron propagation. The conclusion is that, in contrast to the drift mobility, the *Hall mobility is determined by scattering processes only.* The reason is that, within this conceptual framework, for a carrier to be acted upon by **B** it *must be in motion; it must have a velocity different from zero.*

Summarizing: in samples exhibiting a high electron mobility, a mobility ratio significantly larger than unity and, in particular, a decrease in μ_D accompanied by a relatively constant μ_H that takes place over certain temperature range, can be interpreted as evidence of traps. This is the case of NP close to the critical point (Figure 11). On the contrary, a mobility ratio coinciding with one of the ratios discussed above over a wide temperature range, can be used as a powerful tool for identifying the scattering mechanisms at work. Such is the case of TMS, where the mobility ratio stays constant at 1.0 ± 10% between 22 and 164°C; this is consistent with an admixture of the scattering mechanism proposed by BC and optical phonon scattering.[13] Molecular vibrations in TMS provide an inelastic electron scattering mechanism that should be the analog of optical phonon scattering in crystalline solids; Raman active molecular vibrations have been invoked to explain the saturation of the electron drift velocity observed in TMS, at fields E >180 kV/cm.[44]

The preceding discussion is justified in liquids where $k\ell$ >1. At room temperature, when m* = m_0, the de Broglie wavelength of a thermal electron, is λ = h/(3m* k_BT)$^{1/2}$ ~ 60 Å. Under the same conditions, the electron mean-free path in a sample where μ = 100 cm^2/V · s, is ℓ = (μ/q) × (3m* k_BT)$^{1/2}$ ~ 70 Å. Therefore, the condition $k\ell$ >1 seems satisfied in NP and TMS, whereas in 2244-TMP, 224-TMP and 22-DMB it is perhaps marginally satisfied. Application of the above-mentioned theoretical framework leads to the identification of intrinsic traps in 224-TMP throughout the temperature range from 22°C to 160°C (Figure 12).

B. COMPOUNDS EXHIBITING A LOW ELECTRON MOBILITY

Although at present μ_H has been measured only in samples exhibiting a relatively high electron mobility (i.e., μ >10 cm^2/V · s), it seems appropriate to complete this discussion by briefly addressing the question concerning the Hall effect in materials that exhibit a low electron mobility. From the experimental point of view, the definition of μ_H and μ_D based on Equation 1, although inspired on the itinerant electron gas picture, may be extended to samples displaying a low electron mobility.

Regarding the theoretical interpretation, in view of the discussion presented above (where the Hall effect is due to the Lorentz force acting on *carriers that are in motion*), and given that the experimental evidence indicates that charge transport in liquids exhibiting a low electron mobility involves localized states (for which a carrier velocity is a questionable concept), we may start by asking *what is the nature of the processes that might produce a Hall voltage*? An answer to a similar question has been found in the case of hopping conduction in molecular crystals;[136] it may be relevant to the present discussion because small polaron hopping conduction is probably one of the mechanisms responsible for charge transport in liquids exhibiting a low electron mobility.

As shown by Friedman and Holstein,[136] the effect of the magnetic field $\mathbf{B}$ can be described (in the weak field limit) as a change of phase of the wave function, so that if $\phi(\mathbf{r} - \mathbf{g})$ stands for the wave function representing the excess electron (localized at the crystal site designated by the lattice vector $\mathbf{g}$) when $\mathbf{B} = 0$ then, when $\mathbf{B} \neq 0$, the corresponding wave function is $\exp\left[(-iq/2\hbar)\,\mathbf{B} \times \mathbf{g} \cdot \mathbf{r}\right]\phi(\mathbf{r} - \mathbf{g})$. The exponential term is the Peierls phase factor. The probability per unit time that an electron localized at site $\mathbf{g}$ performs a jump to site $\mathbf{g} + \mathbf{h}$, is proportional to the transfer integral $J_{\mathbf{g},\mathbf{g}+\mathbf{h}} = \int \phi^*(\mathbf{r} - \mathbf{g})U(\mathbf{r} - \mathbf{g})\phi(\mathbf{r} - \mathbf{g} - \mathbf{h})\,d^3\mathbf{r}$, where $U(\mathbf{r} - \mathbf{g})$ is the contribution of the molecule located at $\mathbf{g}$ to the effective one electron potential. This transition rate is modified by the presence of the magnetic field $\mathbf{B}$, because of the Peierls phase factor.

To first order in $\mathbf{B}$, processes involving direct transitions between two neighboring sites (where the molecular configurations are similar), do not contribute. However, when processes involving direct and indirect transitions between nearest neighbors are added (i.e., processes of the type $(A \rightarrow B \rightarrow C) + (A \rightarrow C)$, where A, B, and C are mutually nearest neighbors *displaying similar molecular configurations*), the result is a net electron drift along the direction of $\mathbf{E} \times \mathbf{B}$, reminiscent of the classical Hall effect due to itinerant electron motion. As stated in Reference 136, this effect stems from "the interference between the amplitude for a direct transition between the initial and final states, and the amplitude for an indirect, second order transition, involving intermediate occupancy of the third site". The resulting μ_H can be described by an Arrhenius-type function involving a preexponential factor that is weakly temperature dependent. The activation energy of μ_H is $\varepsilon = \varepsilon_3 - \varepsilon_2$, where ε_3 is the activation energy associated with the indirect hop involving a triple coincidence of molecular configurations, and ε_2 is that associated with the direct hop (two site coincidence).[136,137] On the other hand, as discussed in Section VI, the effect of the electric field $\mathbf{E}$ alone is to modify the direct transition rates between two neighboring sites, producing a net electron drift in the direction of $\mathbf{E}$ leading, in the weak field limit, to an Arrhenius-type drift mobility, except for a preexponential factor which is weakly temperature dependent;[136,137] the corresponding activation energy is ε_2.[136]

Two interesting consequences follow from mechanisms similar to the one outlined above, involving second or higher order indirect transitions between the initial and the final state. First, a mobility ratio *larger than unity* can occur. Moreover, μ_H may exhibit an Arrhenius-type temperature dependence, with an activation energy *smaller than that corresponding to* μ_D. Second, in spite of the fact that charge transport takes place via *electron hopping*, since electron drift in the direction of $\mathbf{E} \times \mathbf{B}$ turns out to be proportional to a product of transfer integrals, *the sign of the Hall voltage may be opposite to that found in the case of electronic conduction via band motion**.[137,138]

A charge transport mechanism analogous to that outlined above may be expected in compounds that exhibit a low activated (Arrhenius-like) electron mobility. An upper limit for the expected mobility at room temperature may be obtained by considering a vibrational

---

* Recall that the Lorentz force acting on positively and negatively charged carriers produces a Hall voltage of opposite sign.

frequency $\nu_0 \sim 10^{13}$ Hz, and a jumping (nearest neighbor) distance of $\Lambda \sim 5$ Å, which leads to a preexponential factor $\mu = (q/k_BT)\Lambda^2\nu_0 \sim 1$ cm^2/V · s.

VIII. CONCLUDING REMARKS

The preceding sections point to the fact that experiments have evolved considerably ahead of theory. In this section we summarize the evidence concerning some of the fundamental problems that await an answer.

A. HOT ELECTRONS IN THE RARE GAS LIQUIDS

One of the outstanding problems that has yet to be solved is that regarding the nature of the processes that lead to a saturation of the electron drift velocity in the rare gas solids and liquids. If we follow Schockley's model,[139,140] a transition from ohmic to nonohmic behavior (i.e., $v_D(E) \sim E^{+1/2}$), requires that the carriers pick up more energy from the electric field, E, between scattering events than they can lose by essentially elastic collisions with acoustic phonons, hence their average kinetic energy increases above $(3/2)$ k_BT. However, application of this model to describe the sublinear electric field dependence of the electron drift velocity $v_D(E)$ observed in the rare gas solids and liquids, raises some conceptual difficulties.

First, based on the classical case of hot electron behavior in Ge, we would expect that deviations from Ohm's law become important when the kinetic energy associated with the electron drift velocity, is of the order of k_BT. However, an evaluation of the ratio $R = m^* v_s^2/2k_BT$ (where v_s stands for the saturation drift velocity), using for m^* the values determined experimentally via measurements of photoconductivity close to the triple point ($m^* = 0.54$ m_0 in liquid Ar;[141] $m^* = 0.27$ m_0 in liquid Xe),[117] leads to R ~ 0.01 in liquid Ar at 87 K, and R ~ 0.0003 in liquid Xe at 163 K. A similar argument has been used by Ascarelli to challenge the interpretation of the sublinear field dependence of $v_D(E)$ on E as being due to hot electron behavior,[73,142] proposing instead an explanation invoking the effect of the electric field E on the thermal ionization of shallow traps, known as Schottky ionization. This proposition generated a wide polemic.[143-146]

The alternative explanation requires, either (1) the existence of intrinsic traps in liquid Ar, Kr, and Xe close to the triple point, or (2) extrinsic traps (i.e., impurities). In view of the discussion presented in Section VI about the role of disorder, intrinsic traps would imply the existence of an *electron trapping mechanism of unknown nature* that dominates the conduction process in the pure liquid near the triple point; extrinsic traps cannot be entirely excluded. However, measurements of $v_D(E)$ performed in several laboratories produced rather consistent results in the pure liquid. If the sublinear field dependence of $v_D(E)$ on E is caused by Schottky ionization of extrinsic traps then, since the effect depends on the concentration of traps it would imply that, in all of these cases, *the experimenters found similar concentrations of the same kind of impurities,* a peculiar coincidence difficult to explain. Moreover, as pointed out by Schmidt et al.,[145] there is independent experimental evidence that excess electrons in the rare gas liquids behave as if an important fraction of them did have a kinetic energy larger than $(3/2)$ k_BT. The evidence, not readily explained by the model involving Schottky ionization, arises from: (1) measurement of the electron mean energy in liquid Ar, as a function of the electric field strength;[147] (2) electron multiplication by impact ionization observed in liquid Xe;[148] (3) electric field dependence of the luminescence induced by α particles observed in liquid Xe;[149] (4) effect of nonattaching solutes in liquid Ar, Kr, and Xe on the electron drift velocity at high electric field strength,[41] (5) observation of a field dependence of the rate constant for electron scavenging by SF_6, N_2O, and O_2 in liquid Ar and Xe.[150,151]

There is another questionable assumption related to the application of Schockley's model to the rare gas solids and liquids. To explain the actual saturation of the electron drift velocity observed at high fields (as opposed to a monotonous increase, $v_D(E) \sim E^{+1/2}$), the model requires the action of a second *inelastic* scattering mode, capable of absorbing the kinetic energy of the electrons once it exceeds a certain threshold value; such a mechanism is the emission of optical phonons.[139,140] Nevertheless, phonon splitting into acoustic and optical branches (i.e., phonons where neighboring atoms vibrate 180°C out of phase rather than in phase as in the case of acoustic phonons, and that have an energy dispersion curve $\varepsilon(\mathbf{k})$ vs. $\mathbf{k}$ which exhibits a finite or threshold energy $\hbar\omega_0$ at zero wavevector $\mathbf{k} = 0$, rather than a linear dependence on $\mathbf{k}$ characteristic of acoustic phonons),[152] takes place only in crystalline structures that accommodate at least two atoms per primitive unit cell.[153] The rare gases crystallize in a FCC lattice,[154] with only one atom per primitive unit cell.[155] Therefore, there is no optical branch in the rare gas crystals,[156] and it is difficult to justify the existence of such modes in the liquid.

Molecular vibrations with energies in the range of ~ 0.1 eV or below may provide an inelastic electron scattering mechanism in the liquid which should be the analog of optical phonon scattering in crystalline solids; such scattering processes have already been invoked to explain the constant drift velocity observed in TMS at fields E $>$180 kV/cm.[44] This mechanism is responsible for the increase in the electron drift velocity observed in the rare gas liquids (over and above the velocity measured in the pure liquid at the same field strength) when doped with small concentrations of molecular solutes.[41] Nevertheless, such vibrations do not exist in the pure (noble gas) liquids, and the first excited state is in the range of ~ 10 eV, a few eV below the ionization energy.

In the rare gas liquids, the energy sink available that involves the least amount of energy is the first excited state. However, once the electrons have picked up sufficient energy from the field to reach the threshold for excitation of electronic levels, there should be enough electrons of higher energies in the tail of the distribution to produce collisional ionization. Since in these liquids the first excited state (and the threshold for photoionization) is of the order of ~ 10 eV, this would correspond to a saturation velocity of the order of $\sim 10^8$ cm/s, two orders of magnitude larger than observed — unless the electron distribution function becomes skewed and anisotropic at large fields,[97] or unless the effective mass* increases significantly with increasing electron kinetic energy.[156]

In the rare gas solids, deviations from Schockley's theory have been found at intermediate fields, leading to actual saturation of the drift velocity at large fields.[156] A change of effective mass m* as a function of energy, arising from the nonparabolicity of the band, has been suggested as an explanation of these results.[156] Some calculations of hot electron behavior in liquid Ar lead to an electron drift velocity which *decreases* with increasing field, for fields E $\geq$ 40kV/cm,[97,98] contrary to the constant velocity displayed in Figure 7a. The calculation performed by Sakai, Nakamura, and Tagashira,[99] including an inelastic scattering cross section with a threshold energy of about 2 eV, and the agreement between the calculated and the experimental results, suggests that such a mechanism is, indeed, needed to explain the data in the rare gas liquids. However, a derivation of this cross section, starting from the interaction potential between the excess electron and the liquid, is lacking. Thus, the nature of the scattering process responsible for the saturation of the electron drift velocity in the rare gas solids and liquids has yet to be understood; the explanation of this effect in the solid may help to clarify the situation in the liquid.

B. INTRINSIC TRAPS IN NEOPENTANE NEAR THE CRITICAL POINT

A second puzzle yet to be solved is the striking difference observed close to the critical point in the mobility ratio of excess electrons in NP and in TMS, indicating the presence of intrinsic traps in the former but not in the latter. Resonant scattering was suggested as the reason for the behavior observed in NP.[66,67] The idea is simple: resonant scattering should occur whenever the actual potential well is such that the reflection of the wave function produced at the boundary of the well, results in constructive interference leading to a metastable state, so that the entity being scattered remains localized in the neighborhood of the scatterer for a time Δt that is long compared with the natural time scale in the problem, $\Delta t_0 = a/v_0$ (a: radius of the potential well; v_0: incident velocity of the excess electron); i.e., $\Delta t > \Delta t_0$.

To explore this possibility quantitatively, Muñoz proposed a formalism to calculate this time delay Δt (the delay between the absorption and reemission of the excess electron caused by resonant scattering), using a partial wave expansion[157]

$$\Delta t = \frac{\Sigma_\ell \, (2\ell + 1) \, \sin^2 \delta_\ell \, \tau_\ell}{\Sigma_\ell \, (2\ell + 1) \, \sin^2 \delta_\ell} \qquad (28)$$

where $\delta_\ell(\varepsilon)$ is the phase shift characterizing the ℓ-th partial wave of the incident wave packet, ε is the kinetic energy of the projectile, and $\tau_\ell = \hbar \partial \delta_\ell(\varepsilon)/\partial \varepsilon$ is Wigner's time delay.[158] (Equation 28 is a natural extension of Wigner's time delay calculation to the case of wave packets that do not exhibit a well-defined angular momentum). Independent evidence of resonant scattering was recently found in the calculation performed by Ascarelli;[115,116] it manifests itself as some sharp downward spikes that appear when the excess electron mean-free path is plotted as a function of kinetic energy. Nevertheless, the calculation of the time delay according to Equation 28 indicates that the density of states available for producing metastable states (i.e., states such that $\Delta t > \Delta t_0$) is small, in the case of scattering of a particle off a 3-D potential well; the states for which $\Delta t > \Delta t_0$ are all located in narrow regions of momentum space corresponding to resonances dominated by one partial wave (Figure 2 of Reference 157).

Although, in principle, this effect exists (namely, resonant scattering of excess electrons scattered by potential wells produced by density fluctuations, *excluding any coupling between the excess electron and the well*), in practice it is not strong enough to account for the experimental results in NP. In spite of the limitations that may arise from the mathematical representation of the scattering potential as 3-D square barriers and square wells, which imply unrealistic discontinuities in the density of the fluid, the experiment indicates that there is one ingredient that is missing when the BC model is applied to TMS and NP. If resonant scattering produced by density fluctuations would be the origin of the intrinsic traps observed in NP close to the critical point, then they should also manifest themselves in TMS, since (apart from a shift of about 0.1 eV) the density dependence $V_0(N)$ on N is similar in these two liquids (Figure 5b). A similar argument holds if the traps observed in NP close to the critical point are interpreted as arising from a tail of localized states appearing at the bottom of the conduction band owing to disorder caused by critical density fluctuations; if such traps are present in NP, then they should also appear in TMS, since the two molecules are alike.

In order to account for the observed effect in NP, a *coupling between the excess electron and the density fluctuations* is needed, which on the one hand enlarges the density of states available for resonant scattering, but on the other hand does not lead to a deep trap (e.g., a localized state). Furthermore, this coupling must be important in NP but *not* in TMS. A similar problem exists concerning the observation of a fast signal detected at short times in liquid Ne, indicating the existence of a *fast moving carrier that precedes electron localization in the liquid*;[58] in this case the coupling *should lead to a deep trap* (a bubble).

C. ELECTRON MOBILITY IN LIQUID Ar, Kr, AND Xe NEAR THE CRITICAL POINT

Another question that remains unanswered concerns the excess electron mobility observed in the rare gas liquids in the proximity of the critical point. From the arguments presented in Section VI about the role of disorder, we would expect that as the density of the liquid decreases approaching the critical density, the role of disorder becomes dominant.

This would manifest itself as a *tail of localized states* appearing at the bottom of the conduction band that dominates the conduction process, leading to a low electron mobility, and to a drift velocity $v_D(E)$ that at high fields exhibits a superlinear dependence on the electric field strength E. However, $v_D(E)$ at, or in the neighborhood of the critical point, exhibits a behavior that deviates from the one outlined above.

First, the electron drift mobility μ_0 in the limit of zero field in liquid Ar, in the proximity of the critical point, is $\sim$230 cm²/V · s.[45] This unexpectedly large value corresponds to charge transport via extended states (and therefore, suggests a sublinear dependence of $v_D(E)$ on E), rather than transport via localized states (from which a superlinear dependence of $v_D(E)$ on E would be expected). However, under similar conditions, the observed μ_0 in liquid Kr[46,159] and liquid Xe[46] is approximately 5 cm²/V · s, certainly much closer to what would be expected if charge transport did take place via localized states.

Second, at sufficiently low fields, $\mu_D(E)$ is constant as a function of E. As the field increases beyond a certain value, $\mu_D(E)$ begins to *increase* with increasing field, reflecting a superlinear dependence of $v_D(E)$ on E. However, a further increase of electric field strength E *reverses* this dependence, resulting in a *decrease* of $\mu_D(E)$ (corresponding to a sublinear dependence of $v_D(E)$ on E) with increasing field strength. Such behavior, illustrated in the insert in Figure 7a, is observed in all three rare gas liquids.[45,46,159]

The nature of the processes that determine this intriguing behavior remains unclear; the theoretical efforts related to this problem have focused mostly on predicting the low field electron mobility in the neighborhood of the critical point.

Lekner and Bishop (LB)[107] developed a criteria for establishing whether or not localization would take place in liquid Ar near the critical point, using an Ornstein-Zernicke description of critical fluctuations. The localization mechanism they propose is based on the existence of potential wells that are deep enough to accommodate one bound state; these wells are the result of critical density fluctuations, giving rise to changes in the bottom of the conduction band V_0, analogous to the BC deformation potential model. LB conclude that electron localization in liquid Ar in the proximity of the critical point, along the liquid-vapor coexistence line, would take place if $|T - T_C| \leq 0.2$ K, but the prediction of a localized state turns out to be rather sensitive to the choice of the numerical factors used in the calculation. If localization would take place, they predict a mobility of the order of 0.01 cm²/V · s; if, at the critical point, conduction would take place via extended states, they predict $\mu \sim 40$ cm²/V · s.

Freeman[103] modified this theory by assuming that the density fluctuations that need to be considered are four times as large as those used in the LB calculation; the revised theory rules out electron localization at the critical point. The assumption concerning the size of the density fluctuations is based on the introduction of the "radian length", $\lambda = \lambda/2\pi$, instead of the wavelength, λ, as the proper scale of distance that should be used to describe bound states;[103] its theoretical foundations are questionable. Steinberger and Zeitak[104] discussed the applicability of modified versions of the BC theory (the original BC theory predicts $\mu = 0$ at the critical point, since the isothermal compressibility diverges), and updated the LB calculation using experimental data for V_0 in liquid Ar which was not available to LB. The revised value for μ at the critical point is 16 cm²/V · s, if conduction takes place via extended states. Watanabe has calculated the electron mobility in liquid Ar near the critical point using Kubo's linear response theory,[105,106] but agreement with experiment is poor; the mobility maximum observed experimentally at $N/N_C \sim 1.5$ (Figure 4) is not predicted by theory. The same author calculated the change in the electron effective mass m*,[108] and discussed the formation of a localized state in liquid Ar near the critical point.[109]

Recent Monte Carlo simulations of an excess electron in a cluster of Xe atoms predict electron localization in clusters of six Xe atoms or more,[160,161] thus suggesting an interesting possibility to explain qualitatively the low electron mobility observed in fluid Xe close to the critical point. In this simulation, the potential between Xe atoms is modeled by a Lennard-Jones potential, and the electron-Xe atom potential is modeled by a pseudopotential exhibiting a repulsive core and a deep minimum of about -0.43 eV at a distance r = 0.5 σ (where σ is the diameter of the Xe atom). When the Xe atoms are brought together, the overlap of the electron-Xe atom pseudopotentials gives rise to channels of negative potential energy between two sources of positive potential energy; the excess electron is partially confined in these channels. The decrease in potential energy brought about by the formation of the cluster (arising because of the large polarizability of Xe atoms) turns out to be larger than the increase in kinetic energy associated with the confinement of the electron; hence, localization takes place in the cluster.

Finally, among the theoretical problems that await a solution, we mention the intriguing behavior of the electron mobility observed in 22-DMB and 224-TMP under pressure (Figure 8b). A reliable theory of electron mobility applicable to liquids exhibiting intermediate and low electron mobility (e.g., μ <10 cm^2/V · s) is needed. A formalism proposed recently may prove to be a step in the right direction.[162]

From the experimental point of view, measurements of the excess electron effective mass m* as a function of density would contribute to remove the parameter fitting that still plagues the theoretical calculations applicable to liquids where electrons exhibit a high mobility. If the value of the effective mass m* = 0.27 m_0 determined in liquid Xe close to the triple point[117] is used in the BC theory, the calculated mobility turns out to be about *25 times too large.*[116] A similar situation exists for liquid Ar.

Measurements of μ_H in the rare gas solids and liquids are certainly needed, the measurement across the solid-liquid phase transition and in the proximity of the critical point would be particularly enlightening; the results may help to elucidate the nature of the mechanisms that scatter the excess electrons. Measurements of μ_H in liquids exhibiting a low electron mobility would also be extremely interesting. Unfortunately, since the Hall signal (according to Equation 1) depends roughly on the product $\mu_H \mu_D$, an improvement of at least an order of magnitude in the signal to noise ratio quoted in Reference 12 may be required to perform such measurements.

ACKNOWLEDGMENTS

The author wishes to acknowledge fruitful discussions with W. F. Schmidt and with R. A. Holroyd.

POSTSCRIPT

Since the writing of this manuscript, a most interesting work has been published reporting the measurement of μ_H in solid and liquid Ar[163] and in liquid Xe;[164] as a result of this work, the list of fundamental problems that await an answer grows longer. The measurement was performed using a 4-electrode cell; the author does not report a "calibration run" (i. e., a measurement of μ_H performed in a liquid for which μ_H is known, with the purpose of verifying the validity of Equation 3 in the geometry defined by this particular 4-electrode cell).

Recall that the derivation leading to Equation 3 is based on the calculation of the current induced on each electrode of a parallel plate cell by the motion of the excess electrons in the liquid contained between the resistive plates P_1 and P_2 (see Figure 3).[12,19] The introduction of two extra electrodes (say, P_3 and P_4, perpendicular to P_1 and P_2[65]) results in a modification of the currents induced on P_1 and P_2. Consequently, the above-mentioned derivation and its result do not necessarily apply to a 4-electrode cell;[12,19] a modification to the right-hand side (RHS) of Equation 3 cannot be excluded *a priori*. The only liquid for which a measurement of μ_H has been performed with a 2-electrode (parallel plate) cell and with a 4-electrode cell is TMS. The measurement reported in Reference 65 yields μ_H = 125 ± 10 cm^2/V · s (4-

electrode cell); the average of measurements performed with two different 2-electrode cells reported in Reference 13, is $\mu_H = 104 \pm 5 \text{ cm}^2/\text{V} \cdot \text{s}$. Furthermore, a theoretical calculation suggests that the modification to the RHS of Equation 3 in the case of a 4-electrode cell may be strongly dependent on the geometry of the cell, e. g., on the distance between electrodes P_3 and P_4, and electrodes P_1 and P_2.[19]

If the correction to the RHS of Equation 3 applicable to the cell used in the experiment reported in Reference 163 happens to be comparable to the reproducibility of the measurement ($\pm 10\%$), then the conclusion reached by Ascarelli that in liquid Ar, along the liquid-vapor coexistence line, μ_D is not affected by trapping events, implies that *traps do not play a significant role in the charge transport process in liquid Ar.** This conclusion contradicts the hypothesis of Schottky ionization of traps proposed by the same author, discussed in Section VIII.A.

However, the most striking feature reported in Reference 163, is the observation that, at densities close to that at which the mobility maximum takes place, and within the range of fields employed in the experiment (E <35 V/cm; B <1 Tesla), *the Hall field E_H exhibits a sublinear dependence both as a function of electric field E and magnetic field B*. A similar, more pronounced sublinear dependence of the Hall field vs. magnetic field B has been reported in liquid Xe.[164]**

As discussed in Section VI.B.4, the effect of the external fields **E** and **B** can be described as a change in the equilibrium electron distribution function, from a Maxwellian $f_0(\epsilon)$ centered at the origin of momentum space (that therefore carries no momentum, appropriate when $\mathbf{E} = 0$, $\mathbf{B} = 0$), to another distribution function $f(\epsilon)$ which depends on **E** and on **B**. Such field dependence may be calculated by solving the BTE by means of the relaxation time approximation.

Within the context of charge transport in dielectric liquids that exhibit a high electron mobility, the linearized solution of BTE using the relaxation time approximation should furnish a valid description of the motion of the excess electrons in the liquid, provided that: (1) the change in the electron distribution function, f, induced by the external fields **E** and **B** can be considered as a perturbation, in the sense that it is *small* compared to f_0, e.g., $|f - f_0|/f_0 \ll 1$,[78,79] see Section VI.B; (2) the collision operator (Equation 6) can be represented by a relaxation time, as in Equation 9, (3) quantum corrections—due to the rearrangement of the electron states available to the excess electrons (caused by the presence of the magnetic field **B**), e. g., the splitting of the electron states into "Landau levels"— can be ignored, in the sense that the excess electrons are spread out over many Landau levels.[140] Another way of saying this is that the separation between Landau levels is small compared with $k_B T$, i. e., $\hbar\omega_C \ll k_B T$ (here $\omega_C = qB/m^*$ is the cyclotron frequency), a condition that is certainly satisfied in these experiments. Concerning condition (2), in the case of semiconductors, it is satisfied by a rather wide variety of electron scattering mechanisms.[79]

If the two conditions mentioned above are satisfied, then the average velocity $\langle \mathbf{v} \rangle$ of the electron gas should depend *linearly* on **E** and on $\mathbf{E} \times \mathbf{B}$ (Equation 11), except perhaps at large **B**, in which case the average $\langle \tau \rangle$ and $\langle \tau^2 \rangle$ in Equations 11, 24-26 should be replaced by $\langle \dfrac{\tau}{1 + (\omega_C\tau)^2} \rangle$ and $\langle \dfrac{\tau^2}{1 + (\omega_C\tau)^2} \rangle$, respectively, (with the average $\langle \ \rangle$ defined by Equation 12) plus a term proportional to $\mathbf{B}(\mathbf{B} \cdot \mathbf{E})$,[165] which in the present case is zero since **E** and **B** are orthogonal.

---

* Recall that in the language of Reference 163, "the low-field time-of-flight data in the liquid coincide with the low-field drift mobility"; "the drift mobility" of Reference 163 is what we have defined in Section VII.A as μ_F, the microscopic mobility corresponding to the conduction band.

** In the language of Reference 163 and 164, the Hall mobility μ_H—except for a minus sign—is given by (1/B) tan θ, where θ is the Hall angle and B is the magnetic field strength.

Therefore, within this linearized solution of BTE, deviations of the Hall field from a linear dependence on **E** *are not expected*, and deviations from a linear dependence on **B** are to be expected whenever the curvature of the trajectory followed by the excess electrons (between scattering events) caused by the magnetic field **B**, can no longer be ignored; that is, whenever $\omega_c\tau \sim \mu B \sim 1$, since $\mu \sim q\tau/m^*$.

In liquid Ar, the sublinear dependence of the Hall field on the electric field strength E, together with the sublinear dependence of the electron drift velocity $v_D(E)$ on E observed under similar conditions,[39] suggest that *the linearized solution of BTE according to the relaxation time approximation breaks down*. At 144 K, a field of 35 V/cm seems to be large enough to give rise to a change in the electron distribution function $f(\varepsilon)$, such that terms beyond the first order need to be considered in the expansion of $f(\varepsilon)$ in terms of Legendere polynomials (Equation 7), or the leading (spherically symmetric) term $F_0(\varepsilon)$ in Equation 7, when $\mathbf{E} \neq 0$, is no longer the Maxwellian distribution $f_0(\varepsilon)$ characterizing the electron gas when $\mathbf{E} = 0$, as it was assumed in deriving Equation 10. Thus, condition (1) mentioned above appears *not* to be satisfied. This could happen if there was "electron heating", e. g., if at the fields considered, the average electron kinetic energy $(1/2)\,\langle m^*v^2\rangle$ would exceed $(3/2)\,k_BT$, the value expected from equipartition (see Section VIII.A).

Concerning the sublinear dependence of the Hall field on the magnetic field B observed in liquid Ar at 144 K, the highest magnetic field utilized in the experiment was such that $\mu B \sim 0.2$, *an order of magnitude too small* for the aforementioned magnetic field-induced curvature to become dominant.[163]

In liquid Xe, where a more pronounced nonlinearity of the Hall field vs. magnetic field has been reported at 217.4 K, the highest magnetic field used in the experiment was such that condition $\omega_c\tau \sim \mu B \sim 1$ was satisfied. This sublinear dependence was interpreted as "evidence for non-Boltzmann transport".[164] An explanation involving interference between electrons whose wave functions have been phase shifted by π after being elastically scattered several times has been proposed; it is argued that interference between these electrons results in weakly localized states, such that a certain number of interfering electrons ". . .do not contribute to electrical transport. Borrowing an expression from paramagnetic resonance, this can be envisaged as a hole burned into the electron distribution."[164]

The experimental results in liquid Ar and liquid Xe—the sublinear dependence of the Hall field on the magnetic field **B**—do not agree with the prediction based on the solution of BTE according to the relaxation time approximation. A momentum relaxation time $\tau \sim \varepsilon^0$ (energy independent relaxation time) would lead to a linear dependence of the Hall field on magnetic field B (e.g., a Hall mobility μ_H which is a constant independent of B); a BC relaxation time $\tau \sim \varepsilon^{-1/2}$ would lead to a nonlinear dependence much weaker than observed. The experimental findings suggest that *the change in the electron distribution function induced by the electric field* **E** *and magnetic field* **B** *cannot be considered as a perturbation*. Thus, the applicability of the relaxation time approximation (in the sense discussed in Section VI.B.4) and of formulas 10, 11, 24-26 becomes questionable.

The explanation of these discrepancies involving weak electron localization and the "hole burned into the electron distribution" raises some fundamental questions. First, the preceding discussion casts doubts on the validity of the calculation of this effect using a relaxation time, since the very fact that such marked nonlinearities are observed, implies that the applied fields must induce a rather drastic change in the electron distribution function. We may then ask, what is the meaning of the relaxation time τ employed in the calculation? Second, the proposed explanation suggests that this sublinear dependence between the Hall field and the magnetic field B is due to some quantum interference taking place between electrons that have been backscattered following a different sequence of scattering events, the so-called "weak localization".[166,167] Such interference would modify the effect of collisions on the electron distribution function. A calculation of the quantum corrections to BTE arising from electron scattering in a disordered medium has been reported,[168] the interference effects mentioned above were incorporated into the calculation of the transition probability $T(\mathbf{v}, \mathbf{v}')$ entering Equation 6. However, we are not aware of any theoretical study of the quantum corrections to BTE arising from weak localization, that would explain the nonlinearities reported in Reference 164.

The theoretical description of the electron motion leading to weak localization seems far from clear, and the connection (if any) between the sublinear behavior of the Hall field vs. magnetic field B and vs. electric field E has yet to be elucidated. Moreover, as discussed in Sections VIII, current theories (based on a solution of BTE) concerning the electron-fluid interaction in liquids where electrons exhibit a high mobility, suffer from a "classical" omission. The Hamiltonian describing the interaction between the excess electron and the fluid via the potential barriers/potential wells induced by density fluctuations, contains no coupling between the electron and the density fluctuations. Another way of stating this is that the charged particle undergoing scattering does not influence the scattering potential; the density fluctuations considered in the BC theory are those that exist in the liquid *regardless of whether or not excess electrons are present in the fluid*. We do not know if such a coupling term would play a significant role in describing electron motion in liquid Ar or liquid Xe. However, without such a coupling, the description of electron motion in similar fluids remains incomplete, for if the excess electron cannot modify these density fluctuations, then it is difficult to explain why fast carriers preceding electron localization have been reported in liquid Ne,[58] and why high mobility carriers coexisting with low mobility carriers have been observed in dense He gas.[60]

In view of the absence of a coherent theoretical description capable of explaining the several unresolved questions discussed in Section VIII and in this Postscript, the conclusion that the results presented in Reference 164 imply a breakdown of BTE seems premature.

REFERENCES

1. **Tewari, P. H. and Freeman, G. R.**, Dependence of radiation-induced conductance of liquid hydrocarbons on molecular structure, *J. Chem. Phys.*, 49, 4394, 1968.
2. **Schmidt, W. F. and Allen, A. O.**, Mobility of free electrons in dielectric liquids, *J. Chem. Phys.*, 50, 5037, 1969.
3. **Freeman, G. R.**, Ionization and charge separation in irradiated materials, in *Kinetics of Nonhomogeneous Processes*, Freeman, G. R., Ed., Wiley and Sons, New York, 1987, chap. 2.
4. **Meyer, L., Davis, H. T., Rice, S. A., and Donnelly, R. J.**, Mobility of ions in liquid ^{4}He I and ^{3}He as a function of pressure and temperature, *Phys. Rev.*, 126, 1927, 1962.
5. **Levine, J. L. and Sanders, T. M.**, Mobility of electrons in low temperature helium gas, *Phys. Rev.*, 154, 138, 1967.
6. **Bruschi, L., Mazzi, G., and Santini, M.**, Localized electrons in liquid Neon, *Phys. Rev. Lett.*, 28, 1504, 1972.
7. **Loveland, R. J., Le Comber, P. G., and Spear, W. E.**, Experimental evidence for electronic bubble states in liquid Ne, *Phys. Lett.*, 39A, 225, 1972.
8. **Harrison, H. R. and Springett, B. E.**, Electron mobility variation in dense hydrogen gas, *Chem. Phys. Lett.*, 10, 418, 1971.
9. **Springett, B. E., Jortner, J., and Cohen, M. H.**, Stability criterion for the localization of an excess electron in a nonpolar fluid, *J. Chem. Phys.*, 48, 2720, 1968.
10. **Miyakawa, T. and Dexter, D. L.**, Stability of electronic bubbles in liquid Neon and Hydrogen, *Phys. Rev.*, 184, 166, 1969.
11. **Seager, C. H. and Emin, D.**, High temperature measurements of the electron Hall mobility in the alkali halides, *Phys. Rev.*, B2, 3421, 1970.
12. **Muñoz, R. C.**, Measurement of the Hall effect in liquid insulators: challenge and surprises, *Radiat. Phys. Chem.*, 32, 169, 1988.
13. **Muñoz, R. C. and Holroyd, R. A.**, Measurement of the Hall mobility of injected electrons in liquid tetramethyl silane between 22 and 164°C, *Chem. Phys. Lett.*, 137, 250, 1987.

14. **Fuochi, P. G. and Freeman, G. R.,** Molecular structure effects on the free ion yields and reaction kinetics in the radiolysis of the methyl-substituted propanes and liquid Argon: electron and ion mobilities, *J. Chem. Phys.,* 56, 2333, 1972.
15. **Davis, H. T. and Brown, R. G.,** Low energy electrons in nonpolar fluids, in *Advances in Chemical Physics,* Vol. 31, Prigogine, I. and Rice, S. A., Eds., Wiley and Sons, New York, 1975.
16. **Dodelet, J. P., Shinsaka, K., and Freeman, G. R.,** Electron mobilities in liquid olefins: structure effects, *J. Chem. Phys.,* 59, 1293, 1973.
17. **Schmidt, W. F. and Allen, A. O.,** Yields of free ions in irradiated liquids; determination by a clearing field, *J. Phys. Chem.,* 72, 3730, 1968.
18. **Redfield, A. G.,** Electronic Hall effect in diamond, *Phys. Rev.,* 94, 526, 1954.
19. **Muñoz, R. C.,** *Measurement of the Hall Mobility and of the Time of Flight Mobility in Liquid Tetramethyl silane and Neopentane,* Ph. D. Thesis, Purdue University, West Lafayette, 1983.
20. **Holroyd, R. A. and Gangwer, T. E.,** Electron attachment to oxygen and other solutes in nonpolar liquids, *Radiat. Phys. Chem.,* 15, 283, 1980.
21. **Miller, L. S., Howe, S., and Spear, W. E.,** Change transport in solid and liquid Ar, Kr, and Xe, *Phys. Rev.,* 166, 871, 1968.
22. **Nakamura, Y., Shinsaka, K., and Hatano, Y.,** Electron mobilities and electron—ion recombination rate constants in solid, liquid, and gaseous methane, *J. Chem. Phys.,* 78, 5820, 1983.
23. **Shinsaka, K. and Freeman, G. R.,** Electron mobilities and ranges in solid neopentane: effect of liquid-solid phase change, *Can. J. Chem.,* 52, 3556, 1974.
24. **Namba, H., Shinsaka, K., and Hatano, Y.,** Effect of n-butane impurity on electron mobility and electron-ion recombination rate constant in solid neopentane, *J. Chem. Phys.,* 70, 5331, 1979.
25. **Nakamura, Y., Namba, H., Shinsaka, K., and Hatano, Y.,** Electron mobilities in solid tetramethyl silane: effect of liquid-solid phase change, *Chem. Phys. Lett.,* 76, 311, 1980.
26. **Schmidt, W. F.,** Electron mobility in nonpolar liquids: the effect of molecular structure, temperature, and electric field, *Can. J. Chem.,* 55, 2197, 1977.
27. **Stephens, J. A.,** Comment on electron mobility in the liquid isomeric pentanes, *J. Chem. Phys.,* 84, 4721, 1986.
28. **Freeman, G. R.,** Reply to Comment on electron mobility in the liquid isomeric pentanes, *J. Chem. Phys.,* 84, 4723, 1986.
29. **Gyorgy, I., Gee, N., and Freeman, G. R.,** Molecular shape effect on electron transport in fluids: density and temperature studies in mixtures of n-pentane and neopentane, *J. Chem. Phys.,* 79, 2009, 1983.
30. **Gee, N. and Freeman, G. R.,** Effect of molecular properties on electron transport in hydrocarbon fluids, *J. Chem. Phys.,* 78, 1951, 1983.
31. **Gee, N. and Freeman, G. R.,** Electron mobilities, free ion yields, and electron thermalization distances in liquid long-chain hydrocarbons, *J. Chem. Phys.,* 86, 5716, 1987.
32. **Nyikos, L., Zador, E., and Schiller, R.,** Temperature dependence of electron mobility in liquid hydrocarbons, *IV Int. Symp. Radiat. Chem.,* Keszthely, Hungary, 1976.
33. **Schmidt, W. F.,** Electronic conduction processes in dielectric liquids, *IEEE Trans. Elec. Insul.,* EI–19, 389, 1984.
34. **Allen, A. O.,** Drift mobilities and conduction band energies of excess electrons in dielectric liquids, *Nat. Stand. Ref. Data Ser.,* National Bureau of Standards, Report 58, 1976.
35. **Holroyd, R. A.,** The electron: its properties and reactions, in *Radiation Chemistry–Principles and Applications,* Farhataziz, and Rodgers, M. A. J., Eds., V.C.H. Publishers, New York, 1987.
36. **Holroyd, R. A. and Cipollini, N. E.,** Correspondence of conduction band minima and electron mobility maxima in dielectric liquids, *J. Chem. Phys.,* 69, 501, 1978.
37. **Asaf, U., Reininger, R., and Steinberger, I. T.,** The energy V_0 of the quasifree electron in gaseous, liquid and solid methane, *Chem. Phys. Lett.,* 100, 363, 1983.
38. **Engels, J. M. L. and Van Kimmenade, A. J. M.,** The mobility of excess electrons in compressed liquid methane, *Phys. Lett.,* 59A, 43, 1976.
39. **Jahnke, J. A., Meyer, L., and Rice, S. A.,** Zero field mobility of an excess electron in fluid Argon, *Phys. Rev.,* A3, 734, 1971.
40. **Muñoz, R. C. and Holroyd, R. A.,** Pressure induced changes in electron mobility in liquid neopentane and tetramethyl silane near the critical point, *Radiat. Phys. Chem.,* 32, 49, 1988.
41. **Yoshino, K., Sowada, U., and Schmidt, W. F.,** Effect of molecular solutes on the electron drift velocity in liquid Ar, Kr, and Xe, *Phys. Rev.,* A14, 438, 1976.
42. **Schmidt, W. F.,** Excess electron mobility in liquid methane and ethane, *Proc. of the Sixth Int. Congr. of Radiat. Res.,* Tokyo, May 13–19, 1979.
43. **Bakale, G. and Schmidt, W. F.,** On excess electron transport in liquid hydrocarbons, *Chem. Phys. Lett.,* 22, 164, 1973.
44. **Doldissen, W. and Schmidt, W. F.,** Saturation of the electron drift velocity in liquid tetramethyl silane, *Chem. Phys. Lett.,* 68, 527, 1979.

45. **Huang, S. S. S. and Freeman, G. R.,** Electron transport in gaseous and liquid argon: effects of density and temperature, *Phys. Rev.,* A24, 714, 1981.

46. **Huang, S. S. S. and Freeman, G. R.,** Electron mobilities in gaseous, critical and liquid Xenon: density, electric field, and temperature effects: quasilocalization, *J. Chem. Phys.,* 68, 1355, 1978.

47. **Muñoz, R. C., Holroyd, R. A., and Nishikawa, M.,** Effect of high pressure on the electron mobility in liquid n-hexane, 2,2-dimethyl butane, and tetramethyl silane, *J. Phys. Chem.,* 89, 2969, 1985.

48. **Muñoz, R. C. and Holroyd, R. A.,** The effect of temperature and pressure on excess electron mobility in n-hexane, 2,2,4 trimethyl pentane, and tetramethyl silane, *J. Chem. Phys.,* 84, 5810, 1986.

49. **Muñoz, R. C., Holroyd, R. A., Itoh, K., Nakagawa, K., Nishikawa, M., and Fueki, K.,** Excess electron mobility in hydrocarbon liquids at high pressure, *J. Phys. Chem.,* 91, 4639, 1987.

50. **Basak, S. and Cohen, M. H.,** Deformation-potential theory for the mobility of excess electrons in liquid Argon, *Phys. Rev.,* B20, 3404, 1979.

51. **Cipollini, N. E., Holroyd, R. A., and Nishikawa, M.,** Zero field mobility of excess electrons in dense methane gas, *J. Chem. Phys.,* 67, 4636, 1977.

52. **Nishikawa, M., Holroyd, R. A., and Sowada, U.,** Electron mobility in supercritical ethane as a function of density and temperature, *J. Chem. Phys.,* 72, 3081, 1980.

53. **Nishikawa, M. and Holroyd, R. A.,** Electron mobility in supercritical propane as a function of density and temperature, *J. Chem. Phys.,* 77, 4678, 1982.

54. **Itoh, K., Nakagawa, K., and Nishikawa, M.,** Electron mobility in supercritical butanes as a function of density and temperature, *J. Chem. Phys.,* 84, 391, 1986.

55. **Itoh, K., Nakagawa, K., and Nishikawa, M.,** Electron mobility in supercritical pentanes as a function of density and temperature, *Radiat. Phys. Chem.,* 32, 221, 1988.

56. **Schwarz, K. W. and Stark, R. W.,** Phonon-limited drift of the electron bubble in superfluid Helium, *Phys. Rev. Lett.,* 21, 967, 1968.

57. **Schwarz, K. W.,** Charge-carrier mobilities in liquid Helium at the vapor pressure, *Phys Rev.,* A6, 837, 1972.

58. **Sakai, Y., Bottcher, H., and Schmidt, W. F.,** Excess electrons in liquid hydrogen, liquid Neon, and liquid Helium, *J. Electrostatics,* 12, 89, 1982.

59. **Harrison, H. R. and Springett, B. E.,** Electron mobility variation in dense ^{4}He gas, *Phys. Lett.,* 35A, 73, 1971.

60. **Jahnke, J. A. and Silver, M.,** Mobility of electrons in dense Helium gas, *Chem. Phys. Lett.,* 19, 231, 1973.

61. **Woolf, M. A. and Rayfield, G. W.,** Energy of negative ions in liquid Helium by photoelectric injection, *Phys. Rev. Lett.,* 15, 235, 1965.

62. **Bruschi, L., Maraviglia, B., and Moss, F. E.,** Measurement of a barrier for the extraction of excess electrons from liquid Helium, *Phys. Rev. Lett.,* 17, 682, 1966.

63. **Northby, J. A. and Sanders, T. M.,** Photoejection of electrons from bubble states in liquid Helium, *Phys. Rev. Lett.,* 18, 1184, 1967.

64. **Le Comber, P. G., Wilson, J. B., and Loveland, R. J.,** Transport of excess charge carriers in liquid and solid hydrogen, *Solid State Comm.,* 18, 377, 1976.

65. **Muñoz, R. C. and Ascarelli, G.,** Measurement of the room temperature Hall mobility of injected electrons in liquid tetramethyl silane, *Chem. Phys. Lett.,* 94, 235, 1983.

66. **Muñoz, R. C. and Ascarelli, G.,** Hall mobility of electrons injected into fluid neopentane (Dimethylpropane) along the liquid-vapor coexistence line between the triple and the critical points, *Phys. Rev. Lett.,* 51, 215, 1983.

67. **Muñoz, R. C. and Ascarelli, G.,** Motion of electrons in a classical liquid: injected electron Hall mobility in liquid neopentane along the liquid-vapor coexistence line between the triple and the critical points, *J. Phys. Chem.,* 88, 3712, 1984.

68. **Itoh, K., Muñoz, R. C., and Holroyd, R. A.,** The Hall mobility of excess electrons in 2,2-dimethyl butane, 2,2,4-trimethyl pentane and 2,2,4,4-tetramethyl pentane, *J. Chem. Phys.,* 90, 1128, 1989.

69. **Itoh, K., Holroyd, R., and Nishikawa, M.,** The Hall mobility of excess electrons in n-pentane-neopentane mixtures, *J. Chem. Phys.,* 94, 2073 (1991).

70. **Mott, N. F. and Davis, E. A.,** *Electronic Processes in Non-Crystalline Materials,* Clarendon Press, 2nd ed., Oxford, 1979, chap. 1.

71. **Ioffe, A. F. and Regel, A. R.,** Non-crystalline, amorphous, and liquid electronic semiconductors, in *Progress in Semiconductors,* Vol. 4, Gibson, A. F., Ed., Heywood, London, 1960.

72. **Mott, N. F. and Davis, E. A.,** *Electronic Processes in Non-Crystalline Materials,* Clarendon Press, 2nd ed., Oxford, 1979, Chap. 5.

73. **Ascarelli, G.,** The motion of electrons injected in classical nonpolar insulating liquids, *Comments in Solid State Physics,* 11, 179, 1985.

74. **Kevan, L.,** Localization and Solvation of electrons in condensed media. The simplest radical ion? *J. Phys. Chem.,* 82, 1144, 1978.

75. **Kimura, T. and Fueki, K.,** Effect of density on excess electron localization in ethane and propane, *J. Chem. Phys.,* 66, 366, 1977.

76. **Kimura, T., Fueki, K., and Kevan, L.,** Ground state energy of excess electrons in n-hexane, 2,2,4-trimethyl pentane, and tetramethyl silane, *J. Chem. Phys.,* 68, 3945, 1978.

77. **Ziman, J. M.,** *Principles of the Theory of Solids,* 2nd ed. Cambridge, 1972, chap. 7.

78. **Conwell, E. M.,** Transport: the Boltzmann equation, in *Handbook on Semiconductors,* Vol. 1, Paul, W., Ed., North Holland, Amsterdam, 1982.

79. **Conwell, E. M.,** High field transport in semiconductors, *Solid State Physics, Advances in Research and Applications,* Suppl. 9, Academic Press, New York, 1967, chap. 5.

80. **Overhauser, A. W. and Huberman, M.,** Reply to "Scattering of electron by impurities in a weak magnetic field: a comment", *Phys. Rev.,* B27, 7796, 1983.

81. **Kittel, C.,** *Introduction to Solid State Physics,* Wiley and Sons, 4th ed., New York, 1971, chap. 10.

82. **Cohen, M. H. and Lekner, J.,** Theory of hot electrons in gases, liquids, and solids, *Phys. Rev.,* 158, 305, 1967.

83. **Schnyders, H., Rice, S. A., and Meyer, L.,** Electron mobilities in liquid argon, *Phys. Rev. Lett.,* 15, 187, 1965.

84. **Swan, D. W.,** Electron drift velocity in liquid argon and argon-nitrogen mixtures, *Nature,* 196, 977, 1962.

85. **Swan, D. W.,** Drift velocity of electrons in liquid argon, and the influence of molecular impurities, *Proc. Phys. Soc. (London),* 83, 659, 1964.

86. **Lekner, J.,** Motion of electrons in liquid argon, *Phys. Rev.,* 158, 130, 1967.

87. **Schnyders, H., Rice, S. A., and Meyer, L.,** Electron drift velocities in liquified argon and krypton at low electric field strengths, *Phys. Rev.,* 150, 127, 1966.

88. **Jahnke, J. A., Holzwarth, N. A. W., and Rice, S. A.,** Comments on the theory of electron mobility in simple fluids, *Phys. Rev.,* A5, 463, 1972.

89. **Lekner, J.,** Mobility maxima in the rare gas liquids, *Phys. Lett.,* 27A, 341, 1968.

90. **Lekner, J.,** Scattering of waves by an ensemble of fluctuating potentials, *Phil. Mag.,* 18, 1281, 1968.

91. **Gryko, J. and Popielawski, J.,** Comment on the application of the Cohen-Lekner theory to excess electron mobility in liquid krypton, *Phys. Rev.,* A16, 1333, 1976.

92. **Harrison, W. A.,** *Solid State Theory,* McGraw-Hill, New York, 1970, chap. 3.

93. **Berlin, Y. A., Nyikos, L., and Schiller, R.,** Mobility of localized and quasifree excess electrons in liquid hydrocarbons, *J. Chem. Phys.,* 69, 2401, 1978.

94. **Reininger, R., Asaf, U., Steinberger, I. T., and Basak, S.,** Relationship between the energy V_0 of the quasifree electron and its mobility in fluid Argon, Krypton, and Xenon, *Phys. Rev.,* B28, 4426, 1983.

95. **Polischuk, A. Y.,** Concerning a theory of electron mobility in simple fluids, *J. Phys.,* B17, 4789, 1984.

96. **Atrazhev, V. M. and Iakubov, I. T.,** Hot electrons in non-polar liquids, *J. Phys.,* C14, 5139, 1981.

97. **Atrazhev, V. M. and Dmitriev, E. G.,** Heating and diffusion of hot electrons in non-polar liquids, *J. Phys.,* C18, 1205, 1985.

98. **Sakai, Y., Sukegawa, K., Nakamura, S., and Tagashira, H.,** Monte Carlo simulation of hot electrons in liquid argon, Conference Record, *Ninth Int. Conf. on Conduction and Breakdown in Dielectric Liquids,* Salford, U.K., July 27—31, 1987.

99. **Sakai, Y., Nakamura, S., and Tagashira, H.,** Drift velocity of hot electrons in liquid Ar, Kr and Xe, *IEEE Trans. Elec. Insul.,* EI–20, 133, 1985.

100. **Vertes, A.,** Quasifree electron mobility by the method of partial waves in liquid hydrocarbons and in fluid argon, *J. Chem. Phys.,* 79, 5558, 1983.

101. **Vertes, A.,** Electron mobility calculations in liquid Xenon by the method of partial waves, *J. Phys. Chem.,* 88, 3722, 1984.

102. **Nishikawa, M.,** Electron mobility in fluid argon: application of a deformation potential theory, *Chem. Phys. Lett.,* 114, 271, 1985.

103. **Freeman, G. R.,** Estimation of electron mobilities in simple fluids near the critical point, *Phil. Mag. Lett.,* 56, 47, 1987.

104. **Steinberger, I. T. and Zeitak, R.,** Estimation of electron mobilities near the critical point in simple nonpolar fluids, *Phys. Rev.,* B34, 3471, 1986.

105. **Watanabe, S.,** Density dependent electron mobility in nonpolar fluids along critical isotherm, *J. Phys. Soc. Jpn.,* 50, 1049, 1981.

106. **Watanabe, S.,** Critical scattering of excess electrons in nonpolar fluids, *J. Phys. Soc. Jpn.,* 50, 1095, 1981.

107. **Lekner, J. and Bishop, A. R.,** Electron mobility in simple fluids near the critical point, *Phil. Mag.,* 27, 297, 1973.

108. **Watanabe, S.,** Excessive electrons in fluids near critical points, *J. Phys. Soc. Jpn.,* 49, 38, 1980.

109. **Watanabe, S.,** Localization of an excess electron in nonpolar fluids near critical points, *J. Phys. Soc. Jpn.,* 54, 1665, 1985.

110. **Leycuras, A. and Larour, J.,** Model for excess electron transport in dense heavy non-polar fluids, *J. Phys.,* C15, 6765, 1982.

111. **Leycuras, A. and Levesque, D.**, Molecular dynamics simulation of excess electron transport in simple fluids, *Phys. Rev.*, A32, 1180, 1985.

112. **Gee, N. and Freeman, G. R.**, Comment on "Molecular dynamics simulation of excess electron transport in simple fluids", *Phys Rev.*, A35, 4449, 1987.

113. **Leycuras, A., and Levesque, D.**, Reply to "Comment on Molecular dynamics simulation of excess electron transport in simple fluids", *Phys. Rev.*, A35, 4451, 1987.

114. **Boehm, R.**, Excess electron mobility maxima in Argon in the critical region, *Phys. Rev.*, A12, 2189, 1975.

115. **Ascarelli, G.**, Calculation of the mobility of electrons injected in liquid Argon, *Phys. Rev.*, B33, 5825, 1986.

116. **Ascarelli, G.**, Calculation of the mobility of electrons injected in liquid Xenon, *Phys. Rev.*, B34, 4278, 1986.

117. **Asaf, U. and Steinberger, I. T.**, Photoconductivity and electron transport parameters in liquid and solid Xenon, *Phys. Rev.*, B10, 4464, 1974.

118. **Minday, R. M., Schmidt, L. D., and Davis, H. T.**, Excess electrons in liquid hydrocarbons, *J. Chem. Phys.*, 54, 3112, 1971.

119. **Schiller, R.**, Localization probability and mobility of electrons in liquid hydrocarbons, *J. Chem. Phys.*, 57, 2222, 1972.

120. **Schiller, R., Vass, S., and Mandics, J.**, Energy of the quasifree electrons and the probability of electron localization in liquid hydrocarbons, *Int. J. Radiat. Phys. Chem.*, 5, 491, 1973.

121. **Wada, T., Shinsaka, K., Namba, H., and Hatano, Y.**, Electron reactivity in liquid hydrocarbon mixtures, *Can. J. Chem.*, 55, 2144, 1977.

122. **Dodelet, J. P., Shinsaka, K., and Freeman, G. R.**, Molecular structure effects on electron ranges and mobilities in liquid hydrocarbons: chain branching and olefin conjugation: mobility model, *Can. J. Chem.*, 54, 744, 1976.

123. **Dodelet, J. P., Shinsaka, K., and Freeman, G. R.**, On electron mobility models for liquid hydrocarbons, *J. Chem. Phys.*, 63, 2765, 1975.

124. **Kestner, N. R. and Jortner, J.**, Conjecture on electron mobility in liquid hydrocarbons, *J. Chem. Phys.*, 59, 26, 1973.

125. **Schiller, R. and Nyikos, L.**, Percolation model of electron and hole mobility in liquid mixtures, *J. Chem. Phys.*, 72, 2245, 1980.

126. **Schiller, R., Vertes, A., and Nyikos, L.**, Quasipercolation: charge transport in fluctuating systems, *J. Chem. Phys.*, 76, 678, 1982.

127. **Chandler, D., Singh, Y., and Richardson, D. M.**, Excess electrons in simple fluids, I. General equilibrum theory for classical hard sphere solvents, *J. Chem. Phys.*, 81, 1975, 1984.

128. **Nichols, A. L., III, Chandler, D., Singh, Y., and Richardson, D. M.**, Excess electrons in simple fluids. II. Numerical results for the hard sphere solvent, *J. Chem. Phys.*, 81, 5109, 1984.

129. **Nichols, A. L., III and Chandler, D.**, Excess electrons in simple fluids. III. Role of solvent polarization, *J. Chem. Phys.*, 84, 398, 1986.

130. **Nichols, A. L., III and Chandler, D.**, Excess electrons in simple fluids. IV. Real time behaviour, *J. Chem. Phys.*, 87, 6671, 1987.

131. **Feynman, R. P.**, Slow electrons in a polar crystal, *Phys. Rev.*, 97, 660, 1955.

132. **Laria, D. and Chandler, D.**, Comparative study of theory and simulation calculations for excess electrons in simple fluids, *J. Chem. Phys.*, 87, 4088, 1987.

133. **Smith, R. A.**, *Semiconductors*, 2nd ed., Cambridge University Press, 1978, chap. 5.

134. **Smith, R. A.**, *Semiconductors*, 2nd ed., Cambridge University Press, 1978, chap. 8.

135. **Conwell, E. M.**, High field transport in semiconductors, *Solid State Physics, Advances in Research and Applications*, Suppl. 9, Academic Press, New York, 1967, chap. 3.

136. **Friedman, L. and Holstein, T.**, Studies of Polaron Motion. Part III. The Hall mobility of the small polaron, *Ann. Physics*, 21, 494, 1963.

137. **Emin, D.**, The Hall effect in hopping conduction, in *The Hall Effect and its Applications*, Chien, C. L. and Westgate, C. R., Eds., Plenum Press, New York, 1980.

138. **Friedman, L.**, The Hall effect in low mobility and amorphous solids, in *Physics of Disordered Materials*, Adler, D., Fritzsche, H., and Ovshinsky, S. R., Eds., Plenum Press, New York, 1985.

139. **Schockley, W.**, Hot electrons in germanium and Ohm's law, *Bell System Tech. J.*, 30, 990, 1951.

140. **Smith, R. A.**, *Semiconductors*, 2nd ed., Cambridge University Press, 1978, chap. 12.

141. **Reininger, R., Steinberger, I. T., Bernstorff, S., Saile, V., and Laporte, P.**, Extrinsic photoconductivity in xenon doped fluid argon and krypton, *Chem. Phys.*, 86, 189, 1984.

142. **Ascarelli, G.**, The role of shallow traps on the mobility of electrons in liquid Ar, Kr, and Xe, *J. Chem. Phys.*, 71, 5030, 1979.

143. **Freeman, G. R.**, On the role of shallow traps in electron transport in liquid Ar, Kr, and Xe, *J. Chem. Phys.*, 74, 3079, 1981.

144. **Shibamura, E., Takahashi, T., Kubota, S., Doke, T., and Mozumder, A.,** Comments on Ascarelli's papers on electron trapping in liquid Ar, Kr, and Xe, *J. Chem. Phys.*, 77, 3290, 1982.

145. **Schmidt, W. F., Sowada, U., and Yoshino, K.,** Comment on the paper "The role of shallow traps on the mobility of electrons in liquid Ar, Kr, and Xe", by G. Ascarelli, *J. Chem. Phys.*, 71, 5030, 1979, *J. Chem. Phys.*, 74, 3081, 1981.

146. **Ascarelli, G.,** Response to the critique to the paper "The role of shallow traps on the mobility of electrons in liquid Ar, Kr and Xe," *J. Chem. Phys.*, 74, 3082, 1981.

147. **Shibamura, E., Takahashi, T., Kubota, S., and Doke, T.,** Ratio of diffusion coefficient to mobility for electrons in liquid argon, *Phys. Rev.*, A20, 2547, 1979.

148. **Derenzo, S. E., Mast, T. S., Zaklad, H., and Muller, R. A.,** Electron avalanche in liquid xenon, *Phys. Rev.*, A9, 2582, 1974.

149. **Dolgoshein, B. A., Lebedenko, V. N., and Rodionov, B. U.,** Luminescence induced by alpha particles in liquid xenon in an electric field, *JETP Lett.*, 6, 224, 1967.

150. **Sowada, U., Bakale, G., Yoshino, K., and Schmidt, W. F.,** Electric field effect on electron capture by SF_6 in liquid Argon and Xenon, *Chem. Phys. Lett.*, 34, 466, 1975.

151. **Bakale, G., Sowada, U., and Schmidt, W. F.,** Effect of an electric field on electron attachment to SF_6, N_2O and O_2 in liquid Argon and Xenon, *J. Phys. Chem.*, 80, 2556, 1976.

152. **Kittel, C.,** *Introduction to Solid State Physics,* 4th ed., Wiley and Sons, New York, 1971, chap. 5.

153. **Harrison, W. A.,** *Solid State Theory,* McGraw-Hill, New York, 1970, chap. 4.

154. **Kittel, C.,** *Introduction to Solid State Physics,* Wiley and Sons, 4th ed., New York, 1971, chap. 1.

155. **Kittel, C.,** *Introduction to Solid State Physics,* 4th ed., Wiley and Sons, New York, 1971, chap. 2.

156. **Spear, W. E. and Le Comber, P. G.,** A possible explanation of the observed electron drift velocity saturation in solid Ar, Kr, and Xe, *Phys. Rev.*, 178, 1454, 1969.

157. **Muñoz, R. C.,** Delay time experienced by a particle undergoing resonant scattering: the case of the 3D square well, *J. Chem. Phys.*, 83, 6242, 1985.

158. **Wigner, E. P.,** Lower limit for the energy derivative of the scattering phase shift, *Phys. Rev.*, 98, 145, 1955.

159. **Jacobsen, F. M., Gee, N., and Freeman, G. R.,** Electron mobility in liquid Krypton as a function of density, temperature, and electric field strength, *Phys. Rev.*, A34, 2329, 1986.

160. **Martyna, G. J. and Berne, B. J.,** Structure and energetics of Xe_n^-, *J. Chem. Phys.*, 88, 4516, 1988.

161. **Martyna, G. J. and Berne, B. J.,** Structure and energetics of Xe_n^-: many body polarization effects, *J. Chem. Phys.*, 90, 3744, 1989.

162. **Cohen, M. H., Economou, E. N., and Soukoulis, C. M.,** Microscopic mobility, *Phys. Rev.*, B30, 4493, 1984.

163. **Ascarelli, G.,** Hall mobility of electrons in liquid and solid argon, *Phys. Rev.*, B40, 1871, 1989.

164. **Ascarelli, G.,** Motion of electrons in liquid Xenon: evidence for non-Boltzmann Transport, *Phys. Rev. Lett.*, 66, 1906, 1991.

165. **Blatt, F. J.,** Theory of mobility of electrons in solids, in *Solid State Physics, Advances in Research and Applications,* Vol. 4, Seitz, F. and Turnbull, D., Eds., Academic Press, New York, 1957.

166. **Chakravarty, S. and Schmid, A.,** Weak localization: the quasiclassical theory of electrons in a random potential, *Physics Rep.*, 140, 193, 1986.

167. **Bergmann, G.,** Weak localization in thin films, *Physics Rep.*, 107, 1, 1984.

168. **Polischuk, A. Y.,** Quantum corrections to the Boltzmann equation for electrons in a disordered medium, *J. Phys.*, B16, 3845,1983.

169. **Engels, J. M. L. and Van Kimmenade, A. J. M.,** The mobility of excess electrons in liquid methane, *Chem. Phys. Lett.*, 42, 250, 1976.

170. **Schmidt, W. F. and Allen, A. O.,** Mobility of electrons in dielectric liquids, *J. Chem. Phys.*, 52, 4788, 1970.

171. **Kalinowski, I., Rabe, J. G., and Schmidt, W. F.,** On excess electron mobility in liquid and glassy 3-methyl pentane, *Z. Naturforsch.*, 30a, 568, 1975.

172. **Robinson, M. G. and Freeman, G. R.,** Electron mobilities and ranges in liquid $C_1 — C_3$ hydrocarbons and in Xenon: effects of temperature and field strength, *Can. J. Chem.*, 52, 440, 1974.

173. **Reininger, R., Asaf, U., and Steinberger, I. T.,** The density dependence of the quasifree electron state in fluid Xenon and Krypton, *Chem. Phys. Lett.*, 90, 287, 1982.

174. **Cipollini, N. E. and Allen, A. O.,** Electron mobilities in liquid tetramethyl silane at temperatures up to the critical point, *J. Chem. Phys.*, 67, 131, 1977.

175. **Tauchert, W., Jungblut, H., and Schmidt, W. F.,** Photoelectric determination of V_0 values and electron ranges in some cryogenic liquids, *Can. J. Chem.*, 55, 1860, 1977.

Chapter 7

QUANTUM CALCULATIONS ON EXCESS ELECTRONS IN DISORDERED MEDIA

David F. Coker and Bruce J. Berne

TABLE OF CONTENTS

I. INTRODUCTION

With the advent of supercomputers it has become possible to simulate electron solvation in liquids [1-11] and clusters [12-18] to determine the absorption spectrum of solvated electrons, to understand electron transport in fluids, and, finally, to begin to understand basic electron transfer reactions on a molecular level.[19] This chapter is aimed at providing an overview of this rapidly developing field.

In Section II we review the theoretical methods now being applied to these problems including:

1. Path-integral Monte Carlo methods based on the Feynman path-integral formulation of quantum statistical mechanics
2. Spectral methods based on the propagation of wave functions using fast Fourier transform (FFT) algorithms
3. Diffusion Monte Carlo methods and variational basis set approaches using simulated annealing to determine the ground states of excess electrons in clusters and liquids
4. Adiabatic molecular dynamics propagation on ground state electronic potential energy surfaces
5. Propagation methods for nonadiabatic dynamics

Section III is devoted to a discussion of the molecular modeling of electron-solvent interactions. The most common approach is to treat the overall interaction as pairwise additive over electron-atom pseudopotentials, an approximation in which the electron interacts independently with the dipoles it induces on each of the polarizable atoms. This ignores the dipole-dipole interaction between all of the induced dipoles, and can be corrected by treating the electric field at each atom as a superposition of the unscreened Coulomb field arising from the electron plus the fields due to all of the dipoles induced by the electron (see Sections III.B and III.C). For many systems the problem cannot be reduced to that of a single excess electron moving independently of the other molecular electrons, and the many electron nature of the problem must be treated explicitly. Density functional methods have proven useful for this purpose (see Section III.D).

The equilibrium properties of electrons in a variety of different fluids have been studied using the above methods, and some of these studies are reviewed in Section IV. Structural, thermodynamic, and spectroscopic properties obtained from these calculations indicate that electrons in fluids can exist in either very localized or highly delocalized states depending on the nature of the solvent. The simplest situation occurs in solvents, such as helium where the electron ''sees'' an atomic repulsive core that is the same size as the repulsive core seen by another helium atom. In a dense fluid there will be no room for the electron density to pass between the solvent atoms. The electron will then seek out regions of low helium density where its kinetic energy and potential energy will be minimized. Thus, in dense fluid helium, the electron localizes into a bubble-like state. The solvent is excluded from this region and the electron behaves much like a particle trapped in a spherical box.

Similar trapped states are observed in polar solvents such as water and ammonia. In these systems, the trapping results from the anisotropy of the electron-solvent interaction. The trapped electron disrupts the local solvent structure, causing the dipoles of the surrounding molecules to orient in a way that produces a dipolar cage, or polaron, around the excess electron. Detailed simulation studies of the local solvent environment around excess electrons in both water and ammonia have revealed that the local solvent structure is disrupted least when the solvent dipoles closest to the electron are oriented so that, on average, they point approximately parallel to the surface of the bubble, rather than directly toward the center of the excess charge. Studies indicate that an excess electron in a molten salt exists in a localized state surrounded by a shell of positive ions in an F-center-like environment in the fluid. We also discuss spin pairing of several electrons solvated in a molten salt.

A qualitatively different situation is observed in nonpolar but highly polarizable solvents like liquid xenon, where the excess electron can deeply penetrate the polarizable charge distribution of the solvent molecules. The electron-solvent interaction radius is then considerably smaller than that of the solvent-solvent interaction radius; that is, the electron ''sees'' a smaller xenon repulsive core than that seen by another xenon atom. In the condensed phase, this difference in interaction length scales gives rise to attractive channels between the solvent atoms through which the excess electron can readily move. In such polarizable fluids, the excess electron density behaves like a fluid flowing through a porous medium and exists in an extended state. In branched alkanes the same behavior is expected, but in normal alkanes intramolecular local field effects, coupled with topological disorder, should lead to localized states, and corresponding low mobilities are observed. This section concludes with a presentation of the results of calculations on the electron-xenon system, which demonstrate the importance of many body polarization interactions.

If the states of excess electrons are highly localized, the electronic energy levels are widely separated compared to k_BT. Then electronic transport phenomena, to a good approximation, will involve nuclear motion on the ground electronic state potential surface since nonadiabatic transitions to excited states will occur infrequently. In Section II we present the equations of motion of the electron (see, for example, equations beginning with Equation 12), and in Section IV.A we discuss the results of adiabatic dynamics calculations on electrons in molten salts. These results indicate that electronic transport in these system involves sudden hops between localized states. In contrast, the bubble structure of the hydrated electron remains intact and slowly diffuses through the solution, being buffeted along by solvent motions. We also discuss the nonadiabatic transient relaxation phenomena involved when an initially delocalized free electron scatters through the fluid, dissipating energy, till eventually it finds a trap in which it localizes and equilibrates.

There has been considerable interest in the nature of the quantum states of an excess electron interacting with atomic and molecular clusters. What is the smallest cluster that will bind an electron? Does the excess electron get trapped inside the bulk of the cluster and have properties similar to electrons in fluids, or does the electron density accumulate on the surface of the cluster? Section VI summarizes the results of quantum simulations of electron attachment to water and xenon clusters.

The most fundamental chemical process involves the transfer of an electron between a pair of chemical species. When the intervening solvent between the two species reorganizes to some special transition state geometry which makes it no longer favorable for the electron to remain on the donor, the electron will hop to the acceptor species. In section VII, we discuss the first attempts to develop a microscopic understanding of these processes from simulation studies of a model of the hydrated $Fe^{2+}-Fe^{3+}$ electron transfer system. We also discuss some path-integral calculations on electron tunneling pathways through proteins.

This review is intended to display the wide applicability of these very general quantum simulation methods. We have tried to indicate what can currently be done with these powerful techniques and also the aspects of electron solvation problems which cannot yet be addressed by these methods. We offer some speculations on approaches which may prove fruitful in answering these questions at the frontier of our understanding.

II. QUANTUM SIMULATION METHODS

The computer simulation methods for studying excess electons in disordered media all have their formal basis in making a short time approximation to a propagator for the motion of the quantum particle. The solvent, on the other hand, is usually treated using either the classical Monte Carlo method, in which the Boltzman distribution is used to sample configurations of the classical particles, or classical molecular dynamics, in which Newton's equations are solved to give dynamical trajectories of the solvent coordinates.

In this section, we first review the path-integral techniques for calculating thermally averaged properties of excess electron systems. Next, we describe FFT propagator methods, a simulated annealing variational basis set approach, and the diffusion Monte Carlo method. These techniques can be used to calculate the Born-Oppenheimer eigenstates of an excess electron moving in the potential field of a static configuration of solvent molecules. Finally, we outline adiabatic dynamics methods which have been used to study nuclear dynamics on the ground excess electronic surface and briefly discuss modeling nonadiabatic processes.

A. THERMAL AVERAGES AND PATH-INTEGRAL METHODS

The partition function for an electron moving in a potential field $V(r)$ is

$$Z = \text{Tr}[e^{-\beta H}] \tag{1}$$

where $H = -(\hbar^2/2m)\nabla^2 + V(r)$ and $\beta = 1/k_B T$. The partition function can be written as

$$Z = \text{Tr}[(e^{-\beta H/P})^P] \tag{2}$$

and if we insert P complete sets of position representation states $\int dr_1 |r_1\rangle\langle r_1| = 1$, the partition function becomes

$$Z = \int dr_1. \, . \, . \int dr_P \langle r_1|e^{-\beta H/P}|r_2\rangle. \, . \, .\langle r_P|e^{-\beta H/P}|r_1\rangle \tag{3}$$

Equation 3 is still an exact expression for the partition function but it is not computationally useful in this form. We proceed by recognizing that at the high temperature, PT, the density matrix elements appearing in this expression can be approximated by free particle propagation with a perturbation by the potential term, so that

$$\lim_{P\to\infty}\langle r_1|e^{-\beta H/P}|r_2\rangle = \left(\frac{mP}{2\pi\hbar^2\beta}\right)^{3/2} \exp\left[-\frac{mP}{2\hbar^2\beta}(r_1 - r_2)^2\right]\exp\left[-\frac{\beta}{2P}\{V(r_1) + V(r_2)\}\right] \tag{4}$$

Substituting Equation 4 into Equation 3, we find that $Z = \lim_{P\to\infty} Z_P$, where

$$Z_P = \left(\frac{mP}{2\pi\hbar^2\beta}\right)^{3P/2} \int dr_1. \, . \, . \int dr_P \exp[-\beta V_{eff}(r_1,. \, . \, .,r_P)] \tag{5}$$

$$V_{eff}(r_1,. \, . \, .,r_P) = \sum_{i=1}^{P}\left\{\frac{mP}{2\hbar^2\beta^2}(r_i - r_{i+1})^2 + \frac{1}{P}V(r_i)\right\} \tag{6}$$

Since Equation 5 is equivalent to the classical configurational partition function of P particles with potential V_{eff}, the quantum system is said to be isomorphic to a classical P particle cyclic chain polymer in which each particle i interacts with its neighbors $i - 1$ and $i + 1$ through a harmonic potential with force constant $mP/\beta^2\hbar^2$, and each particle i experiences an attenuated potential $V(r_i)/P$.

The classical isomorphism embodied in Equations 5 and 6 has several features worth noting. For a free electron, the rms bond length in the isomorphic chain is proportional to $(\hbar^2\beta/mP)^{1/2}$. The classical isomorphism will be a good approximation only if the potential, $V(r)$, does not vary much over the rms bond length. If σ is a characteristic distance specifying the range of $V(r)$, then P must be chosen such that $(\hbar^2\beta/mP) << \sigma^2$, and we see that the lower the temperature (i.e., the larger the value of β), the larger P must be to maintain this condition. In path-integral simulations one empirically determines a P sufficiently large that the thermodynamic properties cease changing for larger P values.

In most simulations of electrons in liquids, it is assumed that the solvent can be treated classically, but that the electron must be described quantum mechanically as outlined above. This approximation can be justified by recognizing that if all the solvent atoms were treated quantum mechanically as discrete paths we would have an isomorphic classical system of many interacting polymers, one for each solvent atom as well as the one for the electron. The force constants for the harmonic bonds of the isomorphic polymers representing a solvent molecules would be thousands of times larger than the force constant for the electron polymer, since these force constants are proportional to the masses of the particles. Thus, in the discrete path-integral picture, all the beads of an isomorphic polymer representing a given solvent molecule will be essentially localized at a point compared to the diffuse electron polymer. When all the polymer beads collapse to a point, the classical configurational partition function is recovered. In this approximation the electron-liquid partition function is then given by

$$Q_P(\beta) = \left(\frac{mp}{2\pi\hbar^2\beta}\right)^{3P/2} \int dR^N \exp\left[-\beta\Phi(R^N)\right] \int dr_1 \ldots \int dr_P$$

$$\times \exp\left[-\frac{P}{2\lambda^2}\sum_{i=1}^{P}(r_i - r_{i+1})^2\right] \exp\left[-\frac{\beta}{P}\sum_{i=1}^{P} V(r_i, R^N)\right] \tag{7}$$

This is the path-integral version of the Born-Oppenheimer approximation. Here $\Phi(R^N)$ describes how the N classical solvent atoms interact with one another, $V(r,R^N)$ describes the interaction between the excess electron and the solvent, and $\lambda = (\hbar^2\beta/m)^{1/2}$ is the thermal wavelength of the electron. In general, in this adiabatic approximation the average of some property $O(r,R^N)$ of an excess electron in a classical fluid can be written as

$$\langle O \rangle = \left(\frac{mP}{2\pi\hbar^2\beta}\right)^{3P/2} \int dR^N \exp\left[-\beta\Phi(R^N)\right] \int dr_1 \ldots \int dr_P \frac{1}{P}\sum_{i=1}^{P} O(r_i, R^N)$$

$$\times \exp\left[-\frac{P}{2\lambda^2}\sum_{i=1}^{P}(r_i - r_{i+1})^2\right] \exp\left[-\frac{\beta}{P}\sum_{i=1}^{P} V(r_i, R^N)\right] \tag{8}$$

Averages of excess electronic properties over a thermal distribution of quantum states can be calculated very efficiently using this discretized Feynman path-integral approach.

As stated before, the integral in Equation 8 is isomorphic to a classical statistical mechanical average of a property of a P particle cyclic polymer system moving through a classical fluid. Particles which follow one another in the polymer exert harmonic forces on their neighbors, and each polymer particle interacts with all the solvent atoms. Path-integral calculations proceed by generating configurations of the fluid atoms and cyclic polymer chain according to the distribution

$$P(R^N, r_1, \ldots r_P) = \exp\left[-\beta\Phi(R^N)\right] \exp\left[-\sum_{i=1}^{P}\left[\frac{P}{2\lambda^2}(r_i - r_{i+1})^2\right.\right.$$

$$\left.\left. - \frac{\beta}{P} V(r_i, R^N)\right]\right] \tag{9}$$

and averaging the property of interest over the set of configurations. Various importance sampling Monte Carlo methods have been developed to sample this distribution efficiently.[1,20,21] The key feature of these algorithms is their ability to generate acceptable new configurations of the electron path which involve substantial displacement of many particles. This is achieved by directly sampling the Gaussian free particle part of the probability distribution to generate new segments of the path which are accepted on the basis of the change in the interaction potential part of the distribution.

The equilibrium path-integral methods outlined above involve representing the cyclic electron path as a discrete set of points in space. The integral over all possible cyclic paths is then performed using Monte Carlo or molecular dynamics methods to integrate over all possible positions of these discrete points. Alternatively, to represent a path of a quantum particle which is simply a three-dimensional function of time, one could expand the path in terms of some basis set of functions. Path-integrals can then be performed by integrating over all possible values of the expansion coefficients. In obtaining equilibrium properties, the paths of interest are cyclic; thus, a Fourier series provides an appropriate basis set expansion. Fourier path-integral methods have been developed by Freeman and Doll,[22,23,24] who have applied this approach to study the importance of quantum effects in rare gas clusters. However, this approach has not yet been applied to excess electron systems.

In addition to obtaining simple thermodynamic averages of the properties of solvated electron systems, path-integral methods can be used to compute the free energies of electron solvation. Klein and his co-workers[25] have developed a method to obtain an estimate of free energy of solvation of an excess electron in liquid ammonia. Their approach involves charging up the electron from zero to one, then performing a thermodynamic integration of the average electron-solvent interaction energy as the electronic interactions are gradually turned on in the system. The results of these calculations agree well with experimental measurements. These workers have also developed a constant-pressure path-integral method for obtaining enthalpy changes for electron solvation processes.[26] This approach involves sampling volume fluctuations in the electron-solvent system.

B. EIGENSTATES AND FFT PROPAGATOR METHODS

Path-integral simulations provide a record of solvent atom configurations which are in thermal equilibrium with the excess electron at a given temperature. If the system is stripped of the classical isomorphic chain representing the electron, one is left with a set of configurations of the liquid which have been distorted by the presence of the excess electron. It is possible to solve the Schrödinger equation for the excess electron in each of these distorted configurations and thereby to determine eigenenergies and eigenstates of the solvated electron. By averaging over the distribution of distorted configurations, we can then calculate the inhomogeneous absorption spectrum of the excess electron. Thus, working in the Born-Oppenheimer approximation, we can use the perturbed configurations of solvent atoms produced by path-integral calculations as static potential fields through which the quantum electron moves rapidly. The electronic eigenstates of these disordered potential fields have been calculated for a variety of systems using discrete grid propagator methods. These calculations give information about dynamical electronic properties in these disordered systems, such as the absorption spectrum and the density of excess electronic states.

The lowest few localized electronic states in polar solvents like water and ammonia and the bubble-like trapped states observed in fluid helium have been studied by taking solvent configurations from path-integral calculations and using the propagator methods which we shall now outline.

The quantum operator which propagates the state of a system is $\exp[-H\tau]$. If $\tau = it/\hbar$, this operator propagates the state in real time, so that, since $\psi(r,t) = \Sigma_n a_n \exp[-iE_n t/\hbar]\phi_n(r)$, the amplitudes of the contributions of the various eigenstates, $\phi_n(r)$, to the wave function will oscillate in time at the eigenfrequencies $E_n/\hbar$. On the other hand, if τ is a real variable so that t is imaginary, and if the energy scale is chosen so that the eigenvalues are positive, then the various eigenstate contributions will decay with exponential rate constants which are proportional to the eigenvalues.

Algorithms involving both real and imaginary time propagation have been presented. In real time techniques, correlation functions of some arbitrary initial wave function with the propagated solution are calculated. These ocillatory correlation functions are then Fourier analyzed, and the frequencies obtained are the eigenvalues of the component eigenfunctions contained in the initial wave function.[27-29]

Alternatively, in imaginary time algorithms, the contributions from states in the initial wave function which have higher eigenvalues will decay most rapidly leaving, at long times, a wave function which is dominated by the lowest lying eigenstate contribution in the initial wave function. With this approach the eigenvalue can be estimated by monitoring the rate of decay of the wave function normalization at long times. Higher energy states can be determined with this technique by filtering previously determined lower lying states from the propagating wave function by Gram-Schmidt-type orthogonalization.[2,13,30-35]

Central to these methods is a technique for propagating the wave function. For a single quantum particle we set up the initial wave function on a discrete grid in three-dimensional space. Next we apply a short time approximation for the propagator to this wave function to move it through a time increment $\Delta\tau$. Like the short time approximation employed in obtaining the discrete path-integral expression discussed above, the incremental propagation involves free particle motion for a short time followed by perturbation by the potential. The free particle propagator is most easily applied in momentum space, where it has the simple form $\exp[-p^2\Delta\tau/2m]$. Thus, we first Fourier transform the wave function on the grid in position space to a grid in momentum space by FFT techniques, then simply multiply the wave function by the free particle propagator. Perturbation by the potential is most conveniently performed in position space, where the potential part of the short time propagator has the form $\exp[-V(r)\Delta\tau]$. Therefore, to complete the short time propagation cycle we back-transform the partially propagated wave function to position space and multiply by the potential part of the propagator. Many such short time propagation cycles result in accurate propagation to long times.

The FFT propagator method outlined above is a special case of a more general approach for applying the propagator $\epsilon^{-\Delta\tau H}$ so that the wave function may be evolved. If A and B are two noncommuting operators, like the kinetic and potential energy operators, that make up H, then $\lim_{\alpha\to 0} e^{-\alpha(A + B)} = e^{-\alpha A}e^{-\alpha B}$. This result enables us to split the propagation up into separate operations of the kinetic and potential propagators, provided $\Delta\tau$ is sufficiently small. With the FFT, approach the Hamiltonian is naturally split up into its kinetic and potential components, and the wave function is expanded in terms of the complete set of orthogonal eigenfunctions of the kinetic operator, i.e., $e^{ik\cdot r}$. We choose to expand the wave function in terms of this basis so that the kinetic propagator will be diagonal and may be applied to the wave function in this space by simple multiplication. These ideas are completely general in that we can split the Hamiltonian up any way we choose, and if the various parts are noncommuting, we must repeatedly apply the short time approximation. Propagation is most conveniently performed when the orthogonal eigenfunctions of the different pieces of the Hamiltonian are used as basis functions to expand the wave function, resulting in diagonal representations of the various parts of the propagator. Recently Webster et al. have developed a Block Lanczos method for refining the FFT propagation method outlined here to eliminate the short time step error.[63a,b]

The Fourier or plane wave basis has been very useful for representing the wave functions of electrons in fluids, but it is a particularly poor choice to represent the wave functions of electrons attached to certain types of clusters. In rare gas atom clusters containing tens of hundreds of solvent molecules, the attached electron is often found in a very diffuse surface state in which electron density can extend out to tens of cluster diameters. However, these wave functions change rapidly only in the small region near the cluster. The uniform grid of points in space employed with the FFT basis would thus have to extend out to large distances to accurately represent the slowly varying tail of the wave function; and also be densely packed with points to accurately account for the rapid variation of the wave funciton close to the cluster. Use of such a basis for this type of electron-cluster problem is clearly extremely inefficient.

Martyna and Berne[13] choose to write the excess electronic Hamiltonian of the cluster system in spherical polar coordinates and propagate the function $r\psi$. The propagator is thus conveniently split into a potential piece, a radial kinetic component, and an angular kinetic part. With this choice, each component of the short time propagator can be applied by transforming between a sine basis for the radial part and a basis of associated Legendre functions for the $\cos\theta$ part. The position space is defined by equally spaced grid points in r and ϕ and Gaussian quadrature points in $\cos\theta$. These quadrature points are spaced to efficiently reproduce both the angular dependence of the wave function and its radial variation close to the origin.

In some cluster systems such as larger clusters of dipolar molecules for example on the order of ten water or ammonia molecules of in ionic clusters, there is evidence that some of the excess electronic states are localized in the interior of the cluster in states reminiscent of those in the bulk.[34,35] Under these circumstances the simple plane wave basis approach may give accurate results.

C. VARIATIONAL BASIS SET EXPANSION METHODS: SIMULATED ANNEALING OPTIMIZATION OF PARAMETERS

An alternative to the method outlined above for obtaining the eigenstates of an excess electron has been presented by Sprik et al.[36] This approach involves expanding the eigenfunctions in terms of sets of basis functions and performing a variational calculation to determine the parameters in the basis.

With the imaginary time FFT approach discussed above, a basis set of plane waves is employed to represent a wave function, but not for the purposes of a variational calculation. Rather, this basis is used to enable the application of a short time approximation for the propagator so that the Schrödinger equation may be integrated and the wave function evolved in time.

Sprik and Klein employ a distributed Gaussian basis set (which should be useful for representing localized states of excess electrons in fluids) and variationally optimize the amplitudes $\{A_i\}$, widths $\{w_i\}$, and positions $\{r_i\}$ of these Gaussians. Instead of solving the secular equation which results from minimizing the excess electronic energy function $E(\{A_i\}, \{W_i\}, \{r_i\})$ with respect to variations in the Gaussian parameters, these workers employ an efficient optimization procedure developed recently by Car and Parrinello[37] which is generally useful for complicated nonlinear variational problems (see Section III.D). The approach involves treating the variational parameters as fictitious classical dynamical variables whose motion is derived from an artificial Lagrangian (see the discussion surrounding Equation 40).

This approach can be conveniently applied to calculate the adiabatic dynamics of the classical solvent coordinates by obtaining the optimum ground state, and then allowing the nuclei to move. The Lagrangian includes the solvent kinetic and intermolecular potential energy, and the extended set of classical equations of motion which move the solvent and adjust the variational parameters are solved using molecular dynamics. If the masses for the variational parameters are chosen to be sufficiently small, their relaxation times can be made much faster than the solvent motion. Thus by continually rescaling the fictitious velocities of the parameters, the simulated annealing will allow the electronic distribution to respond instantantly to the slow variation of the solvent dynamics, giving adiabatic nuclear motion in which the electronic wave function remains close to the instantaneous Born-Oppenheimer ground state. The adiabatic molecular dynamics approach is discussed in more detail in Section II.E.

D. THE DIFFUSION MONTE CARLO METHOD

The final technique discussed here for studying the quantum states of electrons moving in disordered potential fields is the diffusion Monte Carlo method[38-42] which has been used to solve the imaginary time Schrödinger equation for the ground states of excess electrons moving in the potential field of frozen clusters of solvent molecules.[12] The approach involves simulating the action of each term in the imaginary time dependent Schrödinger equation

$$\frac{\partial \psi(r,\tau)}{\partial \tau} = \left[\frac{\hbar^2}{2m_e} \nabla^2 - (V(r) - V_{ref}) \right] \psi(r,\tau) \tag{10}$$

Here V_{ref} is a reference energy and $V(r)$ is the electron-solvent interaction potential. This equation is simulated by creating an ensemble of replicas of the electron-cluster system. The members of this ensemble are moved through space in such a way that the long time limit of the spatial distribution of replica electrons gives the ground state wave function. By dividing the time up into small increments $\Delta\tau$, we can assume that each term in the Schrödinger equation modifies the ensemble distribution independently. The kinetic term is isomorphic to the diffusion operator, so at each time step each replica electron is given a random displacement chosen from a Gaussian distribution as dictated by this diffusion term. The potential in the Schrödinger equation resembles a spatially varying birth-death or chemical reaction rate constant. Therefore, at each time step, the action of the potential is modeled by allowing each replica system to give birth to more replicas of itself or to die, depending on the value of its potential energy relative to V_{ref}.

The diffusion Monte Carlo method can be made very efficient by the use of importance sampling if we have some approximate information about the shape of the wave function in the form of some trial function ψ_T. The Schrödinger equation can be transformed from an equation for $\psi(r,\tau)$, the wave function, to an equation for $f(r,\tau) = \psi(r,\tau)\psi_T(r)$, where ψ_T is an arbitrary function of the electron coordinates.[38,42,43]

$$\frac{\partial f(r,\tau)}{\partial \tau} = \frac{\hbar^2}{2m_e} \nabla^2 f(r,\tau) - \frac{\hbar^2}{m_e} \vec{\nabla} \left[f(r,\tau) \cdot \vec{\nabla} \log\psi_T(r) \right] - \left[\frac{H\psi_T(r)}{\psi_T(r)} - V_{ref} \right] f(r,\tau) \tag{11}$$

The first and third terms on the right-hand side of this equation are simulated as outlined above, by analogy with diffusion and birth-death processes, respectively. The second term is modeled by analogy with an equation describing the drift of a density under the influence of a concentration gradient. Consequently, at each time step, the replicas are given an additional drift, whose magnitude and direction are determined by ψ_T. The drift term thus acts to push replicas into the regions where ψ_T is largest. If ψ_T is close to the ground state wave function, $H\psi_T/\psi_T$ will be approximately the ground state energy and will be nearly independent of position. This effect reduces fluctuations in the ensemble and greatly decreases the statistical errors of the calculations.[43]

E. ADIABATIC MOLECULAR DYNAMICS METHODS

In principle, path-integral methods like those outlined in Section II.A can be used to obtain dynamical properties of quantum systems. Formally, one simply substitutes $\beta = it/\hbar$ in all the path integral results in section II.A and obtains expressions for the real time quantum propagator. The problem with using these results to calculate dynamical properties is that the weights for the different quantum paths over which we must integrate are now complex. In general, the weighting function can now be a rapidly oscillating function when one makes small changes from one path to the next. This oscillation results in severe Monte Carlo path sampling problems. Various "stationary phase" sampling procedures have been devised by a number of groups[44-50] to search out the most important quantum paths, but considerable development of these methods is still required before it is practical to apply these techniques to study the dynamics of solvated electron systems. Recently a promising new path integral approach to the calculation of real time path integral involving distortion of the integration contours and stationary phase filtering has been proposed.[51]

However, in certain circumstances, it is possible to use the methods outlined above for obtaining the individual quantum states to simulate the real time dynamics of a solvated electron. If the electronic energy levels are widely separated compared to k_BT, the electron will be found in its ground state, and the system is said to be ground state dominated. Excess electrons in high density He, liquid water, fused salts, and liquid n-alkanes are ground state dominated. In this case the electron moves on the ground state potential energy surface, and it is possible to simulate electron solvation dynamics. Adiabatic equations of motion for a coupled classical-quantum system have been written down by several workers.[52] A derivation has recently been presented by Thirumalai et al.,[53] who have employed a path integral approach presented by Pechukas,[54,55] to obtain the equations of motion.

The Hamiltonian for the mixed quantum-classical system is

$$H = H(P^N,R^N) + T + V(r,R^N) \tag{12}$$

or

$$H = H(P^N,R^N) + h(r,R^N) \tag{13}$$

where the first term on the right of Equation 12 is the Hamiltonian of the fluid, the second term is the kinetic energy of the electron, and the third term is the electron-fluid pseudo-potential. The equations of motion for the fluid atoms are

$$M\ddot{R}_j = -\nabla_{R_j} H(P^N,R^N) - \nabla_{R_j} V_{eff}(R^N,t) \tag{14}$$

where

$$V_{eff}(R^N,t) = \langle \psi_o(r,t)|h(r,R^N)|\psi_o(r,t)\rangle \tag{15}$$

and where the ground state wave function $\psi_0(r,t)$ is determined using the various methods described above. The basic approach is

1. Given an initial solvent configuration, use grid methods to calculate $\psi_0(r,0)$
2. Determine the forces $-\nabla_{R_j} V_{eff}(R^N,0)$ and advance the solvent configuration one time step, Δt, using dynamics
3. Compute $\psi_0(r,\Delta t)$ corresponding to the new fluid configuration
4. Repeat the above procedure

Thirumalai et al. solved these equations in the Gaussian wave packet approximation for the evolution of the vibrational state following photoexcitation of a Br_2 molecule in solid argon. More recently, there have been a host of applications of these equations to electron solvation and a wide variety of other problems, such as studying the ground and excited state dynamics of electrons in bulk water,[4,56,57] ammonia,[36] water clusters,[33-35,58] and molten salts.[11] In fact this approach has become so popular that some authors have assigned their own acronyms to the above molecular dynamics procedure and claim priority for it. Unfortunately, this procedure fails unless the system is strongly ground state dominated. In systems where the electron is in an extended state, the energy levels are closely spaced and there are nonadiabatic transitions. Under these conditions, the above equations give a poor approximation to the true dynamics. Surface hopping techniques, which are outlined briefly in Section II.F, must then be applied. It behooves the simulator to demonstrate the validity of the adiabatic equations if they are being used.

In their applications to the problem of electrons in a molten salt, Parrinello and his co-workers[11] use the real time FFT propagator approach to evolve the time dependent Schrödinger equation for the motion of the electronic degrees of freedom and a molecular dynamics approach to evolve the classical nuclear equations of motion. The system thus moves in a mixed quantum state, in which the nuclei experience a potential surface which is a weighted average of the different adiabatic potential curves. If the time evolving wave function is strongly dominated by one state, this approximation is reasonable, but if the states are strongly mixed, the results are more complicated to interpret.

Consequently, these workers must check on the adiabaticity of their dynamics by periodically projecting the evolving electronic wave function onto the instantaneous adiabatic ground state, which they determine variationally. They see short-lived, infrequent deviations from adiabatic motion on the order of 10 to 20%. The reliability of adiabatic dynamics algorithms under these circumstances is questionable, as these methods cannot really say anything about processes. Departures from adiabaticity of this magnitude merely indicate that application of an adiabatic algorithm to this system is probably inappropriate. With adiabatic dynamics algorithms which involve calculating the Ehrenfest nuclear forces by using the instantaneous adiabatic ground state wave function obtained from, say, imaginary time FFT propagation, at least the dynamics is well defined in that it involves motion over a single adiabatic potential energy surface. For real time propagation procedures where significant departures from adiabaticity are observed, the dynamics occasionally involves motion over an ill-defined mixture of potential surfaces, and there is no clear interpretation of such results.

F. NONADIABATIC DYNAMICS METHODS

A semiclassical approach to nonadiabatic transitions given by Nikitin[59] was developed by Tully and Preston[60,61] in their trajectory surface hopping approach to nonadiabatic molecular collisions. Variations on this approach are currently being developed to study nonadiabatic processes in solvated electron systems.

Briefly, within the semiclassical theory, we treat the nuclei classically and the electronic degrees of freedom quantum mechanically. The electronic wave function thus becomes time dependent as a result of the time-dependent potential provided by the classical nuclear motion. The electronic wave function $\psi(r,t)$ satisfies the time dependent Schrödinger equation

$$i\hbar \frac{\partial \psi(r,t)}{\partial t} = H_e(r,R(t))\psi(r,t) \tag{16}$$

where $H_e(r,R(t))$, is the electronic Hamiltonian for the fixed nuclear configuration at time t. We proceed by expanding the time-dependent wave function in the basis of instantaneous adiabatic eigenstates $\phi_j(r,R(t))$, where the instantaneous adiabatic eigenfunctions $\phi_j(r,R(t))$ and eigenvalues $E_j(R(t))$ are parameterized by the nuclear configuration and satisfy

$$H_e(r,R(t))\phi_j(r,R(t)) = E_j(R(t))\phi_j(r,R(t)) \tag{17}$$

thus:

$$\psi(r,t) = \sum_j a_j(t)\phi_j(r,R) \exp\left[-i/\hbar \int^t E_j(R)dt \right] \tag{18}$$

Substituting Equation 18 into Equation 16, we find that the time-dependent expansion coefficients satisfy the following system of coupled equations:

$$\frac{da_j}{dt} = - \sum_i a_i \langle \phi_j \left| \frac{\partial}{\partial t} \right| \phi_i \rangle \exp\left[-i/\hbar \int^t (E_i - E_j)dt \right] \tag{19}$$

The probability $|a_j(t)|^2$ that the electronic degrees of freedom are in state j at time t can be calculated from Equation 19 for any assumed nuclear trajectory R(t). The problem, then, is to find the right semiclassical nuclear trajectory, or equivalently, the time-dependent effective nuclear potential $V_{eff}(R(t))$, from which R(t) can be obtained by solving the classical equations of motion. The time dependence of this potential will be influenced by the evolution of the electronic distribution, which can only be determined if we know the nuclear trajectory. Consequently, to obtain the rigorous solution for the nonadiabatic semiclassical nuclear trajectory, we are faced with a problem that must be solved self-consistently.

The surface hopping trajectory method of Tully and Preston circumvents this difficult self-consistent problem by making assumptions about the strength of the coupling matrix element $\langle \phi_j | {}^{\partial}/_{\partial} | \phi_i \rangle$ between the adiabatic states. Consider the two state situation and suppose the system starts out in State 1. In the adiabatic case, the coupling between the two states remains small and the system stays in Electronic State 1, so the effective potential should be well approximated by $E_1(R(t))$. In the diabatic situation, on the other hand, the coupling between the states is very strong in an extremely narrow region of configuration space. Consequently, probability $|a_2|^2$ of finding the electronic distribution in State 2 changes abruptly from 0 to 1 as the trajectory passes through the transition region. The surface hopping trajectory approach employs an approximate effective potential which is consistent with both the adiabatic and diabatic extreme situations:

$$V_{eff}(R) = |a_1|^2 E_1(R) + |a_2|^2 E_2(R) \tag{20}$$

The nonadiabatic surface hopping trajectory method thus involves simultaneously solving Newton's equations for the classical nuclear coordinates and the expansion coefficient equations of motion (Equation 19). Depending on the probabilities $|a_j|^2$, a stochastic approach is employed to split the trajectory into branches on the different adiabatic electronic surfaces.

The surface hopping trajectory method is an approximate approach to nonadiabatic nuclear dynamics which avoids iterating the classical nuclear, and quantum electronic parts of the problem to self-consistency.

Pechukas[54,55] derived a rigorous self-consistent iterative method for finding the nonadiabatic semiclassical nuclear trajectory by considering a path integral expression for the reduced propagator which gives the probability for the nuclear coordinates to evolve from R' to R'' as the electronic state changes from $\phi_\alpha(r)$ at time t' to $\phi_\beta(r)$ at time t''. Pechukas expands this propagator about the path of stationary phase to obtain an iterative prescription for finding this semiclassical path of nuclear coordinates. The approach involves solving a Newton's equation of the form

$$M_i \ddot{R}_i = -\nabla_{R_i} V(R,t) \tag{21}$$

where the time-dependent potential depends on the classical path and has the form

$$V(R,t) = \frac{\int dr \psi^{\beta *}(r,t,t'') H_e(r,R(t)) \psi^{\alpha}(r,t,t')}{\int dr \psi^{\beta *}(r,t,t'') \psi^{\alpha}(r,t,t')} \tag{22}$$

Here the wave functions $\psi^{\alpha}(r,t,t')$ are the solutions of the time-dependent Schrödinger equation with initial condition $\psi^{\alpha}(r,t',t') = \phi_{\alpha}(r)$ where ϕ_{α} is the initial adiabatic eigenstate.

The iterative procedure involves guessing a nuclear path which specifies $H_e(r,R(t))$ and solving the time-dependent Schrödinger equations for ψ^{α} and $\psi^{\beta *}$, then evaluating $V(R,t)$ and solving Newton's equations to obtain a new nuclear path. The iteration process is repeated till convergence, when the nuclear coordinates no longer change. Another alternative approach to nonadiabatic nuclear dynamics has been presented by Miller and George.[62]

At this stage, very little is known about the sorts of nonadiabatic processes which are to be expected in solvated electron systems. For these systems, many different dynamical situations in which nonadiabatic processes may be important can be envisioned. For example, when we inject an energetic electron in a delocalized plane wave like state into an unperturbed fluid, it will scatter through the system, losing its energy, until it eventually equilibrates, perhaps significantly distorting the fluid and giving rise to a localized equilibrium state. As discussed in Section V.C, nonadiabatic processes may be important in such an equilibration in the hydrated electron system. Alternatively, when an electron can exist in various localized excited states in a bubble in the fluid, if it is excited into one of these states, what sorts of nonradiative nonadiabatic transitions are responsible for the relaxation of the system into its equilibrium ground state? For localized states in molten salts, solvent fluctuations can apparently lead to nonadiabatic transitions out of the ground state. These transitions may have a significant effect on electron transport in these systems.

Recently Space and Coker[63] have developed a nonadiabatic surface hopping dynamics method to study nonadiabatic processes of excess electrons influid helium. Webster, Friesner, and Rossky[63a] have also been working to develop a nonadiabatic dynamics approach for exploring electron hydration dynamics.

III. ELECTRON-SOLVENT INTERACTIONS

A. PAIRWISE ADDITIVE PSEUDOPOTENTIALS

An excess electron interacts with all of the electrons of the solvent molecules through a sum of Coulomb potentials, but both analytical theory and numerical simulation often use simplified electron-molecule pseudopotentials. Pseudopotential theory has a long history in electron-molecule scattering theory.

In the simplest pseudopotentials, positive and negative partial charges are assigned to sites within the solvent molecules and the excess electron interacts through Coulomb forces with these sites. To avoid negative divergences and artificial bound states, the interaction between the electron and the positive partial charges is set equal to a constant for distances closer than a certain cutoff radius.

More elaborate pseudopotentials have been constructed, but most still include a static Coulomb potential to account for the interaction between the electron and the unperturbed charge distribution, which is modeled by partial charges or multipole terms on the target molecule.

At large separations, the asymptotic behavior of the pseudopotential is usually taken to be a spherical interaction of the form $-\alpha_i/2r^4$ centered on each atom of polarizability α_i in the solvent molecule. This term accounts for the polarization of the molecular charge distribution by the electron. At short distances, this term becomes unphysical, due to the neglect of the spatial variation of the molecular charge distribution being polarized. The short range attractive divergence of the polarization term is often removed by replacing r^{-4} by $(r^2 + r^2_{vdw})^{-2}$ where r_{vdw} is the van der Waals radius of a given atom. More elaborate polarization switching functions have been used to account for the spatial variation of the charge density within the target molecule.[5,13,14]

At short distances, the classical electrostatic description of the electron-solvent interaction must be abandoned due to quantum mechanical exchange and correlation effects. The total wave function describing the molecular electrons together with the excess electron must be antisymmetric with respect to interchange of identical electrons. Within the Hartree-Fock self-consistent field formulation of the many electron problem, Slater[64] devised a local approximation to the nonlocal exchange potential by using free electron gas results. In this X_α approximation, the exchange interaction energy of an excess electron at r is proportional to the cube root of the electron density, $\rho(r)$, of the target molecule at that point, so $V_{X_\alpha} = -\alpha e^2(3\pi^2\rho(r))^{1/3}/\pi$. Here the αs are constants (not polarizabilities) which have been tabulated from atomic Hartree-Fock calculations. This is the simplest local approximation for the exchange interaction. Density functional theory, which is briefly described in Section III.D, provides a framework in which more accurate local exchange approximations can be incorporated.

Finally, these more sophisticated pseudopotentials usually include a core repulsion term which accounts for the fact that the spin orbital describing the excess electron must be orthogonal to the spin orbitals containing the molecular electrons. The repulsion is a consequence of the Pauli exclusion principle, which dictates that the excess electron cannot be pushed into the already filled orbitals of the target molecule. The term is usually calculated using only the valence orbitals of the molecule.

To summarize, electron-molecule pseudopotentials often incorporate some or all of the following:

1. Electrostatic interactions with unperturbed atomic charge distribution
2. Exchange potential
3. Repulsive core interactions arising from orthogonality
4. Attractive static polarization terms arising from the electron interaction with the induced dipoles.

B. MANY BODY POLARIZATION INTERACTIONS

Electron-molecule pseudopotentials were formulated according to the electron-molecule scattering theory, in which the electron interacts with a single gas phase molecule. Although it is quite common for investigators to use a pairwise additive superposition of these electron-molecule pseudopotentials to describe electron scattering and solvation in liquids, this approach is not justified, because pairwise additivity ignores very important many body polarization effects in fluids of highly polarizable molecules such as xenon, hydrocarbons, and even water.

The induced dipole at each atom of the fluid is just the polarizability of the given atom times the local electric field at the atom. The local electric field at a given atom is the superposition of the Coulomb field of the excess electron and the electric fields due to all the dipoles induced by this electron. This reasoning gives rise to a set of 3N coupled linear equations for the electric fields that must be solved self-consistently in order to determine the electrostatic polarization energy. For a system with N solvent particles a 3N × 3N matrix equation must be inverted in order to evaluate the electron-solvent interaction potential.

Explicitly,[65,66] the many body polarization energy of an electron at position r_e interacting with an assembly of N polarizable centers at positions R_i can be written in the form

$$V_p^{MB}(r_e, R^{3N}) = -\frac{1}{2} \sum_{i=1}^{N} \mu_i \cdot E_i^{(0)} \tag{23}$$

where $E_i^{(0)}$ is the electric field at the ith polarizable center due to the excess electron:

$$E_i^{(0)} = \frac{e(R_i - r_e)}{|R_i - r_e|^3} [S(|R_i - r_e|)]^{1/2} \tag{24}$$

and μ_i is the induced dipole moment of the i-th center. The function $S(r)$ gradually switches off the electric field arising from the excess electron. This function is needed because as the excess electron penetrates the molecular charge distribution, it polarizes the charge density and causes it to vary spatially.

The induced dipole moment in the ith center is defined as

$$\mu_i = \alpha_i E_i \tag{25}$$

where α_i is the static polarizability of the i-th solvent atom and E_i is the local electric field at the center of this atom, which is the superposition of the field due to the excess charge and the fields associated with the induced dipole moments in the other solvent atoms:

$$F_i = F_i^{(0)} + \alpha \sum_{k \neq i}^{N} T_{i,k} \cdot F_k \tag{26}$$

where $T_{i,k}$ is the second rank dipole propagator

$$T_{i,k} = \left[\frac{3R_{ik}R_{ik}}{R_{ik}^2} - I \right] \frac{1}{R_{ik}^3} \tag{27}$$

Given the expense in calculating these many body forces, it is fortunate that they need only be treated in highly polarizable solvents like xenon and hydrocarbons. In such systems the excess electron can penetrate deeply into the solvent molecules which are held apart by their large van der Waals radii. Under these conditions the excess electron will almost always exist in a highly delocalized state, and will not significantly perturb the solvent structure. In this case it is possible to treat the problem by a mean field theory[65] which requires only the determination of the pure solvent pair distribution function as input. In less polarizable solvents, the electron will be unable to penetrate the solvent cores as deeply, and, as discussed earlier, localized excess electronic states which significantly perturb the local solvent structure will result. This perturbation makes the mean field treatment of the many body polarization interaction inapplicable, but these interactions will be less important because of the smaller polarizability of the solvent. In systems of highly polarizable molecules which exhibit localized excess electronic states and significant local structural perturbation, full many body treatment of the polarization interaction becomes necessary. The hydrated electron system provides an example of a fairly polarizable solvent showing localized states, and detailed studies of the effects of many body polarization in this system indicate these interactions account for about 10% of excess electronic thermodynamic and spectroscopic properties.[5,6]

Including many body polarization in the way described in this section involves making the Born-Oppenheimer on a new level. Usually, we assume that the time scale for nuclear motion is much slower than the time scale for the motion of the molecular electrons so that the molecular electronic distribution can adjust instantaneously to each new nuclear configuration. This approximation enables us to separate the nuclear and molecular electronic coordinates. In the treatment of many body polarization described here we are making a further approximation which involves assuming that in the solvated electron system there are three distinctively different relevant time scales: the molecular electrons move much more rapidly than the excess electron, which in turn moves very quickly on the time scale of the slow nuclear motions. The difference in time scale between the excess electron motions and the molecular electronic degrees of freedom is implicit in our use of the adiabatic polarizability determining the induced dipole in each of the solvent atoms. The molecular electrons instantly reorganize under the influence of the field resulting from the current excess electronic charge distribution to give the new induced dipole. We expect this separation of excess electronic, and molecular electronic time scales to be justified if the excess electronic kinetic energy is much smaller than the kinetic energy of the molecular electrons. This will be a good approximation when the excess electron is delocalized as is the case in fluids of highly polarizable molecules.

C. MEAN FIELD TREATMENT OF MANY BODY POLARIZATION

Calculating the many body polarization interaction discussed above requires much more computer time than does the direct pairwise superposition of electron-molecule pseudopotentials. Lekner[65] has presented a mean field solution to this problem which involves making use of the pair correlation function of the solvent so that the average solvent structure can be incorporated into the self-consistent field. This technique allows one to express the potential as a sum of two-body terms, like the simple electron-molecule pseudopotential, with each term containing a screening function arising from the mean field theory.

Briefly, in the absence of other solvent atoms, the electric field acting on an atom at R due to an electron at the origin is $E^{(0)}(R)$. The presence of other polarizable solvent atoms will modify this field. In an isotropic fluid, the average local electric field on the atom at R still points along R, but its magnitude will be the sum of the direct field $E^{(0)}(R)$ together with the contributions from the fields of the induced dipoles in the surrounding solvent atoms. We assume that the average local field can be written as $E^{(0)}(R)f(R)$, where the function $f(R)$ must be determined. Substitution of this local field into Equation 26 yields an equation for the factor $f(R_i)$:

$$f(R_i) = 1 + \alpha \sum_{k \neq i}^{N} \hat{R}_i \cdot T_{i,k} \cdot \hat{R}_k \left(\frac{R_i^2}{R_k^2}\right)\left(\frac{S(R_k)}{S(R_i)}\right)^{1/2} f(R_k) \tag{28}$$

This matrix equation for $f(R_i)$ is not equivalent to Equation 26, because the system has been assumed to be isotropic. A mean field equation can be derived by assuming that the distribution of atoms around atom i is the average distribution specified by the pair correlation function g(R). Then Equation 28 becomes

$$f(R) = 1 + \alpha\rho \int d^3R' g(|R - R'|)[\hat{R} \cdot T(R - R') \cdot \hat{R}'] \left(\frac{R^2}{R'^2}\right)\left(\frac{S(R')}{S(R)}\right)^{1/2} f(R') \tag{29}$$

Lekner first derived this equation using bipolar coordinates and found that

$$S^{1/2}(R)f(R) = S^{1/2}(R) + \pi\rho\alpha \int_0^\infty ds g(s)s^{-2} \int_{|R-s|}^{|R+s|} dt S^{1/2}(t)f(t)t^{-2}$$

$$\times \left[\frac{3}{2s^2}(s^2 + t^2 - R^2)(s^2 + r^2 - t^2) + (R^2 + t^2 - s^2)\right] \tag{30}$$

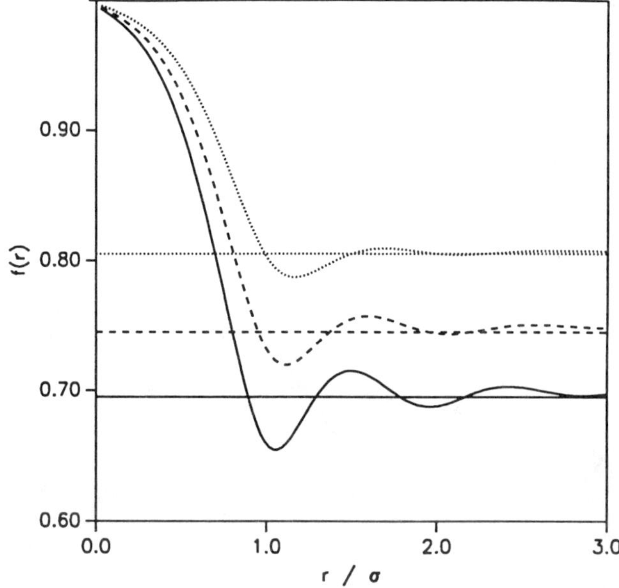

FIGURE 1.   Lekner local field functions at various densities in fluid xenon. Solid curve is $\rho^* = 0.9$, dashed curve is $\rho^* = 0.7$, and dotted curve is $\rho^* = 0.5$.

This equation can be solved numerically, using an iterative procedure, to obtain the average local field function $f(R)$. The dipole moment induced in the atom at R is then $\mu = \alpha e S^{1/2}$ $(R)f(R)/R^2$, and the energy of this dipole when it interacts with the direct electric field $eS^{1/2}$ $(R)/R^2$ gives Lekner's effective mean field pairwise additive approximation to the many body polarization energy:

$$V_p^{Lek}(R) = -\frac{\alpha e^2}{2R^4} S(R)f(R) \tag{31}$$

This polarization term in the pseudopotential is thus density dependent since $f(R)$ involves the solvent pair distribution function. The average local field function $f(R)$ for several different solvent densities in fluid xenon is presented in Figure 1. This function decays from unity to the Lorentz local field factor f_L in an oscillatory fashion. The factor f_L depends on the density of polarizable centers. The total self-consistent mean field pseudopotentials obtained in fluid xenon are displayed and compared with the zero-density pair polarization pseudopotential in Figure 2. This full potential is obtained by subtracting the bare pair polarization term from the pseudopotential, leaving only the exchange, orthogonality, and core repulsion terms, then adding the effective mean field many body polarization term. The self-consistent mean field potentials are more weakly attractive than the pair polarization potential, because the average local field function is generally less than unity for separations where the electron-solvent interaction is attractive.

We expect that the mean field theory will work best for excess electrons in extended states. In these states the electron perturbs the radial distribution of fluid atoms very little, and the mean field equations can be iterated using the radial distribution function of the pure solvent without the added electron. When solvated electrons are localized, the fluid structure can be greatly perturbed, requiring an altered mean field equation. A detailed account of the use of many body polarization interactions in the calculation of the energy of the bottom of the conduction band in noble gas fluids has been presented by Space et al.[95]

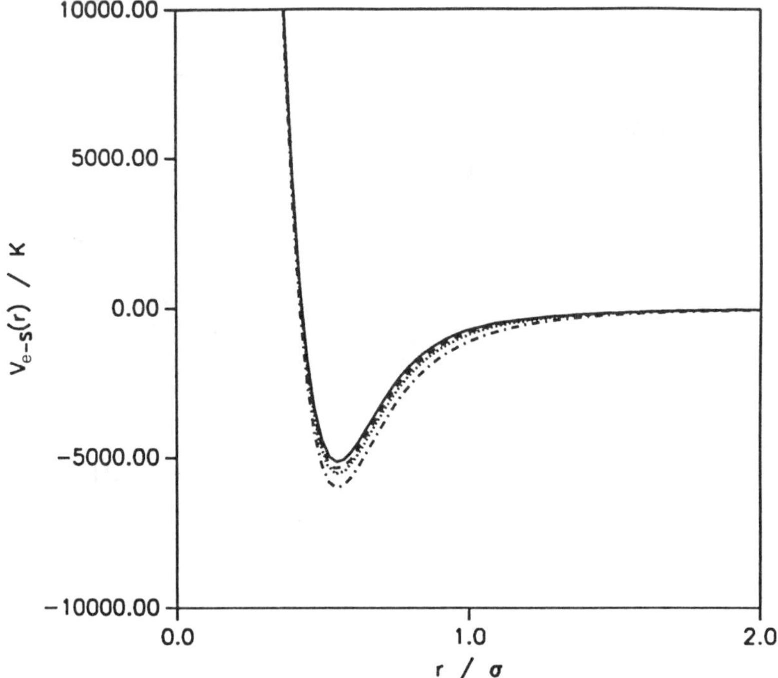

FIGURE 2.   Lekner effective self-consistent polarization pair potential for various densities in fluid xenon. Curves are labeled in Figure 1 and the dash-dot curve is the density independent pair polarization potential.

D. BEYOND PSEUDOPOTENTIALS: DENSITY FUNCTIONAL METHODS FOR MANY ELECTRON SYSTEMS

To go a step beyond the pseudopotential description of the electronic interactions, one uses lower level approximate pseudopotentials to describe the interactions of valence electrons of the surrounding molecules and performs a many electron calculation on these electrons. A very useful tabulation of these valence electron pseudopotentials has been presented by Bachelet et al.[67]

With the development of the powerful new simulated annealing techniques for performing *ab initio* quantum variational calculations which are outlined later in this section, this many electron approach has become feasible. Parrinello and co-workers[37] have pioneered the application of such an approach via the Hohenberg-Kohn-Sham density functional and spin density functional formalism,[68,69] a formalism that provides an accurate approximate way to include electron correlation while retaining an independent particle framework. Because correlation is neglected, the Hartree-Fock description of a molecular system gives too much weight to ionic configurations and results in a poor description of bond breaking and bond formation processes. Even within the lowest levels of approximation of the spin density functional formalism, the inclusion of correlation effects supresses these ionic configurations, enabling the spin density functional approach to bridge the gap between the HF-MO and valence bond descriptions.

Briefly, the spin density functional approach is based on the variational principle.[69] The total energy of the electonic system is expressed as a functional of the electron densities of spin-up, $n_\uparrow(r)$, and spin-down, $n_\downarrow(r)$, electrons which we write in terms of a set of single particle spin orbitals $\psi_{\alpha\sigma}$ as

$$n_\sigma(r) = \sum_\alpha f_{\alpha\sigma}|\psi_{\alpha\sigma}(r)|^2 \tag{32}$$

where $0 \leq f_{\alpha\sigma} \leq 1$ is the occupation number of the spin orbital with quantum number α and spin σ. The total electronic energy in atomic units thus has the form

$$E = \sum_{\alpha\sigma} f_{\alpha\sigma} \langle \psi_{\alpha\sigma} | -\frac{1}{2} \nabla^2 | \psi_{\alpha\sigma} \rangle + \int dr v(r)n(r) + U[n] + E_{xc}[n_\uparrow, n_\downarrow] \tag{33}$$

where $n(r) = n_\uparrow(r) + n_\downarrow(r)$ is the total electron density, and $v(r)$ describes the external electric field experienced by the electrons, including the electron-nuclear interaction. The electronic Coulomb interaction term has the form

$$U[n] = \frac{1}{2} \int dr \int dr' \frac{n(r)n(r')}{|r - r'|} \tag{34}$$

and E_{xc} is the exchange and correlation energy term. The physical interpretation of this final term is that if there is a spin-up electron at r say, then the Pauli exclusion principle requires that the other spin-up electrons avoid the region around r since two electrons with the same quantum numbers cannot occupy the same region of space. This exclusion gives rise to a so-called exchange-correlation hole in the electron density. It is found[68,70] that the exchange-correlation energy, E_{xc}, can thus be written as the sum of the electrostatic interaction of each electron with its positively charged exchange-correlation hole, so that

$$E_{xc} = \frac{1}{2} \int dr \int dr' \frac{n(r)\rho(r,r')}{|r - r'|} \tag{35}$$

where $\rho(r,r')$ is the density at r' of the hole around an electron at r. In general, exact results for E_{xc} are not available, and in practice the local spin density (LSD) approximation is often made, in which

$$E_{xc}^{LSD} = \int dr n(r) \epsilon_{xc}(n_\uparrow(r), n_\downarrow(r)) \tag{36}$$

where $\epsilon_{xc}(n_\uparrow, n_\downarrow)$ is the exchange-correlation energy per particle of an electron gas with uniform spin densities $n_\uparrow$, $n_\downarrow$. In this way, the results from very accurate calculations of the exchange-correlation energy of the homogenous interacting electron gas[71] can be used in studies of systems with inhomogeneous electron densities.

Care must be taken to insure that the exchange-correlation energy of a single, fully occupied spin orbital cancels the Coulomb interaction energy of the electron density of this orbital with itself so that

$$U[n_{\alpha\sigma}] + E_{xc}[n_{\alpha\sigma}, 0] = 0 \tag{37}$$

With an appropriate self-interaction correction scheme[69] the spin density functional approach outlined here can reproduce many experimental results for atomic, molecular, and solid state properties which are very sensitive to the accurate treatment of exchange and correlation, such as binding energies of negative ions.

The usual method for minimizing the variational energy involves functional differentiation of Equation 33 with respect to the spin orbitals, subject to the constant that the orbitals are normalized by using Lagrange multipliers $\{\epsilon_{\alpha\sigma}\}$. Thus,

$$\frac{\delta}{\delta \psi_{\alpha\sigma}(r)} \left(E - \sum_{\alpha'\sigma'} \epsilon_{\alpha'\sigma'} f_{\alpha'\sigma'} \int dr |\psi_{\alpha'\sigma'}(r)|^2 \right) = 0 \tag{38}$$

which leads to a set of one-electron self-consistent Schrödinger-like equations known as the Kohn-Sham equations

$$\left[-\frac{1}{2} \nabla^2 + v(r) + u([n];r) + v_{xc}^{\sigma}([n_\uparrow, n_\downarrow];r) \right] \psi_{\alpha\sigma}(r) = \epsilon_{\alpha\sigma} \psi_{\alpha\sigma}(r) \qquad (39)$$

and the effective one-electron potentials u and v_{xc}^{σ} are functional derivatives of the various energy terms with respect to the spin densities. The spin orbitals are then written in terms of some appropriate basis set expansion containing variational parameters, and the variational problem is solved by finding the set of parameters for which the determinant of the matrix vanishes. The problem with this matrix approach is that as we increase the size of the basis set to get more accurate ground state properties, larger and larger matrix eigenvalue problems must be solved, thereby giving many approximate excited state eigenvalues which are of little interest if only the ground electronic state behavior is required.

An efficient alternative to this matrix approach has recently become popular in condensed matter physics applications. This new technique, known as simulated annealing quantum molecular dynamics, was developed by Car and Parrinello[37] and has been used by them and their co-workers to study the electronic and structural properties of crystalline and amorphous silicon,[37,72] the properties of silicon microclusters,[73] and the dynamical bond breaking and bond formation processes that occur as hydrogen diffuses through silicon.[74] Clusters of alkali metal atoms have also been studied with related techniques.[75]

The approach involves writing the spin orbitals in terms of a basis set expansion and substituting them into the spin density functional expression for the variational energy, which is now a multidimensional function $E(\{\alpha_i\})$ of the variational parameters $\{\alpha_i\}$ contained in the basis set expansion. This function is treated as an effective potential energy surface over which the variational parameters move, and the problem boils down to determining the minimum effective potential energy configuration of the parameters. A particularly efficient way to do this involves introducing a fictitious kinetic energy term for each parameter and defining an artificial classical Lagrangian for the motion of these parameters of the form

$$L = \frac{1}{2} \sum_i m_i \dot{\alpha}_i^2 - E(\{\alpha_i\}) + \sigma(\{\alpha_i\}, \{\epsilon_{mn}\}) \qquad (40)$$

Here a holonomic constraint function σ has the same form as in Equation 38, and the Lagrange multipliers $\{\epsilon_{mn}\}$ correspond, as before, to the constraint that the orbitals be orthogonal. The fictitious classical dynamical system defined by this artificial Lagrangian and the Lagrange equations of motion can be integrated using standard molecular dynamics methods. The average fictitious kinetic energy is proportional to the artificial temperature, associated with the motion of the variational parameters over the effective potential surface. The calculation thus proceeds by heating the fictitious dynamical system up to some elevated artificial temperature then slowly cooling the system down by periodically rescaling the parameter velocities to reduce the average kinetic energy. As the fictitious temperature is reduced, the system will "crystallize" into the minimum variational energy parameter geometry. Once the minimum energy parameters are determined for a fixed nuclear conformation, Car and Parrinello extend the Lagrangian to include the nuclear coordinates and momenta, which are treated classically. By making the fictitious parameter masses small compared to the nuclear masses they adjust the relaxation times of the electronic parameter variables so that they respond very quickly on the time scale of the much slower classical nuclear motion. In this way they simulate nuclear dynamics over an *ab initio* Born-Oppenheimer ground state potential energy surface.

As outlined in Section IV.A, the spin density functional approach has been applied to the study of the behavior of several electrons solvated in a molten salt environment. An intriguing biochemical application of this approach is to the study of intramolecular electron transfer phenomena between electron donor and acceptor sites embedded in proteins. As will be discussed in Section VII.B, in such systems there is considerable ambiguity as to what kind of interaction to use to adequately account for the superexchange processes in a particular system. For example, in some cases it may be appropriate to describe the charge transfer process interms of an excess electron moving through the molecular structure, while in other situations a more accurate treatment may be provided by pseudopotentials describing the interaction of a hole in the electron distribution with its surroundings. Treating selected valence electrons of the environment and the transferring electron with many electron simulated annealing spin density functional techniques, would remove this ambiguity.

IV. EQUILIBRIUM THERMODYNAMICS AND STRUCTURE OF ELECTRONS IN FLUIDS

It is remarkable that solvents can be subdivided into essentially two distinct categories with respect to the solvation properties of electrons. In the first group, which consists of water, fused salts, low polarizability cryogenic fluids like helium and neon, and normal alkanes, the mobility of the excess electron decreases monotonically with the solvent density. In the second group of solvents, which consists of highly polarizable spherical molecules and highly branched hydrocarbons, the mobility first decreases as the density is increased, then reaches a minimum and next rises to a maximum, after which it falls again, as the density is further increased. The minimum mobility occurs not far from the critical point density. What causes these dramatic differences? Can the behavior of the mobility be related to the equilibrium solvation structure, or is it caused by subtle effects in many body scattering? Although there has been very important work on these problems in the past, especially by Lekner,[65] Jortner,[76] Cohen,[77-79] Rice[80,81] and Kestner,[82] it is of interest to apply the methods outlined in the foregoing sections to these problems.

In this section we first review the results of equilibrium calculations of the thermodynamics, structure, and spectroscopy of excess electrons in fluids exhibiting localized electronic states. We compare and contrast the different behaviors of localized states observed in fluids of polar molecules, in molten salts, and in simple cryogenic fluids. These systems have been studied experimentally over the past two decades. In the final subsection, Section IV.B, we discuss fluids in which delocalized excess electronic states are observed and indicate the importance of the self consistent treatment of the polarization interactions in these systems.

A. LOCALIZED STATES IN HELIUM, POLAR FLUIDS, AND MOLTEN SALTS
1. Helium
The simplest fluid which exhibits localized excess electronic states is dense helium. As outlined in the introduction, the mechanism of localization here involves simple packing considerations. At low gas densities the electron is delocalized and scatters through the fluid like a quasi-free particle, making occasional encounters with the hard sphere solvent atoms. As the gas density is increased and the interparticle separation becomes comparable to the electron thermal wavelength, multiple scattering events become important and the electron begins to localize, resulting in a decrease in the mobility. At condensed phase densities the helium atoms are in contact, separated by their hard sphere diameter. Because the excess electron cannot penetrate into the unpolarizable helium atoms, at liquid densities it digs a cavity (or bubble) in the fluid and cannot leak between the solvent atoms forming the bubble walls.

When the electron localizes in the higher density fluid, its behavior becomes ground

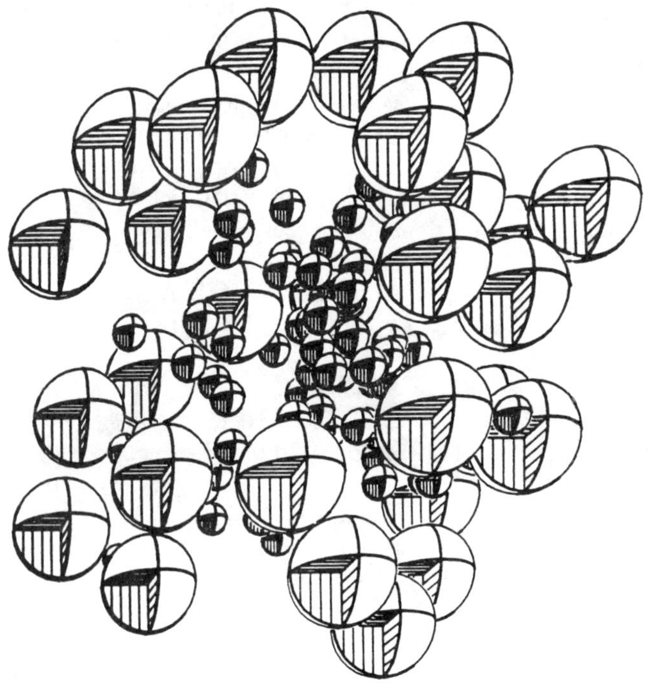

FIGURE 3.   A small portion of a configuration of the electron-helium system with T = 309 K and ρ^{*} = 0.5. Only solvent atoms within an 8 Å radius of the center-of-mass of the electron chain are drawn. Every 15th electron bead in the chain is represented as a 0.5 Å diameter sphere and the solvent sphere and the solvent atoms are drawn with a diameter of 1.5 Å.

state dominated, since under these conditions the excited excess electronic states of the bubble are well spearated from the ground state compared to $k_{B}T$. The electron behaves much like a particle in a spherical box. Increasing the solvent density compresses the bubble and causes the electronic energy to increase. The equilibrium cavity radius is determined by a balance between the ground state electronic energy and the PV and surface tension working to create a cavity in the fluid.[76]

A typical excess electron cavity configuration obtained from path integral calculations on the electron-helium system[1] is presented in Figure 3. The beads of the isomorphic cyclic polymer representing the electron path pack into the bubble and exclude solvent atoms from this region. Finite grid methods have been used[2] to obtain the eigenstates of an excess electron in an ensemble of representative cavities taken from path integral calculations. At equilibrium, the inhomogeneously broadened absorption spectrum and the density of states have been calculated. These studies have revealed that the electronic potential well created by the cavity can support several localized bound states. For example, at a reduced density of $\rho^{*} = \rho\sigma^{3} = 0.9$ (where σ is the hard sphere diameter of the solvent atoms), the solvent cavity can support four bound states: an s-like ground state and three p-like excited states. If the cavity was spherically symmetric, these three excited states would be degenerate but nonspherical fluctuations in the cavity break this degeneracy, giving three broad component bands in the density of states shown in Figure 4. The widths of the bands result from fluctuations in the overall cavity radius from one configuration to the next. For the bound

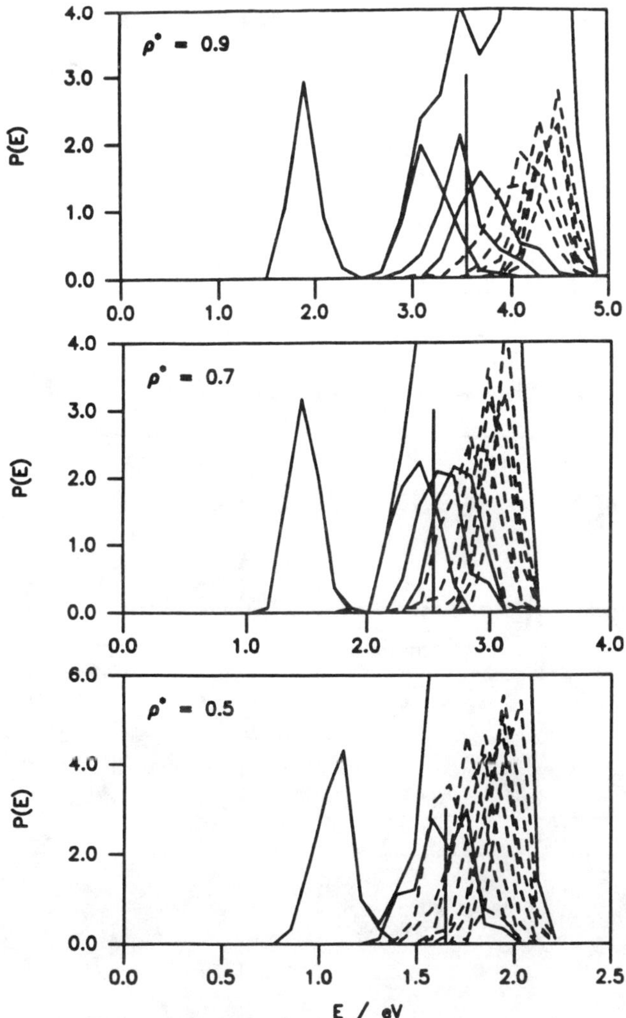

FIGURE 4. Density of excess electronic states in helium. The envelope curve is the total density of states and the component break up is based on the energy. Solid lines are the bound state contributions and dashed lines are continuum state contributions. Note the change in energy scale for the different fluid densities.

states, the electronic wave functions decay rapidly to zero as we move away from the center of the cavity. However, the excited state wave functions, above some threshold energy, show finite probability amplitude at large distances from the center of the bubble, though they are still peaked in the region of the cavity. This behavior is also demonstrated in Figure 5, where we show the average radial projections of the excess electron density. The bound states exhibit an exponential decay outside the cavity, whereas the high energy continuum states have finite electron density at large distances from the center of the bubble. As the density of the fluid is lowered, the cavities can support fewer and fewer bound states, until eventually even the ground state becomes a delocalized continuum state. The threshold energy separating localized and extended electronic states increases with fluid density. At densities lower than $\rho\sigma^3 = 0.3$ no localized states are expected.

Simulations of model electron-helium systems[2,83] have provided a benchmark for theories of electron solvation. For example, Chandler and co-workers [84-86] have devised a RISM

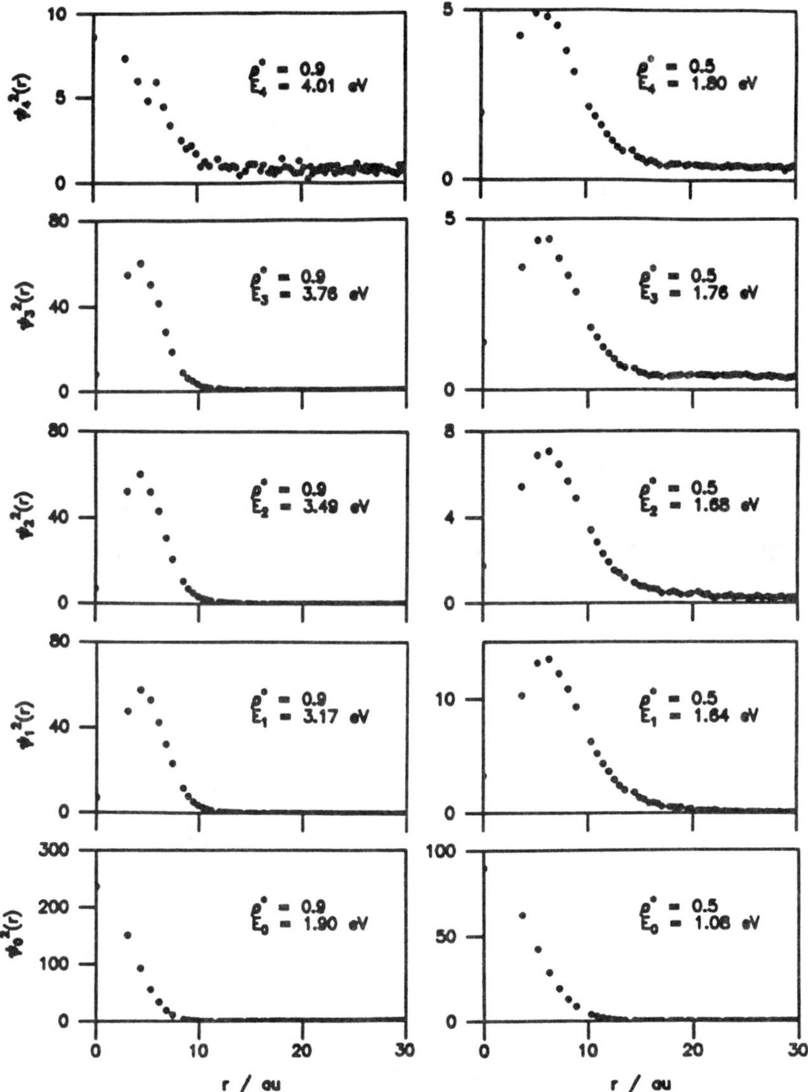

FIGURE 5.   Radial electron densities for different energy states obtained from simulations. Left column is for $\rho^* = 0.9$ and right column is for $\rho^* = 0.5$. The appearance of the nonzero tail at large r indicates the onset of continuum state behavior.

polaron theory of electron solvation which treats the electron solvent atom interaction as a hard sphere interaction with a certain radius which can be taken to be smaller than the solvent-solvent hard sphere radius. This theory does not provide an independent way of determining the radii involved, but if the theory is fit to the helium simulations at one density, it successfully gives agreement with the simulations at all other densities.

2. Polar Molecular Liquids: Water

The interaction between a water molecule and an electron is complex and can be determined only by *ab initio* many body quantum chemical calculations. For the purposes of Monte Carlo and MD simulations, realistic pseudopotentials (see Section III) can be used to simulate electron solvation in water. The pseudopotential can be expressed as

$$U(r,R_i) = V^S + V^E + V^P + V^R \tag{41}$$

where V^s represents the electrostatic interaction between the electron and the electronic and nuclear charge distribution in the unperturbed ground electronic state of the water molecule, V^E represents the exchange potential, V^P represents the interaction between the electron and the dipole it induces on the water, and V^R represents the repulsive interaction which arises because the one-electron wave function of the excess electron must be orthogonal to the wave function representing the core electrons of the water molecule.

A number of realistic pseudopotential models have been used for the electron-water interaction, that of Wallqvist et al.[5,6] that due to Rossky et al.[3] and a potential used by Barnett et al.[15,87,88] are examples. The potential due to Barnett et al. explicitly includes an exchange interaction term but otherwise is very similar to the Rossky potential. Below we shall compare and contrast in detail the potentials of Wallqvist et al. and that of Rossky et al. The other potential terms contained in all these models are a polarization term V^P centererd on the oxygen, a repulsive term $V^R = V^R_{H1} + V^R_{H2} + V^R_O$ consisting of Hartree-Fock repulsions centered both on the two hydrogen atoms and on the oxygen atom, and an electrostatic part V^S consisting of Coulombic terms centered on all atoms. The interaction between an excess electron and the water molecules in liquid water is often treated by adding up all of the electron-single molecule pseudopotentials. Many bond polarizations have been treated only in the electron-oxygen potential of Wallqvist et al. (see Section III), and is ignored by Rossky et al.

The differences in these pseudopotentials can be summarized as follows. The hydrogen core repulsion of Wallqvist et al. is more repulsive than that of Rossky et al.; however, the opposite is true of the core repulsion on the oxygen atoms. When both contributions are considered, the Rossky potential is the more repulsive of the two. The polarization of Wallqvist et al. has two thirds the depth of that of Rossky et al., and the minima are slightly shifted toward the oxygen atom. Finally, the static potential contains different charges, and the water models used have different equilibrium O–H bond lengths and internal angles. This results in differences in the dipole moments of the solvating water molecules in these two models.

Thus the water models used by the two groups are different as are the pseudopotentials. The cavities produced by the Rossky model and that of Wallqvist et al. may also be different. However, as the cavity is basically a void space, it may be assumed for purposes of comparison that the differences are small.

These two pseudopotentials have been used in path-integral simulations to determine the structure and thermodynamics of the hydrated electron. Wallqvist et al. have used Monte Carlo techniques with a convergence-enhancing staging algorithm, whereas Rossky et al. have used molecular dynamics techniques in the pure primitive algorithm without any convergence-enhancing techniques. Despite differences in the potential model it is possible to arrive at the following conclusions:

1. The electron exists in a localized cavity state, reminiscent of a particle in a spherical box. The radius of the cavity ranges between 2.1 to 2.3 Å and the fluctuations in cavity radius are characterized by a Gaussian distribution, with a rms fluctuation of 0.1 Å.
2. The water molecules in the first hydration shell around the electron are coordinated such that one O – H bond points into the electron and the other water O–H bond is hydrogen-bonded to the surrounding waters. The result of this arrangement is that on average the dipole moments of the water molecules in the first hydration shell point tangentially to the surface of the cavity.
3. Compared to the hydration of a classical anion like F^-, which has a well-defined coordination-shell, the hydrated electron is much softer because of quantum fluctuations (dispersion). Because it does not have a hard core radius like F^-, the electron has an

ill-defined coordination shell such that there are, on average, about four water molecules within 3.6 Å of the electron barycenter.

4. The electronic canonical density matrix is ground state dominated, meaning that at thermal energies k_BT the electron is found in the ground state of the cavity.

5. When many body polarization is included, the electron excludes water from a larger region than when it is ignored. This springs from the fact that when many body polarization effects are included, the induced dipoles in water molecules in the first solvation shell of the electron repel each other.

6. In a long simulation run, a large number of cavity configurations are generated. Analysis of the principal moments of inertia of the isomorphic electron polymer chain indicates a small asphericity. Since the electronic distribution is influenced by the solvent structure, on average the cavities are not spherical.

7. The thermodynamic solvation properties are different for the two simulation models. Wallqvist et al. find that $E_{hydr.}^{Berne} = -26.0$ kcal/mol; whereas Rossky et al. find that $E_{hydr.}^{Rossky} = -50$ kcal/mol and experiment gives $E_{hydr.}^{exp} = -38.0$ kcal/mol. Thus both simulations disagree with experiment by 12 kcal/mol.

It should be noted that to obtain accurate absolute hydration energies to compare with experiment we need to include an estimate of the long-range polarization interactions which are necessarily truncated in these calculations which use periodic boundary conditions and minimum imaging to make the computations manageable. There is evidence which suggests that the effect of truncation of long-range polarization interactions on fluid structure are small[89-91] but there are probably substantial errors in estimates of absolute energies.[4] At this point there is need for quantitative investigation of long-range correction schemes in solvated electron systems. It is worth noting that the long-range parts of the electron-solvent interaction are expected to be attractive on average so the corrections to the calculated hydration energies reported above will be negative. From the values reported above, these corrections will push $E_{hydr.}^{Berne}$ in the direction of the experimental value and $E_{hydr.}^{Rossky}$ further from agreement with experiment.

The absorption spectrum of a solvated electron in water is known to be a structureless asymmetric band with a blue tail. The optical spectrum is, strictly speaking, a dynamic property. Path-integral simulation techniques have not yet been developed to determine dynamic properties for real time processes. One approach is to determine the imaginary time correlation functions and to analytically continue these into real time. Unfortunately, there is presently no accurate numerical method to do this. Instead, grid methods (see Section II) have been used to determine the energies and energy eigenstates of the electron for different fluid configurations. We have already seen that simulations suggest that electron solvation in liquid water involves the formation of a bubble-like state. Path-integral techniques allow us to generate a large number of statistically independent cavity configurations in liquid water. Using FFT techniques in the imaginary time domain, we then compute the energies,[2,13,31] corresponding eigenstates, and dipole matrix elements, and thus determine the inhomogeneously broadened absorption spectrum of a hydrated electron. These calculations clearly show that

1. Just as with helium, the ground state of the electron in water is an s-like state, and the lowest three excited states are like p_x, p_y, and p_z states.

2. In a spherical cavity these three p-like states would be degenerate, but in the simulations we find that their degeneracy is lifted because the cavities are slightly aspherical, similar to the situation observed in dense helium. The average splitting between the p-like states is $\Delta E \simeq 0.2$ eV.

3. The oscillator strengths for each of the three s $\rightarrow$ p transitions are essentially equal.

4.　There are more highly excited bound and continuum states, but more than 90% of the oscillator strength is due to the three s → p transitions.

5.　From the equilibrium distribution of cavities, the energy width of each of the three s → p transitions is $\Delta E0.3 \simeq eV$.

6.　Thus, the overall band is dominated by three overlapping bands arising from the three s → p transitions, each with an intrinsic width due to the cavity size fluctuations and each shifted from the other by lifting of degeneracy due to the cavity shape fluctuations.

7.　Many body polarization shifts the s states to the blue by 0.2 eV and the p states to the blue by 0.4 eV compared to the values obtained in using the pair polarization model.[5,6]

8.　The experimental absorption spectrum at 300 K consists of a broad band located at $E^{Peak}_{expt.} = 1.72$ eV, compared to a value of $E^{Peak}_{Berne} = 1.8$ eV obtained from simulations using the many-body polarization potential. Unfortunately, the Rossky pseudopotential gives $E^{Peak}_{Rossky} = 2.25$ eV.

9.　In all cases, the absorption spectra determined from simulations fail to predict the extended high energy tail or a small tail in the low energy region of the spectrum. The latter failure probably arises from the fact that in the thermodynamic limit, liquids can exhibit large cavities albeit with small probability, whereas simulations on small systems will miss these large traps (Lifshitz traps). The failure to reproduce the high energy tail probably springs from the obvious difficulty simulations on finite systems must have in accounting for the extended continuum states expected in bulk fluids.

In concluding this subsection of the comparison of different model potentials for the hydrated electron system it is worth noting some still open questions concerning these models. As discussed above, including the many body nature of the induced polarization interactions can have a significant effect on the hydrated electron absorption spectrum. There is an apparent inconsistency in the implementation of the self-consistent many body potential developed by Berne and co-workers when the solvent molecules have permanent dipoles which are represented by partial point charges. In their approach, the interactions between the permanent partial charges and the induced dipoles are not included. The reasoning behind this choice is that the partial charges in the model water-water intermolecular interaction, in the absence of the electron, already take polarization effects into account by virtue of the effective nature of this potential. A true self-consistent model would not use effective partial charges, but rather the bare charges on the atoms, which would be allowed to interact, through a self-consistent potential, with the induced dipoles. The potential of Berne and co-workers takes an intermediate point of view.

A further complication in calculating spectra of excess electrons which has not yet been addressed in any studies is that, by virtue of the way they are formulated, electron-molecule pseudopotentials depend on the energy of the incoming electron. It is generally assumed in solvated electron systems that the energy is low, and that in this regime the pseudopotential varies slowly with excess electronic energy. When excited excess electronic states are needed in the absorption spectrum calculation, their characteristic higher energies may require the use of a different excited state pseudopotential.

The consistent treatment of many body polarization interactions, the energy dependence of excess electron pseudopotentials and their effect on the spectroscopy of excess electron systems is a subject of continuing research.

3. Molten Salts

Under a wide variety of conditions, the equilibrium states of excess electrons in ionic salts are localized. In solid KCl, for example, an excess electron will become localized at a Cl^- vacancy or F-center in which it is surrounded by an octahedral coordination shell of 6 K^+ ions, with 12 Cl^- ions surrounding this shell, etc. Path-integral calculations by

Parrinello and Rahman[10] have revealed that in molten KCl at T = 1000 K and P = 10.4 kb, that there is a distribution of electron-K^+ coordination numbers peaking at around 3 and extending to larger values giving a fluctuating average tetrahedral environment of 4 K^+ ions. As shall be seen in Section V.A this equilibrium picture actually results from the dynamics of the electron, which frequently hops between localized octahedral sites in the fluid.[11]

Path-integral methods have also been applied to study the properties of electron pairs in molten salts. These pairs result when metal atoms are dissolved in their molten metal halide salts and the metal atoms ionize. Paramagnetic susceptibility measurements indicate that the electrons form spin pairs in the fluid. The problem with applying path-integral methods to study systems of several electrons (or fermions) is that it is necessary to sample cyclic paths for which the two electrons retain their identities after a cycle, so-called direct paths, and paths for which the electrons exchange identities after a cycle.[92] The contributions of direct and exchange paths must be weighted with opposite signs to account for antisymmetry of the wave functions of the two electron system and cancellations due to these opposite signs cause considerable statistical sampling problems.

Selloni et al.[93] have circumvented this problem by applying spin density functional theory combined with the quantum molecular dynamics technique (see Section III.D). In their calculations on the bielectron in a molten salt, Selloni et al. studied both the singlet and triplet states, in which the electron spins are antiparallel and parallel, respectively. For the triplet state, the spin parts of the two spin orbitals are the same, so the Pauli exclusion principle requires that the spatial parts of the orbitals must be orthogonal. This orthogonality is observed in snapshots of the equilibrated triplet state electron density, in which two distinct spatially separated F-centers are found in the fluid. In the singlet state, on the other hand, the spatial parts of the orbitals can overlap appreciably as the spins are orthogonal. Thus, two electrons are found to be localized in a single, slightly extended F-center in the fluid.

In these calculations, the ground state of the bielectron complex in the molten salt environment is found to be the singlet state, and its energy is lower than that of two noninteracting solvated electrons. Consequently, the paired singlet state is actually a bound state with an estimated binding energy of ~0.5 eV. The opposite is true for the triplet state: the energy of the spin parallel system is higher than that for two noninteracting solvated electrons.

Because the electronic density is relatively spread out in the singlet state compared to the two compact F-centers found for the triplet state, the electronic kinetic energy is lower for the singlet state than it is for the triplet. This kinetic contribution, however, is almost exactly balanced by the differences in the Coulomb repulsion energy between the two electrons in these different states. In the singlet state the two electrons are in close proximity to one another, since they sit in the same F-center cavity, so there are strong Coulomb repulsions relative to the triplet state situation, where the two electrons occupy spatially separated F-centers. Further, it is found that the average exchange interactions for the singlet and triplet states are very similar. It is thus the differences in the electron-solvent interactions which are ultimately responsible for making the singlet state bound and the triplet state dissociative. If the number of positive nearest neighbor ions around an F-center is proportional to the surface area of the cavity, which is proportional to the square of the cavity radius, then the radius of the single large F-center need only be $\sqrt{2} = 1.414$ times the radius of the individual F-centers observed in the triplet state to give about the same electron-solvent potential energy. The singlet state F-center must have a slightly larger cavity radius than this, providing, on average, more nearest neighbor cations than for the two smaller, separated, triplet state F-centers.

B. DELOCALIZED STATES IN POLARIZABLE FLUIDS: XENON

Excess electrons in fluids of nonpolar but polarized molecules can exhibit either localized

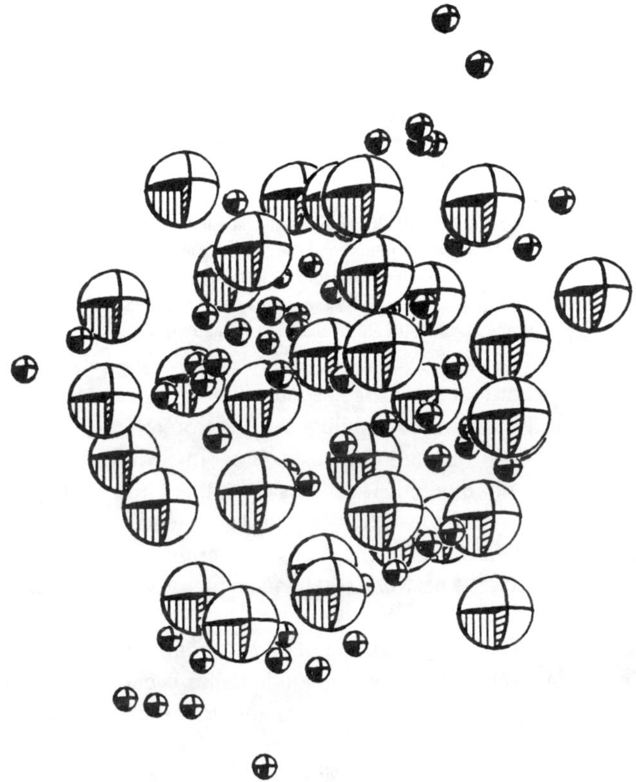

FIGURE 6.   Same as for Figure 3, except here the electron-xenon system is considered.

or extended electronic states. The character of the state generally depends on the molecular shape. For example, in liquids consisting of spherically shaped molecules such as the heavier noble gas atoms or highly branched (or spherical) alkanes, the excess electron is found in very delocalized states with characteristicly high electron mobilities. At low gas densities, the mobility first decreases as the gas density is increased. Around the critical density the mobility begins to increase rather rapidly, and a mobility maximum is observed in the dense liquid.

In contrast, in liquids consisting of polarizable nonspherical molecules with a predominantly chain-like structure, like normal alkanes, the excess electron is found to be in localized states. The electron mobility in these systems has a density dependence much like that observed in fluid helium: the mobility simply decreases as the fluid density is increased from gas to liquid densities.

The reasons for these differences are not entirely clear. Path-integral Monte Carlo studies show[1] that in fluids of highly polarizable spherical molecules, like liquid xenon, the short range repulsive interactions keep the solvent atoms apart, and because the electron can penetrate into the polarizable charge distribution, it sees an open network of attractive potential channels through which it can "percolate". From this it follows that the electron can exist in a delocalized state with high mobility at liquid xenon densities. A typical configuration in liquid xenon is shown in Figure 6. At subcritical temperatures, and at densities around the critical density, the mobility of an excess electron reaches its minimum

TABLE 1
Comparison of Electronic Potential
Energies[a]

ρ^*	$\langle U_c^{Pol}/k_B T \rangle$	$\langle U_c^{Lek}/k_B T \rangle$	$\Delta\%$
0.9	-45.45	-49.47	8.8
0.7	-39.50	-42.57	7.8
0.5	-32.50	-34.71	6.8

[a]   Data obtained with the full self-consistent many body
polarization potential and the Lekner mean field polarization potential. Results have been averaged over
100 representative Monte Carlo configurations and
include an estimate of the long range correction.

value. It has been suggested[1] that this arises because the strong attractive polarization forces
between an electron and the xenon atoms allow the electron to attach to either a preexisting
xenon cluster or to nucleate it through electrostriction. In Section VI, electron attachment
to clusters is considered in more detail. It follows from this cluster effect that slow electron
transport involves break-up of the clusters, which act as extended potential wells, combined
with hopping between these disconnected density fluctuations. In the dense fluid, the clusters
condense, providing a connected network of channels along which the electron can readily
transport.

A structural explanation for why fluids of polarizable chain-like molecules should support
localized excess electronic states rather than extended ones is currently under investigation[94]
with path-integral simulation techniques. It is possible that the excess electron cannot penetrate deeply into the bonds of a molecule where there is a high electron density region. In
chain-like molecules these bonds are ''exposed'', so we can envision disordered fluid configurations in which these repulsive bond regions give rise to narrower and less attractive
channels in which the electron percolates. For spherically shaped branched molecules, the
bonds are buried within the core of the molecule. Such molecules will pack with a network
of strong attractive channels, like liquid xenon. In disordered configurations of chain-like
molecules, the exposed bonds close off these channel structures.

In fluids of highly polarizable solvent molecules, the many body nature of the electron-
solvent polarization interaction (see Section III) may be extremely important. As indicated
in Section III, it is extremely expensive to solve the full self-consistent matrix equation to
determine the induced dipoles each time it is necessary to calculate the many body polarization
interaction. Because the mean field, self-consistent pairwise potential described in Section
III requires much less computer time, it would be useful to determine whether it is accurate.
To test the mean field approximation, the total electron-solvent interaction energy was
calculated[95] for a number of representative configurations of a discrete electron path and
solvent system sampled from path-integral Monte Carlo calculations. For comparison, the
energy calculations were performed with both the approximate mean field potential and by
solution of the full set of self-consistent field equations. Energies averaged over a hundred
different configurations and at a variety of solvent densities are presented in Table 1. Over
the moderate to high density range considered here, the mean field potential gives energies
which are about 6 to 8% more negative than the results of the full self-consistent many body
calculations.

The slight differences noted above indicate, as might be excpected, that fluctuations in
solvent structure give electron-solvent geometries with slightly higher many body polarization
energies than when the mean solvent structure is used to determine the local fields. The
proportional differences between the many body and mean field polarization energies are

TABLE 2
Electronic Potential and Kinetic Energies[a]

ρ^*	$\langle U_e^{pair}/k_BT \rangle$	$\langle U_e^{pair}/k_BT \rangle \times f_L$	$\langle U_e^{Lek}/k_BT \rangle$	$\langle K_e^{pair}/k_BT \rangle$	$\langle K_e^{Lek}/k_Bt \rangle$
0.9	-67.65	-46.53	-46.48	19.34	18.08
0.7	-53.35	-39.43	-39.48	15.77	11.22
0.5	-41.20	-32.90	-32.16	10.00	10.74

[a] Data obtained from path-integral Monte Carlo calculations with the pair polarization and Lekner self-consistent polarization potentials. Pair polarization results multiplied by the Lorentz local field factors are also included.

quite small and reasonably constant over the density range considered. Thus, for the potential energies reported below, we use the mean field approximation and we include a 7% correction to take into account the solvent fluctuations.

In Table 2, the results of path-integral Monte Carlo calculations employing the mean field many body polarization potential are compared with those calculated using the simple pair polarization model. The average electronic potential energy increases by 30 to 40% when many body polarization effects are included. The potential energy results obtained using the many body polarization potential can be reproduced quite closely at the various densities by simply multiplying the pair polarization results by the Lorentz local field factors, f_L (see Section III.C). Figure 1 shows that the Lekner average local field function tends to the Lorentz value at large separations and that the function decreases from unity and begins oscillating around f_L when the separation is about one solvent particle diameter. From Figure 2, however, it can be seen that the electron-solvent interaction potentials have both positive and negative regions within this radius. Under the conditions considered here, the average spatial variation of the excess electronic distribution around a solvent atom must be such that contributions to the average potential energy from those regions where the Lekner function differs from the Lorentz value cancel one another, leaving contributions only at large distances, where the pair polarization potential and the effective self-consistent mean field polarization potential essentially differ only by the multiplicative Lorentz factor.

In Figure 7 the total electronic energies obtained from path-integral Monte Carlo calculations on the electron-xenon system employing both the pair polarization and the self-consistent many body polarization potentials, are compared with the experimental results of Reininger et al.[96] Clearly, the pair polarization description of the electron-xenon interaction is unable to reproduce the experimental density dependence of the electronic energy, giving results which are about two times too negative. The self-consistent many body polarization potential, on the other hand, accurately reproduces the experimental electronic energies across a wide range of densities from the gas phase to the high density liquid, and even reproduces the upturn in the total energy at about the right density.

Path-integral calculations are currently being used to study the structure of excess electronic states in mixtures of very polarizable atoms combined with hard sphere like atoms to see how the composition of such a mixture affects the transition between localized and delocalized excess electronic states.[97]

V. ADIABATIC DYNAMICS OF LOCALIZED EXCESS ELECTRONIC STATES IN FLUIDS

Recently, the molecular dynamics methods outlined in Section II.E have been applied to study excess electronic transport phenomena in a variety of different solvents. These studies have all involved systems in which the equilibrium state is a localized excess electron. However, qualitatively different characteristic transport mechanisms are observed depending

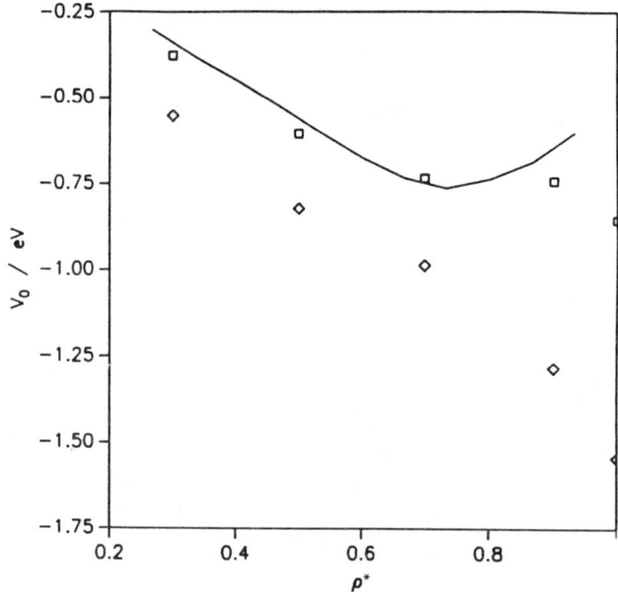

FIGURE 7.   Comparison of ground state excess electronic energies in xenon obtained from experiment (solid curve) and theoretical calculations using the pair polarization potential (diamonds) and the Lekner many body polarization effective pair potential (squares).

on the nature of the solvent. The adiabatic dynamics of localized excess electronic states in molten salts and in liquid water are examined in Sections V.A and V.B. Finally, in Section V.C, recent experimental studies and adiabatic dynamics simulations of the relaxation of injected electrons in liquid water into their localized equilibrium states are discussed. These studies indicate that the electron trapping process in water may involve nonadiabatic relaxation of the electron-solvent system once a precursor quasilocalized state is formed.

Schnitker et al.[4,56,57] and Parrinello et al.[11,93] employ the molecular dynamics techniques outlined in Section II.E to propagate the electronic wave function and the classical degrees of freedom of the liquid. As discussed in Section II.E, these adiabatic dynamics calculations involve moving the nuclei over the instantaneous ground electronic state and ignoring the possibility of nonadiabatic transitions between electronic states, which may occur around particular nuclear conformations where the adiabatic energies of different states come close together.

As pointed out in Section II.E, these adiabatic dynamics calculations are valid only if the behavior of the system is ground state dominated and if nonadiabatic events occur with low probability. We expect these conditions to hold in systems exhibiting strongly localized excess electronic states in which there is a considerable energy gap between the ground state and the excited states. Consequently, the dynamics of the unexcited system may be approximated by nuclear motion only on the electronic ground state potential energy surface.

A. ELECTRONIC TRANSPORT IN MOLTEN SALTS BY LOCALIZATION AND HOPPING

Parrinello[11] and his co-workers used a real time ground state adiabatic dynamics method to study the propagation of an excess electron in liquid KCl (T = 1300 K, ρ = 1.52 g/cm^3). As discussed in Section IV.A the equilibrium state of this system involves a localized excess electron confined in an F-center like environment in the fluid. How does the electron migrate through the liquid? The diffusion coefficient, D_e, of the excess electron was calculated by monitoring the expectation value of the electron position $\langle r(t)\rangle = \langle\psi(t)|r|\psi(t)\rangle$, evaluating the mean square displacement as

$$r^2(t) = \lim_{T\to\infty} \frac{1}{T} \int_0^T dt'[\langle r(t' + t)\rangle - \langle r(t')\rangle]^2 \tag{42}$$

and using the fact that if the process is diffusive, $\lim_{t\to\infty} r^2(t) = 6D_e t$. The calculated value, $D_e \sim 2 \times 10^{-3} cm^2 s^{-1}$ approximately compares with that obtained from experimental mobility measurements, $D_e \sim 3 \times 10^{-3} cm^2 s^{-1}$,[98] indicating that the potential model may well be providing a reasonable description of the dynamics of the real system.

To explore the mechanism of electron transport in this system, Selloni et al.[11] studied the correlation function

$$\chi(t) = \sum_{I^+} \langle |\psi(R_{I+}, t_0 + t)|^2 |\psi(R_{I+}, t_0)|^2\rangle \tag{43}$$

where the overall sum is the cations in the system and $\psi(R_{I+}, t)$ is the value of the electronic wave function at the site of one of the cations at time, t. The decay time of this correlation function gives an estimate of the residence time, τ, of the electron around the region of a given ion. The adiabatic dynamics simulations give a residence time of $\tau \sim 0.3$ ps, which compares with experimental estimates of this time.[99]

The evolution of contour maps of the electron density obtained from these simulations reveals the mechanism of electron transport in the molten salt. The electron makes fairly rapid but infrequent jumps between spatially separated F-center sites in the fluid. These hopping events typically take about 0.05 ps. Monitoring the time evolution of the expectation value of the electronic potential energy shows that during these hopping transitions, the electron climbs over a potential energy barrier separating the initial and final localized states. The height of these barriers is typically about 0.7 eV. During these rapid hopping events, the energy gap between the ground and excited states is found to decrease substantially. When the e^- is trapped in an F-center, the gap between these states is typically about 1.2 eV. During a hopping event, however, the gap narrows to typically about 0.35 eV.

The energy gap between the ground and excited states was generally large enough during these hopping processes so that the probability of a nonadiabatic transition to an excited state remained quite small throughout the calculation. By projecting $\psi(t)$ onto the adiabatic eigenstates of the instantaneous solvent configurations, Selloni et al. monitored the accuracy of the adiabatic dynamics approximation. During their entire trajectory of over a million time steps (a total of about 25 ps), they observed only one occurrence of an appreciable increase in excited state amplitude, when during a hopping event, the square of the expansion coefficient of the adiabatic ground state $|a_0|^2$ fell to 0.76 in about 0.015 ps. Since the adiabatic dynamics approach cannot accurately treat these nonadiabatic events, the propagating wave function was set equal to the instantaneous adiabatic ground state and the adiabatic dynamics calculation was resumed with this electronic distribution. These nonadiabatic events are so rare in this system that it was unnecessary to repeat this *ad hoc* nonadiabatic correction procedure since $|a_0|^2$ never fell below 0.99 for the rest of the trajectory.

We can view the electron transport process in a molten salt in terms of the Marcus ideas of electron transfer.[100,101] The electron remains localized in an F-center like structure until the solvent around it reorganizes so that it is equally favorable for the electron to be found at another F-center some distance away, when the hopping transition occurs. These ideas are discussed further in connection with path integral simulations of electron transfer between Fe^{2+} and Fe^{3+} ions in aqueous solution in Section VII. In this system, the electron remains localized on the Fe^{2+} ion until a fluctuation in the hydration spheres of this ion and a nearby

Fe^{3+} ion make it equally probable for the electron to be at either ionic site, resulting in electron transfer between these species.

B. DIFFUSIVE TRANSPORT OF HYDRATED ELECTRONS

The adiabatic dynamics studies of electrons in water reveal a totally different mechanism of electronic transport from that found in molten salts. Schnitker and Rossky[57] have employed imaginary time FFT propagator methods to obtain the instantaneous adiabatic ground state wave function for the hydrated electron. They use this electronic distribution to determine the forces on the solvent molecules in their adiabatic dynamics calculations. These simulations show no evidence of the hydrated electron hopping between localized sites in the fluid, as the electron does in molten salts. Rather, the hydrated electron remains in a tightly localized state, and the electronic transport occurs by diffusion of this localized electron and its hydration shell. The water molecules in the hydration shell are partially hydrogen bonded to the surrounding solvent, and these interactions make the motion of the hydrated electron complex very slow. Because of this different electronic transport mechanism, the hydrated electron diffusion constant is about two orders of magnitude smaller than the electronic diffusion constant observed in molten salts.

In their study, Schnitker and Rossky compare and contrast the adiabatic dynamics of the hydrated electron with the classical motion of halide ions in aqueous solution. In these simulations, the hydrated electron is found to diffuse at about the same rate as the solvent water molecules, with $D_e = 3.3 \times 10^{-5} cm^2/s$ and $D_{H_2O} = 3.5 \times 10^{-5} cm^2/s$. The hydrated electron is a little smaller than a bromide ion and classical simulation of aqueous Br^- gives the diffusion constant of the hydrated ion as $D_{Br^-} = 0.5 \times 10^{-5} cm^2/s$. Similar qualitative trends are observed experimentally. The electronic diffusion constant obtained from conductivity measurements is $4.9 \times 10^{-5} cm^2/s$, which is about 2.5 to 3.5 times larger than the experimental diffusion coefficients observed for hydrated halide ions. The experimental self-diffusion constant of water is $2.4 \times 10^{-5} cm^2/s$. The quantitative differences between theory and experiment probably arise from the inadequacy of the interaction potentials employed in the calculations. However, the qualitative feature that the electron diffuses more rapidly than an anion of comparable size seems to be reliably reproduced by these adiabatic simulations.

C. TRANSIENT DYNAMICS OF ELECTRON HYDRATION

Rossky and co-workers[4] have applied the adiabatic dynamics approach to study the fast transient relaxation of the hydrated electron absorption spectrum. After the electron is injected into an equilibrium water configuration, they observed the initial broad absorption band undergo a continuous blue shift of over 1000 nm and a narrowing until it coincided with their previously determined equilibrium spectrum centered at about 600 nm. The time scale for this adiabatic equilibration was about 80 fs. The experimental time-resolved optical spectrum[102] shows qualitatively different behavior in that two distinctively different dynamical stages are identifiable. The first stage is consistent with a broad band undergoing a continuous blue shift on a time scale of about 100 fs. However, during this initial stage, which is believed to involve thermalization and initial localization of the electron, the absorption maximum remains above 1250 nm. The experimental spectrum subsequently shows a second stage in its dynamical evolution, when the absorption band which formed during the first stage disappears and, simultaneously, the equilibrium spectrum appears with an absorption maximum at around 700 nm. This second stage, which takes over 200 fs, may involve a nonradiative electronic transition from the precursor state formed in the first stage to the ground electronic state of the equilibrating solvent cavity.[102] The model interactions used in the above simulations give qualitatively reasonable static equilibrium results, and as indicated in the previous section, the model equilibrium dynamics of electronic diffusion shows the same qualitative features observed experimentally. Thus we might expect that calculations employing this model should give at least a qualitative representation of the nonequilibrium electron solvation dynamics. The fact that the adiabatic dynamics calculations cannot reproduce even the qualitative features observed experimentally indicates that nonadiabatic effects may play an important role in determining these spectral transients. Rossky and co-workers have recently reproduced the two phase spectral relaxation of the hydrated electron observed experimentally by using nonadiabatic dynamics methods.[102a]

VI. EQUILIBRIUM PROPERTIES OF ELECTRONS ATTACHED TO CLUSTERS

The attachment of an excess electron to neutral clusters of atoms or molecules has been the subject of considerable experimental interest. Path-integral calculations have been performed to study electron attachment to clusters of ions.[17,18] The attachment of an electron to a large cluster is analogous to the solvation of an excess electron in a bulk fluid, but because of free surfaces the electron may behave very differently in a cluster. For example, in a water cluster, it is very likely that an excess electron will form a diffuse surface state, as opposed to the localized state found in the homogeneous fluid. How does an electron attach to xenon clusters? In liquid xenon the electron is found in an extended state. We expect, therefore, that an electron will not attach to small clusters of xenon. How large must the cluster be before an electron can attach to it, and how will it be distributed if it attaches? The quantum simulation methods outlined in Section II can be used to answer many questions like these.

A. XENON CLUSTERS

The Xe^- ion has never been observed despite the fact that Xe is very polarizable. It is reasonable to assume that an electron will bind to Xe_n for n is sufficiently large. Since accurate pseudopotentials are known for both the e-Xe interaction and the Xe-Xe interaction, this is an ideal system to study. Martyna and Berne[12,13] performed Diffusion Monte Carlo calculations to determine the electron binding energy (ground state energy) and electronic structure of clusters of xenon atoms (for n $<$19) frozen in the minimum energy geometries by Hoare and Pal.[103,104] The calculations were performed using the pair polarization approximation as well as the full many body polarization model. Figure 8 shows the full binding energy, and the increment in binding energy per atom as a function of cluster size. Several things should be noted:

1. Only clusters n $\geqslant$7 bind an electron in the many body polarization case, and n $\geqslant$6 in the pair polarization case.
2. The binding energies corresponding to the pair polarization model are 2.7 times larger than for the many body polarization model. The dipole-induced dipole interaction destabilizes the binding of the excess electron. The binding energies scale almost exactly as

$$E_{pair-pol} = (2.7)E_{many-body-pol} \tag{44}$$

for all cluster sizes studied.
3. The electron exists in a highly diffuse state, with very little density penetrating into the cluster and most of it extending well outside the cluster. For example, in the 19-atom cluster, the electron extends out to 10 Xe diameters, whereas the cluster itself has a radius of about 1.75 diameters. As n increases, more electron density is sucked into the channels in the interior of the cluster, until at n $\simeq$100 very little electron density extends outside the cluster.

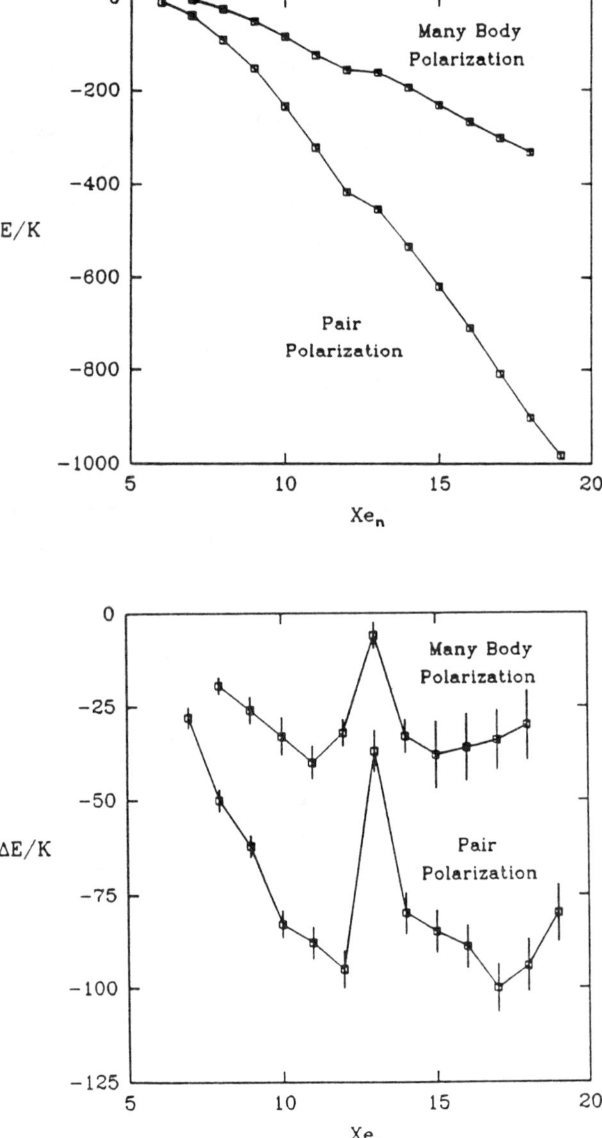

FIGURE 8.   Binding energies of excess electrons attached to xenon clusters as a function of cluster size, obtained from quantum calculations with pairwise additive and many body polarization potentials.

4.   For small clusters, a dielectric continuum model[105] in which the cluster is represented by a sphere of dielectric gives very good agreement with the molecular simulation. Here the electrostatic potential energy of an electron interacting with the dielectric sphere is substituted into the Schrödinger equation, which upon solution gives the ground state energy. A comparison is given in Figure 9. It should be noted that for large clusters, where the electron is found entirely inside the cluster and thus probes the channel network, this continuum model breaks down.

B. WATER CLUSTERS

The attachment of an excess electron to water clusters[106-110] and to ammonia clusters[111]

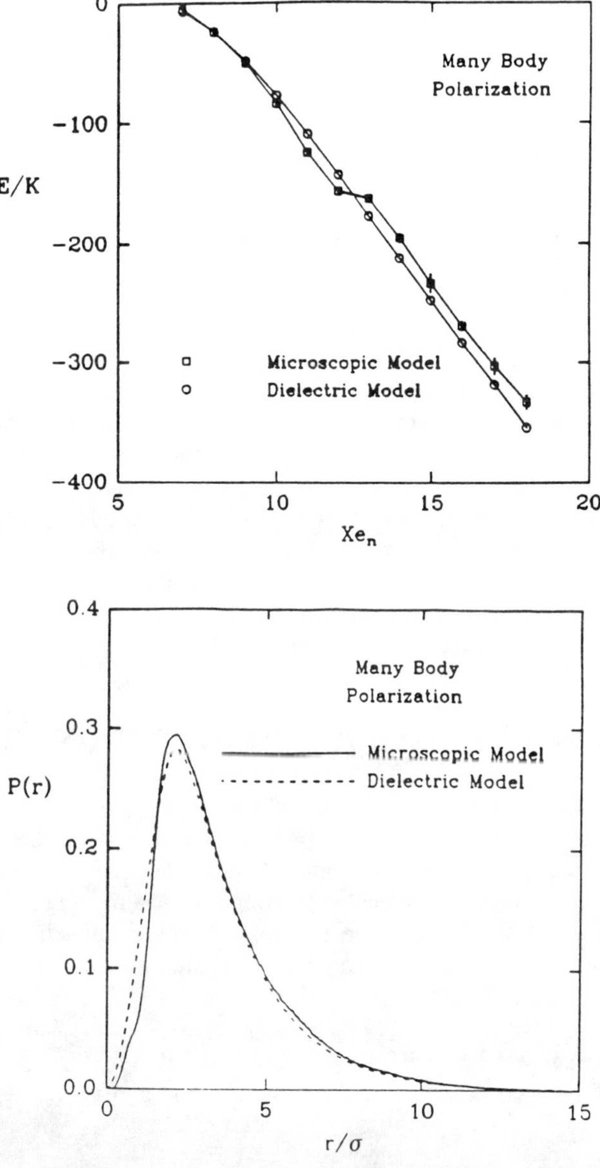

FIGURE 9. Comparison of binding energies obtained from quantum simulation calculations employing the many body polarization potential and the values calculated from the continuum dielectric model.

has been studied experimentally. It is known that for water the cluster anions $(H_2O)_n^-$ exist for n = 2,4,5... Aside from *ab initio* quantum chemical calculations on the water dimer,[112-117] little theoretical work had been done to elucidate the electronic structure and binding energies of the electrons. Wallqvist and Berne[14] used path-integral Monte Carlo methods to study electron attachment to water dimers and trimers. The electron-water-pseudopotential adopted in the studies was a very simple one involving the electron-instantaneous dipole interaction. The water molecules and the electron were all treated quantum mechanically, because at the low temperatures of the simulation, the intramolecular vibrations of the water molecules, as well as the intermolecular vibrations, have substantial zero point amplitudes that could be important in the binding process. These studies found that

1.   The monomer anion H_2O^- does not form.
2.   The dimer anion $(H_2O)_2^-$ forms and has a very weak binding energy. The electron forms a very diffuse state extending out from the center of the cluster to 80 atomic units. The dimer structure is not perturbed very much by the electron.
3.   The electron does not bind to the trimer in its lowest energy geometry. If the geometry is distorted so that the trimer has a sufficiently large dipole moment, the electron will bind, but the overall energy of this distorted $(H_2O)_3^-$ moiety is very high compared to the thermal energy and should thus not be seen at equilibrium.

More recently, Barnett et al.[15,16] have repeated these studies, using a more realistic pseudopotential and nevertheless getting similar results: weak binding and very diffuse electron states. In their dimer calculations these authors used MD simulations to move both the solvent molecules and the isomorphic ring polymer chain. Because the ring polymer representing the electron is stiff, small time steps are required. The simulation was performed for only one half the period of the lowest frequency normal mode of this ring polymer, hardly enough time to be sure that the polymer has equilibrated.

These investigators have gone on to study much larger water clusters and find that for clusters containing around n = 128 molecules, the electron binds in a bulk localized or cavity state much as Berne et al.[5] and Rossky et al.[3] found to occur in liquid water.

There are still many interesting questions involving clusters. This is a new and rapidly growing field for which the theoretical methods outlined here should continue to be very useful.

VII. SIMULATION STUDIES OF ELECTRON TRANSFER

Electron transfer reactions are of paramount importance in chemistry. In the successful theories of Marcus, Levich, and others,[100,101] the solvent is treated phenomenologically. It has now become possible to simulate some of the significant steps in electron transfer reactions, taking into account the molecular structure and dynamics of the solvent. In complex media, like proteins, it is also possible to explore electron tunneling pathways through heterogeneous regions of protein. Although very few studies have been performed to date, two recent studies should provide some insight into how the field is likely to develop.

A. HYDRATED Fe^{2+}–Fe^{3+} ELECTRON TRANSFER

The work of Kuharski et al.[19] focuses on the electron transfer reaction,

$$Fe^{3+} + e \rightleftharpoons Fe^{2+} \tag{45}$$

The model starts with an effective pair potential acting between the water molecules, pseudopotentials describing the electron-water and electron-Fe^{3+} interactions, and an Fe^{3+}-water pair potential which was fitted to give the observed $Fe^{3+} - O$ bond length for the $Fe^{3+}(H_2O)_6$ complex. The model system consists of a pair of classical Fe^{3+} ions and a collection of classical water molecules coupled to an isomorphic ring polymer representing the path integral for the quantal electron (see Section II).

The most important aspect of the electron transfer process in this system is the reorganization of the hydration spheres around the ions which makes the two redox sites degenerate, enabling the electron to resonate between these sites and thus promoting the charge transfer. To better understand this feature of their model these workers performed a path-integral simulation of the solvated Fe^{2+} ion treated as an Fe^{3+} particle interacting with a quantal electron and the system of water molecules. They found that the calculated Fe^{2+}–O bond length agreed well with the experimental value for the $Fe^{2+}(H_2O)_6$ complex and that the isomorphic polymer was strongly localized close to the ion. Identical results were obtained from a fully classical simulation employing a charge of $2+$ instead of $3+$ in the ion-water potential. Thus, the electron appears to be so strongly localized on the ion that quantal charge fluctuations have an insignificant influence on the model ion-water interaction.

The reorganization of the solvent to special transition state conformations is the first step in the phenomenological description of electron transfer processes. It is found that these transition state conformations are characterized by structures of the water ligands in the ion solvation shells, which rearrange on a very slow time scale compared to fluctuations in the bulk fluid. The waters in the ion solvation shells thus exhibit glassy structure.

Next, a calculation was performed in which the barycenter of the electron polymer was held midway between the two Fe^{3+} particles separated by 5.5 Å and surrounded by the 12 water molecules at their first coordination shells. This interionic separation is estimated to be in the region where electron transfer between the ions is optimal.[118-120] The calculation revealed that less than 1% of the electron polymer resided in the tunneling region between the two ions, with the rest of the polymer being tightly bound to the Fe^{3+} particles. Thus, the appropriate model is a two state tight binding model in which the electron is always strongly localized on the ions, allowing them to be treated as classical point charges.

The transition state for electron transfer between the Fe^{2+} and Fe^{3+} species involves the formation of a high energy configuration of the system of solvated ions in which, for example, the coordination shell of the Fe^{2+} ion compresses, and that of the Fe^{3+} ion expands. This change provides environments around the two ions which are degenerate with regard to the interaction potential they provide for the extra electron. Due to their high energies, these important solvent fluctuations for the electron transfer process are very infrequent, so sampling these conformations with the Boltzman distribution will be an extremely impractical way to study their properties.

To sample these infrequent configurations, Kuharski et al. employ an umbrella sampling scheme based on the classical point charge or tight binding model. The aim is to calculate the reversible work required to deform the solvation shells from their stable equilibrium conformation into the less probable transition state configurations. Within the tight binding approximation, the transition state is characterized by having 50% of extra electronic charge localized on each of the two Fe^{3+} centers. The reversible work required to bring the nuclei to this transition state geometry can be calculated by following any reversible path between the equilibrium and transition state conformations. A computationally convenient path is to put the extra electron into its transition state distribution (two $Fe^{2.5+}$ ions within the tight binding approximation) and then allow the surrounding water molecules to equilibrate under the influence of this electronic distribution, sampling the solvent transition state configurations.

In these studies, the free energy difference between the transition and equilibrium states is obtained from the probability distribution $P(E)$ for the difference in the potential fields experienced by the electron when it is localized at the different redox centers. This difference E is obtained from the electron-water pseudopotentials as

$$E(r^N) = \sum_i \sum_a [u_{e-\alpha}(r_i^\alpha - R_A) - u_{e-\alpha}(r_i^\alpha - R_B)] \tag{46}$$

where $u_{e-\alpha}(r_i^\alpha - R_A)$ describes how an electron localized at the position of the Fe^{3+} ion labeled A interacts with the α-th site of the i-th water molecule. The interactions between the electron localized on the ions and the ions themselves cancel when we take the difference, due to the symmetry of this particular ionic system.

The probability of obtaining a solvent conformation for which the two redox centers provide degenerate environments for the transferring electron will be $P(0)$. The average free energy of these conformations is $-k_B T \ln P(0)$. The equilibrium conformations will be characterized by an average difference in the electronic potential environments between the two redox centers, E_{min}, the value of which gives the minimum free energy $-k_B T \ln P(E_{min})$.

Thus the free energy difference between the equilibrium state and the states with average electrical potential difference between redox centers E is

$$\Delta G(E) = -k_B T \ln[P(E)/P(E_{min})] \tag{47}$$

The electron transfer activation free enegy $\Delta G^\dagger$ is simply the average free energy difference between the transition and equilibrium states, so

$$\Delta G^\dagger = -k_B T \ln[P(0)/P(E_{min})] = \Delta G(0) \tag{48}$$

The quantity E can be considered to be reaction coordinate, because the system must move from the equilibrium regions of multidimensional configuration space characterized by an average electrical potential energy difference between the centers $E = E_{min}$, to the transition state region of configuration space where $E = 0$ in order for the electron transfer reaction to occur. By assuming harmonic interactions between the particles, Marcus proposed that the free energy difference should be a parabolic function of such a reaction coordinate. The most important result to emerge from the microscopic model calculations outlined in the section is the verification of this parabolic dependence. Over a wide range of interionic separations the $\Delta G(E)$ functions are well fit by least square parabolas. Perhaps the most convincing evidence for Marcus' ideas is the microscopic verification of the relationship between the activation free energy $\Delta G^\dagger$ and the solvent reogranization energy λ. The rules of intersecting parabolas give

$$\Delta G^\dagger = \lambda/4 \tag{49}$$

Within the tight binding approximation, λ is simply the energy required to take the electron from the Fe^{2+} ion to the Fe^{3+} ion when the solvent is held in the equilibrium configuration around the initial ion pair. It is the vertical transition energy which takes the system from DSA to D^+SA^- with no nuclear rearrangements. This value is nothing more than the electric potential energy difference between the two redox centers in the equilibrium conformation; thus, $\lambda = E_{min}$. The results for $\Delta G^\dagger$ and λ obtained from these simulations verify the Marcus relation in Equation 49 to within the statistical uncertainties of these calculations for a range of interionic separations.

Not only do these microscopic model calculations verify the Marcus phenomenology, they also provide remarkable agreement with experiment considering the crudity of the model interactions employed. The experimental activation free energy is ~ 15 kcal/mol and the calculated value is ~ 20 kcal/mol for an interionic separation of 5.5 Å. This calculated result ignores the tunnel splitting and reversible work to bring the ions together. These two terms are on the order of $k_B T$, and they should approximately cancel each other.

Kuharski et al. have also calculated the tunnel splitting Δ for an electron moving in the pseudopotential field of two Fe^{3+} particles by using a cylindrical discrete grid of points and the methods described in Section II. They solve the one-electron Schrödinger equation and obtain the tunnel splitting as the energy difference between the ground and first excited states. They find that the dependence of Δ on interionic separation is well represented by an exponential decay, and that the splittings and exponential decay constant agree reasonably well with *ab initio* calculations[118] which included the influence of the first shell of water molecules and their valence electrons. Therefore, the intervening solvent molecules may not have a significant influence on the splitting.

At the relevant separations where electron transfer takes place, the tunnel splittings are very small on the scale of the energies and the electronic states, so that dense grids were necessary to give the small energy differences accurately. With a more highly structured intervening environment like that found in a protein, an alternative path-integral approach developed by Kuki and Wolynes[121] described in the following section may be more useful for calculating Δ.

B. ELECTRON TUNNELING PATHWAYS IN PROTEINS

Kuki and Wolynes[121] have employed path-integral methods to study the electron tunneling pathways between redox centers in proteins. Rather than studying the nuclear reorganization of the protein, which must take place in order for electron transfer to occur, these workers concentrate on the other important factor governing electron transfer, namely, electron tunneling. Tunneling determines the magnitude of the coupling between the Born-Oppenheimer potential energy surfaces describing the nuclear motion.

The rate of electron transfer between chromophores embedded in proteins is most strongly influenced by the magnitude of the electron tunneling matrix element, which couples the DSA and D^+SA^- potential energy surfaces. The chromophores are held in place at a relatively large separation by the rigid protein matrix so nuclear reorganization is not as significant a factor as it is with electron transfer between ions in solution. Usually, some nuclear rearrangement is necessary to give the two chromophores electronically degenerate environments. These rearrangements are generally localized around either or both of the chromophores and probably do not involve significant reorganization of the protein. Fixing the donor and acceptor relative to one another results in an electron transfer rate which is determined primarily by the rate at which the electron tunnels through the large region of protein separating the chromophores. Due to electron tunneling, the DSA and D^+SA^- potential energy surfaces are never truly degenerate; rather, they avoid one another and are pulsed apart in energy by an amount 2Δ. Time dependent perturbation theory gives that the tunneling rate is proportional to Δ^2.

Kuki and Wolynes have presented a path-integral approach for calculating the tunnel splitting Δ at the crossing point of the two Born-Oppenheimer surfaces. They show that provided there are no other potential surfaces close in energy to these two nearly degenerate surfaces, the tunnel splitting between them is related to an integral over paths which stretch between the two chromophores. Using the discretized path integral formulation outlined in Section II, we can write

$$\beta\Delta_\lambda = \left(\frac{mP}{2\pi\hbar^2\beta}\right)^{3P/2} \int dr_1 \ldots \int dr_{P-1} \exp[-\beta V_{eff}^\lambda(r_0,\ldots,r_P)] \tag{50}$$

where the end points of the paths are at the positions of the centers of the chromophores r_0 and r_P and

$$V_{eff}^\lambda(r_0,\ldots,r_P) = \sum_{i=0}^{P-1} \left\{ \frac{mP}{2\hbar^2\beta^2}(r_i - r_{i+1})^2 + \frac{1}{P}[V_0(r_i) + \lambda(V(r_i) - V_0(r_i))] \right\} \tag{51}$$

Here we have introduced the mutation parameter λ. As this parameter varies between 0 and 1, the electron-protein interaction potential changes continuously between the reference form $V_0(r)$ and the full potential $V(r)$. The reason for introducing this parameter will be clear shortly. As it stands, Equation 50 has the form of a classical partition function for a P-particle linear polymer with nearest neighbor harmonic bonds and moving over a potential $[V_0 - \lambda(V - V_0)]/P$. The end points of the polymer are tied to the two chromophores.

Importance sampling Monte Carlo techniques are useful for computing averages of quantities over a distribution function. In general, the Monte Carlo method cannot be used to calculate partition function integrals like Equation 50. After some simple maniupulation, however, we find that the partition function ratio $\Delta_\lambda/\Delta_{\lambda'}$ can be written as an average quantity as follows:

$$\frac{\Delta_\lambda}{\Delta_{\lambda'}} = \frac{\int dr_1 \ldots \int dr_{P-1} \, \exp[-\beta(\lambda - \lambda')\sum_{i=0}^{P-1}\{V(r_i) - V_0(r_i)\}]\exp[-\beta V_{\text{eff}}^{\lambda'}(r_0,\ldots,r_P)]}{\int dr_1 \ldots \int dr_{P-1}\exp[-\beta V_{\text{eff}}^{\lambda'}(r_0,\ldots,r_P)]}$$

$$= \langle\exp[-\beta(\lambda - \lambda')\sum_{i=0}^{P-1}\{V(r_i) - V_0(r_i)\}]\rangle_{\lambda'}. \tag{52}$$

Thus, if we can find a reference potential for V_0 for which we can calculate Δ_0, the above result can be used to evaluate Δ_1/Δ_0 and obtain the tunnel splitting for the full potential.

The path-integral approach outlined for computing Δ is designed to handle highly structured potential environments where the interaction potential varies very rapidly, and a discrete grid method like that employed by Kuharski et al. may require a very dense grid and be prohibitively expensive. It should prove very useful to employ discrete grid methods to calculate Δ_0 for a smoothly varying reference potential and then use the path-integral method to calculate Δ_1/Δ_0 to obtain the absolute tunnel splitting for a highly structured potential.

Kuki and Wolynes consider electron transfer between a zinc porphyrin and a ruthenium site in a myoglobin that contains a tryptophan in the through space line of flight.[121] Extensive experimental studies of these systems have been conducted.[122,123] In this system, the transition state results primarily from nuclear reorganization around the ruthenium site, with little reorganization around the zinc porphyrin being necessary. Thus the degenerate electronic environments around the chromophores in the transition state should look like that of the zinc porphyrin which reacts from its triplet state which is 1.8 eV, above the ground state. In this triplet state the zinc porphyrin is a strong electron donating species. This fact led Kuki and Wolynes to construct a detailed pseudopotential describing the interactions experienced by an excess electron in the protein. If the transition state environment was characteristic of electron withdrawing species, pseudopotentials describing the interaction of a hole with the protein would be required. The interaction potential experienced by the electron thus consists of a spherical well of constant potential centered on the $[Ru(N\,H_3)_5$ $-\,His^{12}]^{3+}$ acceptor group, an ellipsoid of constant potential centered on the electron donating porphyrin heme, and the fully structured pseudopotential describing the interaction of the excess electron with the intervening protein. The reference potential V_0 was chosen to be zero everywhere except for the wells representing the chromophores. There are both attractive and repulsive core regions in the fully structured pseudopotential.

The calculations reveal that the tunnel splitting ratio is $\Delta_1/\Delta_0 = 0.56 \pm 0.03$; that is, the tunnel splitting for the fully structured potential is about half that of the zero intervening potential. Since the rates go like Δ^2, the effect of the intervening protein is to slow the electron transfer by about a factor of four over the rate we would expect if there were free space between the chromophores. Thus the effect of the repulsive cores of the intervening potential, which restrict the space available to the electron paths and slow the tunneling, outweighs the influence of the attractive regions of the pseudopotential.

By studying the accumulation of points of the electron path in different regions of space, we can get an indication of the probability that the electron will visit these regions as it tunnels between the chromophores. These model studies suggest that the tryptophan does not give rise to a loosely bound anionic state but only slightly distorts the electron tunneling paths, which sweep out a continuous cylindrical zone with about a 2 Å radius that stretches over 28 Å between the two redox centers. The amino acids outside this region have very little influence on the electron tunneling paths.

VIII. CONCLUSION

In this chapter, we have tried to give an overview of the quantum simulation methods that have emerged in the last few years and to indicate how they have been applied to study the equilibrium and dynamical behavior of excess electrons in disordered media. The future of this field for the next few years lies in the development of these new methods in two important areas.

We have already seen some early developments which extend the methods described in this chapter to systems in which the many electron nature of the interactions are explicitly responsible for the physical phenomena of interest. Dynamical chemical processes which involve the breaking of bonds or the reorganization of the electronic distribution of a molecular system must, in general, be treated within such a many electron framework. The applications of spin density functional methods, which were outlined in Section III.D, represent the first promising attempts at unifying classical molecular dynamics methods for describing nuclear motions and making accurate many electron quantum calculations into a powerful tool for exploring such phenomena.

The second important area of future development of the methods described in this chapter will be their application to study dynamical processes which involve transitions between quantum states. As discussed in Section II.F, the treatment of electronically nonadiabatic classical nuclear dynamics presents a formidable theoretical and computational problem. These new developing methods for many electron quantum calculations, combined with treatment of nonadiabatic nuclear dynamics, will enable the detailed microscopic study of many unexplored chemical phenomena.

ACKNOWLEDGMENT

This work was supported by a grant from the National Science Foundation.

REFERENCES

1. **Coker, D. F., Berne, B. J., and Thirumalai, D.,** Path integral Monte Carlo studies of the behavior of excess electrons in simple fluids, *J. Chem. Phys.,* 86, 5689, 1987.
2. **Coker, D. F. and Berne, B. J.,** Excess electronic states in fluid helium, *J. Chem. Phys.,* 89, 2128, 1988.
3. **Rossky, P. J., Schnitker, J., and Kuharski, R. A.,** Quantum simulations of aqueous systems, *J. Stat. Phys.,* 43, 949, 1986.
4. **Rossky, P. J. and Schnitker, J.,** The hydrated electron: quantum simulation of structure, spectroscopy, and dynamics, *J. Phys. Chem.,* 92, 4277, 1988.
5. **Wallqvist, A., Thirumalai, D., and Berne, B. J.,** Path integral Monte Carlo study of the hydrated electron, *J. Chem. Phys.,* 86, 6404, 1987.
6. **Wallqvist, A., Martyna, G., and Berne, B. J.,** Behavior of the hydrated electron at different temperatures: structure and absorption spectrum, *J. Phys. Chem.,* 92, 1721, 1988.
7. **Sprik, M., Impey, R. W., and Klein, M. L.,** Study of electron solvation in liquid ammonia using quantum path integral Monte Carlo calculations, *J. Chem. Phys.,* 83, 5802, 1985.
8. **Sprik, M., Impey, R. W., and Klein, M. L.,** Study of electron solvation in polar solvents using path integral calculations, *J. Strat. Phys.,* 43, 967, 1986.
9. **Sprik, M. and Klein, M. L.,** Application of path integral simulations to the study of electron solvation in polar fluids, *Comput. Phys. Rep.,* 7, 147, 1988.
10. **Parrinello, M. and Rahman, A.,** Study of an F center in molten KC1, *J. Chem. Phys.,* 80, 860, 1984.
11. **Selloni, A., Carnevali, P., Car, R., and Parrinello, M.,** Localization, hopping, and diffusion of electrons in molten salts, *Phys. Rev. Lett.,* 59, 823, 1987.
12. **Martyna, G. J. and Berne, B. J.,** Structure and energetics of Xe_n^-, *J. Chem. Phys.,* 88, 4516, 1988.

13. **Martyna, G. J. and Berne, B. J.**, Structure and energetics of Xe_n^-: many-body polarization effects, *J. Chem. Phys.*, 90, 3744, 1989.

14. **Wallqvist, A., Thirumalai, D., and Berne, B. J.**, Localization of an excess electron in water clusters, *J. Chem. Phys.*, 85, 1583, 1986.

15. **Barnett, R. N., Landman, U., Cleveland, C. L., and Jortner, J.**, Electron localization in water clusters. I. Electron-water pseudopotential, *J. Chem. Phys.*, 88, 4421, 1988.

16. **Barnett, R. N., Landman, U., Cleveland, C. L., and Jortner, J.**, Electron localization in water clusters. II. Surface and internal states, *J. Chem. Phys.*, 88, 4429, 1988.

17. **Landman, U., Scharf, D., and Jortner, J.**, Electron localization in alkali-halide clusters, *Phys. Rev. Lett.*, 54, 1860, 1985.

18. **Scharf, D., Jortner, J., and Landman, U.**, Cluster isomerization induced by electron attachment, *J. Chem. Phys.*, 87, 2716, 1987.

19. **Kuharski, R. A., Bader, J. S., Chandler, D., Sprik, M., Klein, M. L., and Impey, R. W.**, Molecular model for aqueous ferrous-ferric electron transfer, *J. Chem. Phys.*, 89, 3248, 1988.

20. **Sprik, M., Klein, M. L., and Chandler, D.**, Staging: a sampling technique for the Monte Carlo evaluation of path integrals, *Phys. Rev. B*, 31, 4234, 1985.

21. **Pollock, E. L. and Ceperley, D. M.**, Simulation of quantum many-body systems by path-integral methods, *Phys. Rev. B*, 30, 2555, 1984.

22. **Doll, J. D. and Freeman, D. L.**, A Monte Carlo method for quantum Boltzmann statistical mechanics, *J. Chem. Phys.*, 80, 2239, 1984.

23. **Freeman, D. L. and Doll, J. D.**, A Monte Carlo method for quantum Boltzmann statistical mechanics using Fourier representations of path integrals, *J. Chem. Phys.*, 80, 5709, 1984.

24. **Doll, J. D., Coalson, R. D., and Freeman, D. L.**, Fourier path-integral Monte Carlo methods: partial averaging, *Phys. Rev. Lett.*, 55, 1, 1985.

25. **Marchi, M., Sprik, M., and Klein, M. L.**, Calculation of the free energy of electron solvation in liquid ammonia using a path integral quantum Monte Carlo simulation, *J. Phys. Chem.*, 92, 3625, 1988.

26. **Marchi, M., Sprik, M., and Klein, M. L.**, Calculation of the molar volume of electron solvation in liquid ammonia, *J. Phys. Chem.*, 94, 431, 1990.

27. **Feit, M. D., Fleck, J. A., Jr., and Steiger, A.**, Solution of the Schrödinger equation by a spectral method, *J. Comput. Phys.*, 47, 412, 1982.

28. **Kosloff, D. and Kosloff, R.**, A Fourier method solution for the time dependent Schrödinger equation as a tool in molecular dynamics, *J. Comput. Phys.*, 52, 35, 1983.

29. **Kosloff, R.**, Time-dependent quantum-mechanical methods for molecular dynamics, *J. Phys. Chem.*, 92, 2087, 1988.

30. **Kosloff, R. and Tal-Ezer, H.**, A direct relaxation method for calculating eigenfunctions and eigenvalues of the Schrödinger equation on a grid, *Chem. Phys. Lett.*, 127, 223, 1986.

31. **Schnitker, J., Motakabbir, K., Rossky, P. J., and Friesner, R. A.**, A priori calculation of the optical-absorption spectrum of the hydrated electron, *Phys. Rev. Lett.*, 60, 456, 1988.

32. **Motakabbir, K. A. and Rossky, P. J.**, On the nature of pre-existing states for an excess electron in water, *Chem. Phys.*, 129, 253, 1989.

33. **Barnett, R. B., Landman, U., and Nitzan, A.**, Relaxation dynamics following transition of solvated electrons, *J. Chem. Phys.*, 90, 4413, 1989.

34. **Barnett, R. N., Landman, U., and Nitzan, A.**, Dynamics and spectra of a solvated electron in water clusters, *J. Chem. Phys.*, 89, 2242, 1988.

35. **Barnett, R. N., Landman, U., and Nitzan, A.**, Dynamics of electron localization, solvation, and migration in polar molecular clusters, *Phys. Rev. Lett.*, 62, 106, 1989.

36. **Sprik, M. and Klein, M. L.**, Optimization of a distributed Gaussian basis set using simulated annealing: application to the adiabatic dynamics of the solvated electron, *J. Chem. Phys.*, 89, 1592, 1988.

37. **Car, R. and Parrinello, M.**, Unified approach for molecular dynamics and density-functional theory, *Phys. Rev. Lett.*, 55, 2471, 1985.

38. **Grimm, R. C. and Storer, R. G.**, Monte-Carlo solution of Schrödinger's equation, *J. Comput. Phys.*, 7, 134, 1971.

39. **Anderson, J. B.**, A random-walk simulation of the Schrödinger equation: H_3^+, *J. Chem. Phys.*, 63, 1499, 1975.

40. **Anderson, J. B.**, Quantum chemistry by random walk. H^2P, H_3^+ D_{3h} $^1A_1'$, H_2 $^3\Sigma_u^+$, H_4 $^1\Sigma_g^+$, Be ^1S, *J. Chem. Phys.*, 65, 4121, 1976.

41. **Anderson, J. B.**, Quantum chemistry by random walk: a faster algorithm, *J. Chem. Phys.*, 82, 2662, 1985.

42. **Anderson, J. B.**, Quantum chemistry by random walk: higher accuracy, *J. Chem. Phys.*, 73, 3897, 1980.

43. **Ceperley, D. M. and Kalos, M. H.**, Quantum many-body problems, in *Monte Carlo Methods in Statistical Physics*, Binder, K., Ed., Springer-Verlag, Berlin, 1979, chap. 4, 145.

44. **Doll, J. D., Beck, T. L., and Freeman, D. L.**, Quantum Monte Carlo dynamics: the stationary phase Monte Carlo path integral calculation of finite temperature time correlation functions, *J. Chem. Phys.*, 89, 5753, 1988.

45. **Beck, T. L., Doll, J. D., and Freeman, D. L.**, Locating stationary paths in functional integrals: an optimization method utilizing the stationary phase Monte Carlo sampling function, *J. Chem. Phys.*, 90, 3181, 1989.

46. **Filinov, V. S.**, Calculation of Feynman integrals by means of the Monte Carlo method, *Nucl. Phys. B*, 271, 717, 1986.

47. **Chang, J. and Miller, W. H.**, Monte Carlo path integration in real time via complex coordinates, *J. Chem. Phys.*, 87, 1648, 1987.

48. **Makri, N. and Miller, W. H.**, Monte Carlo integration with oscillatory integrands: implications for Feynman path integration in real time, *Chem. Phys. Lett.*, 139, 10, 1987.

49. **Makri, N. and Miller, W. H.**, Monte Carlo path integration for the real time propagator, *J. Chem. Phys.*, 89, 2170, 1988.

50. **Cline, R. E., Jr. and Wolynes, P. G.**, Monte Carlo methods for real-time quantum dynamics of dissipative systems, *J. Chem. Phys.*, 88, 4334, 1988.

51. **Mak, C. H. and Chandler, D.**, Solving the sign problem in quantum Monte Carlo dynamics, preprint.

52. **Kumamoto, D. and Silbey, R.**, A self-consistent semiclassical approach to the inelastic scattering of atoms from solid surfaces, *J. Chem. Phys.*, 75, 5164, 1981.

53. **Thirumalai, D., Bruskin, E. J., and Berne, B. J.**, On the use of semiclassical dynamics in determining electronic spectra of Br_2 in an Ar matrix, *J. Chem. Phys.*, 83, 230, 1985.

54. **Pechukas, P.**, Time-dependent semiclassical scattering theory. I. Potential scattering, *Phys. Rev.*, 181, 166, 1969.

55. **Pechukas, P.**, Time-dependent semiclassical scattering theory. II. Atomic collisions, *Phys. Rev.*, 181, 174, 1969.

56. **Motakabbir, K. A., Schnitker, J., and Rossky, P. J.**, Transient photophysical hole-burning spectroscopy of the hydrated electron: a quantum dynamical simulation, *J. Chem. Phys.*, 90, 6916, 1989.

57. **Schnitker, J. and Rossky, P. J.**, Excess electron migration in liquid water, *J. Phys. Chem.*, 93, 6965, 1989.

58. **Barnett, R. N., Landman, U., and Nitzan, A.**, Dynamics and excitations of a solvated electron in molecular clusters, *Phys. Rev. A.*, 38, 2178, 1988.

59. **Nikitin, E. E. and Umanskii, S. Ya.**, *Theory of Slow Atomic Collisions*, Springer Series in Chemical Physics, Vol. 30, Springer-Verlag, Berlin, 1984.

60. **Tully, J. C. and Preston, R. K.**, Trajectory surface hopping approach to nonadiabatic molecular collisions: the reaction of H^+ with D_2, *J. Chem. Phys.*, 55, 562, 1971.

61. **Tully, J. C.**, Nonadiabatic processes in molecular collisions, in *Dynamics of Molecular Collisions*, Part B, Miller, W. H., Ed., Plenum, New York, 1976, 217.

62. **Miller, W. H. and George, T. F.**, Semiclassical theory of electronic transitions in low energy atomic and molecular collisions involving several nuclear degrees of freedom, *J. Chem. Phys.*, 56, 5637, 1972.

63a. **Webster, F., Rossky, P. J., and Friesner, R. A.**, Nonadiabatic processes in condensed matter: semiclassical theory and implementation, *Comput. Phys. Commun.*, 25, 780, 1990.

63b. **Space, B. and Coker, D. F.**, Nonadiabatic surface hopping dynamics of excess electrons in dense fluid helium, *J. Chem. Phys.*, 94, 1976, 1991.

64. **Slater, J. C.**, Statistical exchange-correlation in the self-consistent field, *Adv. Quant. Chem.*, 6, 1, 1972.

65. **Lekner, J.**, Motion of electrons in liquid argon, *Phys. Rev.*, 158, 130, 1967.

66. **Stillinger, F. H. and David, C. W.**, Polarization model for water and its ionic dissociation products, *J. Chem. Phys.*, 69, 1473, 1978.

67. **Bachelet, G. B., Hamann, D. R., and Schlüter, M.**, Pseudopotentials that work: from H to Pu, *Phys. Rev. B*, 26, 4199, 1982.

68. **Gunnarsson, O. and Lundqvist, B. I.**, Exchange and correlation in atoms, molecules, and solids by the spin-density-functional formalism, *Phys. Rev. B*, 13, 4274, 1976.

69. **Perdew, J. P. and Zunger, A.**, Self-interaction correction to density-functional approximations for many-electron systems, *Phys. Rev. B*, 23, 5048, 1981.

70. **Langreth, D. C. and Perdew, J. P.**, Exchange-correlation energy of a metallic surface: wave-vector analysis, *Phys. Rev. B*, 15, 2884, 1977.

71. **Ceperley, D. M. and Alder, B. J.**, Ground state of the electron gas by a stochastic method, *Phys. Rev. Lett.*, 45, 566, 1980.

72. **Car, R. and Parrinello, M.**, Structural, dynamical, and electronic properties of amorphous silicon: an *ab initio* molecular-dynamics study, *Phys. Rev. Lett.*, 60, 204, 1988.

73. **Ballone, P., Andreoni, W., Car, R., and Parrinello, M.**, Equilibrium structures and finite temperature properties of silicon microclusters from *ab initio* molecular-dynamics calculations, *Phys. Rev. Lett.*, 60, 271, 1988.

74. **Chiarotti, M.**, Private communication.

75. **Martins, J. L., Buttet, J., and Car, R.,** Electronic and structural properties of sodium clusters, *Phys. Rev. B,* 31, 1804, 1985.

76. **Springett, B. E., Cohen, M. H., and Jortner, J.,** Properties of an excess electron in liquid helium: the effect of pressure on the properties of the negative ion, *Phys. Rev.,* 159, 183, 1967.

77. **Cohen, M. H.,** Electrons in fluids: the role of disorder, *Can. J. Chem.,* 55, 1906, 1977.

78. **Cohen, M. H. and Jortner, J.,** Electron cavity formation in solid helium, *Phys. Rev.,* 180, 238, 1969.

79. **Basak, S. and Cohen, M. H.,** Deformation-potential theory for the mobility of excess electrons in liquid argon, *Phys. Rev. B,* 20, 3404, 1979.

80. **Jahnke, J. A., Meyer, L., and Rice, S. A.,** Zero-field mobility of an excess electron in fluid argon, *Phys. Rev. A,* 3, 734, 1971.

81. **Schnyders, H., Rice, S. A., and Meyer, L.,** Electron drift velocities in liquefied argon and krypton at low electric field strengths, *Phys. Rev.,* 150, 127, 1966.

82. **Kestner, N. R., Jortner, J., Cohen, M. H., and Rice, S. A.,** Low-energy elastic scattering of electrons and positrons from helium atoms, *Phys. Rev.,* 140, A56, 1965.

83. **Sprik, M., Klein, M. L., and Chandler, D.,** Simulation of an excess electron in a hard sphere fluid, *J. Chem. Phys.,* 83, 3042, 1985.

84. **Nichols, A. L., III, Chandler, D., Singh, Y., and Richardson, D. M.,** Excess electrons in simple fluids. II. Numerical results for the hard sphere solvent, *J. Chem. Phys,.* 81, 5109, 1984.

85. **Nichols, A. L., III and Chandler, D.,** Excess electrons in simple fluids. IV. Real time behavior, *J. Chem. Phys.,* 87, 6671, 1987.

86. **Laria, D. and Chandler, D.,** Comparative study of theory and simulation calculations for excess electrons in simple fluids, *J. Chem. Phys.,* 87, 4088, 1987.

87. **Barnett, R. N., Landman, U., Cleveland, C. L., and Jortner, J.,** Surface states of excess electrons on water clusters, *Phys. Rev. Lett.,* 59, 811, 1987.

88. **Landman, U., Barnett, R. N., Cleveland, C. L., Scharf, D., and Jortner, J.,** Electron excitation dynamics, localization, and solvation in small clusters, *J. Phys. Chem.,* 91, 4890, 1987.

89. **Andrea, T. A., Swope, W. C., and Andersen, H. C.,** The role of long ranged forces in determining the structure and properties of liquid water, *J. Chem. Phys.,* 79, 4576, 1983.

90. **Linse, P. and Anderson, H. C.,** Truncation of coulombic interactions in computer simulations of liquids, *J. Chem. Phys.,* 85, 3027, 1986.

91. **Brooks, C. L., III,** The influence of long-range force truncation on the thermodynamics of aqueous ionic solutions, *J. Chem. Phys.,* 86, 5156, 1987.

92. **Chandler, D. and Wolynes, P. G.,** Exploiting the isomorphism between quantum theory and classical statistical mechanics of polyatomic fluids, *J. Chem. Phys.,* 74, 4078, 1981.

93. **Selloni, A., Car, R., Parrinello, M., and Carnevali, P.,** Electron pairing in dilute liquid metal-metal halide solutions, *J. Phys. Chem.,* 91, 4947, 1987.

94. **Berne, B. J. et al.,** Studies of electron solvation in liquid alkanes, work in progress.

95. **Space, B., Coker, D. F., Lev, H., Berne, B. J., and Martyner, G.,** The nature of excess electronic states in fluids of polarizable atoms, submitted.

96. **Reininger, R., Asaf, U., and Steinberger, I. T.,** The density dependence of the quasi-free electron state in fluid xenon and krypton, *Chem. Phys. Lett.,* 90, 287, 1982.

97. **Coker, D. F. and Berne, B. J.,** Path integral Monte Carlo studies of excess electron solvation in atomic liquid mixtures, work in progress.

98. **Bredig, M. A.,** Mixtures of metals with molten salts, in *Molten Salt Chemistry,* Blander, M., Ed., Interscience, New York, 1964, 367.

99. **Warren, W. W., Jr., Sotier, S. and Brennert, G. F.,** Localization of electrons in ionic liquids: nuclear magnetic resonance in Cs-CsI and CsI-I solutions, *Phys. Rev. B,* 30, 65, 1984.

100. **Marcus, R. A.,** Chemical and electrochemical electron-transfer theory, *Annu. Rev. Phys. Chem.,* 15, 155, 1964.

101. **Levich, V. G.,** Present state of the theory of oxidation-reduction in solution (bulk and electrode reactions), *Adv. Electrochem. Electrochem. Eng.,* 4, 249, 1966.

102. **Migus, A., Gauduel, Y., Martin, J. L., and Antonetti, A.,** Excess electrons in liquid water: first evidence of a prehydrated state with femtosecond lifetime, *Phys. Rev. Lett.,* 58, 1559, 1987.

102a. **Webster, F. J., Schnitker, J., Friedrichs, M. S., Friesner, R. A., and Rossky, P. J.,** Solvation dynamics of the hydrated electrons: a nonadiabatic quantum simulation, submitted.

103. **Hoare, M. R. and Pal, P.,** Physical cluster mechanics: statics and energy surfaces for monatomic systems, *Adv. Phys.,* 20, 161, 1971.

104. **Hoare, M. R. and Pal, P.,** Physical cluster mechanics: statistical thermodynamics and nucleation theory for monatomic systems, *Adv. Phys.,* 24, 645, 1975.

105. **Stampfli, P. and Bennemann, K. H.,** Theory of the electron affinity of clusters of rare-gas atoms, *Phys. Rev. A,* 38, 4431, 1988.

106. **Haberland, H., Ludewigt, C., Schindler, H.-G., and Worsnop, D. R.,** Clusters of water and ammonia with excess electrons, *Surf. Sci.,* 156, 157, 1985.

107. **Haberland, H., Ludewigt, C., Schindler, H.-G., and Worsnop, D. R.,** Experimental observation of the negatively charged water dimer and other small $(H_2O)_n^-$ clusters, *J. Chem. Phys.,* 81, 3742, 1984.

108. **Knapp, M., Echt, O., Kreisle, D., and Recknagel, E.,** Trapping of low energy electrons at preexisting, cold water clusters, *J. Chem. Phys.,* 85, 636, 1986.

109. **Knapp, M., Echt, O., Kreisle, D., and Recknagel, E.,** Electron attachment to water clusters under collision-free conditions, *J. Phys. Chem.,* 91, 2601, 1987.

110. **Posey, L. A., Campagnola, P. J., Johnson, M. A., Lee, G. H., Eaton, J. G., and Bowen, K. H.,** On the origin of the competition between photofragmentation and photodetachment in hydrated electron clusters, $(H_2O)_n^-$, *J. Chem. Phys.,* 91, 6536, 1989.

111. **Haberland, H., Schindler, H.-G., and Worsnop, D. R.,** Mass spectra of negatively charged water and ammonia clusters, *Ber. Bunsenges. Phys. Chem.,* 88, 270, 1984.

112. **Newton, M. D.,** The role of *ab initio* calculations in elucidating properties of hydrated and ammoniated electrons, *J. Phys. Chem.,* 79, 2795, 1975.

113. **Chipman, D. M.,** Effect of molecular geometry on the electron affinity of water, *J. Phys. Chem.,* 82, 1080, 1978.

114. **Chipman, D. M.,** Theoretical study on the electron affinity of the water dimer, *J. Phys. Chem.,* 83, 1657, 1979.

115. **Kestner, N. R. and Jortner, J.,** Studies of the stability of negatively charged water clusters, *J. Phys. Chem.,* 88, 3818, 1984.

116. **Rao, B. K. and Kestner, N. R.,** *Ab initio* calculations on negatively charged water clusters, *J. Chem. Phys.,* 80, 1587, 1984.

117. **Jortner, J.,** Level structure and dynamics of clusters, *Ber. Bunsenges. Phys. Chem.,* 88, 188, 1984.

118. **Logan, J. and Newton, M. D.,** *Ab initio* study of electronic coupling in the aqueous $Fe^{2+} - Fe^{3+}$ electron exchange process, *J. Chem. Phys.,* 78, 4086, 1983.

119. **Jafri, J. A., Logan, J., and Newton, M. D.,** *Ab initio* study of inner solvent shell reorganization in the $Fe^{2+} - Fe^{3+}$ aqueous electron exchange reaction, *Isr. J. Chem.,* 19, 340, 1980.

120. **Tembe, B. L., Friedman, H. L., and Newton, M. D.,** The theory of the $Fe^{2+} - Fe^{3+}$ electron exchange in water, *J. Chem. Phys.,* 76, 1490, 1982.

121. **Kuki, A. and Wolynes, P. G.,** Electron tunneling paths in proteins, *Science,* 236, 1647, 1987.

122. **Mayo, S. L., Ellis, W. R., Jr., Crutchley, R. J., and Gray, H. B.,** Long-range electron transfer in heme proteins, *Science,* 233, 948, 1986.

123. **Cowan, J. A. and Gray, H. B.,** Long-range electron transfer in metal-substituted myoglobins, *Chem. Scripta,* 28A, 21, 1988.

Chapter 8

ELECTRON SOLVATION IN POLAR LIQUIDS

Jean-Paul Jay-Gerin and Christiane Ferradini

TABLE OF CONTENTS

I. EVIDENCE FOR THE SOLVATED ELECTRON

An excess electron in a polar liquid can occupy a localized state as a result of its interaction with a certain number of neighboring molecules of the solvent. When this interaction leads to a relaxed state of the electron with respect to the medium, the electron is said to be "solvated". The ensemble of electron and solvent molecules thus formed shows all the characteristics of an authentic chemical entity, i.e., its thermodynamic and kinetic properties can be defined. Depending on the nature of the solvent, the lifetime of the solvated electron may either be very long, as is the case for liquid ammonia, or very short as is the case for water. The lifetime of the solvated electron is reduced as a result of its reactivity with the solvent or with other solvated electrons, or with free radicals or molecules that may be present in the solution. It was the particular stability of the electron in liquid ammonia that led to the earliest evidence for the existence of the solvated electron. In fact, it has been known for a long time[1] that alkali metals dissolve in liquid ammonia. Solutions obtained in this way are blue and exhibit an important ionic conductivity. As far back as 1908, Kraus[2] attributed these characteristics to the presence of anions consisting of "electrons surrounded with an envelope of ammonia". The solvated electron, as defined above, is indeed present in these solutions. In the case of potassium, for example, the overall formation process may be written as

$$K + n\,NH_3 \rightarrow K^+ + e_{am}^-$$

The ammoniated electron, which is stable, is responsible for both the blue color and the ionic character of the solution. The absorption spectrum,[3] at room temperature, shows a maximum located at $\lambda = 1850$ nm.

The same spectrum has been obtained by γ-radiolysis of liquid NH_3 at 20°C under basic conditions (i.e., containing KNH_2) and in the presence of hydrogen.[4] The reactions of the ammoniated electron with other radiolysis products (NH_4^+ ions and NH_2 radicals) are inhibited, and the ammoniated electron is stable.

It was only in 1962, when the first pulse radiolysis systems were put into operation, that the transient formation of the solvated electron in water, and afterward in a variety of other solvents, was demonstrated. The initial radiation action being ionization, it is clear that high energy (Compton or accelerated) electrons produce other electrons with greater than thermal energies. These electrons are then slowed down and thermalize in the irradiated medium. In the first interpretations of water radiolysis it was supposed that the thermalized electrons disappeared very rapidly, before any other reaction, leading to the production of hydrogen atoms:

$$e^- + H_2O \rightarrow H + OH^-$$

Obviously, the identification of the reducing entity with "H atoms" in neutral media raised a number of difficulties.

However, as early as 1952, Stein,[5] in explaining the radiolysis of aqueous solutions of methylene blue in the presence of CO_2, suggested that the principal reducing entity in neutral medium could be the hydrated electron, for which a description was given later.[6] Weiss[7,8] justified the stabilization of the negative charge by identifying it as a "polaron".

In the course of an informal conference held in 1953, Platzman[9] developed a more elaborate theoretical model for the hydrated electron and predicted some of its properties. However, he assigned a very short lifetime (10^{-9} s) to it, considering the reaction:

$$e_{aq}^- + H_2O \rightarrow H + OH^-$$

to be fast, while it is in fact very slow.

Various experimental results[10-13] using competition methods showed that the principal reducing entity seemed not to have the same kinetic properties in neutral solution and in acid solution. We now know that this is because the hydrated electron rapidly reacts with H^+:

$$e_{aq}^- + H^+ \rightarrow H + H_2O$$

In acid solution one thus has to deal with an H atom, the conjugate acid of e_{aq}^-. In 1962, two publications[14,15] described the effect of the ionic strength on the rate constant of certain reactions of the principal reducing entity with charged species, and thus established that it carries a unit negative charge in neutral solution.

On the other hand, in 1960,[16] Keene[17,18] reported the existence of a strong transient absorption in the visible region in irradiated water. Matheson[19] suggested that this absorption might be due to the hydrated electron. In 1962, Hart and Boag[20,21] observed a short-lived optically absorbing species in water irradiated with 2 μs pulses of 1.8 MeV electrons. These authors indicated that the corresponding absorption spectrum had a peak near 700 nm, and concluded that the species involved was, indeed, the hydrated electron. Numerous works rapidly confirmed the identification and determined the kinetic and thermodynamic properties of this new radical species,[22,23] while various theoretical models were proposed in an attempt to describe its structure.[24] Moreover, the development of pulse radiolysis methods permitted the demonstration of the existence of the solvated electron in about a hundred pure polar solvents (see Table 1, in Section III). Several review articles have recently appeared in the literature on this subject.[24-26]

II. METHODS OF PRODUCTION AND DETECTION OF THE SOLVATED ELECTRON

Among the methods that are currently used to produce and detect the solvated electron, pulse radiolysis, combined with fast spectrophotometric detection, remains the most widely used technique.[27,28] The method consists of irradiating the solvent of interest with beams of high energy electrons (~1 to 30 MeV) in very short pulses, and observing the subsequent evolution of the system. The pulse duration is most frequently a few nanoseconds but, since 1968, picosecond facilities[29-33] have also been used.

The absorbed radiation gives rise, through Coulombic interactions, to the ionization and excitation of the solvent molecules. Each high energy electron produced by ionization further excites or ionizes a large number of molecules of the solvent until it falls below the energy threshold for electronic excitation: it is then said to be a "subexcitation electron".[34] The value of this energy threshold is a few electronvolts and depends on the nature of the solvent. The electron then thermalizes, losing its excess energy through momentum transfer in elastic collisions, molecular vibrational and rotational excitations, and the excitation of phonons. During this stage, it has a certain probability of recombining with the geminate positive ion. This probability strongly depends on the characteristics of the medium. This ensemble of processes has given rise to a great deal of theoretical and experimental work,[35-45] allowing us to obtain a satisfactory quantitative approach to the phenomena involved with the slowing down of electrons in condensed media and their recombination.

Once thermalized, the electron may solvate according to a complex kinetic law that has been followed for the case of alcohols at low temperature[46-51] by using pulse radiolysis. However, the pulse radiolysis technique is primarily used either to obtain a good spectral characterization of the electron after solvation (see Table 1 below) or to follow its subsequent evolution directly, in order to describe its kinetic behavior.[23]

Ionization processes may also be produced photochemically.[52] Depending on experimental conditions, these photoionizations may either be mono- or multi-photonic. Rare-gas lamps can reach ultraviolet photon energies up to about 7 eV, while synchrotron radiation[53,54] offers a large energy range (typically, 0.1 to 1000 eV). It is easy to select the initial energy and thus to predetermine the transition to be considered. At the present time, pulsed lasers can be used to generate ultrashort pulses with durations of a few tens of femtoseconds[55,56] or even of a few femtoseconds.[57]

Photoionization studies in the liquid phase are numerous: they deal either with solutions (ionic or neutral), or with pure liquids. The formation of solvated electrons has thus been demonstrated in the photolysis of aqueous solutions of inorganic (such as I^{-}[58,59] or $Fe(CN)_6^{4-}$)[60] or organic (formate,[61] acetate,[61] phenolate,[62] etc.) anions. In liquid ammonia, electron photodetachment can be observed from amide ions[63] or ethanolate.[64] Among the neutral solutes that are most frequently used, we may cite tetramethyl-*para*-phenylenediamine, phenol, pyrene, and molecules of biological interest, such as tryptophan and indole. These studies have shown that the value of the ionization threshold is not only a function of the nature of the absorbing molecule, but also depends on the structure of the solvent.[65,66] For the case of pure liquid water, the ionization threshold value is near 6.5 eV.[67-69] Moreover, femtosecond spectroscopic studies have demonstrated the existence in pure water of a precursor state of the hydrated electron, which mainly absorbs in the infrared[70] and whose lifetime is reported as 240[70] or 540[71] fs, depending on the authors.

For the case of photoionization in micellar media, the reader is referred to Chapter 12.

The solvated electron can also be generated electrochemically, either by cathodic reduction or by photoinjection from an electrode. These methods are described in Chapter 11.

We have already seen that a solvated electron can be generated through chemical processes by the dissolution of alkali metals in liquid ammonia or in some other solvents. It is also possible to obtain solutions of hydrated electrons by introducing H atoms into basic aqueous media.[72]

In the above-mentioned experiments, the detection of the solvated electron has most often been carried out either by using suitable scavengers, or principally in the case of flash photolysis or pulse radiolysis, by spectrophotometry. Other methods may also be used depending on the experimental conditions: we mention here electrical conductivity[73] and electron spin resonance[74-76] measurements, provided that they are time resolved.

III. PROPERTIES OF THE SOLVATED ELECTRON IN VARIOUS POLAR SOLVENTS

In Table 1 we have compiled some of the kinetic and optical properties of the solvated electron (e_s^-) in 86 different pure polar solvents (arranged alphabetically by name) and various characteristics of the host liquids whose knowledge seems to be of importance in understanding the phenomenon of electron solvation. Values of G_{fi}, the number of free ion pairs produced per 100 eV of absorbed energy, are first reported. G_{fi} is defined as the yield of electrons that become homogeneously distributed through the bulk of the liquid after spur expansion, which corresponds to times of about 10^{-7} s after passage of the radiation.[204] Values of μ_e and the mobility of e_s^- measured in the considered solvents are listed next followed by those of λ_{max}, the wavelength of the maximum of their optical absorption spectra, ϵ_{max}, the molar extinction coefficient at this maximum, and $W_{1/2}$, the width of the absorption peak at half height. As far as the solvents are concerned, values of the viscosity (η), the static dielectric constant (ϵ_s), and the Kirkwood parameter (g_K) are presented for each of them. We should note that each solvent is given a symbol that is used in the various figures of the chapter with the exception of Figure 1.

TABLE 1

Characteristics of the Solvated Electron and of the Solvent for a Variety of Solvents of Different Polarities.[a]

Solvent	Symbol	η[b]	ϵ_s[c]	g_K[d]	G_n[e]	μ_e[f]	λ_{max}[g]	ϵ_{max}[h]	$W_{1/2}$[i]	Ref.
Ammonia	A	0.135	16	1.4	1.9	14[j,k]	1830 (0.68)	4.8	0.4[j]	4,63,77—85
2-Aminoethanol	B	19.3	37.7	2.2	1.9[j]	—	900 (1.38)	1.85[j]	1.16	78,86—91
1,2-Butanediol	C	52.6	—	—	—	—	560 (2.21)	—	1.42	92
1,3-Butanediol	D	99.7	28.8	2.45	1.55[j]	—	595 (2.08)	0.7[j]	1.46[j]	78,86,90,92—94
1,4-Butanediol	E	1.6	32	2.25	—	—	600 (2.06)	0.7[j]	1.54	78,86,90,92,93
2,3-Butanediol	F	40.6	30[j]	3.3	1.6[j]	—	585 (2.12)	0.85[j]	1.6	78,86,92,93,95
1-Butanol	G	2.56	17.5	2.35	1.2	0.75	635 (1.95)	1.2	1.57	78,86,89,96—105
2-Butanol	H	3.04	16.6	2.6	0.8[j]	0.7[j]	780 (1.59)	1.45[j]	1.57	50,78,86,90,102, 106—108
2-Butanone[l]	I	0.38	18.1	1.5	0.84	1.5	900 (1.38)	2.5	0.5	78,86,87,96, 99,109
n-Butylamine	J	0.53	4.9	1.75	0.27	2.7	>1300	—	—	78,86,99,109,110
Cyclohexanol	K	49[j]	15	2.4	—	—	750 (1.65)	—	1.5	78,86,103,111
Cyclopentanol	L	—	18	3.05	—	—	820 (1.51)	—	1.4[j]	78,94,103,111
1-Decanol	M	11	7.8	2.85	—	—	650 (1.9)	1.4	1.5	78,97,103, 105,111
2-Decanol	N	—	—	—	—	—	800 (1.55)	—	1.3[j]	112
Deuterium oxide	O	1.1	78.3	2.6	3	1.47	705 (1.76)	1.99	0.8	78,113—123
Dibutyl ether	P	0.74	3.1	1.45	0.3	3.7	(0.5)[j]	2.6[j]	0.6[j]	78,86,101,110, 124—127
N,N-Diethylacetamide	Q	—	32.1	—	1.7[j]	—	1700 (0.73)	—	0.4[j]	90,128,129
Diethylamine	R	0.31	3.6	1.5	—	—	1970 (0.63)	3.2	0.5[j]	78,86,110, 130—133
Diethyl ether	S	0.23	4.3	1.15	0.41	8.6[k]	2140 (0.58)	3.5	0.5[j]	78,81,86,101, 110,124—127, 130,132—135
N,N-Diethylformamide	T	—	29.6	1.85	—	—	1775 (0.7)	—	0.5[j]	78,131
Diglyme	U	0.99	7.2[j]	1.4	0.4	—	1915 (0.65)	3.4	0.58	78,86,133, 135,136
1,2-Dimethoxyethane	V	0.455	7.2	1.45	0.4	11[j,k]	1910 (0.65)	3.6	0.5	78,81,99,109, 130,132,133, 135—139

TABLE 1 (continued)
Characteristics of the Solvated Electron and of the Solvent for a Variety of Solvents of Different Polarities.[a]

Solvent	Symbol	η[b]	ϵ_s[c]	g_K[d]	G_{fi}[e]	μ_e[f]	λ_{max}[g]	ϵ_{max}[h]	$W_{1/2}$[i]	Ref.
N,N-Dimethylacetamide	W	0.93	37.8	1.6	1.7	—	1800 (0.69)	—	0.5[j]	78,86,90,131, 140,141
Dimethylamine	X	0.186	5.3	1.9	—	—	1950 (0.64)	—	0.4[j]	86,110,131
Dimethyl ether	Y	0.105	5	1.2	—	10[j]	(0.56)[j]	—	—	78,86,110, 125,142
N,N-Dimethylformamide	Z	0.8	36.7	1.3	2	—	1680 (0.74)	—	0.5[j]	78,86,90,131, 140,141
1,3-Dimethyl-2-imidazolidinone	a	1.94	37.6	1.35	1.9[j]	—	1550 (0.8)	2.3[j]	0.6[j]	78,90,128,143
1,3-Dimethyl-2-oxohexahydropyrimidine	b	2.93	36.1	1.2	1.85[j]	—	1550 (0.8)	2.35[j]	0.6[j]	78,90,128,143
N,N-Dimethylpropionamide	c	—	—	—	—	—	≥2000	—	—	129
Dimethylsulfide	d	0.28	6.2	1.1	—	13.6	>1500	1.2[m]	—	78,80,86, 144—146
Dimethylsulfoxide	e	2	46.4	1.1	1.8	—	(0.56)[j]	1[n]	—	78,86,134, 147—152
Dimethylsulfoxide-d_6	f	—	—	—	2.3[j]	—	>1500	—	—	148
1,4-Dioxane	g	1.2	2.2	1.3	0.1	0.85	(0.74)[j]	—	—	78,101,110,125, 126,134,153,154
N,N-Dipropylacetamide	h	—	—	—	—	—	>1500	—	—	129
Dipropyl ether	i	0.44	3.4	1.3	0.34	4	(0.52)[j]	2.9[j]	0.5[j]	78,101,110, 124—127,155
1,2-Ethanediamine	j	1.39	12.9	1.3	0.5	1.4	1335 (0.93)	1.95	0.88	78,80,81,86,94, 130,132,139, 152,156—159
1,2-Ethanediol	k	17.8	37.7	2.6	1.2	0.28	570 (2.18)	1.3[j]	1.41	78,86,89,92, 93,100,103,111, 160,161
Ethanol	l	1.07	24.5	2.15	1.7	0.37	690 (1.8)	0.99	1.6	78,86,97, 101—105,160, 162—167
2-Ethoxyethanol	m	1.54	29.6	3	1.6[j]	—	740 (1.67)	—	1.5	78,86,90,96,103
Ethylamine	n	0.24	6.5	1.4	0.5	14	1855 (0.67)	3.8	0.65[j]	78,81,83,84, 110,131,132, 156,157,168

Compound										References
3-Ethyl-3-pentanol	o	—	5[j]	—	—	—	≥1500	—	>0.9[j]	103,111
N-Ethylpyrrolidinone	p	—	—	—	—	—	1550 (0.8)	—	0.6[j]	128,169
4-Heptanol	q	—	5.9	1.45	—	—	925 (1.34)	—	1.5[j]	78,103,111
Hexamethylphosphoric triamide	r	3.1	30	1.55	1.9	2.1[k]	2215 (0.56)	3.2	0.45	78,80,81,86, 87,99, 170—173
1,2,6-Hexanetriol	s	1860[j]	—	—	—	—	540 (2.3)	—	1.4[j]	174
1-Hexanol	t	4.59	13.3	3.3	1.1	0.43	675 (1.84)	—	1.5[j]	78,86,100, 103,111
Hydrazine	u	0.93	51.7	2.2	2.35	1.3	980 (1.26)	2	0.76	78,80,81,110, 157,175,176
Hydrazine-d_4	v	—	—	—	2.57	—	1010 (1.23)	2.4	0.78	177
Isobutylamine	w	0.56	4.4	1.3	—	—	(0.8)[j]	—	—	78,86,94,139,178
Isopropylamine	x	0.36	5.1	1.15	0.4	—	2120 (0.58)	3.2	0.56[j]	78,84,86,94, 110,179
Methanol	y	0.55	32.7	2.2	2	0.62	630 (1.96)	1.04	1.43	78,86,89,97, 101—105,160, 162,163, 165—167,180
2-Methoxyethanol	z	1.6	16.9	1.95	—	—	725 (1.71)	1.3[j]	1.63	78,86,103,131, 181
Methylamine	α	0.19	9.4	1.6	2.2	—	1885 (0.66)	3.3	0.6[j]	78,84,86,110, 168,179,182
2-Methyl-1-butanol	β	4.8	15.2	—	—	0.8[j]	—	—	—	86,94,107
2-Methyl-2-butanol	γ	3.53	5.8	1.1	—	—	1250 (0.99)	—	0.8[j]	78,86,103,111, 183
3-Methyl-1-butanol	δ	3.6	14.9	3.05	0.35	0.97	690 (1.79)	—	—	78,86,94, 99—101,103, 109,111,183
4-Methylcyclohexanol[o]	ε	0.3[j]	13.3	2.3	—	—	805 (1.54)	—	1.6[j]	78,86,103
3-Methyl-3-pentanol	ζ	—	5[j]	—	—	—	≥1500	—	>0.9[j]	103,111
2-Methyl-1-propanol	η	3.39	17.9	2.45	≥0.5	0.68	630 (1.97)	1.6[j]	1.61	50,78,86,90,100, 102,103,106,108
2-Methyl-2-propanol	θ	4.47	12.5	2.3	0.7	—	1090 (1.14)	2.15[j]	0.95	78,86,89,100, 103,106,108, 184—187
N-Methyl-2-pyrrolidinone	ι	1.68	32.5	1	0.92	—	1550 (0.8)	3.6	0.5[j]	78,86,128,143, 152,169,188

TABLE 1 (continued)
Characteristics of the Solvated Electron and of the Solvent for a Variety of Solvents of Different Polarities.[a]

Solvent	Symbol	η^b	ϵ_s^c	g_K^d	G_{fi}^e	μ_e^f	λ_{max}^g	ϵ_{max}^h	$W_{1/2}^i$	Ref.
2-Methyltetrahydrofuran	κ	—	7	1.3	0.23	—	2150 (0.58)	3.9	0.45	78,86,133—135, 139
1-Nonanol	λ	9.1	9.1	2.9	—	—	670 (1.85)	—	1.5[j]	78,97,103,111
1,8-Octanediol[p]	μ	—	—	—	—	—	625 (1.98)	—	1.5[j]	189
1-Octanol	ν	7.36	10.3	2.4	0.7[j]	—	630 (1.96)	1.5	1.55	78,86,89,97,102, 103,112,161,190
2-Octanol	ξ	6.49	7.8	2.55	—	—	850 (1.46)	—	1.3[j]	78,86,94, 103,111,112
3-Octanol	π	—	—	—	—	—	800 (1.55)			
1-Pentanol	ρ	3.43	13.9	3.25	0.49	0.5[j]	635 (1.95)	—	1.5[j]	78,86,87,102, 103,183,185
1,2-Propanediamine	σ	—	10.2	—	—	—	1475 (0.84)	—	0.68	94,131,132,191
1,3-Propanediamine	τ	—	9.6	1.1	—	—	1500 (0.83)	2.2[j]	0.7[j]	78,94,131,133
1,2-Propanediol	φ	45	32	2.7	1.7[j]	0.09[j]	570 (2.18)	0.75[j]	1.52	78,86,87,89,90, 92,93,161
1,3-Propanediol	χ	42	35	2.3	1.8[j]	—	575 (2.16)	0.6[j]	1.46	78,86,90,92,93
1,2,3-Propanetriol	ψ	945	42.5	2.5	2.26[j]	0.007[j]	530 (2.33)	0.97[j]	1.45	78,86,89,161, 192,193
1-Propanol	ω	1.92	20.3	2.3	1.25	0.46	640 (1.94)	1.1	1.63	78,87,89,97, 102—104,108, 160,163
2-Propanol	Γ	2.04	19.4	2.2	1.1	0.51	820 (1.51)	1.3	1.45	78,86,87,89,90, 100—103,105, 108,111,134, 160,161,163, 194
n-Propylamine	Δ	0.353	5.1	1.4	0.3	13	1910 (0.65)	3.2	0.75[j]	78,80,81,83,86, 156,157
Tetrahydrofuran	Θ	0.46	7.6	1.2	0.4	6.7[k]	2100 (0.59)	3.8	0.43	78,81,86,101, 125,126,130, 132,135,137, 139,152,168, 182,195—197

		0.764	5.6	1.05	0.51	3.1	$(0.53)^j$	—	—	78,94,101,125
Tetrahydropyran	Λ	0.764	5.6	1.05	0.51	3.1	$(0.53)^j$	—	—	78,94,101,125
Tetraethylurea	Π	—	—	—	—	—	>1700	—	—	129
Tetraglyme	Σ	—	7.4^j	—	—	—	1790 (0.69)	—	0.69	135
Tetrakis (dimethyl-amino) ethylene	Φ	—	2.46	—	0.2	2200	—	—	—	198
Tetramethylurea	Ψ	1.401	23.1	1.22	1.25^j	—	—	—	—	90,128
Tributylamine	Ω	1.33	5	—	0.25	2	>1300	—	—	81,86,94,99
Triglyme	И	—	7.4^j	—	0.4	—	1840 (0.67)	3.8	0.63	135,136,179
Undecanol	э	—	5.9	2.1	—	—	675 (1.84)	—	1.5^j	78,103,111
Water	Я	0.89	78.5	2.85	2.7	1.93	720 (1.73)	1.88	0.85	23,78,89,91,102, 104,114,119— 121, 123,199—202

[a] Selected values at room temperature, assumed to be 298 K.

[b] Viscosity in cP (1 cP = 10^{-3} Pas).

[c] Static dielectric constant.

[d] Kirkwood parameter.

[e] Free ion yield in molecule/100 eV (1 molecule/100 eV $\simeq$ 1.0364 × 10^{-7} mol/J).

[f] Electron mobility in 10^{-3} cm²/Vs (1 cm²/Vs = 10^{-4} m²/Vs).

[g] Wavelength of the maximum of the optical absorption spectrum in nm. Numbers in parentheses correspond to the transition energy $E_{A_{max}}$ [$E_{A_{max}}$(eV) = 1240/λ_{max}(nm)] at the absorption maximum of e_s^-.

[h] Molar extinction coefficient at λ_{max} in 10^4 M^{-1} cm^{-1}.

[i] Width of the absorption peak at half height in eV.

[j] Estimated.

[k] Average of values obtained from e_s^- scavenging rate constants and direct conductivity measurements.

[l] Perhaps, 2-butanone captures electrons to form anions.[153,203]

[m] At 1440 nm.[146]

[n] At 1275 nm.[151]

[o] Mixed isomers.

[p] Spectrum determined at 100°C.

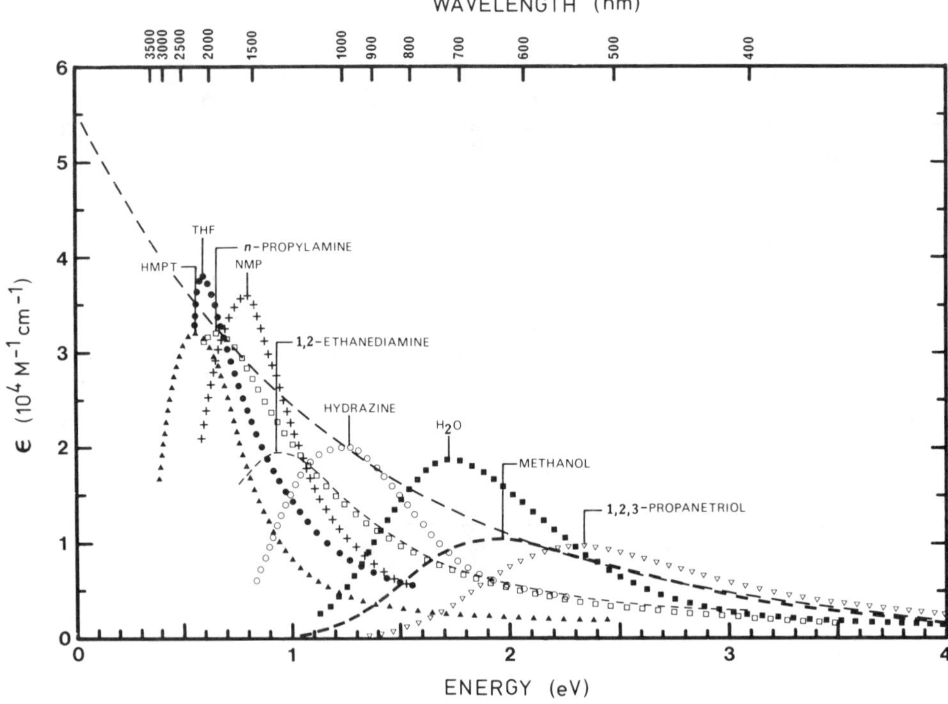

FIGURE 1.   Optical absorption spectra of solvated electrons in various pure solvents at room temperature. The dashed line shows the exponential curve $\epsilon_{max}(E_{A_{max}})$ as calculated from Equations 1 and 2.

IV. INFLUENCE OF VARIOUS PARAMETERS ON THE PROPERTIES OF THE SOLVATED ELECTRON

At the present juncture, it is of interest to establish certain correlations from the various data listed in Table 1. Some of these correlations are presented below.

A. COMPARISON OF SOLVATED ELECTRON SPECTRA IN VARIOUS POLAR SOLVENTS

Figure 1 shows a few typical optical absorption spectra of e_s^- in solvents of various polarities at room temperature. On the basis of the currently available data, it is possible to correlate ϵ_{max} and $W_{1/2}$, defined above, with $E_{A_{max}}$, the energy of the absorption maximum $[E_{A_{max}}(eV) = 1240/\lambda_{max}(nm)]$. This is of obvious practical interest since such correlations would allow us to get an *a priori* estimation of the shape of the e_s^- spectrum in any polar solvent from the experimental determination of λ_{max} only.

As can be seen from a semilogarithmic plot of ϵ_{max} against $E_{A_{max}}$ using the data of Table 1 (Figure 2), a linear correlation is found to describe the experimental results satisfactorily. The relationship between these two parameters can thus be written in the form

$$\epsilon_{max} = \epsilon_{max}^o \exp\left(-\alpha\, E_{A_{max}}\right), \tag{1}$$

where

$$\epsilon_{max}^o = 5.54\,(\pm0.89) \times 10^4\ M^{-1}\ cm^{-1}$$

and $\tag{2}$

$$\alpha = 0.82\,(\pm0.11)\ eV^{-1}$$

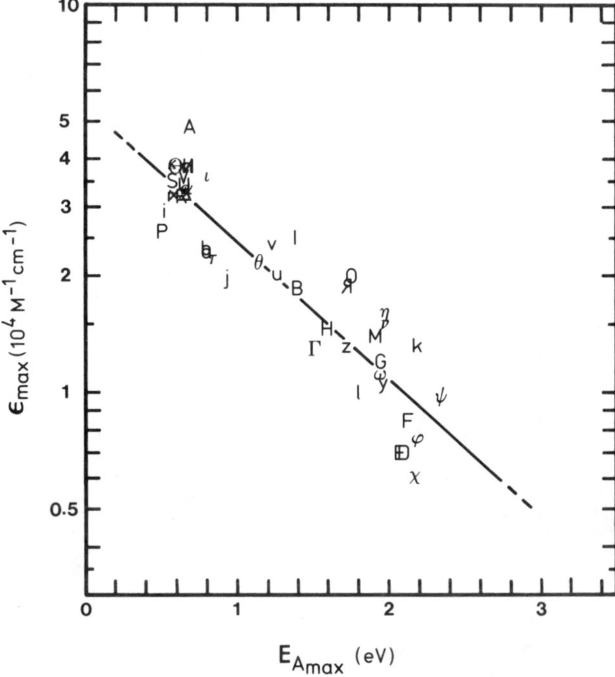

FIGURE 2. Semilog plot showing the variation of ϵ_{max} as a function of $E_{A_{max}}$. Symbols correspond to liquids listed in Table 1. The straight line was obtained from a least-squares fit of ln ϵ_{max} vs. $E_{A_{max}}$ (see Equations 1 and 2) to the data for 44 solvents of various polarities at room temperature.

the values in parentheses indicating the 95% confidence limits of the estimated parameters. The corresponding $\epsilon_{max}(E_{A_{max}})$ curves shown in Figures 1 and 2 are calculated from Equation 1 with these parameter values. As an interesting consequence of Equation 1, the experimental knowledge of λ_{max} in a given polar solvent allows us to predict the value of ϵ_{max}, and thus to deduce from the measured e_s^- spectrum an estimated value of the yield of e_s^- in this solvent.

Figure 3 shows a plot of ϵ_{max} against $1/W_{1/2}$ using the data given in Table 1. A linear correlation is observed, and extrapolation to small values of $1/W_{1/2}$ gives a line which almost passes through the origin. Such a relationship can be written as

$$\epsilon_{max} = \beta \left(\frac{1}{W_{1/2}}\right) + \gamma \tag{3}$$

with

$$\beta = 1.68\,(\pm 0.23) \times 10^4\ M^{-1}\ cm^{-1}\ eV$$

and

$$\gamma = 0.065\,(\pm 0.325) \times 10^4\ M^{-1}\ cm^{-1} \tag{4}$$

where the values in parentheses give the 95% confidence limits of the two estimated parameters. This linear correlation could have been predicted. In fact, a number of authors, and in particular Freeman and co-workers,[89,104,126,162] have indicated that the value of the oscillator strength, f, of the e_s^- absorption spectrum varies only little as one passes from one solvent to another. The usual formula of f is given by[205]

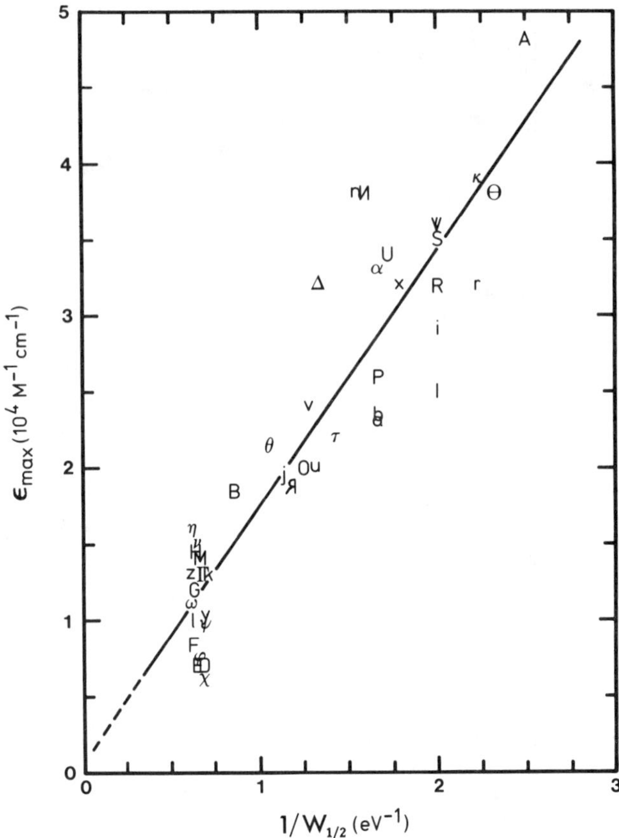

FIGURE 3.   Plot of ϵ_{max} against $1/W_{1/2}$. Symbols correspond to liquids listed in Table 1. The straight line was obtained from a least-squares fit of Equation 3 to the data for 44 solvents of various polarities at room temperature.

$$f = 3.49 \times 10^{-5} \int_0^\infty \epsilon(E) \, dE \tag{5}$$

where $\epsilon(E)$ is the molar extinction coefficient at photon energy $E(eV)$. The integral in Equation 5, which represents the area under the absorption curve $\epsilon(E)$, can be estimated from the approximation

$$\int_0^\infty \epsilon(E) \, dE \approx \epsilon_{max} W_{1/2} \tag{6}$$

According to Equations 5 and 6, the product $\epsilon_{max}W_{1/2}$ should not change much from one solvent to another. ϵ_{max} is thus expected to be nearly proportional to $1/W_{1/2}$. Using Equations 3 to 6 with the data of Table 1, an average value $f \approx 0.61$ can be obtained. This oscillator strength is in good agreement with the values reported in the literature for different solvents.

The existence of a correlation between ϵ_{max} and $E_{A_{max}}$, on the one hand, and between ϵ_{max} and $W_{1/2}$, on the other hand, implies as a consequence a correlation of $W_{1/2}$ with $E_{A_{max}}$. In fact, Equations 1 and 3 give

$$W_{1/2} = \frac{1}{a \exp(-\alpha E_{A_{max}}) - b} \tag{7}$$

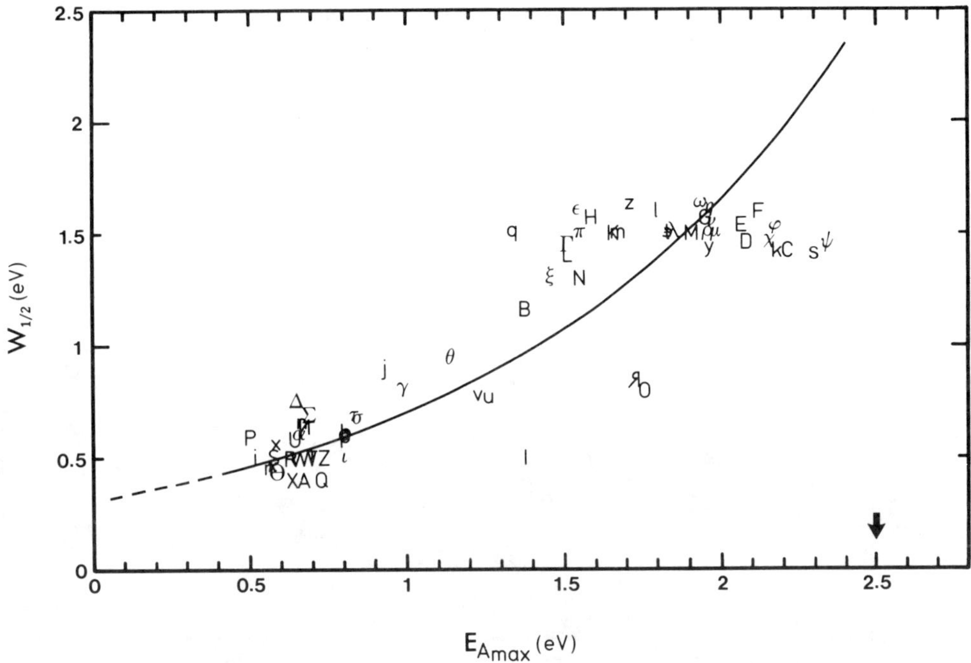

FIGURE 4. Plot of $W_{1/2}$ against $E_{A_{max}}$. The solid line illustrates the variation $W_{1/2}$ ($E_{A_{max}}$) given in Equations 7 and 8. Symbols correspond to experimental data taken from Table 1. The arrow indicates the upper limit for $E_{A_{max}}$ in liquids ($\sim$2.5 eV) as suggested by Freeman.[206]

with

$$a = \epsilon^{\circ}_{max}/\beta = 3.29 \ eV^{-1}$$

and

(8)

$$b = \gamma/\beta = 0.039 \ eV^{-1}$$

where $W_{1/2}$ and $E_{A_{max}}$ are expressed in eV. Equation 7 is illustrated by the curve in Figure 4. In order to test the validity of this relation, we have also plotted in Figure 4 the corresponding experimental data of $W_{1/2}$ ($E_{A_{max}}$) taken from Table 1. The agreement appears to be satisfactory.

The preceding correlations,[95] established between ϵ_{max}, $W_{1/2}$, and $E_{A_{max}}$ (and thus λ_{max}) for a variety of polar solvents, are interesting not only because of their practical importance in allowing certain estimations, but also because they may help to better understand the nature of the interaction of the solvated electron with its stabilizing medium.

B. TENTATIVE CORRELATION BETWEEN $E_{A_{max}}$ AND g_K

It is well known that the Kirkwood parameter, g_K, indicates the extent to which molecules align themselves with respect to their neighbors, and is a useful index to the degree of nonrandom structure in polar liquids. It was therefore tempting to examine whether a correlation exists between this parameter and $E_{A_{max}}$, although the existence of such a correlation has been questioned.[180] A few attempts have been made in this respect.[78,87,152,206-208]

Values of g_K are listed in Table 1 for each of the considered solvents. They are calculated from the Kirkwood-Fröhlich equation[209-211] (in CGS units)

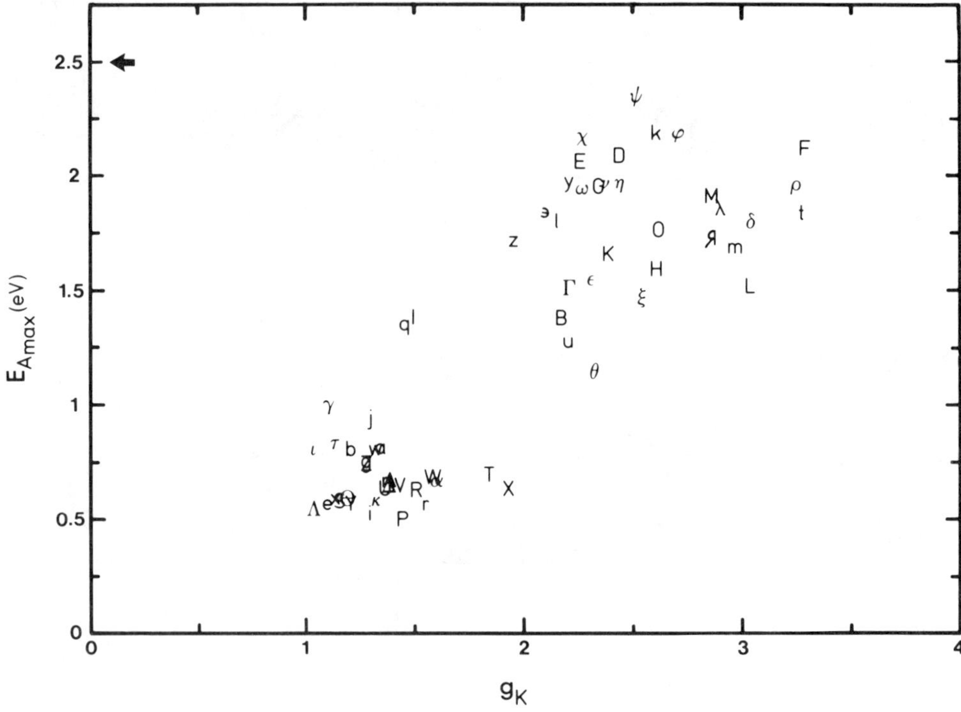

FIGURE 5.   Plot of $E_{A_{max}}$ against g_K for 63 liquids listed in Table 1. The arrow indicates the upper limit for $E_{A_{max}}$ in liquids (~2.5 eV) as suggested by Freeman.[206]

$$g_K = \frac{3k_B T}{\mu_\ell^2} \left\{ \frac{3}{4\pi\, N_{Av}} \left(\frac{M}{d}\right) \left[\frac{(\epsilon_s - 1)(2\epsilon_s + 1)}{9\epsilon_s} \right] - \alpha_m \right\} \tag{9}$$

In this formulation, M is the molecular weight, N_{Av} is Avogadro's number, α_m is the molecular polarizability, ϵ_s and d are the static dielectric constant and the density of the liquid, respectively, μ_ℓ is the permanent dipole moment of a molecule in the liquid phase, and $k_B T$ is the product of Boltzmann's constant and the absolute temperature. Values of μ_ℓ can be estimated from the Onsager's formula[212]

$$\mu_\ell = \frac{(n^2 + 2)(2\epsilon_s + 1)}{3(2\epsilon_s + n^2)} \mu_v \tag{10}$$

where μ_v is the vapor-phase molecular dipole moment and n is the optical refractive index of the liquid. Finally, α_m is evaluated by assuming the validity of the Clausius-Mossotti formula[108,213]

$$\frac{n^2 - 1}{n^2 + 2} \left(\frac{M}{d}\right) = \frac{4\pi}{3} N_{Av}\, \alpha_m \tag{11}$$

Figure 5 shows a plot of $E_{A_{max}}$ against g_K for 63 liquids listed in Table 1. Examination of the data presented in this figure seems to indicate two distinct regions: one corresponding to liquids with g_K values in the range 2 to 3.3 and values of $E_{A_{max}}$ between about 1.2 and 2.3 eV, and the other corresponding to g_K in the range 1 to 1.9 and values of $E_{A_{max}}$ between

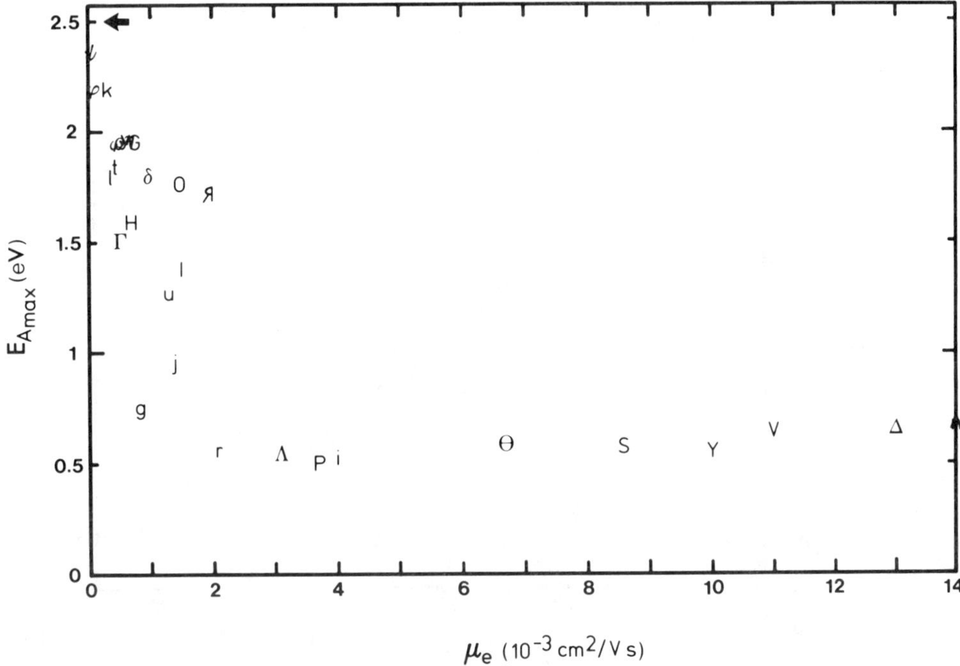

FIGURE 6. Plot of $E_{A_{max}}$ against μ_e for 30 solvents listed in Table 1. The arrow indicates the upper limit for $F_{A_{max}}$ in liquids (~2.5 eV) as suggested by Freeman.[206]

0.5 and 0.9 eV. The former region includes solvents such as water, alcohols, diols, and triols, while the latter is mainly composed of ammonia, amines, amides, and ethers.

Even if a clear trend exists showing that $E_{A_{max}}$ increases with increasing g_K, the scatter of the data indicates that there is no single valued relationship between these two parameters. Such results suggest that, for a given chemical function, the molecular structure of the solvent intervenes per se in a significant way to influence not only the energy distribution of the preexisting electron traps in the liquid, but also the relaxation processes following electron trapping.[78]

C. RELATIONSHIP BETWEEN $E_{A_{max}}$ AND μ_e

As mentioned above, we have listed in Table 1 the values of the mobility (μ_e) of e_s^- measured for the various considered solvents. This parameter is interesting to consider since it provides a direct insight into the mechanisms of electron transport in polar liquids.

The possibility of a correlation between $E_{A_{max}}$ and μ_e has been pointed out by several authors.[80,81,125,127,214,215] Figure 6 shows the variation of $E_{A_{max}}$ with μ_e for the liquids listed in Table 1. This variation can be divided into two distinct regions: (1) for $\mu_e < 3 \times 10^{-3}$ cm²/Vs, $E_{A_{max}}$ decreases approximately linearly with μ_e from about 2.3 to 0.5 to 0.6 eV, and (2) for $\mu_e > 3 \times 10^{-3}$ cm²/Vs, and up to the highest mobility values reported (14 × 10^{-3} cm²/Vs), $E_{A_{max}}$ is nearly constant in the vicinity of 0.5 to 0.6 eV. The existence of these two distinct electron-mobility regions strongly suggests that the interactions of e_s^- with these solvents are of two different types. The region defined by low electron mobility values involves deep trap states in which the electron appears to be quasi-permanently bound to a solvating molecular shell. By contrast, in the high electron mobility region, the electron resides in weakly localized states which would favor its migration through the liquid. However, at present the mechanisms of e_s^- transport involved in these two regions are not yet fully understood.

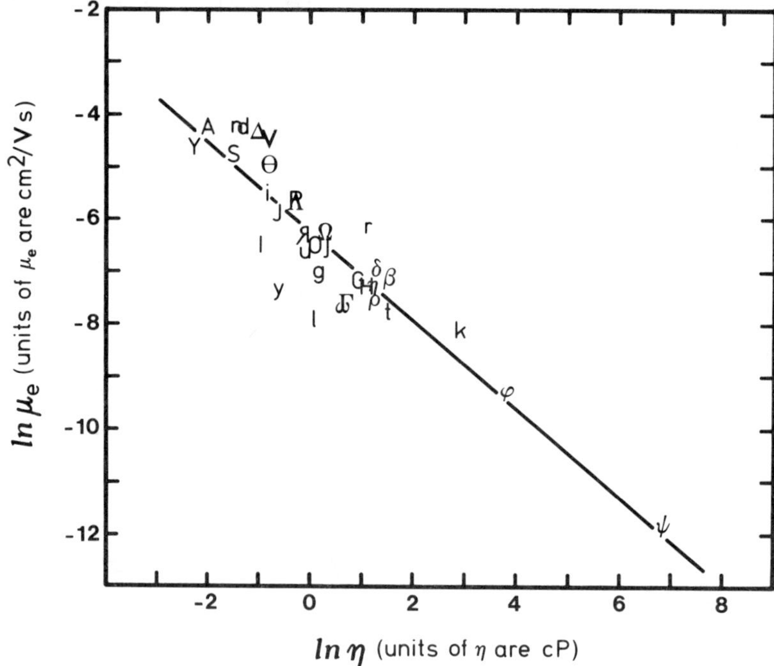

FIGURE 7.   Plot of ln μ_e against ln η. Symbols correspond to liquids listed in Table 1. The straight line was obtained from a least-squares fit of ln μ_e vs. ln η to the data for 34 solvents of various polarities at room temperature (see Equation 12).

D. INFLUENCE OF THE VISCOSITY ON THE MOBILITY OF THE SOLVATED ELECTRON

It has been suggested, that a correlation between the mobility of an excess electron in a polar solvent and the viscosity (η) of the corresponding solvent could provide further information on the nature of the e_s^--solvent interaction.[87,125,167,216,217] For this purpose values of ln μ_e have been plotted against ln η in Figure 7 for the 34 solvents listed in Table 1 for which these two quantities have been measured. Examination of the data presented in this figure shows that there is a linear relationship between ln μ_e and ln η, which indicates that the variation of μ_e (in cm²/Vs) with η (in centipoise) can be represented as

$$\mu_e = C \, \eta^{-P} \tag{12}$$

where p = 0.84 and C = 1.92 $\times$ 10^{-3}, and where the 99% confidence interval of the exponent actually contains the value p = 1.[167] A thorough statistical analysis of the same data has recently indicated the validity of the hypothesis that μ_e is proportional to η^{-1}.[218] This observed viscosity dependence seems to favor a diffusive mechanism for the migration of excess electrons in polar liquids. Such a conclusion is consistent with the results of recent quantum simulation calculations of electron transport by Schnitker and Rossky[219] in liquid water, and by Barnett et al.[220] in water and ammonia clusters. In fact, these authors showed no evidence for electron hopping and quantum mechanical tunneling between neighboring solvent traps, contrary to what was generally proposed in the past.[121,216,221-223] An account of these results is given in Chapter 7.

V. DYNAMICS OF ELECTRON SOLVATION

In this section we report only a brief summary of the dynamics of electron solvation in polar liquids, since detailed information on this question has been given repeatedly in the recent past. The reader is referred to these excellent sources,[52,224-226] and to Chapters 4 and 7.

A number of experiments concern the dynamics of electron solvation in water and alcohols. They have been performed by means of pulse radiolysis or laser photolysis techniques. The experimental data suggest that the solvation process can be described by a two-state model. At short times, an absorption in the infrared is observed, which is assigned to the localization of the electron in preexisting trapping sites of the solvent. This precursor of e_s^- is the so-called incompletely relaxed electron (e_{ir}^-).[101] The e_{ir}^- absorption then decays concurrently with the growth of the characteristic e_s^- absorption spectrum which peaks in the visible region. These absorption spectral changes are attributed to the process of electron solvation.

The first direct observations of the electron solvation process were reported in alcohols. Baxendale and Wardman[46,47] observed this phenomenon by pulse radiolysis as early as 1971 by cooling the alcohols to temperatures near their freezing points and slowing the solvation down to the nanosecond timescale. Subsequently, several pulse radiolysis studies were reported on the time dependence of the e_s^- formation in various liquid alcohols at low temperatures.[49-51,227,228] Since then, room temperature picosecond pulse radiolysis[48,229-233] and laser photolysis[60,233-239] data have been published showing the rapid appearance of the e_s^- absorption spectrum in a range of alcohols and in water. For instance, at 22°C, the formation time of e_s^- ranges from about 18 ps in ethanol to 55 ps in 2-octanol.[224,239]

Recently, Migus et al.[70,240] and Long et al.[71] reported the first time-resolved studies of the solvation dynamics of electrons in liquid water using femtosecond absorption spectroscopy. These experiments indicate that the electron solvation time is considerably shorter than that for the alcohols, the values currently proposed being either 240 or 540 fs, depending on the authors.[70,71,240]

Ultrafast solvation of electrons has also been found in other polar liquids, including ammonia,[241,242] methylamine,[242,243] and ethers,[243] and a competition between solvation and electron attachment seems to exist in liquid dimethylsulfide.[80]

An alternative approach to the dynamics of electron solvation is provided by numerical simulations and a number of such studies have now been published. Such theoretical models make use of the Feynman path-integral formulation of quantum statistical mechanics in conjunction with Monte Carlo and molecular dynamics computer simulations. Details of these methods and their applications to the study of the dynamics of electron localization, solvation, and migration are given in Chapter 7. Calculations of this type have primarily been used to study electron solvation in water and the properties of the hydrated electron. In particular, the study of the structural characteristics of e_{aq}^- has received a great deal of attention.[244-247] Recently, these simulation techniques have been applied to calculate the optical absorption spectrum of e_{aq}^-[248-251] and to study its migration.[219,220] Other studies have dealt with the dynamics of solvent relaxation following sudden excitations of the solvated electron in water and in D_2O,[252] and the simulation of the transient optical hole-burning spectroscopy of e_{aq}^-.[253] Applications of these theoretical models to the electron-liquid ammonia system have also been reported.[254-257] In general, discrepancies exist between experimental results and those obtained using these simulation techniques. These are likely due to inadequacies of the simple model potentials that are used in the simulations to describe both the solvent and the electron-solvent interactions.

At present, the early time dynamics of the electron solvation process in polar liquids and the exact nature of the incompletely relaxed species e_{ir}^- remain far from being well understood. Further experiments are clearly needed. In this respect, advances in ultrafast time-resolved laser spectroscopy have opened new avenues for decisive investigations in this area. They now make it possible to provide important new insights into the molecular nature of the mechanism of electron solvation.

VI. IS THERE A CHEMICAL FATE FOR THE INCOMPLETELY RELAXED ELECTRON?

It is generally believed that the chemical stage begins after the excess electron has reached a fully relaxed, solvated state. However, it should be noted that as soon as the electron becomes trapped in the liquid its localized nature may allow it to undergo chemical reactions before settling into the e_s^- state. Taking into account that the incompletely relaxed localized electron e_{ir}^- has a direct interaction with the solvent itself, it has recently been suggested[258] that it could react in water according to

$$e_{ir}^- + H_2O \rightarrow H + OH^- \tag{13}$$

Reaction 13 would occur on a timescale comparable to the duration of electron hydration, and would be in competition with the latter. It is well known that the corresponding reaction for hydrated electrons is very slow[23]

$$e_{aq}^- + H_2O \rightarrow H + OH^- \quad (k = 19 \text{ M}^{-1}\text{ s}^{-1} \text{ at room temperature}) \tag{14}$$

but the instability of e_{ir}^- with respect to e_{aq}^- would favor the energetic balance of Reaction 13. Although the primary source of atomic hydrogen in water radiolysis has generally been considered at these early times as arising from the dissociation of electronically excited water molecules,[22,259] Reaction 13 offers another possible explanation for the existence of an "initial" H yield.

It has been proposed[258] that a reaction similar to Reaction 13 may also occur in irradiated liquid alcohols. These latter are known to exhibit an "unscavengeable" molecular hydrogen yield[260] which could, at least partly, be assigned to the reaction

$$e_{ir}^- + RCH_2OH \rightarrow H + RCH_2O^- \tag{15}$$

followed by

$$H + RCH_2OH \rightarrow H_2 + RCHOH$$

It is also worth noting that Reactions 13 and 15 would involve electrons in shallow traps,[258] that is, those which contribute to the infrared part of the e_{ir}^- absorption spectrum in these solvents. They would cut down the low energy (red) side of the corresponding optical absorption spectra of e_s^- and thus contribute to the asymmetry of these spectra.

Ultrafast kinetic experiments on the dynamics of electron solvation are needed to assess precisely the fate of e_{ir}^- and thus to test this hypothesis.

VII. CONCLUSIONS

The solvated electron has been the subject of much research since its discovery and identification in the 1960s. A wealth of data have accumulated over the years, as evidenced by the large number of solvents in which the solvated electron has been detected and characterized. Despite this, there are several facets of the problem of an electron in polar liquids that are still not well understood. For instance, there is a great deal of controversy, both experimentally and theoretically, about the mechanism of its solvation, and about the structural characteristics of its solvated state. Little agreement about the correct interpretation of the broad and structureless absorption band of e_s^- has been achieved. Clearly, these points need to be addressed in future studies.

With the advent of ultrafast time-resolved spectroscopy whose frontiers have now moved to femtoseconds, and of new quantum simulation techniques based on the availability of faster computers, the spectroscopic and dynamical data base on e_s^- is growing rapidly. Applications of these powerful techniques offer exciting possibilities for future work, and we hope that they will allow a better characterization of this ubiquitous species. Judging from the intense current activity in this field, we can contemplate an improved understanding of electron solvation in the near future.

ACKNOWLEDGMENTS

We are grateful to the Programme de Coopération franco-québécoise en Recherche médicale and to the NATO Scientific Affairs Division (Grant No. 86/0187) for financial support. Special thanks are due to Drs. Annette Bernas, Nicholas J. B. Green, Bernard Hickel, Darel J. Hunting, and Norman V. Klassen for valuable discussions and for a critical reading of the manuscript. J.-P. Jay-Gerin also acknowledges the financial support of the Medical Research Council of Canada and of the INSERM-FRSQ exchange program in health sciences during his stays at the Laboratoire de Chimie-Physique of the Université René Descartes in Paris.

REFERENCES

1. **Weyl, W.**, Ueber Metallammonium-verbindungen, *Annal. Physik Chemie,* 121, 601, 1864.
2. **Kraus, C. A.**, Solutions of metals in non-metallic solvents. IV. Material effects accompanying the passage of an electrical current through solutions of metals in liquid ammonia. Migration experiments, *J. Am. Chem. Soc.,* 30, 1323, 1908.
3. **Corset, J. and Lepoutre, G.**, Absorption spectra and kinetics of decomposition of potassium-ammonia solutions at room temperature, in *Metal-Ammonia Solutions, Physicochemical Properties,* Lepoutre, G. and Sienko, M. J., Eds., Benjamin, New York, 1964, 186.
4. **Belloni, J. and Fradin de la Renaudière, J.**, Radiolytic formation of long-lived ammoniated electrons at room temperature, *Nature,* 232, 173, 1971.
5. **Stein, G.**, Some aspects of the radiation chemistry of organic solutes, *Discuss. Faraday Soc.,* 12, 227, 1952; see also General Discussion, *Discuss. Faraday Soc.,* 12, 289, 1952.
6. **Jortner, J. and Stein, G.**, Electrons in solutions, *Nature,* 175, 893, 1955.
7. **Weiss, J.**, Radiation chemistry, *Ann. Rev. Phys. Chem.,* 4, 143, 1953.
8. **Weiss, J.**, Primary processes in the action of ionizing radiations on water: formation and reactivity of self-trapped electrons ("polarons"), *Nature,* 186, 751, 1960.
9. **Platzman, R. L.**, Energy transfer from secondary electrons to matter, in *Physical and Chemical Aspects of Basic Mechanisms in Radiobiology,* Magee, J. L., Kamen, M. D., and Platzman, R. L., Eds., Publication 305, Nat. Acad. Sci.-Nat. Res. Council, Washington, D.C., 1953, 22.
10. **Baxendale, J. H. and Hughes, G.**, The X-irradiation of aqueous methanol solutions. Part II. Reaction in D_2O, *Z. Physik. Chem. Neue Folge,* 14, 323, 1958.
11. **Barr, N. F. and Allen, A. O.**, Hydrogen atoms in the radiolysis of water, *J. Phys. Chem.,* 63, 928, 1959.
12. **Hayon, E. and Allen, A. O.**, Evidence for two kinds of "H atoms" in the radiation chemistry of water, *J. Phys. Chem.,* 65, 2181, 1961.
13. **Czapski, G. and Allen, A. O.**, The reducing radicals produced in water radiolysis: solutions of oxygen-hydrogen peroxide-hydrogen ion, *J. Phys. Chem.,* 66, 262, 1962.
14. **Collinson, E., Dainton, F. S., Smith, D. R., and Tazuké, S.**, Evidence for the unit negative charge on the "hydrogen atom" formed by the action of ionizing radiation on aqueous systems, *Proc. Chem. Soc.,* 140, April, 1962.
15. **Czapski, G. and Schwarz, H. A.**, The nature of the reducing radical in water radiolysis, *J. Phys. Chem.,* 66, 471, 1962.
16. **Gilbert, C. W., Keene, J. P., Browne, P. F., and Davy, T. J.**, Radiation Research, in *British Empire Cancer Campaign, 38th Annual Report,* Part II, 1960, 498.
17. **Keene, J. P.**, Kinetics of radiation-induced chemical reactions, *Nature,* 188, 843, 1960.

18. **Keene, J. P.,** Optical absorptions in irradiated water, *Nature,* 197, 47, 1963.

19. **Matheson, M. S.,** Radiation chemistry, *Ann. Rev. Phys. Chem.,* 13, 77, 1962.

20. **Hart, E. J. and Boag, J. W.,** Absorption spectrum of the hydrated electron in water and in aqueous solutions, *J. Am. Chem. Soc.,* 84, 4090, 1962.

21. **Boag, J. W. and Hart, E. J.,** Absorption spectra of "hydrated" electron, *Nature,* 197, 45, 1963.

22. **Klassen, N. V.,** Primary products in radiation chemistry, in *Radiation Chemistry: Principles and Applications,* Farhataziz and Rodgers, M. A. J., Eds., VCH Publishers, New York, 1987, 29.

23. **Buxton, G. V., Greenstock, C. L., Helman, W. P., and Ross, A. B.,** Critical review of rate constants for reactions of hydrated electrons, hydrogen atoms and hydroxyl radicals (OH/O^-) in aqueous solutions, *J. Phys. Chem. Ref. Data,* 17, 513, 1988.

24. **Kestner, N. R.,** Theories of the solvated electron, in *Radiation Chemistry: Principles and Applications,* Farhataziz and Rodgers, M. A. J., Eds., VCH Publishers, New York, 1987, 237.

25. **Brodsky, A. M. and Tsarevsky, A. V.,** Development of the physical chemistry of the solvated electron, *Usp. Khim.,* 56, 1693, 1987 (English transl., *Russian Chem. Rev.,* 56, 969, 1987).

26. **Itskovitch, E. M., Kuznetsov, A. M., and Ulstrup, J.,** The solvated electron, in *The Chemical Physics of Solvation, Part C: Solvation Phenomena in Specific Physical, Chemical, and Biological Systems,* Dogonadze, R. R., Kálmán, E., Kornyshev, A. A., and Ulstrup, J., Eds., Elsevier, Amsterdam, 1988, 275.

27. **Sauer, M. C., Jr.,** Sources of pulsed radiation, in *The Study of Fast Processes and Transient Species by Electron Pulse Radiolysis,* Baxendale, J. H. and Busi, F., Eds., D. Reidel, Dordrecht, 1982, 35.

28. **Roffi, G.,** Optical monitoring techniques, in *The Study of Fast Processes and Transient Species by Electron Pulse Radiolysis,* Baxendale, J. H. and Busi, F., Eds., D. Reidel, Dordrecht, 1982, 63.

29. **Bronskill, M. J. and Hunt, J. W.,** A pulse-radiolysis system for the observation of short-lived transients, *J. Phys. Chem.,* 72, 3762, 1968.

30. **Jonah, C. D.,** A wide-time range pulse radiolysis system of picosecond time resolution, *Rev. Sci. Instrum.,* 46, 62, 1975.

31. **Tabata, Y., Kobayashi, H., Washio, M., Tagawa, S., and Yoshida, Y.,** Pulse radiolysis with picosecond time resolution, *Radiat. Phys. Chem.,* 26, 473, 1985.

32. **Sauer, M. C., Jr., Romero, C., and Schmidt, K. H.,** Dynamics of geminate ions in the pulse radiolysis of cyclohexane and *n*-hexane containing aromatic solutes studied by streak camera measurements of fluorescence and absorption, *Radiat. Phys. Chem.,* 29, 261, 1987.

33. **Grigoryants, V. M., Lozovoy, V. V., Chernousov, Yu. D., Shebolaev, I. V., Arutyunov, V. A., Anisimov, O. A., and Molin, Yu. N.,** Pulse radiolysis system with picosecond time resolution referred to Cherenkov radiation, *Radiat. Phys. Chem.,* 34, 349, 1989.

34. **Platzman, R. L.,** Subexcitation electrons, *Radiat. Res.,* 2, 1, 1955.

35. **Spencer, L. V. and Fano, U.,** Energy spectrum resulting from electron slowing down, *Phys. Rev.,* 93, 1172, 1954.

36. **Dillon, M. A., Inokuti, M., and Kimura, M.,** Time-dependent aspects of electron degradation. I. Subexcitation electrons in helium or neon admixed with nitrogen, *Radiat. Phys. Chem.,* 32, 43, 1988.

37. **Inokuti, M., Kimura, M., and Dillon, M. A.,** Time-dependent aspects of electron degradation. II. General theory, *Phys. Rev. A,* 38, 1217, 1988.

38. **Ashley, J. C.,** Interaction of low-energy electrons with condensed matter: stopping powers and inelastic mean free paths from optical data, *J. Electron Spectrosc. Relat. Phenom.,* 46, 199, 1988.

39. **LaVerne, J. A. and Mozumder, A.,** Energy loss and thermalization of low-energy electrons, *Radiat. Phys. Chem.,* 23, 637, 1984.

40. **Garin, B. M. and Byakov, V. M.,** Energy losses by nonionizing electrons, *Khim. Vys. Énerg.,* 22, 195, 1988, (English transl., *High Energy Chem.,* 22, 165, 1988).

41. **Bader, G., Chiasson, J., Caron, L. G., Michaud, M., Perluzzo, G., and Sanche, L.,** Absolute scattering probabilities for subexcitation electrons in condensed H_2O, *Radiat. Res.,* 114, 467, 1988.

42. **Konovalov, V. V., Raitsimring, A. M., and Tsvetkov, Yu. D.,** Thermalization lengths of "subexcitation electrons" in water determined by photoinjection from metals into electrolyte solutions, *Radiat. Phys. Chem.,* 32, 623, 1988.

43. **Goulet, T. and Jay-Gerin, J.-P.,** Thermalization of subexcitation electrons in solid water, *Radiat. Res.,* 118, 46, 1989.

44. **Goulet, T., Patau, J. P., and Jay-Gerin, J.-P.,** Influence of the parent cation on the thermalization of subexcitation electrons in solid water, *J. Phys. Chem.,* 94, 7312, 1990.

45. **Sanche, L.,** Investigation of ultra-fast events in radiation chemistry with low-energy electrons, *Radiat. Phys. Chem.,* 34, 15, 1989.

46. **Baxendale, J. H. and Wardman, P.,** Direct observation of solvation of the electron in liquid alcohols by pulse radiolysis, *Nature,* 230, 449, 1971.

47. **Baxendale, J. H. and Wardman, P.,** Electrons in liquid alcohols at low temperatures, *J. Chem. Soc., Faraday Trans. I,* 69, 584, 1973.

48. **Chase, W. J. and Hunt, J. W.,** Solvation time of the electron in polar liquids. Water and alcohols, *J. Phys. Chem.,* 79, 2835, 1975.
49. **Gilles, L., Bono, M. R., and Schmidt, M.,** Electron in cold alcohols: a pulse radiolysis study in ethanol, *Can. J. Chem.,* 55, 2003, 1977.
50. **Okazaki, K. and Freeman, G. R.,** Concerning the spectra and solvation process of electrons in liquid butanols, *Can. J. Chem.,* 56, 2305, 1978.
51. **Okazaki, K. and Freeman, G. R.,** Scavenging of electrons prior to solvation in liquid alcohols, *Can. J. Chem.,* 56, 2313, 1978.
52. **Mialocq, J. C.,** La formation de l'électron solvaté en photochimie, *J. Chim. Phys.,* 85, 31, 1988.
53. **Goulon, J.,** Le rayonnement synchrotron: des expériences lourdes!... Mais un potentiel exceptionnel d'applications en chimie!, *J. Chim. Phys.,* 86, 1427, 1989.
54. **Farge, Y.,** Le rayonnement synchrotron, *L'Onde Electrique,* 62, 59, 1982.
55. **Fork, R. L., Shank, C. V., and Yen, R. T.,** Amplification of 70-fs optical pulses to gigawatt powers, *Appl. Phys. Lett.,* 41, 223, 1982.
56. **Migus, A., Antonetti, A., Etchepare, J., Hulin, D., and Orszag, A.,** Femtosecond spectroscopy with high-power tunable optical pulses, *J. Opt. Soc. Am. B,* 2, 584, 1985.
57. **Fork, R. L., Brito Cruz, C. H., Becker, P. C., and Shank, C. V.,** Compression of optical pulses to six femtoseconds by using cubic phase compensation, *Opt. Lett.,* 12, 483, 1987.
58. **Hayon, E.,** The photochemistry of iodide ion in aqueous solutions, *J. Phys. Chem.,* 65, 1937, 1961.
59. **Jortner, J., Ottolenghi, M., and Stein, G.,** Cage effects and scavenging mechanisms in the photochemistry of the iodide ion in aqueous solutions, *J. Phys. Chem.,* 66, 2029, 1962.
60. **Rentzepis, P. M., Jones, R. P., and Jortner, J.,** Dynamics of solvation of an excess electron, *J. Chem. Phys.,* 59, 766, 1973.
61. **Arvis, M., Lustig, H., and Hickel, B.,** Etude par photolyse éclair de la photoionisation des anions formiate, acétate et oxalate dans l'eau, *J. Photochem.,* 13, 223, 1980.
62. **Mialocq, J.-C., Sutton, J., and Goujon, P.,** Picosecond study of electron ejection in aqueous phenol and phenolate solutions, *J. Chem. Phys.,* 72, 6338, 1980.
63. **Belloni, J. and Saito, E.,** Radiolytic and photolytic formation of stable e_{am}^{-} in amide solutions, in *Electrons in Fluids. The Nature of Metal-Ammonia Solutions,* Jortner, J. and Kestner, N. R., Eds., Springer-Verlag, Berlin, 1973, 461.
64. **Belloni, J., Saito, E., and Tissier, F.,** Photodetachment of electrons from alcoholate ions in liquid ammonia, *J. Phys. Chem.,* 79, 308, 1975.
65. **Holroyd, R. A.,** Role of the energy of the excess electron in photoionization of TMPD in solution, *J. Chem. Phys.,* 57, 3007, 1972.
66. **Bernas, A., Grand, D., and Amouyal, E.,** Photoionization of solutes and conduction band edge of solvents. Indole in water and alcohols, *J. Phys. Chem.,* 84, 1259, 1980.
67. **Boyle, J. W., Ghormley, J. A., Hochanadel, C. J., and Riley, J. F.,** Production of hydrated electrons by flash photolysis of liquid water with light in the first continuum, *J. Phys. Chem.,* 73, 2886, 1969.
68. **Nikogosyan, D. N., Oraevsky, A. A., and Rupasov, V. I.,** Two-photon ionization and dissociation of liquid water by powerful laser UV radiation, *Chem. Phys.,* 77, 131, 1983.
69. **Goulet, T., Bernas, A., Ferradini, C., and Jay-Gerin, J.-P.,** On the electronic structure of liquid water: conduction-band tail revealed by photoionization data, *Chem. Phys. Lett.,* 170, 492, 1990.
70. **Migus, A., Gauduel, Y., Martin, J. L., and Antonetti, A.,** Excess electrons in liquid water: first evidence of a prehydrated state with femtosecond lifetime, *Phys. Rev. Lett.,* 58, 1559, 1987.
71. **Long, F. H., Lu, H., and Eisenthal, K. B.,** Femtosecond studies of the presolvated electron: an excited state of the solvated electron? *Phys. Rev. Lett.,* 64, 1469, 1990.
72. **Jortner, J. and Rabani, J.,** The reactivity of hydrogen atoms in alkaline solutions, *J. Am. Chem. Soc.,* 83, 4868, 1961.
73. **Asmus, K.-D. and Janata, E.,** Conductivity monitoring techniques, in *The Study of Fast Processes and Transient Species by Electron Pulse Radiolysis,* Baxendale, J. H. and Busi, F., Eds., D. Reidel, Dordrecht, 1982, 91.
74. **Avery, E. C., Remko, J. R., and Smaller, B.,** EPR detection of the hydrated electron in liquid water, *J. Chem. Phys.,* 49, 951, 1968.
75. **Fessenden, R. W. and Verma, N. C.,** Time resolved electron spin resonance spectroscopy. III. Electron spin resonance emission from the hydrated electron. Possible evidence for reaction to the triplet state, *J. Am. Chem. Soc.,* 98, 243, 1976.
76. **Shiraishi, H., Ishigure, K., and Morokuma, K.,** An ESR study on solvated electrons in water and alcohols: difference in the *g* factor and related analysis of the electronic state by MO calculation, *J. Chem. Phys.,* 88, 4637, 1988.
77. **Perkey, L. M. and Farhataziz,** Specific rates of some reactions of the solvated electron in liquid ammonia measured by pulse radiolysis at 23°C, *Int. J. Radiat. Phys. Chem.,* 7, 719, 1975.

78. **Jay-Gerin, J.-P. and Ferradini, C.,** Solvation of electrons in various polar solvents: can we predict the energy of the optical absorption maximum from the Kirkwood parameter?, *Radiat. Phys. Chem.,* 36, 317, 1990.

79. **Billaud, G. and Demortier, A.,** Dielectric constant of liquid ammonia from -35 to $+50°C$ and its influence on the association between solvated electrons and cation, *J. Phys. Chem.,* 79, 3053, 1975.

80. **Belloni, J. and Marignier, J. L.,** Electron-solvent interaction: attachment solvation competition, *Radiat. Phys. Chem.,* 35, 157, 1989.

81. **Delaire, J. A., Delcourt, M. O., and Belloni, J.,** Mobilities of solvated electrons in polar solvents from scavenging rate constants, *J. Phys. Chem.,* 84, 1186, 1980.

82. **Farhataziz, Perkey, L. M., and Hentz, R. R.,** Pulse radiolysis of liquids at high pressures. V. Absorption spectrum and yield of the solvated electron in liquid ammonia at 23°C and pressures up to 6.7 kbar, *J. Chem. Phys.,* 60, 3483, 1974.

83. **Jou, F.-Y. and Freeman, G. R.,** Effect of alkylation of ammonia on the optical band shape of solvated electrons, *Can. J. Chem.,* 60, 1809, 1982.

84. **Seddon, W. A., Fletcher, J. W., and Sopchyshyn, F. C.,** The effect of temperature on the optical spectra and the yields of solvated electrons and ion-pairs in amines, *Can. J. Chem.,* 56, 839, 1978.

85. **Dewald, R. R. and Roberts, J. H.,** The conductance of dilute solutions of sodium in liquid ammonia at -33.9, -45, and $-65°$, *J. Phys. Chem.,* 72, 4225, 1968.

86. **Riddick, J. A., Bunger, W. B., and Sakano,T. K.,** *Organic Solvents: Physical Properties and Methods of Purification,* Wiley, New York, 1986.

87. **Vannikov, A. V.,** The solvated electron in polar organic liquids, *Usp. Khim.,* 44, 1931, 1975, (English transl., *Russian Chem. Rev.,* 44, 906, 1975).

88. **Vannikov, A. V. and Marevtsev, V. S.,** The absorption spectra of solvated electrons in monoethanolamine and ethanol, *Int. J. Radiat. Phys. Chem.,* 5, 453, 1973.

89. **Jou, F.-Y. and Freeman, G. R.,** Band resolution of optical spectra of solvated electrons in water, alcohols, and tetrahydrofuran, *Can. J. Chem.,* 57, 591, 1979.

90. **Jay-Gerin, J.-P. and Ferradini, C.,** On the variation of the free-ion yield with the static dielectric constant in the radiolysis of liquids, *Radiat. Phys. Chem.,* 33, 251, 1989.

91. **Brodsky, A. M. and Tsarevsky, A. V.,** Determination of the characteristics of solvated electrons from optical absorption data, *Int. J. Radiat. Phys. Chem.,* 8, 455, 1976.

92. **Okazaki, K., Idriss-Ali, K. M., and Freeman, G. R.,** Temperature and molecular structure dependences of optical spectra of electrons in liquid diols, *Can. J. Chem.,* 62, 2223, 1984.

93. **Idriss-Ali, K. M. and Freeman, G. R.,** Solvent effects on solvated electron reaction rates in diols, *Radiat. Phys. Chem.,* 23, 89, 1984.

94. **Dean, J. A.,** *Handbook of Organic Chemistry,* McGraw-Hill, New York, 1987.

95. **Jay-Gerin, J.-P. and Ferradini, C.,** Solvatation des électrons en excès dans les solvants polaires: corrélation entre les valeurs de ϵ_{max}, $W_{1/2}$ et $E_{A_{max}}$, *Can. J. Chem.,* 68, 553, 1990.

96. **Fermeglia, M. and Lapasin, R.,** Excess volumes and viscosities of binary mixtures of organics, *J. Chem. Eng. Data,* 33, 415, 1988.

97. **Rauf, M. A., Stewart, G. H., and Farhataziz,** Viscosities and densities of binary mixtures of 1-alkanols from 15 to 55°C, *J. Chem. Eng. Data,* 28, 324, 1983.

98. **Pikaev, A. K. and Ponomarev, A. V.,** The track and bulk reactions of solvated electrons in irradiated monobasic aliphatic alcohols, *Radiat. Phys. Chem.,* 34, 693, 1989.

99. **Vannikov, A. V., Mal'tzev, E. I., Zolotarevsky, V. I., and Rudney, A. V.,** Transient charged radiolysis products of organic liquids studied by electrical and optical methods, *Int. J. Radiat. Phys. Chem.,* 4, 135, 1972.

100. **Rudnev, A. V., Vannikov, A. V., and Bakh, N. A.,** Optical absorption spectra and mobility of solvated electrons in aliphatic alcohols, *Khim. Vys. Énerg.,* 6, 473, 1972 (English transl., *High Energy Chem.,* 6, 416, 1972).

101. **Freeman, G. R.,** Ionization and charge separation in irradiated materials, in *Kinetics of Nonhomogeneous Processes,* Freeman, G. R., Ed., Wiley, New York, 1987, chap. 2.

102. **Leu, A.-D., Jha, K. N., and Freeman, G. R.,** Solvent structure effects on electron thermalization ranges in water-alcohol mixed solvents; preferred $E_{A_{max}}$ values for e_s^- in alcohols, *Can. J. Chem.,* 61, 1115, 1983.

103. **Hentz, R. R. and Kenney-Wallace, G. A.,** The influence of molecular structure on optical absorption spectra of solvated electrons in alcohols, *J. Phys. Chem.,* 78, 514, 1974.

104. **Jou, F.-Y. and Freeman, G. R.,** Shapes of optical spectra of solvated electrons. Effect of pressure, *J. Phys. Chem.,* 81, 909, 1977.

105. **Hall, G. E. and Kenney-Wallace, G. A.,** Nanosecond laser measurements of optical absorption coefficients of electrons in polar fluids, *Chem. Phys.,* 32, 313, 1978.

106. **Senanayake, P. C., Gee, N., and Freeman, G. R.,** Viscosity and density of isomeric butanol/water mixtures as functions of composition and temperature, *Can. J. Chem.,* 65, 2441, 1987.

107. **Ponomarev, A. V., Makarov, I. E., and Pikaev, A. K.,** Mobility of solvated electrons in alcohols, *Khim. Vys. Énerg.,* 19, 183, 1985 (English transl., *High Energy Chem.,* 19, 150, 1985).

108. **Dannhauser, W. and Bahe, L. W.,** Dielectric constant of hydrogen bonded liquids. III. Superheated alcohols, *J. Chem. Phys.,* 40, 3058, 1964.

109. **Vannikov, A. V., Mal'tsev, E. I., and Rudnev, A. V.,** Investigation of the charged particles in the pulse radiolysis of organic liquids, in *Proc. 3rd Tihany Symp. Radiat. Chem.,* Part 1, Dobó, J. and Hedvig, P., Eds., Akadémiai Kiadó, Budapest, 1972, 157.

110. **Gallant, R. W. and Railey, J. M.,** *Physical Properties of Hydrocarbons,* Vol. 2, 2nd ed., Gulf Publishing, Houston, 1984.

111. **Hentz, R. R. and Kenney-Wallace, G.,** Optical absorption of solvated electrons in alcohols and their mixtures with alkanes, *J. Phys. Chem.,* 76, 2931, 1972.

112. **Kenney-Wallace, G. A.,** Picosecond spectroscopy and dynamics of electron relaxation processes in liquids, *Adv. Chem. Phys.,* 47, 535, 1981.

113. *The Merck Index,* 9th ed., Merck, Rahway, NY, 1976.

114. **Weast, Robert, Ed.,** *Handbook of Chemistry and Physics,* 63rd ed., CRC Press, Boca Raton, FL, 1982.

115. **Hickel, B.,** Activation energy for the recombination of hydrated electrons in H_2O and D_2O, in *Proc. 4th Tihany Symp. Radiat. Chem.,* Hedvig, P. and Schiller, R., Eds., Akadémiai Kiadó, Budapest, 1977, 801.

116. **Fielden, E. M. and Hart, E. J.,** Primary radical yields and some rate constants in heavy water, *Radiat. Res.,* 33, 426, 1968.

117. **Asmus, K.-D. and Fendler, J. H.,** The use of sulfur hexafluoride to determine $G(e^-_{D_2O})$ and relative reaction rate constants in D_2O, *J. Phys. Chem.,* 73, 1583, 1969.

118. **Cygler, J., Klassen, N. V., and Ross, C. K.,** A pulse radiolysis study of solvated electrons in *tert*-butanol/water solutions, *Can. J. Chem.,* 64, 1548, 1986.

119. **Singh, A., Chase, W. J., and Hunt, J. W.,** Reactions in spurs: mechanisms of hydrated electron formation in radiolysis of water, *Faraday Discuss. Chem. Soc.,* 63, 28, 1978.

120. **Schmidt, K. H.,** Electrical conductivity techniques for studying the kinetics of radiation-induced chemical reactions in aqueous solutions, *Int. J. Radiat. Phys. Chem.,* 4, 439, 1972.

121. **Hart, E. J. and Anbar, M.,** *The Hydrated Electron,* Wiley, New York, 1970.

122. **Dixon, R. S. and Lopata, V. J.,** Spectrum of the solvated electron in heavy water up to 445 K, *Radiat. Phys. Chem.,* 11, 135, 1978.

123. **Jou, F.-Y. and Freeman, G. R.,** Temperature and isotope effects on the shape of the optical absorption spectrum of solvated electrons in water, *J. Phys. Chem.,* 83, 2383, 1979.

124. **Gee, N., Shinsaka, K., Dodelet, J.-P., and Freeman, G. R.,** Dielectric constant against temperature for 43 liquids, *J. Chem. Thermodynamics,* 18, 221, 1986.

125. **Dodelet, J.-P. and Freeman, G. R.,** Electron mobilities and ranges in liquid ethers: ion-like and conduction band mobilities, *Can. J. Chem.,* 53, 1263, 1975.

126. **Jou, F.-Y. and Freeman, G. R.,** Optical spectra of electrons solvated in liquid ethers: temperature effects, *Can. J. Chem.,* 54, 3693, 1976.

127. **Dodelet, J.-P., Jou, F.-Y., and Freeman, G. R.,** An electron mobility transition in liquid ethers, *J. Phys. Chem.,* 79, 2876, 1975.

128. **Langan, J. R., Liu, K. J., Salmon, G. A., Edwards, P. P., Ellaboudy, A., and Holton, D. M.,** The radiation chemistry of organic amides I. A pulse radiolysis study of solvated electrons and alkali-metal-electron species in cyclic amides, *Proc. R. Soc. Lond.,* A421, 169, 1989.

129. **Guy, S. C., Edwards, P. P., and Salmon, G. A.,** Solvated electrons and electron-cation aggregates in *N,N*-diethylacetamide, *N,N*-dipropylacetamide, *N,N*-dimethylpropanamide, tetramethylurea, and tetraethylurea, *J. Chem. Soc., Chem. Commun.,* 1257, 1982.

130. **Dorfman, L. M., Jou, F.-Y., and Wageman, R.,** Solvent dependence of the optical absorption spectrum of the solvated electron, *Ber. Bunsenges. Phys. Chem.,* 75, 681, 1971.

131. **Gavlas, J. F., Jou, F. Y., and Dorfman, L. M.,** Optical absorption spectrum of the solvated electron in some liquid amides and amines, *J. Phys. Chem.,* 78, 2631, 1974.

132. **Lok, M. T., Tehan, F. J., and Dye, J. L.,** Spectra of Na^-, K^-, and e^-_{solv} in amines and ethers, *J. Phys. Chem.,* 76, 2975, 1972.

133. **Dorfman, L. M. and Jou, F. Y.,** Optical absorption spectrum of the solvated electron in ethers and in binary liquid systems, in *Electrons in Fluids. The Nature of Metal-Ammonia Solutions,* Jortner, J. and Kestner, N. R., Eds., Springer, Berlin, 1973, 447.

134. **Hayon, E.,** Yield of ions and excited states produced in the radiolysis of polar organic liquids, *J. Chem. Phys.,* 53, 2353, 1970.

135. **Jou, F. Y. and Dorfman, L. M.,** Pulse radiolysis studies. XXI. Optical absorption spectrum of the solvated electron in ethers and in binary solutions of these ethers, *J. Chem. Phys.,* 58, 4715, 1973.

136. **Seddon, W. A., Fletcher, J. W., Sopchyshyn, F. C., and Catterall, R.,** Solvated electrons and the effect of coordination on the optical spectra of alkali metal cation-electron pairs in ethers, *Can. J. Chem.,* 55, 3356, 1977.

137. **Carvajal, C., Tölle, K. J., Smid, J., and Szwarc, M.**, Studies of solvation phenomena of ions and ion pairs in dimethoxyethane and tetrahydrofuran, *J. Am. Chem. Soc.*, 87, 5548, 1965.

138. **Glarum, S. H. and Marshall, J. H.**, Ethereal electrons, *J. Chem. Phys.*, 52, 5555, 1970.

139. **Fox, M. F. and Hayon, E.**, Correlation of the CTTS solvent scale with ν_{max} of the solvated electron. Prediction of band maxima, *Chem. Phys. Lett.*, 25, 511, 1974.

140. **Leader, G. R. and Gormley, J. F.**, The dielectric constant of N-methylamides, *J. Am. Chem. Soc.*, 73, 5731, 1951.

141. **Hayashi, N., Hayon, E., Ibata, T., Lichtin, N. N., and Matsumoto, A.**, Pulse radiolysis of liquid amides, *J. Phys. Chem.*, 75, 2267, 1971.

142. **Gee, N. and Freeman, G. R.**, Electron and cation transport in fluid dimethyl ether: effects of density and temperature, *Can. J. Chem.*, 60, 1034, 1982.

143. **Langan, J. R. and Salmon, G. A.**, Physical properties of *N*-methylpyrrolidinone as functions of temperature, *J. Chem. Eng. Data*, 32, 420, 1987.

144. **Marignier, J.-L. and Belloni, J.**, Solvated electrons in liquid dimethylsulfide, *Chem. Phys. Lett.*, 73, 461, 1980.

145. **Marignier, J.-L.**, Etude des processus radiolytiques de molécules sulfurées à l'état liquide par les méthodes impulsionnelles nanoseconde, *Doctorat 3ème cycle (chimie-physique)*, Université de Paris-Sud, Orsay, 1979.

146. **Marignier, J.-L. and Belloni, J.**, Competition processes between electron attachment and solvation in dimethyl sulfide, *J. Phys. Chem.*, 85, 3100, 1981.

147. **Casteel, J. F. and Sears, P. G.**, Dielectric constants, viscosities, and related physical properties of 10 liquid sulfoxides and sulfones at several temperatures, *J. Chem. Eng. Data*, 19, 196, 1974.

148. **Cooper, T. K., Walker, D. C., Gillis, H. A., and Klassen, N. V.**, The pulse radiolysis of dimethylsulfoxide and binary mixtures with water, *Can. J. Chem.*, 51, 2195, 1973.

149. **Cooper, T. K. and Walker, D. C.**, Free-ion yield in the radiolysis of dimethylsulfoxide, *Can. J. Chem.*, 49, 2248, 1971.

150. **Walker, D. C., Klassen, N. V., and Gillis, H. A.**, Solvated electrons in dimethylsulphoxide, *Chem. Phys. Lett.*, 10, 636, 1971.

151. **Bensasson, R. and Land, E. J.**, Transient species in the pulse radiolysis of dimethyl sulphoxide, *Chem. Phys. Lett.*, 15, 195, 1972.

152. **Tran-Thi, T. H., Koulkès-Pujo, A. M., Sutton, J., and Anitoff, O.**, Pulse radiolysis of amides and their aqueous organic mixtures. Effect of the environment on the reactivity of the presolvated and solvated electrons, *Radiat. Phys. Chem.*, 23, 77, 1984.

153. **Freeman, G. R.**, Personal communication, 1989.

154. **Baxendale, J. H. and Rodgers, M. A. J.**, Electrons, ions, and excited states in the pulse radiolysis of dioxane, *J. Phys. Chem.*, 72, 3849, 1968.

155. **Vermeer, R. A. and Freeman, G. R.**, Radiolysis of liquid di-*n*-propyl ether: alcohol formation and solvated electrons, *Can. J. Chem.*, 52, 1181, 1974.

156. **Delaire, J. A. and Bazouin, J. R.**, Primary mechanisms in the radiolysis of amines: pulse and γ-radiolysis of neutral and acidic ethylamine, *n*-propylamine and ethylenediamine, *Can. J. Chem.*, 57, 2013, 1979.

157. **Delaire, J.**, Modèle de recombinaison de l'électron solvaté dans la radiolyse des liquides polaires: application à l'hydrazine et à quelques amines aliphatiques, *Doctorat ès sciences physiques (chimie)*, Université de Paris-Sud, Orsay, 1980.

158. **Dye, J. L., DeBacker, M. G., and Dorfman, L. M.**, Pulse radiolysis studies. XVIII. Spectrum of the solvated electron in the systems ethylenediamine-water and ammonia-water, *J. Chem. Phys.*, 52, 6251, 1970.

159. **Dye, J. L., DeBacker, M. G., Eyre, J. A., and Dorfman, L. M.**, Pulse radiolysis study of the kinetics of formation of Na^- in ethylenediamine by the reaction of solvated electrons with sodium ions, *J. Phys. Chem.*, 76, 839, 1972.

160. **Sauer, M. C., Jr., Arai, S., and Dorfman, L. M.**, Pulse radiolysis studies. VII. The absorption spectra and radiation chemical yields of the solvated electron in the aliphatic alcohols, *J. Chem. Phys.*, 42, 708, 1965.

161. **Jou, F.-Y. and Freeman, G. R.**, Pressure dependence and structural effect of solvated electron spectra in branched chain and multihydric alcohols, *J. Phys. Chem.*, 83, 261, 1979.

162. **Jha, K. N., Bolton, G. L., and Freeman, G. R.**, Temperature shifts in the optical spectra of solvated electrons in methanol and ethanol, *J. Phys. Chem.*, 76, 3876, 1972.

163. **Ponomarev, A. V. and Pikaev, A. K.**, Radiation chemical yields of electrons in alcohols, *Khim. Vys. Énerg.*, 20, 215, 1986, (English transl., *High Energy Chem.*, 20, 162, 1986).

164. **Dixon, R. S. and Lopata, V. J.**, Effect of temperature on the yield of solvated electrons in liquid ethanol, *J. Chem. Phys.*, 63, 3679, 1975.

165. **Fowles, P.**, Pulse radiolytic induced transient electrical conductance in liquid solutions. IV. The radiolysis of methanol, ethanol, 1-propanol and 2-propanol, *Trans. Faraday Soc.*, 67, 428, 1971.

166. **Barat, F., Gilles, L., Hickel, B., and Lesigne, B.**, Effect of the dielectric constant on the reactivity of the solvated electron, *J. Phys. Chem.*, 77, 1711, 1973.

167. **Jay-Gerin, J.-P. and Ferradini, C.**, Migration of excess electrons in polar liquids: dependence on solvent viscosity, *Radiat. Phys. Chem.*, in press.
168. **Seddon, W. A., Fletcher, J. W., and Catterall, R.**, Correlation of optical and electron spin resonance spectra for metal-electron species in alkali metal solutions, *Can. J. Chem.*, 55, 2017, 1977.
169. **Kadhum, A. A. H., Langan, J. R., Salmon, G. A., and Edwards, P. P.**, Application of pulse radiolysis to the study of the chemistry of radical anions, *J. Radioanal. Nucl. Chem., Art.*, 101, 319, 1986.
170. **Nauta, H. and van Huis, C.**, Pulse radiolysis of hexamethylphosphoric triamide, *J. Chem. Soc. Faraday I*, 68, 647, 1972.
171. **Shaede, E. A., Dorfman, L. M., Flynn, G. J., and Walker, D. C.**, Spectrum, kinetics, and radiation chemical yield of solvated electrons in hexamethylphosphoric triamide, *Can. J. Chem.*, 51, 3905, 1973.
172. **Mal'tsev, E. I. and Vannikov, A. V.**, Influence of temperature and electrolyte ions on the properties and spectrum of solvated electrons in irradiated hexamethylphosphortriamide, *Radiat. Effects*, 20, 197, 1973.
173. **Brooks, J. M. and Dewald, R. R.**, Absorption spectra of the alkali metals in hexamethylphosphoramide, *J. Phys. Chem.*, 72, 2655, 1968.
174. **Chevrel, J., Jay-Gerin, J.-P., and Ferradini, C.**, Solvated electrons in liquid 1,2,6-hexanetriol, unpublished.
175. **Seddon, W. A., Fletcher, J. W., and Sopchyshyn, F. C.**, Pulse radiolytic formation of solvated electrons in hydrazine, *Can. J. Chem.*, 54, 2807, 1976.
176. **Delaire, J., Cordier, P., Belloni, J., Billiau, F., and Delcourt, M. O.**, Nanosecond pulse radiolysis of hydrazine, *J. Phys. Chem.*, 80, 1687, 1976.
177. **Khaikin, G. I., Slepneva, L. F., and Zhigunov, V. A.**, Optical properties of solvated electrons in liquid deuteriohydrazine, in *Proc. 5th Tihany Symp. Radiat. Chem.*, Vol. 1, Dobó, J., Hedvig, P., and Schiller, R., Eds., Akadémiai Kiadó, Budapest, 1983, 313.
178. **Farhataziz,** An intrasolvent and intersolvent correlation of characteristics of absorption spectra of the solvated electron in polar solvents, *Radiat. Phys. Chem.*, 15, 503, 1980.
179. **Seddon, W. A. and Fletcher, J. W.**, The contribution of pulse radiolysis to the chemistry of alkali metal solutions, *Radiat. Phys. Chem.*, 15, 247, 1980.
180. **Chubakova, T. A. and Bakh, N. A.**, The solvated electron in methanol in the range 3-220°C, *Khim. Vys. Énerg.*, 12, 373, 1978 (English transl., *High Energy Chem.*, 12, 315, 1979).
181. **Fox, M. F. and Hayon, E.**, Further correlations of the CTTS solvent scale for halides with v_{max} for the solvated electron, *J. Chem. Soc. Faraday I*, 72, 1990, 1976.
182. **Fletcher, J. W. and Seddon, W. A.**, Alkali metal species in liquid amines, ammonia, and ethers. Formation by pulse radiolysis, *J. Phys. Chem.*, 79, 3055, 1975.
183. **Riggio, R., Martinez, H. E., and Sólimo, H. N.**, Densities, viscosities, and refractive indexes for the methyl isobutyl ketone + pentanols systems. Measurements and correlations, *J. Chem. Eng. Data*, 31, 235, 1986.
184. **Senanayake, P. C. and Freeman, G. R.**, Effect of solvent structure on electron reactivity: *tert*-butyl alcohol/water mixtures, *J. Phys. Chem.*, 91, 2123, 1987.
185. **Capellos, C. and Allen, A. O.**, Ionization of liquids by radiation studied by the method of pulse radiolysis. III. Solutions of galvinoxyl radical, *J. Phys. Chem.*, 74, 840, 1970.
186. **Teather, G. G. and Klassen, N. V.**, A pulse radiolysis study of the reaction of the hydrated electron with *tert*-butyl alcohol, *Int. J. Radiat. Phys. Chem.*, 7, 475, 1975.
187. **Leu, A.-D., Jha, K. N., and Freeman, G. R.**, Composition effects on optical absorption spectra of solvated electrons in alcohol/water mixed solvents, *Can. J. Chem.*, 60, 2342, 1982.
188. **Langan, J. R., Liu, K. J., Salmon, G. A., and Edwards, P. P.**, The radiation chemistry of organic amides. II. Electron scavenging yields in *N*-methylpyrrolidinone, *Proc. R. Soc. London*, A424, 431, 1989.
189. **Samskog, P.-O., Lund, A., and Nilsson, G.**, Localized electrons in 1,8-octanediol crystals studied by ESR spectroscopy and pulse radiolysis, *Chem. Phys. Lett.*, 79, 447, 1981.
190. **Jou, F.-Y. and Freeman, G. R.**, Spectrum shape changes of solvated electrons in 1-octanol, *J. Phys. Chem.*, 88, 3900, 1984.
191. **Farhataziz and Stewart, G. H.**, Width of the absorption spectrum of the solvated electron, *Radiat. Phys. Chem.*, 17, 145, 1981.
192. **Kajiwara, T. and Thomas, J. K.**, The reactions of electrons in glycerol, *J. Phys. Chem.*, 76, 1700, 1972.
193. **Arai, S. and Sauer, M. C., Jr.**, Absorption spectra of the solvated electron in polar liquids: dependence on temperature and composition of mixtures, *J. Chem. Phys.*, 44, 2297, 1966.
194. **Dixon, R. S. and Lopata, V. J.**, Temperature dependence of the absorption maximum of the solvated electron in 2-propanol, *Radiat. Phys. Chem.*, 19, 495, 1982.
195. **Bockrath, B. and Dorfman, L. M.**, Pulse radiolysis studies. XXII. Spectrum and kinetics of the sodium cation-electron pair in tetrahydrofuran solutions, *J. Phys. Chem.*, 77, 1002, 1973.
196. **Kadhum, A. A. H. and Salmon, G. A.**, Reactivity of solvated electrons in tetrahydrofuran, *J. Chem. Soc., Faraday Trans.*, 1, 82, 2521, 1986.
197. **Kadhum, A. A. H. and Salmon, G. A.**, Electron scavenging in the γ-radiolysis of tetrahydrofuran, *Radiat. Phys. Chem.*, 23, 67, 1984.

198. **Holroyd, R. A., Ehrenson, S., and Preses, J. M.,** Electron mobility, ion yields, and photoconductivity in liquid tetrakis(dimethylamino)ethylene, *J. Phys. Chem.,* 89, 4244, 1985.

199. **Hunt, J. W., Lam, K. Y., and Chase, W. J.,** Electron yields and reaction kinetics in polar liquids, in *Radiation Research: Biomedical, Chemical, and Physical Perspectives,* Nygaard, O. F., Adler, H. I., and Sinclair, W. K., Eds., Academic Press, New York, 1975, 345.

200. **Schmidt, K. H. and Buck, W. L.,** Mobility of the hydrated electron, *Science,* 151, 70, 1966.

201. **Keene, J. P.,** The absorption spectrum and some reaction constants of the hydrated electron, *Radiat. Res.,* 22, 1, 1964.

202. **Michael, B. D., Hart, E. J., and Schmidt, K. H.,** The absorption spectrum of e_{aq}^- in the temperature range -4 to 390°, *J. Phys. Chem.,* 75, 2789, 1971.

203. **Allen, A. O.,** Drift mobilities and conduction band energies of excess electrons in dielectric liquids, *Natl. Stand. Ref. Data Ser.-Nat. Bur. Stand. (U.S.),* 58 (NSRDS-NBS 58), U.S. Government Printing Office, Washington, D.C., 1976.

204. **Buxton, G. V.,** Basic radiation chemistry of liquid water, in *The Study of Fast Processes and Transient Species by Electron Pulse Radiolysis,* Baxendale, J. H. and Busi, F., Eds., D. Reidel, Dordrecht, 1982, 241.

205. **Turro, N. J.,** *Modern Molecular Photochemistry,* Benjamin/Cummings, Menlo Park, 1978, chap. 5.

206. **Freeman, G. R.,** Solvent structure dependence of the optical excitation energy of solvated electrons, *J. Phys. Chem.,* 77, 7, 1973.

207. **Freeman, G. R.,** I. Electrons in fluids. II. Nonhomogeneous kinetics, *Ann. Rev. Phys. Chem.,* 34, 463, 1983.

208. **Koulkès-Pujo, A. M., Gilles, L., Halle, J. C., and Sutton, J.,** Pulse radiolysis of N-methyl acetamide (NMA), *Radiat. Phys. Chem.,* 10, 73, 1977.

209. **Kirkwood, J. G.,** The dielectric polarization of polar liquids, *J. Chem. Phys.,* 7, 911, 1939.

210. **Oster, G. and Kirkwood, J. G.,** The influence of hindered molecular rotation on the dielectric constants of water, alcohols, and other polar liquids, *J. Chem. Phys.,* 11, 175, 1943.

211. **Fröhlich, H.,** *Theory of Dielectrics: Dielectric Constant and Dielectric Loss,* 2nd ed., Oxford University Press, New York, 1958.

212. **Onsager, L.,** Electric moments of molecules in liquids, *J. Am. Chem. Soc.,* 58, 1486, 1936.

213. **Cole, R. H.,** Induced polarization and dielectric constant of polar liquids, *J. Chem. Phys.,* 27, 33, 1957.

214. **Belloni, J., Billiau, F., Delaire, J. A., Delcourt, M. O., and Marignier, J. L.,** Ionizing radiation-liquid interactions: a comparative study of polar liquids, *Radiat. Phys. Chem.,* 21, 177, 1983.

215. **Jay-Gerin, J.-P. and Ferradini, C.,** On the interaction between thermal electrons and polar solvents, *J. Chem. Phys.,* 91, 3275, 1989.

216. **Schindewolf, U.,** Physical and chemical properties of dissolved electrons (excess electrons), *Angew. Chem. Int. Ed. Engl.,* 17, 887, 1978.

217. **Farhataziz,** Viscosity dependence of the specific rates of the diffusion-controlled reactions of the solvated electron in polar solvents, *IEEE Trans. Nucl. Sci.,* NS-28, 1779, 1981.

218. **Keszei, E., Jay-Gerin, J.-P., and Ferradini, C.,** Note on electron transport in polar solvents and liquid long-chain *n*-alkanes, unpublished.

219. **Schnitker, J. and Rossky, P. J.,** Excess electron migration in liquid water, *J. Phys. Chem.,* 93, 6965, 1989.

220. **Barnett, R. N., Landman, U., and Nitzan, A.,** Dynamics of excess electron migration, solvation, and spectra in polar molecular clusters, *J. Chem. Phys.,* 91, 5567, 1989.

221. **Dewald, J. F. and Lepoutre, G.,** Thermoelectric properties of metal-ammonia solutions. III. Theory and interpretation of results, *J. Am. Chem. Soc.,* 78, 2956, 1956.

222. **Kenney-Wallace, G. A.,** Electrons in molecular clusters: microscopic probes of fluids, *Acc. Chem. Res.,* 11, 433, 1978.

223. **Holroyd, R. A.,** The electron: its properties and reactions, in *Radiation Chemistry: Principles and Applications,* Farhataziz and Rodgers, M. A. J., Eds., VCH Publishers, New York, 1987, 201.

224. **Kenney-Wallace, G. A.,** Electron localization and femtosecond nonlinear optical responses in liquids, in *The Liquid State and its Electrical Properties,* Kunhardt, E. E., Christophorou, L. G., and Luessen, L. H., Eds., Plenum Press, New York, 1988, 179.

225. **Gauduel, Y., Pommeret, S., Migus, A., and Antonetti, A.,** Electron reactivity in aqueous media: a femtosecond investigation of the primary species, *Radiat. Phys. Chem.,* 34, 5, 1989.

226. **Jonah, C. D., Bartels, D. M., and Chernovitz, A. C.,** Primary processes in the radiation chemistry of water, *Radiat. Phys. Chem.,* 34, 145, 1989.

227. **Baxendale, J. H. and Sharpe, P. H. G.,** A pulse radiolysis study of electrons in 1-propanol at low temperatures, *Int. J. Radiat. Phys. Chem.,* 8, 621, 1976.

228. **Dixon, R. S., Lopata, V. J., and Roy, C. R.,** Effect of temperature on the solvated electron in 1-propanol, *Int. J. Radiat. Phys. Chem.,* 8, 707, 1976.

229. **Bronskill, M. J., Wolff, R. K., and Hunt, J. W.,** Picosecond pulse radiolysis studies. I. The solvated electron in aqueous and alcohol solutions, *J. Chem. Phys.,* 53, 4201, 1970.

230. **Beck, G. and Thomas, J. K.,** Picosecond observations of some ionic and excited-state processes in liquids, *J. Phys. Chem.,* 76, 3856, 1972.

231. **Gilles, L., Aldrich, J. E., and Hunt, J. W.,** Solvation time of the electron in liquid alcohols and water at room temperature, *Nature (Phys. Sci.),* 243, 70, 1973.

232. **Kenney-Wallace, G. A. and Jonah, C. D.,** Picosecond molecular relaxations during electron solvation in liquid alcohol and alcohol-alkane solutions, *Chem. Phys. Lett.,* 39, 596, 1976.

233. **Kenney-Wallace, G. A. and Jonah, C. D.,** Picosecond spectroscopy and solvation clusters. The dynamics of localizing electrons in polar fluids, *J. Phys. Chem.,* 86, 2572, 1982.

234. **Rentzepis, P. M., Jones, R. P., and Jortner, J.,** Relaxation of excess electrons in a polar solvent, *Chem. Phys. Lett.,* 15, 480, 1972.

235. **Mialocq, J. C., Sutton, J., and Goujon, P.,** Picosecond laser photoionization in polar liquids and the study of the electron solvation process, in *Picosecond Phenomena II,* Hochstrasser, R., Kaiser, W., and Shank, C. V., Eds., Springer-Verlag, Berlin, 1980, 212.

236. **Wiesenfeld, J. M. and Ippen, E. P.,** Dynamics of electron solvation in liquid water, *Chem. Phys. Lett.,* 73, 47, 1980.

237. **Wang, Y., Crawford, M. K., McAuliffe, M. J., and Eisenthal, K. B.,** Picosecond laser studies of electron solvation in alcohols, *Chem. Phys. Lett.,* 74, 160, 1980.

238. **Huppert, D., Kenney-Wallace, G. A., and Rentzepis, P. M.,** Picosecond infrared dynamics of electron trapping in polar liquids, *J. Chem. Phys.,* 75, 2265, 1981.

239. **Miyasaka, H., Masuhara, H., and Mataga, N.,** Picosecond 266-nm multiphoton laser photolysis studies on the solvated electron formation process in water and liquid alcohols, *Laser Chem.,* 7, 119, 1987.

240. **Gauduel, Y., Martin, J. L., Migus, A., Yamada, N., and Antonetti, A.,** Femtosecond study of electron localization and solvation in pure water, in *Ultrafast Phenomena V,* Fleming, G. R. and Siegman, A. E., Eds., Springer-Verlag, Berlin, 1986, 308.

241. **Belloni, J., Clerc, M., Goujon, P., and Saito, E.,** Solvation time of electrons in liquid ammonia, *J. Phys. Chem.,* 79, 2848, 1975.

242. **Huppert, D., Rentzepis, P. M., and Struve, W. S.,** Picosecond dynamics of localized electrons in metal-ammonia and metal-methylamine solutions, *J. Phys. Chem.,* 79, 2850, 1975.

243. **Huppert, D., Avouris, Ph., and Rentzepis, P. M.,** Picosecond absorption studies of excess electrons in ammonia, amines, and ethers, *J. Phys. Chem.,* 82, 2282, 1978.

244. **Jonah, C. D., Romero, C., and Rahman, A.,** Hydrated electron revisited via the Feynman path integral route, *Chem. Phys. Lett.,* 123, 209, 1986.

245. **Sprik, M., Impey, R. W., and Klein, M. L.,** Study of electron solvation in polar solvents using path integral calculations, *J. Stat. Phys.,* 43, 967, 1986.

246. **Schnitker, J. and Rossky, P. J.,** Quantum simulation study of the hydrated electron, *J. Chem. Phys.,* 86, 3471, 1987.

247. **Wallqvist, A., Thirumalai, D., and Berne, B. J.,** Path integral Monte Carlo study of the hydrated electron, *J. Chem. Phys.,* 86, 6404, 1987.

248. **Schnitker, J., Motakabbir, K., Rossky, P. J., and Friesner, R. A.,** *A priori* calculation of the optical-absorption spectrum of the hydrated electron, *Phys. Rev. Lett.,* 60, 456, 1988.

249. **Wallqvist, A., Martyna, G., and Berne, B. J.,** Behavior of the hydrated electron at different temperatures: structure and absorption spectrum, *J. Phys. Chem.,* 92, 1721, 1988.

250. **Rossky, P. J. and Schnitker, J.,** The hydrated electron: quantum simulation of structure, spectroscopy, and dynamics, *J. Phys. Chem.,* 92, 4277, 1988.

251. **Romero, C. and Jonah, C. D.,** Molecular dynamics simulation of the optical absorption spectrum of the hydrated electron, *J. Chem. Phys.,* 90, 1877, 1989.

252. **Barnett, R. B., Landman, U., and Nitzan, A.,** Relaxation dynamics following transition of solvated electrons, *J. Chem. Phys.,* 90, 4413, 1989.

253. **Motakabbir, K. A., Schnitker, J., and Rossky, P. J.,** Transient photophysical hole-burning spectroscopy of the hydrated electron: a quantum dynamical simulation, *J. Chem. Phys.,* 90, 6916, 1989.

254. **Marchi, M., Sprik, M., and Klein, M. L.,** Solvation of electrons, atoms and ions in liquid ammonia, *Faraday Discuss. Chem. Soc.,* 85, 373, 1988.

255. **Marchi, M., Sprik, M., and Klein, M. L.,** Calculation of the free energy of electron solvation in liquid ammonia using a path integral quantum Monte Carlo simulation, *J. Phys. Chem.,* 92, 3625, 1988.

256. **Sprik, M. and Klein, M. L.,** Adiabatic dynamics of the solvated electron in liquid ammonia, *J. Chem. Phys.,* 91, 5665, 1989.

257. **Marchi, M., Sprik, M., and Klein, M. L.,** Calculation of the molar volume of electron solvation in liquid ammonia, *J. Phys. Chem.,* 94, 431, 1990.

258. **Ferradini, C. and Jay-Gerin, J.-P.,** Hypothesis of a possible chemical fate for the incompletely relaxed electron in water and alcohols, *Chem. Phys. Lett.,* 167, 371, 1990.

259. **Han, P. and Bartels, D. M.,** H/D isotope effects in water radiolysis. 2. Dissociation of electronically excited water, *J. Phys. Chem.,* 94, 5824, 1990.

260. **Freeman, G. R.,** The radiolysis of alcohols, in *Actions Chimiques et Biologiques des Radiations,* Vol. 14, Haïssinsky, M., Ed., Masson et Cie, Paris, 1970, 73.

Chapter 9

EXCESS ELECTRONS IN POLAR MATRICES

Masaaki Ogasawara

TABLE OF CONTENTS

I. INTRODUCTION

The importance of excess electrons trapped in solid matrices, irradiated by ionizing radiation, has been recognized for a long time, and we are well acquainted with this species by the name of "trapped electron," e_t^-. In particular, the electrons trapped by polar matrices, such as aqueous glasses and glassy alcohols at low temperature, have been extensively studied using electron spin resonance (ESR), electron spin echo, and optical methods. A large body of experimental data concerning this unique intermediate in radiation chemistry has been accumulated during the last 20 to 25 years. Although it still lacks a definite molecular structure, several more or less accepted models exist.

The studies of excess electrons trapped in frozen polar solvents containing OH groups have relevance to hydrated electrons. The determination of a detailed solvation structure of hydrated electrons in liquid water is practically impossible, since the excess electrons produced in the liquid by ionizing radiation can only persist for a short period, eluding direct and precise physical measurements. However, in polar matrices held at low temperature, e_t^- can be stabilized for a long time, which makes the study of the geometrical structure of the trapping site possible by conventional methods. The relatively long lifetime of e_t^- in matrices is also advantageous for investigating the reactions of excess electrons. Indeed, ESR and pulse radiolysis studies of the reactions induced by thermal- and photo-excitation of e_t^- have contributed to the unraveling of the mechanism of radiation chemistry of irradiated condensed media.

This chapter deals with the excess electrons trapped, stably or temporarily, in representative polar matrices such as crystalline ice, aqueous and alcoholic glasses, and organic crystals having OH groups. Particular emphasis is on the formation and reaction mechanisms. The literature concerned with excess electrons in polar matrices is very extensive, and only a part can be discussed in the space available here. The reader is referred to several excellent reviews which deal with earlier works,[1-3] matrix and solvation structures,[4-6] theoretical treatments,[4,7] and excess electrons in crystals.[8]

II. CRYSTALLINE ICE

A. SINGLE CRYSTALS OF PURE ICE
1. Identification of Trapped Electrons

The excess electrons trapped in single crystals of pure ice were first observed by Shubin et al. in 1966 in pulse radiolysis experiments.[9] The subsequent steady state measurements at 77 K by Eiben and Taub of the samples irradiated by γ-rays disclosed further details of the e_t^- spectra.[10] The absorption is characterized by a single, asymmetric band having a steeper tail on the lower energy side, with the absorption maximum at 670 to 680 nm for the sample near the melting point.[11] The band shifts continuously to the higher energy side with decreasing temperature: the straight line obtained by the plots of λ_{max} in energy scale against temperature gives the slope of dE/dT $= -1.2 \times 10^{-3}$ eV/K, compared to -2.9×10^{-3} eV/K for the liquid.[11] The lines obtained in the solid phase and liquid phase intercept each other near melting point without a sharp discontinuity, inferring that the e_t^- in solid phase and the e_s^- in liquid phase are one and the same species. The extinction coefficient at the absorption maximum has been determined, assuming the same oscillator strength as that of hydrated electrons,[12] to be 1.29×10^4 M^{-1} cm^{-1} at 268 K. Although the yield of e_t^- in irradiated crystalline D$_2$O determined by steady state measurements at 77 K is 10 times higher than in crystalline H$_2$O,[13] pulse radiolysis studies revealed that the initial yields of e_t^- were not much different from each other.[12] The difference was ascribed to the faster decay in the H$_2$O crystals.

Kawabata et al. discovered another absorption band in the infrared of γ-irradiated D$_2$O ice at 77 K,[14] which was later confirmed by Buxton et al. using pulse radiolysis technique.[15]

TABLE 1
Initial Yields of e_{vis}^- and e_{ir}^- in Aqueous Matrices Determined by Pulse Radiolysis

Matrices	Temperature	$G(e_{vis}^-)$	$G\,e_{ir}^-)$	Ref.
Crystalline H_2O	268	>0.3, 0.95	—	11, 12
	242	0.13	—	12
	223	0.05	—	12
	177	0.02	—	12
	138	0.01	—	12
Crystalline D_2O	269	1.10	—	24
	252	0.54	—	24
	220	0.02	—	24
	150	—	0.2	32
	77	0.4	0.1	22, 32
	6	0.35	0.29	28
(with NH_4F)[a]	77	—	0.85	32
	6	—	1.8	32
Aqueous Glasses				
12 M LiCl	73	1.46, 1.67 (72 K)	0.36, 0.28 (72 K)	45, 48
	6	0.76	1.29	47
9.5 M LiCl	90	1.02	0.56	47
	76	1.40, 1.11 (72 K)	1.2, 1.06 (72 K)	40, 48
	6	0.53	1.94	47
6 M LiCl	72	0.42	1.63	48
2.5 M $MgCl_2$	76	0.6	2.1	40
7.5 M BeF_2/D_2O	76	0.13	1.65	43
10.7 M BeF_2/D_2O	76	0.74	1.05	43
15 M BeF_2/D_2O	76	1.50	0.25	43
6 M/$LiBr_2/D_2O$	76	0.39	1.18	43

[a] Crystal doped with $3 \times 10^{-3}\ M\ NH_4F$.

The absorption in the infrared has been assigned to another type of e_t^- associated with relatively shallow potential wells in the matrices. We refer to this type of electron as the "infrared absorbing electron" (e_{ir}^-), and to the electron absorbing visible light as the "visible absorbing electron" (e_{vis}^-). The initial yields of e_{vis}^- and e_{ir}^- determined by pulse radiolysis are summarized in Table 1.

Only a few studies have been reported concerning the ESR spectra of e_t^- in crystalline ice. Bennett et al. detected ESR spectra with resolved hyperfine structure from e_t^- in ice doped with an alkali metal;[16] a narrow singlet line with g = 2.0006 observed in γ-irradiated D_2O crystals doped with $10^{-2}M$ NH_4F[17] has been assigned to e_{vis}^-; a doublet signal, observed by Hase and Kawabata, with a separation of 1.3 Gauss in γ-irradiated pure D_2O crystal at 4 K has tentatively been assigned to e_{vis}^-.[18] Further details have not been reported.

2. Visible Absorbing Electrons

Warman et al.[19] have studied the decay kinetics of highly mobile pseudo-free electrons, e_{qf}^-, produced by an electron pulse in crystalline H_2O; the product of the trap concentration and the rate constant of the reaction

$$e_{qf}^- + T \rightarrow T^-$$

where T denotes an assumed preexisting trap for the excess electron, was determined by means of the microwave absorption method. The halflife of e_{qf}^- was estimated to be 80 ps near the melting point, which was interpreted as if the preexisting traps of concentration 10^{-4} to 10^{-3} M in the ice reacted with the mobile electrons at a rate constant of 10^{14} M^{-1} s^{-1}.

Kawabata has proposed an anion vacancy model for the pre-existing electron trap to explain the e_{vis}^- yield which increases remarkably on doping with NH_4F.[13] If one water molecule is taken away from a crystal lattice, a vacancy surrounded tetrahedrally by four OH groups is formed with only two of them orienting toward the center. On the other hand, if one water molecule is replaced by a fluoride ion, all of the adjacent OH dipoles will orient toward the ion. He suggested that the excess electron might be trapped by the F^- vacancy. The same configuration of molecules around the electron was treated theoretically by Natori and Watanabe,[20] and Fueki.[21] However, it was revealed later that the role of doped NH_4F was not to increase the initial yield of e_{vis}^- but to reduce the decay rate of it.[22]

Buxton et al. have proposed that, at higher temperatures near the melting point, the electron trap giving rise to the visible band in D_2O crystals should be a vacancy type of trap existing at equilibrium concentration before the irradiation of electron pulse.[22] The observed narrowing of the visible absorption with time was thus interpreted as being due to an orientation of two molecules at the corners of the tetrahedron. The reorientation will result in all of the adjacent OH bonds pointing to the center. However, referring to the results of recent picosecond pulse radiolysis measurements,[23] the narrowing should be ascribed to the preferential decay of the electrons absorbing longer wavelength light. Different values have been reported for the formation energy of the vacancy: 0.5, 0.28 ± 0.07, and 0.3 to 0.4 eV.[24] If the value around 0.3 eV is right, the concentration of vacancies in the crystals can be high enough to explain the yield of e_{vis}^- obtained at temperatures near the melting point.

The formation enthalpy of the vacancy has also been estimated from an analysis of temperature dependent initial yields of e_{vis}^-.[24] The value of 0.7 to 0.4 eV, estimated on the assumption of a competition between the capturing process by vacancies and the scavenging process by radiolysis products of e_{qf}^-, was consistent with the above interpretation by Buxton et al.[22] The radiolysis products, D, OD, and D_2O^+ are possible candidates for e_{qf}^- scavenger.

A Bjerrum type of defect has also been assumed as a possible preexisting trap. There are two types of Bjerrum defects: one is a D defect in which two OH groups orient toward each other and another is an L defect which lacks a proton between two O atoms. Many years ago, by *ab initio* calculations, Naleway and Schwartz demonstrated the D defect to be a possible electron trap.[25] More recent calculation by Murphy and Hibbert arrived at the same conclusion.[26] Warman et al. have invoked a "dressed vacancy," which refers to a combined site of Bjerrum D defect and a vacancy defect, in interpreting the observed reaction rate of e_{qf}^- with preexisting traps in H_2O ice[19]. Considering the very large formation energy of a pure D defect, 0.68 eV, which arises from a repulsive interaction between neighboring H atoms,[22] the preferential formation of the D defect near the vacancy site may be reasonable.

The dielectric relaxation time of ice near the melting point is six orders of magnitude higher than that of water.[27] Therefore, the solvation process of electrons after being captured by a vacancy can be detected in ice. Warman and Jonah succeeded in following the solvation process unambiguously in the pulse radiolysis of crystalline H_2O.[23] The observed halflife of the solvation, 200 to 400 ps at 268 K, was 4 orders of magnitude shorter than the dielectric relaxation time of the medium. It follows that the solvation of the excess electron is not controlled by a bulk polarization of the medium but by a microscopic "inside out" motion of the OH bond which is accelerated by the strong electric field of the excess electron. The process may include a proton jump of the one or two water molecules surrounding the lattice occupied by the electron. The mobility data of an L defect suggest a timescale of several hundred picoseconds for the proton jump in accordance with the above observation.

At higher temperatures near the melting point, the initial yield of e_{vis}^- remarkably decreases with decreasing temperature as indicated in Table 1: the G value is 0.6 at 263 K at midpulse (t = 20 ns), while it drops to 0.02 at 220 K.[22] At low temperatures near 77 K or downward, the initial yield increases because of the appearance of another fast decaying component; the G value has been reported as large as 0.4 at 76 K[22] and 0.35 at 6 K.[28] The vacancy concentration at temperatures lower than 120 K is not high enough to explain the yield of e_{vis}^- even if the equilibrium concentration of vacancies at 120 K is assumed to be frozen at lower temperatures. Buxton et al. have invoked the following spur reactions

$$2D_2O \xrightarrow{\hspace{1cm}} (2D_2O^+ + 2e_{qf}^-)$$

$$(2D_2O^+ + 2e_{qf}^-) \rightarrow (D + OD + D_2O^+ + e_{qf}^- + \square)$$

$$\rightarrow 2D_2O^+ + 2e_{qf}^-$$

to interpret the existence of a high concentration of vacancies in the low temperature matrices.[22] Here the reactions in parentheses indicate spur reaction and $\square$ represents a vacancy.

3. Infrared Absorbing Electrons

Until 1981, e_{ir}^- was only observed in irradiated D_2O crystals, since in H_2O crystals e_{ir}^- decayed very quickly. However, careful studies by pulse radiolysis technique confirmed its formation in both D_2O and H_2O crystals.[29] The absorption starts from 1200 nm and increases continuously toward the lower energy side; an absorption maximum is predicted to exist at about 3000 nm from the line shape analysis of the absorption tail;[22] the yield and decay kinetics are affected by temperature, dose per pulse, accumulated dose, and added NH_4F, HF, and ND_3.[22,28]

Most agree that e_{ir}^- is an excess electron trapped in a shallow potential well in the medium, but little is known about the detailed structure around it. One of the plausible trapping sites is a cavity existing naturally in the crystalline ice. In a crystal of ordinary ice the water molecules form layers of six-membered puckered rings perpendicular to the c axis of the crystal, two adjacent rings forming a dodecahedral cavity.[30] In 1972, Nilsson proposed the dodecahedral cavity to be a preexisting trapping center giving rise to infrared absorption.[30] At that time the infrared band was not yet observed, so he explained that the electron captured in this cavity was relaxed very quickly by a reorientation of the surrounding molecules before the observation. As will be discussed later, such a drastic reorganization of molecules is less likely in crystalline ice, but the electrons may accommodate in this type of cavity for a while under certain conditions.

Funabashi et al. have given a different explanation to the origin of infrared band.[31] They assumed that electrons were localized through the coupling with an intramolecular vibrational mode of the molecules. Thus, e_{ir}^- was considered to only interact with low frequency matrix modes, whereas the e_{vis}^- was thought to interact with both high frequency molecular modes and matrix modes.

So far, no positive evidences supporting e_{ir}^- to be the main precursors of e_{vis}^- have been obtained in spite of the prediction by Nilsson.[30] Although Hase and Kawabata have reported an increase in the visible band on keeping the irradiated D_2O crystals at 4 K for a long time, the difference was very small indeed;[18] an increase of the visible band accompanied by the decay of infrared band has not been observed in pulse radiolysis studies; the analysis of decay kinetics of e_{ir}^- by Gillis et al. has given no definite conclusion about this problem.[24] Buxton et al. considered to the assumption that e_{ir}^- and e_{vis}^- are electrons trapped in natural cavities and vacancies, respectively, but they rejected the idea of orientational polarization in the natural cavities for the following reasons:[22] (1) it would have to work against the strong hydrogen bonding of crystal; and (2) there are as many as 12 molecules to be reoriented in the first coordination shell of the natural cavities.

The yields of e_{ir}^- depend strongly on the accumulated dose: the G value was 0.29 at midpulse in the D_2O crystals at 6 K, while it became as large as 1.08 with the samples already γ-irradiated by 300 krad. The increase was reduced when samples were annealed at higher temperatures. This effect has been interpreted as that the H_2O^+, the reaction partner of e_{qf}^-, is scavenged by the radiolysis products, D, OD, and D_2O.[28] The effect of the dopants, such as NH_4F, can also be understood in the same context.

4. Thermal Decay Reactions

Decay curves of e_{vis}^- observed at temperatures above 212 K consist of a spike and a long-lasting tail, the latter being significant at higher temperatures.[22] The spike and the tail have been ascribed to spur reactions and reactions of free electrons, respectively. It was possible to superimpose the normalized decay curves obtained at 73 to 76 K at different doses on each other, which strongly suggested that the decay reactions occurred in spurs. On the other hand, the decay rate somehow depends on the duration time of the pulse used in the measurements: the halflife in H_2O crystal is 25 ns at 73 K and 75 ns at 6 K with a 20 and 40 ns pulse, respectively,[29] while it becomes as short as 8 ns at 77 K with a 40 ps pulse.[33]

Since the effect of temperature on the decay rate of e_{vis}^- was rather weak in the temperature range 6 to 77 K,[29,32] tunneling transfer of the electrons has been assumed for the decay reaction in spurs. The reaction partners were supposed to be radiolysis products such as D_2O^+, D_3O^+, OD, or D. However, the pulse radiolysis experiments with a fast detection system with time resolution of 40 ps have led to a different conclusion.[33] The fast e_{vis}^- decay, following an empirical equation by Hummel for geminate recombination in spurs, was ascribed to the reaction of the pairs of ions H_3O^+ (or D_3O^+) and e_{vis}^- produced within their mutual Coulombic field from very highly excited intermediates

$$H_2O^* \rightarrow H^+ + e^- + OH$$

with the proton being subsequently attached to H_2O

$$H^+ + H_2O \rightarrow H_3O^+$$

It is well known that H_3O^+ has a high mobility in ice, so that migration of H_3O^+ rather than the thermal diffusion of deeply trapped electrons, is a plausible mechanism. Another piece of evidence supporting this mechanism is that the mobility of H_3O^+ is about seven times larger than that of D_3O^+ near the melting point inconsistent with the shorter lifetime of e_{vis}^- in H_2O ice in comparison to that in D_2O ice. Tunneling transfer of protons has also been invoked at very low temperatures where lattice phonons are completely frozen out.

Since the decay curves of e_{ir}^- in D_2O at 76 K fit simple second order kinetics, e_{ir}^- must be distributed at large distances from D_2O^+, probably outside the Onsager radius.[22] In this connection, it is of interest to note that the effect of accumulated dose or added NH_4F on the initial electron yield is much larger for e_{ir}^- than for e_{vis}^-.[28] The scavenging of D_2O^+ by the radiolysis products or additives easily competes with the neutralization reaction of the e_{qf}^- and D_2O^+ pairs with a large separation distance. The electron capture by shallow traps to give e_{ir}^- thus competes with the scavenging reaction. In contrast, the electron trapping giving rise to the formation of e_{vis}^- occurs mostly within spurs, as the trapping sites at low temperatures are radiolysis products, and therefore are distributed only at short distances from the sites of ionization; the neutralization process between e_{qf}^- and H_2O^+ which compete with electron trapping is also very fast and can only be interfered with by large concentrations of scavengers.

B. CLATHRATE HYDRATES

A clathrate hydrate is a crystalline compound which consists of a hydrogen bonded water "host" lattice around one or more species of "guest" molecules or ions. For instance, when chlorine is passed into ice water, green-yellow crystals are formed.[34] This is a well-known clathrate hydrate of chlorine discovered by Faraday as early as 1823. Recently, Zagórski detected e_t^- in the pulsed crystalline clathrate hydrates of tertiary alkylonium ion.[35-38]

The clathrate hydrate of tetra-n-butyl ammonium fluoride is expressed by the chemical formula, $[(n\text{-}C_4H_9)_4NF] \cdot 38H_2O$. The structural feature common to all clathrate hydrates is the pentagonal dodecahedron of water molecules, $H_{40}O_{20}$, in which each vertex is the site of an oxygen atom and each edge is an O–H$\cdots$O hydrogen bond.[34] A fourth external hydrogen bond is formed to another water molecule at each oxygen atom. All polyhedra thus formed are potential clathrate cages with internal voids which can accommodate guest species of the appropriate size and shape. The smallest void is that of pentagonal dodecahedra with a volume of 0.168 nm^3. The guest ions, tetra-n-butyl ammonium cations, are accommodated within the host lattice; while hydrogen bonded fluoride and/or hydroxide ions are built into the part of the host structure.

Zagórski has reported transient spectra of e_t^- with λ_{max} at 620 nm produced in the pulsed polycrystalline hydrates of sodium, potassium, and tetraalkylammonium salts at room temperature.[35] He has observed similar absorptions in the crystals of trimethylamine, diethylamine and tert-butylamine hydrates.[37] The observed $G\epsilon$ is 6000 M^{-1} cm^{-1}(100 eV)$^{-1}$ at the absorption maximum; the yield depends on the kinds of hydrates but the position of λ_{max} always stays the same; the absorption decays away at room temperature in the time domain of milliseconds. Amorphous samples made by melting the clathrates also gave the absorption but with the λ_{max} being shifted to 710 nm.[37]

The structure of the trapping site is not known. Zagórski has proposed, as a working hypothesis, that it may be a specific H_2O vacancy formed by the absence of OH^- or F^-.[37] If only 1% of the anion sites are vacant, it gives a concentration more than sufficient to trap all electrons produced in a 100 Krad pulse of radiation. The formation of similar e_t^- in pulsed amine clathrates has been interpreted to mean that the amines are protonated and OH^- ions are present in solution which produce a clathrate.[37]

Another example is the electron trapped by the clathrate hydrate of tert-butanol. Addition of tert-butanol to water produces clathrates in which each tert-butanol molecule is enclosed within a shell of 21 water molecules up to a concentration of about 4 mol% tert-butanol; above this concentration, the clathrate cages coalesce. Cygler et al. have studied e_t^- produced in tert-butanol water glasses in the concentration range of 4 to 16 mol% tert-butanol at 6 and 72 K.[39] The pulse radiolysis of these glasses revealed the production of both e_{ir}^- and e_{vis}^- with the former greatly dominating the spectrum. They concluded that electrons trapped in the water clathrate shell existed as e_{ir}^-. This study provides interesting information about the environment of e_{ir}^- which we will discuss in detail in the following section.

III. AQUEOUS GLASSES

A. GENERAL REMARKS

Water containing a high concentration of solute such as NaOH forms a clear, transparent glass at 77 K with a high degree of disorder in its structure whereas rapid freezing of pure water to 77 K gives an opaque polycrystalline ice. Solutions of many other compounds such as other alkalis, $NaClO_4$, LiCl, KF, carbonates, bicarbonates, formates, acetates, and ethylene glycol are known to give glasses. The minimum concentration required to form a glass depends on which kind of solute is dissolved: it is 5.0 M for NaOH, for instance.[1] The electron ejected by ionizing radiation is efficiently trapped in these glasses, and, in fact, the first observation of both ESR and optical absorption spectra was made by using NaOH alkaline ices.[1]

As in the crystalline ice, two types of electrons, e_{vis}^- and e_{ir}^-, are produced by ionizing radiation. Steady state optical absorption spectra of e_{vis}^- at 77 K are similar to those obtained for crystalline ice except for the broader linewidth: it has a range 0.72 to 0.95 eV in H_2O glasses[2] and 1.0 to 1.2 eV in D_2O glasses,[40] λ_{max} of the absorption ranges from 500 to 600 nm, being dependent upon which kind of solute is used to form the glass. The absorption maximum at 77 K is at higher energies than that in crystalline ice in most cases.[2] The e_{ir}^- band has been detected by the pulse radiolysis of the glasses containing 2 to 15 M LiCl, $MgCl_2$, and BeF_2.[15,40] At temperatures near 6 K, pulsed aqueous alkali-metal hydroxide glasses also give e_{ir}^-.[41] Although e_{ir}^- had not been detected in H_2O glasses for a long time, a careful study by Trudel et al.[29] and Buxton et al.[41] confirmed their formation in both H_2O and D_2O glasses.

The visible region of the optical absorption comprises bands of several different e_t^-. When alkaline aqueous glasses were irradiated and measured at very low temperatures such as 1.6 K, a broader absorption spectrum of e_{vis}^- was obtained.[42] Dolivo and Kevan have resolved it into a higher energy component and a lower energy one by photobleaching, and assigned them to a "solvated electron" and a "presolvated electron," respectively.[42] Narrowing and red shifting of the band with time due to the selective decay of the lower energy side of the band have also been observed in pulse radiolysis measurements at 77 K.[15] The visible band observed in pulsed BeF_2 glasses at 77 K has been resolved into three components, "green" band, "red" band, and "red" wing.[43]

The observed $G(e_{vis}^-)$ in irradiated 10 M alkali ice at steady state at 77 K is as large as 3.0 ± 0.6, meaning that almost all the thermalized electrons which escape neutralization are stably trapped in this glass. However, the localization is less efficient in D_2O compared with H_2O: efficiency of the e_{qf}^- scavenging by Cd^+ which competes with electron localization is 47% higher in the D_2O matrices.[44] Hamill et al. have interpreted this in terms of the electron localization coupled with an excitation of an OH-vibrational mode of the water molecules.[44a]

The high concentration of ions in the glasses affects the yield of e_t^-. Gillis et al. measured the yields of e_{vis}^- and e_{ir}^- by pulse radiolysis at different LiCl concentrations; the $G(e_{ir}^-)$ was proportional to the free water content in the glasses, although the total yield of e_{ir}^- and e_{vis}^- remained constant.[45] In aqueous solutions, a lithium ion is coordinated by four water molecules and, consequently, the basic structure of the solutions is $(Li \cdot 4H_2O)^+$ Cl^- which is a face centered cubic with Li^+ and Cl^- coordinated to each other in distorted octahedral arrangement.[46] Water in excess of that required to form $(Li \cdot 4H_2O)^+$ does not make further coordination shells but remains essentially in an amorphous state. Gillis et al. have, therefore, proposed that e_{ir}^- is surrounded by an amorphous environment of free water, while e_{vis}^- is trapped by the water molecules associated to the cation.[45] However, Klassen et al. did not agree with this explanation, since the $G(e_{ir}^-)$, determined by pulse radiolysis at 6 K, was as large as 1.2 despite the very high concentration of LiCl (12 to 15 M) in the D_2O glasses.[47,48] They related the difference of e_{ir}^- and e_{vis}^- to the distance between the electron and its reaction partner, Cl_2^-, instead. Bartczak et al. have offered another model; they regard the $LiCl/D_2O$ glass as a distorted crystal lattice and propose that e_{ir}^- and e_{vis}^- are electrons trapped in cation and anion vacancies, respectively.[49]

Efficiency of the conversion from e_{ir}^- to e_{vis}^- is extremely low, and an accurate estimation of the conversion efficiency is rather difficult because of the intrinsic decay processes for each species. The lower limit has been estimated roughly to be 6% in 9.5 and 6 M LiCl/D_2O.[48]

Studies of electrons trapped in ethylene glycol (EG)/water glasses of varied composition provide deeper insight into the nature of e_{ir}^- and e_{vis}^- and the interdependence of the two species. In pure EG, e_{ir}^- is not formed even at 6 K. The presence of water permits the production of e_{ir}^-: both e_{ir}^- and e_{vis}^- are detected in irradiated 50/50(vol/vol) EG/water glasses.[40,50-52] From these and other results it is concluded that e_{ir}^- and e_{vis}^- are trapped in two distinctly different molecular environments. Wu et al. have proposed a model in which

e_{vis}^- is trapped in clusters of EG/nD_2O and e_{ir}^- resides in an environment of only water molecules.[52] Thus, an increased water concentration is believed to increase the concentration of ir-trap thereby leading to a greater e_{ir}^- yield. This picture apparently contradicts the view that e_{ir}^- is the presolvated form of e_{vis}^- with e_{ir}^- having a nonequilibrium orientation of first solvent shell molecules.[50] The effect of an accumulated dose on the electron trapping in EG/ water glasses[53] is also inconsistent with the assumption that e_{ir}^- was converted to e_{vis}^- by a molecular orientation mechanism.[54]

B. GEOMETRICAL STRUCTURE AND SPATIAL DISTRIBUTION

The ESR spectra of e_{vis}^- in irradiated aqueous glasses reveal a sharp Gaussian singlet line with a g factor close to that of the free electron. The signal is easily saturated by microwave power. There have been no reports on the ESR spectra of e_{ir}^-. Substitution of H_2O by D_2O reduces the ESR linewidth to half or even one third of the original one.[2] Since the D nucleus has a smaller magnetic moment compared with that of a proton, the decrease suggests the linewidth is contributed mostly by hyperfine (hf) interaction between the electron and surrounding protons (or deuterons). Essential information about the geometrical structure of e_t^- is thus hidden in the unresolved hf splittings of the single observed line. To determine the number of nearest protons around the electrons and the isotropic Fermi contact hf coupling (hfc) constant of them, various approaches have been made such as resolution enhancement ESR,[55] electron-nuclear double resonance (ENDOR),[56,57] electron-electron double resonance (ELDOR),[58] and second moment analysis at two different microwave frequencies,[59] but none of them have successfully deduced the numbers and the hfc constant uniquely.

Electron spin echo (ESE) studies by Kevan and co-workers have contributed detailed information about the geometrical structure of e_t^-.[6,60] The decay envelope of ESE observed in pulsed electron spin resonance experiments consists of the product of two factors, a relaxation function, and a modulation function.[61] The modulation function, which can be deduced from an analysis of the ESE decay envelope, is compared with the theory based on certain assumed geometries of e_t^-. Assuming two solvation shells around the excess electron (the distance to the outer shell being 0.36 nm), the ESE modulation pattern from the e_{vis}^- produced in 10 M NaOH by x-irradiation was consistent with a first solvation shell of six water molecules with their OH dipoles pointing toward the center of the cavity.[60] The hfc constant and the distance were determined to be 2.1 Gauss and 0.21 nm, respectively, as shown in Figure 1. An alternative model of the first solvation shell consists of three water molecules with their molecular dipoles oriented toward the center. However, Schlick et al. analyzed the second moment of the ESR spectra recorded from samples of oxygen-17 enriched water and excluded this possibility.[62] It is particularly significant that the OH bond orients with the proton toward the electron (see Figure 1). Theoretical works so far have concluded that solvation is due to the molecular dipole orientation. It was suggested by Kevan that a very strong chemical bonding or exchange type of interaction could give rise to OH bond-oriented solvation structure.[6]

The model shown by Figure 1 is consistent with most of the experimental results obtained so far, but several questions still remain. The first question is whether the alkali ions contained in the glasses in a high concentration deform the solid structure. Kevan rejected this possibility because of the fact that the second moment of the ESR spectra obtained in the sodium-water mixture codeposited on a cold finger gave the same results as those obtained in alkaline glasses.[63] But, free water can only rarely be found in solutions where the concentration of NaOH dissolved is as high as 10 M. The second question concerns the sign of the spin density on the protons near the center of the excess electron: the experiments suggest positive spin density in contrast to the negative spin density predicted by theory by Newton et al.[64] According to Kestner,[7] however, this discrepancy is not surprising as the spin density is so sensitive to the detail of wave functions used in calculations. The third question is whether the analysis of the ESE modulation pattern is unique. Recently, Astashkin et al. made a

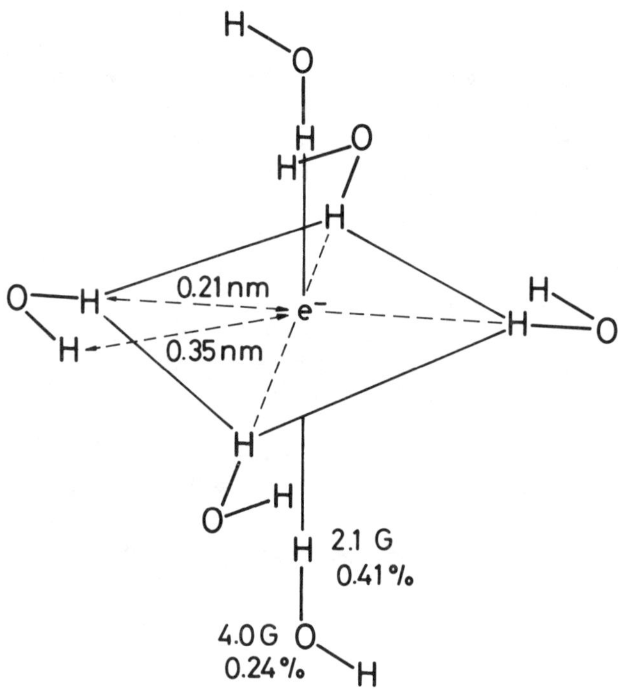

FIGURE 1.   Geometrical structure of e_t^- in 10 M NaOH aqueous glasses determined from electron spin echo modulation analysis. (From Kevan, L., *Acc. Chem. Res.*, 14, 138, 1981. With permission.)

new analysis on the ESE modulation in the frequency and time domain and concluded that the excess electron interacted with only two protons at a distance 0.2 nm away from its center.[65] The disagreement originates from the various ways of subtracting the matrix nuclei contribution from the ESE modulation pattern. The problem is still in dispute.

Little is known about the geometrical structure of e_{ir}^-. Considering the fact that e_{ir}^- was formed when the water content was increased up to 3 water molecules per ethylene glycol molecule [(50/50)(vol/vol)EG/H$_2$O], Van Leeuwen et al. ascribed the e_{ir}^- trap to a water dimer.[51] In this model the electron is trapped in a kind of tetrahedral arrangement of the four hydrogen ions of the two water molecules. They suggested, on the basis of electrostatic calculations, the largest trap depth to be obtained if the oxygen-oxygen distance was ~3.2 Å. The trap depth in the center of the dimer was ~ −2 eV. Recent *ab initio* calculations by Tachikawa and Ogasawara gave slightly different results.[66] The geometry optimization by using the UHF energy gradient method suggests the oxygen-oxygen distance to be 4.290 Å.

The spatial distribution of e_t^- has been studied mainly by utilizing ESR relaxation characteristics. If one denotes T_1 and T_2 as spin-lattice and spin-spin relaxation time, respectively, the product T_1T_2 can be evaluated from the microwave saturation experiments. Assuming a spin-packet model for the ESR lineshape, T_1 and T_2 can be obtained separately, and, from the latter, the average spur radius may be calculated. The spur radius in γ-irradiated 10 M NaOH glasses was estimated to be 4.2 nm,[2] Chernova has reported values of 2.9 to 3.4 nm.[2] Similar measurements on another radical, O⁻, have given slightly smaller values.[67] The size of spurs evaluated from the effect of electron scavengers is consistent with the above results.[2] Samples irradiated with tritium β-particles have a larger value, 7.0 nm.[68] However, according to an ELDOR study by Yoshida et al.,[69] the spin-packet model is not applicable in its strict sense to e_t^- in alkaline ice, so that the results obtained from ESR methods describe only semiquantitatively the spatial distribution of e_t^-.

C. REACTIONS

The reactions of e_t^- have been most extensively studied in LiCl/D_2O glasses.[15,40,45,47,48] The only observed species by pulse radiolysis in this glass are e_{vis}^-(λ_{max} ~600, nm), e_{ir}^- (λ_{max} >3000 nm), and Cl_2^- (λ_{max} ~340 nm). Cl_2^- is produced by the following reactions.

$$D_2O \xrightarrow{} D_2O^+ + e^-$$

$$Cl^- \xrightarrow{} Cl + e^-$$

$$D_2O^+ + Cl^- \rightarrow D_2O + Cl$$

$$Cl + Cl^- \rightarrow Cl_2^-$$

Klassen et al. have studied the time dependent G values of these three species in the time domain of 10^{-7} to 2100 s and concluded that e_{vis}^- reacted with Cl_2^- almost exclusively.[48]

$$e_{vis}^- + Cl_2^- \rightarrow 2Cl^-$$

On the other hand, e_{ir}^- reacts neither with Cl_2^- nor with D_3O^+. Instead, it reacts with an OD radical produced by radiolysis in the same spur to give spontaneous luminescence (λ_{max} = 410 nm) via the excited state formation of OD^-.[47]

$$e_{ir}^- + OD \rightarrow OD^{*-}$$

The effect of temperature on the e_t^- yield and the decay rate is rather strange: going from 75 to 6 K increases the initial yield of e_{ir}^- and decreases its decay rate, while the yield of e_{vis}^- decreases and its decay rate increases.[47] These phenomena have been attributed to fast trap-to-trap tunneling of e_{vis}^- from unrelaxed traps at 6 K and slower tunneling from deeper traps at 75 K. The e_{ir}^- traps are supposed to relax within the duration time of the pulse even at 6 K.

Tunneling electron transfer in aqueous glasses in the presence of electron acceptors is a well-known phenomenon. Miller et al. have demonstrated, in the pulse radiolysis studies of 6 M NaOH glasses containing CrO_4^{-2} or NO_3^- as electron acceptors, that e_{vis}^- decayed according to a tunneling mechanism:[70,72] the observed rate followed linearly with the logarithm of time over a wide range of time from 10^{-8} s to minutes without being dependent on temperature from 77 K to 143 K. Yet, the reaction kinetics between e_{vis}^- and inorganic ions in irradiated LiCl or NaOH glasses could not be rationalized by a direct tunneling mechanism; Buxton et al. have interpreted it in terms of a trap-to-trap tunneling process or hopping mechanism.[45,73]

D. PHOTOEXCITATION

The origin of the optical transition of e_t^- can be probed by comparing the optical absorption spectra to the wavelength dependency of photoconductivity which measures the direct promotion of electrons to the conduction band. If the absorption spectrum and the photoconductivity spectrum are similar to each other, the transition is regarded as a bound-continuum excitation, but if the peak of photoconductivity is shifted to the higher energy side of the optical absorption, a bound-bound transition is more likely. Thus e_{vis}^- in aqueous glasses was classified into two groups: systems like 10 M NaOH, 10 M NaOD/D_2O, 7 M $NaClO_4$, and 7.5 M CH_3COOK exhibiting bound-continuum excitation,[74,75] and systems like 5 to 10 M LiCl/H_2O and LiCl/D_2O and 7.2 M sodium formate having bound-bound transitions.[75-77] It is noted, however, that the e_{vis}^- band in LiCl glasses is of more complex character as it comprises two bands. The higher and lower energy sides are attributed to bound-bound and bound-continuum transitions, respectively.[77]

In 10 M LiCl/H$_2$O and LiCl/D$_2$O aqueous glasses at 4.2 K, the photoconductivity has been correlated with the optical absorption spectra of e_{ir}^-. It was suggested that the e_{ir}^- band was overlapping bound-bound and bound-continuum transitions with the former dominating in the region of the extrapolated optical absorption maximum, whereas the latter causes the high energy tail of the optical absorption spectrum.[76]

Based on the above comparison together with the threshold of the photoconductivity, the depth of the electron trap for e_{vis}^- in 7 M NaClO$_4$/D$_2$O and 5 M K$_2$CO$_3$/D$_2$O glasses has been estimated to be 1.6 eV and 1.0 to 1.2 eV, respectively.[75] The onset of the photoconductivity of e_{vis}^- and e_{ir}^- suggested that e_{vis}^- is bound twice as strongly as e_{ir}^-.[76]

Another interesting approach to examine the origin of the absorption band is to study the consequence of optically exciting e_t^- by laser photolysis. In fact e_{vis}^- produced by γ-radiolysis was excited by the laser light of 694 nm and a transient or a permanent bleaching was observed by monitoring with the light from a different source at the same wavelength or different one.[78] Usually the transient bleaching is much greater at the photolysing wavelength than at any other wavelengths. Furthermore, after the pulse some absorbance returns at 694 nm while it decreases elsewhere across the band. These observations, taken altogether, imply an inhomogeneous broadening of the e_{vis}^- spectra. An upper limit of excited state lifetimes of e_{vis}^- in aqueous glasses has been estimated to be 2 ns from the measured transient bleaching efficiency.[78]

When irradiated 9.5 to 10 M LiCl and 3.1 M MgCl$_2$ glasses are subjected to pulses of laser light, a transient absorption due to e_{ir}^- accompanies an emission at about 410 nm.[79-81] Walker et al. have summarized the processes as follows.[80,81]

$$e_{vis}^- \xrightarrow[\text{(694 nm)}]{h\nu} e_{qf}^- \to e_{ir}^- \xrightarrow[\text{(tunnels)}]{R} \underset{\text{(fast)}}{R^{-*}} \to R^- + h\nu \,(\sim 410 \text{ nm})$$

Photoexcitation of e_{vis}^- leads to the formation of e_{qf}^- directly via bound-continuum transitions or indirectly via bound-bound transitions. Nguyen and Walker have estimated the contribution from bound-bound transition to be approximately 3.7%.[82] Some of these subsequently become e_{ir}^- and the rest return to e_{vis}^- or react with cations or electron scavengers. The newly formed e_{ir}^- tunnel to reaction centers (R) thereby creating short-lived excited states (R^{-*}) that fluoresce at about 410 nm. The kinetics of the luminescence decay obey the relationship derived for geminate recombination, so that the pair, e_{ir}^- and R, is supposed to be in the same spur.[80]

IV. ALCOHOLIC GLASSES

A. FORMATION AND STABILIZATION

Most of the aliphatic alcohols, except for methanol, become transparent glassy matrices by rapid freezing to 77 K. Pure methanol, frozen to 77 K, forms an opaque polycrystalline matrix. If it is mixed with a small amount (>5 vol%) of another solvent, such as water or 2-methyltetrahydrofuran, rapid freezing gives a clear glass. It is well known that electrons are efficiently trapped in these matrices. The G values at 77 K have been reported to be 2.5 to 2.6 for methanol, 2.4 to 3.6 for ethanol, and 1.1 to 3.6 for 2-propanol.[4] Recent experiments by using high-LET (linear energy transfer) ion beams showed a strong LET effect on the e_t^- yield: at 80 K the G values, corrected for the dose effects, were 0.83 for C ion, 0.91 for N ion, and 1.4 for He ion.[83]

The steady state absorption spectra of electrons trapped in alcohols at 77 K are not much different from those obtained in aqueous glasses: they are broad, asymmetric, and singly peaked at 500 to 600 nm. However, the initial spectra detected in pulse radiolyis experiments are completely different. For most aliphatic alcohols, a very strong band which has a maximum at 1600 to 2000 nm is observed at 77 K immediately after the pulse. The absorption shifts to 500 to 600 nm with its rates depending on temperature and size of the matrix molecules. This tremendous spectral shift was first observed in ethanol glass at 77 K by Richards and Thomas in 1970[84,85] and then confirmed by Kevan in other alcohols such as 1-propanol and 2-propanol.[86] Independently, Hase et al. found a similar spectral change of e_t^- in alcohols and in other matrices on warming samples irradiated by γ-rays at 4.2K.[87] In the pulse radiolysis a decrease of the infrared band was accompanied by a prompt increase of the visible band with its extent depending on the kind of alcohol. No loss of the oscillator strength was noted in ethanol glasses before and after the spectral change.

The spectral change has been accounted for by a ''digging'' or ''dipole orientation'' mechanism. Initially the electrons find themselves in preexisting potential wells made up of molecular dipoles which do not have the equilibrium configuration with respect to the field of the electron. But eventually, they induce a relaxation of surrounding molecules by dipole reorientation, thereby leading to a deepening of trapping sites. This interpretation has been supported by various kinds of investigations including theoretical calculations in the framework of a semicontinuum model,[88,89] pulse radiolysis measurements,[86,90,91] ESR measurements,[92] and steady state optical absorption measurements at 4.2 K.[93-95]

However, the mechanism has been disputed by several authors. Baxendale and Sharpe concluded from the studies of the spectra of e_t^- in 1-propanol that the mechanism responsible for the spectral shifts was different depending on temperature.[96] Above 118 K, both the decay in the infrared and the growth in the visible were first order;[97] the temperature dependency of the rate constant was similar to that of the relaxation times obtained from dielectric dispersion measurements and the alcohol viscosity, all in accordance with the idea of molecular reorientation around the electron. However, below 118 K, the kinetics were no longer first order, the changes were much more extended in time and the rate was much faster than expected from an extrapolation of the observation at the higher temperatures. The similar tendency in the spectral change has been found in ethanol solutions.[98] Electron loss during the spectral shift, probably due to ion recombination, was noted in 1-propanol. They proposed that the thermal excitation from shallow traps and subsequent retrapping by deeper ones were responsible for the spectral shift at lower temperatures.[96,99] This explanation is called the ''seekers'' model. Ogasawara et al. disputed the ''digging'' mechanism based on the steady state absorption measurements at 4.2 K.[100-103] It was shown that the spectra of e_t^- in γ-irradiated ethanol and 1-propanol at 4.2 K could be resolved into two components, a visible band and an infrared one.[100,101] By analogy with aqueous matrices, the visible and infrared bands were ascribed to two different species ''unrelaxed e_{vis}^-'' and e_{ir}^-, respectively, although the nature of the latter species was fairly different from that formed in aqueous matrices. The ''unrelaxed e_{vis}^-'' band, which was resolved from the e_t^- band by photobleaching or by adding weak scavengers such as toluene, showed a broad peak at 600 to 800 nm.[101] The analysis of the spectral change observed on warming the samples irradiated at 4.2 to 77 K showed that 20 to 40% of e_{ir}^- converted to e_{vis}^- depending on the alcohol. They concluded that this conversion was due to electron transfer from shallow traps to deeper ones and that the spectral change was essentially caused by the selective decay of the e_{ir}^- band and narrowing and blue shifting of the unrelaxed e_{vis}^- band. Most of the results followed the small-polaron model of Bush and Funabashi:[104] e_{ir}^- are associated with the alkane part and e_{vis}^- with hydrogen bonded OH chains. Experiments on e_t^- in irradiated ethanol/water mixed matrices also supported this view.[105]

Buxton et al.[106-109] pulse irradiated a variety of alcohols, methanol, ethanol, 1-propanol, 1-butanol, and 1-pentanol over the temperature range 6 to 115 K, and found that the end-of-pulse spectra depended on both temperature and alcohol. They interpreted the spectra in terms of a multistep (small-polaron) model: electrons are trapped by OH groups and by at least two other shallow traps associated with the alkane moiety. At 6 K the electrons are initially trapped by CCC mode giving rise to the absorption band in the infrared but at 115

K by the CCO and COH mode through higher activation energy process.[109] All of the initial spectra at 115 K in higher alcohols than 1-propanol showed a maximum at about 800 nm in accordance with this view. Spectral changes with time were ascribed to migration of e_t from shallow to deep traps. Thus the dipole orientation was not invoked to explain the trap deepening observed in glassy alcohols.

However, Klassen et al. reached a completely different conclusion after reexamining the "initial" e_t^- spectra in pulse irradiated 1-propanol, 1-butanol, and 1-pentanol.[110] They measured spectra from 100 ns after the start of a 40 ns pulse. At about 6 K, the 100-ns spectra had maxima at approximately 2000 nm. At 115 K, maxima in the range 1300 to 1700 nm were found with shoulders at about 800 nm being consistent with the previous observations. However, the midpulse spectrum for each alcohol was almost independent of temperature, both in shape and in absolute intensity. Thus, on the 20-ns timescale, the initial spectra of e_t^- were essentially the same between 12 and 115 K. Based on these results, Klassen et al. rejected the different activation energy hypothesis together with the small polaron model.

Briefly, the well known mechanism for electron stabilization in liquid alcohols that the molecular configurations around excess electrons relax by molecular dipole orientation is probably right even in glassy matrices, as it is consistent with most of the experimental results obtained so far. Nevertheless, there still remain several unresolved problems which are not explained by this mechanism.

For instance, the absence of a large spectral shift in irradiated methanol glasses is mysterious; no intense infrared bands have been observed, either in the steady state measurements at 1.6 K[42] or from a 10-ns pulse irradiation at 6 K.[111] Dolivo and Kevan explained that even at 1.6 K the relatively small methanol molecules were able to reorient owing to the strong electric field of the electron before the measurement starts.[42] However, the spectrum of e_t^- with γ_{max} at 625 nm observed at 1.6 or 6 K is not the final one with respect to solvent relaxation: the shift of absorption maximum from this wavelength to the higher energy side was actually observed in pulse radiolysis of methanol glasses at 77 K.[91] According to the pulse radiolysis study by Huddleston and Miller,[112] the solvation of benzophenone anion in methanol glasses at 77 K proceeds in the same time domain.[112] Together, the small shift in the visible is attributed to reorientation of methanol molecules around the electron. It should be emphasized that the rate of this spectral change is not particularly fast compared to that in ethanol; it is clearly demonstrated by the rise curve of the absorbance at 550 nm measured in pulsed methanol at 77 K by Miller et al.[91] The extremely fast shift from infrared to the visible occurring in methanol (it must be occurring because the shift has been observed even in water at room temperature in femtosecond time domain[113]) should be ascribed to a different mechanism. As far as the low temperature matrices is concerned, electron transfer may play a role in the stabilization processes of e_t^-.

Another interesting problem is the origin of an intermediate band at approximately 800 nm, observed in the course of the spectral shift from the infrared to the visible. The intermediate band was found in time dependent spectra obtained by Kevan,[86] and then shown clearly in the work by Klassen et al.;[90] Ogasawara et al. reproduced this band on warming samples from 4.2 to 77 K and called it as the "unrelaxed e_{vis}^-;"[102,103] Buxton et al. observed a shoulder at about 800 nm in the "initial" spectra in higher alcohols at 115 K.[106,109] These suggest the existence of a very fast process in the beginning of the spectral change. It seems that the mechanism of this fast process is different from that of the slow relaxation process of the intermediates itself.[90] Although the origin of the intermediate band is still obscure, the results of the recent study of e_t^- in 2-methyltetrahydrofuran/ethanol mixtures at 77 K is suggestive;[114] the optical absorption spectrum with λ_{max} at about 770 nm observed in the γ-irradiated mixture has been assigned to the electron coordinated by two ethanol molecules. In this analogy, the intermediate band mentioned above may be attributed to e_t^- oriented by two alcohol molecules. This speculation is supported by the fact that alcohol dimers can trap excess electrons in alcohol/alkane mixed solutions.[115] It is not surprising that the alcohol glasses are abundant in preexisting trapping sites of this configuration, because defects of the hydrogen-bonding network may easily become this type of trap. This can explain the prompt formation of so-called unrelaxed e_{vis}^- at 4.2 K in the matrices made up of larger alcohols which are difficult to reorient.[100-103]

Światła et al. have explained the spectral change in a different way.[116-118] They have rejected the small polaron model and returned to the traditional view of e_t^- as localized on a few of the alcohol molecules. They do not admit to the reorientation of dipoles surrounding the trapped electron at very low temperature.[118] The electron capture is treated as a radiationless transition between the conduction state and the trapped state of the electron assisted by a multiple emission of matrix phonons; electrons are primarily captured by excited levels of deep traps and then relax to the ground state by electron-vibron coupling resulting in the shifts in optical absorption spectra. This model explains the optical absorption spectra and their change at low temperatures very well, but it is not verified yet by other experimental methods such as ESR.

B. GEOMETRICAL STRUCTURE

The geometrical structure of e_t^- in glassy alcohols has been extensively studied by means of magnetic resonance. The linewidth of a singlet ESR signal is as large as 16 G in ethanol glasses at 77 K.[4] The deuteration of OH groups causes the reduction of the linewidth to 6.7 G, suggesting that e_t^- is surrounded by at least partially oriented OH groups.[119] The matrix proton ENDOR measurements of e_t^- in methanol glasses by Schwartz et al.[120] indicated that the excess electron was partially localized on OH bonds. The linewidth observed in ethanol irradiated at 4.2 K is only 6.7 G, which suggests a less perfect OH orientation toward the electron compared with that of e_t produced at 77 K. The linewidth increases irreversibly on warming the sample to 77 K in accordance with the large spectral shifts from the infrared to the visible observed in ethanol glasses. Bales and Kevan have evaluated the second moment of the ESR spectra of e_t^- produced in ethanol at 4.2 and 77 K to demonstrate a larger OH proton component and a smaller ethylene proton component at 77 K.[121] Analogous measurements on e_t^- formed in matrices of ethylene glycol/water mixtures revealed that the ethylene protons moved 0.02 to 0.04 nm further away from the electron as solvation became more complete at higher temperatures.[122]

The geometrical structure of e_t^- trapped at 77 K in ethanol matrices was determined in detail by Kevan et al. by using the ESE method.[6,123,124] They fitted the ESE modulation patterns from e_t^- produced at 77 K in partially deuterated compounds, CH_3CH_2OD, CD_3CH_2OH, CH_3CD_2OH to simulations averaged over all angles to approximate a disordered system. Analysis of the deuterium modulation pattern gave the number, n, of nearest protons around the electron, the distance, r, and hfc constant, a; n = 4, r = 0.22 nm and a = 0.7 MHz for the OD deuterons; n = 8, r = 0.33 nm and a = 0.1 MHz for the CD_2 deuterons; and n = 12, r = 0.38 nm, and a = 0 MHz for the CD_3 deuterons. Thus, the number of first solvation shell molecules was four for each specifically deuterated molecule. A striking feature of the model which is consistent with this experiment is that the OH groups do not orient toward the electron as had long been thought.[123,124] The electron is approximately located on the bisector of the COH angle in ethanol, the direction of which is considered to coincide with that of the molecular dipole. This contrasts with the structure of e_t^- in aqueous glasses (Figure 1).

A quantitative analysis of the ESE modulation patterns from e_t^- in methanol glasses was difficult because of a short relaxation time in this system. The structure determination has been made by measuring forbidden proton spin-flip satellites arising from the magnetic dipole-dipole interaction between the unpaired electron and its nearest matrix protons.[123,125] The measured separation between the main and satellite transition gave r = 0.228 ± 0.015 nm with reasonable accuracy, but the spin-flips were too weak to unambiguously determine the number of protons in the nearest shell.[125] Kevan has assumed a geometrical model with the OH bonds of the four solvation shell molecules orienting tetrahedrally toward the electron.[6,125]

C. PHOTO- AND THERMAL-INDUCED REACTIONS

The singlet in the ESR spectrum and the visible absorption band of e_t^- are readily bleached by visible light to produce one hydroxyalkyl radical for each vanishing e_t^-. The electrons are consumed presumably by the following reactions.

$$e^- + RHOH \rightarrow H + RHO^- \tag{1}$$

$$e^- + 2RHOH \rightarrow H_2 + \dot{R}OH + RHO^- \tag{2}$$

The hydroxyalkyl radical may also be formed by the reaction,

$$e^- + RHOH_2^+ \rightarrow \dot{R}OH + H_2, \tag{3}$$

where, $RHOH_2^+$ is formed mostly via a protonation reaction of an alcohol molecule by the primary cation, $RHOH^+$. However, there are several reasons to consider that Reaction 3 does not contribute to the hydroxyalkyl radical formation in glassy 1-propanol. To attain this reaction, the electron has to be detrapped by the action of light.[126] In the presence of electron scavengers, some of the detrapped electrons undergo an electron scavenging reaction, so that the quantum efficiency of photobleaching must markedly depend on the fraction of e_t^- photobleached. However, the quantum efficiency is constant throughout the bleaching process, and the electron decay rate does not change at all.[126]

The following evidence also supports the importance of Reactions 1 and 2, rather than 3, for the formation of hydroxyalkyl radicals in glassy alcohols. The quantitative conversion of photoexcited electrons to hydroxyalkyl radicals are observed even in the absence of alcohol cations: photoexcitation of e_t^- generated by means of the photoionization of aromatic amines in ethanol glasses or by doping glasses with Na metal leads to a one-to-one formation of the radical;[3] HD was formed by the action of light on e_t^- in HROD matrices.[127]

Among Reactions 1 and 2 the latter is preferred, since (1) Reaction 1 is endothermic and Reaction 2 is exothermic, and (2) the following reaction which is required for the formation of hydroxyalkyl radical via Reaction 1 is unlikely at low temperatures because of its high activation energy, 8 to 10 kcal/mol, in gas phase.[3]

$$H + RHOH \rightarrow H_2 + \dot{R}OH$$

The above considerations lead to a mechanism in which the excited e_t^- decompose within their own traps. In this context, Dainton et al.[130] preferred the following expressions instead of 1 and 2.

$$e_t^- + h\nu \rightarrow \dot{H} + RHO^-$$

$$e_t^- + h\nu \rightarrow H_2 + \dot{R}OH + RHO^-$$

Similar reactions occur when the electrons trapped in single crystals of alkanediols at 4.2 K are bleached by visible light.[128] Tachikawa et al.[129] have carried out *ab initio* calculations to examine a possible structure for e_t^- which copes with the quantitative conversion to the hydroxyalkyl radical; details are described in the following section.

The thermal decay reaction of e_t^- seems to depend on the matrix molecules. In methanol glasses, the decay of e_t^- is not accompanied by an increase in the hydroxyalkyl radical concentration. The electron is obviously detrapped by thermal agitation, since the decay rate is accelerated in the presence of scavengers.[130] In contrast, the detrapping of the electron is not observed in 1-propanol and, hence, the decomposition of the trap, i.e., reaction with solvent, is assumed as a possible mechanism for electron decay.[126]

V. ORGANIC CRYSTALS

A. POLYCRYSTALLINE CYCLODEXTRIN

The molecular structure of cyclodextrin consists of six glucose units linked together through glycoside bonds forming a natural cavity with a diameter as large as 0.55 nm. The cavity usually contains two water molecules to form an inclusion complex.[131] The water can be substituted by other molecules such as alcohols and alkanes. According to the X-ray diffraction study by Manor and Saenger,[132] one water molecule is connected to a glucose ring by hydrogen bonds and another is located near the hydrophobic environment of the rings. In 1971, Burdslay et al.[133] reported on a singlet ESR line with a g value of 2.0010 and a linewidth of 12.5 G suggesting the formation of e_t^- in γ-irradiated β-cyclodextrin at 77 K.

Ichikawa and Yoshida made a detailed study on the electrons trapped by polycrystalline α-cyclodextrin.[134] The γ-irradiation of polycrystalline α-cyclodextrin at 77 K resulted in violet coloration of the samples. Steady state optical absorption spectra showed a very broad single band with λ_{max} at about 600 nm, but photobleaching with light of λ at 670 nm resolved it into two components with λ_{max} at 530 and 730 nm. These bands were attributed to the electrons trapped by water molecules in the rings as crystalline water and those by the sites characteristic of the α-cyclodextrin-$(H_2O)_2$ complex, respectively. The electron signal was detected in anhydrous crystals including 1-propanol in the cavity instead of water, but the crystals substituted by neopentane gave no e_t^-. Failure in detecting e_t^- from the latter crystals has been interpreted in terms of the large size of neopentane, leaving no space in the cavity available for an electron to be accommodated. Alternatively, it may suggest the importance of OH groups in the cavity in electron trapping. The details are not yet known.

B. SINGLE CRYSTALS OF POLYHYDROXY COMPOUNDS

The crystal structure of L(+)-arabinose is orthorhombic containing four molecules in each unit cell. In 1978, Box et al. detected an ESR signal with the resolved anisotropic hf structure of e_t^- in this compound.[135] The irradiation and the measurements were performed on samples at 4.2 K. This was a striking finding, because the easily photobleachable singlet line had long been accepted as one of the important characteristics of ESR spectra of radiation-produced e_t^-. They revealed, by means of ENDOR spectroscopy, the hf structure to be due to three interacting hydrogens. The substitution of the OH protons by deuteriums resulted in a collapse of the lines into a singlet, thus indicating hf interactions ascribable to OH protons. The assignment of the ESR spectra to e_t^- is based on the following arguments:[135] (1) The observed ESR hfc could be analyzed as arising from three protons being 0.160 to 0.164 nm away from the center of electron — in the crystal structure, there are indeed three exchangeable hydrogen atoms positioned so that no one hydrogen atom is farther than 0.3146 nm from each other; (2) The g value associated with the ESR signal is nearly isotropic and less than the free electron value (2.0023) — it is difficult to make up a structure of an organic free radical for which all three principle values of the g tensor are less than the free electron value; (3) The x-irradiation, at temperatures as low as 4.2 K, should produce and preserve a stoichiometric amount of free radicals attributable to oxidation and reduction products — actually, the second largest component of the ESR absorption is due to alkoxy radicals, an oxidation product.

The ESR study on e_t^- in arabinose by Box et al. is one of the rare cases in which the geometrical structure can be determined unambiguously.[136,137] We will explain the deter-

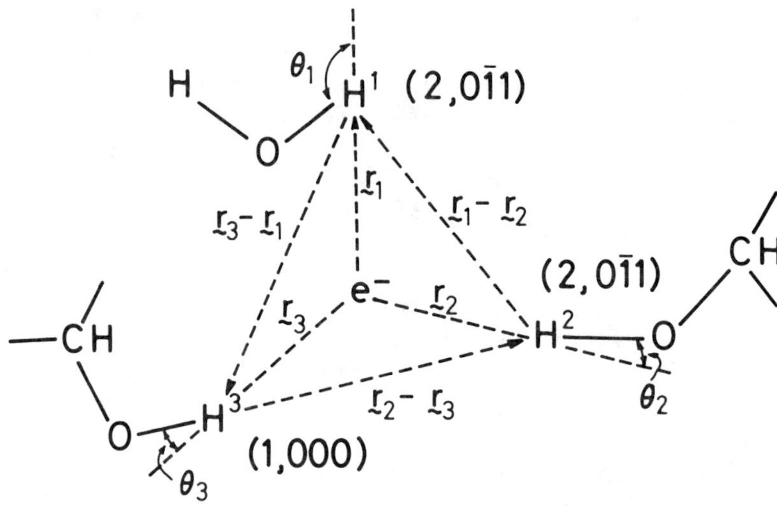

FIGURE 2.   Geometrical structure of e_t^- in single crystal of rhamnose. r_1, r_2, and r_3 are 0.1660, 0.1698, and 0.1718 nm, respectively; r_1–r_2, r_2–r_3, and r_3–r_1 are 0.2631, 0.3060, and 0.2905 nm, respectively; θ_1, θ_2, and θ_3 are 129, 11, and 27°, respectively. (From Samskog, P.-O., Kispert, L. D., and Lund, A., *J. Chem. Phys.*, 79, 635, 1983. With permission.)

mination procedure briefly here. The hfc tensors obtained by measuring the hf structure at various angles of crystals are decomposed into its isotropic and anisotropic components. The isotropic term arising from the Fermi-type contact interactions is proportional to the unpaired spin density on the protons, while the anisotropic term arises from the magnetic dipole-dipole interaction between the electron and the nearby protons. Assuming a point-dipole approximation, the observed hfc constant, a, is related to the distance between the electron and proton by the equation:

$$A_{max} - a = 2g\beta g_N\beta_N/R^3 h$$

Where A_{max} is the principle value of hfc tensor, β is Bohr magneton, g_N is the nuclear g factor, β_N is the nuclear magneton, R is the distance between magnetic dipoles, and h is Planck's constant. The direction cosines of vector R coincide with those of the principal axis of the coupling tensor which corresponds to the maximum hfc. The measured hfc tensors suggest that the excess electron in a single crystal of arabinose is accommodated at a position which is nearly equidistant from three protons as mentioned above.[136] The estimated spin densities on the protons are not so large, mostly less than 1%, and not exceeding 7%. The positions of all OH groups in single crystals of arabinose have been determined with high precision by means of X-ray diffraction and neutron scattering, and it was possible to find a site consistent with the ESR data.[136]

Another interesting example is the electron trapped in rhamnose crystals.[138,139] Figure 2 shows the geometrical structure of the trapping site determined from the observed ESR hf structure, H^1 denotes the proton of a nearby crystal water, and H^2 and H^3 denote the OH protons of rhamnose. It should be noted that two OH dipoles belonging to the hydroxyl groups are pointing approximately toward the site of e_t^-, but the dipole of one of the crystal waters is directed away from the electron. It appears as if the OH proton of crystal water is not contributing to the stabilization of e_t^-. Consequently, only two OH protons are supposed to coordinate to the electron in this crystal.

Table 2 summarizes the characteristics of electrons trapped in single crystals of poly-hydroxy, monosaccharide, and disaccharide compounds.[135-141] The data are considerably

TABLE 2

Excess Electrons Trapped by Single Crystals of Polyhydroxy Compounds

Compounds	Structure	Space Group	Number of proton	R/Å	Spin Density	g value
L (+) Arabinose		$P2_12_12_1$	3	1.60 1.62 1.65	0.035 0.024 0.011	*a* 2.0011 *b* 2.0009 *c* 2.0013
Dulcitol	CH_2OH–$HCOH$–$HOCH$–$HOCH$–$HCOH$–CH_2OH	$P2_1/c$	2	1.59 1.61	0.066 0.066	*a'* 2.0015 *b* 2.0010 *c* 2.0013
Rhamnose		$P2_1$	3	1.66 1.70 1.72	0.006 0.039 0.029	
Sucrose		$P2$	2	1.69 1.71	0.037 0.033	
Glucose-1-phosphate		$P2_1$	2	1.738 1.732	0.039 0.014	*x* 1.9982 *y* 1.9980 *z* 1.9989
Trehalose		$P2_12_12_1$	2	1.708 1.820		2.0012 ~2.0017
Xylitol	CH_2OH–$HCOH$–$HOCH$–$HCOH$–CH_2OH	$P2_12_12_1$	2	1.73 1.75	0.032 0.025	

TABLE 2

Excess Electrons Trapped by Single Crystals of Polyhydroxy Compounds

Compounds	Structure	Space Group	Number of proton	R/Å	Spin Density	g value
D-Sorbitol	CH$_2$OH | HCOH | HOCH | HCOH | HCOH | CH$_2$OH	P2$_1$2$_1$2$_1$	2	1.67 1.68	0.030 0.023	

different from those obtained with amorphous matrices. The distances between the electron and the OH protons are rather short, all of them ranging from 0.16 to 0.18 nm and the number of protons interacting magnetically with e_t^- is only two, except for arabinose and rhamnose as mentioned above.

Not as much is known about the optical absorption spectra of the electrons trapped in single crystals of polyhydroxy compounds. Samskog et al.[138] have reported a transient absorption spectrum with λ_{max} at 500 nm and a half-width of 250 nm measured in pulse-irradiated single crystal of rhamnose at 268 K. Absorption measurements have also been made on irradiated sucrose[142] and arabinose;[143] in both cases λ_{max} is found at about 400 nm.[142,143] Generally speaking, the position of the absorption maximum is considered to reflect the depth of the trapping site and, hence, the trapping sites in crystalline polyhydroxy compounds seem to be deep and stable. However, the steady state spectrum of e_t^- starts to decay at 47 K in dulcitol, at 61 K in sucrose, at 75 K in arabinose.[137] It is apparent that the position of absorption maximum is irrelevant to the thermal stability of the trapping site.

Then, a question arises as to how two or three OH groups, despite the highly restricted orientational motions in the crystal lattice, can trap an electron? In answering this question, Kevan et al.[144] have made calculations on the basis of a semicontinuum model. They have chosen to calculate possible structures of e_t^- in crystalline D-sorbitol. The theoretical model assumes e_t^- interacting with surrounding dipoles in the first solvation shell through the dipole-dipole and charge-induced dipole attractive potentials; the solvent dipoles beyond the first solvation shell are treated as a continuous dielectric which contributes only to the long-range interactions with the electron; Hydrogen-like wave functions are assumed and an optimum configuration is searched for by using a variational method.[88] It has been suggested that the electron in the crystalline D-sorbitol is trapped by two OH protons of the same molecule.[140] They calculated the total energy by changing the distance from the center of the electron to the closest dipole, but no energy minimum could be found. Then, they made another calculation assuming that the OH oxygens are fixed but the protons are allowed to rearrange to point toward the electron. A minimum in the total energy was obtained, but the distances between the protons and the electrons were calculated to be 0.08 to 0.10 nm, deviating very much from the experimentally determined value. Thus, even if one assumes flexible OH bonds, the experimental results cannot be explained in the framework of the semicontinuum model, insofar as the crystal lattice is kept unchanged.

C. SINGLE CRYSTALS OF ALKANEDIOLS

The alkanediols have a simple molecular structure composed of a normal alkane chain as a backbone with the OH groups connected to the end carbons. Single crystals are easily available from the compounds having an even number of carbon atoms in the skeleton such

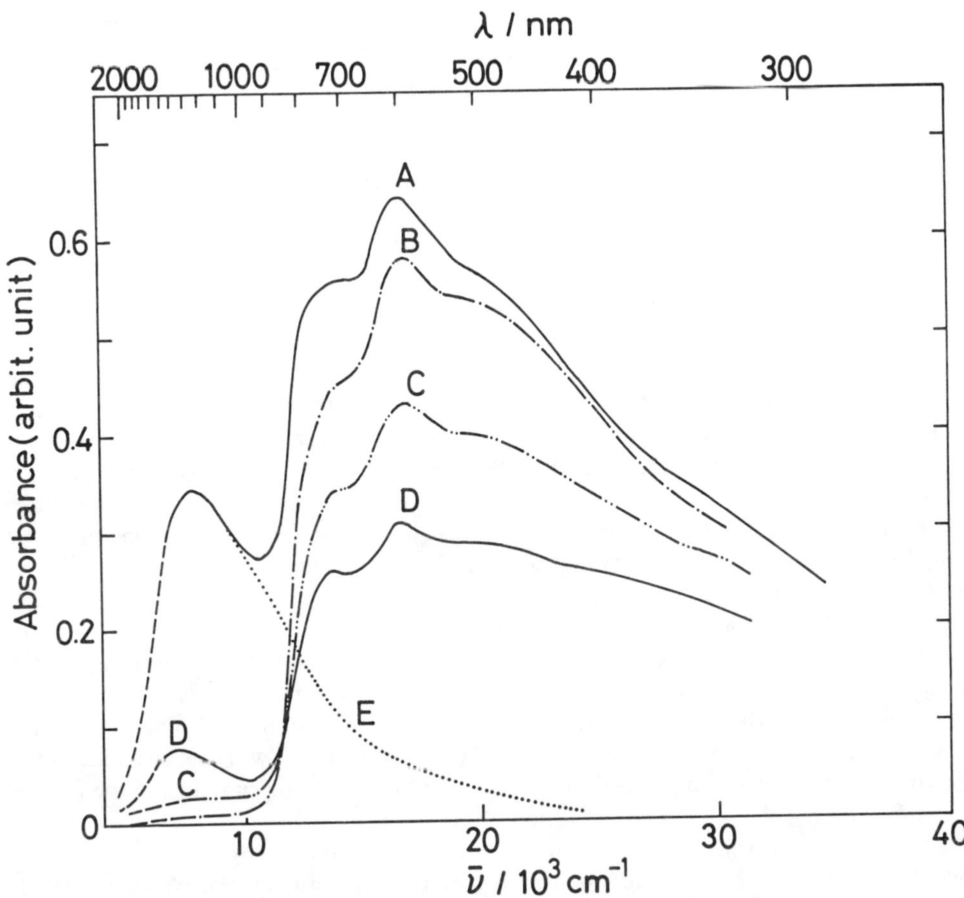

FIGURE 3.   Absorption spectra observed from crystalline 1,8-octanediol γ-irradiated at 4.2 K. (A) Immediately after the γ-irradiation; (B) photobleached by 1000-nm light; (C) photobleached by 730-nm light; (D) photobleached by 590-nm light; (E) the difference spectrum between A and B. (From Ogasawara, M., Claesson, O., Yoshida, H., and Lund, A., *J. Phys. Chem.*, 88, 5004, 1984. With permission.)

as 1,8-octanediol and 1,6-hexanediol. They are colorless, transparent, and belong to the P2$_1$/c space group. In single crystals of 1,6-hexanediol, linear molecules are forming layers in the *bc* plane connected with hydrogen bonds along arrays in the *b* direction. The layers are arranged along the *a* axis so that the arrays of hydrogen bonds form a two-dimensional structure.[145] The crystal structure of 1,8-octanediol has not been examined, but it can be assumed similar to that of 1,6-hexanediol. As the alkanediols resemble alcohols in molecular structure, the electron trapped in crystalline alkanediol may be considered as a kind of model "compound" of e$_t^-$ in the glassy alcohols.

Samskog et al. examined single crystals of 1,8-octanediol by means of pulse radiolysis in 1981 and observed a structureless, broad absorption band with a λ$_{max}$ at 615 nm and a large ε (probably larger than $10^4 M^{-1}$ cm^{-1}).[146] A singlet ESR signal with a negatively shifted g factor was also observed to support the above assignment, but this singlet has not been reproduced. Considering the results obtained later by Ogasawara et al.[147] and Hama et al.,[148] the ESR line may originate from electrons trapped in amorphous parts of the samples.

Claessen et al.[149] and Ogasawara et al.[150] have reported steady state optical absorption spectra of γ-irradiated alkanediols at 4.2 K. As shown in Figure 3, the spectrum obtained from 1,8-octanediol consists of three partially resolved peaks with maxima at 740, 590, and 490 nm, which cannot be separated by photobleaching or annealing. It was concluded, therefore, that they arose from transitions inherent in e_t^-. $G\epsilon$ for e_{vis}^- in 1,8-octanediol was as large as 15,000 M^{-1} cm^{-1} (100 eV)$^{-1}$. In addition to this band, infrared bands with λ_{max} at about 1300 and 1000 nm in 1,6-hexanediol and 1,8-octanediol, respectively, were found. This band was quite sensitive to the light and similar in many respects to those of the electrons trapped in alkane matrices. The effects of temperature, photobleaching, and sample preparation methods altogether indicated that electrons were trapped at two distinctly different trapping sites, shallow and deep. Like those in aqueous matrices, the electrons giving rise to the visible and the infrared bands were called e_{vis}^- and e_{ir}^-, respectively. A small absorption recovery in the near infrared during the photobleaching of e_{vis}^- was also noted as shown in Figure 3. Ogasawara et al. succeeded in detecting an ESR signal with resolved hf structure of the electrons trapped in γ-irradiated crystalline alkanediols at 4.2 K.[147] Interactions with hydrogens showed anisotropy with a maximum splitting as large as 35 G. These results were particularly encouraging with respect to structure determination, but since the complicated ESR signals overlapped with those of free radicals, a complete analysis was not accomplished. However, the effect of deuteration showed that the hyperfine structure was at least partially caused by protons.

Recently, Lindgren et al. investigated thermal- and photo-induced reactions of e_t^- by ESR for the purpose of elucidating the structure of the trapping site from the mechanistic aspects.[128] Their study confirmed a previous suggestion that e_t^- converted to hydroxyalkyl radicals by photoexcitation. The ESR spectra of e_t^- disappeared completely and those of hydroxyalkyl radical increased in intensity on illuminating with light >700 nm on the samples. A similar phenomenon was observed in 1,6-hexanediol. An analysis of integrated signal intensities of the ESR spectra revealed that 93 and 90% of e_t^- in 1,6-hexanediol and 1,8-octanediol, respectively, converted to hydroxyalkyl radicals. It was concluded therefore, that the self-decomposition of photoexcited e_t^- into the radicals occurred as in the case of alcoholic glasses.[130]

It seems that the well-accepted model of e_t^-, the electrons trapped by four or six molecules with their molecular or OH dipoles oriented toward the center of it, cannot explain the results obtained in the crystalline alkanediols. The crystal structure of alkanediols is characterized by a two-dimensional layer of OH groups connected by hydrogen bonds as mentioned earlier. In order to form a trapping site similar to the above model, four or six OH dipoles in the network must be oriented toward the electron against the strong hydrogen bonds. Such a drastic reorganization of molecules is difficult to account for in the crystalline phase at a temperature as low as 4.2 K. Consequently, as a new working hypothesis, they have proposed an electron trap consisting of a hydroxyl group and a methylene group.

Recently, Tachikawa et al. reported *ab initio* MO calculations in an attempt to find possible trapping sites in the alkanediol crystals.[129] In this study, the excess electron together with the environment was considered as a kind of supermolecule: a calculation was made with respect to the stability of the SOMO to determine the metastable state of the super-molecule. They used basis sets added by s,p-type basis as diffuse, Rydberg-like wave functions for the excess electron. Since clusters of octanediol or hexanediol are too large for calculations on this level, a simplified system composed of methanol or ethanol molecules and an excess electron were assumed. Although the calculations were made for the gas phase molecules, the results were discussed assuming a relevance to e_t^- in matrices. The calculated spin density suggested that the hydrogen bonding in a linear form cluster pushed the excess electron toward the end of the cluster. It follows that the excess electron is not stabilized in a perfect crystal. This is consistent with the experimental fact that e_t^- has been detected only in a few cases in molecular crystals. On the other hand, the hydrogen bond plays a role of lowering the total energy of the negatively charged methanol cluster. An excess electron might thus be stabilized in the form of a cluster anion with the electron spin density mainly localizing at the terminal of an array of hydrogen bonds. This conclusion is in accordance with the trapping site proposed by Lindgren et al. which is located at an OH group near a defect in the array of hydrogen bonds along the b axis including a methylene group of another layer.[128] The array of hydrogen bonds exists just above (or below) a methylene group of another layer — the methylene group which abstracts a hydrogen atom to yield the hydroxyalkyl radical.

$$\text{RCH}_2\text{O}^- + \text{H}_2 + \dot{\text{R}}\text{CHOH}$$

Such a trap might thus be associated without a drastic conformational change of the crystal structure.

VI. CONCLUSION

Many agree that our understanding of the excess electrons trapped in polar matrices at low temperatures is reasonably good. Owing to the great progress made in the last two decades, their nature is well established in several matrices. But the story is incomplete in many respects. What is the difference and relation between e_{ir}^- and e_{vis}^-? What is the effect of the phase of media? Are the electrons trapped in crystals like those in amorphous matrices? Is the solvation structure of e_t^- in aqueous matrices really like that of a hydrated electron? All these questions should be answered by further studies, particularly the geometrical structure of the trapping site and its relation to the observed decomposition via excited states to yield specific products.

ACKNOWLEDGMENTS

The author is grateful to Dr. Mikael Lindgren and Mr. Hiroto Tachikawa for comments on, as well as discussions of, the manuscript.

REFERENCES

1. **Kevan, L.,** Radiation chemistry of frozen aqueous solutions, in *Radiation Chemistry of Aqueous Systems,* Stein, G., Ed., The Weizmann Science Press of Israel, Jerusalem, 1968.
2. **Pikaev, A. K.,** The trapped electron in radiation chemistry, in *The Solvated Electron in Radiation Chemistry,* Israel Program for Scientific Translation, Jerusalem, 1971, chap. 6.
3. **Pshezhetskii, S. Ya., Kotov, A. G., Milinchuk, V. K., Roginskii, V. A., and Tupikov, V. I.,** *EPR of Free Radicals in Radiation Chemistry,* John Wiley & Sons, New York, 1974.
4. **Kevan, L.,** Trapped electrons in organic glasses, in *Advances in Radiation Chemistry,* Vol. 4, Burton, M. and Magee, J. L., Eds., John Wiley & Sons, New York, 1974.
5. **Kevan, L.,** Solvated electron structure in glassy matrices, *Acc. Chem. Res.,* 14, 138, 1981.
6. **Kevan, L.,** Geometrical structure of solvated electrons, *Radiat. Phys. Chem.,* 17, 413, 1983.
7. **Kestner, N. R.,** Theories of the solvated electron, in *Radiation Chemistry — Principles and Applications,* Farhataziz and Rodgers, A. J., Eds., VCH Publishers, New York, 1987.
8. **Lund, A. and Schlick, S.,** Trapped electrons in crystalline media, *Res. Chem. Intermed.,* 11, 37, 1989.
9. **Shubin, V. N., Zhigunov, V. A., and Zdotarevsky, V. I.,** Pulse radiolysis of crystalline ice and frozen crystalline aqueous solutions, *Nature,* 212, 1002, 1966.
10. **Eiben, K. and Taub, I. A.,** Solvated electron spectrum in irradiated ice, *Nature,* 216, 782, 1967.
11. **Taub, I. A. and Eiben, K.,** Transient solvated electron, hydroxy, and hydroperoxy radicals in pulse irradiated crystalline ice, *J. Chem. Phys.,* 49, 2499, 1968.

12. **Nilsson, G., Christensen, H., Pagsberg, P., and Nielsen, S. O.,** Transient electron in pulse-irradiated crystalline water, *J. Phys. Chem.,* 76, 1000, 1972.

13. **Kawabata, K.,** Trapped electron in the irradiated single crystal of ice doped with some impurities: its formation and photobleaching at 77K, *J. Chem. Phys.,* 55, 3672, 1971.

14. **Kawabata, K., Horii, H., and Okabe, S.,** Optical absorption spectrum of trapped electron in crystalline ice at −150··. Pulse radiolysis study, *Chem. Phys. Lett.,* 14, 223, 1972.

15. **Buxton, G. V., Gillis, H. A., and Klassen, N. V.,** Pulse radiolysis of aqueous solids at 76 K. An absorption band in the infrared, *Chem. Phys. Lett.,* 32, 533, 1975.

16. **Bennett, J. E., Mile, B., and Thomas, A.,** Trapped electrons produced by the deposition of alkali-metal atoms on ice and solid alcohols at 77 K. Part I. Optical spectra and electron spin resonance spectra, *J. Chem. Soc.(A),* 1967, 1393, 1967.

17. **Kawabata, K.,** Electron traps in irradiated crystalline ice, *J. Chem. Phys.,* 65, 2235, 1976.

18. **Hase, H. and Kawabata, K.,** Trapped electron in crystalline D_2O ice at 4 K, *J. Chem. Phys.,* 65, 64, 1976.

19. **Warman, J. M., de Haas, M. P., and Ververne, J. B.,** Decay kinetics of excess electrons in crystalline ice, *J. Phys. Chem.,* 84, 1240, 1980.

20. **Natori, M. and Watanabe, T.,** Structure model of the hydrated electron, *J. Phys. Soc. Japan,* 21, 1573, 1966.

20a. **Natori, M.,** Structure model of the hydrated electron. II, *J. Phys. Soc. Jpn,* 24, 913, 1968.

21. **Fueki, K.,** Hyperfine interactions in the electron spin resonance of hydrated electrons, *J. Chem. Phys.,* 45, 183, 1966.

22. **Buxton, G. V., Gillis, H. A., and Klassen, N. V.,** Two types of localized excess electrons in crystalline D_2O ice, *Can. J. Chem.,* 55, 2385, 1977.

23. **Warman, J. M. and Jonah, C. D.,** Electron solvation in crystalline ice on a subnanosecond time scale, *Chem. Phys. Lett.,* 79, 43, 1981.

24. **Gillis, H. A., Teather, G. G., and Ross, C. K.,** Mechanism of formation of visible-absorbing excess electrons in crystalline ice near 273 K, *J. Phys. Chem.,* 84, 1248, 1980.

25. **Naleway, C. A. and Schwartz, M. E.,** *Ab initio* studies of the interactions of an electron and two water molecules as a building block for a model of the hydrated electron, *J. Phys. Chem.,* 76, 3905, 1972.

26. **Murphy, S. V. and Hibbert, D. B.,** *Ab initio* study of the transition from trapped to solvated electron in water dimers, *Radiat. Phys. Chem.,* 28, 319, 1986.

27. **Auty, R. P. and Cole, R. H.,** Dielectric properties of Ice and solid D_2O, *J. Chem. Phys.,* 20, 1309, 1952.

28. **Wu, Z., Gillis, H. A., Klassen, N. V., and Teather, G. G.,** Pulse radiolysis of crystalline D_2O at 6 K, *J. Chem. Phys.,* 78, 2449, 1983.

29. **Trudel, G. J., Gillis, H. A., Klassen, N. V., and Teather, G. G.,** Localized excess electrons in H_2O glasses and ice, *Can. J. Chem.,* 59, 1235, 1981.

30. **Nilsson, G.,** Localized excess electrons in water and ice. An ice cavity model for the localized electron and the microscopic relaxation time, *J. Chem. Phys.,* 56, 3427, 1972.

31. **Funabashi, K., Carmichael, I., and Hamill, W. H.,** Photon-induced electron transfer transitions of the solvated electrons, *J. Chem. Phys.,* 69, 2652, 1978.

32. **Kawabata, K., Buxton, G. V., and Salmon, G. A.,** Temperature dependence of trapped electrons in crystalline D_2O ice from 77 to 6 K, *Chem. Phys. Lett.,* 64, 487, 1979.

33. **Kawabata, K., Nagata, Y., Okabe, S., Kimura, N., Tsumori, K., Kawanishi, M., Buxton, G. V., and Salmon, G. A.,** Fast decay of the visible band electron in e-irradiated crystalline ice at low temperature: the isotope effects and the role of a mobile proton in the decay, *J. Chem. Phys.,* 77, 3884, 1982.

34. **Jeffrey, G. A. and McMullan, R. K.,** The clathrate hydrates, in *Progress in Inorganic Chemistry,* Vol. 8, Cotton, F. A., Ed., Interscience Publishers, New York, 1967.

35. **Zagórski, Z. P., Grodkowski, J., and Bobrowski, K.,** Trapped electron spectra in hydrates of sodium, potassium and tetraalkylammonium hydroxides of varying H_2O content, *Radiat. Phys. Chem.,* 54, 343, 1980.

36. **Zagórski, Z. P.,** Electron trapped in aqueous clathrates solid at room temperature, *Chem. Phys. Lett.,* 115, 507, 1985.

37. **Zagórski, Z. P.,** Pulse radiolysis study on electrons trapped in aqueous solid clathrates, *J. Phys. Chem.,* 90, 957, 1986.

38. **Zagórski, Z. P.,** Thermal effects on the behavior of electron trapped in aqueous solid clathrates, *J. Phys. Chem.,* 91, 734, 1987.

39. **Cygler, J., Klassen, N. V., and Teather, G. G.,** Trapped electrons in ter-butanol-water glasses, *Radiat. Phys. Chem.,* 27, 47, 1986.

40. **Buxton, G. V., Gillis, H. A., and Klassen, N. V.,** Evidence for a second kind of trapped electron in some deuterated aqueous glasses at low temperatures. A pulse radiolysis study, *Can. J. Chem.,* 54, 367, 1976.

41. **Buxton, G. B., Medley, M., and Salmon, G. A.,** Evidence for infrared absorbing electrons in alkaline aqueous glasses — a pulse radiolysis study at temperature down to 6 K, *J. Chem. Soc., Faraday Trans. 1,* 81, 3067, 1985.

42. **Dolivo, G. and Kevan, L.,** Optical absorption spectra of localized electrons generated at 1.6 K in polar matrices: evidence for presolvated electron, *J. Chem. Phys.,* 70, 2599, 1979.

43. **Nguyen, T. Q. and Walker, D. C.,** Evolution of the spectrum of the solvated electron in BeF_2 aqueous glasses at 76 K, *J. Chem. Phys.,* 69, 1038, 1978.

44. **Stradowski, Cz. and Hamill, W. H.,** The role of H_2O^- and D_2O^- in competitive electron trapping by the matrix and by cadmium (II) in sodium perchlorate aqueous glasses, *J. Phys. Chem.,* 80, 1431, 1976.

44a. **Ražem, D. and Hamill, W. H.,** Electron scavenging in ethanol and in water, *J. Phys. Chem.,* 81, 1625, 1977.

45. **Gillis, H. A., Teather, G. G., and Buxton, G. V.,** Pulse radiolysis of deuterated aqueous LiCl glasses. Dependence of the yields and rates of reaction of visible and infrared absorbing trapped electrons on LiCl concentration, *Can. J. Chem.,* 56, 1889, 1978.

46. **Baianu, I. C., Boden, N., Lightowlers, D., and Mortimer, M.,** A new approach to the structure of concentrated aqueous electrolyte solutions using pulsed nmr methods, *Chem. Phys. Lett.,* 54, 169, 1978.

47. **Klassen, N. V., Teather, G. G., and Kieffer, F.,** A study of trapped electrons in $LiCl/D_2O$ and other aqueous glasses at temperatures down to 2 K by radiolysis, pulse radiolysis, photolysis, and stimulated luminescence, *Can. J. Chem.,* 57, 1488, 1979.

48. **Klassen, N. V., Adams, R. J., Teather, G. G., and Ross, C. K.,** The reaction of e_{vis}^- and e_{ir}^- in $LiCl/D_2O$ Glasses as measured from 10^{-7} to 2100s, *J. Phys. Chem.,* 84, 3609, 1980.

49. **Bartczak, W. M., Hilczer, M., and Kroh, J.,** Trapped electron in frozen ionic solutions — a theoretical model, *Radiat. Phys. Chem.,* 17, 431, 1981.

50. **Rice, S. A., Dolivo, G., and Kevan, L.,** Comparison of photoconductivity and optical spectra for trapped electrons in aqueous and alcoholic glasses at 4.2 K, *J. Chem. Phys.,* 70, 18, 1979.

51. **Van Leeuwen, J. W., Heijman, M. G. J., and Nauta, H.,** A study of the temperature dependence of the visible absorption band of trapped electrons in ethylene glycol-water and LiCl glasses between 4 and 100 K, *Radiat. Phys. Chem.,* 17, 367, 1981.

52. **Wu, Z., Klassen, N. V., Gillis, H. A., and Teather, G. G.,** A pulse radiolysis study of trapped electrons in aqueous ethylene glycol glasses, *Can. J. Chem.,* 61, 189, 1983.

53. **Cygler, J., Gillis, H. A., Klassen, N. V., and Teather, G. G.,** The effect of accumulated dose at 6 and 72 K on electron trapping in ethylene glycol-water glass, *Radiat. Phys. Chem.,* 17, 379, 1981.

54. **Hase, H. and Higashimura, T.,** Electron trapping and solvation in ethylene glycol-water glass: effect of irradiation at 77 K prior to irradiation at 4 K, *Radiat. Phys. Chem.,* 12, 161, 1978.

55. **Ohno, K., Takemura, I., and Sohma, J.,** Resolution enhanced ESR spectra of the trapped electrons in an alkaline ice, *J. Chem. Phys.,* 56, 1202, 1972.

56. **Helbert, J., Kevan, L., and Bales, B. L.,** Matrix ENDOR linewidths of trapped electrons in glassy matrices at 77 K, *J. Chem. Phys.,* 57, 723, 1972.

57. **Bales, B. L., Helbert, J., and Kevan, L.,** Electron paramagnetic resonance and electron-nuclear double resonance line shape studies of trapped electrons in γ-irradiated deuterium-substituted 10 M sodium hydroxide alkaline ice glass, *J. Phys. Chem.,* 78, 221, 1974.

58. **Yoshida, H., Feng, D. F., and Kevan, L.,** Electron-electron double resonance study of trapped electrons in 10M NaOH alkaline ice glass, *J. Chem. Phys.,* 58, 3411, 1973.

59. **Bales, B. L., Bowman, M. K., Kevan, L., and Schwartz, R. N.,** Separation of isotropic and anisotropic hyperfine constants in disordered systems by analysis of electron paramagnetic resonance lines at two different microwave frequencies: application to the molecular structure around excess electrons in γ-irradiated 10 *M* sodium hydroxide alkaline ice glass, *J. Chem. Phys.,* 63, 3008, 1975.

60. **Narayana, P. A., Bowman, M. K., Kevan, L., Yudanov, V. F., and Tsvetkov, Yu. D.,** Electron spin echo envelope modulation of trapped radicals in disordered glassy systems: application to the molecular structure around excess electrons in γ-irradiated 10 *M* sodium hydroxide alkaline ice glass, *J. Chem. Phys.,* 63, 3365, 1975.

61. **Kevan, L. and Schwartz, R. N.,** *Time domain electron spin resonance,* Wiley-Interscience, New York, 1979.

62. **Schlick, S., Narayana, P. A., and Kevan, L.,** ESR line shape studies of trapped electrons in γ-irradiated ^{17}O enriched 10M NaOH alkaline ice glass: model for the geometrical structure of the trapped electron, *J. Chem. Phys.,* 64, 3153, 1976.

63. **Lin, D. P. and Kevan, L.,** Second moment studies of the electron spin resonance line shape of trapped electrons in sodium-ice condensates. Relation to the molecular structure around trapped electron, *J. Phys. Chem.,* 81, 1498, 1977.

64. **Newton, M.,** The role of *ab initio* calculations in elucidating properties of hydrated and ammoniated electrons, *J. Phys. Chem.,* 79, 2795, 1975.

65. **Astashkin, A. V., Dikanov, S. A., and Tsvetkov, Yu. D.,** Hyperfine interactions of deuterium nuclei in the nearest surroundings of trapped electrons in alkaline glass, *Chem. Phys. Lett.,* 144, 258, 1988.

66. **Tachikawa, M. and Ogasawara, M.,** *Ab initio* molecular orbital study on water dimer anions, *J. Phys. Chem.,* 94, 1746, 1990.

67. **Lin, D. P., Hamlet, P., and Kevan, L.,** Paramagnetic relaxation study of spatial distribution of trapped radicals in γ-irradiated alkaline ice and organic glasses at 4.2 K, *J. Phys. Chem.,* 76, 1226, 1972.

68. **Hase, H. and Kevan, L.,** EPR studies on trapped species produced by tritium β particles in alkaline and acid ices at 77 K, *J. Chem. Phys.,* 52, 3183, 1970.

69. **Yoshida, H., Feng, D. F., and Kevan, L.,** Electron-electron double resonance study of trapped electrons in 10 *M* NaOH alkaline ice glass, *J. Chem. Phys.,* 58, 3411, 1973.

70. **Miller, J.,** Fast electron transfer reactions in a rigid matrix; further evidence for quantum mechanical tunneling, *Chem. Phys. Lett.,* 22, 180, 1973.

71. **Zamaraev, K. I., Khairutdinov, R. F., and Miller, J. R.,** Experimental determination of the electron tunneling probability as a function of distance for chemical reactions, *Chem. Phys. Lett.,* 57, 311, 1978.

72. **Miller, J. R.,** Reactions of trapped electrons by quantum mechanical tunneling observed by pulse radiolysis of an aqueous glass, *J. Phys. Chem.,* 79, 1070, 1975.

73. **Buxton, G. V. and Kemsley, K. G.,** A rate equation for the reaction of trapped electrons in aqueous 6M NaOH glass at 77 K based on trap-to-trap migration and a fixed reaction distance, *Radiat. Phys. Chem.,* 13, 151, 1979.

74. **Eisele, I. and Kevan, L.,** Photoconductivity in γ-irraidated alkaline ice, *J. Chem. Phys.,* 53, 1867, 1970.

75. **Rice, S. A. and Kevan, L.,** Comparison of photoconductivity and optical spectra of the trapped electron in polar aqueous and alcoholic glasses, *J. Phys. Chem.,* 81, 847, 1977.

76. **Rice, S. A., Dolivo, G., and Kevan, L.,** Optical absorption and photoconductivity spectra of localized electrons in x-irradiated aqueous 10 *M* lithium chloride glasses at 4.2 K: support for two types of localized electrons each with overlapping bound-bound and bound-continuum transitions, *J. Chem. Phys.,* 68, 4864, 1978.

77. **Noda, S. and Kevan, L.,** Energy level structure and mobility of excess electrons in γ-irradiated 5 *M* potassium carbonate aqueous glasses. Effect of ions on trapped and mobile electrons in aqueous glasses, *J. Phys. Chem.,* 78, 2454, 1974.

78. **May, R. and Walker, D. C.,** Inhomogeneous bleaching of trapped electron absorption bands in aqueous glasses during laser excitation, *J. Chem. Soc. Faraday Trans. II,* 74, 1833, 1978.

79. **May, R. and Walker, D. C.,** Infrared absorption produced during laser bleaching of trapped electrons, *Chem. Phys. Lett.,* 30, 69, 1975.

80. **Gillis, H. A. and Walker, D. C.,** Luminescence and infrared absorption from laser excitation of trapped electrons in aqueous glasses, *J. Chem. Phys.,* 65, 4590, 1976.

81. **Nguyen, T. Q. and Walker, D. C.,** Luminescence following laser excitation of trapped electrons in mixed-solute aqueous glasses, *J. Chem. Soc., Faraday Trans. I,* 73, 1958, 1977.

82. **Nguyen, T. Q. and Walker, D. C.,** The percentage of bound-continuum character in the optical transition of trapped electron absorption bands, and the fate of quasifree electrons, *J. Chem Phys.,* 67, 2399, 1977.

83. **Kira, A., Imamura, M., and Matsui, M.,** Trapped electron yield in ethanol glasses at 80 K irradiated with high-LET ion beams, *Radiat. Phys. Chem.,* 27, 425, 1986.

84. **Richards, J. T. and Thomas, J. K.,** Trapping of electrons in low temperature glasses. A pulse radiolysis study, *J. Chem. Phys.,* 53, 218, 1970.

85. **Richards, J. T. and Thomas, J. K.,** Photoionization of molecules in low temperature glasses, *Chem. Phys. Lett.,* 8, 13, 1971.

86. **Kevan, L.,** Pulse radiolysis of alcohol glasses at 77 K: Mechanism of electron trapping, *J. Chem. Phys.,* 56, 838, 1972.

87. **Hase, H., Noda, M. and Higashimura, T.,** Electronic spectra of trapped electrons in organic glasses at 4° K, *J. Chem. Phys.,* 54, 2975, 1971.

88. **Fueki, K., Feng, D. F., and Kevan, L.,** Application of the semicontinuum model to the effect of dipole reorientation on trapped electron spectra in glassy ethanol, *J. Chem. Phys.,* 56, 5351, 1972.

89. **Fueki, K., Feng, D. F., and Kevan, L.,** Application of the semicontinuum model to temperature effects on solvated electron spectra and relaxation rates of dipole orientation around an excess electron in liquid alcohols, *J. Phys. Chem.,* 78, 393, 1974.

90. **Klassen, N. V., Gillis, H. A., Teather, G. G., and Kevan, L.,** Pulse radiolysis studies of time dependent spectral shifts of the solvated electron in ethanol and deuterated ethanol glasses at 76 K, *J. Chem. Phys.,* 62, 2474, 1975.

91. **Miller, J. R., Clifft, B. E., Hines, J. J., Runowski, R. F., and Johnson, K. W.,** Solvation of electrons in alcohol glasses from 10^{-6} to 10^2 s after pulse radiolysis at 77 K, *J. Phys. Chem.,* 80, 457, 1976.

92. **Higashimura, T., Noda, M., Warashina, T., and Yoshida, H.,** Electron traps in irradiated water-ethylene glycol glass at 4 and 77° K, *J. Chem. Phys.,* 53, 1152, 1970.

93. **Hase, H., Warashina, T., Noda, M., Namiki, A., and Higashimura, T.,** Trapped electrons produced in ethanol glass at 4° K, *J. Chem. Phys.,* 57, 1039, 1972.

94. **Namiki, A., Noda, M., and Higashimura, T.,** Electron spectra of trapped electrons in organic glasses at 4 K. VI. Photobleaching in ethanol, *Chem. Phys. Lett.,* 23, 402, 1973.

95. **Hase, H., Noda, M., Higashimura, T., and Fueki, K.,** Electronic spectra of trapped electrons in organic glasses at 4 K. II. Ethanol-methanol mixtures, *J. Chem. Phys.,* 55, 5411, 1971.

96. **Baxendale, J. H. and Sharpe, P. H. G.,** A pulse radiolysis study of electrons in 1-propanol at low temperatures, *Int. J. Radiat. Phys. Chem.,* 8, 621, 1976.

97. **Baxendale, J. H. and Wardman, P.,** Electrons in liquid alcohols at low temperatures, *J. Chem. Soc. Faraday Trans. I,* 69, 584, 1973.

98. **Ogasawara, M. and Kevan, L.,** Laser photoionization study of time dependent spectral shifts of localized electrons in ethanol glass. Temperature effects, *J. Phys. Chem.,* 82, 378, 1978.

99. **Baxendale, J. H. and Sharpe, P. H. G.,** Electron solvation in alcohols at 77 K after pulse radiolysis, *Chem. Phys. Lett.,* 39, 401, 1976.

100. **Ogasawara, M., Shimizu, K., Yoshida, K., Kroh, J., and Yoshida, H.,** On the spectral shift of trapped electrons in glassy alcohols irradiated at 4.2 K, *Chem. Phys. Lett.,* 64, 43, 1979.

101. **Noda, S., Yoshida, K., Ogasawara, M., and Yoshida, H.,** Effect of an inefficient electron scavenger on infrared- and visible-absorbing electrons in an ethanol matrix, *J. Phys. Chem.,* 84, 57, 1980.

102. **Ogasawara, M., Shimizu, K., and Yoshida, H.,** Isothermal change of the spectrum of localized electrons in glassy 1-propanol irradiated at 4 K, *Chem. Lett.,* 1201, 1980.

103. **Ogasawara, M., Shimizu, K., and Yoshida, H.,** Evolution of the spectra of the localized electrons in glassy alcohols irradiated at 4 K, *Radiat. Phys. Chem.,* 17, 331, 1981.

104. **Bush, R. L. and Funabashi, K.,** Small polaron model for the absorption spectrum of solvated electrons in alcohols, *J. Chem. Soc., Faraday Trans. II,* 274, 1977.

105. **Stradowski, Cz., Wolszczak, M., and Kroh, J.,** Spectra of electrons trapped in ethanol-water mixtures irradiated at 4.2 K, *Radiat. Phys Chem.,* 16, 465, 1980.

106. **Buxton, G. V., Kroh, J., and Salmon, G. A.,** On electron trapping in glassy alcohols. A pulse radiolysis study of the effect of irradiation temperature, *Chem. Phys. Lett.,* 68, 554, 1979.

107. **Buxton, G. V. and Salmon, G. A.,** Effect of temperature and composition on the absorption spectrum of trapped electrons in glassy mixtures of methanol and n-butanol, *Radiat. Phys. Chem.,* 17, 355, 1981.

108. **Buxton, G. V. and Salmon, G. A.,** Electrons trapped in ethanol-water glasses at ultra-low temperatures, *Radiat. Phys. Chem.,* 17, 361, 1981.

109. **Buxton, G. V., Kroh, J., and Salmon, G. A.,** Electron trapping in glassy normal alcohols. A pulse radiolysis study at temperatures down to 6 K, *J. Phys. Chem.,* 85, 2021, 1981.

110. **Klassen, N. V. and Teather, G. G.,** "Initial" spectra of trapped electrons in alcohol glasses from 12 to 115 K, *J. Phys. Chem.,* 87, 3894, 1983.

111. **Perkey, L. M. and Smalley, J. F.,** Very low temperature pulse radiolysis. Protiated and perdeuterated methanol glasses at 6 K, *J. Phys. Chem.,* 83, 2959, 1979.

112. **Huddleston, R. K. and Miller, J. R.,** Spectral relaxation of the benzophenone negative ion in alcohol glasses: evidence for solvent reorientation, *Radiat. Phys. Chem.,* 17, 383, 1981.

113. **Gauduel, Y., Migus, A., Chambaret, J. P., and Antonetti, A.,** Femtosecond reactivity of electron in aqueous solutions, *Rev. Phys. Appl.,* 22, 1755, 1987.

114. **Ichikawa, T. and Yoshida, H.,** Solvation structure of excess electrons in ethanol and ethanol/2-methyl-tetrahydrofuran mixtures, *J. Phys. Chem.,* 93, 5943, 1989.

115. **Baxendale, J. H. and Sharpe, P. H. G.,** Electron solvation in alcohol-alkane mixtures, *Chem. Phys. Lett.,* 41, 440, 1976.

116. **Bartczak, W. M., Światła, D., and Kroh, J.,** Model of electron capture in low-temperature glasses, *Radiat. Phys. Chem.,* 21, 469, 1983.

117. **Światła, D., Bartczak, W. M., and Kroh, J.,** Electron localization in low-temperature alcohol glasses. A theoretical model, *J. Phys. Chem.,* 89, 3006, 1985.

118. **Kroh, J.,** Mechanism and kinetics of primary processes induced by irradiation of noncrystalline solids, *Radiat. Phys. Chem.,* 28, 415, 1986.

119. **Hase, H. and Warashina, T.,** Trapped electrons produced in deuterated ethanol glasses at 4° K, *J. Chem. Phys.,* 59, 2152, 1973.

120. **Schwartz, R. N., Bowman, M. K., and Kevan, L.,** Matrix proton ENDOR studies of the overlap of trapped electron wave functions with first solvation shell molecules in methanol glass at 77 K, *J. Chem. Phys.,* 60, 1690, 1974.

121. **Bales, B. L. and Kevan, L.,** On the solvation of the trapped electron in γ-irradiated deuterated ethanol glass at 4 K, *J. Chem. Phys.,* 60, 710, 1974.

122. **Hase, H., Ngo, F. Q. H., and Kevan, L.,** Study of solvation of excess electrons in ethylene glycol/water and methyltetrahydrofuran matrices by matrix ENDOR, *J. Chem. Phys.,* 62, 985, 1975.

123. **Narayana, M. and Kevan, L.,** Electron spin echo modulation study of the geometry of solvated electrons in ethanol glass: an example of a molecular dipole oriented solvation shell, *J. Chem. Phys.,* 72, 2891, 1980.

124. **Narayana, M. and Kevan, L.,** Geometrical structure of solvated electrons in ethanol glass determined from electron spin echo modulation analysis, *J. Am. Chem. Soc.,* 103, 1618, 1981.

125. **Kevan, L.,** Forbidden matrix proton spin flip satellites in 70 GHz ESR spectra of solvated electrons. A geometrical model for the solvated electron in methanol glass, *Chem. Phys. Lett.,* 66, 578, 1979.

126. **Dainton, F. S., Salmon, G. A., and Zucker, U. F.,** The radiation chemistry of glassy n-propanol, *Proc. R. Soc. London,* A320, 1, 1970.

127. **Schulte-Frohlinde, D., Lang, D., and von Sonntag, C.,** Strahlenchemie von Alkoholen. IV. Einfluß der Kettenlänge auf die Wasserstoffbildung bei der γ-Radiolyse von gesättigten n-alkoholen, *Ber. Bunsengesellsch. Phys. Chem.,* 72, 63, 1968.

128. **Lindgren, M., Lund, A., Ogasawara, M., and Samskog, P.-O.,** Localized electron to radical conversion in x-irradiated single crystals of 1,6-hexanediol and 1,8-octanediol, *Radiat. Phys. Chem.,* 29, 439, 1987.

129. **Tachikawa, H., Ogasawara, M., Lindgren, M., and Lund, A.,** *Ab initio* calculations on localized electrons in alcoholic matrices: hydrogen-bond defect model, *J. Phys. Chem.,* 92, 1712, 1988.

130. **Dainton, F. S., Salmon, G. A., and Teplý, J.,** The radiation chemistry of low temperature methanolic glasses. *Proc. R. Soc.,* A286, 27, 1965.

131. **Hingerty, B. and Saenger, W.,** Topography of cyclodextrin inclusion complexes. 8. Crystal and molecular structure of the α-cyclodextrin-methanol-pentahydrate complex. Disorder in a hydrophobic cage, *J. Am. Chem. Soc.,* 98, 3357, 1976.

132. **Manor, P. C. and Saenger, W.,** Water molecule in hydrophobic surroundings: structure of α-cyclodextrin-hexahydrate $(C_6H_{10}O_5) \cdot 6H_2O$, *Nature,* 237, 392, 1972.

133. **Bardsley, J., Baugh, P. J., and Phillips, G. O.,** Trapping of radiation-induced electrons in cycloamylose hydrates: electron spin resonance study, *J. Chem. Soc. Chem. Comm.,* 1335, 1972.

134. **Ichikawa, T. and Yoshida, H.,** Trapped electrons in crystalline cyclodextrin matrices, *Radiat. Phys. Chem.,* 11, 173, 1978.

135. **Box, H. C., Budzinski, E. E., and Freund, H. G.,** Electron trapping in irradiated single crystals of organic compounds, *J. Chem. Phys.,* 69, 1309, 1978.

136. **Box, H. C., Budzinski, E. E., Freund, H. G., and Potter, W. R.,** Trapped electrons in irradiated single crystals of polyhydroxy compounds, *J. Chem. Phys.,* 70, 1320, 1979.

137. **Budzinski, E. E., Potter, W. R., Potienko, G., and Box, H. C.,** Characteristics of trapped electrons and electron traps in single crystals, *J. Chem. Phys.,* 70, 5040, 1979.

138. **Samskog, P.-O., Lund, A., Nilsson, G., and Symons, M. C. R.,** Primary reactions of localized electrons in rhamnose crystals studied by pulse radiolysis and ESR spectroscopy, *J. Chem. Phys.,* 73, 4862, 1980.

139. **Samskog, P.-O., Kispert, L. D., and Lund, A.,** Geometric model of trapped electrons in x-ray irradiated single crystals of rhamnose, *J. Chem. Phys.,* 79, 635, 1983.

140. **Budzinski, E. E., Potter, W. R., and Box, H. C.,** Radiation effects in x-irradiated hydroxy compounds, *J. Chem. Phys.,* 72, 972, 1980.

141. **Locher, S. E. and Box, H. C.,** ESR-ENDOR studies of x-irradiated glucose-1-phosphate dipotassium salt, *J. Chem. Phys.,* 72, 828, 1980.

142. **Buxton, G. V. and Salmon, G. A.,** Absorption spectrum of electrons trapped in single crystals of sucrose. A pulse-radiolysis study in the temperature range 6-270 K, *Chem. Phys. Lett.,* 73, 304, 1980.

143. **Li, A. S. W., Kevan, L., and Fujimura, T.,** Absorption spectrum of trapped electrons in rhamnose single crystals x-irradiated at 4 K, *J. Chem. Phys.,* 76, 5647, 1982.

144. **Kevan, L., Schlick, S., Narayana, P. A., and Feng, D. F.,** Application of the semicontinuum potential model to deduce localized electron trapping sites in single crystals of D-sorbitol, *J. Chem. Phys.,* 75, 1980, 1981.

145. **Lindgren, M., Gustafsson, T., Westerling, J., and Lund, A.,** ESR characterization of the hydroxyalkyl radical in single crystals of 1,6-hexanediol and 1,8-octanediol and crystal structure of 1,6-hexanediol, *Chem. Phys.,* 106, 441, 1986.

146. **Samskog, P.-O., Lund, A., and Nilsson, G.,** Localized electrons in 1,8-octanediol crystals studied by ESR spectroscopy and pulse radiolysis, *Chem. Phys. Lett.,* 79, 447, 1981.

147. **Ogasawara, M., Lindgren, M., Lund, A., and Nilsson, G.,** Anisotropic ESR hyperfine interaction of electrons trapped in crystals of 1,6-hexanediol and 1,8-octanediol, *Chem. Phys. Lett.,* 117, 254, 1985.

148. **Hama, Y., et al.,** Private communication.

149. **Claesson, O., Ogasawara, M., Yoshida, H., and Lund, A.,** The absorption spectrum of localized electrons in crystalline 1,8-octanediol, γ-irradiated at 4.2 K, *Chem. Phys. Lett.,* 94, 408, 1983.

150. **Ogasawara, M., Claesson, O., Yoshida, H., and Lund, A.,** Localized electrons in crystalline 1,6-hexanediol and 1,8-octanediol studied by absorption spectroscopy at 4.2 K and 77 K, *J. Phys. Chem.,* 88, 5004, 1984.

Chapter 10

HOT ELECTRON TRANSPORT AND TRAPPING IN SILICON DIOXIDE

Donelli J. DiMaria and Massimo V. Fischetti

TABLE OF CONTENTS

I. INTRODUCTION

Recently, transport studies in wide bandgap insulators have generated much interest for both scientific and technological reasons. Experimentally, work has been performed on materials such as SiO_2, Si_3N_4, Si-rich SiO_2, and SiO_xN_y which are used in the electronics industry,[1-9] polymer films used in the high power insulation industry,[10,11] and rare gas solids.[11] Theoretically, this experimental work has generated rethinking of the hot carrier transport in insulators, with most of the calculations to date performed for SiO_2.[5-8] However, the fundamental picture for electron transport that has evolved appears to apply universally to all these different wide bandgap dielectrics. The discussion here will extensively be about SiO_2 on which most of our work has been performed.

Previous studies of SiO_2 were performed mostly on the amorphous phase sandwiched between metal or semiconductor electrodes in a configuration known in the semiconductor industry as a metal-oxide-semiconductor (MOS) structure. The amorphous SiO_2 layer is either thermally grown from the Si substrate at temperatures $\gtrsim 800°C$ or deposited at lower temperatures from the gaseous phase. Typically, these films have fewer than $\approx 1 \times 10^{12}$ impurities/cm^2 of such ions, atoms, or molecules as Na^+, K^+, Li^+, Al, OH, and H_2O than either bulk quartz or fused silica and can be made uniformly as thin as 35 Å. The MOS structure is schematically shown in Figure 1 in an energy band configuration where both the metal "gate" electrode and semiconductor substrate are kept at ground potential. The interface energy barriers formed by the ≈ 9 eV bandgap SiO_2 layer with contacting metal or semiconductor electrodes are in the range of 2 to 4 eV for electrons and 7 to 5 eV for holes. This energy barrier asymmetry and the minimization of mobile alkaline impurities makes electronic conduction dominant in these films over the electric field range of interest. This and other generally known background information about MOS structures and SiO_2 can be found in the literature.[12]

When a bias is applied to the MOS capacitor structure, an electron current can be observed if an ammeter is attached to the external circuit. Typically, high electric fields in excess of 6 MV/cm across the oxide layer are needed to observe any measurable current flow. This current is interface limited and controlled by Fowler-Nordheim tunneling through the triangular energy barrier formed at the interface which is at a negative potential.[13] On emerging into the oxide conduction band, these electrons are quickly swept to the anode in times $\lesssim 1 \times 10^{-12}$ s for films thinner than 1000 Å.[14] The interface of SiO_2 with Si and some metal electrodes is abrupt (to within angstroms), and there are few localized states near the conduction band edge due to disorder even though the oxide is amorphous.[15] Deep electron trapping states in the forbidden bandgap of the oxide layer, which are initially present in as-fabricated films, are due to residual water or OH-related impurities.[16] These can be minimized with current high temperature processing conditions so that less than one in every 1×10^7 of the electrons traversing the oxide layer is permanently captured into these deep sites. However, other trapping sites can be created by the passage of hot electrons.[17] It is the motion of these hot electrons, while in the SiO_2 conduction band, that will be discussed here.

To investigate the interaction of conduction band electrons with the surrounding atoms comprising the SiO_2 layer requires experiments which measure either the energy, velocity, and/or transit time of the charge carriers. Most standard measurements such as current-voltage (I-V) or capacitance-voltage (C-V) cannot directly yield a measure of carrier motion or energetics. I-V and C-V techniques can sense trapped charge build-up in the oxide or at its interfaces with contacting electrodes via the internal electric field that is generated. Although having yielded some information on the lattice interaction, transit time measurements are limited to fields below the threshold for electron heating. This is due to the thick oxide films that must be used to resolve the very short times involved in carrier transport in silicon dioxide.[14] The two techniques that will be extensively discussed here, carrier

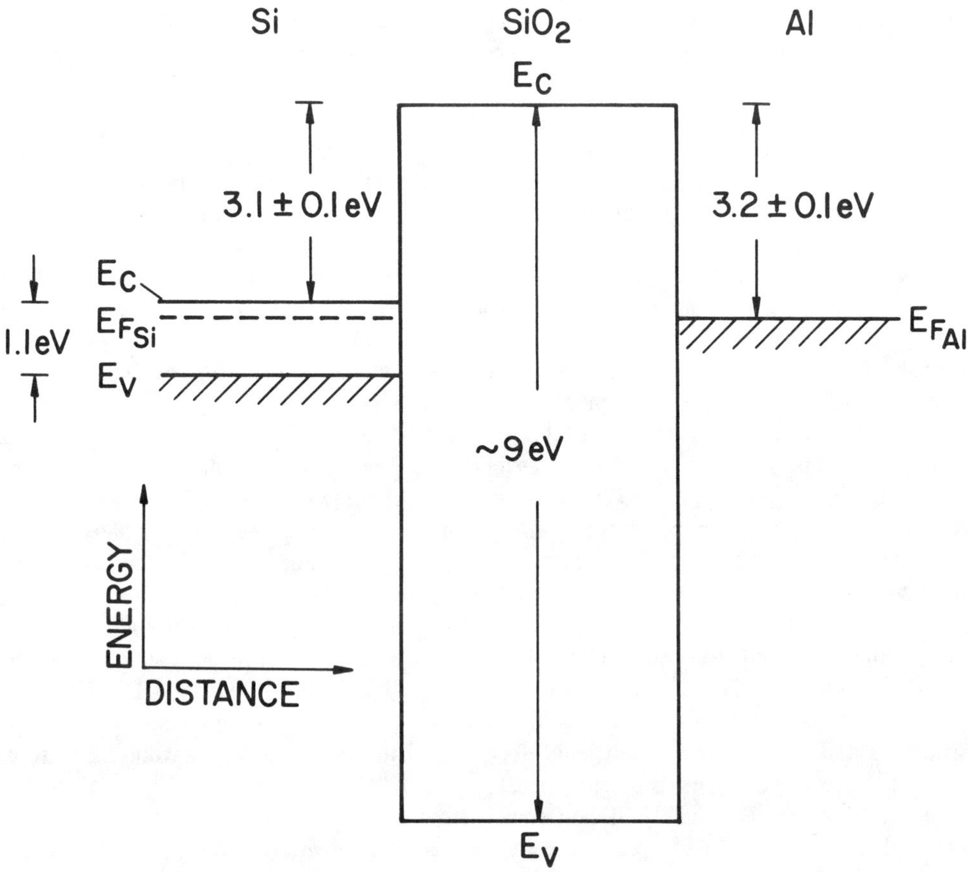

FIGURE 1. Schematic zero field energy band diagram of an MOS structure.

separation and vacuum emission, can give direct information on the emerging energy distribution of the carriers at the SiO_2-anode interface over a wide range of fields and oxide thickness.[1-7,9]

Until 1982, SiO_2 transport was believed to be dominated by interaction of the electrons with only the polar phonon modes of the lattice atoms through inelastic scattering.[18,19] These electrons were thought to lose large amounts of energy (0.153 and 0.063 eV for the dominant optical phonon modes in SiO_2) gained from the applied field every 1 to 2 Å through the creation of these oscillations of the lattice atoms. This takes place through the Coulomb interaction of the electronic carriers with the partially charged electron clouds surrounding the lattice atoms. Electrons were believed to stay at thermal energies, near the bottom of the conduction band, and to gain sufficient amounts of energy from the applied field only near destructive breakdown of the SiO_2 which typically occurred at $\gtrsim 8$ MV/cm. Destructive breakdown then occurred when the electrons gained enough energy to cause impact ionization. The holes created during ionization were thought to enhance the electron injection at the cathode, thereby leading to an unstable feedback condition where current densities became large enough to cause melting and vaporization of the oxide.

Even in the 1970s, however, there were some problems with this picture for electron transport in SiO_2 which had primarily evolved from the earlier theoretical work of Fröhlich and experimental work on alkali halide crystals.[20] In the early 1970s, internal photoemission of hot electrons over an interfacial energy barrier into the SiO_2 layer of an MOS structure was interpreted to give scattering lengths on the order of 30 Å and not the 1 to 2 Å anticipated.[21] Lewicki and Maserjian, as early as 1975, tried to explain oscillations in dark current as a function of voltage characteristics on very thin SiO_2 layers by ballistic transport of electrons where no scattering events had occurred.[22] Their data analysis led to a mean-free path of ≈ 10 Å which was an order of magnitude larger than that assumed by most others during this time period. Also in the mid-1970s, pronounced decreases in Coulomb capture cross sections and occupancy of certain purposely introduced trapping sites above ≈ 1.5 MV/cm were argued to be due to the onset of strong electron heating in the oxide.[12] In the early 1980s, theoretical Monte Carlo simulations by the group at Rostock University also predicted the onset for electron heating to be about 1 to 2 MV/cm, again almost an order of magnitude less than what was believed by others.[23] As will be discussed here, these early observations which contradicted the "old" model for transport in SiO_2 were not artifacts, could be easily reproduced, and were consistent with measured energy distributions and theoretical calculations performed by us after 1982.

After 1982, our understanding of high field steady-state electron transport in SiO_2 began to change as energetics of the charge carriers were directly studied and improved Monte Carlo techniques were applied to study the transport of these electrons. This led to a much different picture from what had been believed previously. These new results showed that the motion of electrons in the oxide layer is completely controlled by their interaction with the lattice and not by electronic excitation or their interaction with defects and/or trapping states in the forbidden gap.[1-9] Both polar (inelastic) and nonpolar (quasi-elastic) collisions occur. Energy gained from the applied electric field is mostly lost in large amounts to the polar modes (in particular, LO phonon emission at 0.153 and 0.063 eV). The nonpolar phonons mostly act to stabilize the electron energy at high fields. They cause large angle scattering and therefore long path lengths characteristic of dispersive transport. Electrons have near thermal energies at electric fields of $\lesssim 1.5$ MV/cm. For fields greater than this value, the carriers can gain significant energy from the applied field. The average energy of the distribution increases up to a steady-state condition where values as large as 6 eV can be attained at the highest fields before destructive breakdown occurs. The scattering of the hot carriers by the nonpolar modes is essential for the stabilization of the distribution above this 1.5 MV/cm threshold. Prior to our work, the importance of the nonpolar phonon modes at high fields in SiO_2 had been neglected.[18,19,23] This resulted in the incorrect idea that most electrons in SiO_2 were near thermal energies until breakdown where the electrons accelerated to energies greater than that of the bandgap (≈ 9 eV) and caused impact ionization. This older model for high field transport has now been conclusively eliminated for SiO_2 films $\gtrsim 500$ Å in thickness.

Other insulators, such as Si_3N_4, have effective low field mobilities which are much smaller than SiO_2 and more strongly temperature dependent (dropping with decreasing temperature). In SiO_2, the mobility is very large for an amorphous material (≈ 30 $cm^2/V-s$) and increases slightly with decreasing temperature, consistent with lattice scattering and not with the influence of disorder or trapping states.[14] In materials like Si_3N_4, however, transport is dominated by defects or traps.[9] In these materials, most electrons move only small distances (10 to 30 Å) before being captured into trapping sites. Subsequently, these carriers can be reemitted thermally, by tunneling, or through a combination of these two processes. However, a small fraction of the electrons can behave in a manner similar to the hot carrier distributions observed in SiO_2.[9] Hole transport in SiO_2, which is very dispersive and has a low mobility, behaves similarly to that observed for trap-modulated electron conduction.[12]

In this chapter, we will be concerned mainly with our work which was used to directly study the energy distributions of hot electrons in SiO_2 films. Theoretical Monte Carlo simulations will be compared to these results. After briefly reviewing the measurement techniques and the original work on steady-state transport in thicker oxide films, we will discuss more recent results on ballistic transport in thin SiO_2 layers, trap creation by hot carriers, and trapping kinetics when carrier heating is important.

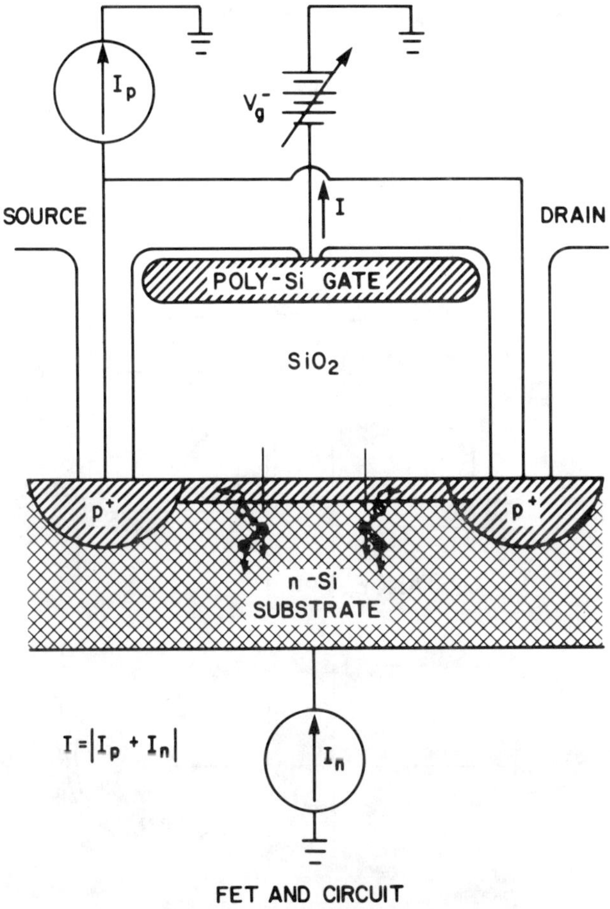

FIGURE 2. Schematic representation depicting the carrier separation technique on a p-channel MOSFET with the device in the measuring circuit configuration.

II. EXPERIMENTAL TECHNIQUES

Three different experimental techniques have been extensively used to study carrier heating. These techniques in chronological order of use are electroluminescence, carrier separation, and vacuum emission.[1-4,9] The latter two, depicted in Figures 2 and 3 and used to obtain the data presented here, will be reviewed. Carrier separation measures only the average energy of the hot electron distribution while vacuum emission measures the energy distribution itself. Each technique has its limitations, but their combination yields a quantitatively consistent picture for transport in SiO$_2$.

Carrier separation measures the average energy of the electron distribution by counting the number of electron/hole pairs produced in single-crystal Si by hot electrons injected from the SiO$_2$. For these experiments, a metal-oxide-semiconductor field effect transistor (MOSFET) with a p-channel is used under negative gate voltage bias to separate the electron/hole pairs produced in the Si before they recombine.[1] The hole current is measured from the p$^+$ source/drain contacts to the channel while electron current is measured from the substrate contact. The number of pairs produced is converted to the average energy of the incoming hot electron distribution using a suitable theory.[1] One of the main advantages of this tech-

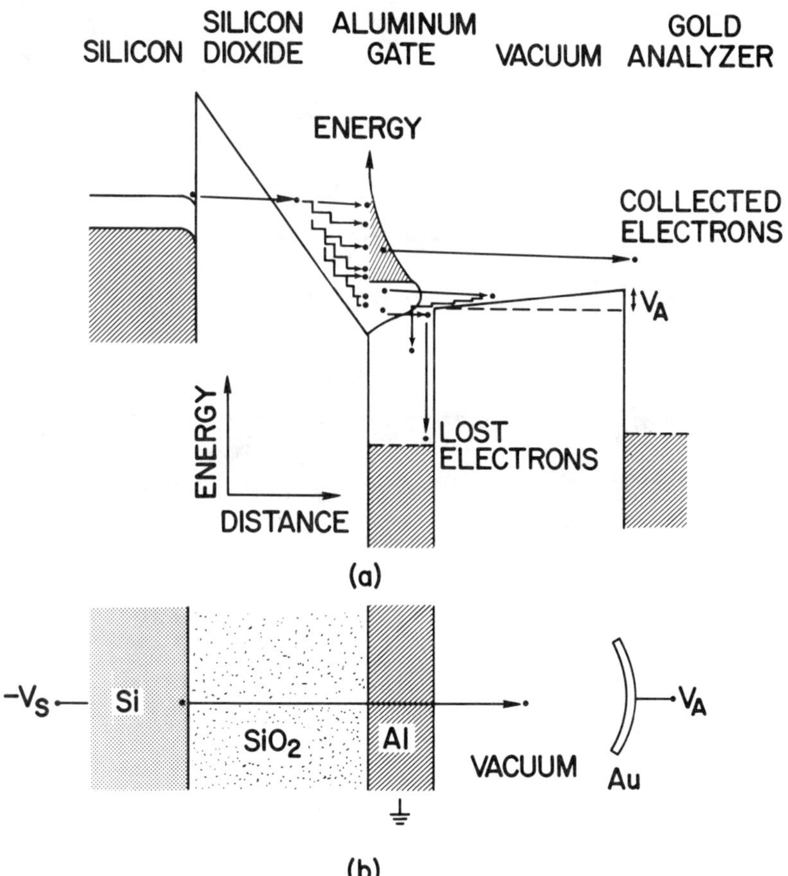

FIGURE 3.   Energy band diagram: (a) Depicting vacuum emission on thin-metal gate MOS and schematic; and (b) showing device in energy analyzing configuration.

nique, as compared to vacuum emission and electroluminescence, is that all carriers entering the contacting anode from the oxide conduction band are counted in determining the average electron energy.

Vacuum emission directly measures the hot electron energy distribution ejected into vacuum through the thin metal gate of an MOS capacitor. An Einzel lens configuration is used to focus and analyze the energy of these hot electrons.[3] Data are obtained by counting the total number of hot electrons emitted into vacuum as a function of the analyzer potential for different oxide fields. The derivative of these data with respect to the analyzer voltage gives the energy distribution. Unless cesiation is used, the first 0.9 eV of information about the distribution is lost due to the small positive electron affinity of SiO$_2$. After extensive studies with different metal gate compositions and thicknesses, we have concluded that most of the emitted electrons we detect escape through thin metallic regions near the edges of the polycrystalline grains. These few electrons ($\approx$1 in 10^5 to 10^6 of those collected by the metal gate) are believed to move through the metal layer without scattering, as demonstrated in some of our later ballistic studies.[4] Typically, we have used thin Al films (150 to 300 Å) for the metal gate electrode. The void space in these films can be up to 2% of the area with grains ($\approx$500 Å in size) missing from the polycrystalline metal film.

For many of the steady-state experiments on thick oxide films, Si-rich SiO$_2$ injectors

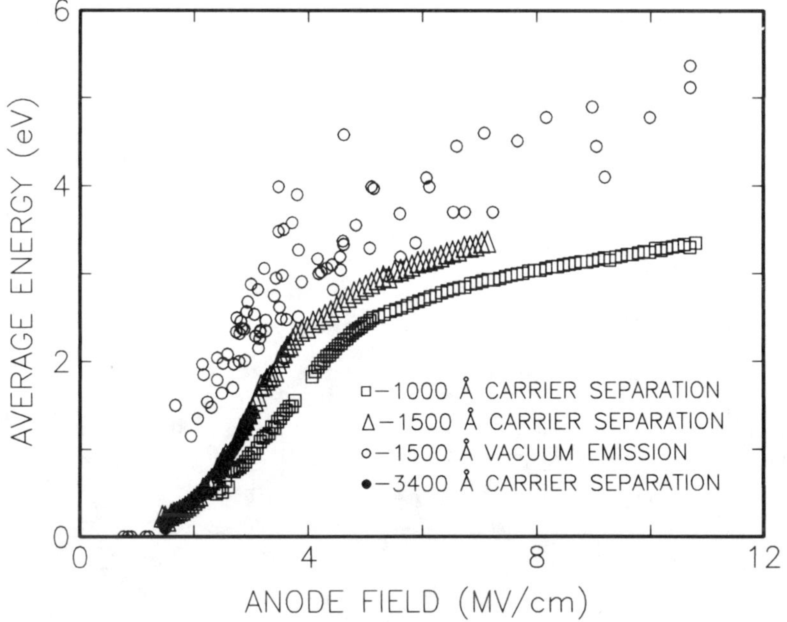

FIGURE 4. Comparison of average energy vs. oxide field data from vacuum emission and carrier separation experiments at room temperature for steady-state hot electron transport in thick SiO_2 layers. This figure shows that significant carrier heating begins at an electric field of ≈1.5 MV/cm which is much lower than that necessary for destructive breakdown of the oxide.

were used.[1-4] The use of Si-rich SiO_2 injectors for cathodic field enhancement and breakdown reduction made many of our early experiments possible. Since the energy relaxation length for electrons in SiO_2 is ≈30 Å, the cathodic injection mode is not important in determining the electron energy in thick oxides.[1] The oxide field near the anode interface determines the average energy of the electron distribution. For cases where structures with and without injectors were compared, identical results were obtained. For all the ballistic transport studies on films ≤100 Å in thickness, injectors were not used because of the difficulty in correcting for the small voltage drop across this layer. Also in many experiments on thicker oxides, the electric field range was extended by using the additive internal field associated with electron charge trapping on energetically deep sites in the oxide film.

Trapping studies were performed using standard electric field sensing techniques of permanent charge build-up in the SiO_2 layer. These measurements, involving variations and combinations of I-V and C-V characterization of MOSs, have been extensively discussed in the literature.[12] Carrier injection from the contacts was done using Fowler-Nordheim tunneling through the interfacial energy barrier[13] or by hot carrier emission over the barrier using light (internal photoemission[21]) or high fields in the Si substrate (avalanche injection[16] or optically induced hot carrier injection[17]).

III. STEADY-STATE TRANSPORT

If the SiO_2 layer thickness is large compared to the characteristic scattering length of the hot carrier distribution, the carriers are expected to be in a steady-state condition for all electric fields, except possibly near breakdown. For oxide films with thicknesses in the range from 100 to 5,000 Å, this has been shown to be true.[1-8] In Figure 4, the average energy characteristics at room temperature for steady-state transport in thick SiO_2 layers are

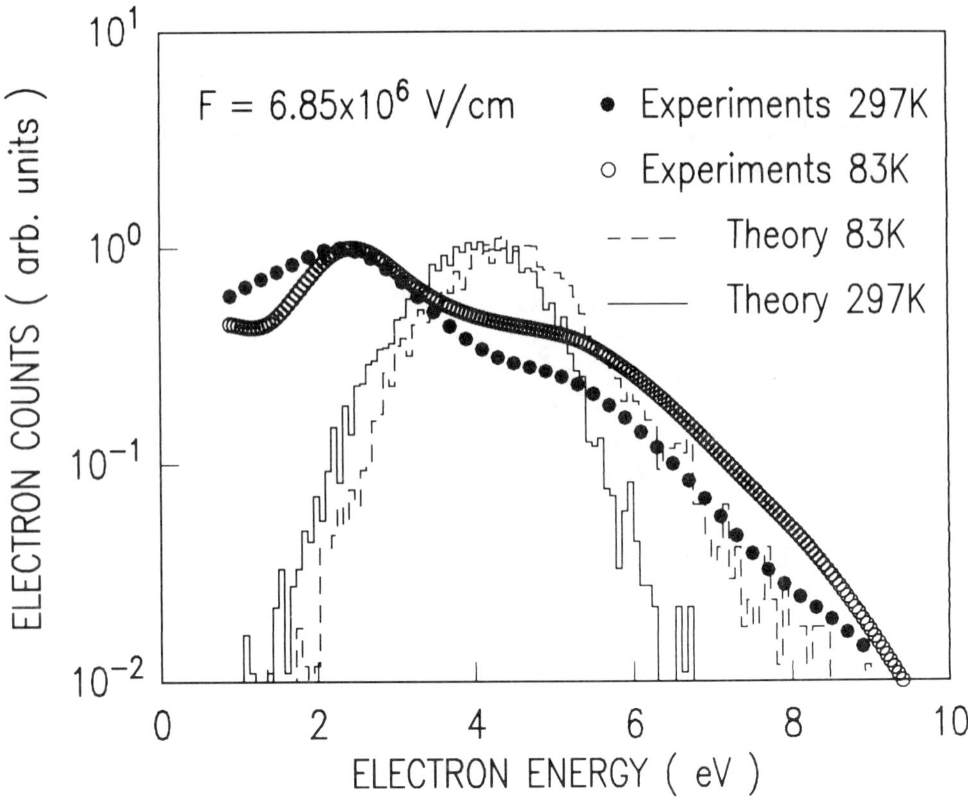

FIGURE 5.   Comparison of energy distributions on thick oxide MOS (similar to those in Figure 1) to Monte Carlo simulations near room and liquid nitrogen temperature at a field well above the heating threshold. These data were obtained using vacuum emission measurements.

shown using both the carrier separation and vacuum emission techniques. As can be seen from this data, runaway from the polar LO phonon modes occurs at a very low field (about 1.5 MV/cm) as initially predicted by Fitting and Frieman.[23] However, the runaway of most of the electrons does not proceed to cause impact ionization even up to fields as high as 16 MV/cm. From 4 to 16 MV/cm, the average energy of the distribution increases very slowly from about 3 to 6 eV with respect to the bottom of the oxide conduction band. The scattering process responsible for stabilizing the distribution at the higher fields is the nonpolar collision (acoustic phonons). From the slope of these data in this regime, an energy relaxation length of ≈ 30 Å can be determined. This quantity corresponds to the distance over which an electron of the distribution must lose an amount of energy equivalent to its average energy value so as to remain in a steady-state condition.

The average energy vs. electric field characteristics, such as that shown in Figure 4, give us a first order view of electron transport and energetics in SiO_2. To get more information on the details of the distributions (such as broadening and the behavior of the high energy tail), vacuum emission must be used. Figure 5 shows typical energy distributions obtained from vacuum emission experiments at two different temperatures which have been used to determine some of the average energies in Figure 4. A comparison of the distribution to a classical Monte Carlo simulation is included. These data have been displayed on a logarithmic scale to show the behavior of the main peak and high energy tail of the distributions. This figure shows a broad distribution with a small temperature dependence over the range from 83 to 297 K, consistent, with the expected behavior for energy loss dominated by LO phonon emission of the 0.153 and 0.063 eV modes in SiO_2 whose energy is large compared to k_BT

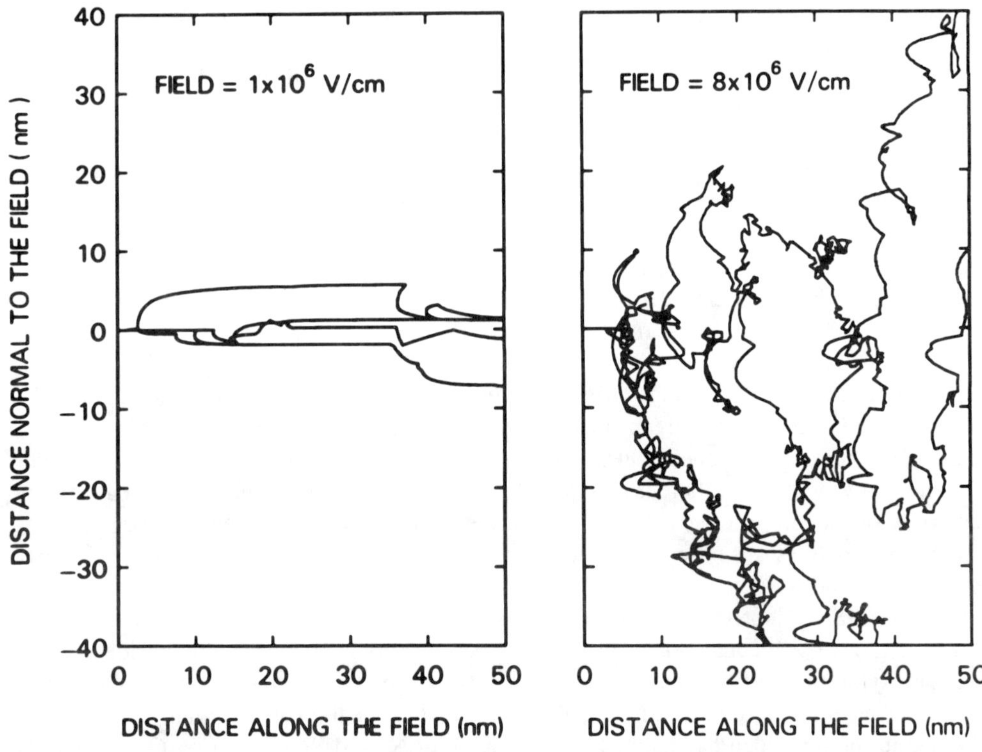

FIGURE 6. Trajectories of several electrons in real space at two values of the electric field. The low field (1 MV/cm) trajectories show that the transport is dominated by forward, polar scattering with LO phonons. At 8 MV/cm, the nonpolar scattering at large angles with acoustic phonons randomizes the direction of motion and effectively increases the actual path length of the electrons. However, the large energy losses of the hot electrons are still dominated by polar, inelastic scattering with LO phonons.

(k_B being Boltzmann's constant and T the absolute temperature). The apparent discrepancy between theory and experiment for the main peak is due to an underestimation of the scattering rate. Better agreement at high electric fields is obtained if a quantum Monte Carlo treatment is used. This will be reviewed in a later section.

These data totally contradict the old model of transport in which LO phonon scattering supposedly keep electrons near thermal energies until runaway occurs near destructive breakdown.[18,19] Using Ridley's earlier suggestion that acoustic or nonpolar phonon scattering should also be considered in SiO_2[24] and the theoretical treatment of this scattering mode by Sparks et al. in alkali halides,[25] we have extended the Monte Carlo calculation of Fitting and Frieman[23] to include it.[5-7] Our initial results using a classical treatment at room temperature compared well with average energy characteristics, like those in Figure 4.[5,6] Later, other groups performed similar Monte Carlo simulations, assuming that nonpolar intervalley scattering to a second conduction band of SiO_2 was limiting the hot electrons to the observed steady-state condition.[8] We will discuss these calculations in Section VI, in light of the available experimental information on the density of states of the SiO_2 conduction band.[26]

The important physical aspect of the acoustic or nonpolar phonon interaction is large angle scattering (momentum reversal) which gives the electrons much larger path lengths than the actual length of material traversed. LO phonon scattering still accounts for most of the energy loss. This is depicted in Figure 6 using classical Monte Carlo simulations for the trajectories of several electrons.[6] At low fields, "streaming-type" transport is observed where most of the electronic paths are in the direction of the applied field. At fields significantly above the threshold for electron heating, dispersive transport dominates due to the large angle scattering caused by the acoustic phonons. Although the scattering rate for the polar phonon modes is decreasing with increasing electron energy, the distribution is stabilized against velocity runaway by the rapid decrease in the path length between collisions due to the increasing scattering rate of the nonpolar phonon modes.

As deduced from average energy field characteristics, a strong dependence on oxide thickness for electron transport in films $\gtrsim 100$ Å has not been observed from the main portion of the electron energy distribution. As shown in Figure 7, however, the high energy tail of the distribution does extend to higher energies with increasing SiO_2 thickness. This dependence approximately scales with oxide thickness and shows that a small number of electrons can gain energies exceeding the SiO_2 bandgap, at least for films thicker than 500 Å.[3] The conventional Monte Carlo approach, which gives the best agreement with low field data near the threshold, cannot predict these high energy tails or their SiO_2 thickness dependence.[5,6] Recently, calculations using a quantum Monte Carlo approach which will be discussed later give better agreement with the measured energy distributions at high fields, particularly in predicting these tails.[7] Although very hot carriers can exist in thicker oxides from the data in Figure 7, there is still little direct evidence that they trigger impact ionization and consequently destructive breakdown of these films.[3] However, average breakdown fields for MOS structures are observed to decrease with increasing oxide thickness, particularly on films $\gtrsim 1000$ Å.[1]

Besides SiO_2, hot electron transport has been studied in other insulators which are important to the semiconductor industry.[3,9,27] These materials include silicon nitride (Si_3N_4), silicon oxynitride (SiO_xN_y), and Si-rich SiO_2 (SRO). The electron transport in all these wide bandgap materials is dominated through the interaction of the carriers with trapping sites formed in the forbidden bandgap by defects (as in Si_3N_4 and SiO_xN_y) or by tiny Si islands (as in Si-rich SiO_2). The electrons spend a significant fraction of time in the traps before thermalizing or tunneling to the conduction band or a neighboring trapping site. This type of trapping/detrapping motion of the carriers leads to the very low electron mobilities observed in these insulators and generally associated with electrons or holes in most amorphous materials besides SiO_2. However, some fraction of the distribution which can escape trapping over a distance comparable to the energy relaxation length could be expected to display transport characteristics similar to those seen for hot electrons in SiO_2, at least over a limited field range. As shown in Figure 8, this situation is realized in silicon nitride and silicon oxynitride MOSs where vacuum emission has been used.[9] This figure clearly shows that the hot electron distributions in these materials are similar to that of SiO_2 even though very few carriers are energetic enough to be emitted into vacuum. The oxynitrides have fewer traps due to decreased nitrogen incorporation and show more hot electrons. Similar results have also been reported for Si-rich SiO_2 layers where each Si island forms a deep potential well in the oxide bandgap.[3,27] Fewer hot carriers are seen in the oxide conduction band with increasing silicon content in these films.

IV. BALLISTIC TRANSPORT

As the SiO_2 thickness is further reduced, a nonsteady-state condition is reached where some of the electrons of the distribution are not energetic enough to be scattered by acoustic or nonpolar phonon modes.[1,4] This occurs for oxides under 100 Å in thickness. (Nonpolar scattering becomes effective above ≈ 2 eV.) The characteristic length at which this occurs is a measure of the average distance through which an electron of the distribution must move, starting from thermal energy, in order to reach a steady-state condition. This length, which is ≈ 30 Å, can be regarded as the "heat-up" distance.[1] Similarly, if hot electrons are injected into an oxide at a low electric field conditon (≤ 1.5 MV/cm) one would expect to measure a characteristic "cool-down" distance of similar magnitude. This distance has also been deduced to be ≈ 25 to 35 Å from internal photoemission measurements where photons

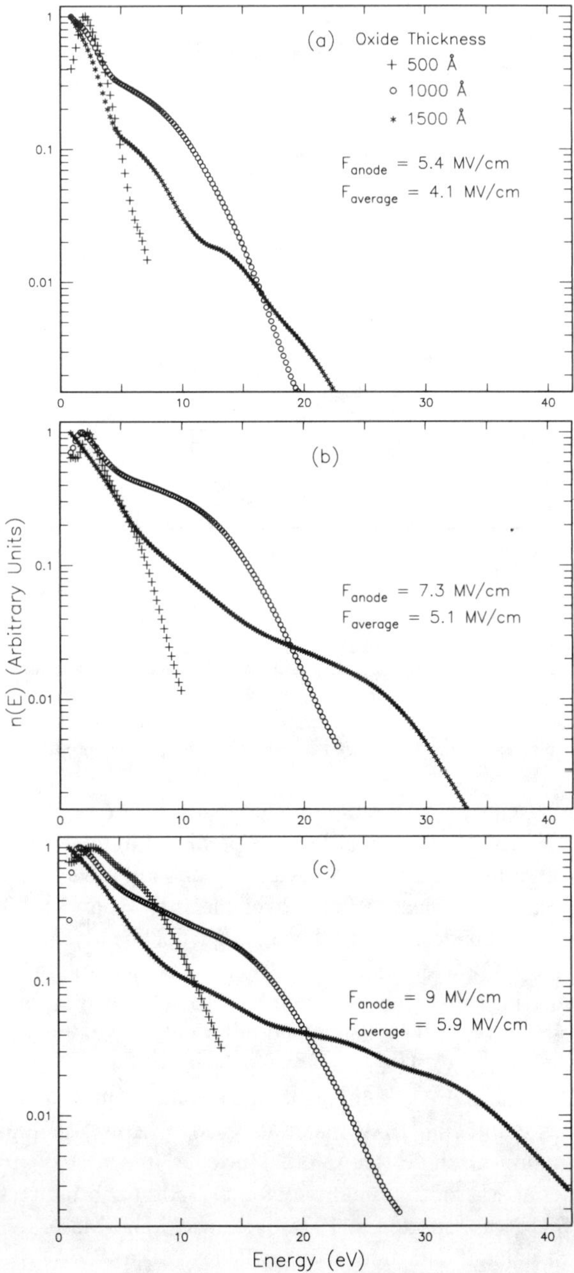

FIGURE 7. Comparison of electron energy distributions for varying oxide thicknesses (500 to 1500 Å) determined using vacuum emission at room temperature for several electric fields. Although the average energies deduced from these distributions are not strongly dependent on oxide thickness, a few ''lucky'' electrons in the high energy tail are not stabilized against runaway.

are used to inject hot electrons.[21] For the determination of both of these lengths, the distribution is in a nonsteady-state condition and at energies low enough so that LO phonon modes dominate the hot electron scattering.

Decreasing the oxide thickness even more results in a condition where distances are so short that some electrons will not collide with any phonon mode, either polar or nonpolar.

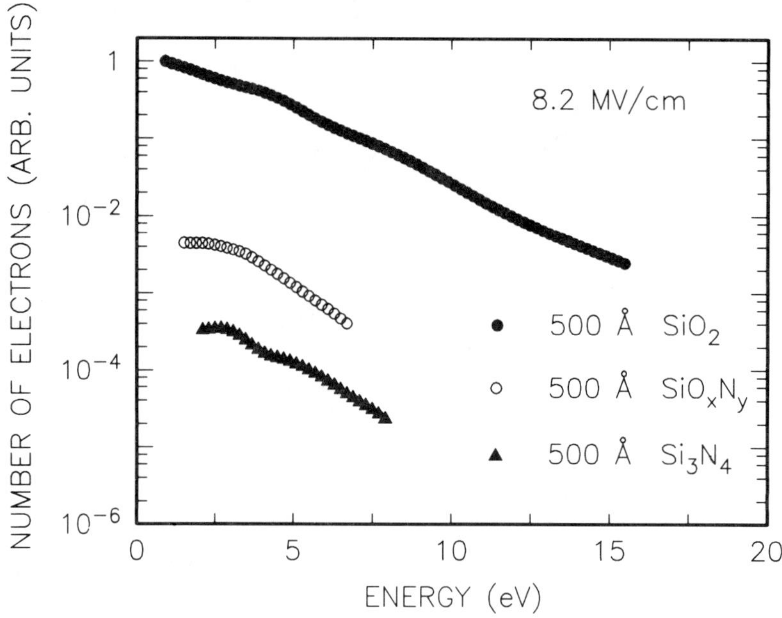

FIGURE 8.   Comparison of the electron energy distributions for varying insulators (SiO_2, Si_3N_4, and SiO_xN_y with 39% atomic nitrogen) determined using vacuum emission at room temperature and high electric fields. These data show a strong decrease in the number of hot carriers as trapping in the insulator increases and controls transport through the film.

These electrons, which can gain kinetic energies equal to the applied potential they fall through, are called "ballistic". This situation was predicted in the Monte Carlo simulations of Figure 6 where scattering does not occur with any phonon mode (as indicated by a lack of displacement normal to the electric field) over the first 10 to 30 Å distance along the field direction. When the oxide layer is thin enough that nearly all the electrons in the distribution traverse the SiO_2 layer ballistically, a characteristic length called the mean-free path can be measured with values ranging from 7 to 15 Å depending on the oxide field.[4,22] The remainder of this section will be concerned only with transport in the ballistic or near ballistic regime in very thin SiO_2 layers.

Currently, ballistic transport is a topic of much interest in potentially fast switching semiconductor devices employing III-V material based systems[28] such as gallium arsenide (GaAs)/gallium aluminum arsenide (GaAlAs). However, the first observation of ballistic electrons in any material was in thin amorphous SiO_2. Although indirect, this first experimental evidence in SiO_2 was reported in 1975 by Lewicki and Maserjian.[22] They observed oscillations in current-voltage data for very thin layers (30 to 75 Å) of SiO_2 incorporated into MOS capacitors. These oscillations were interpreted by them as the result of quantum reflection and interface of ballistic electron wave functions at an abrupt anode interface which was formed by the single-crystal Si substrate and the thin thermally grown SiO_2 layer. They did not observe similar oscillations when the chromium gate electrode interface with the SiO_2 layer was used for the anode.

Data showing oscillations in I-V characteristics due to quantum reflection of ballistic electrons at either interface of a MOS structure are shown in Figure 9. This structure is similar to that used by Lewicki and Maserjian except for the gate electrode material which is degenerately doped n-type polycrystalline silicon. Other gate materials, such as aluminum, which are known to react with the oxide layer and give a less abrupt transition region with SiO_2 show greatly reduced amplitudes for the oscillations when that interface is used as the anode.[4] These I-V oscillations should not show a strong temperature dependence if the

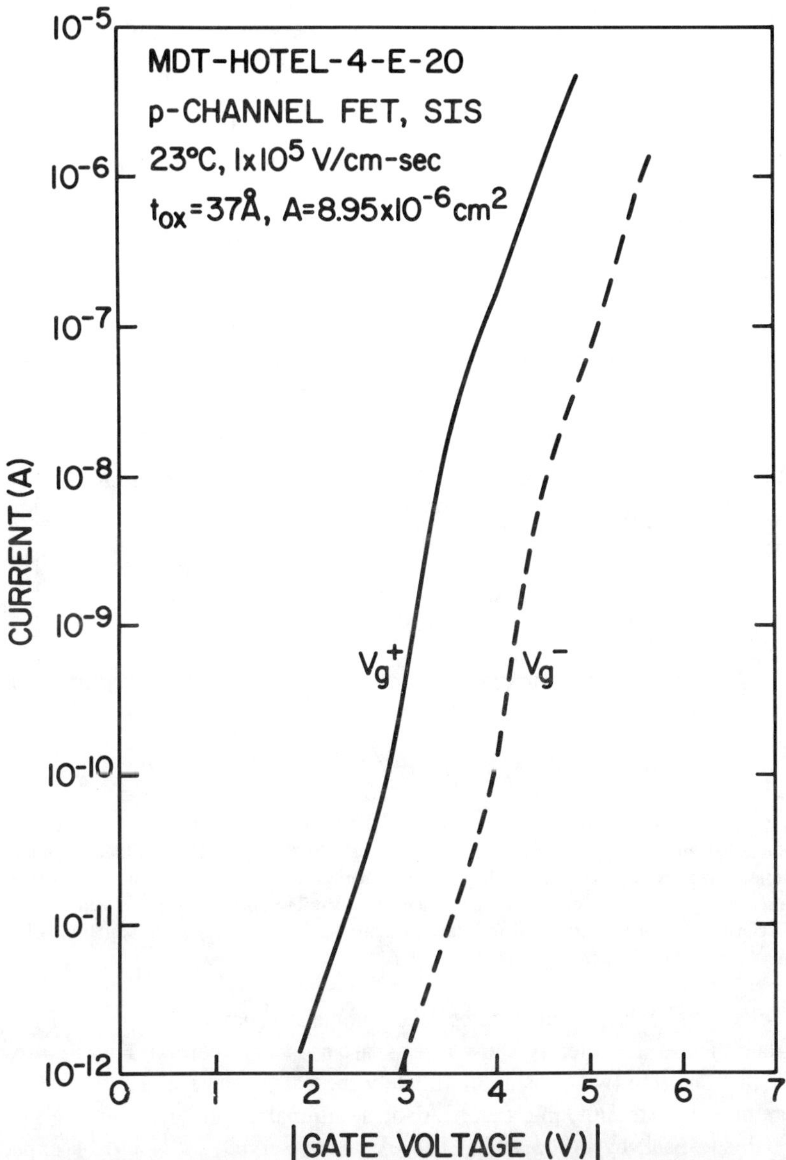

FIGURE 9. Current-voltage characteristics at room temperature of very thin oxide MOS showing oscillations due to quantum interference (reflection at the anode/SiO$_2$ interface) of ballistic-electron wave functions. Oscillations in these tunneling characteristics are observed for either voltage polarity due to the abrupt nature of the anode interface of either the single-crystal substrate (V_g^-) or polycrystalline gate (V_g^+) silicon with the SiO$_2$ layer.

dominant scattering mechanism (which destroys phase coherence or randomizes the path lengths of ballistic electrons) is unaffected over the temperature range of interest. Figure 10 shows this on normalized I-V data for temperatures of 4.2 and 300 K and compares these data to Monte Carlo simulations using the assumptions for lattice scattering previously discussed. The key assumption again is that energy loss is dominated by emission of an LO phonon with energies of 0.153 or 0.063 eV whose population is not strongly affected over this temperature range.

With the use of vacuum emission and carrier separation techniques, ballistic transport can be directly measured in SiO$_2$.[4,29] Hot electron distributions emitted from a thin oxide

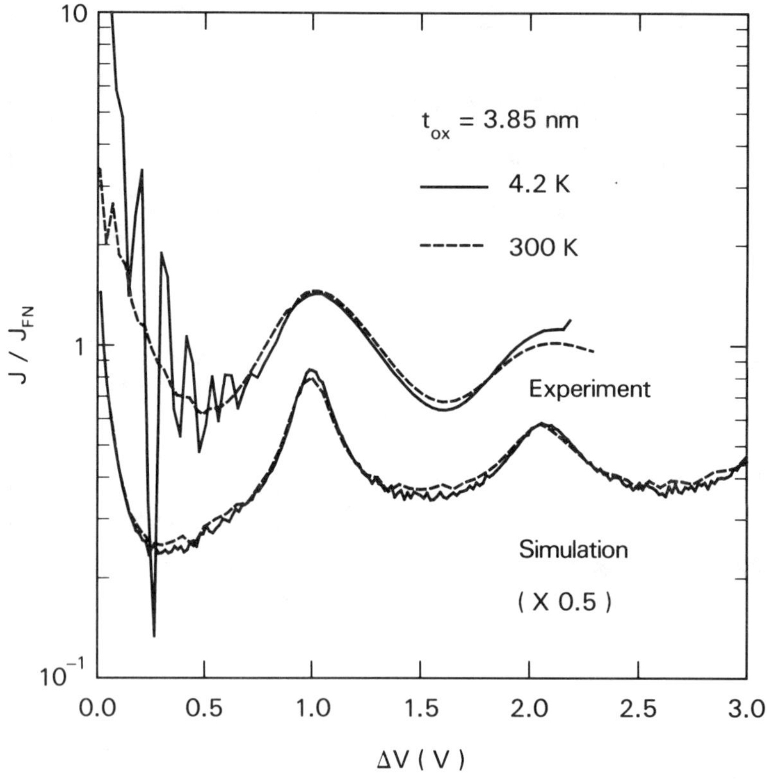

FIGURE 10.   Comparison of data (from I-Vs) to Monte Carlo simulations for the ratio of the actual current density to that predicted by Fowler-Nordheim tunneling (without quantum reflection) as a function of the voltage dropped across the remaining portion of the oxide after tunneling at room and liquid helium temperatures. The small temperature dependence is consistent with that predicted for electron transport dominated by polar scattering with high energy LO phonons.

(51 Å) MOS at 83 and 297 K with a fixed gate voltage are shown in Figure 11 from vacuum emission data. These data clearly show a peak at an energy (here, 1.1 eV) corresponding to the ballistic condition (which is approximately the gate voltage minus the 3.1 eV energy barrier for the substrate/SiO_2 interface). Also, additional peaks at lower energies shifted from the ballistic peak by 0.063 and 0.153 eV are observed, as would be expected, for single phonon scattering by the two dominant LO phonon modes. This is confirmed in Figure 12 by the comparison of vacuum emission data on an MOS structure with a 56-Å thick SiO_2 layer to Monte Carlo simulations. Peaks at energies less than the ballistic peak correspond to single phonon *emission* while those at larger energies, but with greatly reduced amplitudes, correspond to the less probable *absorption* of single LO phonons by the hot electrons. An example of this latter type of phenomenon is demonstrated by the shoulder observed in both the data and simulation in Figure 12 which is shifted up in energy from the ballistic peak by 0.153 eV, particularly at 297 K.

The temperature dependence of the slope of the data at energies above the ballistic peak in Figure 12 is shown to replicate the source function of tunneling electrons at the substrate Si/SiO_2 interface.[29] As shown by the agreement with the simulations which will be discussed later, this source function is mainly due to the Fermi distribution of electrons at this interface multiplied by their tunneling probability through the interfacial barrier. Consistent with the Monte Carlo simulations, the structure, due to the ballistic peak and phonon-induced sidebands, becomes less pronounced with increasing temperature broadening due to the increase of nonpolar phonon modes.

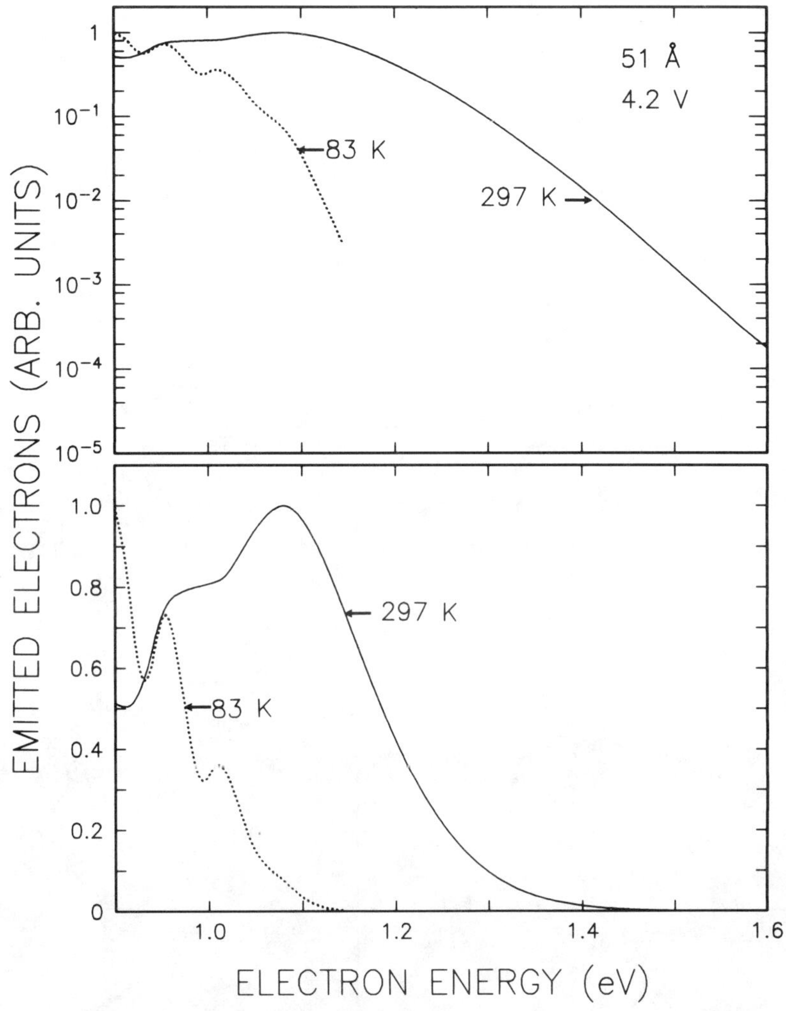

FIGURE 11.   Comparison of electron energy distributions on a very thin oxide MOS at two different temperatures in the ballistic regime determined using vacuum emission at a fixed gate voltage.

Other trends consistent with the Monte Carlo simulations are observed in data for the distributions when the voltage bias is changed. As shown in Figure 13, for fixed lattice temperature the structure, attributed to the ballistic peak and phonon-induced sidebands, shifts by the change in applied voltage. With thinner oxides or lower voltages, the amplitude of the ballistic peak increases with respect to that of the sidebands while the opposite trend is observed with thicker films or higher voltages (see Figures 11 through 13). Modulation of the ballistic peak due to quantum reflection at the SiO_2/gate interface has also been observed as predicted in our simulations and seen in the I-V data.[4] A gradual disappearance of the structure and broadening in the energy distribution, as would be expected for the ballistic to steady-state transition, is seen as the oxide thickness approaches 100 Å.

The only clear discrepancy between the data and theory in Figures 12 and 13 occurs at low temperature and electron energy.[29] The calculations used in the Monte Carlo simulation solve exactly for the three-dimensional electron energy distribution at the Si/SiO_2 interface and the electron tunneling probability into the SiO_2 layer. Although corrections for the quantum reflections in the oxide layer are taken into account, we have made no attempt to

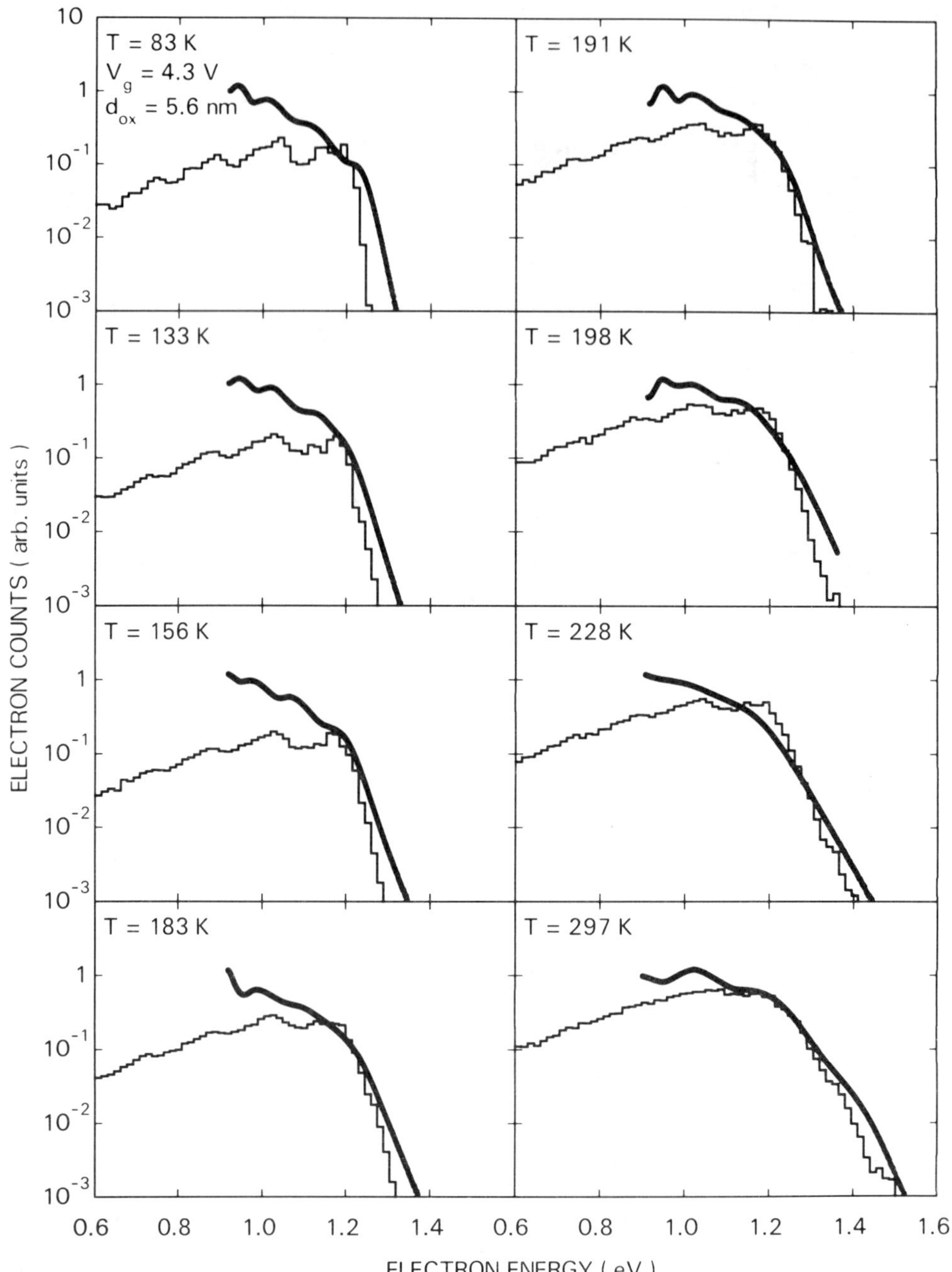

FIGURE 12.   Comparison of vacuum emission data with Monte Carlo simulations for the energy distribution of a very thin oxide MOS in the near ballistic regime at various temperatures under fixed gate voltage. These data show the temperature broadening of the ballistic peak, phonon-induced sidebands, and Fermi source function at the Si/SiO$_2$ interface.

include contributions to the electron source function from lower lying, two-dimensional quantum levels in the Si accumulation layer. Qualitatively, the low energy structure in the hot electron distribution at low temperature behaves in a manner consistent with the contributions expected from these quantum states.

Unless the thin top gate electrode is cesiated, vacuum emission cannot be used to study

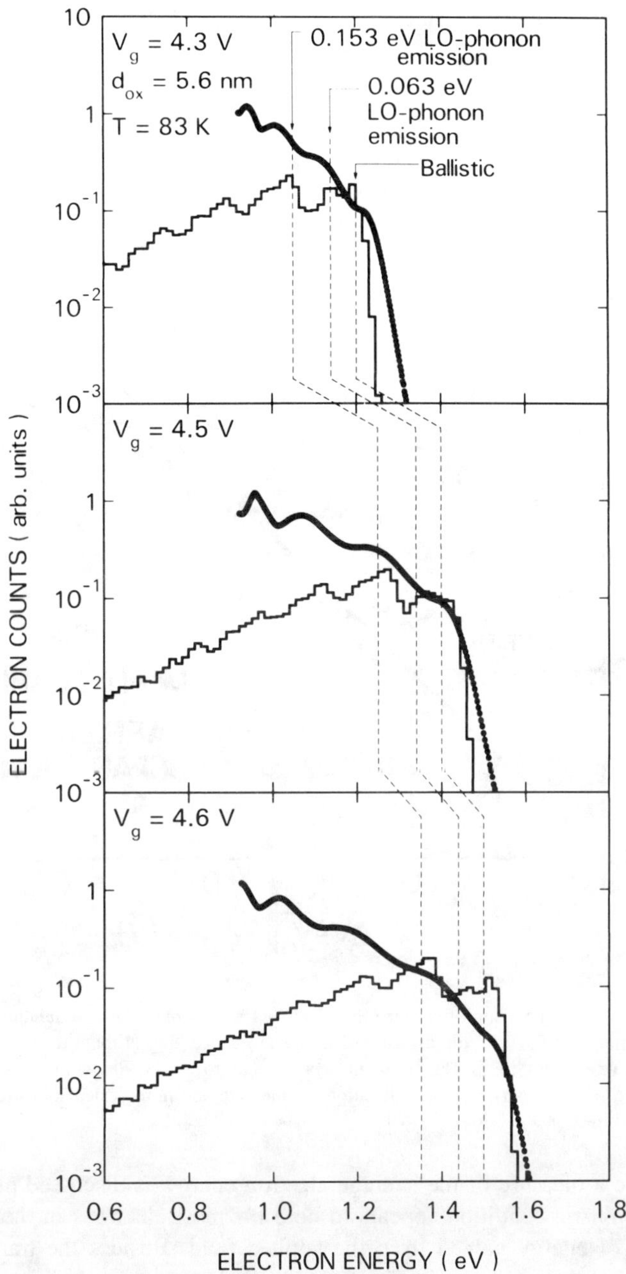

FIGURE 13. Comparison of vacuum emission data with Monte Carlo simulations for the electron energy distributions of a very thin oxide MOS in the near ballistic regime near liquid nitrogen temperature. The ballistic peak and phonon-induced sidebands are shown to track the applied gate voltage.

hot electron behavior at voltages dropped across the oxide layer of $\gtrsim 0.9$ eV (which is the electron affinity of SiO_2). However, carrier separation can be used to detect ballistic behavior even at voltages where electrons tunnel from electrode to electrode without emerging in the oxide conduction band[4] (called direct tunneling). These measurements, such as those shown in Figure 14, also further confirm the existence of ballistic transport in very thin SiO_2 layers,

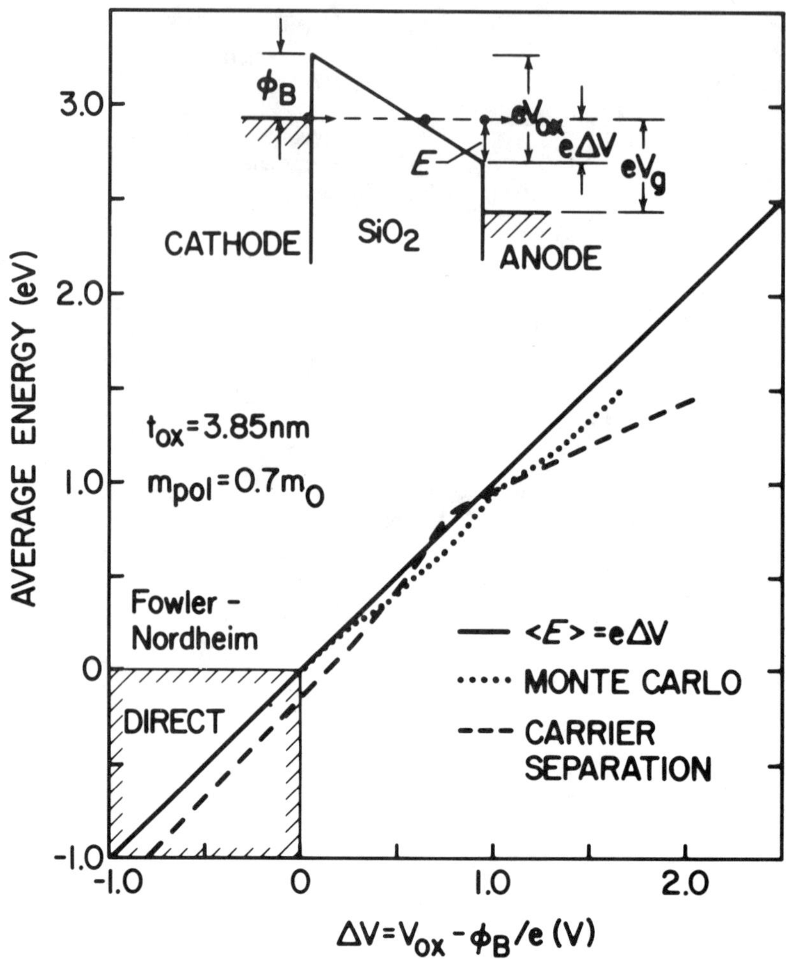

FIGURE 14.   Comparison of carrier separation data with Monte Carlo simulations for the average energy of the electron distribution vs. oxide voltage drop at room temperature on a very thin oxide MOSFET. The solid line corresponds to the condition for pure ballistic transport of the hot carriers. The undulations in the data and in the calculations are due to quantum reflection.

but can only give a measure of the average electron energy as discussed previously. From these data, the ballistic condition appears to hold for most electrons in the distribution for the first 0.8 eV of energy gained from the applied field. Besides the transition from the ballistic to the steady-state regime in Figure 14, ballistic electron behavior is observed from this data in the transition region from direct tunneling (gate into Si substrate) to Fowler-Nordheim tunneling (gate into SiO_2 conduction band). Undulations in the energy voltage data, which appear in the Monte Carlo simulations, are due to quantum reflection at the contact/oxide interfaces as previously described.[4]

V. TRAPPING AND TRAP GENERATION

Electron heating should have an influence on the trapping/detrapping kinetics of as fabricated, preexisting defect or impurity sites in SiO_2. Hot electrons would be more difficult to capture than thermal carriers if their path lengths remained unchanged. However, dispersive transport caused by large angle scattering of hot carriers should increase the number

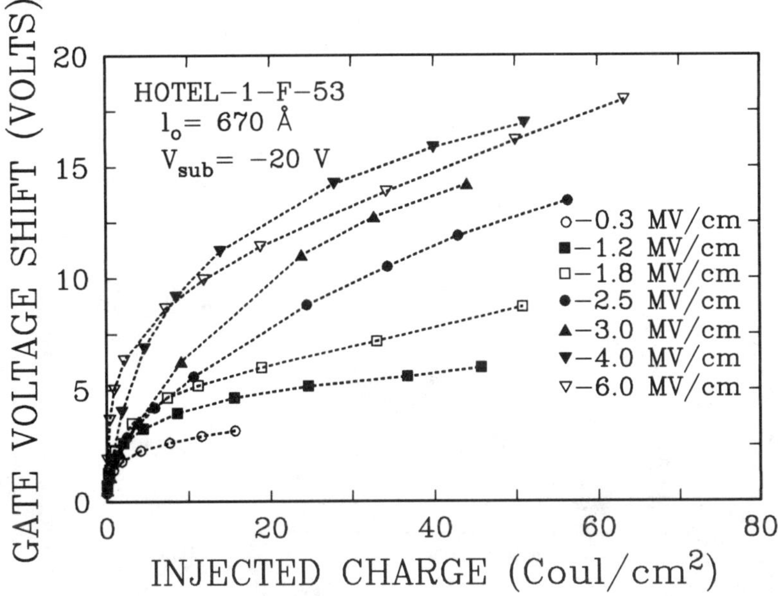

FIGURE 15. Electron trapping characteristics at room temperature on MOSFETs with a thick oxide layer for several different average electric fields. These data show a sudden increase in trapped electron build-up for fields above the heating threshold of ≈ 1.5 MV/cm.

of passes they make through the potential of a trapping site, thereby increasing their capture probability. Hot carriers would have more tendency to impact ionize trapped charges or produce new trapping sites by network modification through their energy loss during scattering. Destructive breakdown of SiO_2 after current passage at high electric fields could be triggered by this trap creation and/or network modification. In this section, these technology related issues concerning the presence of hot electrons in oxides will be reviewed.

Although many of the results were controversial, several groups during the 1980s have claimed to observe trap generation in SiO_2 at high electric fields.[17] In some cases, background trapping in the as fabricated oxides was not separated out from trapping in the generated sites. In other cases, interface charging (where the charge state can be either positive or negative depending on the Si/SiO_2 interface Fermi level position) was not clearly separated from electron trapping in the oxide bulk. Since many of these researchers looked at only a limited range of high electric fields, the effects due to the generated traps were not large. Although some of these researchers discussed the possibility of trap creation by hot electrons, heating characteristics like that shown in Figure 4 were not available until 1986.

Recently, a direct correlation of trap generation (both interface and bulk) with hot electrons in SiO_2 has been demonstrated.[17] This is shown in Figures 15 through 17 where the charge build-up is studied from voltage shifts of both current-voltage and capacitance voltage characteristics. At the threshold field for heating (≈ 1.5 MV/cm), there is a sudden increase in electron trapping, particularly for injected electron fluence $\gtrsim 1$ C/cm^2. From data like that shown in Figure 15, the number of generated traps as a function of an electric field can be shown to follow a trend similar to the heating results in Figure 4. Positive interfacial charges and interface states can be separated from bulk electron charging by using both C-V and I-V characteristics.[12] As shown in Figure 16 for interface states, their behavior as a function of field is similar to that in Figure 15 for bulk electron traps and therefore, also similar to the heating characteristic in Figure 4.

The data in Figure 4 on thick oxides do not unequivocally determine whether the electric field (for example, by polarization) or the hot electron energy is directly responsible for trap

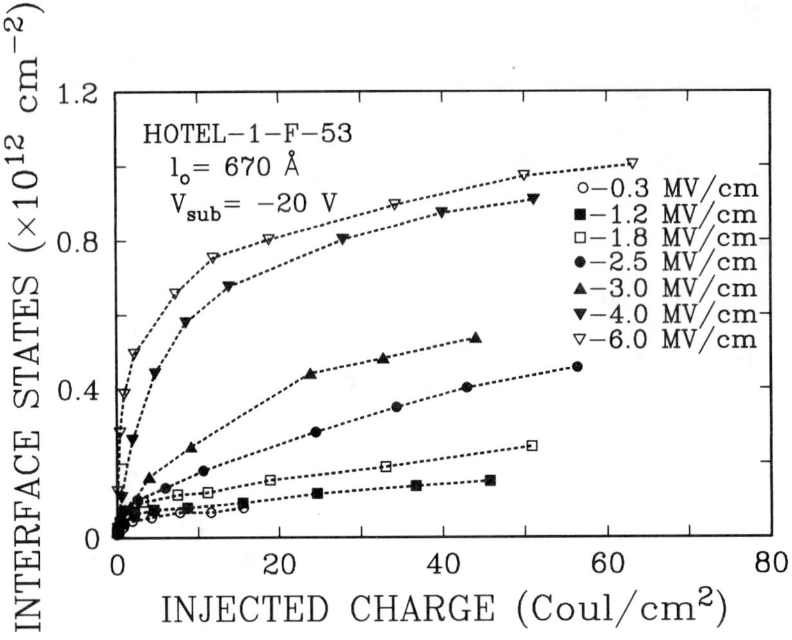

FIGURE 16.   Number of Si/SiO$_2$ interface states at room temperature as a function of injected charge on the same devices as in Figure 15 for fields from 0.3 to 6 MV/cm.

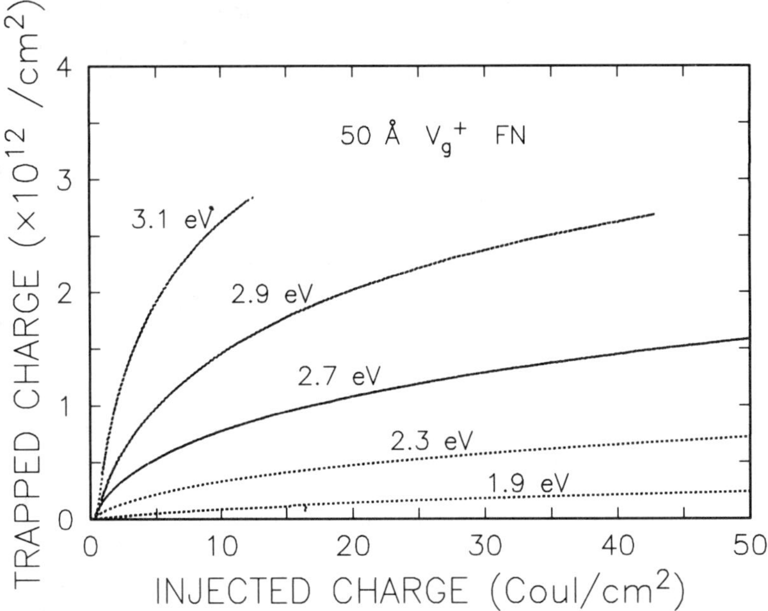

FIGURE 17.   Number of trapped electrons in bulk sites created by hot electrons (injected by Fowler-Nordheim tunneling) in devices with a very thin SiO$_2$ gate insulator as a function of injected charge at room temperature for various average energies of the electron distribution. This figure shows that trap creation starts to occur at average hot electron energies $\geq$2 eV.

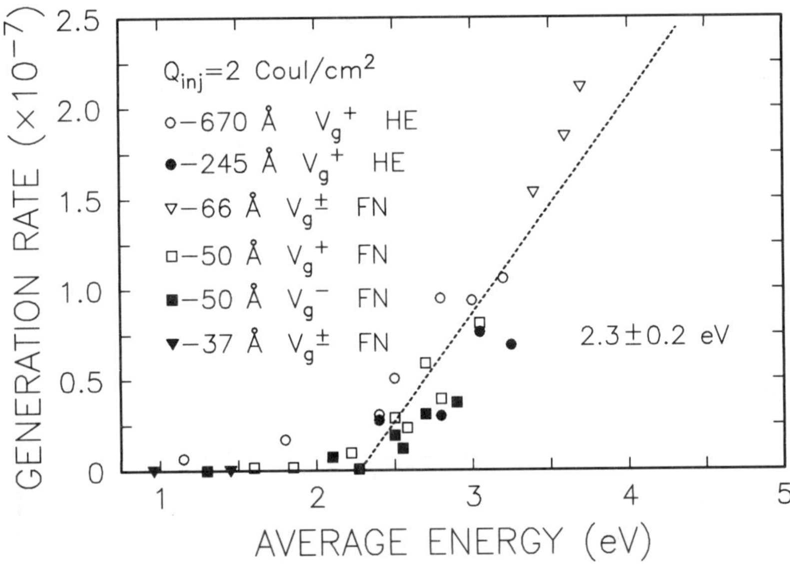

FIGURE 18.   Trap generation rate (traps created per injected electron) in SiO_2 as a function of average electron energy using Fowler-Nordheim tunneling or optically induced hot electron injection. These data show a minimum energy of about 2.3 ± 0.2 eV is necessary for site creation, approximately independent of voltage polarity, means of injection, and oxide thickness.

creation. By additionally studying charge build-up in very thin oxide layers, this separation can be made. As previously discussed in the section on ballistic transport, electrons injected into thin oxides at high electric fields still require a distance of ≈30 Å to reach a steady-state condition. By varying the oxide thickness, the average energy of a near ballistic distribution impinging on the anode/SiO_2 interface can be changed from 0 to 4 eV almost independently of the field. Using devices with 50-Å thick gate oxides, Figure 17 shows that a threshold energy near 2 eV is needed for the observation of trap creation. Trap generation rate as a function of electron energy is plotted for several oxide thicknesses, injection conditions, and both voltage polarities in Figure 18. From a straight line fit of these data above 2 eV, a minimum electron energy of ≈2.3 ± 0.2 eV for trap creation can be deduced. These results in Figure 18 imply that the bulk of the SiO_2 layer is not necessary for the generation of traps or interface states, at least for the oxide layers which have undergone extensive high temperature processing associated with poly-Si gate technologies.

Consistent with our results, other recent studies of hydrogen evolution during electron injection of SiO_2 layers,[30] suggest that a possible cause of trap creation in SiO_2 is the cracking of hydrogen bound to impurity sites near the anode interface. Some form of mobile hydrogen can then diffuse into the oxide layer to produce the sites causing the observed charging phenomenon. Similar arguments have been made for trap creation in polymer films exposed to hot electrons.[11] In Figure 19, we have tested the idea that a mobile species released by hot electrons causes trap generation. These data show that the generation rate is greatly inhibited at lattice temperatures below 150 K independent of oxide thickness and voltage polarity. These experiments and similar ones (where a device is sequentially cycled between high and low temperatures) are consistent with the release of a mobile species from the anode/SiO_2 interfacial region limiting the observed temperature dependence, once activated or cracked by hot electron impact.

Hot electrons would be expected to have different capture kinetics than thermal carriers which move at lattice temperatures near the bottom of the oxide conduction band edge.

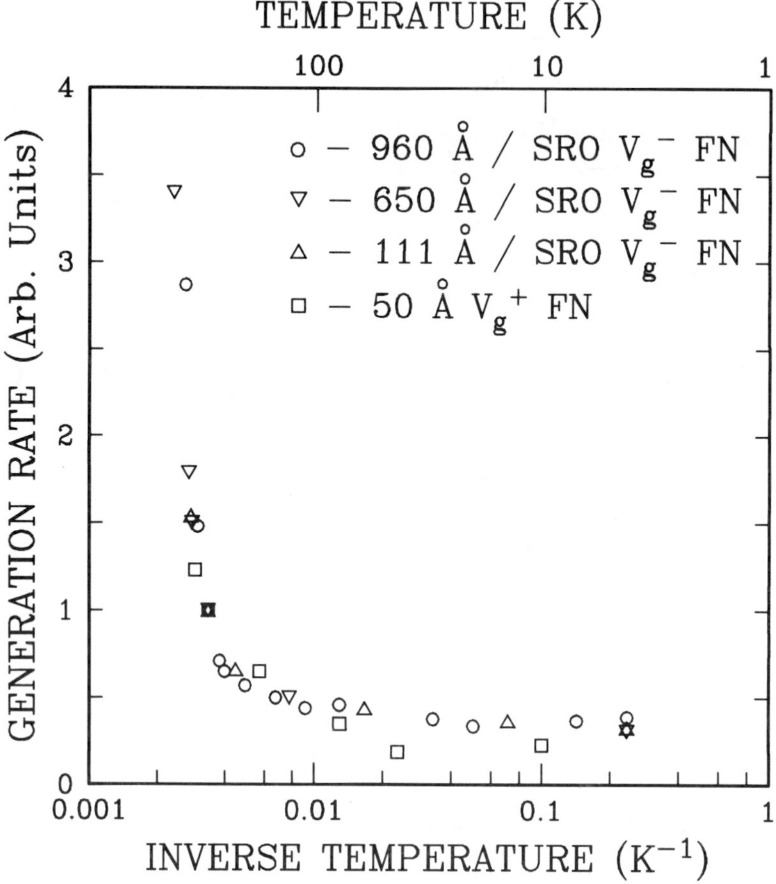

FIGURE 19.   Arrhenius plot (trap generation rate vs. inverse temperature) which was produced from the slope of trapping data such as that in Figure 17 where Fowler-Nordheim tunneling was used to inject carriers. These data show a universal temperature dependence which is not dependent on oxide thickness or voltage polarity, similar to the observations for the cracking energy characteristic in Figure 18. From this curve, it can be inferred after the liberation by hot electrons that the onset of motion away from the anode/SiO$_2$ interface of one or more mobile species occurs at temperatures above 150 K.

Since capture potentials in SiO$_2$, depending on the site, can be Coulomb attractive, neutral, or Coulomb repulsive, each specific case is different.[12,31] For example, a Coulomb attractive site can be created by initially trapping a positively charged hole on a neutral bulk or interfacial site. Capture of electrons subsequently injected into the oxide to annihilate these trapped hole sites can then be studied. This has been done experimentally where an inflection point in the capture cross section vs. electric field data (where the slope changes from ≈ -1.5 to -3.0) has been seen at a field magnitude comparable to the 1.5 MV/cm threshold for electron heating.[12,31] Capture into initially neutral sites (covering the same field range) has also been studied experimentally and does not show a strong dependence on electric field or a pronounced change near the threshold for heating.[12,31]

Preliminary comparisons to Monte Carlo simulations for Coulomb attractive and neutral site capture show reasonable agreement with the experiments.[31] These calculations take into account the heating, increase in path length due to nonpolar scattering above the heating threshold, Coulomb volume shrinkage, creation of bound states in the potential wells of the sites, and field/thermal ionization out of these centers during capture. These calculations do not take into account impact ionization of captured electrons by the hot carriers. There is some experimental evidence for this less probable phenomenon, but for specific energetically shallow trapping sites.[12] For Coulomb repulsive sites, an increase in the magnitude of the capture cross section with increasing electron energy followed by relative insensitivity to further increases in the electric field is expected. This is consistent with the idea of an energy barrier for near thermal electrons with increasing tunneling probability into the potential well of the site as the electronic carriers heat-up. At electron energies greater than the barrier energy, the hot electron capture should be similar to capture into a neutral site, particularly for a dipole-like center.

Over the past 20 years, there have been numerous researchers trying to connect destructive breakdown of SiO_2 with electron heating. Most of this work has been indirect with few attempts, until recently, to correlate heating characteristics with oxide deterioration. One of the most popular of these "breakdown models" involves impact ionization of valence electrons of the lattice atoms.[18,19] As previously discussed, this would require electronic carriers to gain at least >9 eV with respect to the bottom of the conduction band. This is not predicted theoretically and has only been observed experimentally in some of the high energy tails of hot electron distributions in oxides thicker than those currently of technological interest. More recently, cumulative trap creation by hot electrons and the subsequent charging of these generated sites have been argued to be responsible for causing destructive breakdown.[32] Again, the evidence is not conclusive and certainly cannot explain why very thin oxide films where there is no trap creation still breakdown at comparable electric field strengths.[1] The status of destructive breakdown in SiO_2 is still an open question and may simply be related to hot electron induced and/or electric field enhanced chemical or physical changes occurring locally at a defect or impurity site in the oxide bulk or more likely near an interface.

VI. THEORY OF HIGH FIELD ELECTRON TRANSPORT

In previous sections, we have described the experimental evidence supporting our "new" understanding of electron transport in SiO_2 or, even more generally, in large bandgap materials. In this section we will discuss in more detail the physics of the electron lattice scattering processes which we have included in our Monte Carlo simulations.

A. POLAR COLLISIONS

Since insulators are usually ionic materials, the polar electron-phonon interaction is one of the dominant processes by which the electrons lose energy to the lattice. The interaction is due to the coupling between the electron and the dipole field associated with the LO phonons of the crystal, the weaker field associated with the transverse optical (TO) phonons being of negligible importance. In SiO_2, two LO phonons are coupled to the electrons: a "high energy" phonon of energy $\hbar\omega_{LO,1} = 0.153$ eV and a "low energy" phonon of energy $\hbar\omega_{LO,2} = 0.063$ eV. The frequency at which an electron of wave vector of magnitude k scatters with each LO phonon of type i, $1/\tau_{LO,i}(k)$, can be computed from the Fröhlich-Hamiltonian matrix element:[20,33]

$$M_{i,polar}(\mathbf{q}) = \frac{ie(\hbar\omega_{LO,i})^{1/2}}{2q} \left(\frac{1}{\epsilon_{high,i}} - \frac{1}{\epsilon_{low,i}} \right)^{1/2} \left(n_{LO,i} + \frac{1}{2} \pm \frac{1}{2} \right)^{1/2} \tag{1}$$

where $\mathbf{q}$ is the wave vector of the emitted/absorbed phonon, e the magnitude of the electron charge, $n_{LO,i}$ is the occupation number of the LO phonon of type i involved in the process, $\epsilon_{high,i}$ ($\epsilon_{low,i}$) is the dielectric constant at a frequency larger (smaller) than the phonon frequency $\omega_{LO,i}$, and $\hbar$ the reduced Planck's constant. The key characteristics of this coupling (due to the Coulomb nature of the interaction) is the phonon wave vector in the denominator. This implies that the electron lattice collisions will occur most likely at small angles. Since $\mathbf{q} =$

$\mathbf{k}_{before} - \mathbf{k}_{after}$ is also the change of the electron wave vector, $\mathbf{k}$, before and after the collision, small $\mathbf{q}$s imply small changes of the electron direction. Moreover, even accounting for the widening of the region of the phase space available to the final states as the electron energy increases, the coupling between electrons and dipole models decreases. As we have mentioned before, this will eventually lead to the intrinsic instability associated with this interaction (runaway, breakdown, etc.). This is clearly seen by computing the scattering rate derived from the matrix element (Equation 1) using a spherical band model:

$$\frac{1}{\tau_{LO,i}(k)} = \frac{e^2 m^* \omega_{LO,i}}{4\pi\hbar^2 k} \left(\frac{1}{\epsilon_{high,i}} - \frac{1}{\epsilon_{low,i}} \right) \log \left(\frac{q_{max,i}}{q_{min,i}} \right) \left(\frac{1}{2} \pm \frac{1}{2} + n_{LO,i} \right) \tag{2}$$

where $q_{max,i}$ and $q_{min,i}$ are the maximum and minimum phonon wave vectors allowed by momentum and energy conservation (i.e., $q_{max,i} = k(w) + k(w \mp \hbar\omega_{LO,i})$, $q_{min,i} = |k(w) - k(w \mp \hbar\omega_{LO,i})|$, where $k(w)$ is the wave vector of an electron of energy w), and m^* is the electron effective mass. The upper and lower signs refer to phonon emission and absorption, respectively.

We show in Figures 20a and 20b the total (polar and nonpolar) electron phonon scattering rates at two different lattice temperatures. The coupling parameters chosen for the polar rate, Equation 2, are given by:[4-7] $1/\epsilon_{high} - 1/\epsilon_{low} = 0.143 \; \epsilon_0^{-1}$ for the 0.153 eV LO-phonon, $= 0.063\epsilon_0^{-1}$ for the 0.063 eV LO-phonon. The value employed for the effective mass m^* will be discussed below. As seen in these figures, at electron energies above a few tenths of an eV, the polar scattering rate begins to decrease as $1/k$ since the interaction between the fast electron and the dipolar molecules of the crystal is driven off the resonant response of the molecules themselves. This feature has been the starting point for breakdown models in polar insulators since the pioneering work by Fröhlich:[20] If the field, F, applied to the insulator is small enough so that between two successive collisions the electrons gain less kinetic energy than the LO phonon energy (that is, in a simplified picture, $F \lesssim F_c = \hbar\omega_{LO}/eu\tau_{LO}$, with u the average group velocity), transport would occur in steady state with the electrons being kept at energies below a few tenths of an eV. But at fields larger than the critical field F_c, the electrons can be driven above the energy corresponding to the maximum of the collision rate $1/\tau_{LO}(w)$. In this regime, between successive collisions the electrons gain from the external field kinetic energy in excess of the phonon energy. As they are driven to even higher energies, the energy loss rate decreases even further, and the resulting instability (referred to as *velocity runaway*) develops.

B. NONPOLAR COLLISIONS

As we stressed quite a few times above, the essential ingredient to explain our experimental data is the introduction of nonpolar electron-phonon collisions. This is particularly true in the case of SiO_2 due to the moderate ionicity of the Si–O bond as compared, for instance, to the fully ionic bonds of the alkali halides. To understand the essential difference between polar and nonpolar scattering, it is useful to note that the nonpolar coupling between the flying electron and the lattice ions displaced from their equilibrium position is described by the matrix element:

$$M_{nonpolar}(\mathbf{q}) = \frac{i\hbar^{1/2} S_q}{(2\rho\omega_q)^{1/2}} \left(n_q + \frac{1}{2} \pm \frac{1}{2} \right)^{1/2} \tag{3}$$

where ρ is the density of the material, and ω_q and n_q are the frequency and occupation number of the (acoustic or optical) phonons. Notice the difference between Equations 1 and 3. In the latter, the phonon wave vector, $\mathbf{q}$, enters via the coupling constant S_q. At low energy (deformation potential approximation[34]), it is customary to consider the electron as

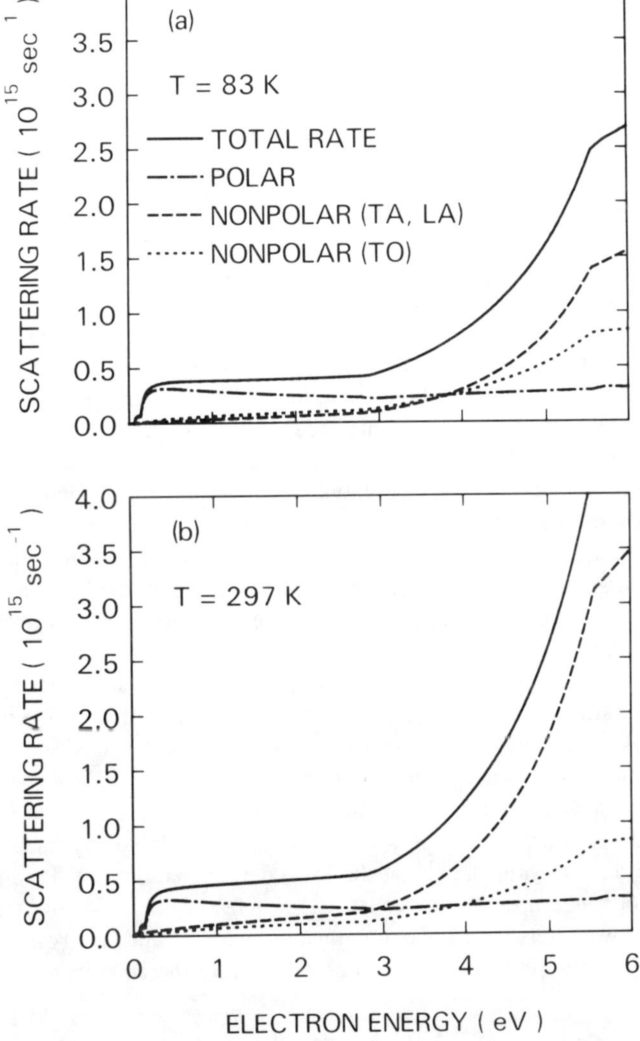

FIGURE 20. Electron-phonon scattering rates in SiO$_2$ as a function of electron energy at two temperatures. The parameters employed are described in the text.

scattered by the energy variation of the bottom of the conduction band as acoustic waves change the interionic distance. The shorter the wavelength, the larger the change of energy going from one unit cell to the other, so that $S_q \simeq C_{ac}q$, with C_{ac} being the deformation potential (i.e., S_q is the energy variation of the bottom of the conduction band as we change the lattice constant by a unit length). At higher energies we must rely on some amount of "guessing", since little is known about the high energy electron-phonon interaction. In our original Monte Carlo simulation, we have followed the ideas of Sparks et al.[25] and Ziman[35], suitably modified for the case of SiO$_2$. We have expressed the coupling constant S_q in terms of the differential scattering cross section σ_q related to the form factor of the lattice:[35]

$$|S_q|^2 = \frac{\pi \hbar^4 N^2 q^2}{m^{*2}} \sigma_q \tag{4}$$

where N is the atomic density. We have further assumed that the scattering cross section

σ_q is independent of $\mathbf{q}$ and is dominated by the larger oxygen ions. A value of 3.5×10^{-15} cm^2 has been assumed, yielding $S_q/q \simeq 3.5$ eV. The same value was originally assumed for the low energy deformation potential, C_{ac}. An alternative way to view this approximation, as suggested by Blatter and Baeriswyl,[36] is to approximate the short-range potential, V_q, of the displaced ions as seen by the high energy electron by a screened Coulomb potential. For a small enough screening length, $1/q_s$, the potential $V_q \propto (q^2 - q_s^2)^{-1}$ is constant, beginning to show a q^{-2} asymptotic behavior only at very high electron energies (i.e., for electrons energetic enough that the short wavelength of the envelope function describing them can probe the ionic potential at distances smaller than the screening length). Therefore, the "form factor", $S_q \propto V_q (\mathbf{q} \cdot \mathbf{e}_\lambda) \approx V_q q$ ($\mathbf{e}_\lambda$ being the phonon polarization vector), increases linearly with the momentum transfer, q, up to high electron energies and the matrix element (Equation 3) shows the same linear dependence on q exhibited in the low energy deformation potential regime. (We should recall that the failure of the adiabatic approximation used to treat the electron-lattice coupling may also help in reducing the matrix element at high electron energies).[7] Therefore, up to energies of the order of 10 eV or so, we take for the coupling constant a form similar to the low energy regime, assuming that displaced ions behave as incoherent scattering centers with a potential deep enough to efficiently scatter at short wavelengths,[35] yielding a matrix element growing with q. Thus, in contrast to the polar coupling given by Equation 1, the nonpolar coupling, (Equation 3), favors large momentum transfer, q (i.e., large-angle collisions). The interaction gains strength as the electron energy (and $\mathbf{k}$-vector) increases as higher q phonons will be emitted more and more efficiently while the phase space opens up.

In order to obtain a monotonically increasing scattering rate at increasing electron energies, an alternative suggestion has been made by Porod and Ferry[8] by postulating the existence of satellite valleys in the conduction band of SiO$_2$. The increased density of final states does indeed compensate in this way for the decreasing form factor S_q assigned to a standard intervalley transition assisted by optical phonons. While we cannot rule out the possible existence of satellite valleys, the experimental evidence available at present[26] suggests that no additional conduction valleys exist at the low energies required by Porod and Ferry.[8]

Sticking to our assumption of a constant form factor, the rate $1/\tau_{ac}(w)$ at which an electron of energy w scatters with a phonon of wave vector, q, and energy $\hbar\omega(q)$ via nonpolar interaction with acoustic phonons has been obtained from the equation:

$$\frac{1}{\tau_{ac}(w)} = \frac{m^* C_{ac}^2}{4\pi\rho\hbar k(w)} \int_{q_c}^{q_{max}} dq \, \frac{q^3}{\hbar\omega_q} \left(\frac{1}{2} \pm \frac{1}{2} + n_{ac,q} \right) \tag{5}$$

where $q_c = 2m^* c_s$, $c_s = 4.6 \times 10^5$ cm/s is an effective sound velocity obtained by averaging over the longitudinal and transverse polarizations,[5,6] $q_{max} = 2k - q_c$, C_{ac} is the nonpolar electron/acoustic phonon coupling constant discussed above, m* is the electron effective mass, k the electron wave vector, and $n_{ac,q}$ the Bose function of acoustic phonons at wave vector q and temperature T. The upper/lower sign in Equation 5 corresponds to emission/absorption processes. The acoustic phonon dispersion has been approximated as

$$\hbar\omega_q \begin{cases} = \dfrac{2}{\pi} \hbar k_{BZ} c_s \left[1 - \cos\left(\dfrac{\pi q}{2k_{BZ}} \right) \right]^{1/2} & (q < k_{BZ}) \\[2ex] = \dfrac{2}{\pi} \hbar k_{BZ} c_s & (q \geq k_{BZ}) \end{cases} \tag{6}$$

where $k_{BZ} = 1.208 \times 10^8$ cm^{-1} is the wave vector at the edge of the Brillouin Zone (BZ).[5,6] Due to the lack of information about the structure of the conduction bands of SiO$_2$, we have approximated the BZ with a sphere having the same **k**-space volume of the BZ of α-quartz. Note that the approximation of Equation 6 treats correctly the high **q** dispersion, but it underestimates it by a factor $2^{1/2}$ the low **q** behavior. The use of Equation 6 will obviously yield a value for the coupling constant C_{ac} somewhat different from those employed in earlier simulations[5,6] employing different approximations to $\hbar\omega_q$. The electron effective mass has been assumed to be half the free electron mass, m* = 0.5 m_0, for electron energies below half the energy at the edge of the BZ, $w_{BZ} \simeq 5.52$ eV. At electron energies above w_{BZ}, the free electron mass has been used. In intermediate regions, we have employed an algebraic interpolation. (This effective mass has been used also to obtain the polar electron-LO phonon interaction. In this case, however, a "polaron corrected" effective mass of 0.7 m_0 has been used at low energy[4]). A similar cut-off between high energy (w > w_{BZ}) and low energy (w < 0.5 w_{BZ}) has been employed also for the "equivalent" density factor ρ in the past.[5] The integration over the q vector is performed numerically in the case of Equation 5 to account for the temperature dependence of the collision rate. It should be noted that the upper limit of integration in Equation 5 has been taken up to a value $q_{max} = 2k - q_c$ which may extend outside the BZ. This accounts for the possibility of Umklapp processes which constitute a crucial ingredient to account correctly for the entire strength of the nonpolar interaction. In particular, Umklapp processes are very effective in randomizing the electron direction after a collision, thus acting as those *catalytic* agents whose importance we stressed in previous sections. These considerations were first suggested by Sparks et al.,[25] later stressed by us,[5] and are now almost generally recognized as the basis to understand high energy electron transport in organic and inorganic insulators.[10,36,37] This characteristic unifying feature of electron transport in large bandgap material is probably the most significant — and largely unexpected — result of our investigations. Except for the already mentioned possibility of trap-assisted conduction in a few materials, the basic physics controlling transport in insulators appears to be the momentum randomizing action of nonpolar collisions with displaced ions. Most of the energy is absorbed by polar vibration. But without the action of the nonpolar processes, no steady-state transport would result in any insulator. The value of the coupling constant C_{ac} has been used as an empirical parameter, determined from comparison with experimental data. A value between 3.5 and 7.0 eV is required to fit the experimental range, the uncertainty due to the scattering of the experimental values of the electron average energy as a function of electric field in thick films. Results for a value of 5 eV are shown in Figures 20 and 21. This value has been consistently used in all simulations to provide a good fit to the vacuum emission data.

We have also included the nonpolar interaction between electrons and transverse optical (TO) phonons in the usual Harrison approach.[38] In this case, the electron-lattice coupling is viewed as proportional to the absolute displacement of the ions from their equilibrium position. The factor S_q is then a constant deformation potential, Ξ_{TO}. The collision rate, $1/\tau_{TO}(w)$, for this process as a function of lattice temperature and electron energy has been obtained as:

$$\frac{1}{\tau_{TO}(w)} = \frac{m^*\Xi_{TO}^2}{4\pi\rho\hbar^2\omega_{TO}k} \int_{q_{min}}^{q_{max}} dqq \left(\frac{1}{2} \pm \frac{1}{2} + n_{TO}\right) \tag{7}$$

where $q_{min/max} = k(\mp 1 \pm \sqrt{1 \mp \hbar\omega_{TO}/w})$. A value of 2×10^9 eV/cm[39] has been chosen for the coupling constant Ξ_{TO}, while the TO phonon energy used was 0.132 eV.[39]

We show in Figure 20 the total electron-phonon scattering rates obtained at two temperatures with our choice of parameters. It can be seen that the choice of the coupling constant Ξ_{TO} is not a critical one: the nonpolar electron/TO-phonon rate is not very large and its effects are minor corrections to the results obtained by turning this interaction off. We may expect that its role would be more significant in controlling the low field electron drift velocities, due to its effect in increasing the momentum relaxation rate.[39]

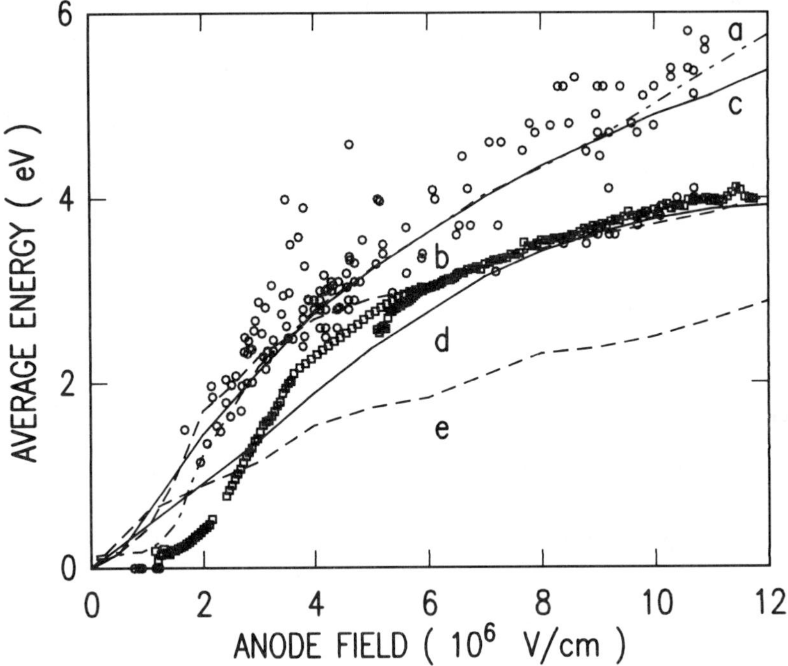

FIGURE 21.   Average electron energy vs. anode field from the experimental data: squares — carrier separation;[1] and circles — vacuum emission;[1-4] and from various Monte Carlo simulations — (a) semiclassical;[5] (b) including collision-broadening (overlapping the carrier separation data,[6] (c) Quantum Monte Carlo simulation,[7] (d) semiclassical with intervalley scattering and (e) with collision-broadening and intracollisional field effects.[8]

C. MONTE CARLO RESULTS

When a *conventional* Monte Carlo (CMC) technique is used,[40,41] together with the scattering rates we have just discussed, to simulate electron transport at high fields in SiO_2, an excellent agreement can be obtained with the experimental data. We show in Figure 21 a selection of experimental data compared to the results of simulations performed by us[5,6] (using semiclassical rates with a simplified acoustic phonon dispersion, a high temperature limit for the high energy nonpolar scattering rates, and a free electron mass at all energies) and by the group at Arizona State University[8] (accounting for intervalley scattering to hypothetical satellite valleys). The conventional Monte Carlo simulation, with the proper choice of the fitting parameter C_{ac}, gives excellent agreement with the data. In particular, the threshold for electron heating and the quasi-linear relation of average energy vs. field at fields above the heating threshold are reproduced in a remarkable way. In the former case, the simulation actually *predicted* the existence of the threshold as well as the magnitude of the threshold field.

However, there are a few fundamental open problems which have somehow been by-passed. First, we had to *guess* the strength of the high energy electron-lattice coupling. We can only justify our guess on the grounds that it seems to well explain the various transport features investigated experimentally. Second, we lack a good knowledge of the band structure of SiO_2, and are actually forced to talk about band structure of an amorphous material to employ the wealth of theoretical formalism available to simplify the description of electrons in ordered materials. Fortunately, it has been common wisdom in the past that a medium-range order exists in the structure of thermally grown SiO_2, so that symmetry can be recovered at the short wavelength exhibited by high-energy carriers. Finally, and more importantly, the use of a CMC technique (i.e., a solution of the semiclassical Boltzmann Transport

Equation, BTE) to treat electron transport at very high fields and energy presents quite a few problems. The BTE makes use of the assumption that the electrons interact only weakly with the scatterers (phonons, impurities, valence or conduction electrons) so that first order perturbation theory can be employed to compute the scattering rates. Moreover, in this weak coupling limit, the electron mean-free path is large compared to the electron wavelength, so that the collision processes can be treated as independent, occurring instantaneously, and the electrons become quasi-classical particles. These approximations can be easily seen in a CMC simulation scheme, since this technique assigns definite positions and momenta to the electrons, as allowed by classical mechanics.

The presence of very high fields and hot electrons definitely requires a revision of the semiclassical picture associated with the BTE. Several attempts have already been made in this direction. Some particular problems, such as the calculation of the mass[42] and drift velocity[43] of polarons in polar insulators, or the effect of the finite duration of the collision processes[44,45] have been successfully treated. However, we are still missing a general technique which enables us to treat, at a full quantum level, the transport problem in realistic instances.

A possible failure of the semiclassical BTE can be seen in Figure 5: at high electron energies, the Monte Carlo results show a lower tail than seen experimentally. While some of the disagreement could be attributed to our poor knowledge of the density of states (DOS), at high energies, our semiclassical approach must be regarded with suspicion. To somehow correct this deficiency, we have included the so-called collison-broadening effects in our Monte Carlo simulations. Originally, this had been done by Chang et al.[46] by using an approximate solution of the Dyson equation. The incorporation of this idea has somewhat improved the picture,[6] but has opened the new basic problem of how to merge the conventional BTE (*requiring* the weak coupling limit) with higher order corrections to the scattering rates. Therefore, we have taken steps in two different directions:[7] first, we have obtained the electron-phonon scattering rates by a direct solution of the Dyson equation — at all orders in the perturbation series — as previously suggested by Chang et al.[46] These are inputs in a Quantum Monte Carlo (QMC) scheme which is viewed as a sampling technique to stochastically evaluate the electron's Green function expressed as a Feynman integral over paths[47] rather than as a simple random generation of classical trajectories. The major difference between the CMC and our QMC technique consists of the weighting procedure for the various electron paths. The CMC assigns unit weight to all electron trajectories and adds them without interference. On the contrary, the QMC assigns a phase to every path and accounts for the interference of different trajectories ending with the same classical parameters (energy, momentum, position, or time).

In addition, we have introduced two major modifications in the band structure of SiO_2 and in the approximation employed to guess the matrix form factor S_q: The conduction band has been assumed to be spherical, but nonparabolic, and tailored in order to reproduce the calculated DOS characterized by a conduction band peaked at about 3 and 7 eV above its bottom at the Γ point.[26] We have taken for the electron effective mass at this point one half of the free electron mass and used a free electron dispersion at energies above 10 eV. Notice that we include a minimum of the DOS at about 9 eV which has been observed experimentally.[26] This contrasts the choice of a much higher DOS made by Porod and Ferry.[8] We have also assumed that at a very high electron energy a rigid ion approximation[35] better represents the matrix element for the nonpolar electron-phonon interaction. This amounts to a decrease of the momentum transfer cross section of the ions as the conduction electron acquires enough velocity to prevent the valence electrons from following its motion adiabatically. This behavior is observed in the collisions between electrons and nonpolar molecules[48] and is consistent with the screened Coulomb matrix element.[36] The critical energy we have assumed for the transition between the (isotropic scattering) Umklapp dominated and the (forward scattering) rigid ion response is 25 eV.

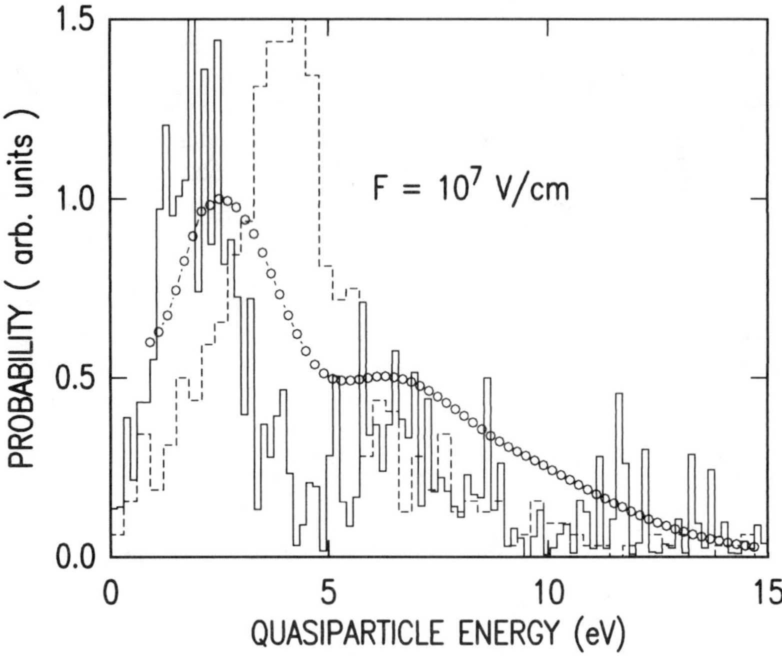

FIGURE 22.   Electron energy distribution obtained from the vacuum emission experiments of Reference 3 compared to the semiclassical[6] (dashed line) and Quantum[7] (solid line) Monte Carlo simulations.

The average electron energy as a function of the external electric field obtained from our quantum simulation is very close to the results of our previous semiclassical simulations, as shown in Figure 21, curve c. Note how the sharp threshold for electron heating is "washed-out" by the collisional broadening effect accounted for by the QMC. However, the major difference between quantum and semiclassical results concerns the actual electron energy distribution. In Figure 22, we show the quantum and semiclassical (no broadening) distributions at a field of 10^7 V/cm. A comparison between the quantum distribution and the experimental data indicates that a satisfactory agreement is now obtained not only in terms of average energies, but also in terms of the distribution of electrons at the high energies at which quantum effects must be included. In particular, the results of the QMC technique differ from those obtained from the CMC because of two somewhat competing effects. First, the peak of the distribution is shifted to lower energies, an effect already known to be due to the collisional broadening.[6,8] Second, the high energy tails are more pronounced, since the low energy paths (farther away from the stationary action trajectory) are most likely to destructively interfere. This second effect is typical of the present QMC technique and can be attributed to both intracollisional field effects and interference among partial waves in multiple collision events.

The rather crude approximations made (for the DOS and matrix element, and, most remarkably, the fact that the phonons are not allowed to back react on the electron system in our approximation of the correct path-integral technique) can only make us suggest that, while the *average* electron seems to be correctly described by the semiclassical BTE, high energy tails can only be treated with more sophisticated approaches. Our approximated QMC points in the right direction. Recent progress toward a complete formulation of the Feynman path quantization[49] makes us hopeful about our ability to treat the problem correctly in the foreseeable future.

D. ELECTRON DISTRIBUTION IN Si

Finally, we conclude this theoretical section by providing a few details on the inclusion of the electron distribution of the source function (the Si substrate) on the CMC simulation of the electron distributions shown in Figures 12 and 13.

We have already shown that the vacuum emission data in thin SiO_2 films rather strongly reflect the distribution of the electrons in the cathode. Actually, in the limit of fully ballistic transport in the oxide, the vacuum emission technique appears to be a viable tool to map the electron distribution in the "source" contact, the intervening oxide layer acting as an almost ideal transparent medium. Therefore, it is necessary to use some care in choosing the initial electron distribution in Monte Carlo simulation of electron transport in very thin SiO_2 films.

The possibility of accounting for quantization effects in the Si accumulation layers (the most common cathodic contact in our experiments) is a rather formidable task that we shall leave for further study. Presently, we shall content ourselves with a semiclassical description of the electron population in the Si accumulation layer.

For the purpose of the Monte Carlo simulation, we have gone through the following steps: first, we have started by solving the Poisson equation numerically for the particular MOS structure and bias conditions to be simulated. This guarantees that, except for the aforementioned quantum effects, we can properly account for the band bending in the accumulated Si region. We then proceed to generate electrons in k space. We employ a parabolic, ellipsoidal approximation to the 6 lower lying conduction valleys. We select a random valley and a random k vector. From its component normal to the Si-SiO_2 interface k_{normal}, we evaluate the energy $E_{normal} = \hbar^2 k^2_{normal}/(2m_{normal})$, where m_{normal} is either the longitudinal or the transverse effective mass in Si, depending on which of the 6 valleys is considered. Finally, we evaluate the tunneling probability for such an electron in the Wenzel-Kramers-Brillouin (WKB) approximation, $T_{WKB}(E_{normal})$, and the occupation probability, $f(\mathbf{k})$, given by the Fermi function at temperature T for the selected k state. The Fermi level in the Si accumulation layer is obviously known from the numerical solution of Poisson equation. The injection of one electron in the selected $\mathbf{k}$ state from Si into SiO_2 is then rejected or accepted according to the probability $Af(\mathbf{k})T_{WKB}(E_{normal})$, where A is a normalization constant equal to the tunneling probability evaluated at $E_{normal} = E_F + 20k_BT$, k_B is the Boltzmann constant, and E_F the Fermi energy in Si. This procedure correctly accounts for the DOS in the Si electrode, for the proper angular distribution of velocity at the interface, and for the tunneling probability (within the approximations we described). Once the injection energy E_{normal} is known, the "launching" position in the oxide (i.e., the initial position used in the Monte Carlo simulation) can be obtained as $(\phi_B - E_{normal})/F_{ox}$, where $\phi_B = 3.15$ eV is the barrier height at the Si-SiO_2 interface, and F_{ox} is the field applied to the SiO_2. The electron is assigned zero kinetic energy and its motion in the oxide is followed with a CMC algorithm. In order to obtain the energy distribution, ballistic electrons are assigned a weight proportional to the modulation of the tunneling probability due to the reflection of the electron wave function at the sharp SiO_2-Al interface at the anode.[22]

VII. CONCLUSIONS

This chapter has summarized both the experimental and theoretical work peformed on electron heating over the past 6 years, mostly at IBM. Electrons injected into SiO_2 can gain significant energy from the applied field for $\gtrsim 1.5$ MV/cm but do not undergo velocity runaway. Stabilization of the hot electron distribution is attained by interaction with the surrounding oxide lattice via polar and nonpolar phonons. There is no evidence at any oxide thickness that impact ionization in the SiO_2 layer can occur for an "average" electron of the distribution, even out to fields of 16 MV/cm. From vacuum emission studies, a small fraction of the hot electrons can be found in the high energy tail of the distribution. Some of these very hot electrons are predicted by quantum Monte Carlo simulations where the scattering rate can significantly decrease with increasing energy.

On very thin oxides, ballistic transport has been observed. SiO_2 is the only material other than III-V systems (for example, GaAs/GaAlAs) where a direct observation of this important phenomenon has been reported. Also, phonon induced side bands in the hot electron distributions caused by single LO phonon scattering have been seen. This latter observation was made possible by minimizing electron-electron interactions with control of the number of electrons introduced into the oxide conduction band during these experiments. Also, the vacuum emission experiments have allowed us to gain information on the electrons at the substrate Si/SiO_2 interface that can subsequently tunnel into these thin oxides under the appropriate applied voltage. The expected behavior of the Fermi function with temperature has been observed in these studies.

Trap creation in SiO_2 has been unequivocally related to the energy of the hot electrons. A minimum energy of ≈ 2 eV is necessary to generate both interface and bulk oxide traps. On very thin SiO_2 devices, this problem is minimized because injected electrons may not have a large enough distance in the oxide to heat-up to ≥ 2 eV before they enter the anode contact, depending on the gate voltage condition. In future MOSFET technologies where hot electron injection into the oxide from near the drain contact is a problem, lower power supply voltages or thinner SiO_2 layers will be necessary to circumvent trap generation.

Future trapping and breakdown studies in SiO_2 must be analyzed in terms of the electron heating phenomenon. Although these hot electrons will be harder to capture, their dispersive motion will give longer path lengths; therefore, more passes at a given trapping site. Also, the hot electrons will be capable of imparting their excess kinetic energy to a trapped carrier, defect, and/or impurity site which in certain cases could produce the observed bulk traps and interface states generated near the cathode.

Although all the oxides we studied were amorphous, we have still been able to observe ballistic electron transport, at least on the scale of the characteristic transport lengths (≈ 10 to 30 Å). This observation could be giving us information on the short-range order of the SiO_2 films. Our results are consistent with the idea that amorphous SiO_2 may have ordered regions which are nearly crystalline on a scale of 10 to 30 Å. As suggested by Mott, ballistic transport and the large value of the low field mobility for electrons in SiO_2 are also consistent with a lack of disorder scattering caused by a negligible number of localized states near the conduction band edge.[50] This latter explanation would be expected for s-like wave functions (which are believed to mainly form the bottom of the SiO_2 conduction band) and small variations of Si–O and O–O interatomic distances even with the large variation in the Si–O–Si bond angle.

REFERENCES

1. **DiMaria, D. J., Theis, T. N., Kirtley, J. R., Pesavento, F. L., Dong, D. W., and Brorson, S. D.,** Electron heating in silicon dioxide films and off-stoichiometric silicon dioxide films, *J. Appl. Phys.,* 57, 1214, 1985.
2. **DiMaria, D. J., Fischetti, M. V., Arienzo, M., and Tierney, E.,** Electron heating studies in silicon dioxide: low fields and thick films, *J. Appl. Phys.,* 60, 1719, 1986.
3. **Brorson, S. D., DiMaria, D. J., Fischetti, M. V., Pesavento, F. L., Solomon, P. M., and Dong, D. W.,** Direct measurement of the energy distribution of hot electrons in silicon dioxide, *J. Appl. Phys.,* 58, 1302, 1985.
4. **Fischetti, M. V., DiMaria, D. J., Dori, L., Batey, J., Tierney, E., and Stasiak, J.,** Ballistic electron transport in thin silicon dioxide films, *Phys. Rev. B,* 35, 4404, 1987.
5. **Fischetti, M. V.,** Monte-Carlo solution to the problem of high-field electron heating in SiO_2, *Phys. Rev. Lett.,* 53, 1755, 1984.

6. **Fischetti, M. V., DiMaria, D. J., Brorson, S. D., Theis, T. N., and Kirtley, J. R.,** Theory of high-field electron transport in silicon dioxide, *Phys. Rev. B,* 31, 8124, 1985.

7. **Fischetti, M. V. and DiMaria, D. J.,** Quantum Monte Carlo simulation of high-field electron transport: an application to silicon dioxide, *Phys. Rev. Lett.,* 55, 2455, 1985.

8. **Porod, W. and Ferry, D. K.,** Monte-Carlo study of high-energy electrons in silicon dioxide, *Phys. Rev. Lett.,* 54, 1189, 1985.

9. **DiMaria, D. J. and Abernathey, J. R.,** Electron heating in silicon nitride and silicon oxynitride films, *J. Appl. Phys.,* 60, 1727, 1986.

10. **Cartier, E. and Pfluger, P.,** Transport and relaxation of hot conduction electrons in an organic dielectric, *Phys. Rev. B,* 34, 8822, 1986.

11. **Cartier, E. and Pfluger, P.,** Experimental determination of energy dependent inelastic and elastic scattering rates of hot electrons in large bandgap insulators, *Physica Scripta,* T23, 235, 1988.

12. **DiMaria, D. J.,** The properties of electron and hole traps in thermal silicon dioxide layers grown on silicon, in *The Physics of SiO$_2$ and Its Interfaces,* Pantelides, S. T., Ed., Pergamon, New York, 1978, 160.

13. **Lenzlinger, M. and Snow, E. H.,** Fowler-Nordheim tunneling into thermally grown SiO$_2$, *J. Appl. Phys.,* 40, 278, 1969.

14. **Hughes, R. C.,** Hot electrons in SiO$_2$, *Phys. Rev. Lett.,* 35, 449, 1975.

15. **Grunthaner, F. J. and Maserjian, J.,** Chemical structure of the transitional region of the SiO$_2$/Si interface, in *The Physics of SiO$_2$ and Its Interfaces,* Pantelides. S. T., Ed., Pergamon, New York, 1978, 389.

16. **Feigl, F. J., Young, D. R., DiMaria, D. J., Lai, S., and Calise, J.,** The effects of water on oxide and interface trapped charge generation in thermal SiO$_2$ films, *J. Appl. Phys.,* 52, 5665, 1981.

17. **DiMaria, D. J.,** Correlation of trap creation with electron heating in silicon dioxide, *Appl. Phys. Lett.,* 51, 655, 1987.

18. **DiStefano, T. H. and Shatzkes, M.,** Dielectric instability and breakdown in SiO$_2$ thin films, *J. Vac. Sci. Technol.,* 13, 50, 1976.

19. **Solomon, P.,** Breakdown in silicon oxide—a review, *J. Vac. Sci. Technol.,* 14, 1122, 1977.

20. **Fröhlich, H.,** Theory of electrical breakdown in ionic crystals, *Proc. R. Soc. London,* Ser. A 160, 230, 1937.

21. **Berglund, C. N. and Powell, R. J.,** Photoinjection into SiO$_2$: electron scattering in the image force potential well, *J. Appl. Phys.,* 42, 573, 1971.

22. **Lewicki, G. and Maserjian, J.,** Oscillations in MOS tunneling, *J. Appl. Phys.,* 46, 3032, 1975.

23. **Fitting, H. H. and Frieman, J. U.,** Monte-Carlo studies of the electron mobility in SiO$_2$, *Phys. Status Solidi A,* 69, 349, 1982.

24. **Ridley, B. K.,** Mechanism of electrical breakdown in SiO$_2$ films, *J. Appl. Phys.,* 46, 998, 1975.

25. **Sparks, M., Mills, D. L., Warren, R., Holstein, T., Maradudin, A. A., Sham, L. J., Loh, E., Jr., and King, D. F.,** Theory of electron-avalanche breakdown in solids, *Phys. Rev. B,* 24, 3519, 1981.

26. **Himpsel, F. J., Fauster, Th., and Straub, D.,** Inverse photoemission in the UV, *J. Lumin.,* 31, 920, 1984.

27. **DiMaria, D. J., Dong, D. W., Falcony, C., Theis, T. N., Kirtley, J. R., Tsang, J. C., Young, D. R., Pesavento, F. L., and Brorson, S. D.,** Charge transport and trapping phenomena in off-stoichiometric silicon dioxide films, *J. Appl. Phys.,* 54, 5801, 1983.

28. **Heiblum, M., Nathan, M. I., Thomas, D. C., and Knoedler, C. M.,** Direct ballistic tunneling hot-electron transfer observation of ballistic transport in GaAs, *Phys. Rev. Lett.,* 55, 2200, 1986.

29. **DiMaria, D. J. and Fischetti, M. V.,** Vacuum emission of hot electrons from silicon dioxide at low temperatures, *J. Appl. Phys.,* 64, 4683, 1988.

30. **Gale, R., Feigl, F. J., Magee, C. W., and Young, D. R.,** Hydrogen migration under avalanche injection of electrons in Si metal-oxide-semiconductor capacitors, *J. Appl. Phys.,* 54, 6938, 1983.

31. **Buchanan, D. A., Fischetti, M. V., and DiMaria, D. J.,** Coulombic and neutral trapping centers in silicon dioxide, *Phys. Rev. B,* 43, 1471, 1991.

32. **Harari, E.,** Dielectric breakdown in electrically stressed thin films of thermal SiO$_2$, *J. Appl. Phys.,* 49, 2478, 1978.

33. **Kittel, C.,** *Quantum Theory of Solids,* Wiley, New York, 1963, 137.

34. **Bardeen, J. and Shockley, W.,** Deformation potentials and mobilities in non-polar crystals, *Phys. Rev.,* 80, 72, 1950.

35. **Ziman, J. M.,** *Principles of the Theory of Solids,* Cambridge University Press, 1972, 205.

36. **Blatter, G. and Baeriswyl, D.,** High-field transport phenomenology: hot-electron generation at semiconductor interfaces, *Phys. Rev. B,* 36, 6446, 1987.

37. **Zakharov, S. I. and Fiveisky, Yu. D.,** The influence of Umklapp processes on the hot carriers' relaxation, *Solid State Commun.,* 66, 1251, 1988.

38. **Harrison, W. A.,** Scattering of electrons by lattice vibrations in nonpolar crystals, *Phys. Rev.* 104, 1281, 1956.

39. **Koester, H., Jr. and Huebner, K.,** Statistics and transport behaviour of electrons in SiO$_2$, *Phys. Status Solid B,* 118, 293, 1983.

40. **Jacoboni, C. and Reggiani, L.,** The Monte Carlo method for the solution of charge transport in semiconductors with applications to covalent materials, *Rev. Mod. Phys.,* 55, 645, 1983.
41. **Price, P. J.,** Monte Carlo calculation of electron transport in solids, *Semiconduct. Semimet,* 14, 249, 1979.
42. **Feynman, R. P.,** Slow electrons in a polar crystal, *Phys. Rev.,* 97, 660, 1955.
43. **Feynman, R. P., Hellwarth, R. W., Iddings, C. K., and Platzman P. M.,** Mobility of slow electrons in a polar crystal, *Phys. Rev. B,* 4, 1004, 1962.
44. **Thornber, K. K.,** High-field electronic conduction in insulators, *Solid-State Electron,* 21, 259, 1978.
45. **Barker, J. R. and Ferry, D. K.,** Self-scattering path-variable formulation of high-field, time dependent, quantum kinetic equations for semiconductor transport in the finite-collision-duration regime, *Phys. Rev. Lett.,* 42, 1779, 1979.
46. **Chang, Y. C., Ting, D. Z.-Y., Tang, J. Y., and Hess, K.,** Monte Carlo simulation of impact ionization in GaAs including quantum effects, *Appl. Phys. Lett.,* 42, 76, 1983.
47. **Feynman, R. P. and Hibbs, A. R.,** *Quantum Mechanics and Path Integrals,* McGraw-Hill, New York, 1965.
48. **Lane, N. F.,** The theory of electron-molecule collisions, *Rev. Mod. Phys.,* 52, 29, 1980.
49. **Mason, B. A. and Hess, K.,** Quantum Monte Carlo calculations of electron dynamics in dissipative solid state systems using real-time path integrals, *Phys. Rev. B,* 39, 5051, 1989.
50. **Mott, N. F. and Davis, E. A.,** *Electronic Processes in Non-Crystalline Materials,* 2nd ed., Clarendon Press, Oxford, 1979, 512.

Chapter 11

THE FATE OF EXCESS ELECTRONS GENERATED BY ELECTROCHEMICAL METHODS

Anatol M. Brodsky

TABLE OF CONTENTS

I. INTRODUCTION

In recent years a substantial body of information on the physical characteristics of excess electrons in liquid polar dielectrics has been gained by electrochemical methods. The greatest body of data has been obtained with the use of the methods of photoelectrochemical and cathodic generation of such electrons in electrolyte solutions. This chapter summarizes the results obtained by the above-mentioned methods.

The foremost attention has been given to the fate of electrons generated by the photoelectrochemical methods. Systematic studies on light-induced electron transitions at the metal/electrolyte solution interface began only recently, though the history of the investigation of the so-called photovoltaic phenomena dates back 150 years. As early as 1839, i.e., long before the discovery of the photoeffect by H. Hertz, E. Becquerel found that the potential of some metal electrodes immersed in solutions of salts, acids, and bases changed on illumination of the electrochemical cell.

The modern stage of photoelectrochemical studies began in 1963 with the work of Barker and co-workers,[1] who suggested, and experimentally proved, that in a certain range of light frequencies and various electrode potentials a photocurrent is observed caused by electron photoemission from the metal electrode into the electrolyte solution, where electrons initially exist in a delocalized state. Barker et al.[1] tried to quantitatively explain the obtained data by Fowler's law of photoemission from metal into vacuum. As they themselves noted, it is not possible in this way to achieve satisfactory agreement with experiment. Their failure to do this is due to the fact that they did not take into account the specific features of photoemission into electrolyte solutions which were first considered in the theoretical works of Brodsky and co-workers.[2,3] It was found that, in the threshold range of light frequencies, the photoemission law is defined by the structure of the final state of photoelectrons,[2-4] so that it becomes possible to determine experimentally the spectrum of the excess electron states in electrolyte solutions. The photoemission law[2] was experimentally checked in subsequent studies. The relations theoretically predicted were confirmed and it was revealed[4-6] that the current-voltage characteristic of photocurrent contains quantitative information not only on interface and bulk electronic structure but also on the kinetics of reactions of solvated electrons and other intermediate species. This is accounted for by the fact that the value of the photocurrent observed, in addition to the photoemission process proper, is also determined by the transformation processes of the emitted nonequilibrium electrons occurring in the bulk of electrolyte solutions outside the range of action of surface forces. These processes are:

1. Thermalization of photoelectrons in the delocalized state resulting from inelastic and incoherent electron scattering processes
2. Solvation of electrons in the liquid, i.e., formation of localized (solvated) electrons
3. Physical and chemical transformations of solvated electrons: their diffusion and migration to the electrode, chemical reactions in the bulk and on the electrode, etc.

The solvated electrons e_s^- formed in the solution enter into chemical reactions with the substances present which possess the properties of scavenging electrons, i.e., electron acceptors, A:

$$e_s^- + A \rightarrow [eA]^-$$

Just as in radiation chemistry, the following species and molecules can act as acceptors, A: H_3O^+, NO_3^-, NO_2^- ions, metal cations, O_2, N_2O, CO, acetone, acrylonitrile molecules, etc. The formation of $[eA]^-$ can be measured by either the change of electrode potential or the photocurrent. The potential of the electrode from which electrons are emitted is usually such that in the absence of acceptors the majority of the solvated electrons return to the electrode and are captured by its surface. This results in a decrease of the current being recorded in the electrochemical cells. Therefore, unlike the electrode-vacuum system, where emitted electrons are transferred from photocathode to anode under the action of the applied potential difference, in electrolyte solutions steady photocurrents can be easily investigated experimentally at small light intensities only in the presence of acceptors. At large light intensities, when hydrated electrons are formed in aqueous solutions in large concentrations, due to the bimolecular reactions responsible for the disappearance of electrons, a photocurrent is observed even in the absence of acceptors.

Cathodic generation of solvated electrons was first observed in 1896 by Cady.[7] The information on cathodic generation remained on a qualitative level for over half a century until Laitinen and Nyman[8] made the first attempt to quantitatively investigate this process. This work, however, remained isolated for a long time, and only in the early 70s the systematic investigations of cathodic generation started[9-11] after photoemission into electrolyte solutions had been thoroughly studied.

The outline of the chapter is as follows. The second section is concerned with the studies on the primary act of photoemission at the metal-electrolyte solution interface. The fundamental interest in this phenomenon arises from the fact that it confirms the existence in the bulk solution of delocalized (''dry'') electrons. These conclusions agree with those deduced from the optical absorption spectra of solvated electrons obtained in radiation chemistry.[12]

The third section considers the results of the studies on the thermalization and solvation of delocalized electrons. These results agree qualitatively with the conclusions of the theory of Mozumder and Magee[13,14] on the thermalization of low energy electrons in condensed media during radiolysis.

The fourth section is devoted to the photoelectrochemical method of investigation of the kinetics of the reactions of solvated electrons.

The fifth section treats the process of cathodic generation of solvated electrons. The results of these studies supplement, in many respects, those of photoelectrochemical measurements.

The conclusion summarizes the main points considered in the review. Special attention is given to the energy characteristics of solvated electrons obtained in electrochemical and related measurements.

II. PHOTOEMISSION AT THE METAL-ELECTROLYTE INTERFACE AND THE STRUCTURE OF THE SPECTRUM OF EXCESS ELECTRONS IN ELECTROLYTES

As it was pointed out in the introduction, the specific features of photoemission from electrodes into polar media at low final electron energies constitute an important evidence in favor of the validity of the concepts of the existence in polar media, along with localized solvated electrons, of quasi-free, or dry, electrons. Quasi-free electrons are understood to be excess electrons in delocalized states, in many respects similar to the conduction electrons in metal.

Introduction in this case of the concept of the conduction band or mobility band edge in a polar solution — a disordered system with a strong electron-phonon (electron-exciton) interaction — requires a special discussion.[12] Being in a delocalized state, a mobile excess electron *has no time* to interact with the slow orientation polarization of the medium. One should take into account only the interactions of the delocalized electron with the almost inertialess electronic polarization of the medium, i.e., the interaction with high frequency phonons and excitons, which may be assumed to be comparatively weak. With certain restrictions,* it is then possible to use the concepts of the conduction band developed for ordered media.

In the low energy (threshold) range the functional form of the electron emission law is determined by the spectrum structure of final states. As explained in References 2 and 3, in the case of emission from electrodes into polar media with a fixed bottom of a wide conduction band, the expression for photocurrent J at the temperature T has the form

$$J = A\ T^{5/2}\ I \int_o^\infty du\ \left[\frac{u^{3/2}}{e^{(u-\beta)} + 1}\right] \tag{1}$$

where $\beta = \hbar(\omega - \omega_0)/kT$, I and ω are the intensity and frequency of light, respectively, and the constant A in the threshold approximation does not depend on ω. At T = 0 we have from Equation 1

$$J = \begin{cases} A\ I\ (\omega - \omega_o)^{5/2} & \text{at } \omega \geq \omega_o \\[2ex] 0 & \text{at } \omega < \omega_o \end{cases} \tag{2}$$

The dependence of J on temperature, which, according to Equation 1 is significant in the range $\hbar(\omega - \omega_0) \lesssim kT$ decreases with increasing frequency, and the photoemission current at $\beta > 1$ is well described by Equation 2.**

To find the dependence of the photocurrent on the electrode potential shift φ during electron emission into concentrated electrolyte solutions it is possible[4-6] to replace ω_0 by the quantity $\omega_o(\varphi)$ determined by the relation

$$\hbar\omega_o(\varphi) = \hbar\omega_o + e\varphi \tag{3}$$

where $\hbar\omega_o$ is the work function at a certain value of the potential shift φ, taken to be zero. According to Equation 3, application of a potential is equivalent to a shift of the photoeffect work function by the quantity $e\varphi$ at the fixed frequency ω. Substituting Equation 3 into Equation 2 leads, in the threshold frequency range, to the so-called 5/2-power law relating the photocurrent to frequency and potential[2-6]

$$J = A\ I\ \Delta^{5/2}, \qquad \Delta = \hbar\omega - e\varphi - \hbar\omega_o \tag{4}$$

Since it is important for the determination of the behavior of excess electrons in polar media to confirm that the laws given by Equations 2 and 4 are effectively obeyed and in view of a certain controversy in the literature,[15] we shall consider the experimental verification of these laws. Such a verification was carried out in a number of studies. At first, only a graphical analysis was made with the use of the least-squares method.[2,4,16] The first verification using quantitative criteria of statistical processing was made for mercury electrode.[17] The experimental curves in Reference 17 for a wide range of final energies (β = 3 to 40) agree well with Equation 4 except in the limiting threshold range $\beta < 5$, where the measured

---

* It should be expected that just below the bottom of a conduction band there is a continuous exponentially decreasing "tail" of localized states.

** The laws of photoemission into vacuum or dielectrics and into polar media differ considerably;[2-4] in particular in the former case, in a wide range of ω we have to do with Fowler's square law and in the latter case with the 5/2 law of Equation 2. The reason of this dissimilarity is the absence, in the case of emission into concentrated electrolyte solutions, of the influence of long-range mirror forces.

current is 2 to 2.5 times greater than that expected from Equation 4. The reasons for this difference will be considered below. Babenko et al.[17] reported the determination of the exponent n in the empirical dependence $J \sim \Delta^n$ using statistical processing of experimental data at a large enough φ, where Equation 4 should be valid. The values of n found by minimizing the experimental current voltage characteristic for a dropping mercury electrode with respect to three parameters were n = 2.46 and n = 2.40 for two experimental runs with different ω in solutions saturated with the electron acceptor N_2O at atmospheric pressure. Allowing for the correction for the dependence of the drop size on potential, these values become n = 2.58 and n = 2.50. The confidence range is the same in the two cases and equals 0.2. Though the values of n for the current voltage curves corresponding to different ω are not quite the same, the value of n = 2.5 always lies within the confidence range for n. It follows from the data obtained in Reference 17 that the maximum confidence range for n at different ω is 2.30 to 2.75 which may be considered an experimental corroboration of the 5/2-power law corresponding to Equations 2 and 4. Besides, the threshold potentials found by linear extrapolation of $J^{0.4}$ to J = 0 depend linearly on ω and there exists a relation between the shift of the frequency $(\delta\omega)$ and that of the threshold potential $(\delta\varphi_0)$

$$\frac{e\ \delta\varphi_o}{\hbar\ \delta\omega} = 1.03 \pm 0.05$$

which once more confirms the validity of Equation 4. The threshold potential value at n = 2.5 proved to be independent of the acceptor concentration in the solution.

The work function from mercury into water at the potentials of zero charge, found from the experimental photocurrent values at n = 2.5,[6,17] is $\hbar\omega_o = 3.04 \pm 0.05$ eV. Using this value one can determine the position of the conduction band of water relative to the Fermi level of electrons in the metal.

The validity of the 5/2-power law was also confirmed for photoemission into solutions with a static dielectric constant ϵ lower than that of water, for example in water-dioxane mixtures where ϵ decreased from 80 to 15.[17]

Measurements of the current-voltage characteristics of photocurrent at the metal-electrolyte interface for different metal electrodes showed that the 5/2-power law is valid for all metals whose surface can be made sufficiently smooth and cleaned by electrochemical methods and for which there exists the necessary range of ideal polarizability in electrolyte solution. In particular, measurements are described for Bi, Ag, Ga, Pb, Zn, Sb, In, Tl, Cu,[6] and it was demonstrated that for these metals the differences of the work functions into vacuum and into electrolyte are constant to within 0.2 eV (corresponding to the spread in the work functions for different faces of single crystals and the accuracy of their determination).

More recent works have also confirmed the validity of the 5/2-power law for photoemission into electrolytes in the threshold range of final energies of emitted electrons.[18-22] At the same time, in the immediate vicinity of and below the formal threshold, considerable departures from this law were revealed just as in Reference 17. This behavior led some authors[22-24] to conclude that in solution two continuous bands with their photoemission thresholds exist — a conclusion in which it is difficult to agree. The insufficient accuracy in the values of ω_o and n in the works of References 23 and 24 is apparently connected with the technical difficulties of investigation in the system under consideration, as well as with the influence[5,6] of the contribution of the ψ' potential, due to the low electrolyte concentration, and to the necessity of taking into account the complex light reflection effects upon the surface of the silver electrode.[57]

Thus, the experimental photoemission data may be considered to prove the conclusions of the theory reliable, namely the universality of the linear dependence of the red boundary of photoeffect on electrode potential and the validity of the 5/2-power law (Equation 4) in photoemission from metals into electrolyte solutions in the range of final electron energies $\Delta = 0.2$ to 2 eV. This confirms the reasonableness of using the concept of a conduction band in polar solutions. This is of primary importance in the examination of the low energy excess electron behavior in dielectrics.

We shall now consider the interpretation of the departure of photoemission law from the threshold power law at very low final electron energies. As noted above, the results of experiments on photoemission from metals into polar media indicate that in the immediate vicinity of the formal threshold (determined by extrapolation to zero values of the photocurrent from the range of final energies $\Delta \gtrsim 0.5$ eV) and in the final energy range of the order of some tenths of an electron volt, photoemission deviates from the simple threshold 5/2-power law. Photoemission falls off more smoothly toward lower light frequencies (more positive potential shifts) than according to the threshold law and does not become zero directly at the threshold. This deviation occurs in the energy range close to the threshold greater than kT and can be explained neither by the influence of the statistical energy distribution of initial electrons in metal at the temperature T nor by the contribution of a two-photon photoeffect.[57] The behavior of photocurrent in the immediate vicinity of the threshold for the mercury electrode was studied in great detail and most accurately in Reference 25.

To explain the phenomena at low final energies (in the case of emission into electrolytes), use was made of the concept of optical transitions to the levels lying below the conduction band of the external medium (transitions to the localized levels on the surface and in the bulk solution),[25,26] and of the concept of the influence of the potential fluctuations near the surface.[27,28] The localized levels in the bulk are understood to be the excited states of the solvated electron. This creates the possibility of investigating these levels without resorting to more complicated methods.

Experimental studies of photoemission from solid metal electrodes as close as possible to the threshold, and below it, were carried out in References 24 through 26. It was proved that near and below the formal photoemission threshold the temperature dependence of photocurrent is very strong, whereas in the range where the 5/2-power law is valid it is slight. The temperature dependence of photoemission was theoretically considered and experimentally compared.[28]

III. THERMALIZATION AND SOLVATION OF PHOTOELECTRONS

As it has been pointed out in the introduction, the electrons emitted into the solution react with the solvent and solutes. To quantitatively describe the kinetics of the processes occurring here, it is necessary to identify the main paths of the transformation of these excess electrons. There is a certain similarity here between photoemission and radiation chemical studies since both deal with solvated and dry electrons. But in one respect photoemission differs from radiolysis. In photoemission, a solvated electron arises in a unique manner as the result of solvation of a dry electron. In radiolysis, however, solvated electrons are formed in the course of a complex multistage process. The question as to whether the formation of solvated electrons during radiolysis is necessarily preceded by the dry electron stage is still a subject of discussion. Therefore, the photoemission studies afford a direct approach to the investigation of the mutual transformation of dry and solvated electrons and to the determination of the dry electron properties and reactivity in liquids.

A complete scheme of transformations following photoemission should, in principle, include both thermalization and solvation and should also direct interaction of thermalized, and possibly also "hot", dry electrons with acceptors.* It is evidenced by radiation chemistry experiments[29] that the effect of direct interaction of dry electrons with acceptors becomes appreciable only at sufficiently large concentrations of the latter where their mole fraction in solution is not much smaller than unity. Even the most effective acceptors of dry electrons — Cd^{2+}, NO_3^-, acetone — compete successfully with water for the capture of dry electrons only at concentrations greater than 1 mol/l. But photoemission studies can effectively be carried out at much lower acceptor concentrations. Besides, most frequently H_3O and NO_3^- are used as acceptors in photoelectrochemical experiments and they practically do not interact with dry electrons. Therefore, in processing experimental data one may assume that only solvated electrons interact with acceptors.

As a result of thermalization and solvation near the electrode, which may be assumed to be flat, solvated electrons are formed and distributed in space in a definite manner. Their distribution is given by the source function referred to unit time $F(x)$, proportional to the photoemission current J, according to

$$F(x) = J \, f(x), \quad \int_0^\infty f(x) \, dx = 1 \tag{5}$$

where x is the distance to the electrode surface. The form of $f(x)$ characterizes the law of thermalization and solvation of low energy electrons. The photocurrent j, observed in steady photoelectrochemical measurements at characteristic measurement times greater than 10^{-5} s, is[5,6]

$$j = J - j_e + j_a \tag{6}$$

where j_e is the current of the return of solvated electrons to the electrode and j_a is the current due to the products of the electron capture by acceptors. The sign of j_a is determined by that of the electrode reaction of the capture products. Specifically, for nitrous oxide and hydrogen ions, the most commonly used electron acceptors, the capture products of the hydrated electron (e_{aq}^-) are OH radicals and H atoms, respectively, which are reduced on the electrode (reaction with e_m^-) as follows

$$H_3O^+ + e_{aq}^- \rightarrow H^{\cdot}$$

$$H^{\cdot} + H_3O^+ + e_m^- \rightarrow H_2$$

$$N_2O + H_2O + e_{aq}^- \rightarrow N_2 + OH^- + OH^{\cdot}$$

$$OH^{\cdot} + e_m^- \rightarrow OH^- \tag{7}$$

Here, the steady current is different from zero, which would be the case if the capture products were oxidized on the electrode in the absence of their secondary reactions in solutions.

With short light pulses it becomes possible in experiments like those described in References 6, 30, and 31, to directly measure the return current j_e.

The diffusion description of the currents of solvated electrons based on the introduction of the source function $f(x)$ was given for the first time by Barker,[1] who took $f(x)$ in the form of a δ-function: $f(x) = \delta(x - x_o)$, where x_o is an adjustable parameter. In the diffusion description, the change in the solvated electron concentration $n_e(x,t)$ in space and time is given by the equation

---

* It cannot be said that delocalized electrons completely disappear after solvation. Thermal motions maintain an equilibrium between localized and delocalized electrons.

$$\frac{\partial n_e}{\partial t} = D_e \frac{\partial^2 n_e}{\partial x^2} - K_a N_{ac} n_e + \phi(t) F(x) \tag{8}$$

where K_a is the rate constant of the capture of solvated electrons by acceptors whose concentration is N_{ac}, $\phi(t)$ is the form of the exciting light pulse, and D_e is the diffusion coefficient of solvated electrons. The boundary conditions at the electrode at $x = 0$ and $x = \infty$ are of the form

$$-D_e \left(\frac{\partial n_e}{\partial x} \right)_{x=0} = \kappa_e n_e(0,t), \; n_e(\infty,t) = 0 \tag{9}$$

where κ_e is the is the rate constant of the electrode reaction of solvated electrons.

For the concentration of the products of the capture of solvated electrons by acceptors, $n_a(x,t)$, we have

$$\frac{\partial n_a}{\partial t} = D_a \frac{\partial^2 n_a}{\partial x^2} - \frac{n_a}{\tau_a} + K_a N_{ac} n_e,$$

$$-D_a \left(\frac{\partial n_a}{\partial x} \right)_{x=0} = \kappa_a n_a(0,t), \; n_a(\infty,t) = 0 \tag{10}$$

where τ_a is the annihilation time of the capture product in solution, and κ_a is the rate constant of its electrode reaction. There are similar equations for subsequent transformation products. It should be noted that Equations 9 and 10 are valid if $n_a < N_{ac}$ and if there is no noticeable contribution of bimolecular reactions of e_s^-. The currents figuring in Equation 6 are determined by the equations

$$j_e = -D_e \left(\frac{\partial n_e}{\partial x} \right)_{x=0}, \qquad j_a = -D_a \left(\frac{\partial n_a}{\partial x} \right)_{x=0} \tag{11}$$

In the simplest case of the steady photocurrent in the absence of homogeneous reactions of capture products (when $\tau_a^{-1} = 0$), we have,[6] from Equations 6, 8, 9, and 11, the following expression for the current j

$$j = 2 J \left[1 - \left(\frac{s}{s + 1} \right) \int_o^\infty e^{-\left(\frac{x}{l_d} \right)} f(x) \, dx \right]$$

where

$$l_d = \left(\frac{D_e}{K_a N_{ac}} \right)^{1/2}, \; s = l_a \frac{\kappa_e}{D_e} \tag{12}$$

In References 5 and 6, we can find solutions of Equation 8 in more complex cases for nonsteady light sources both in the absence and presence of the reactions of the solvated electron capture products. In the monographs cited,[5,6] the influence of the electric field of the double layer on the diffusion of solvated electrons and thus on photocurrent is considered, and the kinetics of the transformations of solvated electrons are described under the conditions where the reactions of their bimolecular disappearance cannot be ignored.

However, even from the simplest formula 12, it ensues that, by photoelectrochemical measurements, it is possible to obtain important characteristics of excess electrons in polar

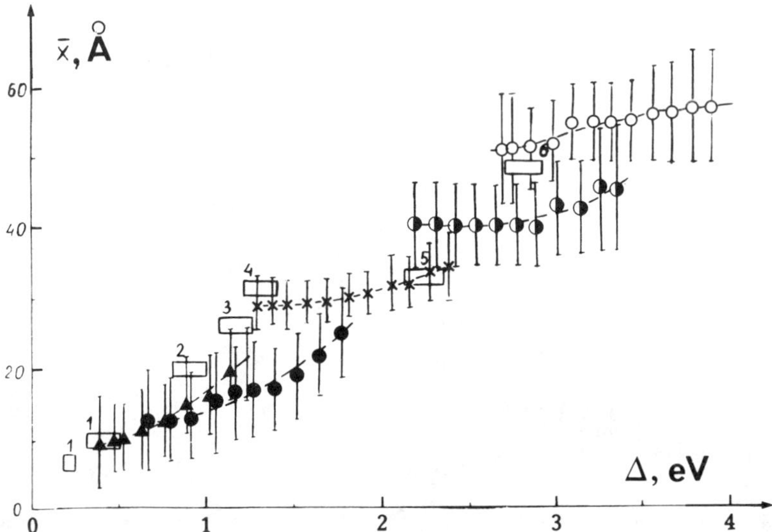

FIGURE 1.   Experimental dependences of $\bar{x}$ for electrons in water on final energy Δ obtained with the use of different acceptors at different light wavelengths.[32,34] Points (N_2O): ($\blacktriangle$), λ = 427 nm; ($\bullet$), 351 nm; (x), 308 nm; ($\circleddash$), 248 nm; ($\circ$), 223 nm. Squares (H_3O^+): 1, λ = 427 nm; 2, 351 nm; 3, 337 nm; 4, 308 nm; 5, 248 nm; 6, 223 nm.

media. In particular, from the experimental dependence of j on the acceptor concentration N_{ac}, one can, in principle, find the moments x_n of the function $f(x)$

$$x_1 = \bar{x} = \int_o^\infty x\, f(x)\, dx = \frac{1}{2J} \left(\frac{j}{N_{ac}^{1/2}}\right)_{N_{ac}\to 0} \left(\frac{D_e}{K_a}\right)^{1/2},$$

$$x_2 = \int_o^\infty x^2\, f(x)\, dx,$$

$$x_n = \int_o^\infty x^n\, f(x)\, dx \tag{13}$$

the first of which is equal to the mean solvation length, $\bar{x}$, of photoelectrons. In writing the last equality in the first relation of Equation 13, which follows from Equation 12, it was assumed that the rate constant of the electrode reaction κ_e contained in the expression of s is large. Therefore, one can put

$$\frac{s}{s + 1} \cong 1$$

Using Equation 13 to determine $\bar{x}$ from the steady-state measurements at different N_{ac} one must know the constants D_e and K_a, which in the initial studies were taken from pulse radiolysis data. The values of $\bar{x}$ may be obtained directly without any additional information in photoelectrochemical measurements with pulsed light sources. The most accurate measurements of $\bar{x}$ with a detailed error analysis for emission into aqueous solutions were reported in References 18 and 32 through 34. The results of these measurements are given in Figures 1 and 2. One should also note the results of earlier studies.[35,36] As is evident from Figures 1 and 2, $\bar{x}$ increases in the range of final electron energies 0.2 to 1.3 eV, while in the range of 1.3 to 2.2 eV $\bar{x}$ remains approximately constant. Comparison of the $\bar{x}(\Delta)$ dependences

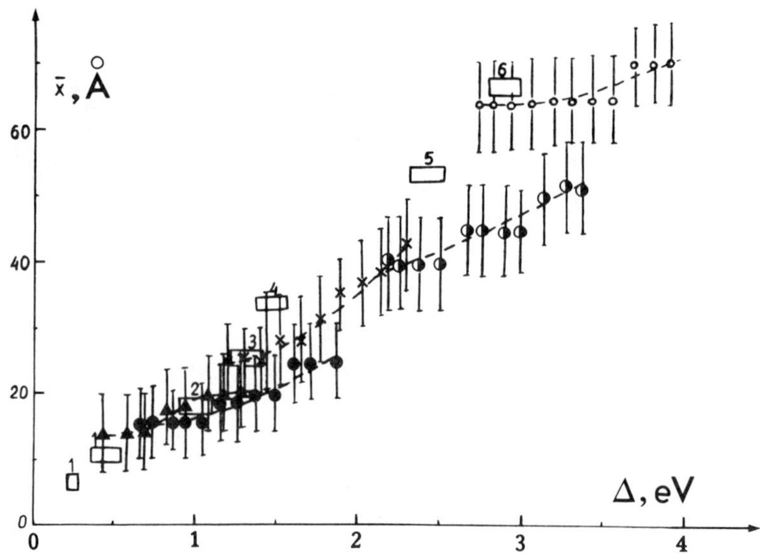

FIGURE 2.   Experimental dependences of $\bar{x}$ for electrons in D_2O.[33,34] Designations are the same as in Fig. 1.

in ordinary and deuterated water shows that at $\Delta < 2.5$ eV they practically coincide. At the same time, for $\Delta > 2.5$ eV, $\bar{x}$ differs for H_2O and D_2O. At $\Delta = 4$ eV, when $\bar{x}$ approaches 100 Å, this difference is as large as about 30%. This fact is an indication of an appreciable contribution of energy transfer to vibrational degrees of freedom when electrons undergo thermalization in water.

Figures 1 and 2 show noticeable differences between values of $\bar{x}$ corresponding to the same excess energy and different ω and φ. This could be explained as the effect of the potential barrier shape in the double layer and nearby (which depends on the potential φ) on the slowing down of electrons.

It is significant that steady and pulse photoelectrochemical measurements give consistent values of $\bar{x}$. This circumstance points to the inherent concordance of the models used and to the coincidence of the values of D_e and K_a obtained in radiation-chemical and photo-electrochemical measurements. The value of $\bar{x}$ is some tens of angstrom points in favor of theoretical models of the type described in References 13, 14, and 37 for thermalization of electrons in water. From photoelectrochemical data it is difficult to reconstruct the detailed form of the distribution function f(x). If the relative error of $x_1 = \bar{x}$ is as little as 5%, then in determining x_2 it becomes as large as 20% and grows rapidly with further increase of the number n of the moment of the function f(x).[32-34] A somewhat higher accuracy in the determination of x_2 was achieved from measurements at high light intensities without acceptors where the bimolecular reactions of solvated electrons were the controlling factors.[34]

In the simplest models,[6] the function f(x) at $x > \bar{x}$ is approximated by the exponential $f(x) = \exp(-x/x_1)$. To fit the form of f(x) to the value of the observed second moment x_2 one has to assume that f(x) has a maximum at a distance of 5 to 20 Å from the electrode.

To conclude this section, we wish to note that the occurrence of bimolecular reactions for the solvated electron disappearance after their photoemission from metals into electrolyte solutions was employed for a quantitative treatment of the extremely interesting phenomenon of a giant enhancement of the effect of electromagnetic field in the case of reflection of light from the rough surface of certain metals, silver in particular. Owing to this enhancement, in photochemical experiments in the absence of acceptors, extremely large photocurrents, up to 10^6 kA/cm^2, were achieved on a rough silver electrode.[38] Due to the nonlinear nature of bimolecular recombination, from the dependence of photocurrent on intensity, which is also nonlinear, it proved possible to estimate the dimensions of the areas of enhanced emission from a nonuniform electrode surface. According to results of the data processing,[38] the fraction of this photoactive surface is as little as 10^{-6} of total surface.

It should be specially noted that with the exceedingly high concentrations of solvated electrons obtained under the conditions of the experiments in Reference 38, it is possible to follow the formation of bielectrons. The most promising is the search for bielectrons in the case of photoemission in the rough silver electrode-ammonia solution system. Unlike solutions of alkali metals, large concentrations of solvated electrons can be obtained in this system in the absence of counterions.

IV. PHOTOELECTROCHEMICAL INVESTIGATIONS OF THE REACTIONS OF SOLVATED ELECTRONS

Even from a qualitative analysis of Equation 10, use of photoemission measurements with pulse or periodic photocurrent sources makes it possible to determine the diffusion coefficient and the absolute rate constants of the reactions of solvated electrons and products of their interaction with acceptors. This method, called the kinetic method, was applied by Barker[39] (see also Reference 6).

In the case of steady experiments it is possible to determine rate constants by the method of competing acceptors which is often used in radiation chemistry. As is known, in this method several reference rate constants are chosen which are measured by independent pulse methods and the ratio of the reaction rates to the reference constant is determined. In photoelectrochemical measurements the competing acceptor method was first used by Barker[1] to determine the rate constants of the capture of hydrated electrons by acceptors. The theory and applications of the two methods for investigation of the kinetic characteristics of solvated electrons mentioned above are considered in Reference 6. The method for investigation of reactions with a periodic light source is described in detail in Reference 40.

Production of various organic compounds with the use of solvated electrons as a strong reducing species is described in various reviews.[41,42]

Bendersky and co-workers[6,31,43] developed procedures for photoelectrochemical studies with nanosecond light pulses and thus have been able to radically improve the kinetic method for investigation of the kinetics of reactions of solvated electrons and products of their interactions. With the use of nanosecond pulses it became possible to study the reduction and oxidation reactions which constitute primary stages of photosynthesis.

V. CATHODIC GENERATION OF SOLVATED ELECTRONS

Some polar solutions are known to exist in which solvated electrons are long lived or even stable. To these belong solutions of metals in ammonia and also such aprotic solvents as hexamethylphosphoric triamide (HMPT), dimethylsulphoxide, diglyme, etc. The method of electrochemical generation of solvated electrons in the above-mentioned systems is of considerable interest since it allows investigation of solvated electrons under steady-state conditions with the chemical transformations in the bulk solution and on the electrode controlled. It is evident that the processes of photoemission and electrochemical generation of solvated electrons are, in many respects, essentially similar and differ only in detailed mechanisms of electron excitation from the Fermi level of the metal: in the former case optical, and in the latter, thermal excitation in the external field takes place. Ammonia was the first solvent for which it was shown by examinations of optical and ESR spectra that the properties of solvated electrons obtained by different methods (by dissolving alkali metals, by pulse radiolysis, by cathodic generation, and by photoelectrochemical methods) were identical.[42]

It should be noted that a number of authors expressed the view that many of the known electrochemical reduction reactions proceed according to the mechanism of solvated electron generation followed by the interaction (in the solution layer adjacent to the electrode) of solvated electrons and positive ions.[44,45] Such a mechanism was discussed in the literature even for an ordinary discharge of H^+ ions in water electrolysis. A detailed study showed, however, that in the case of aqueous electrolyte solutions an ordinary classical discharge of H^+ ions occurs at the electrode.[46-48]

Theoretically, at large enough negative cathode potentials, cathodic generation of electrons should take place. However, for cathodic generation to be realized it is necessary that:

1.   The solution should be stable with respect to cathodic reduction of positive ions at large enough negative cathode potential
2.   The reaction of electrons in the bulk with positive ions should be at least delayed. In the opposite case, instead of cathodic generation of solvated electrons, reduction of counterions occurs either on the electrode (metal deposition) or in the region immediately adjacent to the electrode.

Among all solutions, the above-mentioned aprotic systems (HMPT, diglyme, etc.) satisfy the required conditions best. It is not surprising, therefore, that the greatest amount of data on solvated electrons were obtained precisely with HMPT, which, under normal conditions, is a liquid.

Two reaction paths are possible, in principle, in the cathodic generation of solvated electrons. In the first path metal electrons are thermally excited above the level corresponding to the electrochemical potential of the delocalized dry electron at the bottom of the solution conduction band. The second reaction path consists in a direct tunneling transfer of metal electrons to the level E_s of solvated electrons in solutions. Here electrons must tunnel some distance into the solution. The activation energy of this many particle process, greater than $E_s - E_F$ (E_F is the Fermi level of the metal), is less than the activation energy in the first path, but kinetic difficulties arise here, connected with the low probability of tunneling for comparatively large distances. At different potentials and states of the cathode surface different mechanisms of cathodic generation of solvated electrons may be realized.

Experimental evidence for the realization of both generation mechanisms is available.[42] A possible change in the generation mechanism with the modification of the electrode surface state (surface passivation) is suggested.[42] As a matter of fact, at a fixed electrode potential, the formation of a passivating film on its surface leads to increased tunneling distances. With sufficient film thicknesses, the process of electron thermal excitation into the conduction band of the solution is more likely to occur when the electron passes through the surface barrier.

Investigation of the anodic oxidation reactions of solvated electrons showed that they are controlled by diffusion and that the oxidation potential depends on the state of the electrode surface. Besides the discharge of solvated electrons at the anode, in the case of HMPT one more limiting current plateau was observed in the polarization curves corresponding to the discharge of nonparamagnetic complexes containing two electrons and an alkali metal ion (singly charged bielectrons). With respect to discharge at the anode, monoelectrons are much more active than nonparamagnetic complexes; their oxidation potentials differ by about 0.5 eV. Investigation of anodic reactions of solvated electrons allowed an exact estimation of the concentration of solvated electrons in the solution from the limiting discharge current value and, from the electron concentration, the concentration dependence of the equilibrium "electronic" potential was established and the standard potential of the reversible electronic electrode in HMPT and ammonia was calculated.[42]

Investigation of bulk reactions of cathodically generated electrons holds great promise. Since solvated electrons belong to "fast moving" particles, i.e, they have large diffusion

TABLE 1
Parameters of Excess Electrons Determined in Different Experiments

Type of experiment	The quantity being determined
Photoconductivity	Photoconductivity threshold, energies of autoionization states
Photoemission from metal into liquid	Position for conduction band relative to Fermi level (if emission is into "the band"), level energy (if tunneling occurs to a bound state level)
Cathodic generation	Thermodynamic parameters of the solvated electrons
Photoemission spectroscopy from solution into vacuum (vapor)	Photoemission threshold of solute, connection between bottoms of conduction bands in solution and vapor phase; solvent reorganization energy
Electron thermoemission from solvated electron solution to vapor phase	Activation energy of thermoemission current, heat of electron solvation
Dissolution of metal	Free energy of solvation

coefficients, in many cases it is more advantageous to carry out the reaction in the bulk transferring electrons than to transport the species to be reduced to the electrode. Moreover, the change, observed in many solvents, in the density of the medium near the sites where electrons are localized aids in stirring the solution due to convection, which also accelerates the reaction in the bulk. Solvated electrons, whose standard potential is close to those of alkali metal ions, belong to the strongest reducing agents and by being able to generate them in large concentrations, one can carry out numerous reactions with organic compounds, which are difficult to reduce. At present there is a certain number of examples of such processes.[47,49]

VI. CONCLUSION

As it is evident from the results presented in this chapter, the electrochemical studies of the behavior of solvated electrons described above now form an important part of the investigations of excess electrons in liquid polar dielectrics.

A complete picture of the physical characteristics of these electrons can naturally be obtained only by combining various available methods, especially optical. The experiments described in this chapter are most closely related to those on photon and thermal electron emission spectroscopy from a solution into the vapor phase.[6,50-53] The photo (thermal) emission from a solution of the ions and solvated electrons (in solvents such as hexamethylphosphoric triamide and liquid ammonia where solvated electrons are stable) proceeds in three stages: (1) photo (thermal) ionization of the ions or solvated electrons, (2) movement of generated delocalized electrons to the surface of the solution, and (3) transition of electrons to the vapor phase where they are transferred from the cathode surface to the anode by the external electric field.

Table 1 lists the energy and electrodynamic characteristics of solvated electrons, which can be obtained in experiments of the kind considered above. Photoconductivity measurements in solutions* and coulometric studies of metal dissolution have also been added.

The energy levels for excess electrons in the metal-solution-vapor system are illustrated in Figure 3. As a reference energy level the electron energy in vacuum (vapor) is taken $E^v = O$. This level corresponds to the energy of an electron which is in the neighborhood of the surface of the solution beyond the range of action of pure surface forces. The local levels of Figure 3 could be connected with the solvated electron precursor states found recently in femtosecond spectroscopy experiments.[37,56]

---

* The photoconductivity in polar solutions can be measured with the use of electrochemical cells.

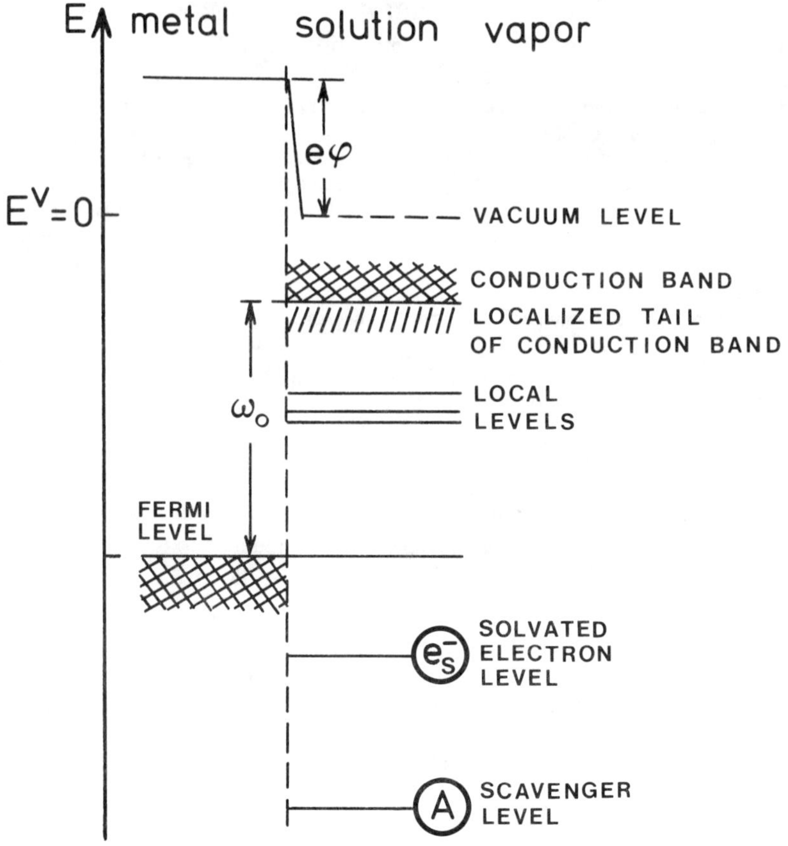

FIGURE 3.   Electron levels in the metal-solution-vapor system.

One may say with confidence that combined measurements of the absorption spectra, photoconductivity, photoemission from metals into solutions and from solutions into vacuum could give all the energy characteristics of the ground and excited states of solvated electrons. Unfortunately, data on photoconductivity are not presently available and the other data were obtained for different systems and are scattered. The data on thermal emission from solutions[50] were criticized[5] for the lack of information on the temperature dependence of the vapor phase density (and accordingly on the energy characteristics of electrons in it).

Experiments on electrochemical generation of solvated electrons may be assumed to give the most exact thermodynamic data. Regrettably, most experiments on cathodic generation were carried out in HMPT and data of photoelectron spectroscopy from solutions are available only for electrons in ammonia. Generally speaking, there is now a large amount of scattered experimental data on thermodynamic characteristics for different systems, and only for one system (electrons in ammonia) is fairly complete information available.[54,55] Table 2 gives the values of the energy of interaction between a delocalized electron at the bottom of the conduction band and different calculated solvents[26,42,57] taking into account available photoelectrochemical data.

Finally, we wish to stress that photoelectrochemical methods afford unique possibilities for investigating the kinetic characteristics of the reactions of solvated electrons and of the products of their interaction with acceptors. With the use of picosecond light pulses, it is possible to quantitatively study the fastest chemical reactions in condensed media that lend themselves to investigation in general. There are also good prospects for investigation of

TABLE 2
The Energy V_o of Interaction
Between a Delocalized Electron
and Different Solvents[26,42,57]

Solvent	V_o (eV)
Water	-1.12
Liquid ammonia	-0.95
Methanol	-0.39
Ethanol	-0.40
Isopropanol	-0.47
Formamide	-1.05
N-Methylformamide	-0.44
HMPT	-0.38
Acetonitrile	-0.26
Formic acid	-2.20

the specific features of the thermalization of low energy electrons in a medium. In particular, considering the derivatives of the mean thermalization length $\bar{x}$ with respect to the final energy of electrons Δ and taking into account the energy distribution law of primary electrons which can be theoretically evaluated, one can, in principle, find the dependence of the scattering cross section of the excess electrons on energy. The value of Δ can be continuously changed in this case by varying the potential shift at constant light frequency ω.

Development of photoelectrochemical studies on the generation of electrons in high concentrations without counterions[38] holds great promise. Under conditions such as these it is possible to carry out reactions in which many electrons participate in one elementary act, e.g., binding of nitrogen. These conditions are also very suitable for looking into the possibilities of formation of the bielectron (a system with a double charge, possessing a specific reactivity) without counterions. Synthesis of such systems aroused special interest in connection with the hypothetical possibility of constructing superconductors of a new type on the basis of their Bose condensate.

REFERENCES

1. **Barker, G. C., Gardner, A. W., and Sammon, D. C.,** Photocurrents produced by ultraviolet irradiation of mercury electrodes, *J. Electrochem. Soc.,* 113, 1182, 1966.
2. **Brodsky, A. M. and Gurevitch, Yu. J.,** Theory of the surface photoemission, *Zh. Eksp. Teor. Fiz.,* 54, 499, 1968.
3. **Brodsky, A. M. and Tsarevsky, A. V.,** Theory of volume photoemission from condensed media, *Zh. Eksp. Teor. Fiz.,* 69, 936, 1975.
4. **Brodsky, A. M. and Pleskov, Yu. V.,** Electron photoemission at a metal-electrolyte solution interface, in *Progress in Surface Science,* Vol. 2, Davison, S., Ed., Pergamon Press, Oxford, 1972, 1.
5. **Brodsky, A. M., Gurevitch, Yu. J., Pleskov, Yu. J., and Rotenberg, Z. A.,** *The Modern Photoelectrochemistry* (in Russian), Nauka, Moscow, 1974.
6. **Bendersky, V. A. and Brodsky, A. M.,** *The Photoemission from Metals into the Electrolyte Solutions* (in Russian), Nauka, Moscow, 1977.
7. **Cady, H. P.,** The electrolysis and electrolytic conductivity of certain substances dissolved in liquid ammonia, *J. Phys. Chem.,* 1, 707, 1896.
8. **Laitinen, H. A. and Nyman, C. J.,** Polarography in liquid ammonia. The electron electrode, *J. Am. Chem. Soc.,* 70, 3002, 1948.
9. **Kanzaki, Y. and Aoyagui, S.,** Electrode kinetics of solvated electrons in hexamethylphosphoric triamide, *J. Electroanal. Chem.,* 41, 395, 1973.
10. **Krishtalik, L. J. and Alpatova, N. M.,** Electrochemical generation of solvated electrons, *J. Electroanal. Chem.,* 65, 219, 1975.

11. **Avake, L. A. and Bewick, A.,** The cathodic generation of solvated electrons in the LiCl, *J. Electroanal. Chem.*, 41, 395, 1973.
12. **Brodsky, A. M. and Tsarevsky, A. V.,** On a theoretical description of solvated electrons, in *Advances in Chemical Physics*, Vol. 4, Prigogine, I. and Rice, S. A., Eds., John Wiley, New York, 1980, 483.
13. **Magee, J. L.,** Electron energy loss processes at subelectronic excitation energies in liquids, *Can. J. Chem.*, 55, 1847, 1977.
14. **LaVerne, J. A. and Mozumder, A.,** Energy loss and thermalization of low-energy electrons, *Radiat. Phys. Chem.*, 23, 637, 1984.
15. **Sass, J. K. and Gerisher, H.,** Photoemission in the electrolyte solutions, in *Photoemission and the Electronic Properties of Surfaces*, Feuerbacher, B., Fitton, B., and Willis, R., Eds., Wiley, New York, 1978, 469.
16. **de Levie, R. and Kreuser, J. C.,** On the measurement of small photocurrents on dropping mercure electrode, *J. Electroanalyt. Chem.*, 21, 221, 1969.
17. **Babenko, S. D., Rudenko, T. S., and Zolotovitsky, Ya. M.,** Experimental investigation of the frequency dependence of photoemission from metals, *Fiz. Tverd. Tela* (in Russian), 14, 3501, 1972.
18. **Konovalov, V. V., Tregub, V. V., and Raitsimring, A. M.,** Photoemission from metal into electrolyte, *Electrokhimiya*, 20, 470, 1984.
19. **Kreitus, J. V., Bendersky, V. A., and Tiliks, Yu. E.,** Photoelectron emission from metal into concentrated electrolyte, Part I, *J. Electroanal. Chem.*, 133, 345, 1982.
20. **Richardson, J. H., Deutscher, S. B., and Maddux, A. S.,** Laser induced photoelectrochemistry, *J. Electroanal. Chem.*, 109, 95, 1980.
21. **Bard, A. I., Itaya, K., Maplas, R. E., and Feherani, T.,** Electrochemical and photoelectrochemical studies of excess electrons in liquid ammonia, *J. Phys. Chem.*, 84, 1262, 1980.
22. **Uribe, F. A., Sawada, T., and Bard, A. I.,** Photoelectron emission from a platinum electrode into liquid ammonia solutions, *Chem. Phys. Lett.*, 97, 143, 1983.
23. **Krebs, P., Bukowski, K., Girard, V., and Heintz, M.,** Energy levels for photoinjection of excess electrons into ammonia vapor at various densities, *Ber. Bunsenges. Phys. Chem.*, 86, 879, 1982.
24. **Krohn, C. E., Antoniewicz, P. R., and Thompson, J. C.,** Energetics for photoemission of electrons into NH_3 and H_2O, *Surf. Sci.*, 101, 241, 1980.
25. **Babenko, S. D., Bendersky, V. A., Brodsky, A. M., and Sorokin, S. B.,** The study of presolvated states of electrons in water by method of photoelectron emission, *Chim. Vys. Energy* (in Russian), 17, 493, 1983.
26. **Bennett, G. T. and Thompson, J. C.,** A model for photoinjection into polar fluids. *J. Chem. Phys.*, 84, 1901, 1986; **Bennett, G. T., Coffman, R. B., and Thompson, J. C.,** *J. Chem. Phys.*, 87, 7242, 1987.
27. **Brodsky, A. M.,** About the connection between potential fluctuations and the threshold laws of photoemission, *Electrokhimiya*, 17, 1687, 1981.
28. **Brodsky, A. M.,** The threshold laws of photoemission from solid metals into electrolytes, *Electrokhimiya*, 20, 1404, 1986.
29. **Hunt, J. W.,** Early events in radiation chemistry, in *Advances in Radiation Chemistry*, Vol. 5, Burton, M. and Magee, L., Eds., Wiley, New York, 1976, 185.
30. **Delahay, P. and Srinivasan, V. S.,** Photocurrents at flash-irradiated mercury electrode interface, *J. Phys. Chem.*, 70, 420, 1966.
31. **Bendersky, V. A., Krivenko, A. G., and Kurmaz, V. A.,** Kinetics of the emitted charge in the laser emission of photoelectrons from metal into solution, *Electrokhimiya*, 22, 728, 1986.
32. **Shokhirev, N. B., Konovalov, V. V., Pusep, A. Yu., and Raitsimring, A. M.,** Determination of the distribution function for e_{aq}^- in the photoemission experiments, *Theor. Exper. Ch.* (in Russian), 20, 336, 1984.
33. **Konovalov, V. V., Puser, A. Yu., Raitsimring, A. M., and Tsvetkov, Yu. P.,** Pulsed photoemission from metals into electrolyte solutions: studies on slow-electron thermalization lengths by the method of e_{aq}^- recombination, *Chem. Phys. Lett.*, 126, 472, 1986.
34. **Konovalov, V. V., Raitsimring, A. M., and Tsvetkov, Yu. D.,** Thermalization lengths of "subexcitation electrons" in water determined by photoinjection from metals into electrolyte solutions, *Radiat. Phys. Chem.*, 32, 623, 1988.
35. **Kreitus, I. V., Bendersky, V. A., and Tiliks, Yu. E.,** Photoelectron emission from metal into concentrated electrolytes. Part II. Thermalization path-length, *J. Electronal. Chem.*, 140, 311, 1982.
36. **Neff, H., Sass, J. K., Lewerenz, H. J., and Ibach, H.,** Photoemission studies of electron localization at very low excess energies, *J. Phys. Chem.*, 84, 1135, 1980.
37. **Mozumder, A.,** Conjecture on electron trapping in liquid water, *Radiat. Phys. Chem.*, 32, 287, 1988.
38. **Funtikov, A. M., Sigalaev, S. K., and Kazarinov, V. E.,** Surface enhanced Raman scattering and local photoemission currents on the freshly prepared surface of a silver electrode, *J. Electroanal. Chem.*, 228, 197, 1987.

39. **Barker, G. C.,** Electrochemical effects produced by light-induced electron emission, *Ber. Bunsenges. Phys. Chem.,* 75, 728, 1971.

40. **Rottenberg, Z. A. and Kazarinov, V. E.,** A new method of modulation of concentration of reactant particles near electrode by photoemission, *J. Electroanal. Chem.,* 180, 337, 1984.

41. **Lund, H.,** Use of solvated electrons, in *Organic Electrochemistry,* Baizer, M. M. and Lund, H., Eds., Marcel Dekker, New York, 1983, 873.

42. **Alpatova, N. M., Krishtalik, L. J., and Pleskov, Yu. V.,** Electrochemistry of solvated electrons, in *Topics in Current Chemistry,* Vol. 138, Springer, Berlin, 1987, 149.

43. **Bendersky, V. A, Krivenko, A. G., Kurmaz, V. A., and Simbirzeva, G. V.,** *Electrokhimiya,* 22, 728, 1986.

44. **Walker, D. C.,** Spectroscopic evidence for hydrated electrons in the electrolysis of water, *Can. J. Chem.,* 45, 807, 1967.

45. **Pyle, T. and Roberts, C.,** The reaction of solvated electrons at metal-electrolyte interfaces, *J. Electrochem. Soc.,* 115, 247, 1968.

46. **Brodsky, A. M. and Frumkin, A. N.,** About the thermoemission of electrons during the cathodic evolution, *Electrokhimiya,* 6, 658, 1970.

47. **Parsons, R.,** The pressure coefficient of hydrogen electrode reaction, *J. Electrochem. Soc.,* 113, 1118, 1966.

48. **Bewick, A., Conway, B. E., and Tuxford, A. M.,** Electrochemically generated hydrated electrons, *J. Electroanal. Chem.,* 42, 11, 1973.

49. **Hart, E. J. and Anbar, M.,** *The Hydrated Electron,* Wiley, New York, 1970.

50. **Baron, B., Delahay, P., and Lugo, R.,** Thermoionic emission by solutions of solvated electrons, *J. Chem. Phys.,* 53, 1399, 1970.

51. **Delahay, P.,** Photoelectron emission spectroscopy of liquids and solutions, in *Electron Spectroscopy: Theory, Techniques and Applications,* Baner, A. D., Ed., Academic Press, London, 1984, 123.

52. **Watanabe, J., Maya, K., Yabuhara, Y., and Ineda, S.,** Photoionization threshold energies of amine solutions studied by photoelectron emission, *Bull. Chem. Soc. Jpn.,* 59, 907, 1986.

53. **Brodsky, A. M. and Tsarevsky, A. V.,** Emission of electrons from solutions, *J. Chem. Soc., Faraday Trans. II,* 72, 1781, 1976

54. **Thompson, J. C.,** *Electrons in Liquid Ammonia,* Clarendon Press, Oxford, 1976.

55. **Brodsky, A. M., and Tsarevsky, A. V.,** The development of physical chemistry of solvated electrons, *Usp. Khim.* (in Russian), 66, 1693, 1987.

56. **Migus, A., Gauduel, Y., Martin, J. L., and Antonetti, A.,** Excess electrons in liquid water: first evidence of a prehydrated state with femtosecond lifetime, *Phys. Rev. Lett.,* 58, 1559, 1987.

57. **Brodsky, A. M. and Urbakh, M. I.,** *Electrodynamics of Metal-Electrolyte Interface* (in Russian), Nauka, Moscow, 1989.

Chapter 12

EXCESS ELECTRONS IN MICROHETEROGENEOUS SYSTEMS

Annette Bernas, Dora Grand, and Simone Hautecloque

TABLE OF CONTENTS

I. INTRODUCTION

The role played by surfaces or interfaces in chemical or photochemical reactions has long been a source of interest and endeavor in various fields of chemistry, e.g., catalysis, polymer and colloid chemistry, biochemistry, etc.[1,2]

In such research areas, the microheterogeneous systems investigated were typically *inorganic colloids* (e.g., CdS, ZnO, TiO_2, etc., various clays) or aqueous solutions of *organic aggregates*. Such a clear-cut picture is somewhat blurred, however, by the recourse to more sophisticated samples. Mixed systems have, in effect, been prepared and examined: e.g., surfactant molecules adsorbed on colloidal clays[3] or semiconductor particles entrapped in natural or synthetic vesicles.[4] At any rate, and in spite of one common goal — the photochemical conversion and storage of solar energy[5] — the studies performed during the last decade or so on either inorganic or organic microheterogeneous systems have experienced such a rapid expansion that they definitely require distinct presentations.

In this chapter only organic microheterogeneous systems will be considered: micelles (Mi), vesicles (Ve), and microemulsions (μE) often referred to a closed organized media (COM), i.e., assemblies of surfactant (or detergent) molecules endowed with an amphiphilic character.

Valuable specific properties characterize aqueous solutions of these organic aggregates, so that surface processes may take the place of bulk reactions and transients and/or reaction products may become localized in two distinct phases. Moreover, the surface electric potential associated with charged Mi or Ve can obviously become a key parameter when charge transfer and particularly electron ejection and transfer processes are involved. It can assist or inhibit the forward and backward excess electron reactions.

Colloid chemistry and radiation physics and chemistry have both known a long historical development but it is only in the 70s that some of their interests have merged.

While investigations of a large number of radiation-induced reactions in aqueous homogeneous systems had been prompted to a large extent by their relevance to biological processes, it became quite clear that extrapolation of dilute homogeneous solution data to complex heterogeneous biosystems was questionable.

Synthetic micellar or vesicular assemblies then appeared as good candidates to bridge the gap and mimic bioaggregates. The first observation of a catalytic effect of micelles in radiation chemistry dates back to 1969.[6] It was followed by the pioneering studies of Fendler,[7] Grätzel, Henglein, and Thomas.[8-12]

Regarding the attractive properties just mentioned, studies dealing with excess electron production or electron transfer (ET) reactions in closed organized media have developed with motivations of a fundamental or more or less applied character. Most importantly, the following can be mentioned:

1. To control the course of chemical or photochemical reactions, particularly the successive stages of the charge separation and recombination processes, this being related to the hope of a better understanding of the charge separation process itself in liquid homogeneous solutions
2. To increase the efficiency of charge separation, attempt related to the problem of the photochemical conversion and storage of solar energy
3. To determine the characteristics of thermal or photoinduced electron transfer reactions inside or across Mi or Ve boundaries, this being related to the frequently underlined analogy, (structural and functional) between artificial Ve and natural membranes
4. Conversely (as was applied repeatedly in homogeneous solutions) the excess electrons can serve as probes to characterize the system itself: this strategy has been adopted, for example, in reversed micelles

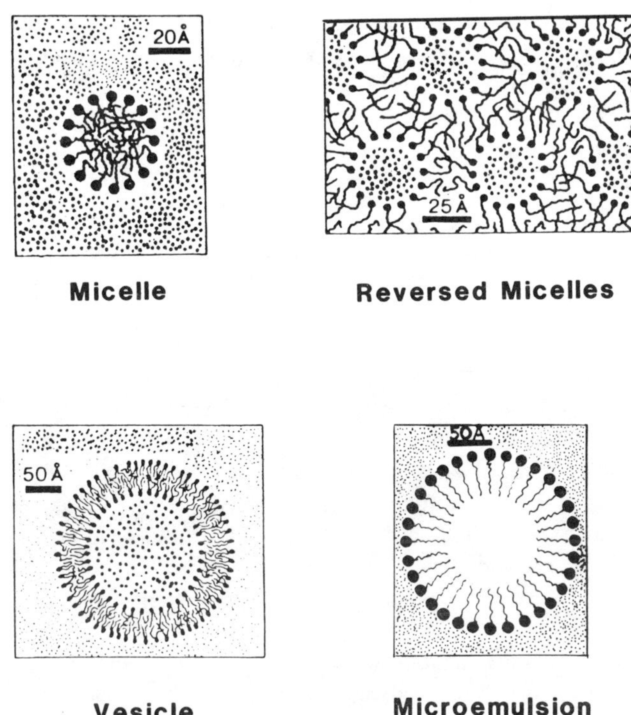

Micelle   Reversed Micelles

Vesicle   Microemulsion

FIGURE 1.   Schematic representation of various closed organized media.

Roughly speaking, the studies concerned with Mi, Ve, μE aggregates fall into two more or less distinct categories. They deal either with; (1) the structure and dynamics of the aggregates themselves; or (2) the chemistry and photochemistry occurring in Mi, Ve, μE solutions. Studies of the second type which clearly include excess electron production usually rest on relatively well-defined systems whose main physical properties have been determined from studies of the first type. Thus, a brief survey of the conditions which control the formation of COM, their main characteristics and potentials will first be summarized. In view of the present context, special reference will be made to the surface electric potential, particularly to the different methods advocated for its determination. The formation and decay of the excess electron will then be presented. Even though these two stages have often been treated in the same reports, they will now be considered in succession. Finally, the conditions for excess electron production and decay which are specific to microheterogeneous systems will be emphasized.

II. FORMATION AND SPECIFIC PROPERTIES OF CLOSED ORGANIZED MEDIA (COM)

A. FORMATION AND STRUCTURE

Amphiphilic molecules have distinct hydrophobic and hydrophilic moieties and can undergo self-association to give various organized assemblies in water or apolar solvents (Figure 1).

The formation of these aggregates has been extensively studied, particularly the energetics of the process.[13-15] In spite of some uncertainties concerning their structural and dynamic properties, such microheterogeneous systems — or at least typical ones — are now well characterized.[16,17]

Depending on their functional groups, surfactants can be neutral, cationic, anionic or zwitterionic (Table 1). As to the hydrophobic part of the amphiphile, it can be of different length and consist of one, two, or more linear hydrocarbon chains. The surfactant structure and concentration, as well as the nature and temperature of the solvent govern the aggregation process, but the preparative technique also intervenes.[18-21]

1. Direct Micelles

Direct micelles are among the simplest organized molecular systems. Micellization occurs spontaneously when the surfactant concentration is larger than a certain "critical micellar concentration" (cmc) and the temperature higher than a critical threshold which depends on the surfactant.

The hydrophobic character of the surfactant hydrocarbon chain which typically contains 8 to 18 carbon atoms plays an important role in the micellization process. The latter may be viewed as a thermodynamic analog of a phase separation, the cmc corresponding to a saturation concentration. The micelles, however, do not coalesce but are uniformly distributed in the bulk aqueous phase, forming a "pseudophase".

A hydrophobic energy and the competition energy-entropy, have been considered.[22] In accordance with the second principle of thermodynamics, micellization is accompanied by an increase of entropy:[23] the chains would be more free to move inside the Mi than in contact with water.[24]

In short, the repulsion between the polar headgroups, and association of the hydrophobic chains are responsible for the formation and stability of Mi.[24]

When the surfactant concentration is slightly above the cmc, the Mi are roughly spherical (hydrodynamic diameter 30 to 100 Å) and hold 50 to 100 molecules (the aggregation number). Only a substantial increase of surfactant or added electrolyte concentration leads to the formation of rod-like Mi which may then reach very large sizes.[25]

The micellar core is filled exclusively by chain segments which are not radially but randomly distributed; it is virtually devoid of internal water but substantial water-hydrocarbon contact occurs at the Mi interface which contains the polar head groups[26-28] (Figure 1).

It should be stressed that the interfacial area is fairly large: e.g., it amounts to 5m^2 for 1 cc of 0.1 M solution of NaLS, while the hydrocarbon phase represents only 2% of the total volume.

In the case of charged Mi, the effective charge density depends on the dissociation degree of the headgroups. The boundary or Stern layer (a few Å) consists of the detergent headgroups and bound counterions, whereas the immediate outside or Gouy-Chapman region (up to several hundred Å) contains unbound counterions.

The surfactant monomers are in dynamic equilibrium between the Mi and the bulk. Aqueous Mi rapidly break up and reform with two relaxation times: the release of a single surfactant molecule and its reincorporation occurs on a microsecond timescale; a complete disruption of Mi requires milliseconds.

2. Vesicles

Vesicles, like Mi, are binary systems, but unlike Mi, they are bilayers with two lipid-water interfaces, the lipid bilayer separating an inner water compartment from the external bulk water (Figure 1).

Ve may be formed from biological materials (liposomes) or from synthetic surfactants. Double chain surfactants aggregate as Ve rather than Mi when the volume ratio of the hydrophobic part over that of the headgroups is large, as discussed and rationalized by Nagarajan and Ruckenstein.[14]

Sonic dispersal of surfactants is the most general method for Ve preparation; it has to be carried out above the phase transition temperature, a temperature at which they are

TABLE 1
Typical Surfactants

Closed organized media	COM charge	Surfactant	Chemical formula	Critical concentration at 25°C
Direct micelle	Nonionic	Polyoxyethylene (10) p-1,1,3,3, tetramethyl butyl-phenol (Triton® X 100)	$C_8H_{17}C_6H_4(OCH_2CH_2)_{10}$ OH	0.54 Vol %
		Polyoxyethylene (23) dodecanol (Brij 35)	$C_{12}H_{25}(OCH_2CH_2)_{23}$ OH	6 to 9.1×10^{-5} M
	Anionic	Sodium dodecylsulfate (NaLS)	$CH_3(CH_2)_{11}$ O SO_3^- Na^+	8.1×10^{-3} M
		Sodium 4-n-dodecylbenzosulfonate	$CH_3(CH_2)_{11}$ $C_6H_4SO_3^-$ Na^+	1.6×10^{-3} M
	Cationic	Dodecyltrimethylammonium chloride (DTAC)	$CH_3(CH_2)_{11}N^+$ $(CH_3)_3$ Cl^-	2.1×10^{-2} M
		Hexadecyltrimethylammonium bromide (CTAB)	$CH_3(CH_2)_{15}N^+$ $(CH_3)_3$ Br^-	9.2×10^{-4} M
		Hexadecylpyridinium chloride (CPC)	$CH_3(CH_2)_{15}$ $(C_5H_5$ $N^+)$ Cl^-	9×10^{-4} M
Synthetic vesicle	Nonionic	DL α dipalmitoylphosphatidylcholine (DPPC)	$[CH_3(CH_2)_{14}CO$ $O]_2$ $C_3H_5PO_4^-$ $C_2H_4N^+$ $(CH_3)_3$	
	Anionic	Dihexadecylphosphate (DHP)	$[CH_3$ $(CH_2)_{15}$ $O]_2$ P O O^- H^+	
	Cationic	Dioctadecyldimethylammonium chloride (DODAC)	$[CH_3$ $(CH_2)_{17}]_2$ N^+ $(CH_3)_2$ Cl^-	$\sim 10^{-8}$ M
Reversed micelle	Anionic	Bis-(2 ethylhexyl)-sodium sulfosuccinate (AOT)	$C_8H_{17}CO_2CH_2CH$ SO_3^- Na^+ CO_2 C_8H_{17}	5×10^{-4} M in isooctane 6×10^{-4} M in CCl_4 10^{-3} M in cyclohexane
	Nonionic	Dodecylammonium propionate (DAP)	$CH_3(CH_2)_{11}$ $\overset{+}{N}H_3$ $\overset{-}{O}_2$ C CH_2 CH_3	2.1 to 2.5×10^{-2} M in CCl_4 $\sim 5 \times 10^{-3}$ M in benzene

transformed from a gel to a liquid. Ultrasonication leads first to the formation of onion-like multilamellar Ve; then with a controlled ultrasonic power, fairly uniform single compartment bilayer Ve are finally obtained.

Ve aggregates are significantly larger than Mi, with aggregation numbers of the order of 20,000 to 50,000 and diameters in the range 300 to 600 Å (Figure 1).

Once formed, Ve cannot be destroyed by dilution and their dynamic stability, greater than that of Mi, can be further enhanced by polymerization at different sites of the surfactant chain.[18]

The Ve size and shape are greatly influenced by the ionic environment and the inclusion of solutes. The three space regions, outer aqueous phase, lipid membrane, and inner aqueous phase, may contain a large number of guest molecules, a situation which is profitable in a few studies or applications. Nevertheless, ion migrations from the outside to the inside or vice versa occur until an equilibrium is reached.

3. Reversed Micelles and Microemulsions

Unlike direct Mi and Ve, reversed (or inverse) Mi are *ternary systems*. Some surfactants with appropriate lipophilic properties undergo self-association in apolar solvents.[29] This is the case, for example, of sodium bis-2-ethylhexyl-sulfosuccinate (aerosol OT or AOT) (Table 1) selected in most studies. Such surfactants are able to entrap water droplets in a hydrocarbon solvent, usually heptane or isooctane (Figure 1).

The size of the Mi increases with the concentration of the surfactant entrapped water (diameter 40 to 100 Å) and also the structure and state of the water molecules depend on w_o, the molar ratio H_2O/AOT, as repeatedly shown by various physical techniques: RMN spectroscopy,[30] fluorescence of probes,[31-33] photon correlation,[34] and Raman spectroscopy.[35]

As the water concentration increases, there is a progressive transition from reverse Mi to the so-called "water in oil" microemulsion. The term microemulsion designates isotropic, monophasic, thermodynamically stable systems consisting of the three basic constituents surfactant, oil, and water to which a cosurfactant is often added (typically butanol or pentanol) to increase the flexibility of the layers and the stability range of the μE.[36]

Their formation is spontaneous if the constituents are present in adequate proportions.[37] Phase diagrams allow selection of a composition which will provide a μE with desired properties.

As for Mi and Ve, ionic surfactants give charged interfaces which play an important role in the reactivity of solubilized species, especially in ET reactions. μE, of larger dimensions than Mi or Ve, are commonly used for extraction processes and in preparative chemistry.

B. INTERFACIAL AND TRANSMEMBRANE ELECTRIC POTENTIAL

Selection of an ionic surfactant allows the aggregate to possess one or two positively or negatively charged interfaces. The electrical potential ψ at the surface can then differ by a few hundred millivolts from that of the bulk. Such interfacial potential is known to affect a large number of physicochemical properties (e.g., colloid stability) and of biological phenomena (membrane permeability, cell fusion, etc.)[38] and obviously charge separation and recombination processes taking place across the interface of synthetic aggregates.

Unlike externally applied electric potentials, interfacial potentials are highly localized.

1. Theoretical Determination

For large colloid particles — and provided the surface charge density is known — ψ can be evaluated from the Gouy-Chapman theory of diffuse double layers extrapolated to curved interfaces.

Such extrapolation is, however, invalid for aggregates of micellar size. In the latter case, potential distance functions can be derived from the nonlinearized Poisson-Boltzmann equation which combines the Poisson equation with the assumption of a Boltzmann distribution of ions:

$$\frac{1}{r^2}\frac{d}{dr}\left(r^2\frac{d\psi}{dr}\right) = (8\pi en_{\pm}z_{\pm}/\epsilon)\ \sinh\ (z_{\pm}e\psi/kT) \tag{1}$$

where r is the distance from the center of the aggregate, $n_{\pm}$ is the number of ions of charge $z_{\pm}$ per cc in the bulk, and ϵ is the dielectric constant. While Equation 1 cannot be solved analytically for a spherical geometry, numerical[39,40] or approximate[41] solutions have been proposed.

A representation of $\psi(r)$ given in Reference 10 for the two typical sets of parameters given is shown in Figure 3.

2. Experimental Determinations

Among the various methods proposed, two are the most widespread.

Electrokinetic measurements — the electrophoretic mobility u of the aggregate is used to calculate ζ potentials from the Helmholtz-Smoluchovski equation:

$$\zeta = \frac{6\pi\eta_0}{RFD_0}\ u \tag{2}$$

where η_o and D_o represent the bulk viscosity and dielectric constant, respectively. However, by essence, the method is unable to take into account the Stern layer of the aggregate and the corresponding potential drop. The ratio ζ/ψ is thus a function of the surface charge density σ: it decreases when σ increases.

pK of lipoid pH indicators — Such pK is related to the interfacial potential and pK measurements have long been used to estimate ψ.[42,43] In effect, the apparent shift of pK detectable in Mi (Ve) solution as compared to pure aqueous solutions is due to a change in the proton activity a_{H+}^i at the interface of charged Mi (Ve) as compared to that of bulk water a_{H+}^w. The pH shift is itself related to ψ by Boltzmann's law:

$$a_{H+}^i = a_{H+}^W\ e^{-F\psi/RT} \tag{3}$$

For a given Mi (Ve), ψ can be modulated by the addition of variable concentrations of different electrolytes.[44,45] The results show that if the obtaining of absolute values is questionable, the relative values reported by different authors for different indicators agree satisfactorily.

Besides an interfacial potential, Ve bilayers may also exhibit a transmembrane electric potential.[46,47] The latter develops when different concentrations of a given ion or different ionic species at the same concentration exist in the two water compartments. Unlike ψ, which is a stationary state potential, the transmembrane potential is a dynamic quantity: it vanishes when the ion concentrations on both sides of the bilayer have reached an equilibrium.

C. SPECIFIC PROPERTIES OF AND REACTIVITY IN COM

Reactivity studies in COM must take into account some characteristic properties of the microheterogeneous system, namely its size and structure,[48] the charge density at the lipid-water interface, the viscosity, and stability of the aggregate (Table 2). One may note at this point that a control of the monodispersity of the surfactant assemblies can be gained from quasi-elastic light scattering methods.

In Mi, Ve, and μE the solubilization site can vary widely: it depends on the degree of

TABLE 2
Some Properties of Closed Organized Media

Media	Method of preparation	Shape hydrodynamic diameter	Viscosity, η, in c.pois (CP)	Interfacial electric constant, ϵ_{eff}
Direct micelles	Dissolution of the surfactant above the critical concentration (CMC) in water	Spheric diameter 40 to 100 Å rodlike: length $\gg$ 100 Å	A few tens of CP in the lipidic phase	$30 < \epsilon_{eff} < 60$
Vesicles	Ultrasonication of surfactant in water Alcohol injection Surfactant dialysis	Spheric with two aqueous phases. Diameter: 300 to 10,000 Å	$100 < \eta > 200$ CP in the lipidic phase	$30 < \epsilon_{eff} < 55$
Reversed micelles	Dissolution of the surfactant above the critical concentration in an apolar solvent and in presence of small amounts of H_2O	Spheric diameter: 40 to 100 Å	Depending on the water pool size: $\eta \searrow$ with (H_2O) $\nearrow$ $70 < \eta < 10$ CP for DAP	Depending on the water pool size $2 < \epsilon_{eff} < 5$ at AOT hydrocarbon interface
Microemulsions	Dissolution of appropriate concentrations of the surfactant and co-surfactant (alcohols) in water or in oil μemulsion, oil in water (o/w); μemulsion, water in oil (w/o)	Spheric diameter: 50 to 5000 Å	Depending on the ratio of the constituents. Generally about or less than 10 CP for (o/w) in the lipidic phase	$20 < \epsilon_{eff} < 40$

hydrophobicity and charge of the guest molecule. Among others, NMR spectroscopy,[49] ESR and spin echo techniques[50,51] have proved particularly valuable tools to locate the neutral solute and its cation.

In the case of Mi, the solubilization process is a dynamic one and solubilized molecules are in dynamic equilibrium between the water bulk and the aggregate.[52] In Ve, the solute local concentration may be quite high allowing bimolecular reactions which are otherwise infrequent in Mi under usual conditions. Distribution of solubilizates in Mi has been carefully analyzed with the conclusion that it is well described by a Poisson distribution.[53-55]

Diffusion of the reactants in the aggregate greatly affects the course of the reaction. Viscosity of the lipid phase depends on the nature of the surfactant and on the structure of the aggregate.

In a bilayer, larger viscosity values η and a more stable structure than in Mi or μE usually result in a very low mobility for the guest molecule (e.g., η which is of the order of 10 to 20 cp in spherical Mi[56,57] is about 150 cp in DODAC Ve[58]). In Ve, substrates are entrapped and permeabilities are low.

Reactivity studies must also take into account the existence of the ionic layer near the charged interfaces. The interfacial dielectric constant, always lower than that of the bulk, is an increasing function of the ion concentration.[59,60]

Finally, it may be mentioned that peculiar properties can be obtained from functionalized synthetic surfactants.[58]

In conclusion, and in spite of the numerous parameters involved, COM offer specific advantages which overbalance their complexity. They can be summarized as follows:

1. The reactants can be organized at a molecular level in Mi, and hydrophobic compounds can be studied in an aqueous environment in Mi or Ve
2. Surface reactions may replace bulk reactions, which carry some analogy with enzymatic catalysis
3. Transients (and reaction products) may be localized in distinct phases whereby transient lifetime may be drastically increased
4. The restricted space may confer specific interesting properties to the reaction products (e.g., polymerization in μE)
5. An electrical potential, high and local, may be interposed between the ion-pair partners, enhancing the efficiency of the charge separation process

Various studies of fundamental character as well as multiple applications have turned the above properties to account.

III. EXCESS ELECTRON PRODUCTION IN COM

As for homogeneous solutions, the excess electrons can be generated either by radiolysis or through the photoionization of suitable chromophores.

In the case of direct Mi, the micellar critical concentration usually corresponds to dilute surfactant solutions (10^{-4} to $10^{-2} M$). The radiolytically produced electrons are thus generated in the aqueous phase, in the absence of direct effects on the Mi aggregates, and scavenging from the spurs. The first stages of the electron production, localization, and solvation are therefore expected to be identical for homogeneous or direct Mi (Ve) solutions. The main purpose of pulse radiolysis experiments on such systems was, therefore, to investigate the catalytic effect of Mi (Ve) on the reactivity of e_{aq}^{-} with the surfactant molecules themselves or with additives dissolved within or at the periphery of the aggregate (see Section IV.B).

On the other hand, numerous studies of chromophores embedded in Mi (Ve) and photoionized under pulsed or continuous excitations have provided a wealth of informations concerning the dynamics and energetics of the excess electron production in COM.

At last, as in homogeneous solutions, the hydrated electron spectroscopic properties can be used to probe the surrounding water structure and state. This is specially illustrated by studies of the water pools of inverted Mi, based on both radiolysis and photolysis experiments.

A. DYNAMICS OF EXCESS ELECTRON PRODUCTION

Various photoionizable chromophores have been selected, mainly for their hydrophobic character and for the low values of their gas phase ionization potentials, typically in the range 6 to 8 eV. These are mostly polyarenes and substituted polyarenes: pyrene (Py) and aminopyrene, perylene (Pe) and aminoperylene, as well as phenothiazine (PTH) and methylphenothiazine (MPTH), tetramethylbenzidine (TMB), chlorophyll and bacteriochlorophyll.

Other miscellaneous electron donor-acceptor pairs have been employed in various electron transfer studies. In the present context, however, ET reactions will not be dealt with since excess electrons do not appear as separate distinct entities in such processes.

In direct Mi (Ve), the charge separation initial step is intralipidic. While the bulky solute cation remains confined in the hydrocarbon phase for a relatively long time, the photoelectron diffuses to the water and is spectroscopically or chemically characterized as e_{aq}^-.

In early studies, laser pulse excitations of COM were restricted to nanosecond resolution times and the attention mostly focused on the ionization mechanism and yields, on the influence of the Mi (Ve) structure and charge on the latter rather than on the dynamics of the solvated electron formation.[9,10,61]

Recent technological advances have now been achieved in obtaining ultrashort laser pulses,[62,63] leading to resolution times in the femtosecond domain and to correlative significant forward steps. Especially the existence of a precursor of the solvated electron (e_{presol}^-), asserted in the case of alcohols,[64,65] was directly demonstrated for the first time in photoionized pure water.[66] The evolution of the IR absorbing species to the classical blue e_{aq}^- was monitored and shown not to be accompanied by a continuous spectral shift, suggesting that the initial localized and the fully hydrated electron indeed correspond to two distinct electronic states.[66b]

The dynamics of the electron localization and hydration has also been investigated in organic assemblies using femtosecond time resolved absorption spectroscopy.[67-69] Hydrophobic PTH was incorporated in either anionic Mi of NaLS,[67-69] cationic Mi of CTAB[69] or dissolved in the hydrocarbon phase of reversed Mi solutions with various water/AOT molar ratios ($w_o \leqslant 60$).[67] Micellized PTH chromophore was photoionized by pulses of typically 100 fs duration, 5 to 20 μJ at 310 nm, and a resolution time of 50 fs.

Whatever the Mi system, a transient species absorbing in the IR ($\lambda \leqslant 1250$ nm) was observed and assigned, as in pure water, to a localized state of the electron, designated as e_{presol}^- (Figure 2). Note that in the micellar case the apparent shift of the e_{aq}^- 720 nm band, if real, might be explained by a partial overlap of the e_{aq}^- and PTH$^+$ absorption spectra.

Whereas the PTH$^+$ cation spectrum, peaking around 520 nm, appears within the laser pulse, i.e., in less than 100 fs, the rise time of the 1250 nm induced absorption is $T_1 = 250 \pm 20$ fs for PTH ionized in either positively or negatively charged direct Mi, and $T_1 = 140 \pm 20$ fs in the case of reversed Mi ($w_o \geqslant 30$).[67b] The latter figure is close to the value reported for pure water.[66] As suggested by Gauduel et al.,[67a] the difference in the trapping time observed in direct or reversed Mi, respectively, can likely be ascribed to a different localization of the chromophore in the hydrocarbon phase.

In the three micellar systems investigated (as in pure water) e_{presol}^- then relaxes to the fully hydrated state with a single time constant $T_2 = 270 \pm 30$ fs.

All these data led the authors to draw the following conclusions:[68,69]

1.   The presence of the Mi aggregate introduces a delay in the electron solvation process; such a time interval would be related to "the existence of an interfacial layer which

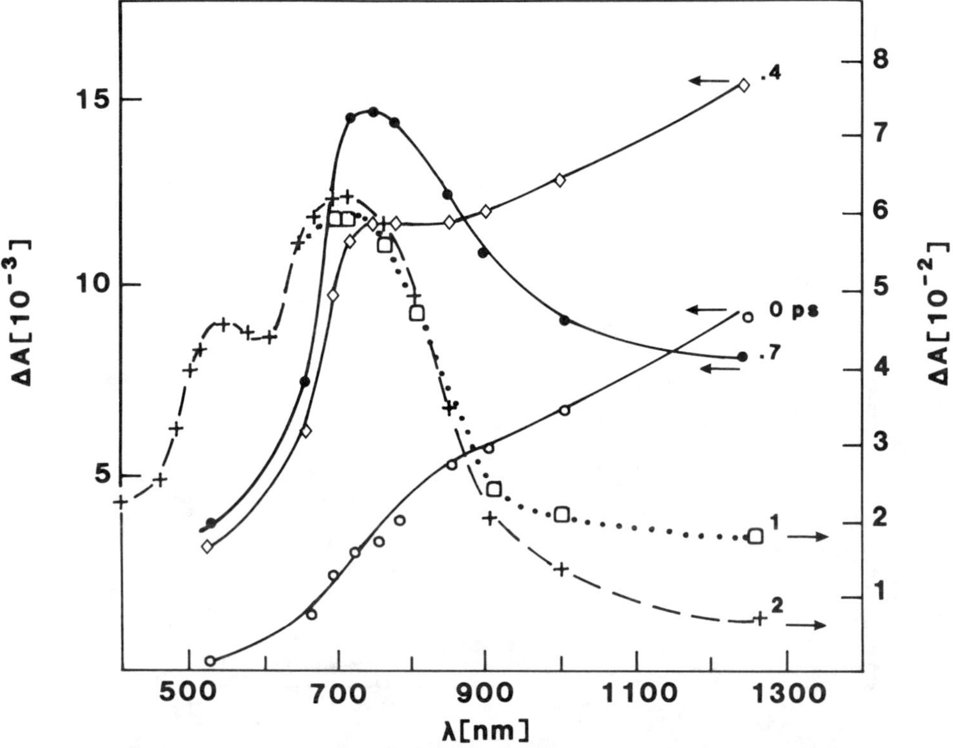

FIGURE 2. Transient absorption spectra from femtosecond laser experiments at different delays after photoionization. In pure water — solid line; [66b] in PTH-NaLS aqueous solution — dashed line.[68]

is able to modify the thermalization distance of e^- or the primary energy loss rate of epithermal electrons before trapping.'' Moreover, the observation of such a time delay seems to rule out the assumption of a predominant role of a high density of preexisting deep traps into which the electron would necessarily fall upon photoejection.

2. The dynamics of the primary stages of photoionization, including the intramicellar charge separation and the electron escape in the aqueous phase are similar, whatever the sign of the interfacial potential.

3. The similarity between the T_2 values measured in pure water on one hand and ionic Mi on the other, suggests that in Mi aggregates (and plausibly by analogy in Ve systems) the electron localization and solvation mainly occur in a space region where the local dielectric constant and ionic strength are tantamount to that of bulk water.

B. THERMALIZATION DISTANCE OF e^- AND DISTRIBUTION FUNCTIONS OF e_{aq}^-

To estimate the mean thermalization distance r_{th} of photoelectrons in Mi aqueous solutions, the e_{aq}^- lifetimes determined in either pulse radiolysis or laser photolysis experiments have been compared.[10] The former technique produces a homogeneous distribution of e_{aq}^-, whereas in the laser photolysis experiments e_{aq}^- are initially formed within the Gouy-Chapman layer where NO_3^- scavenger concentration is lower than in the bulk in the case of Mi^- (NaLS). A longer lifetime ensues which depends on the thermalization distance.

The data finally indicate that the thermalization range of the photoelectron is larger than the extension of the Gouy-Chapman layer, i.e., thermalization would occur at 20 to 40 Å from the Mi surface.

Such a conclusion is qualitatively substantiated by the data mentioned above, namely

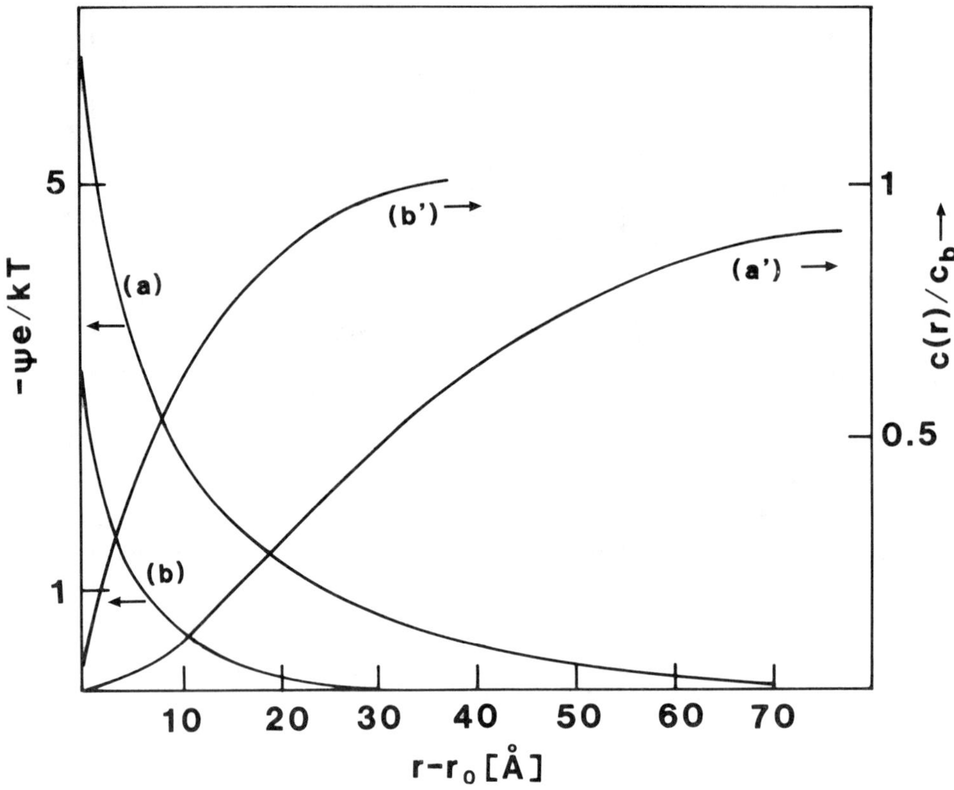

FIGURE 3.   Left ordinate — potential distance function in the diffuse double layer of NaLS micelles for different ionic strengths μ: (a) $\mu = 0.008\ M$; (b) $\mu = 0.1\ M$. Right ordinate — distribution function for monovalent negative ions: (a') $\mu = 0.008\ M$; (b') $\mu = 0.1\ M$.[10]

that the same time constant T_2 is required for e^-_{presol} to relax to e^-_{aq} in either pure water or various micellar solutions.

It may also be remarked that the above estimate of r_{th} is quite similar to the thermalization distance values characterizing low energy electrons (0.1 to 1 eV) and deduced from electrode to electrolyte photoemission electrode techniques.[70,72]

In the case of charged Mi (Ve), the electrostatic potential ψ in the Gouy-Chapman layer deviates from the potential in the water bulk as shown in Figure 3.

A gradient of ion concentration in the vicinity of Mi (Ve) is thus induced which is described by Boltzmann's law:

$$c(r) = c_b\,\exp(\mp Z_\pm\ e\psi/kT) \tag{4}$$

where $c(r)$ denotes the ion concentration at a distance $r - r_o$ from the Mi (Ve) surface, c_b that in the bulk, and $Z_\pm$ the charge of the ions.

Distribution functions deduced from the potential distance curves by means of the above equation[10] are also displayed in Figure 3 for monovalent negative ions and negatively charged Mi. They illustrate the expected inhomogeneous spatial distribution of e^-_{aq} around charged Mi or Ve.

C. ENERGETICS OF EXCESS ELECTRON PRODUCTION

Thermodynamical considerations lead to the assertion that the interfacial potential ψ of charged Mi or Ve should intervene in the energetics of the charge separation process.[11] That is, besides the adiabatic polarization P_+ of the medium by the micellized (Ve) cation and the fundamental energy of the ejected electron in a quasi-free (V_o) or localized state, negative values of ψ should contribute to lower the solute ionization potentials ($E_{i,Mi}$, $E_{i,Ve}$) as compared to corresponding gas phase values ($E_{i,gas}$).

Indeed the interest in the magnitude of the difference $\Delta E_i = E_{i,gas} - E_{i,Mi}$ was not only academic but stemmed from the prospect of inducing monophotonic charge separation with light in the visible part of the sun spectrum.

With the latter objective in mind, aminoperylene was photoionized in negatively charged Mi of NaLS by 20 ns neodymium laser pulses ($\lambda_{exc} = 530$ nm). The excess electron production was considered to be a one photon process, hence a difference ΔE_i amounting to 4.6 eV, explained not by an $e\psi$ contribution but by an anomalously high value of the polarization term.[73a]

It is well known, however, that discriminating between mono- and biphotonic processes may not be an easy task. The occurrence of a monophotonic ionization in the above experimental conditions has thus been questioned and a hindered ion recombination suggested to account for the results.[74] In turn, the arguments presented by Hall have been refuted.[73b]

Similarly, N_2 laser pulses (337 nm) were used to induce the photoionization of Py incorporated in negatively charged Ve (Ve^-) of DHP. The e^- ejection was again interpreted as a one photon process resulting in $\Delta E_i = 2.39$ eV, difference interpreted in terms of P_+ and V_o.[61] Later on however, the monophotonic character of the ionization process was also questioned.[58]

Conversely, the biphotonic character of the e^- ejection process under laser pulse excitation was turned to account for the estimate molecular photoionization thresholds.[75] Pyrene (Py) molecules occluded in Mi^- (NaLS) were first excited by a fixed frequency laser pulse, then 20 ns later the relaxed $^1Py^*$ state was excited by a tunable laser. The ionization potential was located by tuning the frequency of the second laser pulse until the appearance of the e_{aq}^- absorption and the depletion of the fluorescence from the $^1P_y^*$ state.

An extrapolated estimate of the ionization threshold of Py in NaLS was thus $E_{i,Mi} = 5.4 \pm 0.1$ eV, corresponding to $\Delta E_i = 2.3$ eV. Such a substantial lowering was then attributed to the electric field gradient at the Mi surface which would lower the barrier to formation of ions, analogous to Stark ionization of atomic Rydberg states.[75]

To locate the ionization thresholds of micellized solutes (perylene, tetracene, chlorophyll) a somewhat different strategy was adopted.[76,77] Continuous illuminations were performed with a xenon source fitted to a monochromator in which conditions the low light fluxes to totally disregard multiphoton processes and a large span of λ_{exc} is available. The excess electron formation was ascertained by means of negatively or positively charged electron scavengers (NO_3^-, H^+, Eu^{3+}).

Photoionization efficiency curves φe_{aq}^- vs. λ_{exc} were thus determined for Pe in Me_4Si and methanol homogeneous solutions or otherwise occluded in neutral (Mi^0) or negative $Mi(Mi^-)$.

The results are displayed in Figure 4 which calls for the following remarks:

1. Comparing Pe ionization in either Mi^0 or Mi^-, the electric potential at the Mi surface does not appear to appreciably alter the energetics of the charge separation process. This is not in contradiction, however, with the theoretical considerations developed;[11] in effect the expected electrostatic contribution of Mi^- is of the same order (~ 0.1 eV) as the experimental uncertainty attained in References 76 and 77. Moreover, such an $e\psi$ term is about one order of magnitude smaller than the estimated values of the P_+ and $V_{o,aq}$ terms, respectively.

2. The ΔE_i values obtained for Pe, Py, tetracene in Mi^- or Py in Ve^-,[61] are equal or very close which would imply similar values of the polarization terms in all these cases.

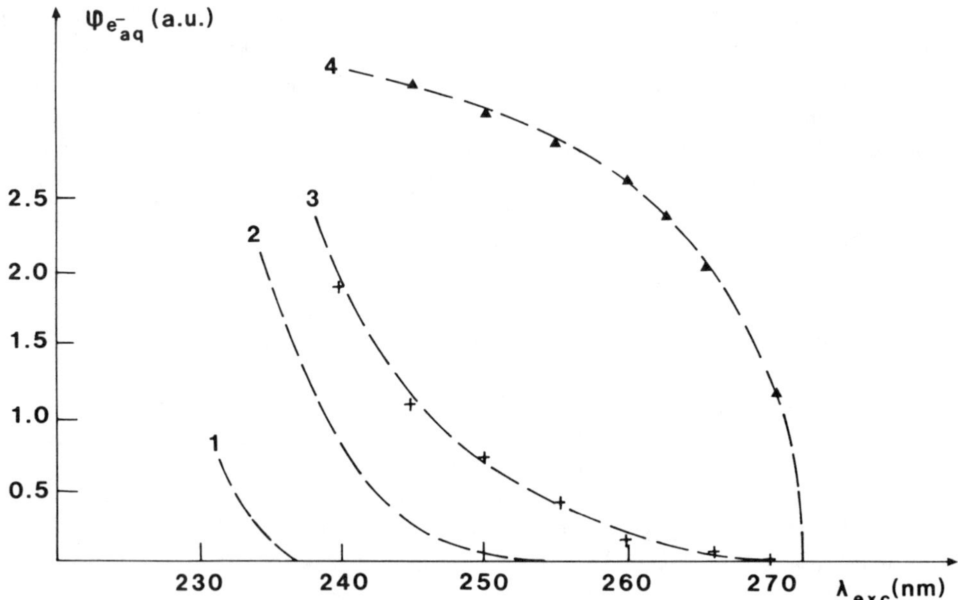

FIGURE 4. Photoionization yields of perylene vs. excitation light wavelength: (1) in tetramethylsilane; (2) in methanol; (3) in neutral micelles (Brij 35, 2.7×10^{-3} M); (4) in anionic micelles (NaLS 0.1 M).[86]

3.   The different shapes of the ionization yield curves for Mi^- on one hand, and Mi^0 or homogeneous solutions on the other, might reflect the increased probability for the electron to escape from geminate recombination in the former case.

4.   At a fixed λ_{exc} in the threshold region the overall ionization yields are larger the more negative the interfacial potential ψ is,[78] the difference vanishing when the photon energy exceeds the threshold energy by about 1 eV.

As argued,[75] the Mi electric field might interfere with the Rydberg-like states, precursors of the ion-pairs. The effect of the local field would also be to distort the Coulomb potential experienced by the photoelectron and thus to lower the barrier to the formation of ions.

In conclusion, the relatively low E_i and high ΔE_i values estimated for micellized Pe, Py, and tetracene would mainly result from the mere fact that one combines hydrophobic polyacenes with low gas phase ionization potentials with the low V_o value of water.[79]

D. EXCESS ELECTRONS AS PROBES OF THE WATER STATE IN REVERSED Mi

In addition to the physical techniques mentioned in Section II.A.3, data on the production and spectroscopic properties of the hydrated electron can be correlated with the water molecular organization in the water bubbles of inverse micelles.

e^-_{aq} may derive from the photoionization of aqueous electron donors,[80] hydrophobic solutes,[67b,81] or else from the solvent radiolysis.[31a,82,83] In the latter case, primary ionization obviously occurs in both phases, the excess electron initially formed in the hydrocarbon solvent being ultimately captured by the water pools which act as sinks for the ionic species.

Some of the reported results are contradictory as underlined and discussed in Reference 80. However there seems to be a general agreement on the following points:

1.   In radiolysis experiments, the electron capture efficiency by the water pools increases with w_o and the kinetics does not become diffusion controlled until $w_o \geqslant 30$.[82]

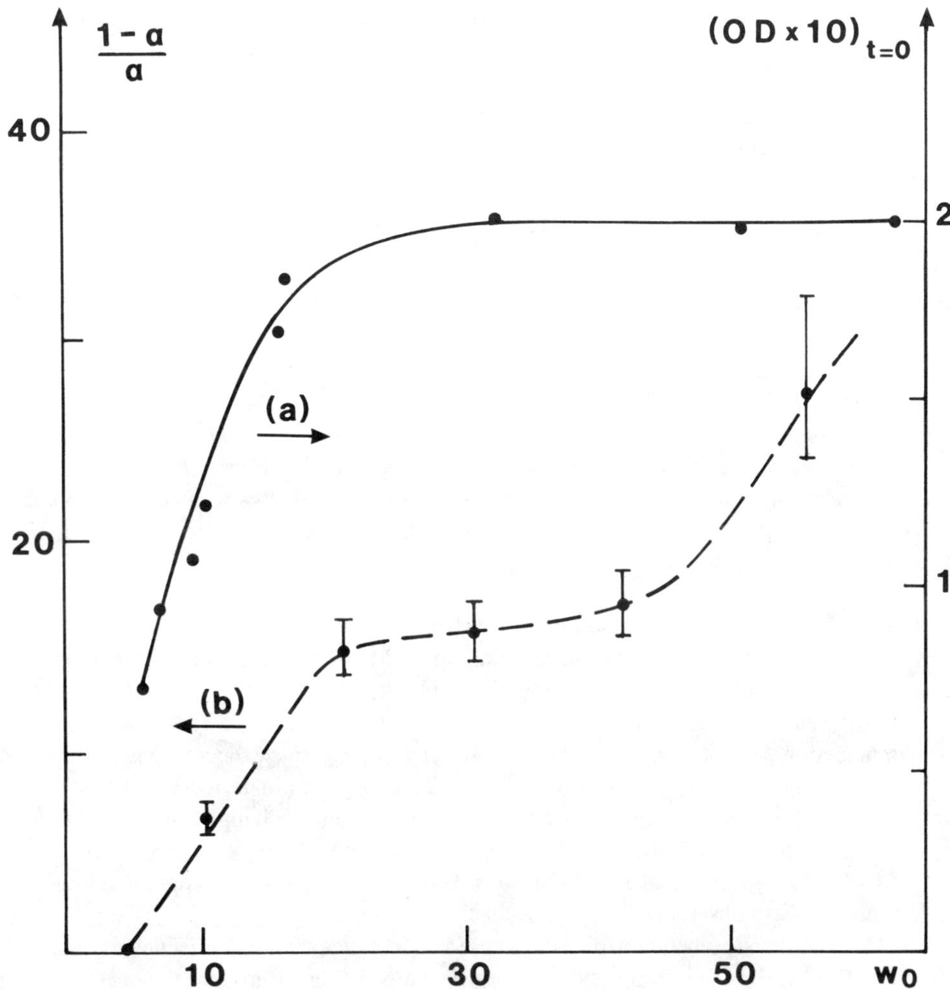

FIGURE 5. e_{aq}^- yields and water molecular organization in reversed micelles (AOT) of variable size: (a) optical density at 720 nm vs. H_2O/AOT molar ratio w_o;[83] (b) Raman contributions ratio of "bulk" $(1-\alpha)$ and bonded water (α) vs. w_o.[35]

2. e_{aq}^- are not observed below a certain critical size of the water pool ($w_o \simeq 5$ to 8).[31,67b,83] This finding correlates well with the observation that at such low H_2O concentration, the water is highly structured:[30,34] all the water molecules would be engaged in the solvation shells of the Na^+ counterions, hence unable to accommodate the excess electron. Such an interpretation is further substantiated by recent results[35] showing that the Raman spectra can be partitioned into two contributions linked, respectively, to the "bonded" and "bulk" water, the latter vanishing precisely for $w_o \simeq 5$ (Figure 5).

3.   For $w_o > 20$, the e_{aq}^- spectrum appears similar to that obtained in pure water[83] and by lowering w_o a blue shift is observed[31,67b,80,83] which could be associated with the high local concentration of Na^+ in the micellar core.[83]

4.   The band width of the e_{aq}^- absorption band has been reported to decrease with w_o[31,81] and also with the temperature; such narrowing has been taken to indicate a more ordered, ice-like environment.[81]

Altogether, even if the reported figures — and sometimes even the trend of their variation with w_o are somewhat discordant — the interpretation always associates the e_{aq}^- absorption characteristics with the state of aggregation of the water molecules in the Mi pools.

Hence, not only is there a concentration gradient of e_{aq}^- at the periphery of charged COM (Figure 3), but their spectroscopic properties depend on the state of organization of the water in the interfacial region. Such remarks should obviously be extended to biological membranes and biopolymers.

IV. DECAY OF EXCESS ELECTRONS IN COM

A. INTRA- OR EXTRAMICELLAR (Ve) DECAY OF DRY ELECTRONS

As compared to the wealth of information regarding the types and rate constants of e_{aq}^- reactions, the data relative to "dry" electron chemical behavior are rather scarce. In the present context, dry electron will denote any transitory state of the photoelectron prior to the solvated state.

As indicated above, when the photoionization primary act occurs in direct Mi or Ve, e_{dry}^- is produced in the lipid phase of the aggregate before it diffuses to water and solvates. Hence, COM may *a priori* constitute promising media to discriminate between e_{aq}^- and e_{dry}^- reactions and help to a better characterization of the latter.

1. Reactions of e_{dry}^- with Adsorbed or Solubilized Scavengers

A thorough study of dry electron scavenging in Mi has been reported,[10] the selected scavengers S differing mainly in their solubilization site.

Upon the photoionization of Py in the micellar core of Mi^-(NaLS), the efficiency of the photoelectron scavenging is reflected by the yield of e_{aq}^- since the scavenging reaction competes with the electron exit and hydration. Neglecting geminate recombination of the ion pair, the cation concentration is itself expected to remain unchanged upon the addition of S, except if the neutralization reaction, $S^- + Py^+ \rightarrow S + Py$, is taking place. In the case of Cu^{2+}, Cd^{2+} ions adsorbed in the Stern layer of Mi^-, both scavenging and neutralization reactions are observed.

On the other hand, when benzoic acid is used as a scavenger, the molecule arrangement is such that the carboxylic group resides close to the NaLS head groups, the aromatic ring protruding into the palisade layer. The dry electron capture then appears very efficient and the Py^+ decay kinetics seems to indicate that the intramicellar neutralization reaction is taking place.

A somewhat different situation prevails when the scavenger is solubilized in the inner core region of Mi, for example biphenyl or benzene. In such cases, a decrease of the e_{aq}^- yield is not observed, a tentative interpretation being that the scavenger encounters non-thermalized electrons, hence a correlative decrease of the electron capture cross section.

The last striking case is that of methylene iodide which shows a very efficient dry electron capture: the e_{aq}^- yield drops sharply at an average concentration of only one CH_2I_2 per micelle. The Py^+ yield first remains constant, then decreases as CH_2I_2 concentration increases.

The general conclusion of this series of experiments is that the scavenger molecules residing in the periphery region of Mi scavenge e_{dry}^- more efficiently than those incorporated in the Mi core.

Also, in reversed Mi solutions (AOT/heptane), dry electron scavenging has been ascertained. From experiments performed in the presence of SF_6 dissolved almost exclusively in the hydrocarbon phase, one can derive the relative proportion of excess electrons initially formed in each phase.[84] The results also demonstrate the competition between the dry electron scavenging by SF_6 and by AOT micelles.

2. Geminate Recombinations Involving Dry Electrons

Such recombinations, which may occur in the lipid phase of COM in the first 100 fs following the photoionization primary step, are not directly observable. They are not revealed by the above scavenging experiments; even for highly efficient e_{dry}^- capture an increase of Py^+ concentration was not observed.

Ultrafast geminate recombination may, however, be inferred from either their inhibition or their promotion. Stearic acid nitroxide spin labels (SASL) of different chain length incorporated in NaLS micelles have been shown to act both as an e_{dry}^- scavenger and as a cosurfactant: the carboxyl group would be anchored near the lipid-water interface, with the alkyl chain extending into the Mi interior.[85]

The photoionization of TMB was then evaluated from the intensity of the ESR signals of both TMB^+ and SASL with the result that upon the addition of spin labels of suitable chainlength, the TMB^+ yield increases while the SASL ESR signal disappears. Such observations clearly imply the inhibition of the dry electron geminate recombination by the spin label.

Such recombinations involving e_{dry}^- may also be deduced from experiments performed in the absence of scavengers, in Mi of increasing diameter.

Triphenylene, Py, or Pe have thus been photoionized in Mi^- whose surfactant chains contained 8, 12, or 16 carbon atoms with a radius of 11, 16.6, and 21 Å,[9] respectively. In such conditions, the e_{aq}^- yield was found lower for the longer hydrocarbon chain Mi, the effect being more pronounced when the photoelectron possessed low excess kinetic energy.

Along the same lines, linear or branched hydrocarbons have been used to swell NaLS micelles and the photoionization yield of micellized TMB evaluated from TMB^+ optical absorption.[86] When n-hexane is the swelling agent, one observes a regular decrease of TMB^+ yield as n-hexane concentration is increased, which is directly related to an increase in the Mi diameter and a concurrent repelling of TMB toward the micellar core.

For the other three hydrocarbons, a minimum is observed in the curves TMB^+ yield vs. hydrocarbon concentration. This was interpreted in the following way: for the relatively low hydrocarbon concentrations, an influence of the Mi dimension and TMB localization would favor the ion-pair geminate recombination, as in the n-hexane case; on the other hand, in the high concentration range, the electron mobility factor would prevail and assist the electron ejection out of the Mi. Indeed, in the ascending part of the curves, TMB^+ yields increase with the electron mobility value characterizing the hydrocarbon.

At any rate, the data[9,85,86] were taken to indicate a dry electron geminate recombination efficiently competing with the photoelectron exit and hydration in the aqueous phase.

B. DECAY OF HYDRATED ELECTRON

As in homogeneous solutions, following the fate of e_{aq}^- in COM has provided a means of studying reactions of the simplest model anion. Besides the classical decay reactions occurring in pure water: e_{aq}^- + cation, e_{aq}^- + OH, e_{aq}^- + e_{aq}^-, e_{aq}^- + O_2 or added electron scavengers, an unusual reactivity is disclosed in COM solutions, by virtue of their specific properties.

1. Reactions of e_{aq}^-

Apart from the scavengers added in the aqueous phase, the species responsible for e_{aq}^- decay are essentially the cations solubilized in either one of the two phases of COM, various micellized (Ve) solutes, the surfactant molecules themselves, H^+ or other ionic species adsorbed at the interface of Mi^- or Ve^-.

Investigations of such reactions (dealing mainly with micellar systems) display the important role of the interfacial electric potential which may either catalyze or inhibit the decay reaction as compared with homogeneous phase conditions (Table 3).

TABLE 3
Rate Constants of e^-_{aq} Decay

			Direct micelles			
	Water or	Anionic	Nonionic	Cationic		Reversed
Reactant	alcohol	NaLS	Triton X 100	CTAB	CPC	micelles AOT
			k $(M^{-1} s^{-1})$			
Pe$^+$			7.6×10^6 [l]	$< 3 \times 10^6$ [a]		
Pyrene	1.1×10^{10}	$< 10^7$ [a]	2×10^9 [a]	1×10^{11} [e]		
Pyrene sulfonic acid				2×10^{11} [b]		
Viologen sulfonate						4×10^{10} [c]
Benzene	1.3×10^7	4×10^6 [d]	6×10^6 [d]	1.1×10^8 [d]		
Biphenyl	5×10^9	1.3×10^8 [d]	4×10^9 [a]	2.4×10^{10} [d]		
Nitroanthracene		1.5×10^9 [e]		9×10^{10} [e]		
Vit K$_1$	1×10^{10} [f]	no reaction				
Duroquinone	3.1×10^{10} [f]	4.2×10^9 [f]				
Acetone	1.4×10^{10}	9×10^9 [g]				
Surfactant						
$< $ c.m.c.					7×10^9 [i]	4.7×10^7 [c]
$>$ c.m.c.		$\leqslant 2 \times 10^5$ [h]	3.8×10^7 [a]	$\leqslant 9 \times 10^5$ [h]	5×10^{12} [i]	
H$^+$	2.3×10^{10}	7.5×10^9 [d]				
Cu^{2+}	3×10^{10}	4.4×10^9 [d]		1.6×10^{10} [j]		
Cd^{2+}	5.4×10^{10}	4.5×10^{10} [j]				
NO$_3^-$	1×10^{10}	1.1×10^{10} [d]		1×10^{10} [d]		
βNAD$^+$	2.5×10^{10}	5.7×10^{10} [k]				

[a]Value from Reference 8; [b]Value from Reference 112; [c]Value from Reference 12; [d]Value from Reference 8; [e]Value from Reference 12; [f]Value from Reference 93; [g]Value from Reference 113; [h]Value from Reference 7; [i]Value from Reference 94; [j]Value from Reference 10; [k]Value from Reference 68; [l]Value from Reference 111.

a. e_{aq}^- + *Micellized Cation*

Whether in negative or positive Mi solutions no appreciable geminate recombination has been observed between $e_{localized}^-$ or e_{aq}^- and PTH$^+$ cation in the first 50 ps,[69] or in the ns time range upon Py photoionization.[10]

It has long been claimed that e_{aq}^- resulting from the photoionization of chromophores occluded in Mi$^-$(Ve$^-$) are prevented from recombining with cations by the repulsive Coulombic interaction with the charged interface. e_{aq}^- decay reactions are then generally the same as those encountered in pure water.

The decay kinetics of Pe$_{Mi}^+$ and e_{aq}^- have been recorded for Pe photoionized in *neutral* (Mi^o) *or anionic* (Mi$^-$) *micelles*[86] (Figure 6). In Mi^o (Triton X$_{100}$ or Brij) e_{aq}^- and Pe$^+$ are found to decay at the same rate and e_{aq}^- scavenging by Eu^{3+} demonstrate the occurrence of mutual recombination. On the contrary, in Mi$^-$ (NaLS), e_{aq}^- decays much faster than Py$_{Mi}^+$ or Pe$_{Mi}^+$ which exhibit relatively long lifetimes.

It should be underlined that very long lifetimes of cations occluded in Mi$^-$ have also been reported in various studies of electron transfer reactions. In this respect, TMB$^+$ offers a particularly striking example: it lives several hours in Mi$^-$,[87,88] instead of about 0.5 μs in homogeneous solution (3 methylpentane). In such a case, the weak reactivity of TMB$^+$ toward electron donors[89] combines with the electrostatic effect of the charged interface. As noted earlier, the interfacial electric potential would not only affect the distribution of charged reactants in the vicinity of the aggregate interface but also the electron transfer rate constant itself.[88,90]

In reversed Mi(AOT), the effect of the electrical interfacial potential has also been evidenced. One observes an easy trapping of the photoelectron by the water pools but the return of e_{aq}^- toward the cation located in the hydrocarbon phase is prevented by the repulsive potential.

On the contrary, when the interface of direct Mi(Ve) is *positively charged,* one may expect an enhanced reactivity of e_{aq}^-. This has been effectively observed in radiolysis experiments when relatively high concentrations of e_{aq}^- are created (see next paragraph).

On the other hand, e_{aq}^--micellized cation recombination cannot be inferred from photoionization data. Onsager distances should be significantly larger than in pure water — 18 Å has been given for Mi$^+$ = DTAC[10] — and ultrafast geminate recombinations very likely occur on a timescale shorter than the resolution time accessible in the laser excitation experiments.

As for e_{aq}^-, they would be attracted by all the surrounding Mi$^+$, i.e., mostly by empty ones ($\sim$99% in typical experimental conditions) with no evidence of geminate or nongeminate recombinations.

b. e_{aq}^- + *Micellized (Ve) Solute*

Reactions of e_{aq}^- with various micellized solutes have been extensively studied by pulse radiolysis techniques.[7,8]

Typically e_{aq}^- decay rates were compared for pyrene incorporated in Mi$^-$, Mi$^+$ or Mi^0.[91] The reaction Py$_{Mi}$ + $e_{aq}^- \rightarrow$ Py$_{Mi}^-$ is inhibited by negatively charged interfaces and favored by positive ones as compared to homogeneous phase conditions (Table 3). For an electrically neutral interface, the reaction rate is smaller than in homogeneous solution, which is ascribed to the reluctance of e_{aq}^- to leave the aqueous phase and enter the less polar region of the probe.[92]

In vesicular systems, the reaction rate of e_{aq}^- with the lipophile probe may be lower than in Mi when the probe is able to deeply penetrate in the thick lipid bilayer.[90]

Similarly, in direct Mi, the role of the separation distance e_{aq}^--reactant was evidenced. For example, the highly hydrophobic character of the electron acceptor vitamin K$_1$ provokes its penetration in the micellar core and prevents its reaction with e_{aq}^-, in contrast to the other quinones investigated.[93]

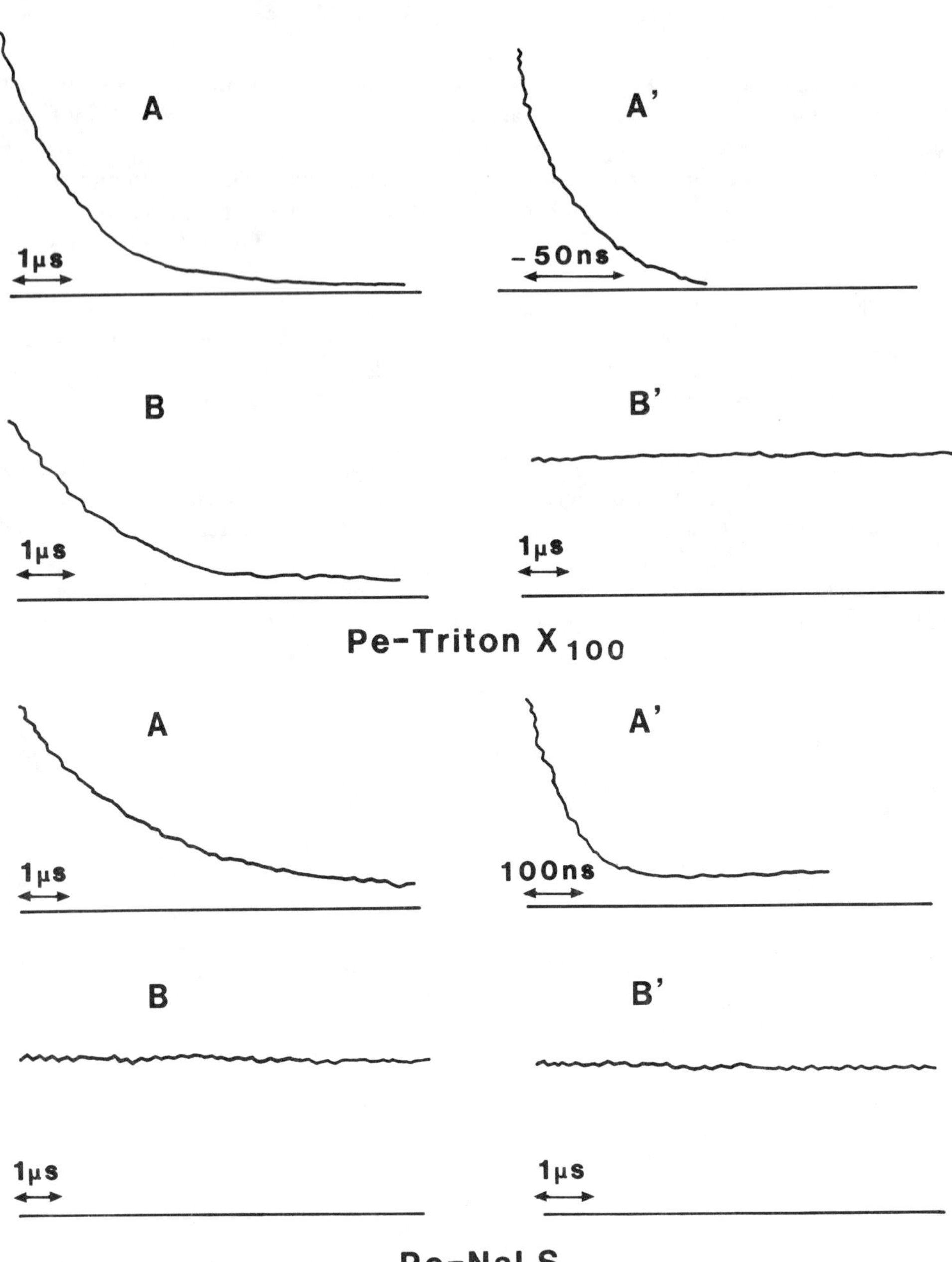

FIGURE 6. e_{aq}^- and Pe_{Mi}^+ decay upon laser photoionization (λ_{exc} = 265 nm) of perylene (Pe) in neutral micelles (Triton X 100) and in anionic micelles (NaLS);[86] (A) e_{aq}^- decay (λ_{an} = 720 nm); (B) Pe_{Mi+} decay (λ_{an} = 540 nm). A' and B' — same as A and B but in presence of Eu^{3+} (5.10^{-4} *M*).

It may then be concluded that the solubilization site of the probe as well as the size and compactness of the aggregate head groups greatly influence the e_{aq}^- decay reactions.

In the water pools of reversed Mi, solutes such as O_2, NO_3^-, and N_2O react with e_{aq}^- produced either by radiolysis or by the photoionization of aminopyrene.[84] Rate constants are found comparable to those determined in bulk water. Nevertheless, e_{aq}^- lifetime in the absence of scavengers was correlated with the viscosity, i.e., with the size of the water pools.[81] In small micelles (w_o <15), e_{aq}^- reacts within the viscous hydration shell of the surfactant counterions. Fewer collisions result in longer lifetimes and smaller rate constants.

c. e_{aq}^- + *Surfactant*

Cosurfactants, such as methylviologen derivatives, or surfactants whose headgroups are recognized as good electron acceptors, lead to very efficient trapping of e_{aq}^-. A typical example is cetylpyridinium chloride (CPCl) as pyridinium cations react very rapidly with e_{aq}^-.[94]

While the rate constant for the reaction of e_{aq}^- with monomeric CPCl is 7×10^9 M^{-1} s^{-1}, it becomes larger than 10^{12} $M^{-1}s^{-1}$ when CPCl is in an aggregated state and the reactant concentration referred to the micellar concentration.

The interpretation of such an increased reactivity is based on the electrical properties of the micellar double layer and the correlative high value of the Onsager distance. In accordance with such a view, an increase of the ionic strength which decreases the surface potential and the reciprocal Debye length was found to result in a significant decrease of the e_{aq}^- scavenging rate.

d. e_{aq}^- + *Adsorbed Ionic Scavengers*

As emphasized above (Section III.B), when the aggregate interface is electrically charged, the distribution of ionic species in the aqueous phase (whether e_{aq}^- or e_{aq}^- scavengers) is clearly quite inhomogeneous and, as expected, the adsorption affinity of ionic scavengers on oppositely charged interfaces increases with their hydrophobic character.

For *negatively charged scavengers* such as NO_3^-, the capture of e_{aq}^- will occur exclusively in the bulk in the case of $Mi^-(Ve^-)$ solutions, while NO_3^-, adsorbed or concentrated at the interface of $Mi^+(Ve^+)$ is expected to catalyze the scavenging reaction.

Conversely when a *positive scavenger* is associated with $Mi^+(Ve^+)$, scavenging should occur predominantly in the bulk; in the case of $Mi^-(Ve^-)$ attention has been focused mainly on H^+ and divalent ions Cu^{2+} and Cd^{2+}.

Various experimental techniques have confirmed that H^+ ions are concentrated or even "incorporated" into negative NaLS micelles.[95] Similarly, a high local concentration of H^+ has been ascertained in the case of Ve^-.[61] The observed molecular hydrogen would then result from concurrent e_{aq}^- scavenging in the bulk and in vesicle interfacial region.

More quantitative experiments on Cu^{2+} and Cd^{2+} ions have shown that such divalent ions are much less reactive toward e_{aq}^- when adsorbed in the Stern layer than when free in the aqueous phase, but a tenfold increase of the rate constant is observed upon the addition of 0.1 M NaCl which leads to a less negative interface potential.[10]

C. DECAY KINETICS

Only in a limited number of situations can the e_{aq}^- decay be expressed in terms of homogeneous kinetics. This is clearly the case when the reaction occurs in the bulk of direct Mi, far from the interface where the reactant diffusion is not influenced by the charged surfactants.

Also in large aggregates like microemulsions (oil in water), diffusion within the lipidic phase is little restricted and the kinetic behavior can approach that encountered in homogeneous solutions. Similarly, in large water pools of reversed micelles, when e_{aq}^- reacts with aqueous scavengers, a second order kinetics is observed and the reaction rate is the same as in homogeneous systems.

Most generally however, reactivities in COM cannot be described in terms of homogeneous kinetics. The structural and dynamic features of the systems as well as the partitioning of the reactants have to be taken into account.[96]

Different treatments of reactivities in COM have been proposed. For reactions slower than the reorganization time of the system investigated, a simple model of partitioning of the reactants and later on a model of pseudophase ion exchange have been elaborated.[97] In the case of micellar system, reactivities have been treated in the framework of stochastic and lattice statistical theories.[98] Along these lines the role of dimensionality and spatial extent in micellar kinetic processes has been considered.[99] For highly simplified systems, the reaction of two molecules or ions in the core of Mi or diffusing on its surface is expected to obey a pseudo-first order law, which qualitatively agrees with experimental findings.

Usually, however, the reaction kinetics are more complex, being influenced by a number of factors inherent in the specific characteristics of COM; so that the kinetics differ from one system to another, and in one particular system from one reaction to another.

Such complexity may be illustrated by the following typical examples dealing with excess electrons. When the electrons are photochemically produced, some concurrent alteration of the aggregate may occur (structure of the aggregate, motion of the cation, etc.) and during a first period of time, a multiexponential decay is frequently observed. Kinetics thus vary with time defining a reaction order which appears somewhat arbitrary.

Some structure modifications can also be induced by a temperature change. Thus, two e_{aq}^- decays are observed in DHP vesicles, one below the transition temperature (46°C) and the other one above. Two pseudo-first order decays are observed successively when the Ve structure is the most rigid. At 46°C, only one pseudo-first order is obtained, indicating a morphological change which influences the reaction rate.[61]

As to an e_{aq}^- decay mechanism, it has been considered in the case of an electron transfer between e_{aq}^- and acceptors (pyrene, 9 nitroanthracene) incorporated in the lipoidic part of Mi^+ or Mi^-.[12] It was thus stressed that the electron must be transferred through the electrical double layer between the two phases. Since the occupied electronic levels of the redox system aq/e_{aq}^- lie much lower than the levels of unoccupied solvent traps and the level of the delocalized electron in hydrocarbons, an electron cannot spontaneously move from water into the lipoidic part of the Mi. A tunneling mechanism through the double layer was thus proposed. In the reverse situation when electron photoinjection occurs in the Mi^- lipid phase, electron tunneling has also been invoked to account for the e_{dry}^- exit out of the micelle.

It may finally be emphasized that electron transfer in bioaggregates may occur via a similar tunneling mechanism and that knowing the detailed ''mechanism of electron transfer between the aqueous and lipoidic phases should be of importance for the understanding of redox processes in biochemical systems''.[12]

V. SPECIFIC CONDITIONS FOR EXCESS ELECTRON PRODUCTION IN COM

A. REDUCED DIMENSIONALITY AND INCREASED VISCOSITY

A priori, such specific properties are expected to mainly affect bimolecular secondary processes. Mi aggregates acting as a ''supercage'' and some chemical selectivity may be achieved.[100] Similarly, the segregation of reactants and the reduced reaction volume inhibit triplet quenching and T–T annihilation, allowing for phosphorescence observation for room temperature fluid solutions.[101,102] As mentioned above, the reactant confinement leads to kinetic equations which differ from the laws describing homogeneous phase bimolecular reaction kinetics.[55,103]

COM may intervene also at an earlier photophysical stage and affect the competition between the various relaxation paths of the electronic excitation energy. Thus, isomerization of transstilbene is hindered in Mi or Ve systems as compared to homogeneous solutions while the fluorescence quantum yield is found to increase.[104] In a similar way, cyanine dyes (which play an important role as photosensitizers in photographic processes) show a remarkable increase of photostability when incorporated in micellar aggregates. The rigidization of the dye related to the medium high viscosity would inhibit the nonradiative decay, thereby increasing the fluorescent state lifetime and fluorescence yield, as well as the molecular stability.[105]

As for electron photoejection from micellized solutes, it does not seem to be notably affected by the chromophore confinement and rigidization, even though monophotonic ionization most generally derives from the first excited singlet state, vibrationally relaxed or not. The fluorescence, triplet state, and ionization yields have been estimated for TMB photoexcited in different environments and the φ_F, φ_T, and φ_+ found nearly equivalent for Mi°, Mi$^+$ and acetonitrile homogeneous solutions.[106]

B. COEXISTENCE OF TWO PHASES

As emphasized above, the juxtaposition in COM of the lipid and aqueous phases permits the combination of hydrophobic chromophores (often characterized by low gas phase ionization potentials) with a photoelectron trapping and solvation in water. This unusual combination results in low ionization threshold energy values E_i of the micellized (Ve) solute and a notable lowering (~ 1 eV) of E_i as compared with homogeneous hydrocarbon solutions.[76]

Such phase coexistence also results in the decrease of the reaction rates when the reactive transients are located in two distinct phases. This has been observed for electron transfer from aqueous nucleophiles to micellized (Ve) radical cations. A striking case is that of micellized TMB$^+$ cation which may live several minutes in neutral Mi, as compared to a few microseconds in 3 MP solutions.[88]

The water-lipid interface should not, however, be viewed as a continuous invariant structure separating two independent phases.

The more or less open structure of COM — particularly of direct Mi — the water penetration inside the aggregate, and the water interaction with the sequestered solute in its neutral or oxidized form have been the subject of various experimental studies and prolonged controversy. In particular, experiments on solid (77 K) Mi or Ve solutions have emphasized that the water-micellized solute interaction would largely contribute to the electron photoejection efficiency.[107]

It should also be pointed out that, when electrically charged Mi or Ve are concerned, the water dipole orientation is anticipated to differ for positive or negative interfaces. Moreover, the free or bound state of water depends on the presence and distribution of counterions. On the analogy of reverse micelle data, it may be expected that for direct Mi$^\pm$ the water molecules engaged in ion hydration shells would not be able to accommodate the excess electrons.

C. INTERFACIAL ELECTRIC POTENTIAL Ψ

The role of the surface electric potential is manifold and seemingly complex. Contrary to expectation, it does not appreciably modify the micellized (Ve) solute *ionization potentials*. Typically, for monophotonic electron ejection from micellized Pe, the threshold was practically unchanged upon the addition of various electrolytes to NaLS solutions at concentrations which allow modulation of ψ without significantly altering the micellar structure.[78] As noted above (Section III.C), such findings can be accounted for when considering that $e\psi$ is of the same order (~ 0.1 eV) as the experimental accuracy attained in[76-78] and at least one order of magnitude smaller than the other energy terms contributing to E_i.

The effect of ψ on the photoionization yields φ_{ion} is still a matter of debate. From *femtosecond* laser pulses (310 nm) one might approach a primary ionization cross section φ_{ip}. On the other hand, a *two photon* ionization of micellized PTH provides an electronic excitation energy which largely exceeds the ionization threshold. In such conditions, the concentration of e_{aq}^-, 2 ps after the photoionization step, was found to be equal for Mi$^+$ and Mi$^-$, suggesting that neither φ_{ip} nor the electron escape probability W are influenced by the surface electric potential.[69]

By contrast, for *nanosecond* laser flash excitation, *monophotonic* ionization of TMB led to much larger values of $\varphi_{e_{aq}^-}$ in Mi$^-$ (NaLS) than in Mi$^+$ (DTAC),[87,106] whereas on the contrary, *biphotonic* ionization of various aromatics resulted in equivalent e_{aq}^- yields in the same Mi^+ and Mi^- as above.[9] Such different behavior was then attributed to the high excess kinetic energy of the photoelectron in the latter case (~ 1 to 2 eV) which would occult the electrostatic contribution of the charged interface.[87] A similar conclusion has also been drawn from steady-state experiments which have disclosed that the difference in $\varphi_{e_{aq}^-}$ noted in the threshold region for negatively charged or uncharged Mi, respectively, can vanish for high enough photon energy (Figure 4).

The various micelle + chromophore relaxation times being several orders of magnitude longer than the laser flash duration,[9,87,106] it seems that the effect of ψ on *overall* ionization yields φ_{ion} is genuine, affecting either a primary ionization yield, φ_{ip}, or the excess electron escape probability, W, or both.

As mentioned above (IVB), following the photoionization of micellized solutes, no e_{aq}^--cation recombinations have been recorded in Mi^+ or Mi^- systems in the 100 fs to 0.1 ns time range[69] nor in the nanosecond domain,[10,86] while mutual recombinations have been shown to occur in Mi^0 solutions.[86]

Considering the regular decrease of φ_{ion} when Mi^- is progressively neutralized, it seems plausible that such a decrease originates from a progressively allowed return of the excess electrons, thermalized or fully hydrated, toward the micellized parent cation, as observed for Mi^0.

One may notice at this point that the linear relationship φ_{ion} vs. ψ obtained from various Mi^- or Ve^- electrolyte solutions[78,90] bears some analogy with Onsager's prediction[108] that the fraction of escaping ions in the incipient stage of the process is proportional to the intensity of an externally applied electric field.

As to the difference in overall yields, $\varphi_{ion,Mi^+} \ll \varphi_{ion,Mi^-}$, it can result not only from more or less efficient $e_{aq}^- + cation_{Mi}$ recombination but also from a different fate, intra-or per-imicellar, of the dry electron.

In $Mi^-(Ve^-)$ the photoelectron would tunnel through the electrostatic barrier[11] and once the barrier crossed, negative potential will assist in the removal of excess electrons. Conversely, one may speculate that in $Mi^+(Ve^+)$ the photoelectron would be temporarily trapped in the Mi(Ve) interface region associating with a surfactant molecule. The competing decay channels would then be either an ultrafast highly probable intramicellar geminate recombination or its exit out of the Mi(Ve) in a free or associated state ultimately dissociating in water.

Last but not least, reaction rates are altered by the existence and variation of ψ, particularly when the reaction partners are located on each side of the charged interface. As evoked above (Section IV.B.1.b) a typical situation is that of e_{aq}^- radiolytically produced in the water bulk and reacting with micellized pyrene.

The reaction $Py_{Mi} + e_{aq}^-$ was thus shown to be inhibited by negative interfaces and favored by positive ones as compared to homogeneous solutions.[91]

Similarly, the rate of the electron transfer from the triplet state of PTH to metal ions (Cu^{2+}, Eu^{3+}) was compared for PTH dissolved in methanol or confined in Mi^- aggregates, in which case the positive ions are strongly adsorbed on the Mi surface. The specific rate of the transfer reaction was then found considerably larger in the Mi^- case than in methanolic homogeneous solutions. This finding was taken to support the occurrence of an electron tunneling and the assertion that a negative potential assists the excess electron escape from the aggregate.[109]

The effect of ψ on reaction rates, in fact, arises from the combination of two distinct factors:

1. The first factor is the spatial distribution of charge species, especially e_{aq}^-, in the vicinity of $Mi^\pm$, $Ve^\pm$ interfaces, this aspect has been considered above (Section III.B) and displayed in Figure 3.

2. The rate constants themselves are affected by the interfacial potential. This has been demonstrated again from the electron transfer process responsible for the decay of TMB^+ cation occluded in either Mi^- or Ve^- of variable interfacial charge density. In such conditions, the decay rate constant was found to be an exponential function of ψ which would reflect the existence of an energy barrier separating $TMB^+_{Mi^-, Ve^-}$ and the electron donating aqueous nucleophile.[88,90] Note that such a law is reminiscent of the exponential Butler-Volmer expression which governs the rate of electron transfer at an electrode-electrolyte interface under the influence of an electric field. It exemplifies the frequently evoked analogy between charged surfactant assemblies and microelectrodes, despite the fact that the former are porous to ions.

On the whole and in spite of some controversial points, the various reported data illustrate the direct or indirect role played by the interfacial electric potential. The latter would not interfere with the valence to ionized state transition but first at the stage of the dry electron transfer from the lipid to the aqueous phase. Furthermore, ψ controls the spatial distribution of charged reactants around the charged aggregates and regulates the subsequent e^-_{aq} reactions.

VI. CONCLUSIONS

Concerning excess electron production and decay, closed organized systems reproduce some of the features characterizing homogeneous solutions but some new ones are introduced as detailed in the preceding paragraph.

As in pure water, photoelectron solvation does not proceed through a single step and, in micellar systems, a presolvated state has also been directly identified. Such presolvation occurs in a space region possessing no specific structure imposed by the interface and the energy loss mechanism of e^-_{presol} appears comparable in liquid water and aqueous Mi.

On the other hand, the dissimilar shapes of the photoionization efficiency curves observed for neutral Mi and homogeneous solutions and negatively charged Mi may be tentatively correlated with hindered geminate recombination in the latter case. As a corollary the shape of ionization yield curves relative to homogeneous liquid solutions would be essentially governed by thermal electron back transfer.

The combined existence of two phases and of an interface electric potential definitely results in an enhanced efficiency of ion-pair separation and a relative stabilization of the ionic products. Mi and Ve (particularly when built up from functionalized surfactants) have thus appeared as good candidates for solar energy harvesting and storage. However, such aggregates nowadays seem superseded by more sophisticated hybrid systems e.g., colloidal semiconductors loaded with various catalysts or entrapped in natural or synthetic vesicles.

On the other hand, in the rapidly developing field of molecular electronics, COM may constitute beneficial matrices for molecular devices and electron transfer reactions. Thus, guest molecules incorporated in a bilayer membrane may function as a continuous transmembrane electron channel, i.e., as a molecular wire.[110]

The use of Mi and Ve to mimic bioaggregates still appears as a wide promising research area. Artificial aggregates serve as model systems to study excess electron ejection or electron transfer reactions across, on, or near neutral or charged biomembranes. Also, encapsulation of a hydrophobic electron donor permits study of the reduction of aqueous compounds of biological interest by e^-_{aq} or one of its precursors, with no interference of the oxidized form of the donor, at least in short time range kinetic studies.

The structural complexity and dynamic character of COM are such that experimental investigations so far have largely outnumbered theoretical treatments. But the wealth of reported data as well as the present computer facilities make the field ripe for theoreticians to bring forth their much needed contributions.

Otherwise, the complexity and variability of surfactant assemblies may be turned to account for and permit a controlled modulation or "tailoring" of structural and chemical properties.

Finally, we wish to stress that closed organized media illustrate the paradoxical situation in which *complex systems* may lead to *simplified processes:* they can control the competition between different decay channels and inhibit reverse reactions.

ACKNOWLEDGMENT

Our gratitude goes to Sydney Leach without whom this chapter would probably not have been written.

REFERENCES

1. **Somorjai, G. A.,** *Chemistry in two dimensions: surfaces,* Cornell University Press, Ithaca, New York, 1981.
2. **Adamson, A. W.,** *Physical chemistry of surfaces,* Wiley, New York, 1982.
3. **Thomas, J. K.,** Characterization of surfaces by excited states, *J. Phys. Chem.,* 91, 267, 1987.
4a. **Fendler, J. H.,** Photochemical solar energy conversion; an assessment of scientific accomplishments, *J. Phys. Chem.,* 89, 2730, 1985.
4b. **Fendler, J. H.,** Atomic and molecular clusters in membrane mimetic chemistry, *Chem. Rev.,* 87, 877, 1987.
5. **Grätzel, M.,** Artificial photosynthesis: water cleavage into hydrogen and oxygen by visible light, *Acc. Chem. Res.,* 14, 376, 1981.
6. **Gebicki, J. M. and Allen, A. O.,** Relationship between critical micellar concentration and rate of radiolysis of aqueous sodium linoleate, *J. Phys. Chem.,* 73, 2443, 1969.
7. **Bansal, K. M., Patterson, L. K., Fendler, E. J., and Fendler, J. H.,** Reactions of hydrated electrons, hydrogen atoms and hydroxyl radicals in micellar systems, *Int. J. Radiat. Phys. Chem.,* 3, 321, 1971.
8. **Wallace, S. C. and Thomas, J. K.,** Reactions in micellar systems, *Rad. Res.,* 54, 49, 1973.
9. **Wallace, S. C., Grätzel, M., and Thomas, J. K.,** Laser photoionization of aromatic hydrocarbons in micellar solutions, *Chem. Phys. Lett.,* 23, 359, 1973.
10. **Grätzel, M. and Thomas, J. K.,** Laser photoionization in micellar solutions. The fate of photoelectrons, *J. Phys. Chem.,* 78, 2248, 1974.
11. **Alkaitis, S. A., Grätzel, M., and Henglein, A.,** Laser photoionization of phenothiazine in micellar solutions. Mechanism and light induced redox reactions with quinones, *Ber. Bunsenges. Physik. Chem.,* 79, 541, 1975.
12. **Grätzel, M., Henglein, A., and Janata, E.,** Mechanism of electron transfer form e_{aq}^- to acceptors in micelles, *Ber. Bunsenges. Physik. Chem.,* 79, 475, 1975.
13. **Fromherz, P.,** Micelle structure: a surfactant-block model, *Chem. Phys. Lett.,* 77, 460, 1981.
14. **Nagarajan, R. and Ruckenstein, E.,** Aggregation of amphiphiles as micelles or vesicles in aqueous media, *J. Coll. Interface Sci.,* 71, 580, 1979.
15. **Stigter, D.,** Micelle formation by ionic surfactants. IV. Electrostatic and hydrophobic free energy from Stern-Gouy ionic double layer, *J. Phys. Chem.,* 79, 1015, 1975.
16. **Menger, F. M. and Doll, D. W.,** On the structure of micelles, *J. Am. Chem. Soc.,* 106, 1109, 1984.
17. **Degorgio, V., Lang, J., Realto, L., and Leng, C.,** *Micelles, Vesicles and Microemulsions in Physics of Amphiphiles,* Degorgio, V. and Cozti, M., Eds., North Holland, Amsterdam, 1985.
18. **Fendler, J. H.,** *Membrane Mimetic Chemistry,* Wiley Interscience, New York, 1982.
19. **Mittal, K. L.,** *Solution Chemistry of Surfactants,* Plenum, New York, 1979.
20. **Mittal, K. L. and Fendler, E. J.,** *Solution Behaviour of Surfactants. Theoretical and Applied Aspects,* Plenum, New York, 1982.
21. **Mittal, K. L. and Lindman, B., Eds.,** *Surfactants in Solution,* Plenum, New York, 1984.
22. **Israelachvili, J. N., Mitchell, D. J., and Ninham, B. W.,** Theory of self-assembly of hydrocarbon amphiphiles into micelles and bilayers, *Trans. Far. Soc. II,* 72, 1525, 1976.
23. **Shaw, J. D.,** *Introduction to Colloid and Surface Chemistry,* Butterworths, London, 1980.
24. **Tanford, C.,** *The Hydrophobic Effect Formation of Micelles and Biological Membranes,* Wiley, Interscience, New York, 1980.

25. **Berthod, A.,** Mise au point. Structures physicochimiques des milieux dispersés, micelles, émulsions et microemulsions, *J. Chim. Phys.,* 30, 407, 1983.

26. **Gruen, D. W. R.,** A model for the chains in amphiphilic aggregates. Thermodynamic and experimental comparisons for aggregates at different shape and size, *J. Phys. Chem.,* 89, 153, 1985.

27. **Cabane, B., Duplessix, R., and Zemb, T.,** High resolution neutron scattering on ionic surfactant micelles: SDS in water, *J. Phys. (Paris),* 46, 2161, 1985.

28. **Dill, K. A., Koppel, D. E., Cantor, R. S., Dill, J. D., Bendedouch, D., and Chen, S. H.,** Molecular conformations in surfactant micelles, *Nature,* 309, 42, 1984.

29. **Luisi, P.,** *Biological and Technical Relevance of Reversed Micelles,* Plenum, New York, 1983.

30. **Wong, M., Thomas, J. K., and Nowak, T.,** Structure and state of H_2O in reversed micelles, *J. Am. Chem. Soc.,* 99, 4730, 1977.

31a. **Wong, M., Grätzel, M., and Thomas, J. K.,** On the nature of solubilized water clusters in aerosol O. T. alkane solutions. A study of the formation of hydrated electron and 1.8. anilinonaphthalen sulfonate fluorescence, *Chem. Phys. Lett.,* 30, 329, 1975.

31b. **Wong, M., Thomas, J. K., and Grätzel, M.,** Fluorescence probing of inverted micelles. The state of solubilized water clusters in alkane/Diisoctyl sulfosuccinate (A.O.T.) solution, *J. Am. Chem. Soc.,* 98, 2391, 1976.

32. **Keh, E. and Valeur, B.,** Determination of the hydrodynamic volume of inverted micelles containing water by the fluorescence polarization technique, *J. Phys. Chem.,* 83, 3305, 1979.

33. **Zinsli, P. E.,** Inhomogeneous interior of aerosol O.T. microemulsions probed by fluorescence and polarization decay, *J. Phys. Chem.,* 83, 3223, 1979.

34. **Zulauf, M. and Eicke, H. F.,** Inverted micelles and microemulsions in the ternary system H_2O/AOT/Isooctane as studied by photon correlation spectroscopy, *J. Phys. Chem.,* 83, 480, 1979.

35. **D'Aprano, A., Lizzio, A., Turco Liveri, V., Aliotta, F., Vasi, C., and Migliardo, P.,** Aggregation states of water in reverse AOT micelles: Raman evidence, *J. Phys. Chem.,* 92, 4436, 1988.

36. **De Gennes, P. G. and Taupin, C.,** Microemulsions and the flexibility of oil/water interfaces, *J. Phys. Chem.,* 86, 2294, 1982.

37. **Mittal, K. L.,** *Micellization, Solubilization and Microemulsions,* Plenum, New York, 1977.

38. **Barber, J.,** Membrane surface charges and potentials in relation to photosynthesis, *Biochim. Biophys. Acta,* 594, 253, 1980.

39. **Loeb, A. L., Overbeeck, J. Th., and Wiersema, P. H.,** *The Electrical Double Layer Around a Spherical Colloid Particle,* MIT Press, Cambridge, MA, 1961.

40. **Hoskin, N. E.,** The interaction of two identical spherical colloidal particules. I. Potential distribution, *Trans. R. Soc. Sec. A,* 248, 433, 1956.

41. **Brenner, S. L. and Roberts, R. A.,** A variational solution of the Poisson-Boltzmann equation for a spherical colloid particle, *J. Phys. Chem.,* 77, 2367, 1973.

42. **Hartley, G. S. and Roe, J. W.,** Ionic concentrations at interfaces, *Trans. Far. Soc.,* 36, 101, 1940.

43. **Fernandez, M. S. and Fromherz, P.,** Lipoid pH indicators as probes of electrical potential and polarity in micelles, *J. Phys. Chem.,* 81, 1755, 1977.

44. **Lukac, S.,** Surface potential at surfactant and phospholipid vesicles as determined by amphiphilic pH indicators, *J. Phys. Chem.,* 87, 5045, 1983.

45. **Charbit, G., Dorion, F., and Gaboriaud, R.,** Influence comparée de divers alkylsulfates sur la protonation de quelques indicateurs, *J. Chim. Phys.,* 81, 187, 1984.

46. **Ohki, S.,** Membrane potential of phospholipid bilayers: ion concentration and pH difference, *Biochim. Biophys. Acta,* 282, 55, 1972.

47. **Mille, M. and Vanderkooi, G.,** Electrochemical properties of spherical polyelectrolytes. II. Hollow sphere model for membranous vesicles, *J. Coll. Interface Sci.,* 61, 455, 1977.

48a. **Russel, J. C. and Whitten, D. G.,** Solute partitioning in aqueous surfactant assemblies: comparison of hydrophobic-hydrophilic interactions in micelles, alcohol swollen micelles, microemulsions and synthetic vesicles, *J. Am. Chem. Soc.,* 104, 5937, 1982.

48b. **Shin, D. M., Schanze, K. S., and Whitten, D. G.,** Solubilization sites and orientations in microheterogeneous media. Studies using donor-acceptor-substituted azobenzenes and bichromophoric solvatochromic molecules, *J. Am. Chem. Soc.,* 111, 8494, 1989.

49. **Grätzel, M., Kalyanasundaram, K., and Thomas, J. K.,** Proton nuclear magnetic resonance and laser photolysis studies of pyrene derivatives in aqueous and micellar solutions, *J. Am. Chem. Soc.,* 96, 7869, 1974.

50. **Narayana, P. A., Li, A. S. W., and Kevan, L.,** Electron spin resonance and spin echo studies of the photoionization of tetramethylbenzidine in frozen aqueous anionic, cationic and non ionic micellar solutions. Effect of micelle type and anionic micelle size, *J. Am. Chem. Soc.,* 104, 6502, 1982.

51. **Li, A. S. W. and Kevan, L.,** Optical absorption, electron spin resonance and electron spin echo studies of the photoionization of tetramethylbenzidine in cationic and anionic synthetic vesicles: comparison with analogous micellar systems, *J. Am. Chem. Soc.,* 105, 5752, 1983.

52. **Zana, R.,** Micelles et microemulsions: revue de résultats structuraux et dynamiques récents en relation avec la réactivité, *J. Chim. Phys.,* 83, 603, 1986.

53. **Almgren, M., Grieser, F., and Thomas, J. K.,** Solubilization of small organic molecules by ionic micelles: Poisson distribution, *J. Am. Chem. Soc.,* 101, 279, 1979.

54. **Infelta, P. P. and Grätzel, M.,** Statistics of solubilizate distribution and its application to pyrene fluorescence in micellar systems. A concise kinetic model, *J. Chem. Phys.,* 70, 179, 1979.

55. **Tachiya, M.,** Stochastic and diffusion models of reactions in micelles and vesicles in *Kinetics of nonhomogeneous processes,* G. R. Freeman, Ed., John Wiley & Sons, 1987.

56. **Müller, N., Wolff, T., and von Bünau, G.,** Light induced viscosity changes of aqueous solutions containing 9-substituted anthracene solubilized in cetylammonium micelles, *J. Photochem.,* 24, 37, 1984.

57. **Turro, N. J., Grätzel, M., and Braun, A.,** Photophysical and photochemical processes in micellar systems, *Angew. Chem. Int. Ed. Engl.,* 19, 675, 1980.

58. **Fendler, J. H.,** Surfactant vesicles as membrane mimetic agents: characterization and utilization, *Acc. Chem. Res.,* 13, 7, 1980.

59. **Zachariasse, K. A., Nguyen, V. P., and Kozankiewicz, B.,** Investigation of micelles, microemulsions and phospholipic bilayers with the pyridinium N-Phenolbetaïne E_T (30); a polarity probe for aqueous interfaces, *J. Phys. Chem.,* 85, 2676, 1981.

60. **Cevc, G. and Marsh, D.,** Properties of the electrical double layer near the interface between a charged bilayer membrane and electrolyte solution: experiment vs theory, *J. Phys. Chem.,* 87, 376, 1983.

61. **Escabi-Perez, J. R., Romero, A., Lukac, S., and Fendler, J. H.,** Aspects of artificial photosynthesis. Photoionization and electron transfer in dihexadecyl phosphate vesicles, *J. Am. Chem. Soc.,* 101, 2231, 1979.

62. **Fork, R. L., Shank, C. V., and Yen, R. T.,** Amplification of 70 fs optical pulses to gigawatt powers, *Appl. Phys. Lett.,* 41, 223, 1982.

63. **Migus, A., Antonetti, A., Etchepare, J., Hulin, D., and Orzag, A.,** Femtosecond spectroscopy with high power tunable optical pulses, *J. Opt. Soc. Am.,* B2, 584, 1985.

64. **Chase, W. J. and Hunt, J. W.,** Solvation time of electrons in polar liquids. Water and alcohols, *J. Phys. Chem.,* 79, 2835, 1975.

65. **Lewis, M. A. and Jonah, C. D.,** Evidence for two electron states in solvation and scavenging processes in alcohols, *J. Phys. Chem.,* 90, 5367, 1987.

66a. **Gauduel, Y., Martin, J. L., Migus, A., Yamada, N., and Antonetti, A.,** Femtosecond study of electron localization and solvation in pure water, in "Ultrafast Phenomena V", *Chemical Physics,* Fleming, G. R. and Sigman, A. E., Ed., Springer-Verlag, Berlin, 308, 1986.

66b. **Migus, A., Gauduel, Y., Martin, J. L., and Antonetti, A.,** Excess electrons in liquid water: first evidence of a prehydrated state with femtosecond lifetime, *Phys. Rev. Lett.,* 58, 1559, 1987.

67a. **Gauduel, Y., Migus, A., Martin, J. L., and Antonetti, A.,** Femtosecond and picosecond time-resolved electron solvation in aqueous and reversed micelles, *Chem. Phys. Lett.,* 108, 319, 1984.

67b. **Gauduel, Y., Pommeret, J., Yamada, N., Migus, A., and Antonetti, A.,** Femtosecond electron attachment of excess electron to water pool of AOT reversed micelles, *J. Am. Chem. Soc.,* 111, 4974, 1989.

68. **Gauduel, Y., Berrod, S., Migus, A., Yamada, N., and Antonetti, A.,** Femtosecond charge separation in organized assemblies: free radical reactions with pyridine nucleotide in micelles, *Biochemistry,* 27, 2509, 1988.

69. **Gauduel, Y., Migus, A., and Antonetti, A.,** Electron reactivity in aqueous media: a femtosecond investigation of the primary species, *Rad. Phys. Chem.,* 34, 5, 1989.

70. **Neff, H., Sass, J. K., Lewerenz, H. J., and Ibach, H.,** Photoemission studies of electron localization at very low excess energies, *J. Phys. Chem.,* 84, 1135, 1980.

71. **Pleskov, Yu V. and Rotenberg, Z. A.,** Photoemission studies of the behaviour of solvated electrons and unstable species, *High Energy Chem.,* 8, 83, 1974.

72a. **Konovalov, V. V., Raitsimring, A. M., Tsvetkov, Yu D., and Benderskii, V. A.,** The thermalization length of low energy electrons determined by nanosecond photoemission into aqueous electrolyte solutions, *Chem. Phys.,* 93, 163, 1985.

72b. **Konovalov, V. V., Raitsimring, A. M., and Tsvetkov, Yu D.,** Thermalization lengths of "subexcitation electrons" in water determined by photoinjection from metals into electrolyte solutions, *Rad. Phys. Chem.,* 32, 623, 1988.

73a. **Thomas, J. K. and Piciulo, P.,** Photoionization by green light in micellar solution, *J. Am. Chem. Soc.,* 100, 3239, 1978.

73b. **Thomas, J. K. and Piciulo, P.,** Photionization by green light, *J. Am. Chem. Soc.,* 101, 2502, 1979.

74. **Hall, G. E.,** Comment on the communication "Photoionization by greenlight in micellar solution", *J. Am. Chem. Soc.,* 100, 8263, 1978.

75. **Wallace, S. C., Hall, G. E., and Kenney-Wallace, C. A.,** Influence of micellar interface on molecular ionization potentials: a tunable laser spectroscopy study of photoionization of pyrene, *Chem. Phys.,* 49, 279, 1980.

76. **Bernas, A., Grand, D., Hautecloque, S., and Chambaudet, A.,** Ionization potentials of polyacene molecules in micellar systems or liquid homogeneous solutions, *J. Phys. Chem.,* 85, 3684, 1981.
77. **Bernas, A., Grand, D., Hautecloque, S., and Myasoedova, T.,** On the ionization potential of chlorophyll or bacteriochlorophyll in an aqueous environment, *Chem. Phys. Lett.,* 104, 105, 1984.
78. **Hautecloque, S., Grand, D., and Bernas, A.,** Salt effects on photoionization yields in micellar systems, *J. Phys. Chem.,* 89, 2705, 1985.
79. **Grand, D., Bernas, A., and Amouyal, E.,** Photoionization of aqueous indole; conduction band edge and energy gap in liquid water, *Chem. Phys.,* 44, 73, 1979.
80. **Abd-el Kader, M. H. and Krebs, P.,** Localized excess electrons in solubilized water clusters in aerosol OT-n-heptane solutions, *J. Chem. Soc. Faraday Trans. I,* 84, 2241, 1988.
81. **Calvo-Perez, V., Beddard, G. S., and Fendler, J. H.,** Hydrated electrons in surfactant solubilized water pools in heptane, *J. Phys. Chem.,* 85, 2316, 1981.
82. **Bakale, G., Beck, G., and Thomas, J. K.,** Electron capture in water pools of reversed micelles, *J. Phys. Chem.,* 85, 1062, 1981.
83. **Pileni, M. P., Hickel, B., Ferradini, C., and Pucheault, J.,** Hydrated electrons in reverse micelles, *Chem. Phys. Lett.,* 92, 308, 1982.
84. **Thomas, J. K., Grieser, F., and Wong, M.,** Fast reaction in micelles, *Ber. Bunsenges. Phys. Chem.,* 82, 937, 1978.
85. **Bales, B. L. and Kevan, L.,** Spatial dependent electron scavenging by stearic acid nitroxide spin-labels in the photoionization of N,N,N′,N′ tetramethylbenzidine in sodium dodecyl sulfate micelles, *J. Phys. chem.,* 90, 3836, 1982.
86. **Bernas, A., Grand, D., and Hautecloque, S.,** On ion pair photoseparation and recombination in micellar aggregates, *Radiat. Phys. Chem.,* 32, 309, 1988.
87. **Alkaitis, S. A. and Grätzel, M.,** Laser Photoionization and light-initiated redox reactions of tetramethylbenzidine in organic solvents and aqueous micellar solution, *J. Am. Chem. Soc.,* 98, 3549, 1976.
88. **Bernas, A., Grand, D., Hautecloque, S., and Giannotti, C.,** Interfacial electrical potential in micelles: Its influence on N,N,N′,N′ tetramethylbenzidine cation decay, *J. Phys. Chem.,* 90, 6189, 1986.
89. **Saito, T., Haida, K., Sano, M., Hirata, Y., and Mataga, N.,** Monophotonic ionization of N-N-N′,N′, tetramethylbenzidine in various polar solvents. Dynamics of long-lived ion pairs, *J. Phys. Chem.,* 90, 4017, 1986.
90. **Ben Chaabane, T., Bernas, A., Grand, D., and Hautecloque, S.,** Electron transfer reactions in artificial vesicle solutions, *J. Phys. Chem.,* 91, 6055, 1987.
91. **Thomas, J. K.,** Effect of structure and charge on radiation induced reactions in micellar systems, *Acc. Chem. Res.,* 10, 133, 1977.
92. **Kalyanasundaram, K. and Thomas, J. K.,** Radiation induced reactions in non ionic micelles, *Proc. Micelle Symp., Albany,* NY, 569, 1976.
93. **Hubig, S. M., Dionne, B. C., and Rodgers, A. J.,** Effect of micellar media on the electron transfer reaction between benzylviologen and quinones, *J. Phys. Chem.,* 90, 5873, 1986.
94. **Grätzel, M., Thomas, J. K., and Patterson, L. K.,** Trapping of hydrated electrons by cetylpyridinium chloride micelles, *Chem. Phys. Lett.,* 29, 393, 1974.
95. **Bunton, C. A., Ohmenzetter, K., and Sepulveda, L.,** Binding of hydrogen ions to anionic micelles, *J. Phys. Chem.,* 81, 2000, 1977.
96. **Fendler, J. H.,** Interactions and kinetics in membrane mimetic systems, *Ann. Rev. Phys. Chem.,* 35, 137, 1984.
97. **Quina, F. H. and Chaimovitch, H.,** Ion exchange in micellar solutions. I. Conceptual framework for ion exchange in micellar solutions, *J. Phys. Chem.,* 83, 1844, 1979.
98. **Hatlee, M. D. and Kozak, J.,** A stochastic approach to the theory of intramicellar kinetics, *J. Chem. Phys.,* 72, 4358, 1980.
98a. **Hatlee, M. D. and Kozak, J.,** *J. Chem. Phys.,* 74, 1098, 1981.
98b. **Hatlee, M. D. and Kozak, J.,** *J. Chem. Phys.,* 74, 5627, 1981.
99. **Hatlee, M. D., Kozak, J. J., Rothenberger, G., Infelta, P. P., and Grätzel, M.,** Role of dimensionality and spatial extent in influencing intramicellar kinetic processes, *J. Phys. Chem.,* 84, 1508, 1980.
100. **Turro, N. J. and Cherry, W. R.,** Photoreactions in detergent solutions. Enhancement of regio selectivity resulting from the reduced dimensionality of substrates sequestered in a micelle, *J. Am. Chem. Soc.,* 100, 7431, 1978.
101. **Kalyanasundaram, K., Grieser, F., and Thomas, J. K.,** Room temperature phosphorescence of aromatic hydrocarbons in aqueous micellar solutions, *Chem. Phys. Lett.,* 5, 501, 1977.
102. **Turro, N. J., Liu, K. C., Chow, M. F., and Lee, P.,** Convenient and simple methods for the observation of phosphorescence in fluid solutions. Internal and external heavy atom and micellar effects, *Photochem. Photobiol.,* 27, 523, 1978.
103. **Rothenberger, G. and Grätzel, M.,** Effects of spatial confinement on the rate of bimolecular reactions in organized liquid media, *Chem. Phys. Lett.,* 154, 165, 1989.

104. **Russel, J. C., Costa, S. B., Seiders, R. P., and Whitten, D. G.,** Photochemistry of a surfactant stilbene in organized media: a probe for hydrophobic sites in micelles, vesicles and other assemblies, *J. Am. Chem. Soc.,* 102, 5678, 1980.

105. **Humphry-Baker, R., Grätzel, M., and Steiger, R.,** Drastic fluorescence enhancement and photochemical stabilization of cyanine dyes through micellar systems, *J. Am. Chem. Soc.,* 102, 847, 1980.

106. **Hashimoto, S. and Thomas, J. K.,** Effect of environment on decay pathways of the singlet excited state of N,N,N',N' tetramethylbenzidine, *J. Phys. Chem.,* 88, 4044, 1984.

107. **Szajdzinska-Pielek, E., Maldono, R., Kevan, L., and Jones, R. R. M.,** Electron spin resonance and electron spin echo modulation studies of N,N,N',N'-tetramethylbenzidine photoionization in anionic micelles: structural effects of tetramethylammonium cation counterion substitution for sodium cation in dodecyl sulfate micelles, *J. Am. Chem. Soc.,* 106, 4675, 1984.

108. **Onsager, L.,** Initial recombination of ions, *Phys. Rev.,* 54, 554, 1938.

109. **Alkaitis, S. A., Beck, G., and Grätzel, M.,** Laser photoionization of phenothiazine in alcoholic and micellar solutions. Electron transfer from triplet states to metal ion acceptors. *J. Am. Chem. Soc.,* 97, 5723, 1975.

110a. **Arrhenius, T. S., Blanchard-Desce, M., Dvolaitsky, M., Lehn, J. M., and Malthete, J.,** Molecular devices: caroviologens as an approach to molecular wires, synthesis and incorporation into vesicle membranes, *Proc. Natl. Acad. Sci. U.S.A.,* 83, 5355, 1986.

110b. **Johansson, L. B. A., Blanchard-Desce, M., Almgren, M., and Lehn, J. M.,** Orientation of caroviologens in model membranes, *J. Phys. Chem.,* 93, 6751, 1989.

111. **Grand, D. and Hautecloque, S.,** Electron transfer from nucleophilic species to N, N, N, N tetramethyl-beuzidine cation in micellar media: effect of interfacial electric potential on cation decay, *J. Phys. Chem.,* 94, 837, 1990.

112. **Grätzel, M., Henglein, A., and Thomas, J. K.,** Electron reactions and electron transfer reactions catalyzed by micellar systems, *J. Chem. Phys.,* 62, 1632, 1975.

113. **Gauduel, Y., Migus, A., Martin, J. L., Lecarpentier, Y., and Antonetti, A.,** Femtosecond optical techniques. Application to reaction dynamics in liquids, *Ber. Bunsenges. Physik. Chem.,* 89, 218, 1985.

Chapter 13

INTRAMOLECULAR ELECTRON TRANSFER REACTIONS IN PEPTIDES AND PROTEINS

Moshe Faraggi and Michael H. Klapper

TABLE OF CONTENTS

I. INTRODUCTION

That an electron could migrate within a protein at distances that would preclude direct contact beetween donor and acceptor (r $\geq$ 5 Å) was, until recently, viewed by biochemists as improbable. Hence, when a protein's crystal structure indicated that the electron donor and/or acceptor was isolated from its counterpart, conformational changes preceding electron transfer were generally invoked. For example, it was proposed that electron transfer to and from the redox centers of the respiratory chain components might involve back and forth protein rotations that would significantly shorten the distances between the redox centers involved sequentially in the electron transfers.[1] On the other hand, Winfield, in a well-known speculation regarding cytochrome c, did suggest that an electron might reach the heme redox center of that protein over an apparent noncovalent aromatic pathway leading from the protein's surface to the buried heme group.[2]

In contrast, workers in the field of pulse radiolysis had acquired experimental evidence, both circumstantial and direct, to support the possibility of intramolecular long range electron transfer (LRET) in proteins.[3,4] For example, reactions between proteins and the solvated electron ($e_{aq}^{-}\cdot$) that lead to metal center (ferriheme, copper[II]) reduction occur under diffusion control (k $>10^{10}$ M^{-1} s^{-1}) and in high yields ($\sim$50%). Since the metals are either buried within the protein or located in a tight crevice and not readily accessible to the solvated electron and since there can be only a limited number of reactant collisions in a diffusion controlled reaction, the high reduction yields could be explained by rapid and efficient intramolecular electron migration from the surface of the protein to the metal.[5,6] Faraggi and Pecht found experimental evidence for such a transfer in the reaction of the hydrated electron with three "blue" type Cu(II) proteins; reduction by $e_{aq}^{-}\cdot$ is fast and is followed by a slower first order reduction of the Cu(II).[7,8] They associated the initial fast reduction with the formation of electron adduct intermediates (histidine or disulfide) that then decay over the same time during which Cu(II) is reduced to Cu(I). Additional evidence for LRET was reported by Prutz and co-workers, who observed intramolecular one-electron oxidation of the tyrosine side chain by the tryptophan indolyl radical in proteins and peptides containing these amino acids.[9,10]

The conclusions based on these early pulse radiolysis results are now justified by the numerous reports[11] of LRET in proteins. There is now little doubt that the many redox centers attached to proteins (e.g., metals ions, metalloporphyrins, flavins, and quinones) may be able to accept or donate electrons over long distances. The importance of LRET in biological processes, such as photosynthesis and respiration, and in various pathologic processes, such as radiation damage and oxygen toxicity, are now acknowledged and under investigation. In this chapter we present an account of LRET studies published until December 1988. We shall provide a description of systems and techniques used in these studies, the theoretical basis of LRET, and a discussion of previous results and future trends in experiments with proteins and model systems (peptides), studies performed in an attempt to understand the efficient biological electron transfer processes.

II. MATERIALS AND METHODS

LRET in proteins is most appropriately studied when there are two attached redox centers (an electron donor and an electron acceptor) and an available X-ray crystal structure. As an example, the laboratories of Gray, Isied, and others[11c,12-20] have covalently attached an electron donor (acceptor). (NH$_3$)$_5$Ru$^{3+/2+}$ or py(NH$_3$)$_4$Ru$^{3+/2+}$ (Ru = ruthenium, py = pyridine), to surface histidine side chains in a variety of metalloproteins to form two-metal proteins such as Ru[III]:Fe[III]-cytochrome c, Ru[III]:Fe[III]-cytochrome c^{551} Ru[III]:Fe[III]-myoglobin, Ru[III]:Cu[II]-azurin, Cu[II]:Ru[III]-plastocyanin, and Ru(III):Cu(II)-stellacyanin. LRET was observed in these modified proteins by first reducing the Ru[III]-complex in these modified proteins and then observing the slower electron transfer from the Ru[II] to the oxidized prosthetic group. Variants of these synthetic two-metal proteins in which zinc[21,22] and palladium[23] substitute for the heme iron have also been the subject of experimentation. Upon photoexcitation, the zinc and palladium prophyrins form triplet state electron donors that can reduce the protein bound Ru[III] complexes. Ruthenium is not required as one of the two redox centers; Hoffman and co-workers,[24] and McLendon and co-workers[25] developed systems that contain both photoexcitable Zn[II]-porphyrin as reductant and a Fe(III) heme as oxidant, while Faraggi and Klapper[26] have converted a hemoglobin sulfhydryl group to a mixed disulfide which, when reduced to its radical anion, becomes a reductant of the heme iron.

But structurally simpler systems have also been the subjects of attention. Starting with Prutz and co-workers,[9] latent redox centers separated by a peptide chain have been utilized as models for LRET in proteins. To avoid questions of peptide flexibility, we shall limit our discussion to those cases in which the proline oligomer, a rigid spacer because of the steric constraints,[27,28] separates the two centers: the two liganded metal centers Ru(III)/Ru(II):Co(III)/Co(II) and Os(III)/Os(II):Co(III)/Co(II);[29-31] the polypyridyl Ru[II] complex as the photoexcited electron donor with a p-benzoquinone moiety as the electron acceptor;[32] and the two organic redox sites (phenoxy radical/tyrosine):(indolyl radical/tryptophan) abbreviated as tyrO·/tyrOH:trp·/trpH.[33,34]

The two fast kinetic techniques used in most protein and polypeptide LRET experiments are laser flash photolysis and pulse radiolysis.[4] In flash photolysis a light pulse with known energy ($\leqslant 6$ eV) and duration (t $\geqslant 100$ fs) is delivered to the solution and directly excites one or more *solute* species. The resultant photoexcitation and/or photoionization are monitored most often by spectral changes in the UV-visible. In pulse radiolysis an electron pulse with known energy ($\geqslant 2$ MeV) and duration (100 ps $\leqslant$ t $\leqslant$ 1500 ns) is delivered to the solution and excites the *solvent*. The primary radicals produced because of this solvent excitation are then used to establish the LRET chemistry on the solutes. In this case also, the UV-visible spectrophotometer remains the most popular detection device.

In flash, now usually laser, photolysis the photon is absorbed specifically by those solute molecules with the appropriate absorption band and with an efficiency dependent on the extinction coefficient of the solute at the irradiation wavelength. The electron transfer process is established if the light absorption creates an excited state electron donor (Equation 1) or results in a ground state electron acceptor due to photoejection (Equation 2). Both Reactions 1 and 2, figure in LRET studies, although the second approach is not yet commonly used.

$$\text{Ru(bpy)}_3^{2+} \xrightarrow{\ h\nu\ } {}^*\text{Ru(bpy)}_3^{2+} \tag{1}$$

$$\text{TrpH} \xrightarrow{\ h\nu\ } \text{TrpH}^+ + e_{sol}^-\cdot \tag{2}$$

The amount of oxidant/reductant obtained from a particular solute depends on the characteristic quantum yield Φ, defined as the number of oxidant/reductant molecules produced per photon absorbed into the system. In the case of Reaction 1 with light of wavelength 532 nm, $\Phi = 1$, or every absorbed photon results in the formation of a photoexcited reductant molecule. For Reaction 2, with light of wavelength 266 nm, $\Phi < 1$ since tryptophan triplets are formed in competition with the radical oxidant.

Thus, Gray et al. reported the rapid reduction of the ruthenium in $a_5\text{Ru}^{3+}$(His-33), (Fe^{3+})cyt c by the excited Ru(bpy)$^{2+}$ produced with an 8 ns pulse of 532 nm light. The electron from the Ru(II) attached to the protein then transferred over a distance of about 15 Å for the slower reduction of the ferriheme iron.[13] Gray and co-workers argued that the slower reaction was indeed an intramolecular LRET since the ferriheme iron of the un-modified protein was not reduced by photoexcited $Ru(bpy)^{2+}$, which is known to reduce Ru^{3+} pentaamino complexes such as that attached to the modified cytochrome c.[13,35] This same laboratory took a similar approach in studies with Ru modified azurin.[15] There have also been a number of reports in which photoexcited Zn-porphyrin (or Pd-porphyrin), also produced with 532 nm light served as the reductant in LRET experiments; Hoffman et al. studied LRET between the Zn and Fe of a $[Zn^{2+}, Fe^{3+}]$ hybrid hemoglobin; McLendon and co-workers worked with the hybrid Zn^{2+} hemoglobin/cytochrome b_5 complex; Gray et al. followed LRET between Ru complexed to Zn- or Pd-porphyrin substituted myoglobins; and Winkler et al. investigated LRET in ruthenium modified Zn substituted cytochrome c.[21-23,36,37]

Pulse radiolysis has not been used as frequently for LRET studies in proteins. The accelerator required for the production of the high energy electron pulses is both large and expensive. But the lack of acceptance of this technique may actually be due to unwarranted concerns over the many possible reactions of the "highly reactive" radicals generated by the electron pulse. In order to allay these concerns, we shall describe the pulse radiolysis method in some detail, albeit discussing only the chemistry in aqueous solutions.

High energy electrons, introduced into dilute aqueous solutions, lose their energy pre-dominantly by Coulombic interactions with the *solvent* and not, as in flash photolysis with the solute. The initial products formed are:

$$e^-(>2\,\text{Mev}) \rightarrow H_2O \rightarrow e_{aq}^-\cdot,\ H\cdot,\ OH\cdot,\ H_2,\ H_2O_2,\ H_3O^+ \tag{3}$$

with yields of: hydrated electron $(e_{aq}^-\cdot)$, $G = 2.9$ radicals per 100 eV absorbed by the solution; hydrogen atom $(H\cdot)$, $G = 0.55$; hydroxyl radical $(OH\cdot)$, $G = 2.8$; (H_2O_2), $G = 0.75$; and hydronium ion (H_2O^+), $G = 2.9$. The three nonradical products are generally of little interest; hydrogen peroxide, a minor product, is unreactive over the timescale of most pulse radiolysis experiments, the hydronium ion is controlled with buffering and H_2 is inert. The three primary radicals can be converted to other radicals or removed by scavengers, permitting a choice of reactants for subsequent studies. For example, the hydrated electron is transformed into the hydroxyl radical via:

$$e_{aq}^-\cdot + N_2O + H^+ \rightarrow N_2 + OH\cdot \tag{4}$$

so that $OH\cdot$ is the major radical species in N_2O saturated solutions. In order to obtain e_{aq}^- as the predominant radical, $OH\cdot$ is removed from solution by the hydrogen abstracting reaction:

$$OH\cdot + (CH_3)_3COH \rightarrow (CH_3)_2(\cdot CH_2)COH + H_2O \tag{5}$$

Since the *t*-butanol radical is neither a good hydrogen abstractor nor a good reducing agent, it is effectively invisible over the time range of most pulse radiolysis experiments and is usually lost by radical-radical recombination.

The hydrated electron, a potent reducing agent, with a redox potential of about -2.8 V, broadly absorbs in the visible and near infrared. Even in the absence of added solutes, $e_{aq}^-\cdot$ decays via multiple paths: (1) combination with a proton to form $H\cdot$; (2) combination with itself, or a hydrogen atom to ultimately form H_2; (3) combination with the hydroxyl radical to form water; or (4) by reaction with solution impurities. The overall $e_{aq}^-\cdot$ decay rate and the relative product yields from each of the different decay paths depend on the con-centrations of all the possible reacting species. Under the conditions of the experiments we normally perform, $e_{aq}^-\cdot$ with an initial concentration of 5 μM has a first halflife of approximately 20 μs. Because of its large negative redox potential, the hydrated electron reduces a variety of organic and inorganic compounds; the rate constants of many of these reactions are known and compiled.[38]

The hydroxyl radical, a potent oxidant with a redox potential of approximately $+2$ volts (at pH 7), reacts rapidly with electron donors and also abstracts hydrogen from a variety of organic compounds. Since OH$\cdot$ absorbs light only weakly and in the far UV (220 nm), it is difficult to monitor by absorbance methods. However, the reactions of many compounds with OH$\cdot$ have been studied by competition techniques or by direct observation of oxidation products and there is a large compilation of known OH$\cdot$ reaction rate constants.[38]

A significant number of additional radicals, oxidants or reductants, can be accessed from OH$\cdot$. Among these are the formate radical ($CO_2^-\cdot$), a one-electron reductant obtained from the reactions:

$$OH\cdot \, + \, HCOO^- \rightarrow CO_2^-\cdot \, + \, H_2O \tag{6a}$$

$$H\cdot \, + \, HCOO^- \rightarrow CO_2^-\cdot \, + \, H_2 \tag{6b}$$

Since $e_{aq}^-\cdot$ can be converted to OH$\cdot$ by Reaction 4, $CO_2^-\cdot$ is the sole radical in solutions that contain formate and are saturated with N_2O. $CO_2^-\cdot$, which absorbs strongly in the UV, is a weaker and more specific reductant with a redox potential of about -1.9 V than is $e_{aq}^-\cdot$.[39,40]

Both OH$\cdot$ and H$\cdot$ also serve as precursors of carbon centered alcohol radicals. For example, the isopropanol radical is formed via hydrogen abstraction:

$$OH\cdot \, + \, (CH_3)_2CHOH \rightarrow (CH_3)_2\dot{C}OH \, + \, H_2O \tag{7a}$$

$$H\cdot \, + \, (CH_3)_2CHOH \rightarrow (CH_3)_2\dot{C}OH \, + \, H_2 \tag{7b}$$

as are the analogous ethanol and methanol radicals. These three neutral reductants have potentials of -1.1 V (isopropanol), -1.0 V (ethanol) and -0.8 V (methanol), and constitute a series of neutral reducing species.[40] The hydroxyl radical can also be converted to milder oxidants. Among these are the diatomic halide radicals ($X_2^-\cdot$), products of the reactions:

$$OH\cdot \, + \, X^- \rightarrow X\cdot \, + \, OH^-$$

$$X\cdot \, + \, X^- \rightleftharpoons X_2^-\cdot \tag{8}$$

and the uncharged azide radical formed in the reaction:

$$OH\cdot \, + \, N_3^- \rightarrow N_3\cdot \, + \, OH^- \tag{9}$$

These radicals have oxidation potentials covering the approximate range from 1.0 to 1.5 V.[40] Thus, through various reactions, only some of which we have briefly mentioned here, the pulse radiolysis technique can provide a variety of radical (and nonradical) reagents covering a wide range of redox potentials.

Both the primary and secondary radicals generated in the pulse radiolysis experiment react with many organic and inorganic compounds. To simplify the kinetics of these various reactions, the nonradical reactant is generally held at a concentration tenfold or greater than that of the radical, which itself may be varied from 0.1 to 30 μM. (With some machines 200 μM can be attained). With this arrangement potential, second order reactions of the radicals become pseudo-first order. The second order rate constants are then obtained easily from the dependence of the pseudo-first order constant on the nonradical reactant's concentration.

In a later part of this review we shall deal with the pulse radiolytic study of protein reduction by $e_{aq}^-\cdot$ and secondary radicals. Therefore, it is important to discuss now the perceived and real limitations of pulse radiolysis studies that involve $e_{aq}^-\cdot$. Concerns are often expressed because $e_{aq}^-\cdot$ is so reactive and has the potential to reduce a number of protein sites (disulfides, aromatic amino acids, sulfhydryls, and the peptide bond carbonyl group).[3,4] If all these reactions were to occur simultaneously, there would be a confusing multiplicity of radical reaction sequences difficult to disentangle from one another. In fact, observed reactions have not been unduly complex, and experimental results have been generally open to reasonable interpretation. Why this apparent simplicity?

One, and perhaps the most important, answer to this question is that reactions between proteins and $e_{aq}^-\cdot$ occur at or very near the diffusion controlled limit.[3] Thus, most $e_{aq}^-\cdot$ protein collisions are productive, and at sufficiently high (at least 10) protein to radical concentration ratios the probability of two or more $e_{aq}^-\cdot$ reacting with one protein molecule becomes negligible. This consideration leads to the *one-radical/one-macromolecule principle* upon which one may base most experimental interpretation. At sufficiently large protein to radical concentration ratios a fraction of all the protein molecules in solution will contain only one radical, and few if any protein molecules will contain more than one. Then, because of a protein's large size and resultant small diffusion coefficient, each radical bearing macromolecule serves as an "island" over short timespans in the pulse radiolysis experiment; i.e., the fate of a particular radical on a protein molecule need not be disturbed by events on any of the other protein molecules in the system. Hence, over short times there are few cross reactions between the products formed on the individual protein molecules, and possible multi-radical complications are minimized. However, we must recognize that more than one site on a particular protein molecule may compete for the hydrated electron. As a result, there may well be different radicals formed initially on different protein molecules, with subsequently different reaction sequences occurring simultaneously — one sequence on a set of molecules, second on another set, and so on — and some or all of the reactions under observation at the same time. Fortunately, disentangling these possible simultaneous, albeit independent reactions, has not been a major problem for reasons to be discussed next.

Reaction of the bulky and polar $e_{aq}^-\cdot$ at the protein surface should be fast relative to the time required for that radical to diffuse into the protein matrix. Thus, we would expect the initial reaction to occur primarily on the protein surface, and not at internal loci shielded from the aqueous environment. Moreover, and irrespective of whether the first reaction is on the surface or in the interior, the various potential reaction sites have different relative reactivities toward the radical reactant, and partitioning between the different reaction windows will depend on differences in relative "concentrations" and intrinsic reactivities. Unless present in much greater quantity, a group with a low intrinsic reactivity cannot compete effectively with one of high reactivity. Thus, location, intrinsic reactivity, and relative concentration all tend to reduce the many potential reactions to the few actually observed. Another factor for consideration is the selectivity introduced by the measurement method. In most laboratories the appearance or disappearance of a transient is monitored by absorption changes in the UV-visible region. Whether certain transients are even detected depends, therefore, on their absorbance intensity at the monitored wavelengths, The reaction product that does not absorb might be overlooked. As a result, we lose information about those reactions that we do not "see", but we gain back a somewhat compensatory simplification. This technical selectivity might be the reason that only a limited number of radical transients — the one-electron reduced disulfide, histidine and occasional coenzyme — have been well-documented products in the reactions of $e_{aq}^-\cdot$ with proteins.

But one cannot expect a simple interpretation in all pulse radiolysis studies: (1) the possibility of multiple reactions, a problem that is not unique to just this technique, still faces us in each new study; and (2) slower reactions (e.g., with half lives larger than 1 ms) present a potential difficulty, since the one-radical/one-macromolecule principle may breakdown due to both diffusional cross reactions between proteins and to secondary reactions with accumulated radical side products, such as the *t*-butanol radical produced from OH· scavenging (Equation 5). There is no fail-safe method for avoiding these complications, but there are ways with which to test for the possible artifacts that these reactions might introduce. For example, the reaction of the *t*-butanol radical with a protein can be detected with a control experiment in which all the primary radiolytic radicals are converted to the butanol radical via Reactions 4 and 5. Where cross reactions between protein molecules may be important, we can establish the observed reaction's kinetic order. Diffusional events should be second order and the intramolecular reactions, of special interest in LRET studies, should be first order.

The typical LRET pulse radiolysis experiment begins with the rapid preferential oxidation or reduction of one redox site on a macromolecule followed by the intramolecular electron transfer. Isied et al. rapidly reduced the ruthenium of a_5Ru^{3+}(His-33) (Fe^{3+})cyt c with the formate radical CO_2^-·, and then followed the slower intramolecular reduction of the ferriheme by the bound Ru^{2+}. Sykes et al. utilized similar CO_2^-· reductions of Ru^{3+} to determine LRET in ruthenium modified plastocyanin, HIPIP and cyt-c^{551}. Faraggi and Klapper reacted CO_2^-· with a single mixed disulfide on the modified hemoglobin α-subunit to produce a disulfide radical anion reductant; the observed electron transfer was from the disulfide radical anion to the ferriheme group.[26] The azide radical preferentially oxidizes the indole side chain of tryptophan to the neutral (above pH 6) indolyl radical (Reaction 10) with which Butler et al., Bobrowski et al., and Faraggi et al. determined the LRET oxidation of the tyrosine side chain to its neutral phenoxy radical (Reaction 11).

$$N_3\text{·} + \text{TrpH--(\ )--TyrOH} \rightarrow \text{Trp·--(\ )--TyrOH} + N_3^- + H^+ \tag{10}$$

$$\text{Trp·--(\ )--TyrOH} \rightleftharpoons \text{TrpH--(\ )--TyrO·} \tag{11}$$

III. LRET MECHANISM

Most interpretations of electron transfer reactions in solution are now associated with proposals presented originally by Marcus and Hush.[44-46] These proposals are based on a relatively simple model for electron transfer and, hence, provide simple mathematical expressions for experimental test. Initially, the Marcus (or Marcus-Hush) theory was applied successfully to outer sphere electron transfer reactions between two metal centers that do not share a common ligand; but with the experimental observation of LRET in proteins has come a modification of the original theoretical formulation. We shall briefly introduce the outer sphere electron transfer argument, but spend more time on the long-range formulation.

It is convenient to separate the electron transfer rate constant, k_{et}, between two species not bound to one another into discrete contributions:

$$k_{et} = K_o \kappa_{el} \kappa_{nu} \tag{12a}$$

where K_0 accounts for the diffusional collision of the two reactants, and κ_{el} and κ_{nu} are the electronic and nuclear transmission coefficients, respectively, under the assumption that the nuclear and electronic motions can be separated (the Born-Oppenheimer approximation).

The Marcus theory starts with the assumption that the potential energy/nuclear coordinate diagrams with the electron in either donor or acceptor site are parabolic (Figure 1A). (Since the potential energy of the reactants and surrounding medium depends upon many nuclear coordinates, the true potential energy surface is hyperdimensional. The two-dimensional curves of Figure 1A are thus "projections" of very complex surfaces onto the plane of the

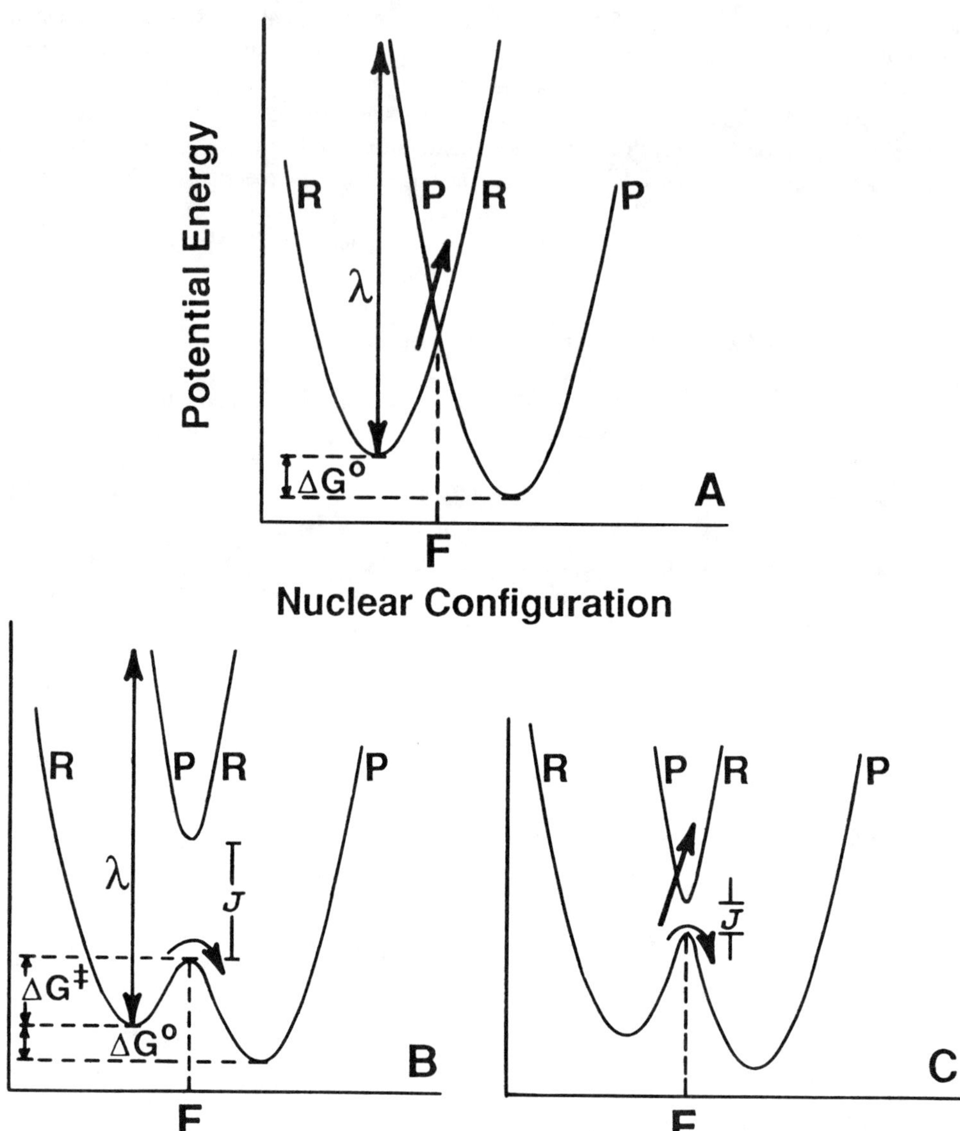

FIGURE 1.   A reaction coordinate diagram and the parameters influencing the electron transfer process according to the theory of Marcus. See text for an explanation of the figure.

paper. (It is these projections that are assumed to be parabolic.) According to classical formalism, the electron located at the donor site (reactants) can transfer to the acceptor site (products) when the potential energy due to nuclear motion acquires the added energy λ (Figure 1A), often called the nuclear reorganization energy. However, quantum theory places the electron with a probability distribution everywhere in space including both donor and acceptor site with no nuclear reorganization required. Hence, transfer could occur by tunneling at a rate governed by the overlap between donor and acceptor wave functions, an overlap indicated by the magnitude of the tunneling matrix element $|V_{av}|^2$, which we shall abbreviate as a coupling factor J. Tunneling has an associated resonance reaching a peak when the energy difference between the electron associated with the donor and the acceptor becomes zero.[47] Hence, the probability of tunneling from donor to acceptor increases as the configurational energy increases along the parabolic reaction coordinate and reaches a maximum at point F in Figure 1A, the point at which the free energies of the two configurations, electron in the donor and electron in the acceptor, are the same. Thus, the actual rate of electron transfer will depend both on the energy required to reach the point F and on the coupling factor J.

When the overlap between reactant and project wave functions of J is sufficiently large, then electron transfer via tunneling becomes instantaneous with respect to nuclear motion ($\kappa_{el} \rightarrow 1$) and we can represent the reaction diagram as in Figure 1B. In this limit the potential energy of the system moving along the reaction coordinate always remains on the lower surface as indicated by the arrow in Figure 1B, or the reaction is *adiabatic*, and Equation 12a becomes:

$$k_{et} = K_o \kappa_{nu} \tag{12b}$$

According to semiclassical theory, the nuclear transmission coefficient is given by:

$$\kappa_{nu} = \Gamma \exp[-\Delta G_{cl}/RT] \tag{13}$$

where ΔG_{cl} is an activation free energy and Γ is a nuclear tunneling correction factor, which for the rest of this discussion will be set to unity (i.e., nuclear tunneling is assumed to be negligible in these reactions). Because of the initial assumption that the two potential energy curves of Figure 1A are parabolic, it is a simple task to calculate the magnitude of the activation free energy from a given λ and ΔG^0, the standard free energy of reaction:

$$\Delta G_{cl} = (\lambda/4)(1 + \Delta G^o/\lambda)^2 \tag{14}$$

As an aside we wish to point out an aspect of this theory, that while not germane to our topic, is of some interest. There is a cross relation that follows from Marcus theory. We consider the three related reactions:

$$A^- + B \xrightarrow{k_{AB}} A + B^-$$

$$A^- + A \xrightarrow{k_{AA}} A + A^-$$

$$B^- + B \xrightarrow{k_{BB}} B + B^- \tag{15}$$

Without going into the argument, it is known that the rate constant k_{AB} for electron transfer between A and B (A $\rightarrow$ B) is related to the two rate constants k_{AA} and k_{BB} for the electron self exchanges A $\rightarrow$ A and B $\rightarrow$ B):[44]

$$k_{AB} = (k_{AA} k_{BB} K_{AB} f_{AB})^{1/2}$$

where K_{AB} is the equilibrium constant for the electron transfer from A to B and f_{AB}, a well-defined function of k_{AB}, k_{AA}, and k_{BB}, is often close to unity. This cross correlation has been shown to be valid for outer sphere electron transfers between metal complexes.[46]

To return to the main thrust of this discussion, Equation 12 applies at the limit of large *J* where nuclear motion dominates the electron transfer process. The opposite limit of small *J* (Figure 1C) is of more interest for this review. In this case the probability of electron transfer becomes sufficiently small that Equation 12 applies and electron transfer, not nuclear reorganization will become rate determining; i.e., the probability of jumping across the gap to the upper limb of the curve so that no electron transfer occurs is greater than the probability of adiabatic reaction along the lower energy surface. Hence, the nuclear term comes to represent the equilibrium between ground state and the point of electron transfer. There are two physical cases in which J could become so small: (1) reactions in which there is little orbital overlap because the distance, r_{DA}, between the donor (D) and acceptor (A) sites is large; and (2) reactions for which the electron transfer process is forbidden, of $J > 0$. We shall discuss here only the first of these two, since transfer between donor and acceptor fixed apart at a "large" distance by an intervening spacer is the LRET phenomenon.

When the coupling is weak due to a large r_{DA}, then the electronic transmission coefficient is thought to have an exponential dependence on r_{DA} so that:[48]

$$\kappa_{el} \propto \exp[-\beta(r_{DA} - r_o)] \tag{16}$$

where r_0 is the distance at which the reaction becomes adiabatic and β is a damping factor characteristic of the wave function overlap between the two redox centers. The magnitude of the constant β, which determines the physical distance scale, is in some dispute. Thus, it was set to 1.44 Å^{-1} by Hopfield,[49] 2.6 Å^{-1} by Jortner,[50] $\leqslant 1$ Å^{-1} by Hush[51] and based on McConnell superexchange model,[52] Beitz and Miller suggested a value of 1.1 Å^{-1}.

Let us now summarize the theoretical conclusions from the discussion until this point. When r_{DA} is small and there is strong electronic coupling between redox and donor sites, then J is large and electron transfer is adiabatic and is described by Equation 14, or as emphasized by Sutin et al.:[54-56]

$$\kappa_{nu} \propto \exp\left[-\frac{(\lambda + \Delta G^o)^2}{4\lambda RT}\right] \tag{17}$$

Since ΔG^0 must be negative for a net reaction in the forward direction and the nuclear reorganization energy, λ, as defined must be positive, Equation 17 predicts the following dependence of k_{et} on ΔG^0, the standard free energy change for the electron transfer. With all other factors kept constant, as ΔG^0 becomes more negative the rate of the reaction will increase to a maximum at which point further decrease in ΔG^0 will cause the reaction rate to drop. This rate behavior is due to the fact that at fixed λ the term $(\lambda + \Delta G^0)$, positive at small ΔG^0 and going through zero to become negative at larger ΔG^0, is raised to the second power. The predicted region at which the rate decreases with decreasing ΔG^0 is called the inverted region. Second, since the nuclear reorganization factor λ should also involve solvent rearrangements around the reacting molecule, κ_{nu} and, hence, k_{et} should also depend on the nature of the solvent. Third, when J is sufficiently small, due perhaps to a large value of r_{DA}, then there should be both an inverted region and, on the basis of Equation 16, exponential decline of the electron transfer rate constant with distance between donor and acceptor sites. All three of these predictions have been met experimentally, as we shall see. But while these verifications of Marcus theory are satisfying, there are still complications that must be considered before we can accept the applicability of Marcus theory to the observation of LRET in proteins.

As originally interpreted, Marcus theory predicts an inverted region that is due solely to nuclear rearrangements (Equation 14) and an exponential decline with increasing distance due solely to the electron tunneling (Equation 16). But these two statements may not be entirely correct. First, it is well known[46] that the probability of tunneling is dependent on the difference in energies of the electron when in the donor and acceptor sites, with the fastest rate predicted to occur when that energy difference is zero. Now, while the classical picture of Figure 1 is intuitively reasonable, it gives us no information concerning the distribution of energy levels in product and reactant states near the point that is assigned as the maximum of the adiabatic surface. Since the relative positions of these various levels are most certainly dependent on the magnitude of ΔG^0, it is conceivable that there is a resonance maximum due to electron tunneling; i.e., an inverted region might also have been predicted on the basis of tunneling. Second, it has recently been argued that nuclear reorganization, the classical contribution to the overall electron transfer process, may also display an exponential dependence with distance due to the sensitivity of λ to distance.[56,57] Neither of these arguments validate or invalidate the applicability of Marcus theory. They do, however, force us: (1) to account for the distance dependence of k_{nu}, the nuclear rearrangement term, when we attempt to isolate the distance dependence of k_{el}, the tunneling term; and (2) to look at observations of the inverted region more critically. Additionally, we should note that other explanations also predict an exponential dependence of k_{et} on distance. For example, both hopping[58,59] and percolation[60] theories have been proposed for LRET, and both these mechanisms can lead to the exponential distance dependence. Hence, the observation of such a distance dependence does not of itself establish Marcus theory as a reliable explanation for LRET.

Before turning to experimental evidence, we should consider one more major concern with present theoretical formulations. Hush and co-workers have been pointing out that when donor and acceptor are connected by a system of nonconjugated bonds, the LRET process has generally been considered as a through-space process over a spacer that fulfills a purely geometrical role,[61] But, are we permitted to consider, as in the first adaptations of Marcus theory to LRET, the spacer as a link effectively invisible to the tunneling electron? Or is LRET modulated by the nonconjugated bridge linking donor and acceptor? This concern has been discussed with two slogans: electron transfer as a through-space (direct spatial overlap of orbitals associated with each redox center) or through-bond (mutual mixing of the redox couple orbitals with those of some communicating framework) process. While there are still few experimental results that can be used to answer this question, there does appear to be a strong interest in the through-bond formulation, for which theories of superexchange coupling have been suggested.[52,62,63] Interestingly, were a mechanism such as superexchange significant, then the distance dependence of the electronic coupling factor need not be exponential.

We turn next to the experiments that test the validity of Marcus theory, or a close variant, as an explanation of LRET in any experimental system, but most particularly in proteins. Until the early 1980s this question could not be answered; for a while there was strong evidence that LRET did occur, at least in proteins,[3] it was not until the first experiments with cytochrome c[13,14] and the construction of redox centers widely separated through a rigid organic spacer,[64-66] that reasonable estimates of the distance between donor and acceptor could be entertained. Since these early papers, a number of LRET reports have been published, and we turn now to the experimental evidence accumulated over approximately the past five years.

IV. LRET IN PEPTIDES AND OTHER MODEL SYSTEMS

There are now a variety of nonpeptide spacers that have been utilized for LRET studies,[57,67-70] with examples presented in Figure 2. Miller and co-workers, with the technique of pulse radiolysis, found an inverted region (decreasing k_{et} with increasing ΔG^0) for both intermolecular electron transfers between a variety of aromatic donors and acceptors dissolved in dilute organic glasses,[71] and intramolecular LRET between organic donor/acceptor center held apart by the rigid framework of a steroidal 5α-androstane skeleton.[67] A similar free energy inversion was subsequently found by Wasielewski et al. in the photoinduced intramolecular LRET between an excited Zn-porphyrin and quinones covalently attached together through a rigid bicyclooctane ring.[68] As had also been proposed, the rate of electron transfer

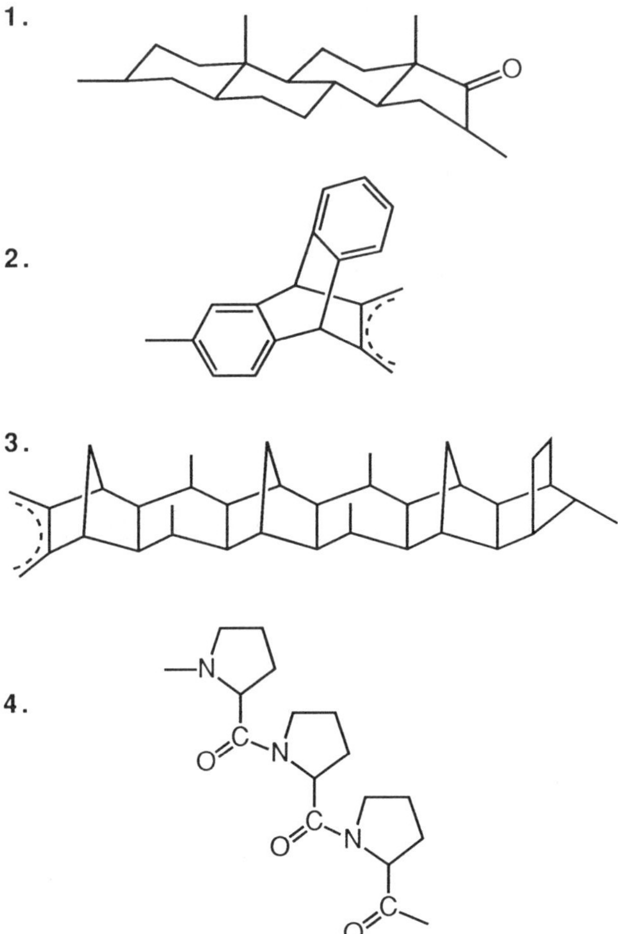

FIGURE 2.   Examples of spacers utilized in LRET studies: (1);[64] (2);[68] (3);[57] (4).[31-34]

is dependent on the solvent,[61,67,71] although Oevering and co-workers report only a small solvent dependence in their series of experiments.[57] Finally, as would have been expected, there is clearly a decline in the LRET rate with increasing distance,[57,69] with Oevering et al. showing that this dependence may be exponential as predicted by theory. What is perhaps most surprising are the magnitudes of the rates when donor and acceptor are linked by a rigid system of saturated carbon atoms. Rate constants $\geq 10^6$ s^{-1} have been reported even to distances of separation close to 20 Å.[57,64,69] Such fast rates may justify the proposal of Oevering et al. for an important through-bond electron transfer channel, at least in these systems.

Early work with peptides was reported by Prutz and co-workers,[9] who used pulse radiolysis to observe electron transfer from the tyrosine side chain phenol to the tryptophan side chain indolyl radical. But, it was with the oligoproline spacer that LRET could be studied in greater detail, for the two ends of an oligoproline chain are held scrupulously apart due to steric constraints on the proline peptide.[27] Moreover, it is known from experimental measurements of fluorescence quenching that the distance between a tyrosine and tryptophan, separated from one another by an oligoproline spacer, increases linearly with the addition of each intervening proline residue.[28] Hence, there is little likelihood of electron transfer through direct redox center collision. Four different laboratories have utilized the oligoproline spacer with different donor/acceptor pairs for LRET studies: Isied et al. have measured pulse radiolytically initiated LRET from Os(II) to Co(III) in the mixed peptide system $(NH_3)_5Os(II)$-X-Co(III)$(NH_3)_5$,[31] where X is the peptide spacer, and from Os(II) to Ru(III) in $(NH_3)_5Os(II)$-X-Ru(III)$(NH_3)_5$;[56] Schanze and Sauer have looked at LRET between a photoexcited polypyridyl Ru(II) complex donor and a p-benzoquinone type electron acceptor;[32] Faraggi et al.[33] and Bobrowski et al.[34] have utilized the tyrosine/tryptophan system. What are the similarities, other than the common -(pro)$_n$-spacer, and differences between the different systems studied in these four laboratories?

Isied and co-workers established LRET by preferentially reducing an osmium center in aqueous solution with the $CO_2^-\cdot$ radical produced through the reactions of Equation 6:

$$\boxed{Os(III)} - X - \boxed{Co(III)} + CO_2^-\cdot \; - - \blacktriangleright \; \boxed{Os(II)} - X - \boxed{Co(III)} + CO_2$$

$$(18)$$

These authors then monitored the intramolecular oxidation of the Os(II) complex by its absorbance decay at 525 nm:

$$\boxed{Os(II)} - X - \boxed{Co(III)} \; - - \blacktriangleright \; \boxed{Os(III)} - X - \boxed{Co(II)} \qquad (19)$$

Electron transfer in the tyrosine/tryptophan system is initiated by the preferential $N_3\cdot$ oxidation of the tryptophan side chain, Reaction 10, followed by a slower trp$\cdot$ oxidation of tyrosine, Reaction 11. The intramolecular electron transfer is easily monitored either from the decrease of the trp$\cdot$ absorption (510 nm) or the increase of the tyrO$\cdot$ absorption (410 nm; see Figure 3). In contrast to Reaction 19, Reaction 11 is formally a hole transfer. But Reactions 11 and 19 are similar in that the electron transfers are from ground state to ground state in aqueous solutions.

Schanze and Sauer[32] worked with a Ru(II) polypyridyl complex linked through a proline chain to a p-quinone. The electron transfer in this system was initiated by excitation of the metal complex with 400 nm light. The excited complex *Ru(II) then returns to the ground state either by light emission or electron transfer:

$$\boxed{*Ru(II)} - NH - (pro)_n - Q \; \Big\langle \; \begin{array}{l} \xrightarrow{\;\text{emission}\;} \boxed{Ru(II)} - NH - (pro)_n - Q \\[2em] \xrightarrow{\;\text{transfer}\;} \boxed{Ru(III)} - NH - (pro)_n - Q^-\cdot \end{array}$$

$$(20)$$

Since LRET is competitive to at least one other process, the LRET rate constant cannot be obtained directly, but must be computed from the enhancement of quenching, emission decay rate increase, due to the presence of the quinone, Q. There are two important differences relative to the other systems discussed: (1) this is a reductive system in which the electron

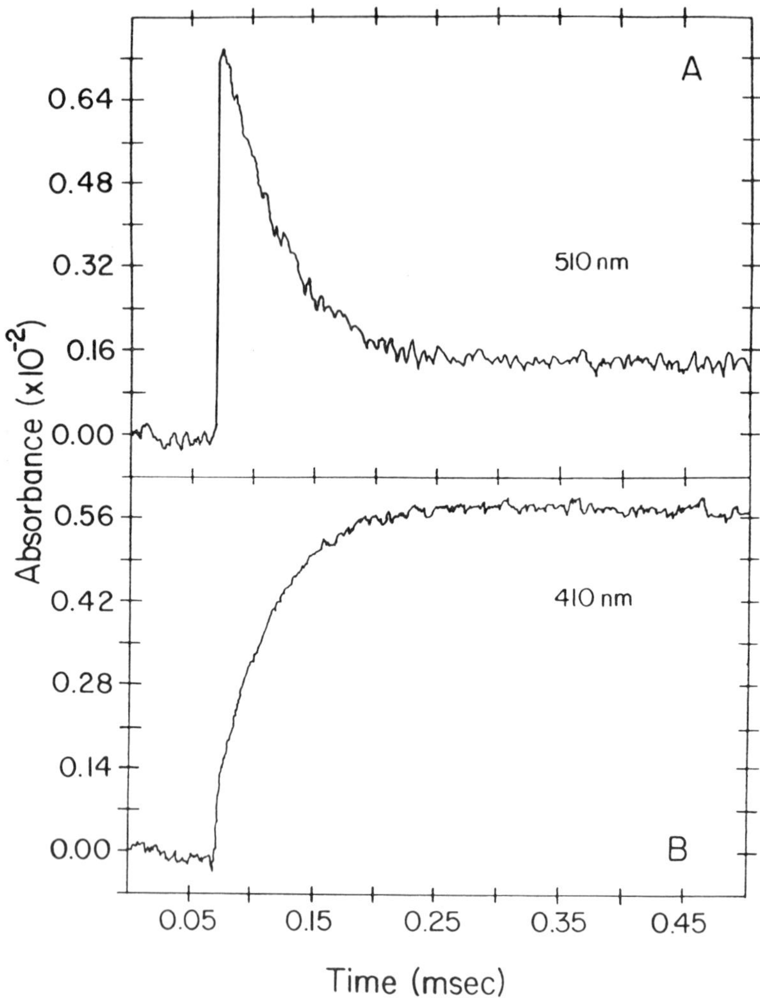

FIGURE 3.   One-electron transfer between tyrosine and tryptophan in the peptide tyr-glu-trp.[33]

is transferred from an excited state to a ground state; and (2) the experiments were performed in CH_2Cl_2 which may be an important difference vis-á-vis the κ_{nu}.

There are clearly important differences between the donor/acceptor chemistry utilized by these four groups, but there are also important similarities. Each system does display an apparent exponential dependence of the rate constant on distance, a dependence predicted by theory. However, in contradiction to theoretical formulations the magnitude of β, the descriptor of that dependence (Equation 16) is not constant (Table 1 and Figure 4). On the one hand the observed variation of β could be construed as due solely to differences of the redox centers, the only structural variant in these experiments. On the other hand, it has been suggested that κ_{nu} (Equation 17) in addition to κ_{el} has an exponential distance dependence,[56,57] and Isied et al.[56] have proposed that the observed β of 1.6 Å^{-1} in the peptide series Os– – –Ru (Table 1) is the sum of a nuclear, $\beta_{nu} \approx 0.9$ Å^{-1}, and an electronic, $\beta_{el} \approx 0.7$ Å^{-1}, contribution. But even accepting this fragmentation of β, the results still argue against a universal value associated with either β_{nu} or β_{el}. As seen from Table 1 both observed β for the two peptide series trp– – –tyr and tyr– – –trp are lower than either the β_{nu} and β_{el} estimated by Isied et al. for the Os– – –Ru series.[52] Hence, we must conclude that β_{nu} and/

TABLE 1

System[a]	$\Delta\epsilon_m$ (V)[b]	ϵ_m [Reductant] (V)	Obs. β (Å^{-1})[c]
Os– – –Co	0.15	0.25	2.1
Os– – –Ru	0.25	0.25	1.6
Ru– – –Quinone	0.35	0.61	0.6
trp– – –tyr	~0.15	0.90	0.4
tyr– – –trp	0.06	0.90	0.2

[a] The LRET systems from top to bottom were reported in References 30, 56, 32—34, 33, respectively.

[b] The driving forces are estimates of uncertain reliability.

[c] The slope of the $\ln(k_{et})$ vs. distance between donor and acceptor were computed on the assumption that each proline inserted into the spacer extends the distance by 3.1 Å.

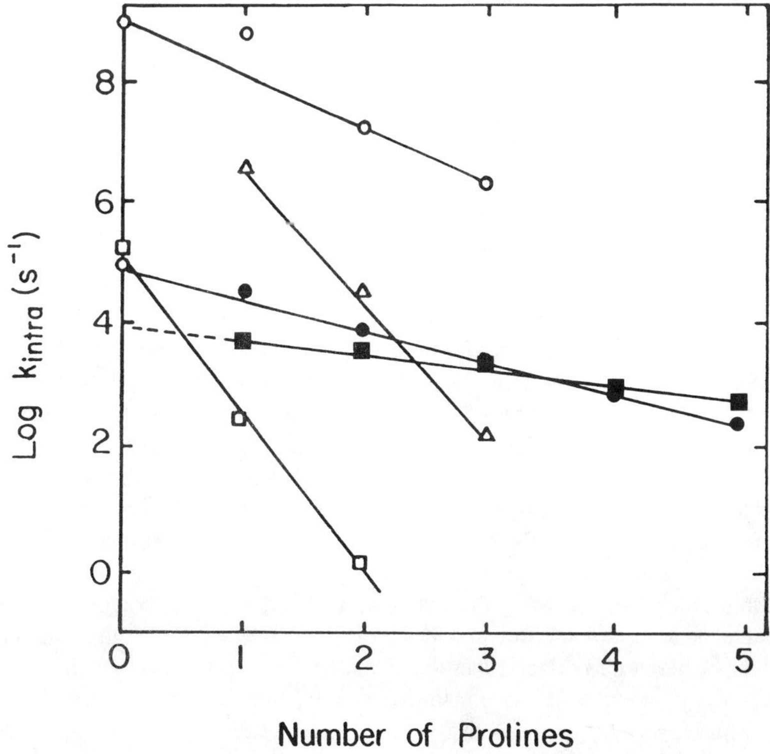

FIGURE 4. Distance dependence of one-electron transfer over the oligoproline spacer: O–O — Ru– – –Q;[32] △–△ — Os– – –Ru;[56] □–□ — Os– – –Co;[31] ●–● — Trp– – –tyr;[33] ■–■ — Tyr– – –trp.[33]

or β_{el} are dependent on the nature of the redox centers, although we cannot yet explain the basis of this variation. Recently, with the determination of the redox potentials (vs. HHE) of the tryptophan/indolyl radical (1.05 ± 0.02 V) and the tyrosine/phenoxy radical (0.93 ± 0.02 V),[72-76] we had suggested that the magnitude of β might be dependent on the driving force, $\Delta\epsilon$, between the redox centers,[33] but with the data now available this suggestion appears incorrect. On the other hand, there may be an inverse correlation between the redox potential of the electron donor and the observed β (Table 1). Such a correlation is still speculative.

There is another remarkable feature to the data summarized in Table 1; the two peptide series, tyr– – –trp and trp– – –tyr, behave differently even though the donor, acceptor, and spacer are the same in both. Clearly, this difference must be due to asymmetry in the polypeptide chain with its amino and carboxyl terminal ends. That the two β do not have the same value may be due directly to the change in amino acid sequence ordering. On the other hand, this difference may be due only indirectly to peptide chain asymmetry, with the difference in $\Delta\epsilon_m$ the direct cause; i.e., under conditions of indentical free energy driving force both orderings of donor and acceptor on the chain may result in the same value of β. There are at least two reasons for attempting to determine whether ordering or potential difference is the primary cause, or even to find other possible causes. First, if the chain order is intrinsically significant in establishing the magnitude of β, then a strong case could be made for a through-bond LRET mechanism. Second, such a directional effect could be a way in which the protein structure might impose a path specificity on intramolecular electron transfer.

Irrespective of the theoretical consequences, the data of Figure 4 and Table 1 lead to an important empirical observation. Even with a very small driving force (e.g., about 60 mV for the series tyr– – –trp) LRET in peptides can occur over large distances and at rates fast enough to be physiologically significant. These results are consistent with the observation of LRET in proteins, and suggest that LRET should be physiologically significant.

V. LRET IN PROTEINS

Experimental evidence for LRET in proteins has been accumulating since the phenomenon was first proposed in the 1970s. Isied et al. and Gray et al. have described LRET between the heme of cyrochrome c and a ruthenium covalently attached to that protein's histidine at position 33;[12-14] Hoffman and co-workers reported an intramolecular electron transfer between the metals of a mixed metal [Zn(II), Fe(III)] heme hybrid hemoglobin;[36] McLendon et al. detected electron transfer between the mixed metal heme groups of a cytochrome b_5/cytochrome c complex;[25] we have reported LRET between the disulfide radical anion and Fe(III) in α-hemoglobin.[26] By now, the occurrence of LRET in proteins is well established,[11] as seen from the sampling of LRET rate constants in Table 2. But despite this rash of activity, much still remains to be learned about the protein LRET mechanism.

A primary goal of much of the research with proteins has been to verify the applicability of the Marcus theory. As discussed earlier, one prediction is that of the "inverted" region: with all other factors fixed, when the reaction ΔG^n becomes more negative, the electron transfer rate constant first increases, reaches a maximum dependent on the magnitude of the nuclear reorganization term λ, and then declines. McLendon has, in fact, reported such an inversion in the photoinduced electron transfer between mixed metal heme centers in the protein complex between cytochrome c and by cytochrome b_5.[11c] Gray et al., while not directly observing the inverted region, do have evidence consistent with an inversion (Figure 5).[31] A second major prediction is that at a fixed ΔG^0 the LRET rate constant should decrease exponentially with the distance r_{DA} when the electronic coupling is weak. While we have seen in Section IV that such a distance dependence appears to hold in small model compounds, in proteins observations of an exponential distance dependence remain ambiguous even with closely related proteins; e.g., Figure 6.[11c,22] Considering accumulated results for a variety of proteins (Table 2), we find no overall correlation between electron transfer rate constants and distance separation. The absence of such a correlation should not, however, be surprising. As we discussed above, both electronic (κ_{el}) and nuclear (κ_{nu}) contributions may have a distance dependence, and the environments around the electron donor and acceptor most surely differ in the various proteins. Until it is possible to account for these structural differences, we may be unable to discern a general relationship between k_{et} and distance in collections of proteins.

TABLE 2
Selected Intramolecular Protein LRET Rate Constants

Donor/acceptor	System	$\Delta\epsilon m(V)$[a]	$d(\text{Å})$[b]	$k(s^{-1})$[c]	Ref.
Zn(II)*/Fe(III)	Zn-Hb/cyt b_5 complex	1.0	7	8,000	37
Fe(II)/Fe(III)	cyt b_2/cyt c complex	0.2	8	150	77
	cyt b_5/yeast cyt c complex	0.2	8	900(?)	25
	cyt b_5/bovine cyt c complex	0.2	8	6,000	11e
Ru(II)/Fe(III)	Ru-HIPIP	0.27	8	18	17
porph*/Fe(III)	porph-cyt-c/cyt b_2 complex	0.4	8	150	77
	porph-cyt c/cyt b_5 complex	0.4	8	50,000	25
Zn(II)*/Fe(III)	Zn-cyt c/cyt b_2 complex	0.8	8	150	77
	Zn-cyt c/cyt b_5 complex	0.8	8	500,000	25
porph$^-$/Fe(III)	porphy-cyt c/cyt b_5 complex	1.1	8	8,000	25
Ru(II)/Fe(III)	Ru-cytochrome c	0.15	12	40	35, 41
Fe(II)/Ru(III)	Ru-cytochrome c	0.18	12	<0.0001	43
Ru(II)/Cu(II)	Ru-Azurin	0.24	12	2	15
S-S·$^-$/Fe(III)	Fe(III)-α-Hb-SSR	0.44	12	200	26
Zn(II)*/Ru(III)	Zn-porph-Ru(III)-(His-33)cytochrome c	0.88	12	770,000	21
Ru(II)/Fe(III)	Ru-myoglobin	−0.02	13	0.02	78
Pd(II)*/Ru(III)	Pd-porph-Ru(III)-(His-48)myoglobin	0.72	13	9,000	23
S-S·$^-$/Fe(III)	Fe(III)-Hr-SSR	0.78	13	15	79
Zn(II)*/Ru(III)	Zn-porph-Ru(III)-(His-48)myoglobin	0.8	13	70,000	22
α-Zn(II)*/β-Fe(III)	Zn/Fe-hemoglogin	0.8	20	100	80
β-Zn(II)$^+$/α-Fe(III)	Zn/Fe-hemoglobin	0.8	20	1,000	80

[a] Driving force of the reaction in volts (at pH 7).

[b] Edge-to-edge through space distance.

[c] Apparent first-order rate constant.

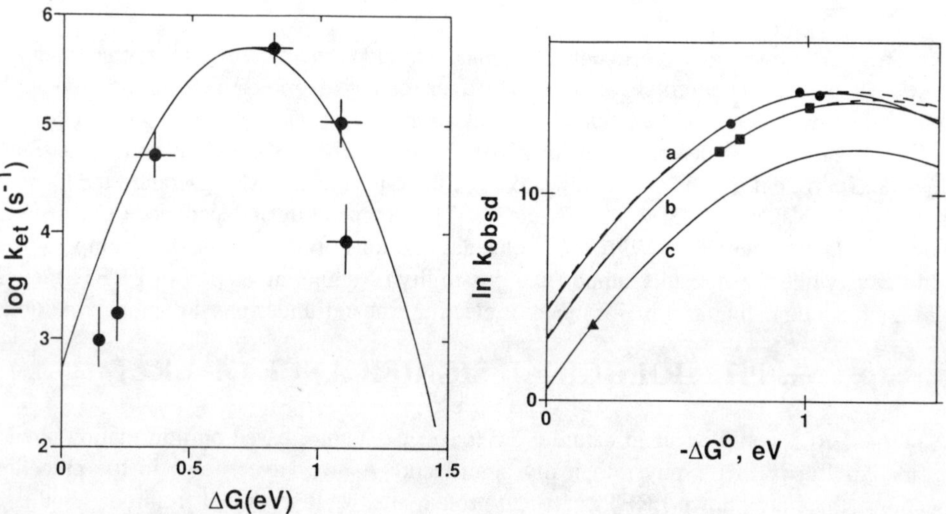

FIGURE 5. Dependence of intramolecular electron transfer rate constants on equilibrium free energy. Left — cytochrome c/cytochrome b_5 complex (solid line calculated assuming α = 0.8 V).[11] Right — ruthenium modified substituted metallo-porphyrin cytochrome c.[81] ● — Zn* to Ru(III); ■ — Ru(II) to Zn$^+$·; ▲ — Ru(II) to Fe(III). Solid lines best fit three different electron transfer models.

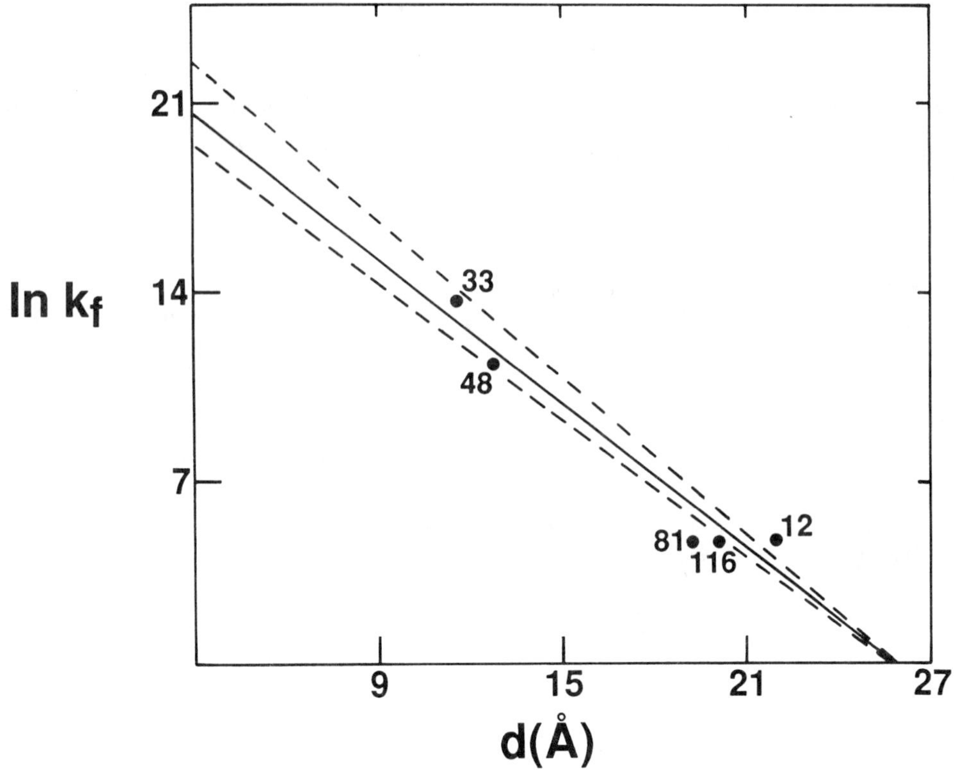

FIGURE 6.   The distance dependence of LRET from the excited Zn*-porphyrin to Ru(III) in ruthenium modified Zn substituted cytochrome *c*.[22]

There is, however, experimental evidence for electron tunneling as a component of LRET in at least one protein system. Hoffman and co-workers have reported that the temperature dependence of the intramolecular electron transfer rate constant in a [Zn(II),Fe(III)] hybrid hemoglobin is described above 253 K by the Arrhenius equation with an assigned $E_a \approx 8$ kJ/mol, but that from 170 down to 77 K the rate constant is temperature independent at a constant value of 9 ± 4 s^{-1} (Figure 7). This temperature independence is consistent with an electron tunneling with the molecular motions frozen out at low temperatures. However, while these results support the possibility of tunneling as part of LRET, they do not establish tunneling as a major mode of electron transfer under physiological conditions.

VI. PHYSIOLOGICAL SIGNIFICANCE OF LRET

That LRET could occur in nature is no longer debatable, based on information already available. But is LRET physiologically significant? As we have shown in the preceding sections, there are known LRETs between protein sites well separated from one another in crystal structures. Moreover, these LRET rates are relatively fast and are driven by free energy differences comparable to those ordinarily found in nature. Thus, while some might still wish to argue that *in vivo* electron transfers occur only by direct collision of redox sites, it is now the simpler premise that LRET does indeed play a role in biological oxidation/reduction phenomena. On the basis of that premise, we shall now speculate on the biological significance of the LRET process and, in particular, whether there is more to LRET than just a mechanism for electrons to exchange between two distant sites. We begin this speculation with a description of an experiment that, while intended to be simple, turned out to have unexpectedly important complications.[82]

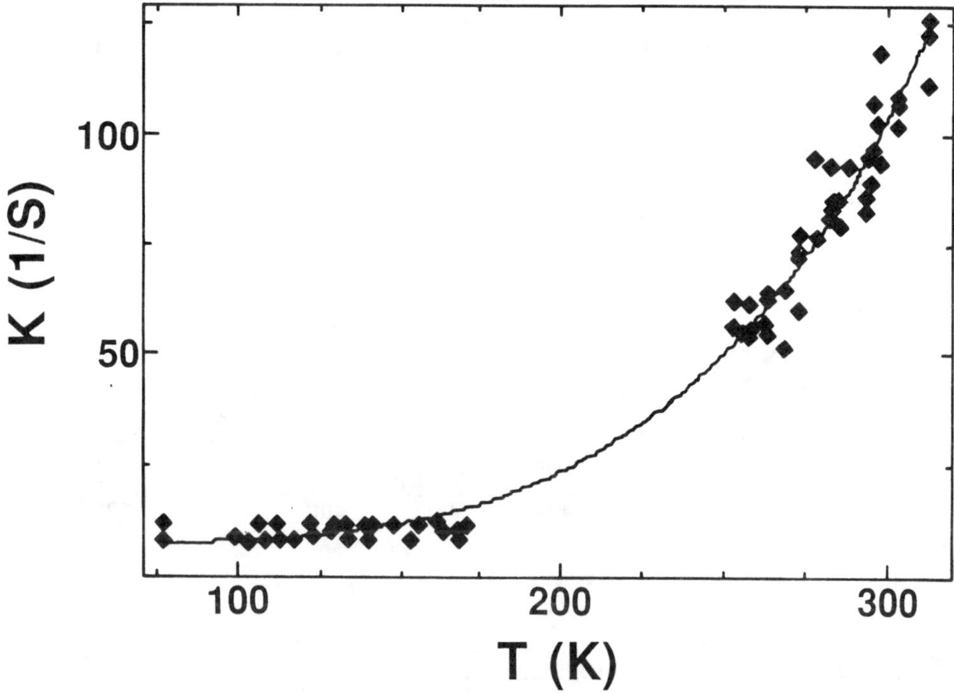

FIGURE 7. Temperature dependence of the intramolecular electron transfer rate constant in $\alpha(Fe^{III}H_2O)$, $\beta(Zn)$ hybrid hemoglobin.[80]

The hydrated electron reduces the histidine imidazole side chain to the electron adduct, (Im·) a free radical with an absorption band centered near 360 nm.[3] Since imidazolium (ImH+), the conjugate acid species, reacts at least 1000-fold faster with e_{aq}^{-}· than does imidazole (Im), the conjugate base, the apparent yield of the imidazolium radical is governed in small peptides by the pK_a of the histidine side chain.[83] Our curiosity pricked by this observation, we wondered if it would be possible to use this Im· pH dependent yield in order to determine the pK_a of histidines bound into proteins. This (perhaps idle) interest in an unconventional experiment was rewarded by success in the first set of expriments; the yield of Im· in pancreatic ribonuclease A was pH dependent (Figure 8).[82] To understand a complicating factor in these results, one must realize that e_{aq}^{-}· reduces not only the imidazole group but also the protein disulfide bond (RSSR) to its disulfide radical anion (RSSR^{-}·), which absorbs intensely near 410 nm. With this additional information one sees from the results of Figure 8 that at low pH, where the protein histidines are all protonated, there is a distinct band centered about 360 nm due to Im·. At higher pH where the protein histidines are deprotonated, the 360 nm band disappears to be replaced by the absorption band of RSSR^{-}·. The pH dependence of absorbance yields for both peaks can be described in terms of a single proton equilibrium with an apparent pK_a of 5.9 to 6.0.

Although gratified with these results, we nonetheless faced the problem that yields of RSSR^{-} from small model disulfides do not exhibit the pH dependence observed in ribonuclease. Because of this discrepancy we searched for Im· in other proteins.[82,84] The results of that search, summarized in Table 3, led us to a totally unexpected conclusion: while the pH dependent e_{aq}^{-}· reduction of histidine side chains in proteins is common, the histidines reduced appear to be located primarily at functional active sites. Furthermore, any modification to an enzyme so as to render it catalytically nonfunctional also resulted (with the exception of the S-methylated papain derivative) in the loss of the imidazole electron adduct. Additionally, while lysozyme, with only one histidine that is not located at the active site,

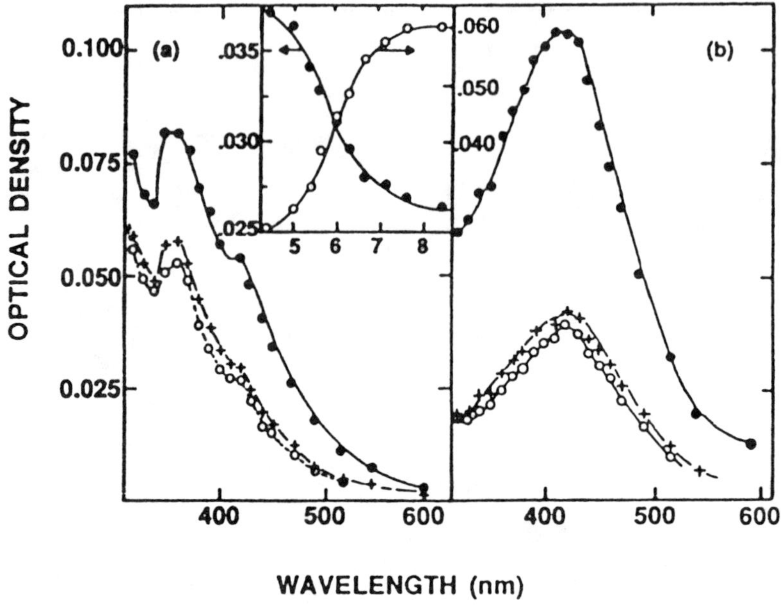

WAVELENGTH (nm)

FIGURE 8.   Transient spectra of argon saturated solution of 0.47 mM RNase containing 0.4 M t-butanol 1 mM phosphate buffer. Obtained at: (a) pH 4.3; and (b) pH. 8.5 and at different times after the pulse. $\circ$ — t = 0; + — t = 350 μs; $\bullet$ — t = 1.2 ms. Inset-effect of pH on the optical density at: $\circ$ — 420; $\bullet$ — 360 nm.[82]

reacts with $e_{aq}^-\cdot$, it does not form Im$\cdot$. Nor is there any experimental evidence for Im$\cdot$ formation from any of the papain histidines that are not at that protein's active site. Whether the observation of $e_{aq}^-\cdot$ specificity toward active site histidines is general over a larger set of proteins will be determined as more data are collected. But irrespective of any additional results it is surprising that $e_{aq}^-\cdot$ with a redox potential of $\leqslant -2.7$ V[40] should appear to have any such specificity in even these proteins; not only did we detect just the active site histidines, but we were able to modulate the histidine response by either covalent or noncovalent modifications at the active site (Table 3).

An explanation as simple as histidine accessibility to $e_{aq}^-\cdot$ cannot explain these results; for the solvent accessible histidine of lysozyme[85] and the histidines of anhydrochymotrypsin and chymotrypsinogen, in which one would expect similar exposure in comparison with chymotrypsin, are not reduced by $e_{aq}^-\cdot$. Moreover, it seems unlikely that $e_{aq}^-\cdot$, which reacts with proteins at or near the diffusion controlled limit, should express any intrinsic specificity toward one or another histidine residue, especially since it is capable of reacting with protein surface groups such as the peptide bond, as known from model studies.[86] For these, among other, reasons we were led to the following proposal.[3,82,84] First, we assumed: (1) that rapid electron transfer over long distances is possible in proteins; and (2) that this "migration" exhibits path specificity due to channels of "least resistance" (a kinetic argument) and/or to potential energy sinks into which an electron would fall during its travels around the protein molecule (a thermodynamic argument). Then, we argued that the primary site(s) of $e_{aq}^-\cdot$ reaction in proteins is(are) not necessarily at a histidine or cystine (in proteins with disulfide bonds), the two reductions so easily detected in the near-UV. Rather, the hydrated electron, embedded within its water shell, must first attach itself at some site on the protein surface, and only after this initial event does it migrate to the active site histidine, the disulfide bond, or any other residue that can be reached and reduced. This argument also accounts for the apparently odd pH dependence noted for the formation of RSSR$^-\cdot$ in RNase (Figure 8). There is a competition between active site imidazolium ion and one or more

TABLE 3
Observed Histidine-$e_{aq}^-\cdot$ Adduct in Proteins

Protein	Inactive derivative[a]	360 nm Band
Trypsin		+
	Trypsinogen	−
Chymotrypsin		+
	Chymotrypsinogen	−
	PMS-chymotrypsin	−
	NMe-chymotrypsin	−
	Anhydrochymotrypsin	−
Subtilisin		+
Thiosubtilisin[b]		+
	PMS-subtilisin	−
	subtilisin + borate[c]	−
Papain		+
	SMe-papain	+
Ribonuclease A		+
Lysozyme		+
Lactalbumin		+

Note: Abbreviations used: NMe-chymotrypsin — α-chymotrypsin in which the active site histidine-67 is methylated at the nitrogen; PMS-chymotrypsin — α-chymotrypsin with its active site serine-195 blocked by a phenylmethane sulfonyl group; anhydrochymotrypsin — α-chymotrypsin in which serine-195 has been converted to dehydroalanine; PMS-subtilisin — subtilisin with its active site serine-22 blocked by phenylmethanesulfonyl group; thiosubtilisin — subtilisin in which the serine-221 has been converted to a cysteine; SMe-papain — papain in which the active site thiol group has been reacted with methyl methanethiosulfonate.

[a] Catalytically inactive proteins

[b] Thiosubtilisn cannot catalyze the hydrolysis of peptide bonds, the natural reaction of subtilisin. It, however, catalyzes the cleavage of esters, as does subtilisin.

[c] The $e_{aq}^-\cdot$ reduction of the enzyme is run in the presence of the competitive inhibitor borate.

disulfide bonds for the intramolecularly migrating electron; at low pH (high imidazolium content) the active site histidine competitively traps the migrating electron; at high pH (low imidazolium content) the disulfide bond(s) trap(s) and migrating electron. However, for this proposal to be correct, the two assumptions upon which it was based must be valid.

In Section V we presented evidence for the occurrence of LRET in proteins, the first assumption in the above argument. Is there any experimental support for the second postulate; path specificity in electron movement through a protein? The answer appears to be yes, since there is, in fact, an important physiological phenomenon in which path specificity has been found. Although one would deduce from the crystal structure of the *Rhodopseudomonas viridis* photosynthetic reaction center protein[87] that there are two possible paths over which photoinduced electron transfer could occur, it has been proposed[88] that only one of the two is actually utilized. The physicochemical basis of this path specificity is not yet known, although various structural features of the protein are now being invoked in explanation.[88,89] A second possible example of path specificity comes from the recent work of Liang et al.,[90] who studied LRET within the complex formed between the cytochrome *c* peroxidase Zn-porphyrin derivative and cytochrome *c*. With the heme iron of the cytochrome oxidized, they obtained a photoinduced electron transfer from the peroxidase Zn-porphyrin to the cytochrome Fe(III). The electron then transferred back from the cytochrome Fe(II) to the Zn-porphyrin$^+$ cation.

$$\text{Zn-porph}^* \cdots \text{Fe(III)} \xrightarrow{\ k_f\ } \text{Zn-porph}^+ \cdots \text{Fe(II)} \qquad (21)$$

$$\text{Zn-porph}^+ \cdots \text{Fe(II)} \xrightarrow{\ k_b\ } \text{Zn-porph} \cdots \text{Fe(III)} \qquad (22)$$

On the basis of the known protein crystal structures, Liang and co-workers postulated that phenylalanine-87 of the cytochrome *c* might facilitate electron transfer within the complex. Substitution, by site-directed mutagenesis, of that phenylalanine with tyrosine resulted in no significant change of forward and reverse rate constants. But substitution of a serine at position 87 did lead to an approximately 10,000-fold reduction in the back rate constant, k_b, although there was no effect on the forward rate constant, k_f. The physical basis of this differential effect between the aromatic amino acids and the serine at position 87 was not explored, but the authors speculated that the mechanism of the electron back transfer — formally a hole transfer — might be different from that of the forward photoinduced process. Whatever the explanation, there is LRET enhancement when serine at position 87 is replaced by an aromatic amino acid, an observation suggesting that proteins might have inherently structured path electron transfer specificities.

VII. CONCLUSIONS AND PROSPECTS

The repeated observances, during the past 5 years, of long-range electron transfer reactions in model compounds (inorganic and organic) free us from the requirement for close physical proximity of redox centers and lead us to accept LRET as a common phenomenon. The observations to date suggest that for a homologous series of compounds the first-order rate constant that characterizes the LRET process (1) decreases exponentially, with a damping constant β, as the distance between the redox centers becomes greater, and (2) increases to a maximum and then falls off again as the driving force of the reaction (ΔG) becomes larger. These are the behaviors that were predicted by Marcus-Hush theory, a semiclassical approach which treats electron tunneling and nuclear motion as separable components of the overall reaction. On the other hand, the observed numerical value of β appears to depend on the nature of the compounds under study, with the variations and the magnitudes of these constants still not fully explained by theory. Of great importance is the fact that LRET can occur over 10 to 20 Å at rates that are still fast, with respect to enzyme catalyzed reactions, even when the driving force is less than 100 mV. Thus LRET is accessible under physiological conditions when potential energy differences are generally small. Finally, there is evidence to suggest that electron transfer within proteins may be directed toward specific target sites.[82,84] Such specificity may imply that either a protein could have preferred paths over which electrons might "migrate", or that it might provide a low potential energy "trap" to favor electron capture at the appropriate site. In other words, the specific structure of a protein may modulate the LRET which it houses.

In spite of the recent and intense flurry of studies on LRET in small molecules and proteins, we are left with a number of fundamental questions that range from more technical concerns to highly speculative notions. Thus, there are known examples of protein redox reactions that involve conformational changes; is any newly uncovered intramolecular LRET within a protein governed by the rate of electron transfer or by the rate of a conformational alteration? How are we to view the electron transfer process? Does the electron jump through space essentially independent of the covalent bond network that lies between the two redox centers, or does it migrate specifically through a covalent bond network? An additional question, dependent on the one just formulated, can also be posed; given either a through-space or a through-bond mechanism, how does the material over which the transfer occurs affect the reaction rate? For example, would an intervening α-helical structure promote or inhibit electron transfer; could a hydrogen bond between two residues promote the transfer; etc.? Can proteins modulate LRET reactions by specifically influencing either the nuclear reorganization energy (medium effect) or the electronic term? And finally, the most speculative question of them all, if it were the case that LRET in proteins could be affected by the intervening matrix, then would it be reasonable to assume that evolutionary pressures have molded protein structures so as to optmize, either by enhancing or diminishing, the rate of electron transfer for a physiologically important LRET?

There are at least three approaches for attempting to answer these and related questions. All three have already been utilized. The first is to work with proteins and to attempt modulation of LRET rates by chemical modification or site specific mutagenesis. The second is to consider small peptides as models of proteins and to synthetically construct molecules in order to isolate one or another possible structural factor that might affect LRET over the peptide. The third is to work with small molecules not related to proteins in the hope that the knowledge so gained can be extrapolated to answer similar questions about proteins. No one approach will suffice, as is already known to chemists and biochemists who have investigated any important physical property of proteins. Thus, the successful solution to the still recently posed questions of LRET in proteins will provide challenges to both experimentalists and theorists of many different disciplines.

ACKNOWLEDGMENTS

This work was supported by Grants No. 85-00217 and 86-00206 from the United States-Israel Binational Science Foundation (BSF), Jerusalem, Israel and by NIH grant GM-35718. The authors are grateful to Dr. J. Holcman for sending us his preprint prior to publication.

REFERENCES

1. **Chance, B. and Williams, G. R.,** The respiratory chain and oxidative phosphorylation, *Adv. Enzymol.,* 17, 65, 1956.
2. **Winfield, M. E.,** Electron transfer within and between haemoprotein molecules, *J. Mol. Biol.,* 12, 600, 1965.
3. **Klapper, M. H. and Faraggi, M.,** Applications of pulse radiolysis to protein chemistry, *Q. Rev. Biophys.,* 12, 465, 1979.
4. **Bensasson, R. V., Land, E. J., and Truscott, T. G.,** *Flash Photolysis and Pulse Radiolysis,* Pergamon Press, New York, 1983.
5. **Faraggi, M. and Pecht, I.,** The reaction of *Pseudomonas* azurin with hydrated electrons, *Biochem. Biophys. Res. Comm.,* 45, 842, 1971.
6. **Land, E. J. and Swallow, A. J.,** One-electron reactions in biochemical systems as studied by pulse radiolysis. V. Cytochrome *c, Arch. Biochem. Biophys.,* 145, 365, 1971.
7. **Faraggi, M. and Pecht, I.,** Elementary steps in the action of electron transfer proteins, *Israel, J. Chem.,* 10, 1021, 1972.
8. **Faraggi, M. and Pecht, I.,** The electron pathway to Cu(II) in ceruloplasmin, *J. Biol. Chem.,* 248, 3146, 1973.
9. **Prutz, W. A., Land, E. J., and Sloper, R. W.,** Charge transfer in peptides, *J. Chem. Soc. Faraday Trans I.,* 77, 281, 1981.
10. **Butler, J., Land, E. J., Prutz, W. A., and Swallow, A. J.,** Charge transfer between tryptophan and tyrosine in proteins, *Biochem. Biophys. Acta,* 705, 150, 1982.
11a. **Isied, S. S.,** Long-range electron transfer in peptides and proteins, *Progr. Inorg. Chem.,* 32, 443, 1984.

11b. **Peterson-Kennedy, S. E., McGourty, J. L., Ho, P. S., Liang, N., Zemel, H., Blough, N. V., Margoliash, E., and Hoffman, B. M.,** Long-range electron transfer at fixed and known distance within protein complexes, *Coord. Chem. Rev.,* 64, 125, 1985.

11c. **Mayo, S. L., Ellis, W. R., Crutchley, R. J., and Gray, H. B.,** Long-range electron transfer in heme proteins, *Science,* 233, 948, 1986.

11d. **Miller, J. R.,** Controlling charge separation through effects of energy, distance and molecular structure on electron transfer rates, *Nouv. J. Chim.,* 11, 83, 1987.

11e. **McLendon, G.,** Long-distance electron transfer in proteins and model systems, *Acc. Chem. Res.,* 21, 160, 1988.

12. **Yocom, K. M., Shelton, J. B., Shelton, J. R., Schroeder, W. A., Worosila, G., Isied, S. S., Bordignon, E., and Gray, H. B.,** Preparation and characterization of a pentaamineruthenium(III) derivative of horse heart cytochrome, *c, Proc. Natl. Acad. Sci. U.S.A.,* 79, 7052, 1982.

13. **Winkler, J. R., Nocera, D. G., Yocom, K. M., Bordignon, E., and Gray, H. B.,** Electron transfer kinetics of pentaaminerutheinum(III) (histidine-33)-ferricytochrome *c.* Measurement of the rate of intramolecular electron transfer between redox centers separated by 15 Å in a protein, *J. Am. Chem. Soc.,* 104, 5798, 1982.

14. **Isied, S. S., Worosila, G., and Atherton, S. J.,** Electron transfer across polypeptides. IV. Intramolecular electron transfer from ruthenium(II) to iron(III) in histidine-33 modified horse heart cytochrome, *c, J. Am. Chem. Soc.,* 104, 7659, 1982.

15. **Kostic, N. M., Margalit, R., Che, C.-M., and Gray, H. B.,** Kinetics of long-distance ruthenium-to-copper electron transfer in [pentaamineruthenium histidine-83]azurin, *J. Am. Chem. Soc.,* 105, 7765, 1985.

16. **Gray, H. B.,** Reversible long-range electron transfer in ruthenium-modified sperm whale myoglobin, *J. Am. Chem. Soc.,* 109, 3778, 1987.

17. **Jackman, M. P., Lim, M.-C., Salmon, G. A., and Sykes, A. G.,** Preparation of pentaamineruthenium(III) derivative of *Chromatium vinosum* HIPIP and the kinetics of intramolecular electron transfer, *J. Chem. Soc. Chem. Comm.,* 179, 1988.

18. **Osvath, P., Salmon, G. A., and Sykes, A. G.,** Preparation, characterization, and intramolecular rate constant for Ru(II) —— Fe(III) electron transfer in the pentaamineruthenium histidine modified cytochrome *c551* from *Pseudomonas stutzeri, J. Am. Chem. Soc.,* 110, 7114, 1988.

19. **Jackman, M. P., McGinnis, P., Powls, R., Salmon, G. A., and Sykes, A. G.,** Preparation and characterization of two his-59 ruthenium-modified algal plastocyanins and an unusually small rate constant for ruthenium(II) —— copper(II) intramolecular electron transfer over 12 Å, *J. Am. Chem. Soc.,* 110, 5880, 1988.

20. **Farver, O. and Pecht, I.,** Long-range intramolecular electron transfer in *Rhus vernicifera* stellacyanin: a pulse radiolysis study, *FEBS Lett.,* 244, 379, 1989.

21. **Horst, E., Chu, M. H., and Winkler, J. R.,** Electron-transfer kinetics of Zn-substituted cytochrome *c* and its $Ru(NH_3)_5$(histidine) derivative, *J. Am. Chem. Soc.,* 110, 429, 1988.

22. **Axup, A. W., Albin, M., Mayo, S. L., Crutchley, R. J., and Gray, H. B.,** Distance dependence of photoinduced long-range electron transfer in zinc/ruthenium-modified myoglobins, *J. Am. Chem. Soc.,* 110, 435, 1988.

23. **Karas, J. L., Lieber, C. M., and Gray, H. B.,** Free energy dependence of the rate of long-range electron transfer in proteins. Reorganization energy in ruthenium modified myoglobins, *J. Am. Chem. Soc.,* 110, 599, 1988.

23. **Zemel, H. and Hoffman, B. M.,** Long-range triplet-triplet energy transfer within metal-substituted hemoglobins, *J. Am. Chem. Soc.,* 103, 1192, 1981.

25. **McLendon, G. and Miller, J. R.,** The dependence of biological electron transfer rates on exothermicity: the cytochrome *c*/cytochrome b_s couple, *J. Am. Chem. Soc.,* 107, 7811, 1985.

26. **Faraggi, M. and Klapper, M. H.,** Intramolecular long-range electron transfer in the α-hemoglobin subunit, *J. Am. Chem. Soc.,* 110, 5753, 1988.

27. **Stryer, L. and Haugland, R. P.,** Energy transfer: a spectroscopic ruler, *Proc. Natl. Acad. Sci., U.S.A.,* 58, 719, 1967.

28. **Chiu, H. C., Berson, R.,** Electronic energy transfer between tyrosine and tryptophan in the peptide trp-(pro)*n*-tyr, *Biopolymers,* 16, 277, 1977.

29. **Isied, S. S. and Vassilian, A.,** Electron transfer across polypeptides. II. Amino acids and flexible dipeptide bridging ligands, *J. Am. Chem. Soc.,* 106, 1726, 1984.

30. **Isied, S.S., Vassilian, A., Magnuson, R. H., and Schwarz, H. A.,** Electron transfer across polypeptides. III. Oligoprolines bridging ligands, *J. Am. Chem. Soc.,* 106, 1732, 1984.

31. **Isied, S. S., Vassilian, A., Magnuson, R. H., and Schwartz, H. A.,** Electron transfer across polypeptides. V. Rapid rates of electron transfer between Os(II) and Co(III) in complexes with bridging oligoprolines and other polypeptides, *J. Am. Chem. Soc.,* 107, 7432, 1985.

32. **Schanze, K. S. and Sauer, K.,** Photoinduced intramolecular electron transfer in peptide-bridged molecules, *J. Am. Chem. Soc.,* 110, 1180, 1988.

33. **Faraggi, M., DeFelippis, M. R., and Klapper, M. H.,** Long range electron transfer between tyrosine and tryptophan in peptides, *J. Am. Chem. Soc.,* 111, 5141, 1989.

34. **Bobrowski, K., Wierzchowski, K. L., Holcman, J., and Cikurak, M.,** Intramolecular electron transfer in peptides containing methionine, tryptophan and tyrosine. A pulse radiolysis study, *Intl. J. Radiat. Biol.,* 57, 919, 1990. See also: *Stud. Biophys.,* 122, 23, 1987.

35. **Winkler, J. R., Nocera, D. G., Yocom, K. M., Bordignon, E., and Gray, H. B.,** Kinetics of intramolecular electron transfer from Ru^{II} to FeIII in ruthenium modified cytochrome c, *J. Am. Chem. Soc.,* 106, 5145, 1984.

36. **McGourty, J. L., Blough, N. V., and Hoffman, B. M.,** Electron transfer at cyrstallographically known long distances (25 Å) in [Zn^{II}, Fe^{III}] hybrid hemoglobin, *J. Am. Chem. Soc.,* 105, 4470, 1983.

37. **Simolo, K. P., McLendon, G. L., Mauk, M. R., and Mauk, A. G.,** Photoinduced electron transfer within a protein-protein complex formed between physiological redox partners: reduction of ferricytochrome b_5 by the hemoglobin derivative $\alpha_2^{Zn}\beta_2^{Fe(III)CN}$, *J. Am. Chem. Soc.,* 106, 5012, 1984.

38. **Buxton, G. V., Greenstock, C. L., Helman, W. P., Ross, A. B.,** Critical review of rate constants for reactions of hydrated electrons, hydrogen atoms and hydroxyl radicals (OH·/O −·) in aqueous solutions, *J. Phys. Chem. Ref. Data,* 17, 513, 1988.

39. **Neta, P., Huie, R. E., and Ross, A. B.,** Rate constants for reactions of inorganic radicals in aqueous solutions, *J. Phys. Chem. Ref. Data,* 17, 1027, 1988.

40. **Wardman, P.,** Reduction potentials of one-electron couples involving free radicals in aqueous solution, *J. Phys. Chem. Ref. Data,* 18, 1637, 1989.

41. **Isied, S. S., Kiehn, C., and Worosila, G.,** Ruthenium modified cytochrome c: temperature dependence of the rate of intramolecular electron transfer, *J. Am. Chem. Soc.,* 106, 1722, 1984.

42. **Bechtold, R., Gardiner, M. B., Kazmi, A., van Hemerlyck, B., and Isied, S. S.,** Ruthenium-modified horse heart cytochrome c: effect of pH and ligation on the rate of intramolecular electron transfer between ruthenium(II) and heme(III), *J. Phys. Chem.,* 90, 3800, 1986.

43. **Bechtold, R., Kuehn, C., Lepre, C., Isied, S. S.,** Directional electron transfer in ruthenium-modified horse heart cytochrome c, *Nature,* 322, 286, 1986.

44. **Marcus, R. A.,** On the theory of oxidation-reduction reactions involving electron transfer. I., *J. Chem. Phys.,* 24, 966, 1956.

45. **Hush, N. S.,** Adiabatic rate processes at electrodes. I. Energy-charge relationships, *J. Chem. Phys.,* 28, 964, 1958.

46. **Marcus, R. A. and Sutin, N.,** Electron transfers in chemistry and biology, *Biochim. Biophys. Acta,* 811, 265, 1985.

47. **Marcus, R. A.,** Exchange reactions and electron transfer reactions including isotopic exchange. Theory of oxidation-reduction reactions involving electron transfer, *Discuss. Faraday Soc. I,* 29, 21, 1960.

48. **Gamow, G.,** Quantum theory of the atomic nucleus, *Zeit. Phys.,* 51, 204, 1928.

49. **Hopfield, J. J.,** Electron transfer between biological molecules by thermally activated tunneling, *Proc. Natl. Acad. Sci. U.S.A.,* 71, 3640, 1974.

50. **Jortner, J.,** Temperature dependent activation energy for electron transfer between biological molecules, *J. Chem. Phys.,* 64, 4860, 1976.

51. **Hush, N. S.,** Distance dependence of electron transfer rates, *Coord. Chem. Rev.,* 64, 135, 1985.

52. **McConnell, H. M.,** Intramolecular charge transfer in aromatic free radicals, *J. Chem. Phys.,* 35, 508, 1961.

53. **Beitz, J. V. and Miller, J.,** Exothermic rate restrictions on electron transfer in a rigid medium, *J. Chem. Phys.,* 71, 4579, 1979.

54. **Brunschwig, B. S., Ehrenson, S., and Sutin, N.,** Distance dependence of electron-transfer reactions: rate maxima and rapid rates of large reactant separations, *J. Am. Chem. Soc.,* 106, 6858, 1984.

55. **Brunschwig, B. S., Ehrenson, S., and Sutin, N.,** Solvent reorganization in optical and thermal electron-transfer processes, *J. Phys. Chem.,* 90, 3657, 1986.

56. **Isied, S. S., Vassilian, A., Wishart, F., Creutz, C., Schwartz, H. A., and Sutin, N.,** The distance dependence of intramolecular electron-transfer rates: importance of the nuclear factor, *J. Am. Chem. Soc.,* 110, 635, 1988.

57. **Oevering, H., Paddon-Row, M. N., Heppener, M., Oliver, A., Costaris, E., VerHoeven, J. W., and Hush, N. S.,** Long-range photoinduced through-bond electron transfer and radiative recombination via rigid nonconjugated bridges: distance and solvent dependence, *J. Am. Chem. Soc.,* 109, 3258, 1987.

58. **Beratan, D. N. and Hopfield, J. J.,** Calculations of electron tunneling matrix elements in rigid systems: mixed-valence dithiaspirocyclobutane molecules, *J. Am. Chem. Soc.,* 106, 1584, 1984.

59. **Beratan, D. N.,** Electron tunneling through rigid molecular bridges: bicyclo[2.2.2]octane, *J. Am. Chem. Soc.,* 108, 4321, 1986.

60. **Jay-Gerin, J.-P., Ferradini, C., Houée-Levin, Lopez-Castillo, J.-M. and Faraggi, M.,** Long-range electron transfer in proteins: a percolation mechanism, *Radiat. Phys. Chem.,* 30, 309, 1987.

61. **Paddon-Row, M. N.,** Some aspects of orbital interactions through bonds: physical and chemical consequences, *Acc. Chem. Res.,* 15, 245, 1982.
62. **Endicott, J. F.,** Modifications of transition-metal reaction patterns through the manipulation of superexchange couplings, *Acc. Chem. Res.,* 221, 59, 1988.
63. **Larsson, S.,** Electron transfer in proteins, *J. Chem. Soc. Faraday Trans 2,* 79, 1375, 1983.
64. **Calcaterra, L. T., Closs, G. L., and Miller, J. R.,** Fast intramolecular electron transfer in radical ions over long distances across rigid saturated hydrocarbon spacers, *J. Am. Chem. Soc.,* 105, 670, 1983.
65. **Hoffman, R., Imamura, A., and Hehre, W. J.,** Benzynes, dehydroconjugated molecules, and the interaction of orbitals separated by a number of intervening σ bonds, *J. Am. Chem. Soc.,* 90, 1499, 1968.
66. **Moebius, D.,** Designed monolayer assemblies, *Ber. Bunsenges, Phys. Chem.,* 82, 848, 1978.
67. **Miller, J. R., Calcaterra, L. T., and Closs, G. L.,** Intramolecular long-distance electron transfer in radical anions. The effects of free energy and solvent on the reaction rates, *J. Am. Chem. Soc.,* 106, 3047, 1984.
68. **Wassielewski, M. R., Niemczyk, M. P., Svec, W. A., and Pewitt, E. B.,** Dependence of rate constants for photoinduced charge separation and dark charge recombination on the free energy of reaction in restricted-distance porphyrin-quinone molecules, *J. Am. Chem. Soc.,* 107, 1080, 1985.
69. **Joran, A.D., Leland, B. A., Geller, G. G., Hopfield, J. J., and Dervan, P. B.,** Models for photochemical electron transfer at fixed distances. Porphyrin-bicyclo[2,2,2] octane-quinone and porphyrin-bibicyclo[2,2,2] octane-quinone, *J. Am. Chem. Soc.,* 106, 6090, 1984.
70. **Pasman, P., Mes, G. F., Koper, N. W., and Verhoeven, J. W.,** Solvent effects on photoinduced electron transfer in rigid bichromophoric systems, *J. Am. Chem. Soc.,* 107, 5839, 1984.
71. **Miller, J. R., Beitz, J. V., and Huddleston, R. K.,** Effect of free energy on rates of electron transfer between molecules, *J. Am. Chem. Soc.,* 106, 5057, 1984.
72. **Butler, J., Land, E. J., Swallow, A. J., and Prutz, W. A.,** Comment on "Electron-transfer reactions of tryptophan and tyrosine derivatives", *J. Phys. Chem.,* 91, 3113, 1987.
73. **Harriman, A.,** Further comments on the redox potentials of tryptophan and tyrosine, *J. Phys. Chem.,* 91, 6102, 1987.
74. **Merenyi, G., Lind, J., Shen, X.,** Electron transfer from indoles, phenol and sulfite (SO_3^{2-}) to Chlorine Dioxide $(ClO_2\cdot)$, *J. Phys. Chem.,* 92, 134, 1988.
75. **Faraggi, M., Weinraub, D., Broitman, F., DeFelippis, M. R., and Klapper, M. H.,** One electron oxydation of ferrocenes: a pulse radiolysis study, *Radiat. Phys. Chem.,* 32, 293, 1988.
76. **DeFelippis, M. R., Murty, C. P., Faraggi, M., and Klapper, M. H.,** Redox potentials of tryptophan and tyrosine. A pulse radiolysis study, *Biochemistry,* 28, 4847, 1989.
77. **McLendon, G., Pardue, K., and Bak, P.,** Electron transfer in the cytochrome c/cytochrome b_2 complex: evidence for "conformational gating", *J. Am. Chem. Soc.,* 109, 7540, 1987.
78. **Crutchley, R. J., Ellis, W. R., Jr., and Gray, H. B.,** Long distance electron transfer in pentaamine-ruthenium (histidine-48)-myoglobin. Reorganizational energetics of a high spin heme, *J. Am. Chem. Soc.,* 107, 5002, 1985.
79. **Klapper, M. H. and Faraggi, M.,** Intramolecular long-range electron transfer in the hemerythrin monomer. A pulse radiolysis study, *Biochem. Biophys. Res. Comm.,* 166, 867, 1990.
80. **Peterson-Kennedy, S. E., McGourty, J. L., and Hoffman, B. M.,** Temperature dependence of long-range electron transfer in [Zn,Fe^{III}] hybrid hemoglobin, *J. Am. Chem. Soc,* 106, 5010, 1984.
81. **Meade, T. J., Gray, H. B., and Winkler, J. R.,** Driving-force effects on the rate of long-range electron transfer in ruthenium-modified cytochrome, *c, J. Am. Chem. Soc.,* 111, 4354, 1989.
82. **Faraggi, M., Klapper, M. H., and Dorfman, L. M.,** Fast reaction kinetics of one-electron transfer in proteins. The histidyl radical. Mode of electron migration, *J. Phys. Soc.,* 82, 508, 1977.
83. **Faraggi, M. and Bettelheim, A.,** The reaction of the hydrated electron with amino acids, peptides, and proteins in aqueous solutions. III. Histidyl peptides, *Radiat. Res.,* 71, 311, 1977.
84. **Steiner, J. P., Faraggi, M., Klapper, M. H., and Dorfman, L. M.,** Reactivity of protein histidines toward the hydrated electron, *Biochemistry,* 24, 2139, 1985.
85. **Blake, C. C. F., Mair, G. A., North, A. C. T., Phillips, D. C., and Sarma, V. R.,** On the conformation of the hen egg-white lysozyne molecule, *Proc. R. Soc. B,* 167, 365, 1967.
86. **Faraggi, M. and Tal, Y.,** The reaction of the hydrated electron with amino acids, peptides, and proteins in aqueous solutions. II. Formation of radicals and electron transfer reactions, *Radiat. Res.,* 62, 347, 1973.
87. **Michel, H., Epp, O., and Diesenhofer, J.,** Pigment-protein interactions in the photosynthetic reaction center from *Rhodopseudomonas viridis, EMBO J.,* 5, 2445, 1986.
88. **Michel-Beyerle, M. E., Plato, M., Deisenhofer, J., Michel, H., Bixon, M., and Jortner, J.,** Unidirectionality of charge separation in reaction centers of photosynthetic bacteria, *Biochim. Biophys. Acta,* 932, 52, 1988.

89. **Barkigia, K. M., Chantranupong, L., Smith, K. M., and Fajer, J.,** Structural and theoretical models of photosynthetic chromophores. Implications for redox, light absorption properties and vectorial electron flow, *J. Am. Chem. Soc.,* 110, 7566, 1988.
90. **Liang, N., Pielak, G., Mauk, A. G., Smith, M., and Hoffman, B. M.,** Yeast cytochrome c with phenylalanine or tyrosine at position 87 transfers electrons to (Zinc cytochrome c peroxidase)$^{+}$ at a rate ten thousand times that of the serine-87 or glycine-87 variants, *Proc. Natl. Acad. Sci. U.S.A.,* 84, 1249, 1987.

INDEX

A

Ab initio calculations, 290
Absolute inelastic cross sections, 12
Absorption band, 276
Absorption events, 62
Absorption of phonons, 52
Absorption rates, 54—55
Acoustic deformation interactions, 100
Acoustic deformation potential, 79, 86—89
Acoustic deformation potential parameter, 87
Acoustic lattice modes, 77
Acoustic lattice phonon modes, 86—90
Acoustic modes, 52, 79—80
Acoustic phonon-electron coupling constant, 88
Acoustic phonon interactions, 97, 99, 323
Acoustic phonon relaxation energy, 88
Acoustic phonon scattering, 56—59, 69, 190—191,
  195
Acoustic phonons, 52, 92—93, 97
Activation energies, 170, 172, 174, 192, 197
Adiabatic approximation, 82—83
Adiabatic compressibility, 190
Adiabatic dynamics, 220—221, 241—245
Adiabatic polarization, 379
Adiabatic reaction, 405—406
Adiabatic separation, 78
Admixture function, 63
Age theory, 77
Aging effects, 54
Aging mechanisms, 44, 51—52
Alcohol radicals, 401
Alcoholic glasses, 298—303
Alcohols, 117, 148, 275
  dynamics of electron solvation in, 275
  electron absorption spectrum, 275
Aliphatic alcohols, 298
Alkane, 54—55
Alkanediols, 306—309
Alkylamines, 146
Ammoniated electron
  generation by dissolution of alkali metals in liquid
    ammonia, 260, 262
Amorphous films, 10, 12, 19
Amorphous materials, 184, 342
Amphiphilic molecules, 369
Anion vacancy model, 290
Anionic micelles, 385—386
Annealing, 307
Angle-resolved transmitted intensity, 49
Angular probability distribution, 56—57, 69—70
Angular velocity, 109
Anisotropic materials, 55
Anthracene, 100, 142
Aprotic solvents, 359—361
Aqueous glasses, 293—298

Arabinose, 306
L(+)-Arabinose, 303
Aromatic molecules, 90, 138—139
Asymmetric stretching mode, 20
Atomic clusters, 6
Atomic hydrogen, 276
Autoionization, 30
Avalanche effects, 53
Avalanche injection, 321
Avalanching, 44
Average energy gain, 44
Azide radical, 401, 403
Azulene, 140
Azurin, 398, 400

B

Ballistic transport, 324—332
Band concept, 183—184, 195
Band dispersion relation, 61
Band structure, 53, 57, 343
Band-to-band transition, 22
Basak-Cohen theory, 188—190
Beam modulation lock-in techniques, 9
Bending mode, 20
Bimolecular reactions, 358
Biphotonic ionization, 139
Bjerrum defects, 290
Bloch's functions, 183
Bohr radius, 145—146
Boltzmann transport equation (BTE), 185—186,
  195, 342—344
Boltzmann's law, 378
Born approximation, 54
Born equation, 110—111
Born formula, 131, 140, 154
Bound-bound transition, 297—298
Bound-continuum excitation, 297—298
Bragg peak, 11
Bragg reflection, 186
Breakdown mechanisms, 44, 51—52
Breakdown models, 337
Breaking bonds, 44
Brownian motion, 118
Brownian particles, 80
1-Butanol, 299—300

C

Carrier mobility, 44
Carrier separation, 316—322, 327, 331—332
Catalytic agents, 341
Cathodic generation of solvated electrons, 351,
  359—361
Cation ESD, 27
Charge carrier, 52, 96, 99, 106, 118, 128